Digital Design with CPLD Applications and VHDL

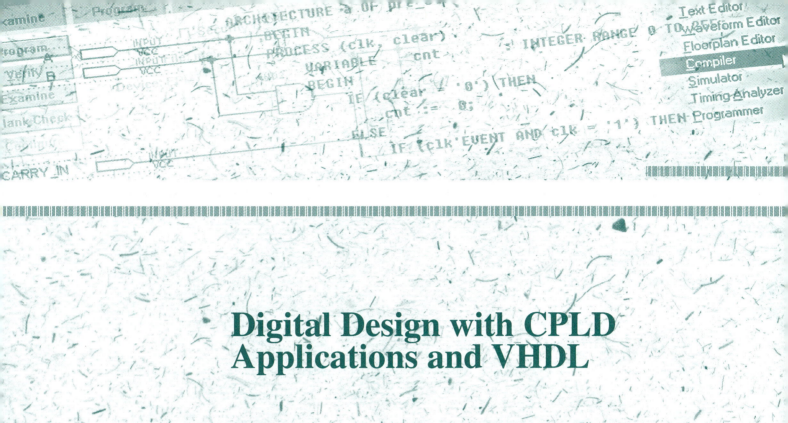

Digital Design with CPLD Applications and VHDL

Robert K. Dueck

Red River College, Winnipeg, Manitoba

Delmar
Thomson Learning™

Africa • Australia • Canada • Denmark • Japan • Mexico
New Zealand • Philippines • Puerto Rico • Singapore
Spain • United Kingdom • United States

NOTICE TO THE READER

Delmar Staff

Business Unit Director: Alar Elken
Executive Editor: Sandy Clark
Acquisitions Editor: Gregory L. Clayton
Developmental Editor: Michelle Ruelos Cannistraci
Editorial Assistant: Jennifer A. Thompson
Executive Marketing Manager: Maura Theriault
Channel Manager: Mona Caron

Marketing Coordinator: Paula Collins
Executive Production Manager: Mary Ellen Black
Production Manager: Larry Main
Senior Project Editor: Christopher Chien
Art & Design Coordinator: David Arsenault
Technology Project Manager: Tom Smith

COPYRIGHT © 2001

Delmar is a division of Thomson Learning. The Thomson Learning logo is a registered trademark used herein under license.

Printer in the United States of America

4 5 6 7 8 9 10 XXX 05 04 03 02

For more information, contact Delmar at 3 Columbia Circle, PO Box 15015, Albany, New York 12212-5015; or find us on the World Wide Web at http://www.delmar.com

Asia
Thomson Learning
60 Albert Street, #15-01
Albert Complex
Singapore 189969

Australia/New Zealand
Nelson/Thomson Learning
102 Dodds Street
South Melbourne, Victoria 3205
Australia

Canada
Nelson/Thomson Learning
1120 Birchmont Road
Scarborough, Ontario
Canada M1K 5G4

Japan
Thomson Learning
Palaceside Building 5F
1-1-1 Hitotsubashi, Chiyoda-ku
Tokyo 100 0003
Japan

Latin America
Thomson Learning
Seneca, 53
Colonia Polanco
11560 Mexico D. F. Mexico

Spain
Thomson Learning
Calle Magellanes, 25
28015-Madrid
Espana

UK/Europe/Middle East
Thomson Learning
Berkshire House
168-173 High Holborn
London
WC1V 7AA United Kingdom

Thomas Nelson & Sons Ltd.
Nelson House
Mayfield Road
Walton-on-Thames
KT 12 5PL United Kingdom

International Headquarters
Thomson Learning
International Division
290 Harbor Drive, 2nd Floor
Stamford, CT 06902-7477
USA

Library of Congress Cataloging-in-Publication Data

Library of Congress Cataloging-in-Publication Data

Dueck, Robert K.
  Digital design with CPLD applications and VHDL / Robert K. Dueck
    p. cm.
  ISBN 0-7668-1160-3
    1. Programming logic devices--Design and construction. 2. Programmable array logic.
3. Logic design. 4. VHDL (Computer hardware description language) I. Title.

TK7872.L64 D84 2000
621.39'5--dc21

                                         00-030838

Contents

Preface

Intended Audience

This book is intended as a textbook for an introductory course in digital electronics in an electronics engineering technologist (EET) or computer engineering technologist (CET) program. There is sufficient material for a second course that expands upon the principles of an introductory course.

No prior knowledge of digital systems is assumed. Prerequisite or corequisite courses in basic DC circuits and introductory college algebra, while not strictly necessary, allow the student to derive maximum benefit from the course of study laid out by this book. A working knowledge of transistors and operational amplifiers is very helpful for the chapters on Logic Gate Circuitry (Chapter 11) and Interfacing Analog and Digital Circuits (Chapter 12). These topics would usually be covered in a second course in digital electronics.

About This Book: Programmable Logic as a Vehicle for Teaching Digital Design

Historically, digital logic or digital design courses at the EET level have primarily focussed on using fixed function TTL and CMOS Small and Medium Scale Integration (SSI and MSI) devices as the vehicle for teaching principles of logic design. However, the digital design field has changed; more and more, digital designs are being implemented in Programmable Logic Devices (PLDs), rendering many of the popular fixed function devices obsolete. The new devices require a different teaching paradigm for which there are few EET-level resources at present.

In the past, many digital components were so-called "fixed function" devices; a manufacturer would have a range of functional offerings (counters, decoders, shift registers, and so on), each of which was manufactured into a single chip. Digital systems were constructed of multiple devices, which sometimes resulted in inefficient designs with large package counts. It was important for technicians and technologists to understand how these devices worked and how to use them in digital circuits. This is the assumption behind many of the classic digital electronics textbooks.

Programmable Logic Devices (PLDs) allow the user to define a function on single or multiple chips, rather than work around standard offerings from device manufacturers. Designs for these devices are developed in concert with sophisticated computer programs that take advantage of the modern PC or workstation environment. The programmable nature and high capacity of these devices result in increased design efficiency and flexibility, as well as decreased time to market.

PLDs have been entering the digital curriculum slowly for a number of years, but have always been treated as one, often optional, topic among many, rather than as an underlying foundation for digital design. Two changes in recent years have made a new approach possible.

First, many of the new Complex PLDs (CPLDs) can be programmed, erased, and re-programmed via a connection to a PC serial or parallel port, without removing them from the circuit in which they are installed. This feature, known as In-System Programmability (ISP), eliminates the need for separate, expensive programming hardware and reduces the mechanical wear-and-tear and risk of damage due to electrostatic discharge (ESD) that always attends multiple insertion-and-removal cycles. Second, the average PC user has access to much more powerful computers than in the past. These computers have sufficient resources to run the design and programming software required by the new CPLDs. Every student with a PC and a CPLD board can now have a complete design and prototyping station at school *and at home,* which has never been possible until now.

This book focuses on the new digital paradigm by using Altera's University Program Laboratory Design Package (UP-1), which includes the Student Edition of MAX+PLUS II, Altera's Programmable Logic Development Software. This Windows-based software allows the student to design, test and program CPLD designs in text-based (VHDL) and graphical (schematic entry) formats. (VHDL, an industry-standard language for PLD design, stands for VHSIC Hardware Description Language. VHSIC is an acronym for Very High Speed Integrated Circuit.) The elementary functions of VHDL and MAX+PLUS II are simple enough to be used successfully in a first course in digital electronics, yet there is enough scope for the software to be useful to students at the senior design project level. A MAX+PLUS II Student Edition CD, which can be legally copied for students' home use, is included with the book. Installation instructions are included on the CD in a file called SE_READ.TXT.

The Altera UP-1 circuit board is available on a donation basis to educational institutions that belong to the Altera University Program and can be purchased by students for $149 (as of May, 2000). It contains two Altera PLDs and a number of standard input and output devices (DIP switches, pushbuttons, LEDs, and seven-segment displays). For more information, see Altera's University Program web page at: http://www.altera.com/html/univ/univ.html

Several lower-cost hardware platforms have been developed or are in the prototyping stage. These are based on the Altera design, but include only one CPLD and some improvements to the interface. One board is manufactured by Intectra (intectra@best.com). All design examples in the book will run on this board, as well as the Altera UP-1 board.

Organization

Although this text is based on CPLD implementation of digital design it does not neglect underlying digital fundamentals. The coverage of binary and hexadecimal numbers, basic logic functions, Boolean algebra, logic minimization, and simple combinational circuits is still present (Chapters 1-3), but application emphasis in later chapters is shifted away from fixed-function SSI and MSI devices and toward CPLDs.

The text introduces CPLDs early in the teaching sequence (Chapter 4) and continues to provide VHDL and MAX+PLUS II applications throughout the book. Students learn the new digital paradigm as an integral part of their training, not just as an optional add-on. Chapters on the MAX+PLUS II design environment (Chapter 4), combinational logic functions (Chapter 5), arithmetic circuits (Chapter 6), latches and flip-flops (Chapter 7), PLD architecture (Chapter 8), counters and shift registers (Chapter 9), and state machines (Chapter 10) are all based primarily on CPLDs and VHDL programming. Sections in Chapter 12 describe CPLD interfacing to digital-to-analog and analog-to-digital converters. Additional topics include electrical characteristics of TTL and CMOS logic, including low-voltage CMOS (Chapter 11), digital-to-analog and analog-to-digital conversion (Chapter 12), and memory devices (Chapter 13). An extensive appendix includes a VHDL language reference with examples, Altera UP-1 User's Guide, and selected data sheets.

Emphasis on CPLDs is not merely descriptive, but experiential. An accompanying lab manual, written by the same author as the text, is tightly integrated with the main text. It contains practical exercises that allow students to design, program, and construct

CPLD-based circuits, rather than just read about them. Also, within the main text, end-of-chapter problems require students to use the MAX+PLUS II design software to create schematic and VHDL design and simulations.

How to Use this Book

Special features of this book include:

Chapter Opener

Each chapter begins with an **Outline** and list of **Objectives** to prepare students for major concepts to be studied and learned.

Key Terms and Notes

Definitions of **Key Terms** are placed at the beginning of each section. All key terms for a chapter are listed in an end-of-chapter Glossary. First use in context of each term is also indicated in boldface. **Notes** are text boxes that contain suggestions, hints, and tips.

VHDL Examples

Numerous solved-problem **Examples** are placed throughout each chapter. Many examples of VHDL programming teach an industry-standard digital hardware design method. Additional examples present **real-world applications**. Examples are stepped in difficulty from basic to more advanced making learning of new material easier.

SECTION
REVIEW
PROBLEMS

CHAPTER
OPENER

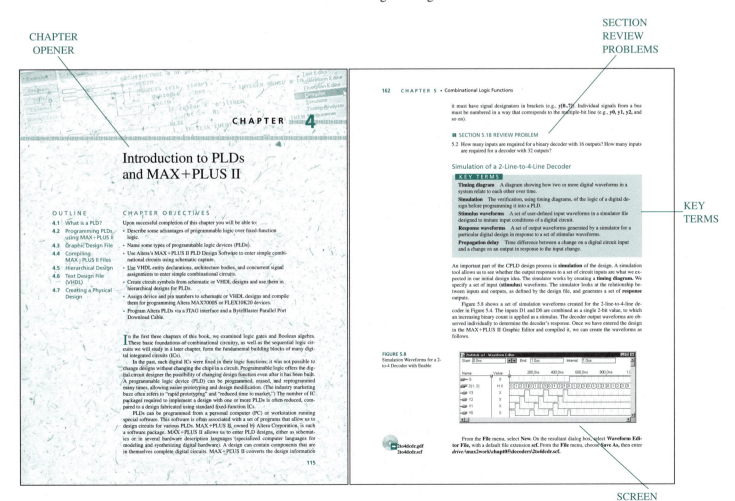

KEY
TERMS

SCREEN
CAPTURES

SCHEMATICS

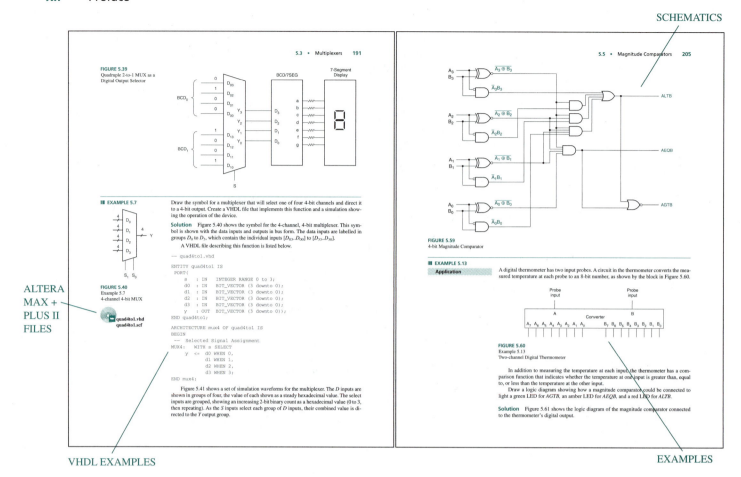

ALTERA
MAX +
PLUS II
FILES

VHDL EXAMPLES

EXAMPLES

Screen Captures and Schematics

Frequent use of schematics, illustrations, tables and screen captures help understanding of the digital principles. Screen captures show exactly what to expect in the Altera™ MAX+PLUS™ II design environment.

Section Review Problems

Numerous section review problems assist in retention of recently learned material. Answers to section review problems are placed at the end of each chapter for quick assessment.

Summary

A chapter Summary provides a recapitulation of the key topics covered in the chapter.

Problem Set

Questions and problems at the end of each chapter are separated into sections. Basic and advanced problems are designated clearly. Answers to selected odd numbered problems are placed at the end of the book.

Altera™ MAX+PLUS™ II Files

The accompanying CD includes **Altera™ Student Edition of MAX+PLUS™ II** CPLD design and programming software along with graphic design files, VHDL files and simulation files from examples. Students can run simulations or program CPLDs with existing error-free design files at home or at school. Students can use existing files as templates for their own modifications. Filenames are identified by a CD icon beside specific examples throughout the text.

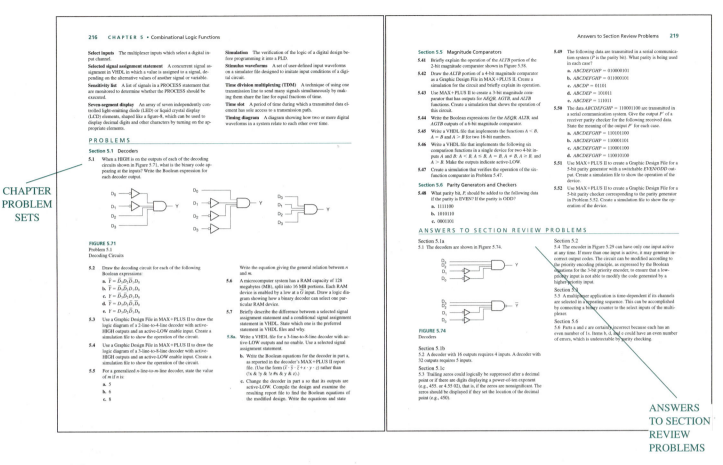

Online Companion™ Web Resources, RealAudio Clips

Delmar's Electronics Website is located at www.electronictech.com. The author has recorded sound clips available to students via **RealAudio**. These sound clips will present more in-depth discussions of difficult topics and will be indicated by a World Wide Web icon in the margin. The Online Companion provides access to text updates, online quizzes, and more.

The Learning Package

In addition to this book, the following materials are available to students and instructors. The complete ancillary package was developed to achieve two goals:

1. To assist students in learning the essential information needed to prepare for the exciting field of electronics.

2. To assist instructors in planning and implementing their instructional programs for the most efficient use of time and other resources. The *Digital Design with CPLD Applications and VHDL* package was created as an integrated whole. Supplements are linked to and integrated with the text to create a comprehensive supplement package that supports students and instructors. The package includes:

Laboratory Manual

This lab manual written by the same author as the textbook, includes introductory and Altera-based lab projects for complex programmable logic devices, design projects, tutorials, introduction to VHDL Programming. An accompanying CD includes a set of design files for the labs and tutorial.
ISBN: 0-7668-1161-1

Instructor's Guide

This comprehensive ancillary contains solutions to all end-of-chapter problems from the textbook.
ISBN: 0-7668-1251-0

e.resource™

This electronic Instructor's Management System is an educational resource that creates a truly electronic classroom. It is a CD-ROM containing tools and instructional resources that enrich your classroom and make your preparation time shorter. The elements of *e.resource* link directly to the text and tie together to provide a unified instructional system.
ISBN: 0-7668-1252-9
Features contained in *e.resource* include:

PowerPoint Presentation Slides

These slides provide the basis for a lecture outline that helps you to present concepts and material. Key points and concepts can be graphically highlighted for student retention. There are over 500 slides, covering every chapter in the text.

ExamView

This computerized testbank includes over 500 questions of varying levels of difficulty provided in multiple formats to assess student comprehension.

Image Library

200 images selected from the textbook allow you to customize power point presentations, or use them as transparency masters. Image Library comes with the ability to browse and search images with key words and allows quick and easy use.

Altera™ MAX+PLUS™ II Files

MAX+PLUS II schematics, VHDL, and simulation files for examples, end-of-chapter problems, and labs. Error-free files can be used as a basis for comparison with student or instructor-developed files.

Electronics Technology Website

Includes Netscape Navigator so you can directly link to the Delmar Electronics Technology website at www.electronictech. com and to the textbook's Online Companion for additional resources.

Acknowledgments

In the writing of this book, I received encouragement, inspiration, and guidance from a great number of people, each of whom deserves my thanks.

I am grateful to Greg Clayton of Delmar Thomson Learning for having the vision to help me start this project and for doing a great deal of ground work to get me a hearing in the electronics teaching community. Thanks to Michelle Ruelos Cannistraci of Delmar, editor, sounding board, and devil's advocate, for her great attention to detail and for keeping me on track during the writing and production of this book. Ben Shriver did a superb job of copyediting, often suggesting finely detailed changes, much to the benefit of the final product. Thanks also to Christopher Chien, David Arsenault, and Larry Main of Delmar for their help in the production process.

Joe Hanson of Altera Corporation has been very helpful in granting permission to use the Student Edition of MAX+PLUS II and in supplying the hardware platform used to develop the material in this book.

Amin Karim of DeVry Institute, graciously invited me to participate in the DeVry Digital Task Force and helped give shape to this book.

Thanks to the following reviewers, whose comments and helpful suggestions greatly contributed to the quality of the final manuscript.

- David G. Delker, Kansas State University, Salina, KS
- Norm Grossman, DeVry Institute, Long Beach, CA
- Robert J. Hofinger, Purdue University, School of Technology, Columbus, IN
- Bruce Johnson, University of Nevada, Reno, NV
- Peter Kerckhoff, DeVry Institute, Kansas City, MO
- Tawfiq Mossadak, Altera Corporation, San Jose, CA
- Vic Quiros, DeVry Institute, Phoenix, AZ
- Arturo Ramirez, DeVry Institute, Addison, IL
- Ken Reid, IUPUI-Indiana University Purdue University Indianapolis, Indianapolis, IN
- Carlo Sapijaszko, DeVry Institute, Calgary, AB
- Gilbert Seah, DeVry Institute, Scarborough, ON
- Lloyd E. Stallkamp, Montana State University-Northern
- Paul Stephanchick, DeVry Institute, Kansas City, MO
- Leslie C. Taylor, DeVry Institute, North Brunswick, NJ
- Jamie Zipay, DeVry Institute, Long Beach, CA

My colleague Mike Gale has been an invaluable collaborator in the development of many of my teaching ideas and has been an advocate for a high standard of education in digital design.

Thanks to my student Ronan Capina for translating a QuickBasic program into C for one of the appendices of this book.

Most especially, thanks to my wife Joan Duerksen who has been more patient than I had any right to expect and has been my greatest encouragement throughout the long writing process.

The author can be contacted by e-mail at bdueck@rrc.mb.ca.

Bob Dueck
Winnipeg, Manitoba, Canada

About the Author

Robert Dueck received an engineering degree from the University of Manitoba and worked for several years as a design engineer at Motorola Canada in Toronto. He began his teaching career in 1986, specializing in digital and microcomputer subjects in the Electronics and Computer Engineering Technology programs at Seneca College in Toronto. His first book, *Fundamentals of Digital Electronics,* was published in 1994. He now teaches digital design at Red River College in Winnipeg. Mr. Dueck is a member of the Association of Professional Engineers of Ontario and the IEEE.

Basic Principles of Digital Systems

CHAPTER OBJECTIVES

Upon successful completion of this chapter, you will be able to:

- Describe some differences between analog and digital electronics.
- Understand the concept of HIGH and LOW logic levels.
- Explain the basic principles of a positional notation number system.
- Translate logic HIGHs and LOWs into binary numbers.
- Count in binary, decimal, or hexadecimal.
- Convert a number in binary, decimal, or hexadecimal to any of the other number bases.
- Calculate the fractional binary equivalent of any decimal number.
- Distinguish between the most significant bit and least significant bit of a binary number.
- Describe the difference between periodic, aperiodic, and pulse waveforms.
- Calculate the frequency, period, and duty cycle of a periodic digital waveform.
- Calculate the pulse width, rise time, and fall time of a digital pulse.

Digital electronics is the branch of electronics based on the combination and switching of voltages called logic levels. Any quantity in the outside world, such as temperature, pressure, or voltage, can be symbolized in a digital circuit by a group of logic voltages that, taken together, represent a binary number. ∎

Each logic level corresponds to a digit in the binary (base 2) number system. The *binary* dig*its*, or bits, 0 and 1, are sufficient to write any number, given enough places. The hexadecimal (base 16) number system is also important in digital systems. Since every combination of four binary digits can be uniquely represented as a hexadecimal digit, this system is often used as a compact way of writing binary information.

Inputs and outputs in digital circuits are not always static. Often they vary with time. Time-varying digital waveforms can have three forms:

1. Periodic waveforms, which repeat a pattern of logic 1s and 0s

2. Aperiodic waveforms, which do not repeat

3. Pulse waveforms, which produce a momentary variation from a constant logic level

1.1 Digital Versus Analog Electronics

The study of electronics often is divided into two basic areas: **analog** and **digital** electronics. Analog electronics has a longer history and can be regarded as the "classical" branch of electronics. Digital electronics, although newer, has achieved greater prominence through the advent of the computer age. The modern revolution in microcomputer chips, as part of everything from personal computers to cars and coffee makers, is founded almost entirely on digital electronics.

The main difference between analog and digital electronics can be stated simply. Analog voltages or currents are **continuously** variable between defined values, and digital voltages or currents can vary only by distinct, or **discrete,** steps.

Some keywords highlight the differences between digital and analog electronics:

Analog	Digital
Continuously variable	Discrete steps
Amplification	Switching
Voltages	Numbers

An example often used to illustrate the difference between analog and digital devices is the comparison between a light dimmer and a light switch. A light dimmer is an analog device, since it can make the light it controls vary in brightness anywhere within a defined range of values. The light can be fully on, fully off, or at some brightness level in between. A light switch is a digital device, since it can turn the light on or off, but there is no value in between those two states.

The light switch/light dimmer analogy, although easy to understand, does not show any particular advantage to the digital device. If anything, it makes the digital device seem limited.

One modern application in which a digital device is clearly superior to an analog one is digital audio reproduction. Compact disc players have achieved their high level of popularity because of the accurate and noise-free way in which they reproduce recorded music. This high quality of sound is possible because the music is stored, not as a magnetic copy of the sound vibrations, as in analog tapes, but as a series of numbers that represent amplitude steps in the sound waves.

Figure 1.1 shows a sound waveform and its representation in both analog and digital forms.

The analog voltage, shown in Figure 1.1b, is a copy of the original waveform and introduces distortion both in the storage and playback processes. (Think of how a photocopy deteriorates in quality if you make a copy of a copy, then a copy of the new copy, and so on. It doesn't take long before you can't read the fine print.)

A digital audio system doesn't make a copy of the waveform, but rather stores a code (a series of amplitude numbers) that tells the compact disc player how to re-create the original sound every time a disc is played. During the recording process, the sound waveform

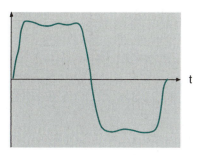
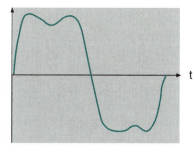
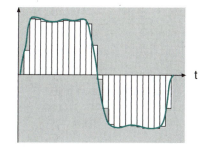

a. Original audio source   **b. Analog reproduction (shows distortion)**   **c. Digital reproduction (simplified)**

FIGURE 1.1
Digital and Analog Sound Reproduction

is "sampled" at precise intervals. The recording transforms each sample into a digital number corresponding to the amplitude of the sound at that point.

The "samples" (the voltages represented by the vertical bars) of the digitized audio waveform shown in Figure 1.1c are much more widely spaced than they would be in a real digital audio system. They are shown this way to give the general idea of a digitized waveform. In real digital audio systems, each amplitude value can be indicated by a number having as many as 16,000 to 65,000 possible values. Such a large number of possible values means the voltage difference between any two consecutive digital numbers is very small. The numbers can thus correspond extremely closely to the actual amplitude of the sound waveform. If the spacing between the samples is made small enough, the reproduced waveform is almost exactly the same as the original.

▐▐ SECTION 1.1 REVIEW PROBLEM

1.1 What is the basic difference between analog and digital audio reproduction?

1.2 Digital Logic Levels

> ### KEY TERMS
>
> **Logic level**   A voltage level that represents a defined digital state in an electronic circuit.
>
> **Logic HIGH** (or **logic 1**)   The higher of two voltages in a digital system with two logic levels.
>
> **Logic LOW** (or **logic 0**)   The lower of two voltages in a digital system with two logic levels.
>
> **Positive logic**   A system in which logic LOW represents binary digit 0 and logic HIGH represents binary digit 1.
>
> **Negative logic**   A system in which logic LOW represents binary digit 1 and logic HIGH represents binary digit 0.

Digitally represented quantities, such as the amplitude of an audio waveform, are usually represented by binary, or base 2, numbers. When we want to describe a digital quantity electronically, we need to have a system that uses voltages or currents to symbolize binary numbers.

The binary number system has only two digits, 0 and 1. Each of these digits can be denoted by a different voltage called a **logic level.** For a system having two logic levels, the lower voltage (usually 0 volts) is called a **logic LOW** or **logic 0** and represents the digit 0. The higher voltage (traditionally 5 V, but in some systems a specific value such as 1.8 V, 2.5 V or 3.3 V) is called a **logic HIGH** or **logic 1,** which symbolizes the digit 1. Except for some allowable tolerance, as shown in Figure 1.2, the range of voltages between HIGH and LOW logic levels is undefined.

FIGURE 1.2

Logic Levels Based on +5 V and 0 V

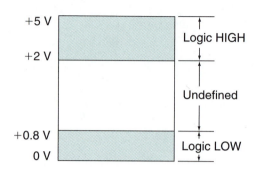

> **NOTE**
>
> For the voltages in Figure 1.2:
>
> $$+5 \text{ V} = \text{Logic HIGH} = 1$$
> $$0 \text{ V} = \text{Logic LOW} = 0$$

The system assigning the digit 1 to a logic HIGH and digit 0 to logic LOW is called **positive logic.** Throughout the remainder of this text, logic levels will be referred to as HIGH/LOW or 1/0 interchangeably.

(A complementary system, called **negative logic,** also exists that makes the assignment the other way around.)

1.3   The Binary Number System

Positional Notation

The **binary number system** is based on the number 2. This means that we can write any number using only two binary digits (or **bits**), 0 and 1. Compare this to the decimal system, which is based on the number 10, where we can write any number with only ten decimal digits, 0 to 9.

The binary and decimal systems are both **positional notation** systems; the value of a digit in either system depends on its placement within a number. In the decimal number 845, the digit 4 really means 40, whereas in the number 9426, the digit 4 really means 400 (845 = 800 + 40 + 5; 9426 = 9000 + 400 + 20 + 6). The value of the digit is determined by *what* the digit is as well as *where* it is.

In the decimal system, a digit in the position immediately to the left of the decimal point is multiplied by 1 (10^0). A digit two positions to the left of the decimal point is multiplied by 10 (10^1). A digit in the next position left is multiplied by 100 (10^2). The positional multipliers, as you move left from the decimal point, are ascending powers of 10.

The same idea applies in the binary system, except that the positional multipliers are powers of 2 ($2^0 = 1, 2^1 = 2, 2^2 = 4, 2^3 = 8, 2^4 = 16, 2^5 = 32, . . .$). For example, the binary number 101 has the decimal equivalent:

$$(1 \times 2^2) + (0 \times 2^1) + (1 \times 2^0)$$
$$= (1 \times 4) + (0 \times 2) + (1 \times 1)$$
$$= \quad 4 \quad + \quad 0 \quad + \quad 1$$
$$= \quad 5$$

▊▊ EXAMPLE 1.1   Calculate the decimal equivalents of the binary numbers 1010, 111, and 10010.

SOLUTIONS
$$1010 = (1 \times 2^3) + (0 \times 2^2) + (1 \times 2^1) + (0 \times 2^0)$$
$$= (1 \times 8) + (0 \times 4) + (1 \times 2) + (0 \times 1)$$
$$= 8 + 2 = 10$$
$$111 = (1 \times 2^2) + (1 \times 2^1) + (1 \times 2^0)$$
$$= (1 \times 4) + (1 \times 2) + (1 \times 1)$$
$$= 4 + 2 + 1 = 7$$
$$10010 = (1 \times 2^4) + (0 \times 2^3) + (0 \times 2^2) + (1 \times 2^1) + (0 \times 2^0)$$
$$= (1 \times 16) + (0 \times 8) + (0 \times 4) + (1 \times 2) + (0 \times 1)$$
$$= 16 + 2 = 18$$

Binary Inputs

KEY TERMS

Most significant bit   The leftmost bit in a binary number. This bit has the number's largest positional multiplier.

Least significant bit   The rightmost bit of a binary number. This bit has the number's smallest positional multiplier.

A major class of digital circuits, called combinational logic, operates by accepting logic levels at one or more input terminals and producing a logic level at an output. In the analysis and design of such circuits, it is frequently necessary to find the output logic level of a circuit for all possible combinations of input logic levels.

The digital circuit in the black box in Figure 1.3 has three inputs. Each input can have two possible states, LOW or HIGH, which can be represented by positive logic as 0 or 1. The number of possible input combinations is $2^3 = 8$. (In general, a circuit with n binary inputs has 2^n input combinations, ranging from 0 to $2^n - 1$.) Table 1.1 shows a list of these combinations, both as logic levels and binary numbers, and their decimal equivalents.

FIGURE 1.3
3-Input Digital Circuit

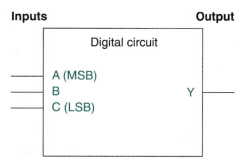

TABLE 1.1 Possible Input Combinations for a 3-Input Digital Circuit

Logic Level			Binary Value			Decimal Equivalent
A	B	C	A	B	C	
L	L	L	0	0	0	0
L	L	H	0	0	1	1
L	H	L	0	1	0	2
L	H	H	0	1	1	3
H	L	L	1	0	0	4
H	L	H	1	0	1	5
H	H	L	1	1	0	6
H	H	H	1	1	1	7

A list of output logic levels corresponding to all possible input combinations, applied in ascending binary order, is called a truth table. This is a standard form for showing the function of a digital circuit.

The input bits on each line of Table 1.1 can be read from left to right as a series of 3-bit binary numbers. The numerical values of these eight input combinations range from 0 to 7 (2^n possible input combinations, having decimal equivalents ranging from 0 to $2^n - 1$) in decimal.

Bit A is called the **most significant bit** (MSB), and bit C is called the **least significant bit** (LSB). As these terms imply, a change in bit A is more significant, since it has the greatest effect on the number of which it is part.

Table 1.2 shows the effect of changing each of these bits in a 3-bit binary number and compares the changed number to the original by showing the difference in magnitude. A change in the MSB of any 3-bit number results in a difference of 4. A change in the LSB of any binary number results in a difference of 1. (Try it with a few different numbers.)

TABLE 1.2 Effect of Changing the LSB and MSB of a Binary Number

	A	B	C	Decimal	
Original	0	1	1	3	
Change MSB	1	1	1	7	Difference = 4
Change LSB	0	1	0	2	Difference = 1

EXAMPLE 1.2

Figure 1.4 shows a 4-input digital circuit. List all the possible binary input combinations to this circuit and their decimal equivalents. What is the value of the MSB?

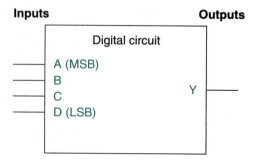

FIGURE 1.4
Example 1.2: 4-Input Digital Circuit

SOLUTION   Since there are four inputs, there will be $2^4 = 16$ possible input combinations, ranging from 0000 to 1111 (0 to 15 in decimal). Table 1.3 shows the list of all possible input combinations.

The MSB has a value of 8 (decimal).

TABLE 1.3   Possible Input
Combinations for a 4-Input Digital Circuit

A	B	C	D	Decimal
0	0	0	0	0
0	0	0	1	1
0	0	1	0	2
0	0	1	1	3
0	1	0	0	4
0	1	0	1	5
0	1	1	0	6
0	1	1	1	7
1	0	0	0	8
1	0	0	1	9
1	0	1	0	10
1	0	1	1	11
1	1	0	0	12
1	1	0	1	13
1	1	1	0	14
1	1	1	1	15

Knowing how to construct a binary sequence is a very important skill when working with digital logic systems. Two ways to do this are:

1. *Learn to count in binary.* You should know all the binary numbers from 0000 to 1111 and their decimal equivalents (0 to 15). *Make this your first goal in learning the basics of digital systems.*

   Each binary number is a unique representation of its decimal equivalent. You can work out the decimal value of a binary number by adding the weighted values of all the bits.

   For instance, the binary equivalent of the decimal sequence 0, 1, 2, 3 can be written using two bits: the 1's bit and the 2's bit. The binary count sequence is:

$$00 \ (= 0 + 0)$$
$$01 \ (= 0 + 1)$$
$$10 \ (= 2 + 0)$$
$$11 \ (= 2 + 1)$$

To count beyond this, you need another bit: the 4's bit. The decimal sequence 4, 5, 6, 7 has the binary equivalents:

$$100 \ (= 4 + 0 + 0)$$
$$101 \ (= 4 + 0 + 1)$$
$$110 \ (= 4 + 2 + 0)$$
$$111 \ (= 4 + 2 + 1)$$

The two least significant bits of this sequence are the same as the bits in the 0 to 3 sequence; a repeating pattern has been generated.

The sequence from 8 to 15 requires yet another bit: the 8's bit. The three LSBs of this sequence repeat the 0 to 7 sequence. The binary equivalents of 8 to 15 are:

$$1000 \ (= 8 + 0 + 0 + 0)$$
$$1001 \ (= 8 + 0 + 0 + 1)$$
$$1010 \ (= 8 + 0 + 2 + 0)$$
$$1011 \ (= 8 + 0 + 2 + 1)$$
$$1100 \ (= 8 + 4 + 0 + 0)$$
$$1101 \ (= 8 + 4 + 0 + 1)$$
$$1110 \ (= 8 + 4 + 2 + 0)$$
$$1111 \ (= 8 + 4 + 2 + 1)$$

Practice writing out the binary sequence until it becomes familiar. In the 0 to 15 sequence, it is standard practice to write each number as a 4-bit value, as in Example 1.2, so that all numbers have the same number of bits. Numbers up to 7 have leading zeros to pad them out to 4 bits.

This convention has developed because each bit has a physical location in a digital circuit; we know a particular bit is logic 0 because we can measure 0 V at a particular point in a circuit. A bit with a value of 0 doesn't go away just because there is not a 1 at a more significant location.

While you are still learning to count in binary, you can use a second method.

2. *Follow a simple repetitive pattern.* Look at Tables 1.1 and 1.3 again. Notice that the least significant bit follows a pattern. The bits alternate with every line, producing the pattern $0, 1, 0, 1, \ldots$. The 2's bit alternates every two lines: $0, 0, 1, 1, 0, 0, 1, 1, \ldots$. The 4's bit alternates every four lines: $0, 0, 0, 0, 1, 1, 1, 1, \ldots$. This pattern can be expanded to cover any number of bits, with the number of lines between alternations doubling with each bit to the left.

Decimal-to-Binary Conversion

There are two methods commonly used to convert decimal numbers to binary: sum of powers of 2 and repeated division by 2.

Sum of Powers of 2

You can convert a decimal number to binary by adding up powers of 2 by inspection, adding bits as you need them to fill up the total value of the number. For example, convert 57_{10} to binary.

$$64_{10} > 57_{10} > 32_{10}$$

- We see that 32 ($=2^5$) is the largest power of two that is smaller than 57. Set the 32's bit to 1 and subtract 32 from the original number, as shown below.

$$57 - 32 = 25$$

- The largest power of two that is less than 25 is 16. Set the 16's bit to 1 and subtract 16 from the accumulated total.

$$25 - 16 = 9$$

- 8 is the largest power of two that is less than 9. Set the 8's bit to 1 and subtract 8 from the total.

$$9 - 8 = 1$$

- 4 is greater than the remaining total. Set the 4's bit to 0.
- 2 is greater than the remaining total. Set the 2's bit to 0.

- 1 is left over. Set the 1's bit to 1 and subtract 1.

$$1 - 1 = 0$$

- Conversion is complete when there is nothing left to subtract. Any remaining bits should be set to 0.

32	16	8	4	2	1
1					

$57 - 32 = 25$

32	16	8	4	2	1
1	1				

$57 - (32 + 16) = 9$

32	16	8	4	2	1
1	1	1			

$57 - (32 + 16 + 8) = 1$

32	16	8	4	2	1
1	1	1	0	0	1

$57 - (32 + 16 + 8 + 1) = 0$

$$57_{10} = 111001_2$$

EXAMPLE 1.3

Convert 92_{10} to binary using the sum-of-powers-of-2 method.

SOLUTION $128 > 92 > 64$

64	32	16	8	4	2	1
1						

$92 - 64 = 28$

64	32	16	8	4	2	1
1	0	1				

$92 - (64 + 16) = 12$

64	32	16	8	4	2	1
1	0	1	1			

$92 - (64 + 16 + 8) = 4$

64	32	16	8	4	2	1
1	0	1	1	1	0	0

$92 - (64 + 16 + 8 + 4) = 0$

$$92_{10} = 1011100_2$$

Repeated Division by 2

Any decimal number divided by 2 will leave a remainder of 0 or 1. Repeated division by 2 will leave a string of 0s and 1s that become the binary equivalent of the decimal number. Let us use this method to convert 46_{10} to binary.

1. Divide the decimal number by 2 and note the remainder.

$$46/2 = 23 + \text{remainder } 0 \text{ (LSB)}$$

The remainder is the least significant bit of the binary equivalent of 46.

2. Divide the quotient from the previous division and note the remainder. The remainder is the second LSB.

$$23/2 = 11 + \text{remainder } 1$$

3. Continue this process until the *quotient* is 0. The last remainder is the most significant bit of the binary number.

$$11/2 = 5 + \text{remainder } 1$$
$$5/2 = 2 + \text{remainder } 1$$
$$2/2 = 1 + \text{remainder } 0$$
$$1/2 = 0 + \text{remainder } 1 \qquad \text{(MSB)}$$

To write the binary equivalent of the decimal number, read the remainders from the bottom up.

$$46_{10} = 101110_2$$

III EXAMPLE 1.4

Use repeated division by 2 to convert 115_{10} to a binary number.

SOLUTION

$$115/2 = 57 + \text{remainder } 1 \text{ (LSB)}$$
$$57/2 = 28 + \text{remainder } 1$$
$$28/2 = 14 + \text{remainder } 0$$
$$14/2 = 7 + \text{remainder } 0$$
$$7/2 = 3 + \text{remainder } 1$$
$$3/2 = 1 + \text{remainder } 1$$
$$1/2 = 0 + \text{remainder } 1 \text{ (MSB)}$$

Read the remainders from bottom to top: 1110011.

$$115_{10} = 1110011_2$$

III

In any decimal-to-binary conversion, the number of bits in the binary number is the exponent of the smallest power of 2 that is larger than the decimal number.

For example, for the numbers 92_{10} and 46_{10},

$$2^7 = 128 > 92 \qquad \text{7 bits: } 1011100$$
$$2^6 = 64 > 46 \qquad \text{6 bits: } 101110$$

Fractional Binary Numbers

KEY TERMS

Radix point   The generalized form of a decimal point. In any positional number system, the radix point marks the dividing line between positional multipliers that are positive and negative powers of the system's number base.

Binary point   A period ("."). that marks the dividing line between positional multipliers that are positive and negative powers of 2 (e.g., first multiplier right of binary point $= 2^{-1}$; first multiplier left of binary point $= 2^0$).

In the decimal system, fractional numbers use the same digits as whole numbers, but the digits are written to the right of the decimal point. The multipliers for these digits are negative powers of 10—10^{-1} (1/10), 10^{-2} (1/100), 10^{-3} (1/1000), and so on.

So it is in the binary system. Digits 0 and 1 are used to write fractional binary numbers, but the digits are to the right of the **binary point**—the binary equivalent of the decimal point. (The decimal point and binary point are special cases of the **radix point,** the general name for any such point in any number system.)

Each digit is multiplied by a positional factor that is a negative power of 2. The first four multipliers on either side of the binary point are:

$$
\begin{array}{cccc|cccc}
 & & & & \text{binary} & & & \\
 & & & & \text{point} & & & \\
2^3 & 2^2 & 2^1 & 2^0 & \cdot & 2^{-1} & 2^{-2} & 2^{-3} & 2^{-4} \\
= 8 & = 4 & = 2 & = 1 & & = 1/2 & = 1/4 & = 1/8 & = 1/16
\end{array}
$$

▉ EXAMPLE 1.5

Write the binary fraction 0.101101 as a decimal fraction.

SOLUTION

$1 \times 1/2 = 1/2$
$0 \times 1/4 = 0$
$1 \times 1/8 = 1/8$
$1 \times 1/16 = 1/16$
$0 \times 1/32 = 0$
$1 \times 1/64 = 1/64$

$1/2 + 1/8 + 1/16 + 1/64 = 32/64 + 8/64 + 4/64 + 1/64$
$= 45/64$
$= 0.703125_{10}$

▉

Fractional-Decimal-to-Fractional-Binary Conversion

Simple decimal fractions such as 0.5, 0.25, and 0.375 can be converted to binary fractions by a sum-of-powers method. The above decimal numbers can also be written 0.5 = 1/2, 0.25 = 1/4, and 0.375 = 3/8 = 1/4 + 1/8. These numbers can all be represented by negative powers of 2. Thus, in binary,

$$0.5_{10} = 0.1_2$$
$$0.25_{10} = 0.01_2$$
$$0.375_{10} = 0.011_2$$

The conversion process becomes more complicated if we try to convert decimal fractions that cannot be broken into powers of 2. For example, the number $1/5 = 0.2_{10}$ cannot be exactly represented by a sum of negative powers of 2. (Try it.) For this type of number, we must use the method of repeated multiplication by 2.

Method:

1. Multiply the decimal fraction by 2 and note the integer part. The integer part is either 0 or 1 for any number between 0 and 0.999. . . . The integer part of the product is the first digit to the left of the binary point.

$0.2 \times 2 = 0.4$ Integer part: 0

2. Discard the integer part of the previous product. Multiply the fractional part of the previous product by 2. Repeat step 1 until the fraction repeats or terminates.

$0.4 \times 2 = 0.8$ Integer part: 0
$0.8 \times 2 = 1.6$ Integer part: 1
$0.6 \times 2 = 1.2$ Integer part: 1
$0.2 \times 2 = 0.4$ Integer part: 0

(Fraction repeats; product is same as in step 1)

Read the above integer parts from top to bottom to obtain the fractional binary number. Thus, $0.2_{10} = 0.00110011 \ldots _2 = 0.0\overline{0011}_2$. The bar shows the portion of the digits that repeats.

▌▌ EXAMPLE 1.6

Convert 0.95_{10} to its binary equivalent.

SOLUTION
$0.95 \times 2 = 1.90$	Integer part: 1
$0.90 \times 2 = 1.80$	Integer part: 1
$0.80 \times 2 = 1.60$	Integer part: 1
$0.60 \times 2 = 1.20$	Integer part: 1
$0.20 \times 2 = 0.40$	Integer part: 0
$0.40 \times 2 = 0.80$	Integer part: 0
$0.80 \times 2 = 1.60$	Fraction repeats last four digits

$$0.95_{10} = 0.1\overline{1110}0_2$$

▌▌

▌▌ SECTION 1.3 REVIEW PROBLEMS

1.2. How many different binary numbers can be written with 6 bits?

1.3. How many can be written with 7 bits?

1.4. Write the sequence of 7-bit numbers from 1010000 to 1010111.

1.5. Write the decimal equivalents of the numbers written for Problem 1.4.

1.4 Hexadecimal Numbers

After binary numbers, hexadecimal (base 16) numbers are the most important numbers in digital applications. Hexadecimal, or hex, numbers are primarily used as a shorthand form of binary notation. Since 16 is a power of 2 ($2^4 = 16$), each hexadecimal digit can be converted directly to four binary digits. Hex numbers can pack more digital information into fewer digits.

Hex numbers have become particularly popular with the advent of small computers, which use binary data having 8, 16, or 32 bits. Such data can be represented by 2, 4, or 8 hexadecimal digits, respectively.

Counting in Hexadecimal

The positional multipliers in the hex system are powers of sixteen: $16^0 = 1$, $16^1 = 16$, $16^2 = 256$, $16^3 = 4096$, and so on.

We need 16 digits to write hex numbers; the decimal digits 0 through 9 are not sufficient. The usual convention is to use the capital letters A through F, each letter representing a number from 10_{10} through 15_{10}. Table 1.4 shows how hexadecimal digits relate to their decimal and binary equivalents.

TABLE 1.4 Hex Digits and Their Binary and Decimal Equivalents

Hex	Decimal	Binary
0	0	0000
1	1	0001
2	2	0010
3	3	0011
4	4	0100
5	5	0101
6	6	0110
7	7	0111
8	8	1000
9	9	1001
A	10	1010
B	11	1011
C	12	1100
D	13	1101
E	14	1110
F	15	1111

NOTE

Counting Rules for Hexadecimal Numbers:

1. Count in sequence from 0 to F in the least significant digit.

2. Add 1 to the next digit to the left and start over.

3. Repeat in all other columns.

For instance, the hex numbers between 19 and 22 are 19, 1A, 1B, 1C, 1D, 1E, 1F, 20, 21, 22. (The decimal equivalents of these numbers are 25_{10} through 34_{10}.)

▌▌ EXAMPLE 1.7

What is the next hexadecimal number after 999? After 99F? After 9FF? After FFF?

SOLUTION The hexadecimal number after 999 is 99A. The number after 99F is 9A0. The number after 9FF is A00. The number after FFF is 1000.

▌▌ EXAMPLE 1.8

List the hexadecimal digits from 190_{16} to 200_{16}, inclusive.

SOLUTION The numbers follow the counting rules: Use all the digits in one position, add 1 to the digit one position left, and start over.

For brevity, we will list only a few of the numbers in the sequence:

190, 191, 192, . . . , 199, 19A, 19B, 19C, 19D, 19E, 19F,

1A0, 1A1, 1A2, . . . , 1A9, 1AA, 1AB, 1AC, 1AD, 1AE, 1AF,

1B0, 1B1, 1B2, . . . , 1B9, 1BA, 1BB, 1BC, 1BD, 1BE, 1BF,

1C0, . . . , 1CF, 1D0, . . . , 1DF, 1E0, . . . , 1EF, 1F0, . . . , 1FF, 200 ▌▌

▌▌ SECTION 1.4A REVIEW PROBLEMS

1.6. List the hexadecimal numbers from FA9 to FB0, inclusive.

1.7. List the hexadecimal numbers from 1F9 to 200, inclusive.

Hexadecimal-to-Decimal Conversion

To convert a number from hex to decimal, multiply each digit by its power-of-16 positional multiplier and add the products. In the following examples, hexadecimal numbers are indicated by a final "H" (e.g., 1F7H), rather than a "16" subscript.

▌▌ EXAMPLE 1.9

Convert 7C6H to decimal.

SOLUTION

$$
\begin{aligned}
7 \times 16^2 &= 7_{10} \times 256_{10} = 1792_{10} \\
C \times 16^1 &= 12_{10} \times 16_{10} = 192_{10} \\
6 \times 16^0 &= 6_{10} \times 1_{10} = \underline{6_{10}} \\
&\qquad\qquad\qquad\qquad\quad 1990_{10}
\end{aligned}
$$

▌▌ EXAMPLE 1.10

Convert 1FD5H to decimal.

SOLUTION

$$
\begin{aligned}
1 \times 16^3 &= 1_{10} \times 4096_{10} = 4096_{10} \\
F \times 16^2 &= 15_{10} \times 256_{10} = 3840_{10} \\
D \times 16^1 &= 13_{10} \times 16_{10} = 208_{10} \\
5 \times 16^0 &= 5_{10} \times 1_{10} = \underline{5_{10}} \\
&\qquad\qquad\qquad\qquad\quad 8149_{10}
\end{aligned}
$$

▌▌

▌▌ SECTION 1.4B REVIEW PROBLEM

1.8 Convert the hexadecimal number A30F to its decimal equivalent.

Decimal-to-Hexadecimal Conversion

Decimal numbers can be converted to hex by the sum-of-weighted-hex-digits method or by repeated division by 16. The main difficulty we encounter in either method is remembering to convert decimal numbers 10 through 15 into the equivalent hex digits, A through F.

Sum of Weighted Hexadecimal Digits

This method is useful for simple conversions (about three digits). For example, the decimal number 35 is easily converted to the hex value 23.

$$35_{10} = 32_{10} + 3_{10} = (2 \times 16) + (3 \times 1) = 23H$$

▌▌ EXAMPLE 1.11   Convert 175_{10} to hexadecimal.

SOLUTION $256_{10} > 175_{10} > 16_{10}$

Since $256 = 16^2$, the hexadecimal number will have two digits.

$$(11 \times 16) > 175 > (10 \times 16)$$

16	1
A	

$175 - (A \times 16) = 175 - 160 = 15$

16	1
A	F

$175 - ((A \times 16) + (F \times 1))$
$= 175 - (160 + 15) = 0$

Repeated Division by 16

Repeated division by 16 is a systematic decimal-to-hexadecimal conversion method that is not limited by the size of the number to be converted.

It is similar to the repeated-division-by-2 method used to convert decimal numbers to binary. Divide the decimal number by 16 and note the remainder, making sure to express it as a hex digit. Repeat the process until the quotient is zero. The last remainder is the most significant digit of the hex number.

▌▌ EXAMPLE 1.12   Convert 31581_{10} to hexadecimal.

SOLUTION
$$31581/16 = 1973 + \text{remainder } 13 \text{ (D) (LSD)}$$
$$1973/16 = 123 + \text{remainder } 5$$
$$123/16 = 7 + \text{remainder } 11 \text{ (B)}$$
$$7/16 = 0 + \text{remainder } 7 \text{ (MSD)}$$
$$31581_{10} = 7B5DH$$

▌▌ SECTION 1.4C REVIEW PROBLEM

1.9   Convert the decimal number 8137 to its hexadecimal equivalent.

Conversions Between Hexadecimal and Binary

Table 1.4 shows all 16 hexadecimal digits and their decimal and binary equivalents. Note that for every possible 4-bit binary number, there is a hexadecimal equivalent.

Binary-to-hex and hex-to-binary conversions simply consist of making a conversion between each hex digit and its binary equivalent.

▌▌ **EXAMPLE 1.13**   Convert 7EF8H to its binary equivalent.

SOLUTION Convert each digit individually to its equivalent value:

$$7H = 0111_2$$
$$EH = 1110_2$$
$$FH = 1111_2$$
$$8H = 1000_2$$

The binary number is all the above binary numbers in sequence:

$$7EF8H = 111111011111000_2$$

The leading zero (the MSB of 0111) has been left out.   ▐▌

▌▌ **SECTION 1.4D REVIEW PROBLEMS**

1.10 Convert the hexadecimal number 934B to binary.

1.11 Convert the binary number 11001000001101001001 to hexadecimal.

1.5 Digital Waveforms

KEY TERM

Digital waveform   A series of logic 1s and 0s plotted as a function of time.

The inputs and outputs of digital circuits often are not fixed logic levels but **digital waveforms,** where the input and output logic levels vary with time. There are three possible types of digital waveform. *Periodic* waveforms repeat the same pattern of logic levels over a specified period of time. *Aperiodic* waveforms do not repeat. *Pulse* waveforms follow a HIGH-LOW-HIGH or LOW-HIGH-LOW pattern and may be periodic or aperiodic.

Periodic Waveforms

KEY TERMS

Periodic waveform   A time-varying sequence of logic HIGHs and LOWs that repeats over a specified period of time.

Period (T)   Time required for a periodic waveform to repeat. Unit: seconds *(s)*.

Frequency (f)   Number of times per second that a periodic waveform repeats. $f = 1/T$ Unit: Hertz (Hz).

Time HIGH (t_h)   Time during one period that a waveform is in the HIGH state. Unit: seconds *(s)*.

Time LOW (t_l)   Time during one period that a waveform is in the LOW state. Unit: seconds *(s)*.

Duty cycle (DC)   Fraction of the total period that a digital waveform is in the HIGH state. $DC = t_h/T$ (often expressed as a percentage: $\%DC = t_h/T \times 100\%$).

Periodic waveforms repeat the same pattern of HIGHs and LOWs over a specified period of time. The waveform may or may not be symmetrical; that is, it may or may not be HIGH and LOW for equal amounts of time.

EXAMPLE 1.14

Calculate the **time LOW, time HIGH, period, frequency,** and **percent duty cycle** for each of the periodic waveforms in Figure 1.5.

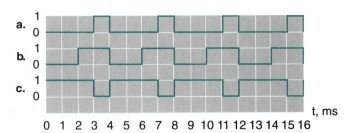

FIGURE 1.5
Example 1.14: Periodic Digital Waveforms

How are the waveforms similar? How do they differ?

SOLUTION

a. Time LOW: t_l = 3 ms

   Time HIGH: t_h = 1 ms

   Period: $T = t_l + t_h$ = 3 ms + 1 ms = 4 ms

   Frequency: $f = 1/T = 1/(4 \text{ ms})$ = 0.25 kHz = 250 Hz

   Duty cycle: $\%DC = (t_h/T) \times 100\% = (1 \text{ ms}/4 \text{ ms}) \times 100\%$
   $$= 25\%$$

   (1 ms = 1/1000 second; 1 kHz = 1000 Hz.)

b. Time LOW: t_l = 2 ms

   Time HIGH: t_h = 2 ms

   Period: $T = t_l + t_h$ = 2 ms + 2 ms = 4 ms

   Frequency: $f = 1/T = 1/(4 \text{ ms})$ = 0.25 kHz = 250 Hz

   Duty cycle: $\%DC = (t_h/T) \times 100\% = (2 \text{ ms}/4 \text{ ms}) \times 100\%$
   $$= 50\%$$

c. Time LOW: t_l = 1 ms

   Time HIGH: t_h = 3 ms

   Period: $T = t_l + t_h$ = 1 ms + 3 ms = 4 ms

   Frequency: $f = 1/T = 1/(4 \text{ ms})$ = 0.25 kHz and 250 Hz

   Duty cycle: $\%DC = (t_h/T) \times 100\% = (3 \text{ ms}/4 \text{ ms}) \times 100\%$
   $$= 75\%$$

The waveforms all have the same period but different duty cycles. A square waveform, shown in Figure 1.5b, has a duty cycle of 50%.

Aperiodic Waveforms

KEY TERM

Aperiodic waveform   A time-varying sequence of logic HIGHs and LOWs that does not repeat.

An **aperiodic waveform** does not repeat a pattern of 0s and 1s. Thus, the parameters of time HIGH, time LOW, frequency, period, and duty cycle have no meaning for an aperiodic waveform. Most waveforms of this type are one-of-a-kind specimens. (It is also worth noting that most digital waveforms are aperiodic.)

Figure 1.6 shows some examples of aperiodic waveforms.

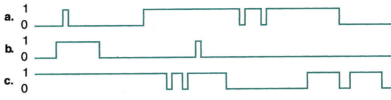

FIGURE 1.6
Aperiodic Digital Waveforms

▋▋ EXAMPLE 1.15

A digital circuit generates the following strings of 0s and 1s:

a. 00111111011010110100000110000

b. 0011001100110011001100110011

c. 00000000111111110000000001111

d. 10111011101110111011101110111011

The time between two bits is always the same. Sketch the resulting digital waveform for each string of bits. Which waveforms are periodic and which are aperiodic?

SOLUTION   Figure 1.7 shows the waveforms corresponding to the strings of bits above. The waveforms are easier to draw if you break up the bit strings into smaller groups of, say, 4 bits each. For instance:

a. 0011 1111 0110 1011 0100 0011 0000

All of the waveforms except Figure 1.7a are periodic.

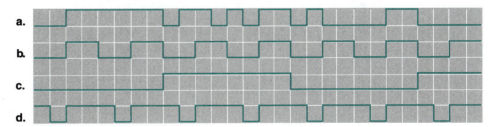

FIGURE 1.7
Example 1.15: Waveforms

Pulse Waveforms

KEY TERMS

Pulse   A momentary variation of voltage from one logic level to the opposite level and back again.

Amplitude   The instantaneous voltage of a waveform. Often used to mean maximum amplitude, or peak voltage, of a pulse.

Edge   The part of the pulse that represents the transition from one logic level to the other.

Rising edge   The part of a pulse where the logic level is in transition from a LOW to a HIGH.

Falling edge   The part of a pulse where the logic level is a transition from a HIGH to a LOW.

Leading edge   The edge of a pulse that occurs earliest in time.

Trailing edge   The edge of a pulse that occurs latest in time.

Pulse width *(t_w)*   Elapsed time from the 50% point of the leading edge of a pulse to the 50% point of the trailing edge.

Rise time *(t_r)*   Elapsed time from the 10% point to the 90% point of the rising edge of a pulse.

Fall time *(t_f)*   Elapsed time from the 90% point to the 10% point of the falling edge of a pulse.

Figure 1.8 shows the forms of both an ideal and a nonideal **pulse.** The **rising and falling edges** of an ideal pulse are vertical. That is, the transitions between logic HIGH and LOW levels are instantaneous. There is no such thing as an ideal pulse in a real digital circuit. Circuit capacitance and other factors make the pulse more like the nonideal pulse in Figure 1.8b.

Pulses can be either positive-going or negative-going, as shown in Figure 1.9. In a positive-going pulse, the measured logic level is normally LOW, goes HIGH for the dura-

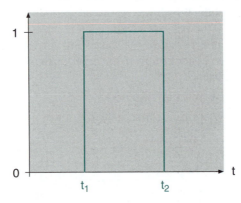

a. Ideal pulse (instantaneous transitions)

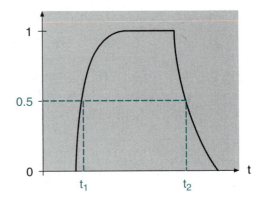

b. Nonideal pulse

FIGURE 1.8
Ideal and Nonideal Pulses

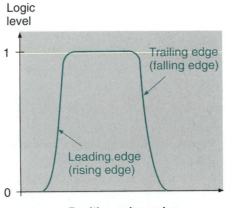

a. Positive-going pulse

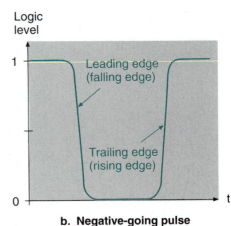

b. Negative-going pulse

FIGURE 1.9
Pulse Edges

tion of the pulse, and returns to the LOW state. A negative-going pulse acts in the opposite direction.

Nonideal pulses are measured in terms of several timing parameters. Figure 1.10 shows the 10%, 50%, and 90% points on the rising and falling edges of a nonideal pulse. (100% is the maximum **amplitude** of the pulse.)

FIGURE 1.10
Pulse Width, Rise Time, Fall Time

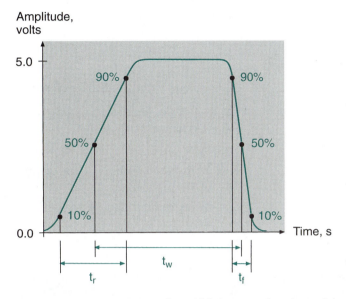

The 50% points are used to measure **pulse width** because the edges of the pulse are not vertical. Without an agreed reference point, the pulse width is indeterminate. The 10% and 90% points are used as references for the **rise and fall times,** since the edges of a nonideal pulse are nonlinear. Most of the nonlinearity is below the 10% or above the 90% point.

▎▎ **EXAMPLE 1.16**   Calculate the pulse width, rise time, and fall time of the pulse shown in Figure 1.11.

FIGURE 1.11
Example 1.16: Pulse

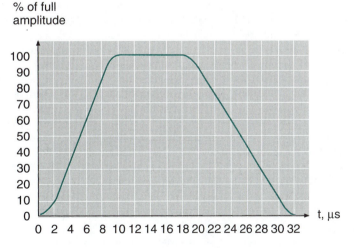

SOLUTION   From the graph in Figure 1.11, read the times corresponding to the 10%, 50%, and 90% values of the pulse on both the **leading and trailing edges.**

Leading edge:	10%:	2 μs	*Trailing edge:*	90%:	20 μs
	50%:	5 μs		50%:	25 μs
	90%:	8 μs		10%:	30 μs

Pulse width: 50% of leading edge to 50% of trailing edge.

$$t_w = 25 \ \mu s - 5 \ \mu s = 20 \ \mu s$$

Rise time: 10% of rising edge to 90% of rising edge.

$$t_r = 8 \ \mu s - 2 \ \mu s = 6 \ \mu s$$

Fall time: 90% of falling edge to 10% of falling edge.

$$t_f = 30 \ \mu s - 20 \ \mu s = 10 \ \mu s$$

Ⅲ SECTION 1.5 REVIEW PROBLEMS

A digital circuit produces a waveform that can be described by the following periodic bit pattern: 0011001100110011.

1.12 What is the duty cycle of the waveform?

1.13 Write the bit pattern of a waveform with the same duty cycle and twice the frequency of the original.

1.14 Write the bit pattern of a waveform having the same frequency as the original and a duty cycle of 75%.

SUMMARY

1. The two basic areas of electronics are analog and digital electronics. Analog electronics deals with continuously variable quantities; digital electronics represents the world in discrete steps.

2. Digital logic uses defined voltage levels, called logic levels, to represent binary numbers within an electronic system.

3. The higher voltage in a digital system represents the binary digit 1 and is called a logic HIGH or logic 1. The lower voltage in a system represents the binary digit 0 and is called a logic LOW or logic 0.

4. The logic levels of multiple locations in a digital circuit can be combined to represent a multibit binary number.

5. Binary is a positional number system (base 2) with two digits, 0 and 1, and positional multipliers that are powers of 2.

6. The bit with the largest positional weight in a binary number is called the most significant bit (MSB); the bit with the smallest positional weight is called the least significant bit (LSB). The MSB is also the leftmost bit in the number; the LSB is the rightmost bit.

7. A decimal number can be converted to binary by sum of powers of 2 (add place values to get a total) or repeated division by 2 (divide by 2 until quotient is 0; remainders are the binary value).

8. The hexadecimal number system is based on 16. It uses 16 digits, from 0–9 and A–F, with power-of-16 multipliers.

9. Each hexadecimal digit uniquely corresponds to a 4-bit binary value. Hex digits can thus be used as shorthand for binary.

10. A digital waveform is a sequence of bits over time. A waveform can be periodic (repetitive), aperiodic (nonrepetitive), or pulsed (a single variation and return between logic levels.)

11. Periodic waveforms are measured by period (T: time for one cycle), time HIGH (t_h), time LOW (t_l), frequency (f: number of cycles per second), and duty cycle (DC or $\%DC$: fraction of cycle in HIGH state).

12. Pulse waveforms are measured by pulse width (t_w: time from 50% of leading edge of 50% of trailing edge), rise time (t_r: time from 10% to 90% of rising edge) and fall time (t_f: time from 90% to 10% of falling edge).

GLOSSARY

Amplitude   The instantaneous voltage of a waveform. Often used to mean maximum amplitude, or peak voltage, of a pulse.

Analog   A way of representing some physical quantity, such as temperature or velocity, by a proportional continuous voltage or current. An analog voltage or current can have any value within a defined range.

Aperiodic waveform   A time-varying sequence of logic HIGHs and LOWs that does not repeat.

Binary number system   A number system used extensively in digital systems, based on the number 2. It uses two digits to write any number.

Bit   *Binary digit.* A 0 or a 1.

Continuous   Smoothly connected. An unbroken series of consecutive values with no instantaneous changes.

Digital   A way of representing a physical quantity by a series of binary numbers. A digital representation can have only specific discrete values.

Digital waveform   A series of logic 1s and 0s plotted as a function of time.

Discrete   Separated into distinct segments or pieces. A series of discontinuous values.

Duty cycle (DC)   Fraction of the total period that a digital waveform is in the HIGH state. $DC = t_h/T$ (often expressed as a percentage: $\%DC = t_h/T \times 100\%$).

Edge   The part of the pulse that represents the transition from one logic level to the other.

Fall time (t_f)   Elapsed time from the 90% point to the 10% point of the falling edge of a pulse.

Falling edge   The part of a pulse where the logic level is in transition from a HIGH to a LOW.

Frequency (f)   Number of times per second that a periodic waveform repeats. $f = 1/T$ Unit: Hertz (Hz).

Hexadecimal number system   Base-16 number system. Hexadecimal numbers are written with sixteen digits, 0–9 and A–F, with power-of-16 positional multipliers.

Leading edge   The edge of a pulse that occurs earliest in time.

Least significant bit (LSB)   The rightmost bit of a binary number. This bit has the number's smallest positional multiplier.

Logic HIGH   The higher of two voltages in a digital system with two logic levels.

Logic level   A voltage level that represents a defined digital state in an electronic circuit.

Logic LOW   The lower of two voltages in a digital system with two logic levels.

Most significant bit (MSB)   The leftmost bit in a binary number. This bit has the number's largest positional multiplier.

Negative logic   A system in which logic LOW represents binary digit 1 and logic HIGH represents binary digit 0.

Period (T)   Time required for a period waveform to repeat. Unit: seconds (s).

Periodic waveform   A time-varying sequence of logic HIGHs and LOWs that repeats over a specified period of time.

Positional notation   A system of writing numbers in which the value of a digit depends not only on the digit, but also on its placement within a number.

Positive logic   A system in which logic LOW represents binary digit 0 and logic HIGH represents binary digit 1.

Pulse   A momentary variation of voltage from one logic level to the opposite level and back again.

Pulse width (t_w)   Elapsed time from the 50% point of the leading edge of a pulse to the 50% point of the trailing edge.

Radix point   The generalized form of a decimal point. In any positional number system, the radix point marks the dividing line between positional multipliers that are positive and negative powers of the system's number base.

Rise time (t_r)   Elapsed time from the 10% point to the 90% point of the rising edge of a pulse.

Rising edge   The part of a pulse where the logic level is in transition from a LOW to a HIGH.

Time HIGH (t_h)   Time during one period that a waveform is in the HIGH state. Unit: seconds (s).

Time LOW (t_l)   Time during one period that a waveform is in the LOW state. Unit: seconds (s).

Trailing edge   The edge of a pulse that occurs latest in time.

PROBLEMS

Problem numbers set in color indicate more difficult problems: those with underlines indicate most difficult problems.

Section 1.1   Digital Versus Analog Electronics

1.1   Which of the following quantities is analog in nature and which digital? Explain your answers.

    a. Water temperature at the beach

    b. Weight of a bucket of sand

    c. Grains of sand in a bucket

    d. Waves hitting the beach in one hour

    e. Height of a wave

    f. People in a square mile

Section 1.2   Digital Logic Levels

1.2   A digital logic system is defined by the voltages 3.3 volts and 0 volts. For a positive logic system, state which voltage corresponds to a logic 0 and which to a logic 1.

Section 1.3   The Binary Number System

1.3   Calculate the decimal values of each of the following binary numbers:

a. 100	**f.** 11101
b. 1000	**g.** 111011
c. 11001	**h.** 1011101
d. 110	**i.** 100001
e. 10101	**j.** 10111001

1.4   Translate each of the following combinations of HIGH (H) and LOW (L) logic levels to binary numbers using positive logic:

a. H H L H	**d.** L L L H
b. L H L H	**e.** H L L L
c. H L H L	

1.5 List the sequence of binary numbers from 101 to 1000.

1.6 List the sequence of binary numbers from 10000 to 11111.

1.7 What is the decimal value of the most significant bit for the numbers in Problem 1.6

1.8 Convert the following decimal numbers to binary. Use the sum-of-powers-of-2 method for parts a, c, e, and g. Use the repeated-division-by-2 method for parts b, d, f, and h.

 a. 75_{10} **e.** 63_{10}

 b. 83_{10} **f.** 64_{10}

 c. 237_{10} **g.** 4087_{10}

 d. 198_{10} **h.** 8193_{10}

1.9 Convert the following fractional binary numbers to their decimal equivalents.

 a. 0.101

 b. 0.011

 c. 0.1101

1.10 Convert the following fractional binary numbers to their decimal equivalents.

 a. 0.01   **c.** 0.010101

 b. 0.0101   **d.** 0.01010101

1.11 The numbers in Problem 1.10 are converging to a closer and closer binary approximation of a simple fraction that can be expressed by decimal integers *a/b*. What is the fraction?

1.12 What is the simple decimal fraction *(a/b)* represented by the repeating binary number 0.101010 . . . ?

1.13 Convert the following decimal numbers to their binary equivalents. If a number has an integer part larger than 0, calculate the integer and fractional parts separately.

 a. 0.75_{10} **e.** 1.75_{10}

 b. 0.625_{10} **f.** 3.95_{10}

 c. 0.1875_{10} **g.** 67.84_{10}

 d. 0.65_{10}

Section 1.4 Hexadecimal Numbers

1.14 Write all the hexadecimal numbers in sequence from 308H to 321H inclusive.

1.15 Write all the hexadecimal numbers in sequence from 9F7H to A03H inclusive.

1.16 Convert the following hexadecimal numbers to their decimal equivalents.

 a. 1A0H   **e.** F3C8H

 b. 10AH   **f.** D3B4H

 c. FFFH   **g.** C000H

 d. 1000H   **h.** 30BAFH

1.17 Convert the following decimal numbers to their hexadecimal equivalents.

 a. 709_{10}

 b. 1889_{10}

 c. 4095_{10}

 d. 4096_{10}

 e. 10128_{10}

 f. 32000_{10}

 g. 32768_{10}

1.18 Convert the following hexadecimal numbers to their binary equivalents.

 a. F3C8H

 b. D3B4H

 c. 8037H

 d. FABDH

 e. 30ACH

 f. 3E7B6H

 g. 743DCFH

1.19 Convert the following binary numbers to their hexadecimal equivalents.

 a. 101111010000110_2

 b. 101101101010_2

 c. 110001011011_2

 d. 110101111000100_2

 e. 101010111100001101_2

 f. 11001100010110111_2

 g. 101000000000000000_2

Section 1.5 Digital Waveforms

1.20 Calculate the time LOW, time HIGH, period, frequency, and percent duty cycle for the waveforms shown in Figure 1.12. How are the waveforms similar? How do they differ?

1.21 Which of the waveforms in Figure 1.13 are periodic and which are aperiodic? Explain your answers.

1.22 Sketch the pulse waveforms represented by the following strings of 0s and 1s. State which waveforms are periodic and which are aperiodic.

 a. 110011110011101100000000110110101

 b. 111000111000111000111000111000111

 c. 111111110000000001111111111111111

 d. 011001100110011001100110011001100110

 e. 011101101001101001011010100111011100

1.23 Calculate the pulse width, rise time, and fall time of the pulse shown in Figure 1.14.

1.24 Repeat Problem 1.23 for the pulse shown in Figure 1.15.

FIGURE 1.12
Problem 1.20: Periodic
Waveforms

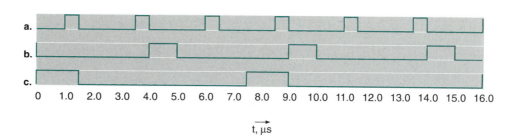

FIGURE 1.13
Problem 1.21: Aperiodic and
Periodic Waveforms

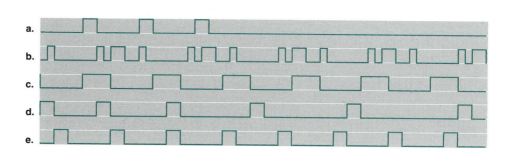

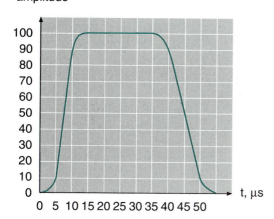

FIGURE 1.14
Problem 1.23: Pulse

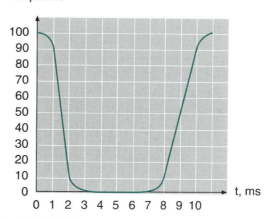

FIGURE 1.15
Problem 1.24: Pulse

ANSWERS TO SECTION REVIEW PROBLEMS

Section 1.1

1.1 An analog audio system makes a direct copy of the recorded sound waves. A digital system stores the sound as a series of binary numbers.

Section 1.3

1.2 64;   **1.3.** 128;   **1.4.** 1010000, 1010001, 1010010, 1010011, 1010100, 1010101, 1010110, 1010111;   **1.5.** 80, 81, 82, 83, 84, 85, 86, 87.

Section 1.4a

1.6 FA9, FAA, FAB, FAC, FAD, FAE, FAF, FB0,   **1.7** 1F9, 1FA, 1FB, 1FC, 1FD, 1FE, 1FF, 200.

Section 1.4b

1.8 41743_{10}.

Section 1.4c

1.9 1FC9.

Section 1.4d

1.10 1001001101001011.   **1.11** C8349.

Section 1.5

1.12 50%;   **1.13** 0101010101010101;
1.14 0111011101110111.

Logic Functions and Gates

CHAPTER OBJECTIVES

Upon successful completion of this chapter, you will be able to:

- Describe the basic logic functions: AND, OR, and NOT

- Draw simple switch circuits to represent AND, OR and Exclusive OR functions.

- Draw simple logic switch circuits for single-pole single-throw (SPST) and normally open and normally closed pushbutton switches.

- Describe the use of light-emitting diodes (LEDs) as indicators of logic HIGH and LOW states.

- Describe those logic functions derived from the basic ones: NAND, NOR, Exclusive OR, and Exclusive NOR.

- Explain the concept of active levels and identify active LOW and HIGH terminals of logic gates.

- Choose appropriate logic functions to solve simple design problems.

- Draw the truth table of any logic gate.

- Draw any logic gate, given its truth table.

- Draw the DeMorgan equivalent form of any logic gate.

- Determine when a logic gate will pass a digital waveform and when it will block the signal.

- Describe several types of integrated circuit packaging for digital logic gates.

All digital logic functions can be synthesized by various combinations of the three basic logic functions: AND, OR, and NOT. These so-called Boolean functions are the basis for all further study of combinational logic circuitry. (Combinational logic circuits are digital circuits whose outputs are functions of their inputs, regardless of the order the inputs are applied.) Standard circuits, called logic gates, have been developed for these and for more complex digital logic functions.

Logic gates can be represented in various forms. A standard set of distinctive-shape symbols has evolved as a universally understandable means of representing the various functions in a circuit. A useful pair of mathematical theorems, called DeMorgan's theorems, enables us to draw these gate symbols in different ways to represent different aspects of the same function. A newer way of representing standard logic gates is outlined in IEEE/ANSI Standard 91-1984, a standard copublished by the Institute of Electrical and

25

Electronic Engineers and the American National Standards Institute. It uses a set of symbols called rectangular-outline symbols.

Logic gates can be used as electronic switches to block or allow passage of digital waveforms. Each logic gate has a different set of properties for enabling (passing) or inhibiting (blocking) digital waveforms. ■

2.1   Basic Logic Functions

> **KEY TERMS**
>
> **Boolean variable**   A variable having only two possible values, such as HIGH/LOW, 1/0, On/Off, or True/False.
>
> **Boolean algebra**   A system of algebra that operates on Boolean variables. The binary (two-state) nature of Boolean algebra makes it useful for analysis, simplification, and design of combinational logic circuits.
>
> **Boolean expression**   An algebraic expression made up of Boolean variables and operators, such as AND, OR, or NOT. Also referred to as a **Boolean function** or a **logic function.**
>
> **Logic gate**   An electronic circuit that performs a Boolean algebraic function.

At its simplest level, a digital circuit works by accepting logic 1s and 0s at one or more inputs and producing 1s or 0s at one or more outputs. A branch of mathematics known as **Boolean algebra** (named after 19th-century mathematician George Boole) describes the relation between inputs and outputs of a digital circuit. We call these input and output values **Boolean variables** and the functions **Boolean expressions, logic functions,** or **Boolean functions.** The distinguishing characteristic of these functions is that they are made up of variables and constants that can have only two possible values: 0 or 1.

All possible operations in Boolean algebra can be created from three basic logic functions: AND, OR, and NOT.[1] Electronic circuits that perform these logic functions are called **logic gates.** When we are analyzing or designing a digital circuit, we usually don't concern ourselves with the actual circuitry of the logic gates, but treat them as black boxes that perform specified logic functions. We can think of each variable in a logic function as a circuit input and the whole function as a circuit output.

In addition to gates for the three basic functions, there are also gates for compound functions that are derived from the basic ones. NAND gates combine the NOT and AND functions in a single circuit. Similarly, NOR gates combine the NOT and OR functions. Gates for more complex functions, such as Exclusive OR and Exclusive NOR, also exist. We will examine all these devices later in the chapter.

NOT, AND, and OR Functions

> **KEY TERMS**
>
> **Truth table**   A list of all possible input values to a digital circuit, listed in ascending binary order, and the output response for each input combination.
>
> **Inverter**   Also called a NOT gate or an inverting buffer. A logic gate that changes its input logic level to the opposite state.
>
> **Bubble**   A small circle indicating logical inversion on a circuit symbol.

---

[1]Words in uppercase letters represent either logic functions (AND, OR, NOT) or logic levels (HIGH, LOW). The same words in lowercase letters represent their conventional nontechnical meanings.

Distinctive-shape symbols   Graphic symbols for logic circuits that show the function of each type of gate by a special shape.

IEEE/ANSI Standard 91-1984   A standard format for drawing logic circuit symbols as rectangles with logic functions shown by a standard notation inside the rectangle for each device.

Rectangular-outline symbols   Rectangular logic gate symbols that conform to IEEE/ANSI Standard 91-1984.

Qualifying symbol   A symbol in IEEE/ANSI logic circuit notation, placed in the top center of a rectangular symbol, that shows the function of a logic gate. Some of the qualifying symbols include: 1 = "buffer"; & = "AND"; ≥1 = "OR"

Buffer   An amplifier that acts as a logic circuit. Its output can be inverting or non-inverting.

NOT Function

The NOT function, the simplest logic function, has one input and one output. The input can be either HIGH or LOW (1 or 0), and the output is always the opposite logic level. We can show these values in a **truth table,** a list of all possible input values and the output resulting from each one. Table 2.1 shows a truth table for a NOT function, where A is the input variable and Y is the output.

The NOT function is represented algebraically by the Boolean expression:

$$Y = \overline{A}$$

This is pronounced *"Y equals NOT A"* or *"Y equals A bar."* We can also say *"Y is the complement of A."*

The circuit that produces the NOT function is called the NOT gate or, more usually, the **inverter.** Several possible symbols for the inverter, all performing the same logic function, are shown in Figure 2.1.

The symbols shown in Figure 2.1a are the standard **distinctive-shape symbols** for the inverter. The triangle represents an amplifier circuit, and the **bubble** (the small circle on the input or output) represents inversion. There are two symbols because sometimes it is convenient to show the inversion at the input and sometimes it is convenient to show it at the output.

Figure 2.1b shows the **rectangular-outline** inverter symbol specified by **IEEE/ANSI Standard 91-1984.** This standard is most useful for specifying the symbols for more complex digital devices. We will show the basic gates in both distinctive-shape and rectangular-outline symbols, although most examples will use the distinctive-shape symbols.

The "1" in the top center of the IEEE symbol is a **qualifying symbol,** indicating the logic gate function. In this case, it shows that the circuit is a **buffer,** an amplifying circuit used as a digital logic element. The arrows at the input and output of the two IEEE symbols show inversion, like the bubbles in the distinctive-shape symbols.

AND Function

AND gate   A logic circuit whose output is HIGH when all inputs (e.g., A AND B AND C) are HIGH.

Logical product   AND function.

The AND function combines two or more input variables so that the output is HIGH only if *all* the inputs are HIGH. The truth table for a 2-input AND function is shown in Table 2.2.

Table 2.1   NOT Function Truth Table

A	Y
0	1
1	0

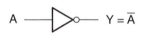

a. Distinctive-shape

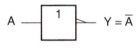

b. Rectangular-outline
(IEEE std. 91-1984)

FIGURE 2.1
Inverter Symbols

Table 2.2   2-input AND Function Truth Table

A	B	Y
0	0	0
0	1	0
1	0	0
1	1	1

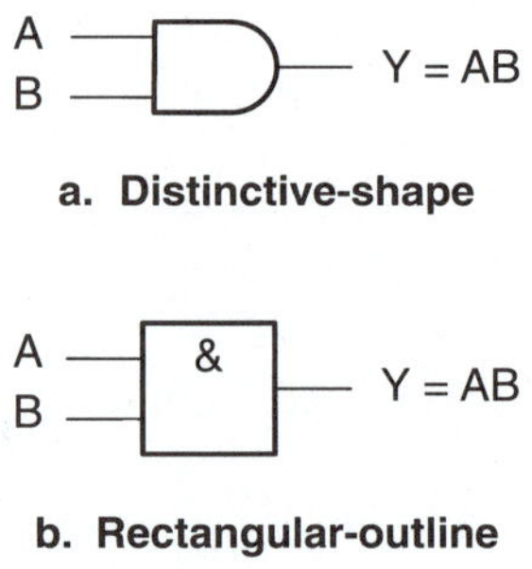

A —⟫— Y = AB
B

a. Distinctive-shape

A —[&]— Y = AB
B

b. Rectangular-outline

FIGURE 2.2
2-Input AND Gate Symbols

Algebraically, this is written:

$$Y = A \cdot B$$

Pronounce this expression *"Y equals A AND B."* The AND function is similar to multiplication in linear algebra and thus is sometimes called the **logical product.** The dot between variables may or may not be written, so it is equally correct to write $Y = AB$. The logic circuit symbol for an **AND gate** is shown in Figure 2.2 in both distinctive-shape and IEEE/ANSI rectangular-outline form. The qualifying symbol in IEEE/ANSI notation is the ampersand (&).

We can also represent the AND function as a set of switches in series, as shown in Figure 2.3. The circuit consists of a voltage source, a lamp, and two series switches. The lamp turns on when switches *A* AND *B* are both closed. For any other condition of the switches, the lamp is off.

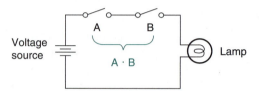

FIGURE 2.3
AND Function Represented by Switches

Table 2.3 3-input AND
Function Truth Table

A	B	C	Y
0	0	0	0
0	0	1	0
0	1	0	0
0	1	1	0
1	0	0	0
1	0	1	0
1	1	0	0
1	1	1	1

Table 2.3 shows the truth table for a 3-input AND function. Each of the three inputs can have two different values, which means the inputs can be combined in $2^3 = 8$ different ways. In general, *n* binary (i.e., two-valued) variables can be combined in 2^n ways.

Figure 2.4 shows the logic symbols for the device. The output is HIGH only when all inputs are HIGH.

FIGURE 2.4
3-Input AND Gate Symbols

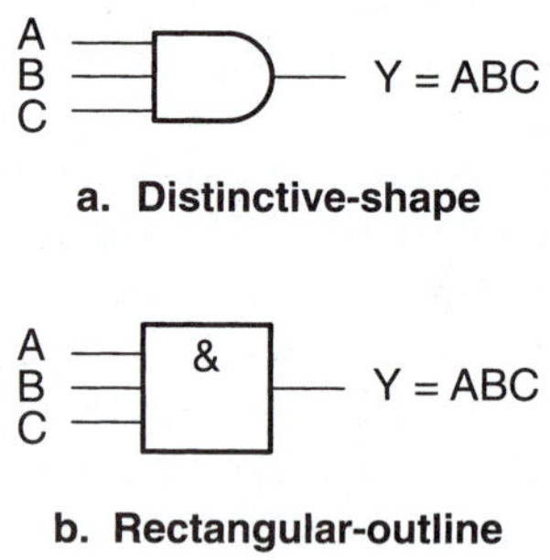

A
B —⟫— Y = ABC
C

a. Distinctive-shape

A
B —[&]— Y = ABC
C

b. Rectangular-outline

OR Function

Table 2.4 2-input OR
Function Truth Table

A	B	Y
0	0	0
0	1	1
1	0	1
1	1	1

KEY TERMS

OR gate   A logic circuit whose output is HIGH when at least one input (e.g., *A* OR *B* OR *C*) is HIGH.

Logical sum   OR function.

The OR function combines two or more input variables in such a way as to make the output variable HIGH if *at least one* input is HIGH. Table 2.4 gives the truth table for the 2-input OR function.

a. Distinctive-shape

A —[≥1]— Y = A + B
B

b. Rectangular-outline

FIGURE 2.5
2-Input OR Gate Symbols

The algebraic expression for the OR function is:

$$Y = A + B$$

which is pronounced *"Y equals A OR B."* This is similar to the arithmetic addition function, but it is not the same. The last line of the truth table tells us that $1 + 1 = 1$ (pronounced "1 OR 1 equals 1"), which is not what we would expect in standard arithmetic. The similarity to the addition function leads to the name **logical sum.** (This is different from the "arithmetic sum," where, of course, $1 + 1$ *does not* equal 1.)

Figure 2.5 shows the logic circuit symbols for an **OR gate.** The qualifying symbol for the OR function in IEEE/ANSI notation is "≥1," which tells us that *one or more* inputs must be HIGH to make the output HIGH.

The OR function can be represented by a set of switches connected in parallel, as in Figure 2.6. The lamp is on when either switch *A* OR switch *B* is closed. (Note that the lamp is also on if *both A* and *B* are closed. This property distinguishes the OR function from the Exclusive OR function, which we will study later in this chapter.)

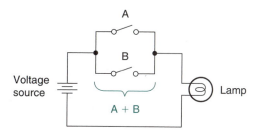

FIGURE 2.6
OR Function Represented by Switches

Table 2.5 3-input OR Function Truth Table

A	B	C	Y
0	0	0	0
0	0	1	1
0	1	0	1
0	1	1	1
1	0	0	1
1	0	1	1
1	1	0	1
1	1	1	1

Like AND gates, OR gates can have several inputs, such as the 3-input OR gates shown in Figure 2.7. Table 2.5 shows the truth table for this gate. Again, three inputs can be combined in eight different ways. The output is HIGH when at least one input is HIGH.

FIGURE 2.7
3-Input OR Gate Symbols

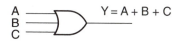

a. Distinctive-shape

b. Rectangular-outline

▌▌ EXAMPLE 2.1

Application

State which logic function is most suitable for the following operations. Draw a set of switches to represent each function.

1. A manager and one other employee both need a key to open a safe.

2. A light comes on in a storeroom when either (or both) of two doors is open. (Assume the switch closes when the door opens.)

3. For safety, a punch press requires two-handed operation.

SOLUTION

1. Both keys are required, so this is an AND function. Figure 2.8a shows a switch representation of the function.

2. One or more switches closed will turn on the lamp. This OR function is shown in Figure 2.8b.

3. Two switches are required to activate a punch press, as shown in Figure 2.8c. This is an AND function.

FIGURE 2.8
Example 2.1

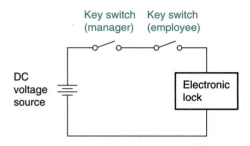

a. Two keys to open a safe (AND)

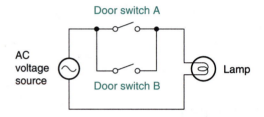

b. One or more switches turn on a lamp (OR)

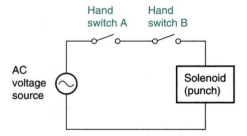

c. Two switches are required to activate a punch press (AND)

Active Levels

An **active level** of a gate input or output is the logic level, either HIGH or LOW, of the terminal when it is performing its designated function. An **active LOW** is shown by a bubble or an arrow symbol on the affected terminal. If there is no bubble or arrow, we assume the terminal is **active HIGH**.

The AND function has active-HIGH inputs and an active-HIGH output. To make the output HIGH, inputs *A* AND *B* must *both* be HIGH. The gate performs its designated function only when *all* inputs are HIGH.

The OR gate requires input *A* OR input *B* to be HIGH for its output to be HIGH. The HIGH active levels are shown by the absence of bubbles or arrows on the terminals.

▌▌ SECTION REVIEW PROBLEM FOR SECTION 2.1

A 4-input gate has input variables *A*, *B*, *C,* and *D* and output *Y.* Write a descriptive sentence for the active output state(s) if the gate is
2.1 AND;

2.2 OR.

2.2   Logic Switches and LED Indicators

Before continuing on, we should examine a few simple circuits that can be used for input or output in a digital circuit. Single-pole single-throw (SPST) and pushbutton switches can be used, in combination with resistors, to generate logic voltages for circuit inputs. Light emitting diodes (LEDs) can be used to monitor outputs of circuits.

Logic Switches

> **KEY TERMS**
>
> V_{CC} The power supply voltage in a transistor-based electronic circuit. The term often refers to the power supply of digital circuits.
>
> **Pull-up resistor**   A resistor connected from a point in an electronic circuit to the power supply of that circuit.

Figure 2.9a shows a single-pole single-throw (SPST) switch connected as a logic switch. An important premise of this circuit is that the input of the digital circuit to which it is connected has a very high resistance to current. When the switch is open, the current flowing through the **pull-up resistor** from V_{CC} to the digital circuit is very small. Since the current is small, Ohm's law states that very little voltage drops across the pull-up resistor; the voltage is about the same at one end as at the other. Therefore, an open switch generates a logic HIGH at point X.

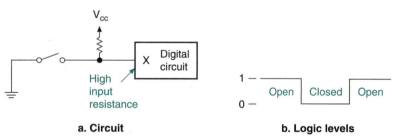

a. Circuit   **b. Logic levels**

FIGURE 2.9
SPST Logic Switch

When the switch is closed, the majority of current flows to ground, limited only by the value of the pull-up resistor. (Since a pull-up resistor is typically between 1 kΩ and 10 kΩ, the LOW-state current in the resistor is about 0.5 mA to 5 mA.) Point X is approximately at ground potential, or logic LOW. Thus the switch generates a HIGH when open and a LOW when closed. The pull-up resistor provides a connection to V_{CC} in the HIGH state and limits power supply current in the LOW state. Figure 2.9b shows the voltage levels when the switch is closed and when it is open.

Figure 2.10 shows how pushbuttons can be used as logic inputs. Figure 2.10a shows a normally open pushbutton and a pull-up resistor. The pushbutton has a spring-loaded plunger that makes a connection between two internal contacts when pressed. When released, the spring returns the plunger to the "normal" (open) state. The logic voltage at X is normally HIGH, but LOW when the button is pressed.

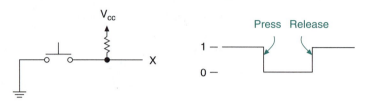

a. Normally open pushbutton

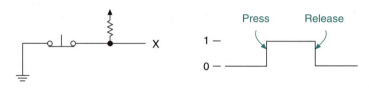

b. Normally closed pushbutton

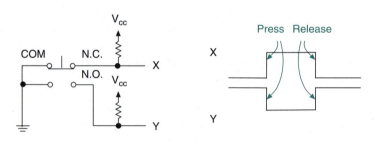

c. Two-pole pushbutton

FIGURE 2.10
Pushbuttons as Logic Switches

Figure 2.10b shows a normally closed pushbutton. The internal spring holds the plunger so that the connection is normally made between the two contacts. When the button is pressed, the connection is broken and the resistor pulls up the voltage at X to a logic HIGH. At rest, X is grounded and the voltage at X is LOW.

It is sometimes desirable to have normally HIGH and normally LOW levels available from the same switch. The two-pole pushbutton in Figure 2.10c provides such a function. The switch has a normally open and a normally closed contact. One contact of each switch is connected to the other, in an internal COMMON connection, allowing the switch to have three terminals rather than four. The circuit has two pull-up resistors, one for X and one for Y. X is normally HIGH and goes LOW when the switch is pressed. Y is opposite.

LED Indicators

KEY TERMS

LED Light-emitting diode. An electronic device that conducts current in one direction only and illuminates when it is conducting.

A device used to indicate the status of a digital output is the **light-emitting diode** or **LED.** This is sometimes pronounced as a word ("led") and sometimes said as separate initials ("ell ee dee"). This device comes in a variety of shapes, sizes, and colors, some of which are shown in the photo of Figure 2.11. The circuit symbol, shown in Figure 2.12, has two terminals, called the anode (positive) and cathode (negative). The arrow coming from the symbol indicates emitted light.

FIGURE 2.11
LEDs

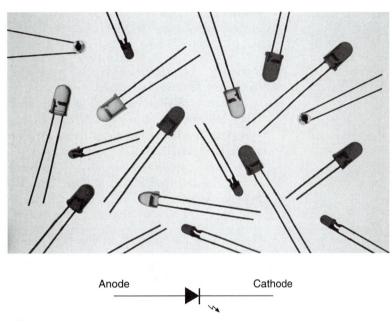

Anode ▷|— Cathode

FIGURE 2.12
Light-Emitting Diode (LED)

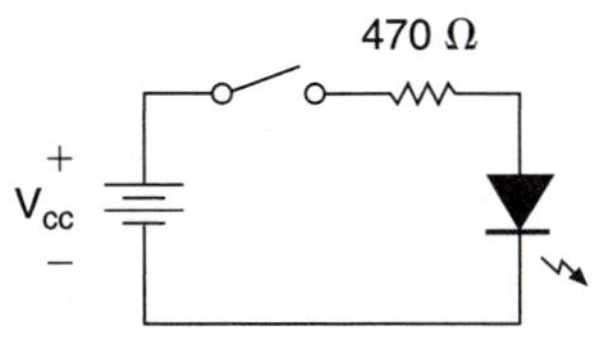

470 Ω

FIGURE 2.13
Condition for LED Illumination

The electrical requirements for the LED are simple: current flows through the LED if the anode is more positive than the cathode by more than a specified value (about 1.5 volts). If enough current flows, the LED illuminates. If more current flows, the illumination is brighter. (If too much flows, the LED burns out, so a series resistor is used to keep the current in the required range.) Figure 2.13 shows a circuit in which an LED illuminates when a switch is closed.

Figure 2.14 shows an AND gate driving an LED. In Figure 2.14a, the LED is on when Y is HIGH (5 volts), since the anode of the LED is more positive than the cathode.

FIGURE 2.14
AND Gate Driving an LED

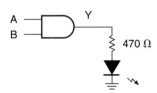

a. LED on when Y is HIGH

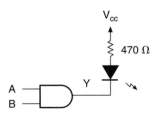

b. LED on when Y is LOW

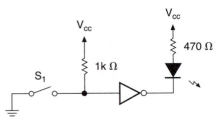

FIGURE 2.15
LED Indicates Status of Switch

In Figure 2.14b, the LED turns on when Y is LOW (0 volts), again since the anode is more positive than the cathode.

Figure 2.15 shows a circuit in which an LED indicates the status of a logic switch. When the switch is open, the 1 kΩ pull-up applies a HIGH to the inverter input. The inverter output is LOW, turning on the LED (anode is more positive than cathode). When the switch is closed, the inverter input is LOW. The inverter output is HIGH (same value as V_{CC}), making anode and cathode voltages equal. No current flows through the LED, and it is therefore off. Thus, the LED is on for a HIGH state at the switch and off for a LOW. Note, however, that the LED is *on* when the inverter output is *LOW.*

▍▍ SECTION 2.2 REVIEW PROBLEM

2.3 A single-pole single-throw switch is connected such that one end is grounded and one end is connected to a 1 kΩ pull-up resistor. The other end of the resistor connects to the circuit power supply, V_{CC}. What logic level does the switch provide when it is open? When it is closed?

2.3 Derived Logic Functions

KEY TERMS

NAND gate   A logic circuit whose output is LOW when all inputs are HIGH.

NOR gate   A logic circuit whose output is LOW when at least one input is HIGH.

Exclusive OR gate   A 2-input logic circuit whose output is HIGH when one input (but not both) is HIGH.

Exclusive NOR gate   A 2-input logic circuit whose output is the complement of an Exclusive OR gate.

Coincidence gate   An Exclusive NOR gate.

The basic logic functions, AND, OR, and NOT, can be combined to make any other logic function. Special logic gates exist for several of the most common of these derived functions. In fact, for reasons we will discover later, two of these derived-function gates, NAND and NOR, are the most common of all gates, and *each* can be used to create any logic function.

NAND and NOR Functions

The names NAND and NOR are contractions of NOT AND and NOT OR, respectively. The NAND is generated by inverting the output of an AND function. The symbols for the **NAND gate** and its equivalent circuit are shown in Figure 2.16.

The algebraic expression for the NAND function is:

$$Y = \overline{A \cdot B}$$

FIGURE 2.16
NAND Gate Symbols

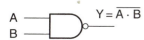 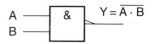 

a. **Distinctive-shape**   b. **Rectangular-outline**   c. **Equivalent circuit**

Table 2.6 NAND Function Truth Table

A	B	Y
0	0	1
0	1	1
1	0	1
1	1	0

The entire function is inverted because the bubble is on the NAND gate output.

Table 2.6 shows the NAND gate truth table. The output is LOW when A AND B are HIGH.

We can generate the NOR function by inverting the output of an OR gate. The NOR function truth table is shown in Table 2.7. The truth table tells us that the output is LOW when A OR B is HIGH.

Figure 2.17 shows the logic symbols for the **NOR gate.**

Table 2.7 NOR Function Truth Table

A	B	Y
0	0	1
0	1	0
1	0	0
1	1	0

 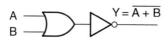

a. **Distinctive-shape**   b. **Rectangular-outline**   c. **Equivalent circuit**

FIGURE 2.17
NOR Gate Symbols

The algebraic expression for the NOR function is:

$$Y = \overline{A + B}$$

The entire function is inverted because the bubble is on the gate output.

We know that the outputs of both gates are active LOW because of the bubbles on the output terminals. The inputs are active HIGH because there are no bubbles on the input terminals.

Multiple-Input NAND and NOR Gates

Table 2.8 shows the truth tables of the 3-input NAND and NOR functions. The logic circuit symbols for these gates are shown in Figure 2.18.

Table 2.8 3-input NAND and NOR Function Truth Tables

A	B	C	$\overline{A \cdot B \cdot C}$	$\overline{A + B + C}$
0	0	0	1	1
0	0	1	1	0
0	1	0	1	0
0	1	1	1	0
1	0	0	1	0
1	0	1	1	0
1	1	0	1	0
1	1	1	0	0

The truth tables of these gates can be generated by understanding the active levels of the gate inputs and outputs. The NAND output is LOW when A AND B AND C are HIGH. This is shown in the last line of the NAND truth table. The NOR output is LOW if one or more of A OR B OR C is HIGH. This describes all lines of the NOR truth table except the first.

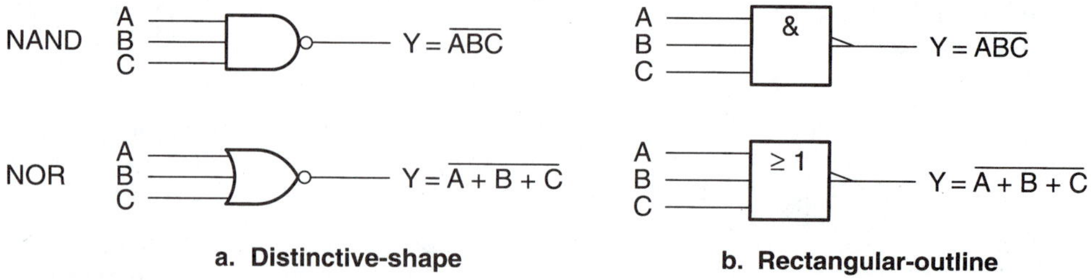

a. Distinctive-shape

b. Rectangular-outline

FIGURE 2.18
3-Input NAND and NOR Gates

Exclusive OR and Exclusive NOR Functions

The Exclusive OR function (abbreviated XOR) is a special case of the OR function. The output of a *2-input* XOR gate is HIGH when *one and only one* of the inputs is HIGH. (Multiple-input XOR circuits do not expand as simply as other functions. As we will see in a later chapter, an XOR output is HIGH when an *odd number* of inputs is HIGH.)

Unlike the OR gate, which is sometimes called an Inclusive OR, a HIGH at both inputs makes the output LOW. (We could say that the case in which both inputs are HIGH is excluded.)

The gate symbol for the **Exclusive OR** gate is shown in Figure 2.19.

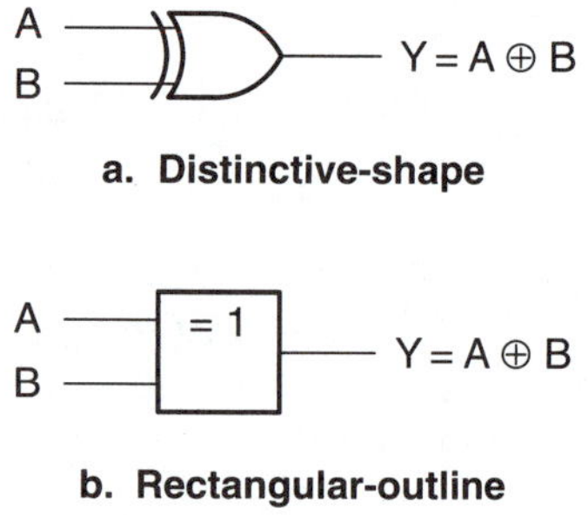

a. Distinctive-shape

b. Rectangular-outline

FIGURE 2.19
Exclusive OR Gate

Table 2.9 Exclusive OR Function Truth Table

A	B	Y
0	0	0
0	1	1
1	0	1
1	1	0

Table 2.9 shows the truth table for the XOR function.

Another way of looking at the Exclusive OR gate is that its output is HIGH when the inputs are different and LOW when they are the same. This is a useful property in some applications, such as error detection in digital communication systems. (Transmitted data can be compared with received data. If they are the same, no error has been detected.)

The XOR function is expressed algebraically as:

$$Y = A \oplus B$$

The Exclusive NOR function is the complement of the Exclusive OR function and shares some of the same properties. The symbol, shown in Figure 2.20, is an XOR gate

FIGURE 2.20
Exclusive NOR Gate

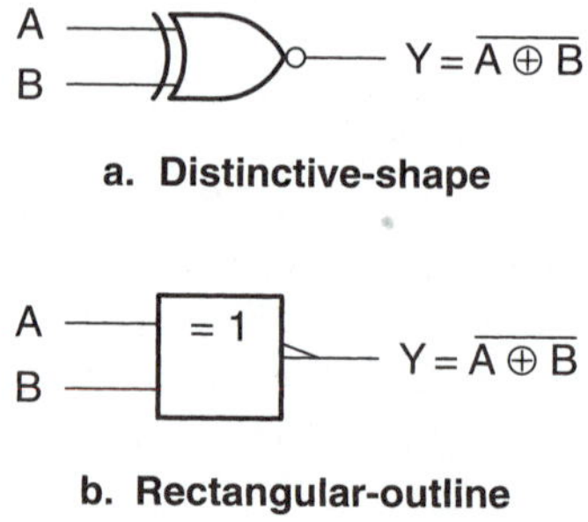

a. Distinctive-shape

b. Rectangular-outline

Table 2.10   Exclusive NOR
Function Truth Table

A	B	Y
0	0	1
0	1	0
1	0	0
1	1	1

with a bubble on the output, implying that the entire function is inverted. Table 2.10 shows the Exclusive NOR truth table.

The algebraic expression for the Exclusive NOR function is:

$$Y = \overline{A \oplus B}$$

The output of the **Exclusive NOR gate** is HIGH when the inputs are the same and LOW when they are different. For this reason, the XNOR gate is also called a **coincidence gate.** This same/different property is similar to that of the Exclusive OR gate, only opposite in sense. Many of the applications that make use of this property can use either the XOR or the XNOR gate.

▌▌ SECTION 2.3 REVIEW PROBLEMS

The output of a logic gate turns on an LED when it is HIGH. The gate has two inputs, each of which is connected to a logic switch, as shown in Figure 2.21.

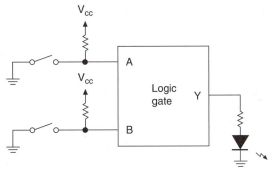

FIGURE 2.21
Section Review Problems: Logic Gate Properties

2.4 What type of gate will turn on the light when the switches are in opposite positions?

2.5 Which gate will turn off the light only when both switches are HIGH?

2.6 What type of gate turns on the light only when both switches are LOW?

2.7 Which gate turns on the light when the switches are in the same position?

2.4   DeMorgan's Theorems and Gate Equivalence

> ### KEY TERMS
>
> **DeMorgan's theorems**   Two theorems in Boolean algebra that allow us to transform any gate from an AND-shaped to an OR-shaped gate and vice versa.
>
> **DeMorgan equivalent forms**   Two gate symbols, one AND-shaped and one OR-shaped, that are equivalent according to DeMorgan's theorems.

Table 2.11   NAND Truth
Table

A	B	Y
0	0	1
0	1	1
1	0	1
1	1	0

Recall the truth table (repeated in Table 2.11) and description of a 2-input NAND gate. "Output Y is LOW if inputs A AND B are HIGH." Or, "Output Y is LOW if *all* inputs are HIGH." The condition of this sentence is satisfied in the last line of Table 2.11.

We could also describe the gate function by saying, "Output Y is HIGH if A OR B (OR both) are LOW," or, "The output is HIGH if *at least one* input is LOW." These conditions are satisfied by the first three lines of Table 2.11.

The gates in Figure 2.22 represent positive- and negative-logic forms of a NAND gate. Figure 2.23 shows the logic equivalents of these gates. In the first case, we combine

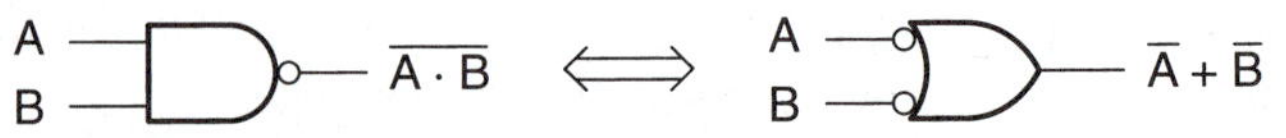

FIGURE 2.22
NAND Gate and DeMorgan Equivalent

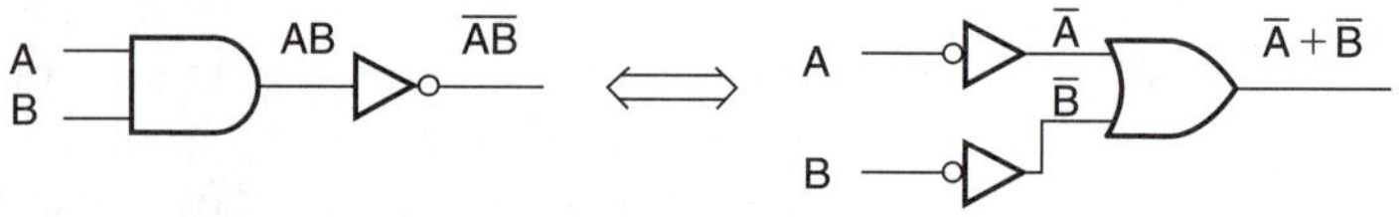

a. AND then invert   b. Invert then OR

FIGURE 2.23
Logic Equivalents of Positive and Negative NAND Gates

the inputs in an AND function, then invert the result. In the second case, we invert the variables, then combine the inverted inputs in an OR function.

The Boolean function for the AND-shaped gate is given by:

$$Y = \overline{A \cdot B}$$

The Boolean expression for the OR-shaped gate is:

$$Y = \overline{A} + \overline{B}$$

The gates shown in Figure 2.22 are called **DeMorgan equivalent forms.** Both gates have the same truth table, but represent different aspects or ways of looking at the NAND function. We can extend this observation to state that *any* gate (except XOR and XNOR) has two equivalent forms, one AND, one OR.

A gate can be categorized by examining three attributes: *shape, input,* and *output.* A question arises from each attribute:

1. What is its shape (AND/OR)?

   AND: *all*

   OR: *at least one*

2. What active level is at the gate inputs (HIGH/LOW)?

3. What active level is at the gate output (HIGH/LOW)?

The answers to these questions characterize any gate and allow us to write a descriptive sentence and a truth table for that gate. The DeMorgan equivalent forms of the gate will yield opposite answers to each of the above questions.

Thus the gates in Figure 2.22 have the following complementary attributes:

	Basic Gate	**DeMorgan Equivalent**
Boolean Expression	$\overline{A \cdot B}$	$\overline{A} + \overline{B}$
Shape	AND	OR
Input Active Level	HIGH	LOW
Output Active Level	LOW	HIGH

▌▌ EXAMPLE 2.2

Analyze the shape, input, and output of the gates shown in Figure 2.24 and write a Boolean expression, a descriptive sentence, and a truth table of each one. Write an asterisk beside the active output level on each truth table. Describe how these gates relate to each other.

FIGURE 2.24
Example 2.2 Logic Gates

a.   b.

SOLUTION

a. **Boolean expression:** $Y = \overline{A + B}$

   Shape: OR *(at least one)*

   Input: HIGH

   Output: LOW

   Descriptive sentence: Output Y is LOW if A OR B is HIGH.

   Truth table:

Table 2.12   Truth Table of Gate in Figure 2.24a.

A	B	Y
0	0	1
0	1	0*
1	0	0*
1	1	0*

b. **Boolean expression:** $Y = \overline{A} \cdot \overline{B}$

   Shape: AND *(all)*

   Input: LOW

   Output: HIGH

   Descriptive sentence: Output Y is HIGH if A AND B are LOW.

   Truth table:

Table 2.13   Truth Table of Gate in Figure 2.124b.

A	B	Y
0	0	1*
0	1	0
1	0	0
1	1	0

Both gates in this example yield the same truth table. Therefore they are DeMorgan equivalents of one another (positive- and negative-NOR gates). ❚❙❚

The gates in Figures 2.22 and 2.24 yield the following algebraic equivalencies:

$$\overline{A \cdot B} = \overline{A} + \overline{B}$$
$$\overline{A + B} = \overline{A} \cdot \overline{B}$$

These equivalencies are known as **DeMorgan's theorems.** (You can remember how to use DeMorgan's theorems by a simple rhyme: "Break the line and change the sign.")

It is tempting to compare the first gate in Figure 2.22 and the second in Figure 2.24 and declare them equivalent. Both gates are AND-shaped, both have inversions. However, the comparison is false. The gates have different truth tables, as we have found in Tables 2.11 and 2.13. Therefore they have different logic functions and are not equivalent. The same is true of the OR-shaped gates in Figures 2.22 and 2.24. The gates may look similar, but since they have different truth tables, they have different logic functions and are therefore not equivalent.

The confusion arises when, after changing the logic input and output levels, you forget to change the shape of the gate. This is a common, but serious, error. These inequalities can be expressed as follows:

$$\overline{A \cdot B} \neq \overline{A} \cdot \overline{B}$$
$$\overline{A + B} \neq \overline{A} + \overline{B}$$

As previously stated, any AND- or OR-shaped gate can be represented in its DeMorgan equivalent form. All we need to do is analyze a gate for its shape, input, and output, then *change everything*.

▌▌ EXAMPLE 2.3

Analyze the gate in Figure 2.25 and write a Boolean expression, descriptive sentence, and truth table for the gate. Mark active output levels on the truth table with asterisks. Find the DeMorgan equivalent form of the gate and write its Boolean expression and description.

FIGURE 2.25
Example 2.3: Logic Gates

SOLUTION

Boolean expression: $Y = \overline{\overline{A} + \overline{B} + \overline{C}}$

Shape: OR *(at least one)*

Input: *LOW*

Output: *LOW*

Descriptive sentence: Output Y is LOW if A OR B OR C is LOW.

Truth table:

Table 2.14   Truth Table of Gate in Figure 2.25

A	B	C	Y
0	0	0	0*
0	0	1	0*
0	1	0	0*
0	1	1	0*
1	0	0	0*
1	0	1	0*
1	1	0	0*
1	1	1	1

Figure 2.26 shows the DeMorgan equivalent form of the gate in Figure 2.25. To create this symbol, we change the shape from OR to AND and invert the logic levels at both input and output.

FIGURE 2.26
Example 2.3: DeMorgan Equivalent of Gate in Figure 2.25

Boolean expression: $Y = ABC$

Descriptive sentence: Output Y is HIGH if A AND B AND C are HIGH.

▌▌ SECTION 2.4 REVIEW PROBLEM

2.8 The output of a gate is described by the following Boolean expression:

$$Y = \overline{A} + \overline{B} + \overline{C} + \overline{D}$$

Write the Boolean expression for the DeMorgan equivalent form of this gate.

2.5   Enable and Inhibit Properties of Logic Gates

In Chapter 1, we saw that a **digital signal** is just a string of bits (0s and 1s) generated over time. A major task of digital circuitry is the direction and control of such signals. Logic gates can be used to **enable** (pass) or **inhibit** (block) these signals. (The word "gate" gives a clue to this function; the gate can "open" to allow a signal through or "close" to block its passage.)

AND and OR Gates

The simplest case of the enable and inhibit properties is that of an AND gate used to pass or block a logic signal. Figure 2.27 shows the output of an AND gate under different conditions of input A when a digital signal (an alternating string of 0s and 1s) is applied to input B.

FIGURE 2.27

Enable/Inhibit Properties of an AND Gate

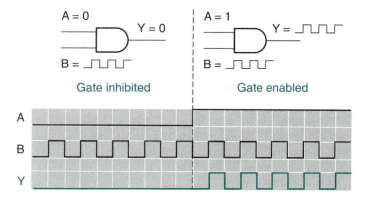

Recall the properties of an AND gate: both inputs must be HIGH to make the output HIGH. Thus, if input *A* is LOW, the output must always be LOW, regardless of the state of input *B*. The digital signal applied to *B* has no effect on the output, and we say that the gate is inhibited or disabled. This is shown in the first half of the timing diagram in Figure 2.27.

If *A* AND *B* are HIGH, the output is HIGH. When *A* is HIGH and *B* is LOW, the output is LOW. Thus, output *Y* is the same as input *B* if input *A* is HIGH; that is, *Y* and *B* are **in phase** with each other. The input waveform is passed to the output in **true form,** and we say the gate is enabled. The last half of the timing diagram in Figure 2.27 shows this waveform.

It is convenient to define terms for the *A* and *B* inputs. Since we apply a digital signal to *B,* we will call it the Signal input. Since input *A* controls whether or not the signal passes to the output, we will call it the Control input. These definitions are illustrated in Figure 2.28.

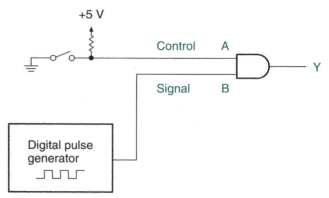

FIGURE 2.28
Control and Signal Inputs of an AND Gate

Each type of logic gate has a particular set of enable/inhibit properties that can be predicted by examining the truth table of the gate. Let us examine the truth table of the AND gate to see how the method works.

Divide the truth table in half, as shown in Table 2.15. Since we have designated A as the Control input, the top half of the truth table shows the inhibit function ($A = 0$), and the bottom half shows the enable function ($A = 1$). To determine the gate properties, we compare input B (the Signal input) to the output in each half of the table.

Inhibit mode: If $A = 0$ and B is pulsing (B is continuously going back and forth between the first and second lines of the truth table), output Y is always 0. Since the Signal input has no effect on the output, we say that the gate is disabled or inhibited.

Enable mode: If $A = 1$ and B is pulsing (B is going continuously between the third and fourth lines of the truth table), the output is the same as the Signal input. Since the Signal input affects the output, we say that the gate is enabled.

Table 2.15   AND Truth Table Showing Enable/Inhibit Properties

A	B	Y	
0	0	0	($Y = 0$)
0	1	0	Inhibit
1	0	0	($Y = B$)
1	1	1	Enable

▮▮ EXAMPLE 2.4

Use the method just described to draw the output waveform of an OR gate if the input waveforms of A and B are the same as in Figure 2.27. Indicate the enable and inhibit portions of the timing diagram.

SOLUTION   Divide the OR gate truth table in half. Designate input A the Control input and input B the Signal input.

As shown in Table 2.16, when $A = 0$ and B is pulsing, the output is the same as B and the gate is enabled. When $A = 1$, the output is always HIGH. (At least one input HIGH makes the output HIGH.) Since B has no effect on the output, the gate is inhibited. This is shown in Figure 2.29 in graphical form.

Table 2.16   OR Truth Table Showing Enable/Inhibit Properties

A	B	Y	
0	0	0	($Y = B$)
0	1	1	Enable
1	0	1	($Y = 1$)
1	1	1	Inhibit

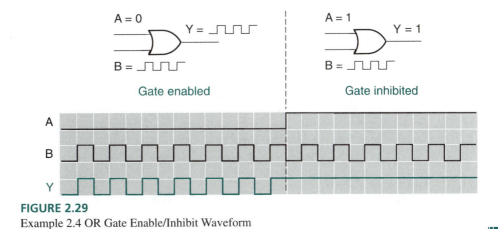

FIGURE 2.29
Example 2.4 OR Gate Enable/Inhibit Waveform

Example 2.4 shows that a gate can be in the inhibit state even if its output is HIGH. It is natural to think of the HIGH state as "ON," but this is not always the case. Enable or inhibit states are determined by the effect the Signal input has on the gate's output. If an input signal does not affect the gate output, the gate is inhibited. If the Signal input does affect the output, the gate is enabled.

NAND and NOR Gates

When inverting gates, such as NAND and NOR, are enabled, they will invert an input signal before passing it to the gate output. In other words, they transmit the signal in **complement form.** Figures 2.30 and 2.31 show the output waveforms of a NAND and a NOR gate when a square waveform is applied to input *B* and input *A* acts as a Control input.

FIGURE 2.30
Enable/Inhibit Properties of a
NAND Gate

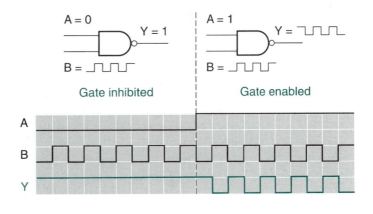

FIGURE 2.31
Enable/Inhibit Properties of a
NOR Gate

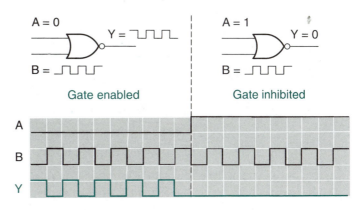

Table 2.17 NAND Truth Table Showing Enable/Inhibit Properties

A	B	Y	
0	0	1	$(Y = 1)$
0	1	1	Inhibit
1	0	1	$(Y = \overline{B})$
1	1	0	Enable

Table 2.18 NOR Truth Table Showing Enable/Inhibit Properties

A	B	Y	
0	0	1	$(Y = \overline{B})$
0	1	0	Enable
1	0	0	$(Y = 0)$
1	1	0	Inhibit

The truth table for the NAND gate is shown in Table 2.17, divided in half to show the enable and inhibit properties of the gate.

Table 2.18 shows the NOR gate truth table, divided in half to show its enable and inhibit properties.

Figures 2.30 and 2.31 show that when the NAND and NOR gates are enabled, the Signal and output waveforms are opposite to one another; we say that they are **out of phase.**

Compare the enable/inhibit waveforms of the AND, OR, NAND, and NOR gates. Gates of the same shape are enabled by the same Control level. AND and NAND gates are enabled by a HIGH on the Control input and inhibited by a LOW. OR and NOR are the opposite. A HIGH Control input inhibits the OR/NOR; a LOW Control input enables the gate.

Exclusive OR and Exclusive NOR Gates

Neither the XOR nor the XNOR gate has an inhibit state. The Control input on both of these gates acts only to determine whether the output waveform will be in or out of phase with the input signal. Figure 2.32 shows the dynamic properties of an XOR gate.

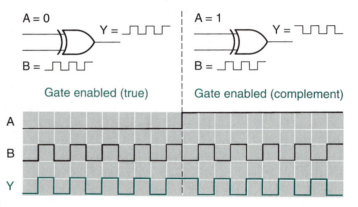

FIGURE 2.32
Dynamic Properties of an Exclusive OR Gate

Table 2.19 XOR Truth Table Showing Dynamic Properties

A	B	Y	
0	0	0	$(Y = B)$
0	1	1	Enable
1	0	1	$(Y = \overline{B})$
1	1	0	Enable

The truth table for the XOR gate, showing the gate's dynamic properties, is given in Table 2.19.

Notice that when $A = 0$, the output is in phase with B and when $A = 1$, the output is out of phase with B. A useful application of this property is to use an XOR gate as a programmable inverter. When $A = 1$, the gate is an inverter; when $A = 0$, it is a noninverting buffer.

The XNOR gate has properties similar to the XOR gate. That is, an XNOR has no inhibit state, and the Control input switches the output in and out of phase with the Signal waveform, although not the same way as an XOR gate does. You will derive these properties in one of the end-of-chapter problems.

Table 2.20 summarizes the enable/inhibit properties of the six gates examined above.

Table 2.20 Summary of Enable/Inhibit Properties

Control	AND	OR	NAND	NOR	XOR	XNOR
$A = 0$	$Y = 0$	$Y = B$	$Y = 1$	$Y = \overline{B}$	$Y = B$	$Y = \overline{B}$
$A = 1$	$Y = B$	$Y = 1$	$Y = \overline{B}$	$Y = 0$	$Y = \overline{B}$	$Y = B$

▌▌ SECTION 2.5 REVIEW PROBLEM

2.9 Briefly explain why an AND gate is inhibited by a LOW Control input and an OR gate is inhibited by a HIGH Control input.

Tristate Buffers

KEY TERMS

Tristate buffer   A gate having three possible output states: logic HIGH, logic LOW, and high-impedance.

High-impedance state   The output state of a tristate buffer that is neither logic HIGH nor logic LOW, but is electrically equivalent to an open circuit.

Bus   A common wire or parallel group of wires connecting multiple circuits.

IN ▷ OUT

$\overline{\text{OE}}$

a. Noninverting

IN ▷ OUT

$\overline{\text{OE}}$

b. Inverting

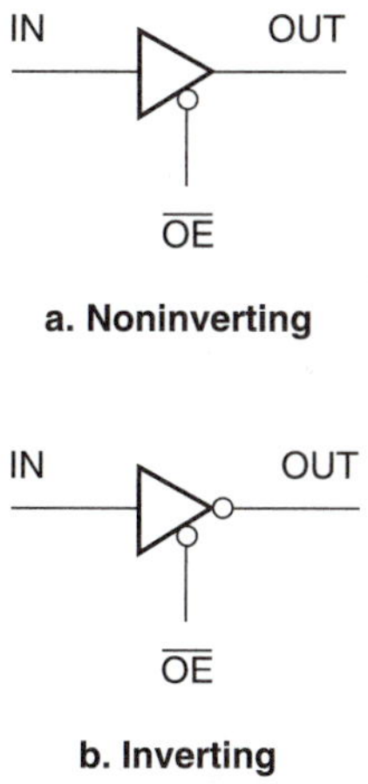

FIGURE 2.33
Tristate Buffers

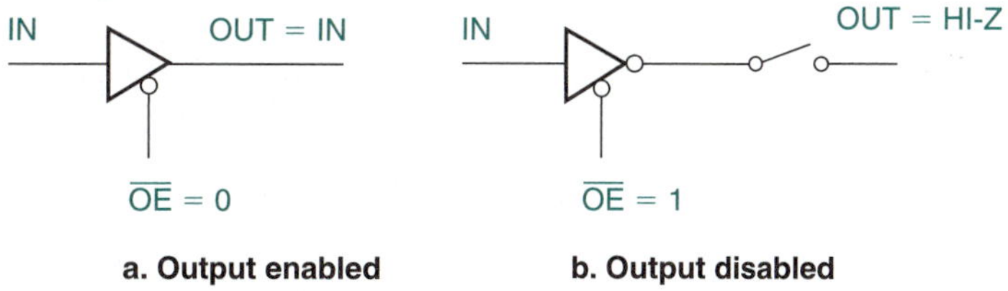

FIGURE 2.34
Electrical Equivalent of Tristate Operation

In the previous section, logic gates were used to enable or inhibit signals in digital circuits. In the AND, NAND, NOR, and OR gates, however, the inhibit state was always logic HIGH or LOW. In some cases, it is desirable to have an output state that is neither HIGH nor LOW, but acts to electrically disconnect the gate output from the circuit. This third state is called the **high-impedance state** and is one of three available states in a class of devices known as **tristate buffers.**

Figure 2.33 shows the logic symbols for two tristate buffers, one with a noninverting output and one with an inverting output. The third input, $\overline{\text{OE}}$ (Output enable), is an active-LOW signal that enables or disables the buffer output.

When $\overline{\text{OE}} = 0$, as shown in Figure 2.34a, the noninverting buffer transfers the input value directly to the output as a logic HIGH or LOW. When $\overline{\text{OE}} = 1$, as in Figure 2.34b, the output is electrically disconnected from any circuit to which it is connected. (The open switch in Figure 2.34b does not literally exist. It is shown as a symbolic representation of the electrical disconnection of the output in the high-impedance state.)

IN ▷ OUT = IN

$\overline{\text{OE}} = 0$

a. Output enabled

IN ▷ OUT = HI-Z

$\overline{\text{OE}} = 1$

b. Output disabled

This type of enable/disable function is particularly useful when digital data are transferred from more than one source to one or more destinations along a common wire (or **bus**), as shown in Figure 2.35. (This is the underlying principle in modern computer systems, where multiple components use the same bus to pass data back and forth.) The destination circuit in Figure 2.35 can receive data from source 1 or source 2. If the source circuits were directly connected to the bus, they could produce contradictory logic levels at the destination. To prevent this, only one source is enabled at a time, with control of this switching left to the two tristate buffers.

FIGURE 2.35
Using Tristate Buffers to Switch Two Sources to a Single Destination

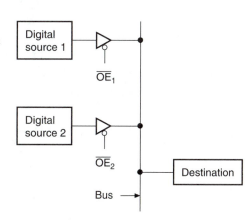

2.6 Integrated Circuit Logic Gates

KEY TERMS

Integrated circuit (IC)   An electronic circuit having many components, such as transistors, diodes, resistors, and capacitors, in a single package.

Small scale integration (SSI)   An integrated circuit having 12 or fewer gates in one package.

Medium scale integration (MSI)   An integrated circuit having the equivalent of 12 to 100 gates in one package.

Large scale integration (LSI)   An integrated circuit having from 100 to 10,000 equivalent gates.

Very large scale integration (VLSI)   An integrated circuit having more than 10,000 equivalent gates.

Transistor-transistor logic (TTL)   A family of digital logic devices whose basic element is the bipolar junction transistor.

Complementary metal-oxide-semiconductor (CMOS)   A family of digital logic devices whose basic element is the metal-oxide-semiconductor field effect transistor (MOSFET).

Chip   An integrated circuit. Specifically, a chip of silicon on which an integrated circuit is constructed.

Dual in-line package (DIP)   A type of IC with two parallel rows of pins for the various circuit inputs and outputs.

Printed circuit board (PCB)   A circuit board in which connections between components are made with lines of copper on the surfaces of the circuit board.

Breadboard   A circuit board for wiring temporary circuits, usually used for prototypes or laboratory work.

Wire-wrap   A circuit construction technique in which the connecting wires are wrapped around the posts of a special chip socket, usually used for prototyping or laboratory work.

Through-hole   A means of mounting DIP ICs on a circuit board by inserting the IC leads through holes in the board and soldering them in place.

Surface-mount technology (SMT)   A system of mounting and soldering integrated circuits on the surface of a circuit board, as opposed to inserting their leads through holes on the board.

Small outline IC (SOIC)   An IC package similar to a DIP, but smaller, which is designed for automatic placement and soldering on the surface of a circuit board. Also called gull-wing, for the shape of the package leads.

Thin shrink small outline package (TSSOP)   A thinner version of an SOIC package.

Plastic leaded chip carrier (PLCC)   A square IC package with leads on all four sides designed for surface mounting on a circuit board. Also called J-lead, for the profile shape of the package leads.

Quad flat pack (QFP)   A square surface-mount IC package with gull-wing leads.

Ball grid array (BGA)   A square surface-mount IC package with rows and columns of spherical leads underneath the package.

Data sheet   A printed specification giving details of the pin configuration, electrical properties, and mechanical profile of an electronic device.

Data book   A bound collection of data sheets. A digital logic data book usually contains data sheets for a specific logic family or families.

Portable document format (PDF)   A format for storing published documents in compressed form.

All the logic gates we have looked at so far are available in **integrated circuit** form. Most of these **small scale integration (SSI)** functions are available either in **transistor-transistor logic (TTL)** or **complementary metal-oxide-semiconductor (CMOS)** technologies. TTL and CMOS devices differ not in their logic functions, but in their construction and electrical characteristics.

TTL and CMOS **chips** are designated by an industry-standard numbering system. TTL devices and the more recent members of the CMOS family are numbered according to the general format 74*XXNN,* where *XX* is a family identifier and *NN* identifies the specific logic function. For example, the number 74ALS00 represents a quadruple 2-input NAND device (indicated by 00) in the advanced low power Schottky (ALS) family of TTL. (Earlier versions of CMOS had a different set of unrelated numbers of the form 4*NNN*B or 4*NNN*UB where *NNN* was the logic function designator. The suffixes B and UB stand for buffered and unbuffered, respectively.)

Table 2.21 lists the quadruple 2-input NAND function as implemented in different logic families. These devices all have the same logic function, but different electrical characteristics.

Table 2.21   Part Numbers for a Quad 2-input NAND Gate in Different Logic Families

Part Number	Logic Family
74LS00	Low-power Schottky TTL
74ALS00	Advanced low-power Schottky TTL
74F00	FAST TTL
74HC00	High-speed CMOS
74HCT00	High-speed CMOS (TTL-compatible inputs)
74LVX00	Low-voltage CMOS
74ABT00	Advanced BiCMOS (TTL/CMOS hybrid)

Table 2.22 lists several logic functions available in the high-speed CMOS family. These devices all have the same electrical characteristics, but different logic functions.

Table 2.22   Part Numbers for Different Functions within a Logic Family (High-Speed CMOS)

Part Number	Function
74HC00	Quadruple 2-input NAND
74HC02	Quadruple 2-input NOR
74HC04	Hex inverter
74HC08	Quadruple 2-input AND
74HC32	Quadruple 2-input OR
74HC86	Quadruple 2-input XOR

Until recently, the most common way to package logic gates has been in a plastic or ceramic **dual in-line package,** or DIP, which has two parallel rows of pins. The standard spacing between pins in one row is 0.1″ (or 100 mil). For packages having fewer than 28 pins, the spacing between rows is 0.3″ (or 300 mil). For larger packages, the rows are spaced by 0.6″ (600 mil).

This type of package is designed to be inserted in a **printed circuit board** in one of two says: (a) the pins are inserted through holes in the circuit board and soldered in place; or (b) a socket is soldered to the circuit board and the IC is placed in the socket. The latter method is more expensive, but makes chip replacement much easier. A socket can occasionally cause its own problems by making a poor connection to the pins of the IC.

The DIP is also convenient for laboratory and prototype work, since it can also be inserted easily into a **breadboard,** a special type of temporary circuit board with internal connections between holes of a standard spacing. It is also convenient for **wire-wrapping,** a technique in which a special tool is used to wrap wires around posts on the underside of special sockets.

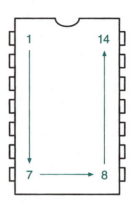

FIGURE 2.36
14-Pin DIP (Top View)

The outline of a 14-pin DIP is shown in Figure 2.36. There is a notch on one end to show the orientation of the pins. When the IC is oriented as shown and viewed from above, pin 1 is at the top left corner and the pins number counterclockwise from that point.

Besides DIP packages, there are numerous other types of packages for digital ICs, including, among others, **small outline IC (SOIC), thin shrink small outline package (TSSOP), plastic leaded chip carrier (PLCC), quad flat pack (QFP),** and **ball grid array (BGA)** packages. They are used mostly in applications where circuit board space is at a premium and in manufacturing processes relying on **surface-mount technology (SMT).** In fact, these devices represent the majority of IC packages found in new designs. Some of these IC packaging options are shown in Figure 2.37.

FIGURE 2.37
Some IC Packaging Options

SMT is a sophisticated technology which relies on automatic placement of chips and soldering of pins onto the surface of a circuit board, not through holes in the circuit board. This technique allows a manufacturer to mount components on both sides of a circuit board.

Primarily due to the great reduction in board space requirements, most new ICs are available only in the newer surface-mount packages and are not being offered at all in the DIP package. However, we will look at DIP offerings in logic gates because they are inexpensive and easy to use with laboratory breadboards and therefore useful as a learning tool.

Logic gates come in packages containing several gates. Common groupings available in DIP packages are six 1-input gates, four 2-input gates, three 3-input gates, or two 4-input gates, although other arrangements are available. The usual way of stating the number of logic gates in a package is to use the numerical prefixes hex (6), quad or quadruple (4), triple (3), or dual (2).

Some common gate packages are listed in Table 2.23.

Table 2.23 Some Common Logic Gate ICs

Gate	Family	Function
74HC00A	High-speed CMOS	Quad 2-input NAND
74HC02	High-speed CMOS	Quad 2-input NOR
74ALS04	Advanced low-power Schottky TTL	Hex inverter
74LS11	Low-power Schottky TTL	Triple 3-input AND
74F20	FAST TTL	Dual 4-input NAND
74HC27	High-speed CMOS	Triple 3-input NOR

Information about pin configurations, electrical characteristics, and mechanical specifications of a part is available in a **data sheet** provided by the chip manufacturer. A collection of data sheets for a particular logic family is often bound together in a **data book.** More recently, device manufacturers have been making data sheets available on their corporate World Wide Web sites in **portable document format (PDF),** readable by a special program such as Adobe Acrobat Reader. Links to some of these manufacturers can be found on the Online Companion Web site for this book. (http://www.electronictech.com)

Figure 2.38 shows the internal diagrams of gates listed in Table 2.23. Notice that the gates can be oriented inside a chip in a number of ways. That is why it is important to confirm pin connections with a data sheet.

In addition to the gate inputs and outputs there are two more connections to be made on every chip: the power (V_{CC}) and ground connections. In TTL, connect V_{CC} to +5 Volts and GND to ground. In CMOS, connect the V_{CC} pin to the supply voltage (+3 V to +6 V) and GND to ground. The gates won't work without these connections.

Every chip requires power and ground. This might seem obvious, but it's surprising how often it is forgotten, especially by students who are new to digital electronics. Probably this is because most digital circuit diagrams don't show the power connections, but assume that you know enough to make them.

The only place a chip gets its required power is through the V_{CC} pin. Even if the power supply is connected to a logic input as a logic HIGH, you still need to connect it to the power supply pin.

Even more important is a good ground connection. A circuit with no power connection will not work at all. A circuit without a ground may appear to work, but it will often produce bizarre errors that are very difficult to detect and repair.

In later chapters, we will work primarily with complex ICs in PLCC packages. The power and ground connections are so important to these chips that they will not be left to chance; they are provided on a specially designed circuit board. Only input and output pins are accessible for connection by the user.

As digital designs become more complex, it is increasingly necessary to follow good practices in board layout and prototyping procedure to ensure even minimal functionality.

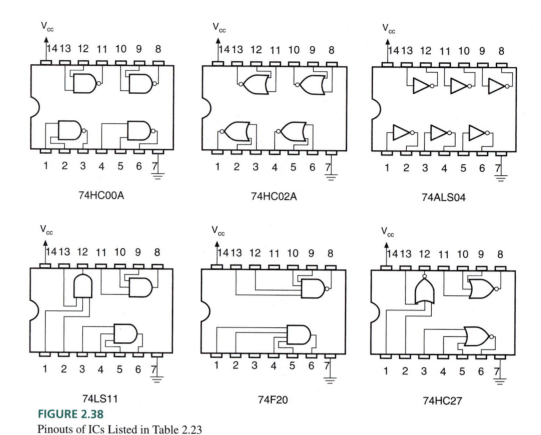

FIGURE 2.38
Pinouts of ICs Listed in Table 2.23

Thus, hardware platforms for prototype and laboratory work will need to be at least partially constructed by the board manufacturer in order to supply the requirements of a stable circuit configuration.

▌▌ SECTION REVIEW PROBLEM FOR SECTION 2.6

2.10.   How are the pins numbered in a dual in-line package?

SUMMARY

1. Digital systems can be analyzed and designed using Boolean algebra, a system of mathematics that operates on variables that have one of two possible values.
2. Any Boolean expression can be constructed from the three simplest logic functions: NOT, AND, and OR.
3. A NOT gate, or inverter, has an output state that is in the opposite logic state of the input.
4. The main 2-input logic functions are described as follows, for inputs A and B and output Y:

    AND: Y is HIGH if A AND B are HIGH. ($Y = A \cdot B$)

    OR: Y is HIGH if A OR B is HIGH. ($Y = A + B$)

    NAND: Y is LOW if A AND B are HIGH. ($Y = \overline{A \cdot B}$)

    NOR: Y is LOW if A OR B is HIGH. ($Y = \overline{A + B}$)

    XOR: Y is HIGH if A OR B is HIGH, but not if both are HIGH. ($Y = A \oplus B$)

    XNOR: Y is LOW if A OR B is HIGH, but not if both are HIGH. ($Y = \overline{A \oplus B}$)

5. The function of a logic gate can be represented by a truth table, a list of all possible inputs in binary order and the output corresponding to each input state.
6. DeMorgan's theorems ($\overline{A \cdot B} = \overline{A} + \overline{B}$ and $\overline{A + B} = \overline{A} \cdot \overline{B}$) allow us to represent any gate in an AND form and an OR form.
7. To change a gate into its DeMorgan equivalent form, change its shape from AND to OR or vice versa and change the active levels of inputs and output.

8. A logic switch can be created from a single-pole single-throw switch by grounding one end and tying the other end to V_{CC} through a pull-up resistor. The logic level is available on the same side of the switch as the resistor. An open switch is HIGH and a closed switch is LOW. A similar circuit can be made with a pushbutton switch.

9. A light emitting diode (LED) can be used to indicate logic HIGH or LOW levels. To indicate a HIGH, ground the cathode through a series resistor (about 470 Ω for a 5-volt power supply) and apply the logic level to the anode. To indicate a LOW, tie the anode to V_{CC} through a series resistor and apply the logic level to the cathode.

10. Logic gates can be used to pass or block digital signals. For example, an AND gate will pass a digital signal applied to input B if the input A is HIGH ($Y = B$). If input A is LOW, the signal is blocked and the gate output is always LOW ($Y = 0$). Similar properties apply to other gates, as summarized in Table 2.20.

11. Tristate buffers have outputs that generate logic HIGH and LOW when enabled and a high-impedance state when disabled. The high-impedance state is electrically equivalent to an open circuit.

12. Logic gates are available as integrated circuits in a variety of packages. Packages that have fewer than 12 gates are called small scale integration (SSI) devices.

13. Many logic functions have an industry-standard part number of the form 74*XXNN*, where *XX* is an alphabetic family designator and *NN* is a numeric function designator (e.g. 74HC02 = Quadruple 2-input NOR gate in the high-speed CMOS family).

14. Some common IC packages include dual in-line package (DIP), small outline IC (SOIC), thin shrink small outline package (TSSOP), plastic leaded chip carrier (PLCC), quad flat pack (QFP), and ball grid array (BGA) packages.

15. Most new IC packages are for surface mounting on a printed circuit board. These have largely replaced DIPs in through-hole circuit boards, due to better use of board space.

16. IC pin connections and functional data can be determined from manufacturers' data sheets, available in paper format or electronically via the Internet.

17. All ICs require power and ground, which must be applied to special power supply pins on the chip.

GLOSSARY

Active HIGH An active-HIGH terminal is considered "ON" when it is in the logic HIGH state. Indicated by the absence of a bubble at the terminal in distinctive-shape symbols.

Active level A logic level defined as the "ON" state for a particular circuit input or output. The active level can be either HIGH or LOW.

Active LOW An active-LOW terminal is considered "ON" when it is in the logic LOW state. Indicated by a bubble at the terminal in distinctive-shape symbols.

AND gate A logic circuit whose output is HIGH when *all* inputs (e.g., *A* AND *B* AND *C*) are HIGH.

Ball grid array (BGA) A square surface-mount IC package with rows and columns of spherical leads underneath the package.

Boolean algebra A system of algebra that operates on Boolean variables. The binary (two-state) nature of Boolean algebra makes it useful for analysis, simplification, and design of combinational logic circuits.

Boolean expression An algebraic expression made up of Boolean variables and operators, such as AND ($\cdot$), OR ($+$), or NOT ($^{-}$). Also referred to as a **Boolean function** or a **logic function.**

Boolean variable A variable having only two possible values, such as HIGH/LOW, 1/0, On/Off, or True/False.

Breadboard A circuit board for wiring temporary circuits, usually used for prototypes or laboratory work.

Bubble A small circle indicating logical inversion on a circuit symbol.

Buffer An amplifier that acts as a logic circuit. Its output can be inverting or noninverting.

Bus A common wire or parallel group of wires connecting multiple circuits.

Chip An integrated circuit. Specifically, a chip of silicon on which an integrated circuit is constructed.

Clock generator A circuit that generates a periodic digital waveform.

Coincidence gate An Exclusive NOR gate.

Complement form Inverted.

Complementary metal-oxide-semiconductor (CMOS) A family of digital logic devices whose basic element is the metal-oxide-semiconductor field effect transistor (MOSFET).

Data book A bound collection of data sheets. A digital logic data book usually contains data sheets for a specific logic family or families.

Data sheet A printed specification giving details of the pin configuration, electrical properties, and mechanical profile of an electronic device.

DeMorgan equivalent forms Two gate symbols, one AND-shaped and one OR-shaped, that are equivalent according to De-Morgan's theorems.

DeMorgan's theorems Two theorems in Boolean algebra that allow us to transform any gate from an AND-shaped to an OR-shaped gate and vice versa.

Digital signal (or pulse waveform) A series of 0s and 1s plotted over time.

Distinctive-shape symbols Graphic symbols for logic circuits that show the function of each type of gate by a special shape.

Dual in-line package (DIP) A type of IC with two parallel rows of pins for the various circuit inputs and outputs.

Enable A logic gate is enabled if it allows a digital signal to pass from an input to the output in either true or complement form.

Exclusive NOR gate A two-input logic circuit whose output is the complement of an Exclusive OR gate.

Exclusive OR gate A two-input logic circuit whose output is HIGH when one input (but not both) is HIGH.

Floating An undefined logic state, neither HIGH nor LOW.

High-impedance state The output state of a tristate buffer that is neither logic HIGH nor logic LOW, but is electrically equivalent to an open circuit.

IEEE/ANSI Standard 91-1984 A standard format for drawing logic circuit symbols as rectangles with logic functions shown by a standard notation inside the rectangle for each device.

In phase Two digital waveforms are in phase if they are always at the same logic level at the same time.

Inhibit (or disable) A logic gate is inhibited if it prevents a digital signal from passing from an input to the output.

Integrated circuit (IC) An electronic circuit having many components, such as transistors, diodes, resistors, and capacitors, in a single package.

Inverter Also called a NOT gate or an inverting buffer. A logic gate that changes its input logic level to the opposite state.

Large scale integration (LSI) An integrated circuit having from 100 to 10,000 equivalent gates.

LED Light emitting diode. An electronic device that conducts current in one direction only and illuminates when it is conducting.

Logic function See Boolean expression.

Logic gate An electronic circuit that performs a Boolean algebraic function.

Logical product AND function.

Logical sum OR function.

Medium scale integration (MSI) An integrated circuit having the equivalent of 12 to 100 gates in one package.

NAND gate A logic circuit whose output is LOW when *all* inputs are HIGH.

NOR gate A logic circuit whose output is LOW when *at least one* input is HIGH.

OR gate A logic circuit whose output is HIGH when *at least one* input (e.g., A OR B OR C) is HIGH.

Out of phase Two digital waveforms are out of phase if they are always at opposite logic levels at any given time.

Plastic leaded chip carrier (PLCC) A square IC package with leads on all four sides designed for surface mounting on a circuit board. Also called J-lead, for the profile shape of the package leads.

Portable document format (PDF) A format for storing published documents in compressed form.

Printed circuit board (PCB) A circuit board in which connections between components are made with lines of copper on the surfaces of the circuit board.

Pull-up resistor A resistor connected from a point in an electronic circuit to the power supply of that circuit. In a digital circuit it supplies the required logic level in a HIGH state and limits current from the power supply in the LOW state.

Quad flat pack (QFP) A square surface-mount IC package with gull-wing leads.

Qualifying symbol A symbol in IEEE/ANSI logic circuit notation, placed in the top center of a rectangular symbol, that shows the function of a logic gate. Some qualifying symbols include: 1 = "buffer"; & = "AND"; ≥1 = "OR"

Rectangular-outline symbols Rectangular logic gate symbols that conform to IEEE/ANSI Standard 91-1984.

Small outline IC (SOIC) An IC package similar to a DIP, but smaller, which is designed for automatic placement and soldering on the surface of a circuit board. Also called gull-wing, for the shape of the package leads.

Small-scale integration (SSI) An integrated circuit having 12 or fewer gates in one package.

Surface-mount technology (SMT) A system of mounting and soldering integrated circuits on the surface of a circuit board, as opposed to inserting their leads through holes on the board.

Thin shrink small outline package (TSSOP) A thinner version of an SOIC package.

Through-hole A means of mounting DIP ICs on a circuit board by inserting the IC leads through holes in the board and soldering them in place.

Transistor-transistor logic (TTL) A family of digital logic devices whose basic element is the bipolar junction transistor.

Tristate buffer A gate having three possible output states: logic HIGH, logic LOW, and high-impedance.

True form Not inverted.

Truth table A list of all possible input values to a digital circuit, listed in ascending binary order, and the output response for each input combination.

V_{CC} The power supply voltage in a transistor-based electronic circuit. The term often refers to the power supply of digital circuits.

Very large scale integration (VLSI) An integrated circuit having more than 10,000 equivalent gates.

PROBLEMS

Problem numbers set in color indicate more difficult problems: those with underlines indicate most difficult problems.

Section 2.1 Basic Logic Functions

2.1 Draw the symbol for the NOT gate (inverter) in both rectangular-outline and distinctive-shape forms.

2.2 Draw the distinctive-shape and rectangular-outline symbols for a 3-input AND gate.

2.3 Draw the distinctive-shape and rectangular-outline symbols for a 3-input OR gate.

2.4 Write a sentence that describes the operation of a 4-input AND gate that has inputs P, Q, R, and S and output T. Make the truth table of this gate and draw an asterisk beside the line(s) of the truth table indicating when the gate output is in its active state.

2.5 Write a sentence that describes the operation of a 4-input OR gate with inputs *J*, *K*, *L*, and *M* and output *N*. Make the truth table of this gate and draw an asterisk beside the line(s) of the truth table indicating when the gate output is in its active state.

2.6 State how three switches must be connected to represent a 3-input AND function. Draw a circuit diagram showing how this function can control a lamp.

2.7 State how four switches must be connected to represent a 4-input OR function. Draw a circuit diagram showing how this function can control a lamp.

Section 2.2 Logic Switches and LED Indicators

2.8 Sketch the circuit of a single-pole single-throw (SPST) switch used as a logic switch. Briefly explain how it works.

2.9 Refer to Figure 2.10 (logic pushbuttons). Should the normally open pushbutton be considered an active HIGH or active LOW device? Briefly explain your choice.

2.10 Should the normally closed pushbutton be considered an active HIGH or active LOW device? Why?

2.11 Briefly state what is required for an LED to illuminate.

2.12 Briefly state the relationship between the brightness of an LED and the current flowing through it. Why is a series resistor required?

2.13 Draw a circuit showing how an OR-gate output will illuminate an LED when the gate output is LOW. Assume the required series resistor is 470 Ω.

Section 2.3 Derived Logic Functions

2.14 For a 4-input NAND gate with inputs *A*, *B*, *C*, and *D* and output *Y*:

   a. Write the truth table and a descriptive sentence.

   b. Write the Boolean expression.

   c. Draw the logic circuit symbol in both distinctive-shape and rectangular-outline symbols.

2.15 Repeat Problem 2.14 for a 4-input NOR gate.

2.16 State the active levels of the inputs and outputs of a NAND gate and a NOR gate.

2.17 Write a descriptive sentence of the operation of a 5-input NAND gate with inputs *A*, *B*, *C*, *D*, and *E* and output *Y*. How many lines would the truth table of this gate have?

2.18 Repeat Problem 2.17 for a 5-input NOR gate.

2.19 A pump motor in an industrial plant will start only if the temperature and pressure of liquid in a tank exceed a certain level. The temperature sensor and pressure sensor, shown in Figure 2.39 each produce a logic HIGH if the measured quantities exceed this value. The logic circuit interface produces a HIGH output to turn on the motor. Draw the symbol and truth table of the gate that corresponds to the action of the logic circuit.

2.20 Repeat Problem 2.19 for the case in which the motor is activated by a logic LOW.

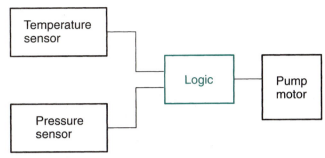

FIGURE 2.39
Problem 2.19: Temperature and Pressure Sensors

2.21 Figure 2.40 shows a circuit for a two-way switch for a stairwell. This is a common circuit that allows you to turn

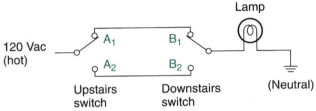

FIGURE 2.40
Problem 2.21: Circuit for Two-Way Switch

on a light from either the top or the bottom of the stairwell and off at the other end. The circuit also allows anyone coming along after you to do the same thing, no matter which direction they are coming from.

The lamp is ON when the switches are in the same positions and OFF when they are in opposite positions. What logic function does this represent? Draw the truth table of the function and use it to explain your reasoning.

2.22 Find the truth table for the logic circuit shown in Figure 2.41.

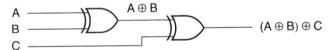

FIGURE 2.41
Problem 2.22: Logic Circuit

2.23 Recall the description of a 2-input Exclusive OR gate: "Output is HIGH if one input is HIGH, but not both." This is not the best statement of the operation of a multiple-input XOR gate. Look at the truth table derived in Problem 2.22 and write a more accurate description of *n*-input XOR operation.

Section 2.4 DeMorgan's Theorems and Gate Equivalence

2.24 For each of the gates in Figure 2.42:

   a. Write the truth table.

   b. Indicate with an * which lines on the truth table show the gate output in its active state.

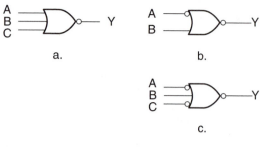

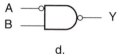

FIGURE 2.42
Problem 2.24: Logic Gates

    c. Convert the gate to its DeMorgan equivalent form.

    d. Rewrite the truth table and indicate which lines on the truth table show output active states for the DeMorgan equivalent form of the gate.

2.25 Refer to Figure 2.43. State which two gates of the three shown are DeMorgan equivalents of each other. Explain your choice.

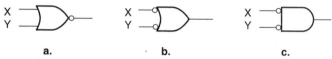

FIGURE 2.43
Problem 2.25: Logic Gates

Section 2.5 Enable and Inhibit Properties of Logic Gates

2.26 Draw the output waveform of the Exclusive NOR gate when a square waveform is applied to one input and

    a. The other input is held LOW

    b. The other input is held HIGH

How does this compare to the waveform that would appear at the output of an Exclusive OR gate under the same conditions?

2.27 Sketch the input waveforms represented by the following 32-bit sequences (use 1/4-inch graph paper, 1 square per bit):

    A: 0000 0000 0000 1111 1111 1111 1111 0000

    B: 1010 0111 0010 1011 0101 0011 1001 1011

Assume that these waveforms represent inputs to a logic gate. (Spaces are provided for readability only.) Sketch the waveform for gate output Y if the gate function is:

    a. AND

    b. OR

    c. NAND

    d. NOR

    e. XOR

    f. XNOR

2.28 Repeat Problem 2.27 for the waveforms shown in Figure 2.44.

2.29 The A and B waveforms shown in Figure 2.45 are inputs to an OR gate. Complete the sketch by drawing the waveform for output Y.

2.30 Repeat Problem 2.29 for a NOR gate.

2.31 Figure 2.46 shows a circuit that will make a lamp flash at 3 Hz when the gasoline level in a car's gas tank drops below a certain point. A float switch in the tank monitors the level of gasoline. What logic level must the float switch produce to make the light flash when the tank is approaching empty? Why?

2.32 Repeat Problem 2.31 for the case where the AND gate is replaced by a NOR gate.

2.33 Will the circuit in Figure 2.46 work properly if the AND gate is replaced by an Exclusive OR gate? Why or why not?

2.34 Make a truth table for the tristate buffers shown in Figure 2.33. Indicate the high-impedance state by the notation

FIGURE 2.44
Problem 2.28: Input Waveforms

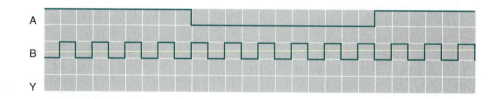

FIGURE 2.45
Problem 2.29: Waveforms

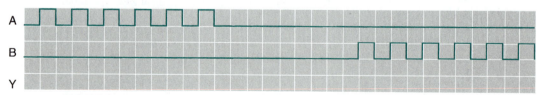

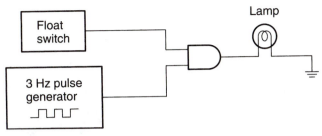

FIGURE 2.46
Problem 2.31: Gasoline Level Circuit

"Hi-Z" How do the enable properties of these gates differ from gates such as AND and NAND?

Section 2.6   Integrated Circuit Logic Gates

2.35 Name two logic families used to implement digital logic functions. How do they differ?

2.36 List the industry-standard numbers for a quadruple 2-input NAND gate in low power Schottky TTL, CMOS, and high-speed CMOS technologies.

2.37 Repeat Problem 2.36 for a quadruple 2-input NOR gate. How does each numbering system differentiate between the NAND and NOR functions?

2.38 List six types of packaging that a logic gate could come in.

ANSWERS TO SECTION REVIEW PROBLEMS

Section 2.1

2.1 AND: "*A* AND *B* AND *C* AND *D* must be HIGH to make *Y* HIGH."   **2.2.** OR: "*A* OR *B* OR *C* OR *D* must be HIGH to make *Y* HIGH."

Section 2.2

2.3 When the switch is open, it provides a logic HIGH because of the pull-up resistor. A closed switch is LOW, due to the connection to ground.

Section 2.3

2.4 XOR;   **2.5** NAND;   **2.6** NOR;   **2.7** XNOR.

Section 2.4

2.8 $Y = \overline{A\,B\,C\,D}$

Section 2.5

2.9 An AND needs two HIGH inputs to make a HIGH output. If the Control input is LOW, the output can never be HIGH; the output remains LOW. An OR output is HIGH if one input is HIGH. If the Control input is HIGH, the output is always HIGH, regardless of the level at the Signal input. In both cases, the output is "stuck" at one level, signifying that the gate is inhibited.

Section 2.6

2.10 Viewed from above, with the notch in the package away from you, pin 1 is on the left side at the far end. The pins are numbered counterclockwise from that point.

Boolean Algebra and Combinational Logic

CHAPTER OBJECTIVES

Upon successful completion of this chapter you will be able to:

- Explain the relationship between the Boolean expression, logic diagram, and truth table of a logic gate network and be able to derive any one from either of the other two.

- Draw logic gate networks in such a way as to cancel out internal inversions automatically (bubble-to-bubble convention).

- Write the sum of products (SOP) or product of sums (POS) forms of a Boolean equation.

- Use rules of Boolean algebra to simplify the Boolean expressions derived from logic diagrams and truth tables.

- Apply the Karnaugh map method to reduce Boolean expressions and logic circuits to their simplest forms.

In Chapter 3, we will examine the rudiments of **combinational logic.** A combinational logic circuit is one in which two or more gates are connected together to combine several Boolean inputs. These circuits can be represented several ways, as a logic diagram, truth table, or Boolean expression.

A Boolean expression for a network of logic gates is often not in its simplest form. In such a case, we may be using more components than would be required for the job, so it is of benefit to us if we can simplify the Boolean expression. Several tools are available to us, such as Boolean algebra and a graphical technique known as Karnaugh mapping. We can also simplify the Boolean expression by taking care to draw the logic diagrams in such a way as to automatically eliminate inverting functions within the circuit. ∎

3.1 Boolean Expressions, Logic Diagrams and Truth Tables

KEY TERMS

Logic gate network Two or more logic gates connected together.

Logic diagram A diagram, similar to a schematic, showing the connection of logic gates.

Combinational logic Digital circuitry in which an output is derived from the combination of inputs, independent of the order in which they are applied.

Combinatorial logic Another name for combinational logic.

In Chapter 2, we examined the functions of single logic gates. However, most digital circuits require multiple gates. When two or more gates are connected together, they form a **logic gate network.** These networks can be described by a truth table, a **logic diagram** (i.e., a circuit diagram), or a Boolean expression. Any one of these can be derived from any other.

A digital circuit built from gates is called a **combinational** (or **combinatorial**) **logic** circuit. The output of a combinational circuit depends on the *combination* of inputs. The inputs can be applied in any sequence and still produce the same result. For example, an AND gate output will always be HIGH if all inputs are HIGH, regardless of the order in which they became HIGH. This is in contrast to **sequential logic,** in which sequence matters; a sequential logic output may have a different value with two identical sets of inputs if those inputs were applied in a different order. We will study sequential logic in a later chapter.

Boolean Expressions from Logic Diagrams

KEY TERMS

Bubble-to-bubble convention The practice of drawing gates in a logic diagram so that inverting outputs connect to inverting inputs and noninverting outputs connect to noninverting inputs.

Order of precedence The sequence in which Boolean functions are performed, unless otherwise specified by parentheses.

Writing the Boolean expression of a logic gate network is similar to finding the expression for a single gate. The difference is that in a multiple gate network, the inputs will usually not consist of single variables, but compound expressions that represent outputs of previous gates.

These compound expressions are combined according to the same rules as single variables. In an OR gate, with inputs x and y, the output will always be $x + y$ regardless of whether x and y are single variables (e.g., $x = A$, $y = B$, output $= A + B$) or compound expressions (e.g., $x = AB$, $y = AC$, output $= AB + AC$).

Figure 3.1 shows a simple logic gate network, consisting of a single AND and a single OR gate. The AND gate combines inputs A and B to give the output expression AB. The OR combines the AND function and input C to yield the compound expression $AB + C$.

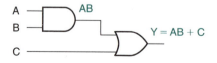

FIGURE 3.1
Boolean Expression from a Gate Network

EXAMPLE 3.1

Derive the Boolean expression of the logic gate network shown in Figure 3.2a.

FIGURE 3.2
Example 3.1

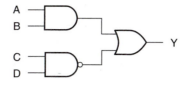

a. Logic gate network

Solution

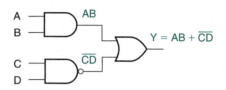

b. Boolean expression from logic gate network

Figure 3.2b shows the gate network with the output terms indicated for each gate. The AND and NAND functions are combined in an OR function to yield the output expression:

$$Y = AB + \overline{CD}$$

The Boolean expression in Example 3.1 includes a NAND function. It is possible to draw the NAND in its DeMorgan equivalent form. If we choose the gate symbols so that outputs with bubbles connect to inputs with bubbles, we will not have bars over groups of variables, except possibly one bar over the entire function. In a circuit with many inverting functions (NANDs and NORs), this results in a cleaner notation and often a clearer idea of the function of the circuit. We will follow this notation, which we will refer to as the **bubble-to-bubble convention,** as much as possible.

EXAMPLE 3.2

Redraw the circuit in Figure 3.2 to conform to the bubble-to-bubble convention. Write the Boolean expression of the new logic diagram.

Solution

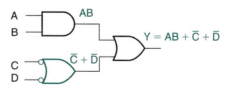

FIGURE 3.3
Example 3.2
Using DeMorgan Equivalents to Simplify a Circuit

Figure 3.3 shows the new circuit. The NAND has been converted to its DeMorgan equivalent so that its active-HIGH output drives an active-HIGH input on the OR gate. The new Boolean expression is $Y = AB + \overline{C} + \overline{D}$.

Boolean functions are governed by an **order of precedence.** Unless otherwise specified, AND functions are performed first, followed by ORs. This order results in a form similar to that of linear algebra, where multiplication is performed before addition, unless otherwise specified.

Figure 3.4 shows two logic diagrams, one whose Boolean expression requires parentheses and one that does not.

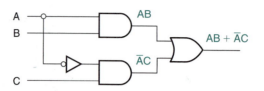

a. No parentheses required (AND, then OR)

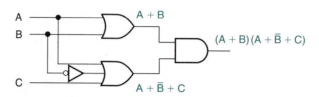

b. Parentheses required (OR, then AND)

FIGURE 3.4
Order of Precedence

The AND functions in Figure 3.4a are evaluated first, eliminating the need for parentheses in the output expression. The expression for Figure 3.4b requires parentheses since the ORs are evaluated first.

▌▌ EXAMPLE 3.3

Write the Boolean expression for the logic diagrams in Figure 3.5.

FIGURE 3.5
Example 3.5
Order of Precedence

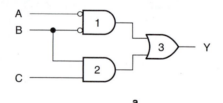

a.

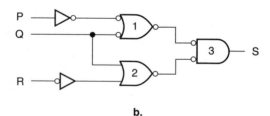

b.

Solution Examine the output of each gate and combine the resultant terms as required.

Figure 3.5a: Gate 1: $\overline{A} \cdot \overline{B}$

Gate 2: $B \cdot C$

Gate 3: $Y =$ Gate 1 + Gate2 $= \overline{A} \cdot \overline{B} + B \cdot C$

Figure 3.5b: Gate 1: $\overline{\overline{P} + \overline{Q}} = P + \overline{Q}$

Gate 2: $\overline{Q + R}$

Gate 3: $S = \overline{\text{Gate1} \cdot \text{Gate2}} = \overline{\overline{(P + \overline{Q})}\overline{(Q + R)}} = (P + \overline{Q})(Q + \overline{R})$

Note that when two bubbles touch, they cancel out, as in the doubly inverted *P* input or the connection between the outputs of gates 1 and 2 and the inputs of gate 3. *In the resultant Boolean expression, bars of the same length cancel; bars of unequal length do not.*

▌▌ SECTION 3.1A REVIEW PROBLEM

3.1 Write the Boolean expression for the logic diagrams in Figure 3.6, paying attention to the rules of order of precedence.

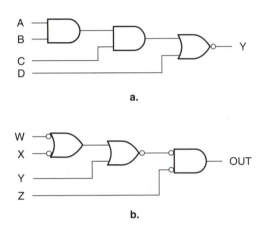

a.

b.

FIGURE 3.6
Section Review Problem 3.1

Logic Diagrams from Boolean Expressions

> **KEY TERMS**
>
> **Levels of gating**   The number of gates through which a signal must pass from input to output of a logic gate network.
>
> **Double-rail inputs**   Boolean input variables that are available to a circuit in both true and complement form.
>
> **Synthesis**   The process of creating a logic circuit from a description such as a Boolean equation or truth table.

We can derive a logic diagram from a Boolean expression by applying the order of precedence rules. We examine an expression to create the first **level of gating** from the circuit inputs, then combine the output functions of the first level in the second level gates, and so forth. Input inverters are often not counted as a gating level, as we usually assume that each variable is available in both true (noninverted) and complement (inverted) form. When input variables are available to a circuit in true and complement form, we refer to them as **double-rail inputs.**

The first level usually will be AND gates if no parentheses are present, OR gates if parentheses are used. (Not always, however; parentheses merely tell us which functions to **synthesize** first.) Although we will try to eliminate bars over groups of variables by use of DeMorgan's theorems and the bubble-to-bubble convention, we should recognize that a bar over a group of variables is the same as having those variables in parentheses.

Let us examine the Boolean expression $Y = AC + BD + AD$. Order of precedence tells us that we synthesize the AND functions first. This yields three 2-input AND gates, with outputs *AC, BD,* and *AD,* as shown in Figure 3.7a. In the next step, we combine these AND functions in a 3-input OR gate, as shown in Figure 3.7b.

FIGURE 3.7
Logic Diagram for
$Y = AC + BD + AD$

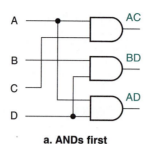

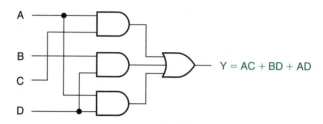

a. ANDs first

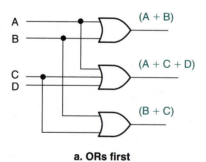

b. Combine ANDs in an OR gate

When the expression has OR functions in parentheses, we synthesize the ORs first, as for the expression $Y = (A + B)(A + C + D)(B + C)$. Figure 3.8 shows this process. In the first step, we synthesize three OR gates for the terms $(A + B)$, $(A + C + D)$, and $(B + C)$. We then combine these terms in a 3-input AND gate.

FIGURE 3.8
Logic Diagram for $Y = (A + B)$
$(A + C + D) (B + C)$

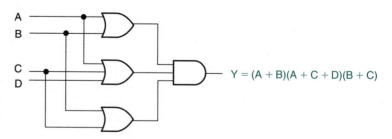

a. ORs first

b. Combine ORs in an AND gate

■■ EXAMPLE 3.4

Synthesize the logic diagrams for the following Boolean expressions:

1. $P = Q\overline{RS} + \overline{S}T$
2. $X = (W + Z + Y)\overline{V} + (\overline{W} + V)\overline{Y}$

Solution

1. Recall that a bar over two variables acts like parentheses. Thus the $Q\overline{RS}$ term is synthesized from a NAND, then an AND, as shown in Figure 3.9a. Also shown is the second AND term, $\overline{S}T$.

Figure 3.9b shows the terms combined in an OR gate.

FIGURE 3.9

Example 3.4
Logic Diagram of
$P = Q\overline{RS} + \overline{S}T$

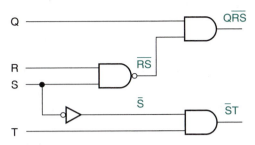

a. Combine inputs (NAND, then AND)

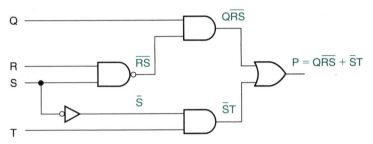

b. First and second level gates combined in and OR

2. Figure 3.10 shows the synthesis of the second logic diagram in three stages. Figure 3.10a shows how the circuit inputs are first combined in two OR gates. We do this first because the ORs are in parentheses. In Figure 3.10b, each of these functions is combined in an AND gate, according to the normal order of precedence. The AND outputs are combined in a final OR function, as shown in Figure 3.10c.

FIGURE 3.10

Example 3.4
Logic Diagram for $X =$
$(W + Z + Y)\overline{V} + (\overline{W} + V)\overline{Y}$

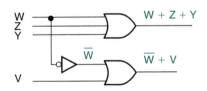

a. ORs first (parentheses)

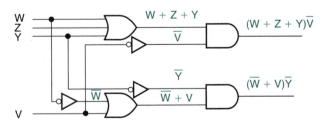

b. Combine with ANDs (order of precedence)

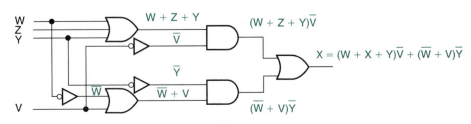

c. Find output (OR)

▌▌ EXAMPLE 3.5

Use DeMorgan's theorem to modify the Boolean equation in part 1 of Example 3.4 so that there is no bar over any group of variables. Redraw Figure 3.9b to reflect the change.

Solution

$$P = Q\overline{RS} + \overline{ST} = Q(\overline{R} + \overline{S}) + \overline{ST}$$

Figure 3.11a shows the modified logic diagram. The levels of gating could be further reduced from three to two (not counting input inverters) by "multiplying through" the parentheses to yield the expression:

$$P = Q\overline{R} + Q\overline{S} + \overline{ST}$$

Figure 3.11b shows the logic diagram for this form. We will examine this simplification procedure more formally in a later section of this chapter.

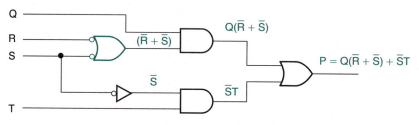

a. Logic diagram of P = Q($\overline{R}$ + $\overline{S}$) + $\overline{ST}$

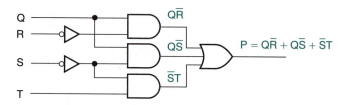

b. Logic diagram of P = Q$\overline{R}$ + Q$\overline{S}$ + $\overline{ST}$

FIGURE 3.11
Example 3.5: Reworking Figure 3.9b

Truth Tables from Logic Diagrams or Boolean Expressions

There are two basic ways to find a truth table from a logic diagram. We can examine the output of each gate in the circuit and develop its truth table. We then use our knowledge of gate properties to combine these intermediate truth tables into the final output truth table. Alternatively, we can develop a Boolean expression for the logic diagram and by examining the expression fill in the truth table in a single step. The former method is more thorough and probably easier to understand when you are learning the technique. The latter method is more efficient, but requires some practice and experience. We will look at both.

Examine the logic diagram in Figure 3.12. Since there are three binary inputs, there will be eight ways those inputs can be combined. Thus, we start by making an 8-line truth table, as in Table 3.1.

FIGURE 3.12
Logic Diagram for AB + C

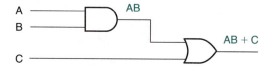

The OR gate output will describe the function of the whole circuit. In order to assess the OR function, we must first evaluate the AND output. We add a column to the truth table for the AND gate and look for the lines in the table where both *A* AND *B* equal logic 1 (in this case, the last two rows). For these lines, we write a 1 in the *AB* column. Next, we look at the values in column *C* and the *AB* column. If there is a 1 in either column, we write a 1 in the column for the final output.

Table 3.1   Truth Table for Figure 3.12

A	*B*	*C*	*AB*	*AB* + *C*
0	0	0	0	0
0	0	1	0	1
0	1	0	0	0
0	1	1	0	1
1	0	0	0	0
1	0	1	0	1
1	1	0	1	1
1	1	1	1	1

▌▌ EXAMPLE 3.6

Derive the truth table for the logic diagram shown in Figure 3.13.

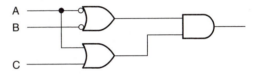

FIGURE 3.13
Example 3.6
Logic Diagram

Solution   The Boolean equation for Figure 3.13 is $(\overline{A} + \overline{B})(A + C)$. We will create a column for each input variable and for each term in parentheses, as well as a column for the final output. Table 3.2 shows the result. For the lines where *A* OR *B* is 0, we write a 1 in the $(\overline{A} + \overline{B})$ column. Where *A* OR *C* is 1, we write a 1 in the $(A + C)$ column. For the lines where there is a 1 in both the $(\overline{A} + \overline{B})$ AND $(A + C)$ columns, we write a 1 in the final output column.

Table 3.2   Truth Table for Figure 3.13

A	*B*	*C*	$(\overline{A} + \overline{B})$	$(A + C)$	$(\overline{A} + \overline{B})(A + C)$
0	0	0	1	0	0
0	0	1	1	1	1
0	1	0	1	0	0
0	1	1	1	1	1
1	0	0	1	1	1
1	0	1	1	1	1
1	1	0	0	1	0
1	1	1	0	1	0

Another approach to finding a truth table involves analysis of the Boolean expression of a logic diagram. The logic diagram in Figure 3.14 can be described by the Boolean expression $Y = \overline{A}BC + \overline{A}\,\overline{C} + \overline{B}\,\overline{D}$.

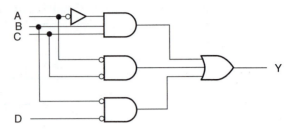

We can examine the Boolean expression to determine that the final output of the circuit will be HIGH under one of the following conditions:

1. $A = 0$ AND $B = 1$ AND $C = 1$;

2. $A = 0$ AND $C = 0$;

3. $B = 0$ AND $D = 0$.

All we have to do is look for these conditions in the truth table and write a 1 in the output column whenever a condition is satisfied. Table 3.3 shows the result of this analysis with each line indicating which term, or terms, contribute to the HIGH output.

Table 3.3 Truth Table for Figure 3.14

A	B	C	D	Y	terms
0	0	0	0	1	$\overline{A}\,\overline{C}$, $\overline{B}\,\overline{D}$
0	0	0	1	1	$\overline{A}\,\overline{C}$
0	0	1	0	1	$\overline{B}\,\overline{D}$
0	0	1	1	0	
0	1	0	0	1	$\overline{A}\,\overline{C}$
0	1	0	1	1	$\overline{A}\,\overline{C}$
0	1	1	0	1	$\overline{A}BC$
0	1	1	1	1	$\overline{A}BC$
1	0	0	0	1	$\overline{B}\,\overline{D}$
1	0	0	1	0	
1	0	1	0	1	$\overline{B}\,\overline{D}$
1	0	1	1	0	
1	1	0	0	0	
1	1	0	1	0	
1	1	1	0	0	
1	1	1	1	0	

▌▌ SECTION 3.16 REVIEW PROBLEM

3.2 Find the truth table for the logic diagram shown in Figure 3.15.

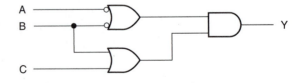

3.2   Sum-of-Products and Product-of-Sums Forms

KEY TERMS

Product term   A term in a Boolean expression where one or more true or comple-ment variables are ANDed (e.g., $A\,\overline{C}$).

Minterm   A product term in a Boolean expression where all possible variables appear once in true or complement form (e.g., $\overline{A}\,\overline{B}\,\overline{C}; A\,B\,\overline{C}$).

Sum term   A term in a Boolean expression where one or more true or comple-ment variables are ORed (e.g., $\overline{A} + B + D$).

Maxterm   A sum term in a Boolean expression where all possible variables appear once, in true or complement form (e.g., $(\overline{A} + \overline{B} + C); (A + \overline{B} + C)$).

Sum-of-products (SOP)   A type of Boolean expression where several product terms are summed (ORed) together (e.g., $\overline{A}\,B\,\overline{C} + \overline{A}\,\overline{B}\,C + A\,B\,C$).

Product-of-sums (POS)   A type of Boolean expression where several sum terms are multiplied (ANDed) together (e.g., $(\overline{A} + \overline{B} + C)(A + \overline{B} + \overline{C})(\overline{A} + B + \overline{C})$).

Bus form   A way of drawing a logic diagram so that each true and complement input variable is available along a continuous conductor called a bus.

Suppose we have an unknown digital circuit, represented by the block in Figure 3.16. All we know is which terminals are inputs, which are outputs, and how to connect the power supply. Given only that information, we can find the Boolean expression of the output.

The first thing to do is find the truth table by applying all possible input combinations in binary order and reading the output for each one. Suppose the unknown circuit in Figure 3.16 yields the truth table shown in Table 3.4.

The truth table output is HIGH for three conditions:

1. When *A* AND *B* AND *C* are all LOW, OR

2. When *A* is LOW AND *B* AND *C* are HIGH, OR

3. When *A* is HIGH AND *B* AND *C* are LOW.

FIGURE 3.16
Digital Circuit with
Unknown Function

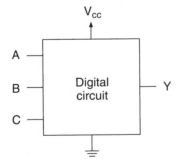

Table 3.4   Truth Table for Figure 3.19

A	B	C	Y
0	0	0	1
0	0	1	0
0	1	0	0
0	1	1	1
1	0	0	1
1	0	1	0
1	1	0	0
1	1	1	0

Each of those conditions represents a **minterm** in the output Boolean expression. (A minterm is a **product term** (AND term) that includes all variables (*A, B, C*) in true or complement form.) The minterms are:

1. $\overline{A}\,\overline{B}\,\overline{C}$

2. $\overline{A}\,B\,C$

3. $A\,\overline{B}\,\overline{C}$

Since condition 1 OR condition 2 OR condition 3 produces a HIGH output from the circuit, the Boolean function Y consists of all three minterms summed (ORed) together, as follows:

$$Y = \overline{A}\,\overline{B}\,\overline{C} + \overline{A}\,B\,C + A\,\overline{B}\,\overline{C}$$

This expression is in a standard form called **sum-of-products** (SOP) form. Figure 3.17 shows the equivalent logic circuit.

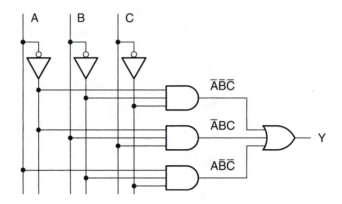

FIGURE 3.17
Logic Circuit for $Y = \overline{A}\,\overline{B}\,\overline{C} + \overline{A}BC + A\,\overline{B}\,\overline{C}$

The inputs A, B, and C and their complements are shown in **bus form.** Each variable is available, in true or complement form, at any point along a conductor. This is a useful, uncluttered notation for circuits that require several of the input variables more than once.

> **NOTE**
>
> We can derive an SOP expression from a truth table as follows:
>
> 1. Every line on the truth table that has a HIGH output corresponds to a minterm in the truth table's Boolean expression.
>
> 2. Write all truth table variables for every minterm in true or complement form. If a variable is 0, write it in complement form (with a bar over it); if it is 1, write it in true form (no bar).
>
> 3. Combine all minterms in an OR function.

▥ EXAMPLE 3.7

Tables 3.5 and 3.6 show the truth tables for the Exclusive OR and the Exclusive NOR functions. Derive the sum-of-products expression for each of these functions and draw the logic diagram for each one.

Table 3.5 XOR
Truth Table

A	B	$A \oplus B$
0	0	0
0	1	1
1	0	1
1	1	0

Table 3.6 XNOR
Truth Table

A	B	$\overline{A \oplus B}$
0	0	1
0	1	0
1	0	0
1	1	1

Solution

XOR: The truth table yields two product terms: $\overline{A}B$ and $A\overline{B}$. Thus, the SOP form of the XOR function is $A \oplus B = \overline{A}B + A\overline{B}$. Figure 3.18 shows the logic diagram for this equation.

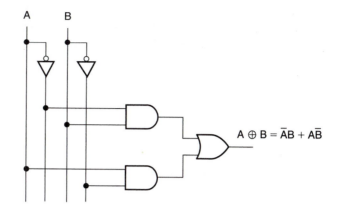

FIGURE 3.18
Example 3.7
SOP Form of XOR Function

XNOR: The product terms for this function are: $\overline{A}\,\overline{B}$ and AB. The SOP form of the XNOR function is $\overline{A \oplus B} = \overline{A}\,\overline{B} + AB$. The logic diagram in Figure 3.19 represents the XNOR function.

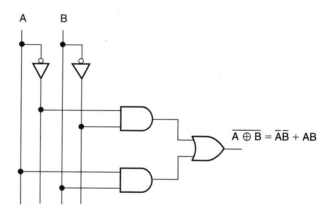

FIGURE 3.19
Example 3.7
SOP Form of XNOR Function

We can also find the Boolean function of a truth table in **product-of-sums** (POS) form. The product-of-sums form of a Boolean expression consists of a number of **maxterms** (i.e., **sum terms** (OR terms) containing all variables in true or complement form) that are ANDed together. To find the POS form of Y, we will find the SOP expression for $\overline{Y}$ and apply DeMorgan's theorems.

Recall DeMorgan's theorems:

$$\overline{x + y + z} = \overline{x}\,\overline{y}\,\overline{z}$$
$$\overline{x\,y\,z} = \overline{x} + \overline{y} + \overline{z}$$

When the theorems were introduced, they were presented as two-variable theorems, but in fact they are valid for any number of variables.

Let's reexamine Table 3.4. To find the sum-of-products expression for Y, we wrote a minterm for each line where $Y = 1$. To find the SOP expression for $\overline{Y}$, *we must write a minterm for each line where* $Y = 0$. Variables A, B, and C must appear in each minterm, in true or complement form. A variable is in complement form (with a bar over the top) if its value is 0 in that minterm, and it is in true form (no bar) if its value is 1.

We get the following minterms for $\overline{Y}$:

$\overline{A}\,\overline{B}\,C$

$\overline{A}\,B\,\overline{C}$

$A\,\overline{B}\,C$

$A\,B\,\overline{C}$

$A\,B\,C$

Thus, the SOP form of $\overline{Y}$ *is*

$$\overline{Y} = \overline{A}\,\overline{B}\,C + \overline{A}\,B\,\overline{C} + A\,\overline{B}\,C + A\,B\,\overline{C} + A\,B\,C$$

To get Y in POS form, we must invert both sides of the above expression and apply DeMorgan's theorems to the righthand side.

$$Y = \overline{\overline{Y}} = \overline{\overline{A}\,\overline{B}\,C + \overline{A}\,B\,\overline{C} + A\,\overline{B}\,C + A\,B\,\overline{C} + A\,B\,C}$$
$$= (\overline{\overline{A}\,\overline{B}\,C})(\overline{\overline{A}\,B\,\overline{C}})(\overline{A\,\overline{B}\,C})(\overline{A\,B\,\overline{C}})(\overline{A\,B\,C})$$
$$= (A + B + \overline{C})(A + \overline{B} + C)(\overline{A} + B + \overline{C})(\overline{A} + \overline{B} + C)(\overline{A} + \overline{B} + \overline{C})$$

This Boolean expression can be implemented by the logic circuit in Figure 3.20.

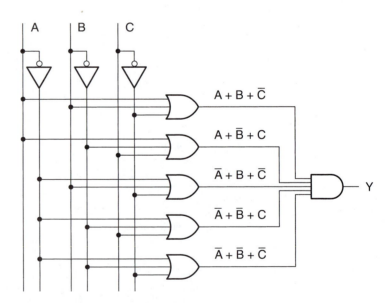

FIGURE 3.20
Logic Circuit for Y = (A + B + $\overline{C}$) (A + $\overline{B}$ + C)($\overline{A}$ + B + $\overline{C}$) ($\overline{A}$ + $\overline{B}$ + C)($\overline{A}$ + $\overline{B}$ + $\overline{C}$)

We don't have to go through the whole process outlined above every time we want to find the POS form of a function. We can find it directly from the truth table, following the procedure summarized below. Use this procedure to find the POS form of the expression given by Table 3.4. The terms in this expression are the same as those derived by DeMorgan's theorem.

> **NOTE**
>
> Deriving a POS expression from a truth table:
>
> 1. Every line on the truth table that has a LOW output corresponds to a maxterm in the truth table's Boolean expression.
> 2. Write all truth table variables for every maxterm in true or complement form. If a variable is 1, write it in complement form (with a bar over it); if it is 0, write it in true form (no bar).
> 3. Combine all maxterms in an AND function.
>
>   Note that these steps are all opposite to those used to find the SOP form of the Boolean expression.

▌▌ EXAMPLE 3.8

Find the Boolean expression, in both SOP and POS forms, for the logic function represented by Table 3.7. Draw the logic circuit for each form.

Table 3.7 Truth Table for Example 3.8 (with minterms and maxterms)

A	B	C	D	Y	Minterms	Maxterms
0	0	0	0	1	$\overline{A}\,\overline{B}\,\overline{C}\,\overline{D}$	
0	0	0	1	1	$\overline{A}\,\overline{B}\,\overline{C}\,D$	
0	0	1	0	0		$A + B + \overline{C} + D$
0	0	1	1	1	$\overline{A}\,\overline{B}\,C\,D$	
0	1	0	0	0		$A + \overline{B} + C + D$
0	1	0	1	0		$A + \overline{B} + C + \overline{D}$
0	1	1	0	0		$A + \overline{B} + \overline{C} + D$
0	1	1	1	0		$A + \overline{B} + \overline{C} + \overline{D}$
1	0	0	0	1	$A\,\overline{B}\,\overline{C}\,\overline{D}$	
1	0	0	1	0		$\overline{A} + B + C + \overline{D}$
1	0	1	0	1	$A\,\overline{B}\,C\,\overline{D}$	
1	0	1	1	0		$\overline{A} + B + \overline{C} + \overline{D}$
1	1	0	0	1	$A\,B\,\overline{C}\,\overline{D}$	
1	1	0	1	1	$A\,B\,\overline{C}\,D$	
1	1	1	0	1	$A\,B\,C\,\overline{D}$	
1	1	1	1	0		$\overline{A} + \overline{B} + \overline{C} + \overline{D}$

Solution   All minterms (for SOP form) and maxterms (for POS form) are shown in the last two columns of Table 3.5.

Boolean Expressions:

SOP form:

$$Y = \overline{A}\,\overline{B}\,\overline{C}\,\overline{D} + \overline{A}\,\overline{B}\,\overline{C}\,D + \overline{A}\,\overline{B}\,C\,D + A\,\overline{B}\,\overline{C}\,\overline{D} + A\,\overline{B}\,C\,\overline{D} + A\,B\,\overline{C}\,\overline{D}$$
$$+ A\,B\,\overline{C}\,D + A\,B\,C\,\overline{D}$$

POS form:

$$Y = (A + B + \overline{C} + D)(A + \overline{B} + C + D)(A + \overline{B} + C + \overline{D})(A + \overline{B} + \overline{C} + D)$$
$$(A + \overline{B} + \overline{C} + \overline{D})(\overline{A} + B + C + \overline{D})(\overline{A} + B + \overline{C} + \overline{D})$$
$$(\overline{A} + \overline{B} + \overline{C} + \overline{D})$$

The logic circuits are shown in Figures 3.21 and 3.22.

FIGURE 3.21
Example 3.8
SOP Form

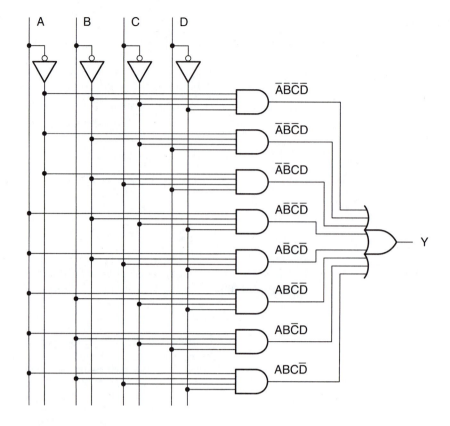

FIGURE 3.22
Example 3.8
POS Form

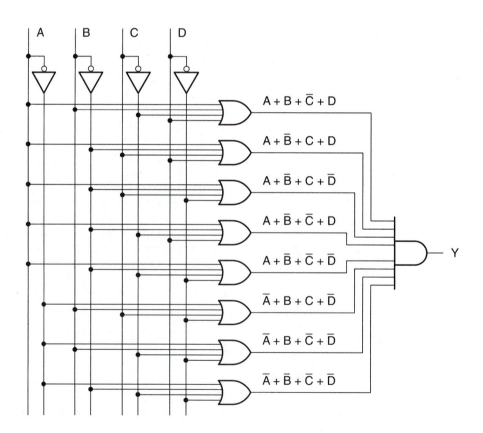

▌▌▌ SECTION 3.2 REVIEW PROBLEM

3.3 Find the SOP and POS forms of the Boolean functions represented by the following truth tables.

a.

A	B	C	Y
0	0	0	0
0	0	1	0
0	1	0	0
0	1	1	0
1	0	0	1
1	0	1	1
1	1	0	0
1	1	1	0

b.

A	B	C	Y
0	0	0	1
0	0	1	0
0	1	0	0
0	1	1	0
1	0	0	1
1	0	1	1
1	1	0	1
1	1	1	0

3.3 Theorems of Boolean Algebra

The main reason to learn Boolean algebra is to learn how to minimize the number of logic gates in a network. Boolean expressions with many terms, such as those represented by the logic diagrams in Figures 3.21 and 3.22, are seldom in their simplest form. It is often possible to apply some techniques of Boolean algebra to derive a simpler form of expression that requires fewer gates to implement.

For example, the logic circuit in Figure 3.21 requires eight 4-input AND gates and an 8-input OR gate. Using Boolean algebra, we can reduce its Boolean expression to $Y = AD + \overline{A}\,\overline{B}\,\overline{C} + \overline{A}\,B\,D + A\,B\,\overline{C}$. This form can be implemented with 4 AND gates and a 4-input OR. You will use a simplification technique for this example in an end-of-chapter problem. In the meantime, let us examine some basic rules of Boolean algebra.

Commutative, Associative, and Distributive Properties

> **KEY TERMS**
>
> **Commutative property**   A mathematical operation is commutative if it can be applied to its operands in any order without affecting the result. For example, addition is commutative $(a + b = b + a)$, but subtraction is not $(a - b \neq b - a)$.
>
> **Associative property**   A mathematical function is associative if its operands can be grouped in any order without affecting the result. For example, addition is associative $((a + b) + c = a + (b + c))$, but subtraction is not $((a - b) - c \neq a - (b - c))$.
>
> **Distributive property**   Full name: distributive property of multiplication over addition. The property that allows us to distribute ("multiply through") an AND across several OR functions. For example, $a(b + c) = ab + ac$.

AND and OR functions are both **commutative** and **associative.** The commutative property states that AND and OR operations are independent of input order. For inputs x and y,

Theorem 1: $xy = yx$

and

Theorem 2: $x + y = y + x$

The associative property allows us to perform several two-input AND or OR functions in any order. In other words,

Theorem 3: $(xy)z = x(yz) = (xz)y$

and

Theorem 4: $(x + y) + z = x + (y + z) = (x + z) + y$

The **distributive** property allows us to "multiply through" an AND function across several OR functions. For example,

Theorem 5: $x(y + z) = xy + xz$

and

Theorem 6: $(x + y)(w + z) = xw + xz + yw + yz$

Figure 3.23 shows the logic gate equivalents of these theorems.

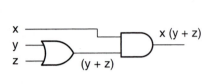

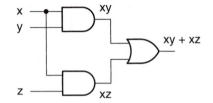

a. x (y + z) = xy + xz

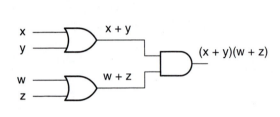

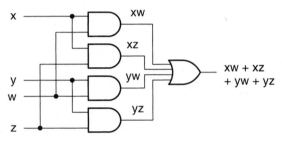

b. (x + y)(w + z) = xw + xz + yw + yz

FIGURE 3.23
Distributive Properties

■■ **EXAMPLE 3.9**

Find the Boolean expression of the POS circuit in Figure 3.24a. Apply the distributive property to transform the circuit to an SOP form.

FIGURE 3.24
Example 3.9
Distributive Property

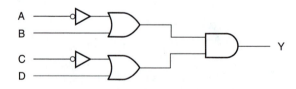

a. POS form

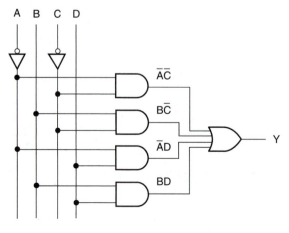

b. SOP form

Solution The Boolean expression for Figure 3.24a is $Y = (\overline{A} + B)(\overline{C} + D)$. Using the distributive property, we get the expression $Y = \overline{A}\,\overline{C} + B\overline{C} + \overline{A}D + BD$. The logic diagram for this expression is shown in Figure 3.24b.

In Example 3.9, we see that the distributive property can be used to convert a POS circuit to SOP or vice versa. In this case, the circuit was not simplified, just transformed.

■ EXAMPLE 3.10

Write the Boolean expression for the circuit in Figure 3.25a. Use the distributive property to convert this to an SOP circuit.

FIGURE 3.25
Example 3.10
Distributive Property

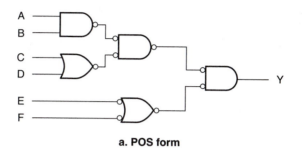

a. POS form

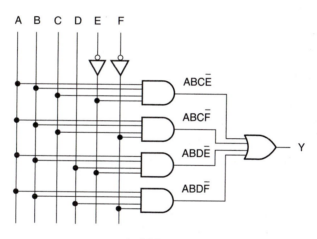

b. SOP form

Solution The Boolean expression for Figure 3.25a is $AB(C + D)(\overline{E} + \overline{F})$. The distributive property can be applied in two stages:

$$Y = (ABC + ABD)(\overline{E} + \overline{F})$$
$$= ABC\overline{E} + ABC\overline{F} + ABD\overline{E} + ABD\overline{F}$$

The logic diagram for this equation is shown in Figure 3.25b. This results in a network that is "wider" (more gates on one level), but also "flatter" (fewer levels). The advantage of the second circuit is that signals would pass through the network faster, since it has fewer levels of gating.

Single-Variable Theorems

There are thirteen theorems that can be used to manipulate a single variable in a Boolean expression. An easy way to remember these theorems is to divide them into three groups:

1. Six theorems: x AND/OR/XOR 0/1
2. Six theorems: x AND/OR/XOR $x/\overline{x}$
3. One theorem: Double Inversion

x AND/OR/XOR 0/1

The theorems in the first group can be generated by asking what happens when *x*, a Boolean variable or expression, is at one input of an AND, an OR, or an XOR gate and a 0 or a 1 is at the other.

Examine the truth table of the gate in question. Hold one input of the gate constant and find the effect of the other on the output. This is the same procedure we used in Chapter 2 to examine the enable/inhibit properties of logic gates.

Each of these six theorems can be represented by a logic gate, as shown in Figure 3.26.

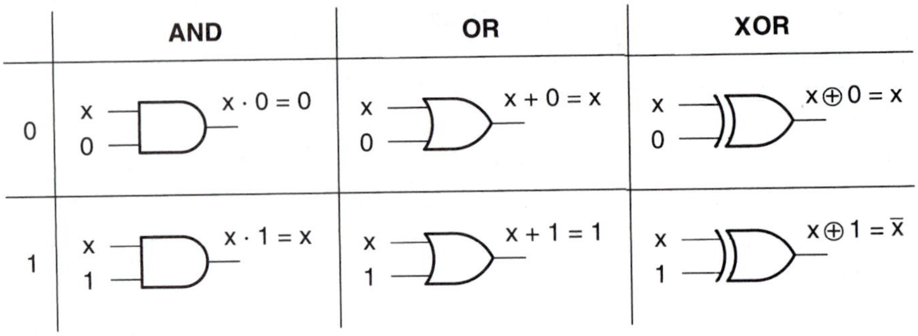

FIGURE 3.26

X AND/OR/XOR 0/1

	AND	**OR**	**XOR**
0	$x \cdot 0 = 0$	$x + 0 = x$	$x \oplus 0 = x$
1	$x \cdot 1 = x$	$x + 1 = 1$	$x \oplus 1 = \overline{x}$

$x \cdot 0$:

A	x	Y
0	0	0
0	1	0
~~1~~	~~0~~	~~0~~
~~1~~	~~1~~	~~1~~

If $x = 0$, $Y = 0$

If $x = 1$, $Y = 0$

(Can never have both inputs HIGH, therefore output is always LOW.)

Theorem 7: $x \cdot 0 = 0$

$x + 0$:

A	x	Y
0	0	0
0	1	1
~~1~~	~~0~~	~~1~~
~~1~~	~~1~~	~~1~~

If $x = 0$, $Y = 0$

If $x = 1$, $Y = 1$

(LOW input enables OR gate.)

Theorem 8: $x + 0 = x$

$x \oplus 0$:

A	x	Y
0	0	0
0	1	1
~~1~~	~~0~~	~~1~~
~~1~~	~~1~~	~~0~~

If $x = 0$, $Y = 0$

If $x = 1$, $Y = 1$

(XOR acts as a noninverting buffer.)

Theorem 9: $x \oplus 0 = x$

$x \cdot 1$:

A	x	Y
0	0	0
0	1	0
1	0	0
1	1	1

If $x = 0$, $Y = 0$

If $x = 1$, $Y = 1$

(HIGH input enables AND gate.)

Theorem 10: $x \cdot 1 = x$

$x + 1$:

A	x	Y
0	0	0
0	1	1
1	0	1
1	1	1

If $x = 0$, $Y = 1$

If $x = 1$, $Y = 1$

(One input always HIGH, therefore output is always HIGH.)

Theorem 11: $x + 1 = 1$

$x \oplus 1$:

A	x	Y
0	0	0
0	1	1
1	0	1
1	1	0

If $x = 0$, $Y = 1$

If $x = 1$, $Y = 0$

(XOR acts as an inverting buffer.)

Theorem 12 $x \oplus 1 = \overline{x}$

x AND/OR/XOR $x/\overline{x}$

Six theorems are generated by combining a Boolean variable or expression, x, with itself or its complement in an AND, an OR, or an XOR function.

Again, we can use the AND, OR, and XOR truth tables. For the first three theorems, we look only at the lines where both inputs are the same. For the other three, we use the lines where the inputs are different.

Figure 3.27 shows the logic gates that represent these theorems.

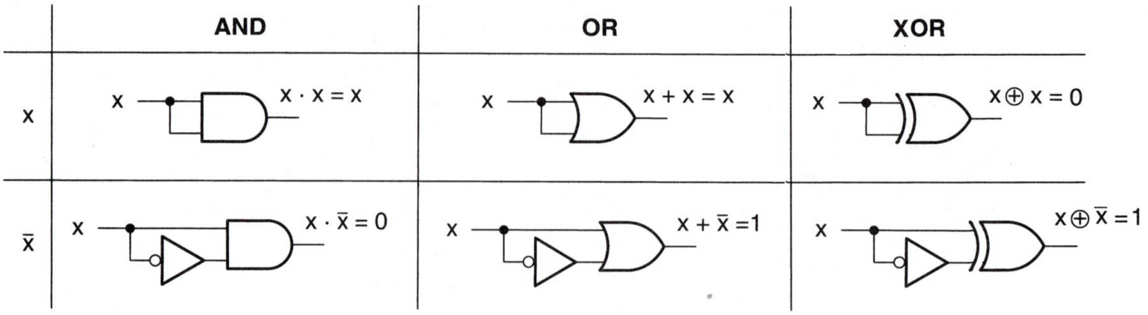

FIGURE 3.27
X AND/OR/XOR X/$\overline{X}$

$x \cdot x$:

A	x	Y
0	0	0
0	1	0
1	0	0
1	1	1

If $x = 0$, $Y = 0$

If $x = 1$, $Y = 1$

Theorem 13: $x \cdot x = x$

$x + x$:

A	x	Y
0	0	0
0	1	1
1	0	1
1	1	1

If $x = 0$, $Y = 0$

If $x = 1$, $Y = 1$

Theorem 14: $x + x = x$

$x \oplus x$:

A	x	Y
0	0	0
0	1	1
1	0	1
1	1	0

If $x = 0$, $Y = 0$

If $x = 1$, $Y = 0$

(Output is LOW if neither input is HIGH or if both are.)

Theorem 15: $x \oplus x = 0$

$x \cdot \overline{x}$:

A	x	Y
0	0	0
0	1	0
1	0	0
1	1	1

If $x = 0$, $Y = 0$

If $x = 1$, $Y = 0$

(Since inputs are opposite, can never have both HIGH. Output always LOW.)

Theorem 16: $x \cdot \overline{x} = 0$

$x + \overline{x}$:

A	x	Y
0	0	0
0	1	1
1	0	1
1	1	1

If $x = 0$, $Y = 1$

If $x = 1$, $Y = 1$

(Since inputs are opposite, one input always HIGH. Therefore, output is always HIGH.)

Theorem 17: $x + \overline{x} = 1$

$x \oplus \overline{x}$:

A	x	Y
0	0	0
0	1	1
1	0	1
1	1	0

If $x = 0$, $Y = 1$

If $x = 1$, $Y = 1$

(One input HIGH, but not both.)

Theorem 18: $x \oplus \overline{x} = 1$

Double Inversion

The final single-variable theorem is just common sense. It states that a variable or expression inverted twice is the same as the original variable or expression. It is given by:

Theorem 19: $\overline{\overline{x}} = x$

This theorem is illustrated by the two inverters in Figure 3.28.

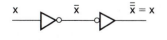

FIGURE 3.28
Double Inversion

Multivariable Theorems

There are numerous multivariable theorems we could learn, but we will look only at five of the most useful.

DeMorgan's Theorems

We have already seen DeMorgan's theorems. We will list them again, but will not comment further on them at this time.

Theorem 20: $\overline{xy} = \overline{x} + \overline{y}$

Theorem 21: $\overline{x + y} = \overline{x}\,\overline{y}$

Other Multivariable Theorems

Theorem 22: $x + xy = x$

Proof:

$$x + xy = x\,(1 + y) \quad \text{(Distributive property)}$$
$$= x \cdot 1 \quad\quad (1 + y = 1; \text{Theorem})$$
$$= x$$

Figure 3.29 illustrates the circuit in this theorem. Note that the equivalent is not a circuit at all, but a single, unmodified variable. Thus, the circuit shown need never be built.

FIGURE 3.29
Theorem 22

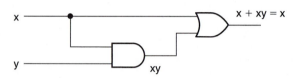

▌▌ EXAMPLE 3.11

Simplify the following Boolean expressions, using Theorem 22 and other rules of Boolean algebra. Draw the logic circuits of the unsimplified and simplified expressions.

a. $H = K\overline{L} + K$

b. $Y = \overline{(A + B)}CD + \overline{(A + B)}$

c. $W = (PQR + \overline{P}\,\overline{Q})(S + T) + (\overline{P} + \overline{Q})(S + T) + (S + T)$

Solution   Figure 3.30 shows the logic circuits for the unsimplified and simplified versions of the above expressions.

a. Let $x = K$, let $y = \overline{L}$:

$$H = x + xy = K + K\overline{L}$$

Theorem 22 states $x + xy = x$. Therefore $K + K\overline{L} = K$.

b. Let $x = \overline{(A + B)}$, let $y = CD$:

$$Y = x + xy = x = \overline{A + B}$$

c. Let $x = S + T$, *let* $y = (\overline{P} + \overline{Q})$:

Since $x + xy = x$, $(\overline{P} + \overline{Q})(S + T) + (S + T) = (S + T)$.

$$W = (PQR + \overline{P}\,\overline{Q})(S + T) + (S + T)$$

Let $x = S + T$, let $y = (PQR + \overline{P}\,\overline{Q})$

$$W = x + xy = x = S + T$$

Alternate method:

$$W = (PQR + \overline{P}\,\overline{Q})(S + T) + (\overline{P} + \overline{Q})(S + T) + (S + T)$$

By the distributive property:

$$W = ((PQR + \overline{P}\,\overline{Q}) + (\overline{P} + \overline{Q}))(S + T) + (S + T)$$

Let $x = S + T$, let $y = ((PQR + \overline{P}\,\overline{Q}) + (\overline{P} + \overline{Q}))$:

$$W = x + xy = x = S + T$$

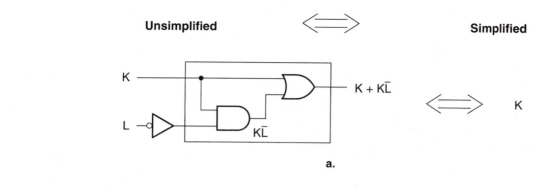

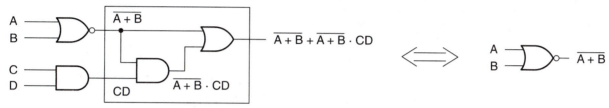

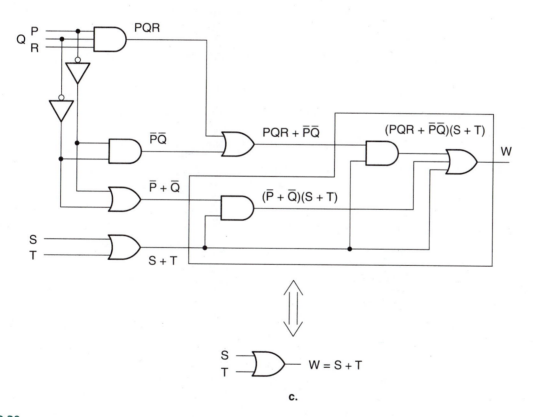

FIGURE 3.30
Example 3.11
Logic Circuits for Unsimplified and Simplified Expressions

Theorem 23: $(x + y)(x + z) = x + yz$

Proof: $(x + y)(x + z) = xx + xz + xy + yz$ (Distributive property)

$$= (x + xy) + xz + yz \quad (xx = x;\ \text{Associative property})$$
$$= x + xz + yz \quad\quad (x + xy = x \text{ (Theorem 22))}$$
$$= (x + xz) + yz \quad\quad (\text{Associative property})$$
$$= x + yz \quad\quad\quad\quad (\text{Theorem 22})$$

Figure 3.31 shows the logic circuits for the left and right sides of the equation for Theorem 23. This theorem is a special case of one of the distributive properties, Theorem 6, where $w = x$.

FIGURE 3.31
Theorem 23

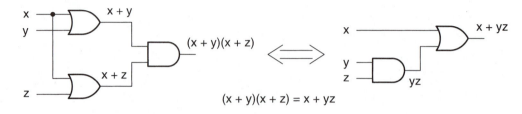

$(x + y)(x + z) = x + yz$

▌▌ **EXAMPLE 3.12**

Simplify the following Boolean expressions, using Theorem 23 and other rules of Boolean algebra. Draw the logic circuits of the unsimplified and simplified expressions.

a. $L = (M + \overline{N})(M + \overline{P})$
b. $Y = \overline{(A + B + AB)}(A + B + C)$

Solution   Figure 3.32 shows the logic circuits for the unsimplified and simplified versions of the above expressions.

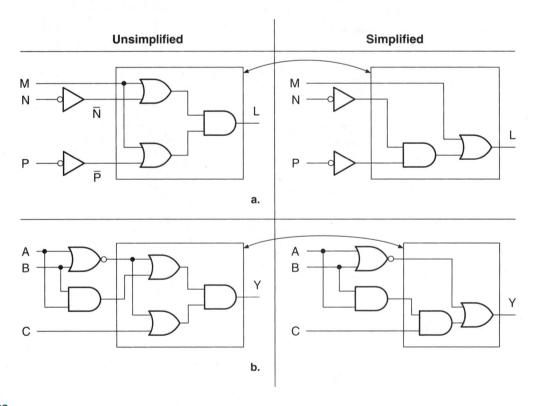

FIGURE 3.32
Example 3.12
Logic Circuits for Unsimplified and Simplified Expressions

Theorem 23: $(x + y)(x + z) = x + yz$

a. Let $x = M$, let $y = \overline{N}$, let $z = \overline{P}$:

$$L = (x + y)(x + z) = x + yz = M + \overline{N}\,\overline{P}$$

b. Let $x = \overline{A} + B$, let $y = AB$, let $z = C$:

$$Y = (x + y)(x + z) = x + yz = \overline{A + B} + ABC = \overline{A}\,\overline{B} + ABC$$

Theorem 24: $x + \overline{x}y = x + y$

Proof: Since $(x + y)(x + z) = x + yz$, then for $y = \overline{x}$:

$$
\begin{aligned}
x + \overline{x}y &= (x + \overline{x})(x + y) \\
&= 1 \cdot (x + y) \qquad (x + \overline{x} = 1) \\
&= x + y
\end{aligned}
$$

Figure 3.33 illustrates Theorem 24 with a logic circuit.

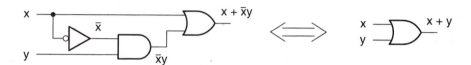

FIGURE 3.33
Theorem 24

> **NOTE**
>
> Here is another way to remember Theorem 24:
> If a variable (x) is ORed with a term consisting of a different variable (y) AND the first variable's complement $(\overline{x})$, the complement disappears.
>
> $$x + \overline{x}y = x + y$$

▌▌ EXAMPLE 3.13

Simplify the following Boolean expressions, using Theorem 24 and other rules of Boolean algebra. Draw the logic circuits of the unsimplified and simplified forms of the expressions.

a. $W = \overline{U} + U\overline{V}$

b. $P = Q\overline{R}S + (\overline{Q} + R + \overline{S})\,T$
   $J = \overline{KM}\,(\overline{K} + L + M) + KM$

Solution   Figure 3.34 shows the circuits for the unsimplified and simplified expressions.

Theorem 24: $x + \overline{x}y = x + y$

a. Let $x = \overline{U}$, let $y = \overline{V}$:

$$W = x + \overline{x}y = x + y = \overline{U} + \overline{V}$$

b.

$$
\begin{aligned}
P &= Q\overline{R}S + (\overline{Q} + R + \overline{S})\,T \\
&= Q\overline{R}S + \overline{Q\,\overline{R}\,S}\;T \qquad \text{(DeMorgan's theorem)}
\end{aligned}
$$

Let $x = Q\overline{R}S$, let $y = T$:

$$P = x + \overline{x}y = x + y = Q\overline{R}S + T$$

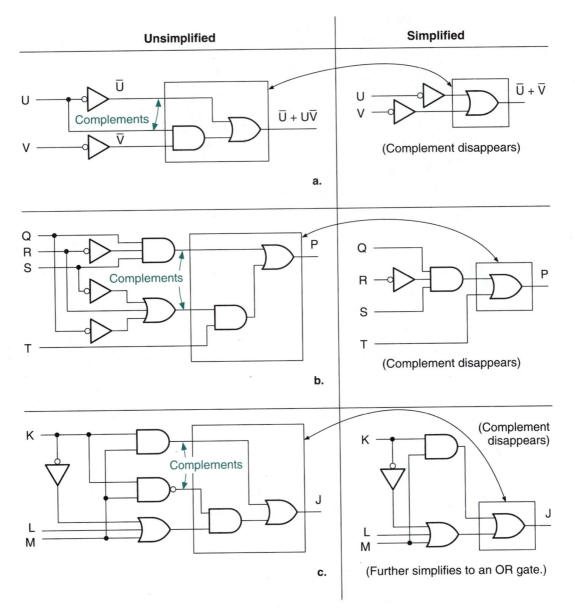

FIGURE 3.34
Example 3.13
Logic Circuits for Unsimplified and Simplified Expressions

c. Let $x = KM$, let $y = (\overline{K} + L + M)$:

$$J = x + \overline{x}y = x + y = KM + \overline{K} + L + M$$
$$= \overline{K} + L + (M + KM) \quad \text{(Associative property)}$$
$$= \overline{K} + L + M \quad \quad \text{(Theorem 22)}$$

The rules of Boolean algebra are summarized in Table 3.8. Don't try to memorize all these rules. The commutative, associative, and distributive properties are the same as their counterparts in ordinary algebra. The single-variable theorems can be reasoned out by your knowledge of logic gate operation. That leaves only five multivariable theorems.

Table 3.8 Theorems of Boolean Algebra

Commutative Properties

1. $x + y = y + x$
2. $x \cdot y = y \cdot x$

Associative Properties

3. $x + (y + z) = (x + y) + z$
4. $x(yz) = (xy)z$

Distributive Properties

5. $x(y + z) = xy + xz$
6. $(x + y)(w + z) = xw + xz + yw + yz$

x AND/OR/XOR 0/1

7. $x \cdot 0 = 0$
8. $x + 0 = x$
9. $x \oplus 0 = x$
10. $x \cdot 1 = x$
11. $x + 1 = 1$
12. $x \oplus 1 = \bar{x}$

x AND/OR/XOR $x/\bar{x}$

13. $x \cdot x = x$
14. $x + x = x$
15. $x \oplus x = 0$
16. $x \cdot \bar{x} = 0$
17. $x + \bar{x} = 1$
18. $x \oplus \bar{x} = 1$

Double Inversion

19. $\bar{\bar{x}} = x$

DeMorgan's Theorems

20. $\overline{xy} = \bar{x} + \bar{y}$
21. $\overline{x + y} = \bar{x}\bar{y}$

Other Multivariable Theorems

22. $x + xy = x$
23. $(x + y)(x + z) = x + yz$
24. $x + \bar{x}y = x + y$

▐▌ SECTION 3.3 REVIEW PROBLEMS

3.4 Use theorems of Boolean algebra to simplify the following Boolean expressions.

a. $Y = \overline{AC} + (\bar{A} + \bar{C})D$

b. $Y = \bar{A} + \bar{C} + ACD$

c. $Y = (A\bar{B} + \overline{BC})(A\bar{B} + \bar{C})$

3.4 Simplifying SOP and POS Expressions

Maximum SOP simplification The form of an SOP Boolean expression that cannot be further simplified by canceling variables in the product terms. It may be possible to get a POS form of the expression with fewer terms or variables.

Maximum POS simplification The form of a POS Boolean expression that cannot be further simplified by canceling variables in the sum terms. It may be possible to get an SOP form of the expression with fewer terms or variables.

Table 3.9 Truth Table for the SOP and POS Networks in Figure 3.35

A	B	C	Y
0	0	0	1
0	0	1	0
0	1	0	0
0	1	1	1
1	0	0	1
1	0	1	0
1	1	0	0
1	1	1	0

Earlier in this chapter, we discovered that we can generate a Boolean equation from a truth table and express it in sum-of-products (SOP) or product-of-sums (POS) form. From this equation, we can develop a logic circuit diagram. The next step in the design or analysis of a circuit is to simplify its Boolean expression as much as possible, with the ultimate aim of producing a circuit that has fewer physical components than the unsimplified circuit.

In Section 3.2, we found the SOP and POS forms of the Boolean expression represented by Table 3.9. These forms yield the logic diagrams shown in Figures 3.17 and 3.20. For convenience, the circuits are illustrated again in Figure 3.35. The corresponding algebraic expressions can be simplified by the rules of Boolean algebra to give us a simpler circuit in each case.

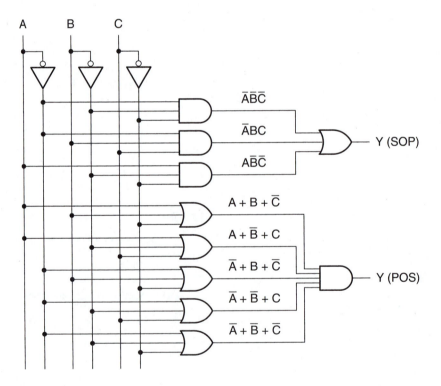

FIGURE 3.35
Unsimplified SOP and POS Networks

The sum-of-products and product-of-sums expressions represented by Table 3.9 are:

$$Y = \overline{A}\,\overline{B}\,\overline{C} + \overline{A}\,B\,C + A\,\overline{B}\,\overline{C} \quad \text{(SOP)}$$

and

$$Y = (A + B + \overline{C})(A + \overline{B} + C)(\overline{A} + B + \overline{C})(\overline{A} + \overline{B} + C)(\overline{A} + \overline{B} + \overline{C}) \quad \text{(POS)}$$

The SOP form is fairly easy to simplify:

$$Y = \overline{A}\,\overline{B}\,\overline{C} + \overline{A}\,B\,C + A\,\overline{B}\,\overline{C}$$
$$= (\overline{A} + A)\,\overline{B}\,\overline{C} + \overline{A}\,B\,C \qquad \text{(Distributive property)}$$
$$= 1 \cdot \overline{B}\,\overline{C} + \overline{A}\,B\,C \qquad (x + \overline{x} = 1)$$
$$= \overline{B}\,\overline{C} + \overline{A}\,B\,C \qquad (x \cdot 1 = x)$$

Since we cannot cancel any more SOP terms, we can call this final form the **maximum SOP simplification.** The logic diagram for the simplified expression is shown in Figure 3.36.

FIGURE 3.36

Simplified SOP Circuit

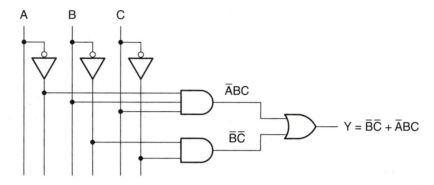

NOTE

Two terms in an SOP expression can be reduced to one if they are identical except for one variable that is in true form in one term and complement form in the other. Such a grouping of a variable and its complement always cancels.

$$x y \overline{z} + x y z = x y(\overline{z} + z) = x y$$

There is a similar procedure for the POS form. Examine the following expression:

$$Y = (A + B + \overline{C})(A + B + C)$$

Recall Theorem 23: $(x + y)(x + z) = x + yz$.
  Let $x = A + B$, let $y = \overline{C}$, let $z = C$.

$$Y = (A + B) + \overline{C}C \quad \text{(Theorem 23)}$$
$$= (A + B) + 0 \qquad (x\overline{x} = 0)$$
$$= (A + B) \qquad (x + 0 = x)$$

NOTE

A POS expression can be simplified by grouping two terms that are identical except for one variable that is in true form in one term and complement form in the other.

$$(x + y + \overline{z})(x + y + z) = (x + y) + \overline{z}z = x + y$$

Let us use this procedure to simplify the POS form of the previous Boolean expression, shown again below with the terms numbered for our reference. The numbered value of each term corresponds to the binary value of the line in the truth table from which it is derived.

$$\overset{(1)}{} \qquad \overset{(2)}{} \qquad \overset{(5)}{} \qquad \overset{(6)}{} \qquad \overset{(7)}{}$$
$$Y = (A + B + \overline{C})\,(A + \overline{B} + C)\,(\overline{A} + B + \overline{C})\,(\overline{A} + \overline{B} + C)\,(\overline{A} + \overline{B} + \overline{C})$$

There can be more than one way to simplify an expression. The following grouping of the numbered POS terms is one possibility.

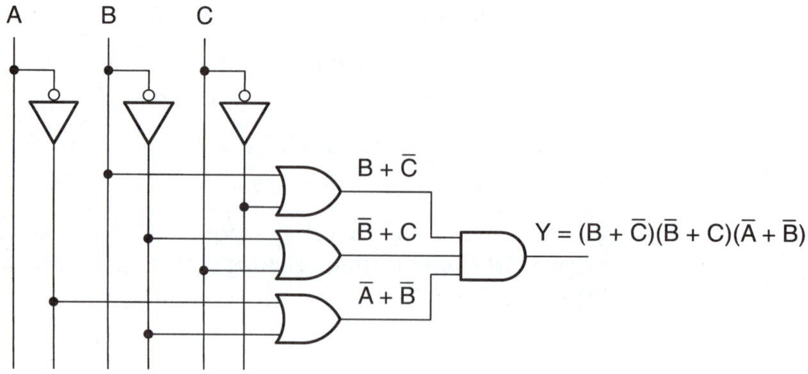

FIGURE 3.37
Simplified POS Circuit

$$(1)(5): (A + B + \overline{C})\,(\overline{A} + B + \overline{C}) = B + \overline{C}$$
$$(2)(6): (A + \overline{B} + C)\,(\overline{A} + \overline{B} + C) = \overline{B} + C$$
$$(6)(7): (\overline{A} + \overline{B} + C)\,(\overline{A} + \overline{B} + \overline{C}) = \overline{A} + \overline{B}$$

Combining the above terms, we get the expression:

$$Y = (B + \overline{C})(\overline{B} + C)(\overline{A} + \overline{B})$$

Figure 3.37 shows the logic diagram for this expression. Compare this logic diagram and that of Figure 3.36 with the unsimplified circuits of Figure 3.35. Since there are no more cancellations of POS terms possible, we can call this the **maximum POS simplification.** We can, however, apply other rules of Boolean algebra and simplify further.

$$
\begin{aligned}
Y &= (B + \overline{C})(\overline{B} + C)(\overline{A} + \overline{B}) \\
&= (\overline{B} + \overline{A}C)(B + \overline{C}) && \text{(Theorem 23)} \\
&= \overline{B}\,B + \overline{B}\,\overline{C} + \overline{A}\,B\,C + \overline{A}\,C\,\overline{C} && \text{(Distributive property)} \\
&= \overline{B}\,\overline{C} + \overline{A}\,B\,C && (x \cdot \overline{x} = 0)
\end{aligned}
$$

This is the same result we got when we simplified the SOP form of the expression.

To be sure you are getting the maximum SOP or POS simplification, you should be aware of the following guidelines:

1. Each term must be grouped with another, if possible.

2. When attempting to group all terms, it is permissible to group a term more than once, such as term (6) above. The theorems $x \cdot x = x$ (POS forms) and $x + x = x$ (SOP forms) imply that using a term more than once does not change the Boolean expression.

3. Each pair of terms should have at least one term that appears only in that pair. Otherwise, you will have redundant terms that will need to be canceled later. For example, another possible group in the POS simplification above is terms (5) and (7). But since both these terms are in other groups, this pair is unnecessary and would yield a term you would have to cancel.

▮▮ EXAMPLE 3.14

Find the maximum SOP simplification for the Boolean function represented by Table 3.10. Draw the logic diagram for the simplified expression.

Solution   SOP form:

$$
\underset{\displaystyle (8)}{} \quad \underset{\displaystyle (9)}{} \quad \underset{\displaystyle (10)}{} \quad \underset{\displaystyle (11)}{} \quad \underset{\displaystyle (12)}{} \quad \underset{\displaystyle (14)}{}
$$
$$Y = A\,\overline{B}\,\overline{C}\,\overline{D} + A\,\overline{B}\,\overline{C}\,D + A\,\overline{B}\,C\,\overline{D} + A\,\overline{B}\,C\,D + A\,B\,\overline{C}\,\overline{D} + A\,B\,C\,\overline{D}$$

Table 3.10 Truth Table for Example 3.14

A	B	C	D	Y
0	0	0	0	0
0	0	0	1	0
0	0	1	0	0
0	0	1	1	0
0	1	0	0	0
0	1	0	1	0
0	1	1	0	0
0	1	1	1	0
1	0	0	0	1
1	0	0	1	1
1	0	1	0	1
1	0	1	1	1
1	1	0	0	1
1	1	0	1	0
1	1	1	0	1
1	1	1	1	0

Group the terms as follows:

$$(8) + (9): \quad A\,\overline{B}\,\overline{C}\,\overline{D} + A\,\overline{B}\,\overline{C}\,D = A\,\overline{B}\,\overline{C}$$
$$(10) + (11): \quad A\,\overline{B}\,C\,\overline{D} + A\,\overline{B}\,C\,D = A\,\overline{B}\,C$$
$$(12) + (14): \quad A\,B\,\overline{C}\,\overline{D} + A\,B\,C\,\overline{D} = A\,B\,\overline{D}$$

Combine the simplified groups and apply techniques of Boolean algebra to simplify further:

$$\begin{aligned}
Y &= A\,\overline{B}\,\overline{C} + A\,\overline{B}\,C + A\,B\,\overline{D} \\
&= A\,\overline{B}(\overline{C} + C) + A\,B\,\overline{D} \\
&= A\,\overline{B} + A\,B\,\overline{D} \\
&= A(\overline{B} + B\,\overline{D}) \qquad \text{Distributive property} \\
&= A(\overline{B} + \overline{D}) \qquad \text{Theorem 24: } x + \overline{x}y = x + y \\
&= A\,\overline{B} + A\,\overline{D}
\end{aligned}$$

Figure 3.38 Shows the logic diagram of the simplified expression.

Table 3.11 Truth Table for Section Review Problem

A	B	C	Y
0	0	0	0
0	0	1	1
0	1	0	1
0	1	1	1
1	0	0	0
1	0	1	0
1	1	0	1
1	1	1	0

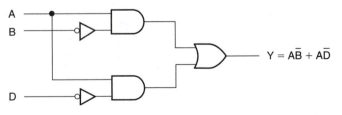

FIGURE 3.38
Example 3.14
Simplified SOP Circuit

$$Y = A\overline{B} + A\overline{D}$$

■■ SECTION 3.4 REVIEW PROBLEM

3.5 Find the maximum SOP and POS simplifications for the function represented by Table 3.11.

3.5 Simplification by the Karnaugh Map Method

Karnaugh map A graphical tool for finding the maximum SOP or POS simplification of a Boolean expression. A Karnaugh map works by arranging the terms of an expression in such a way that variables can be canceled by grouping minterms or maxterms.

Cell The smallest unit of a Karnaugh map, corresponding to one line of a truth table. The input variables are the cell's coordinates, and the output variable is the cell's contents.

Adjacent cell Two cells are adjacent if there is only one variable that is different between the coordinates of the two cells. For example, the cells for minterms ABC and $A\overline{B}C$ are adjacent.

Pair A group of two adjacent cells in a Karnaugh map. A pair cancels one variable in a K-map simplification.

Quad A group of four adjacent cells in a Karnaugh map. A quad cancels two variables in a K-map simplification.

Octet A group of eight adjacent cells in a Karnaugh map. An octet cancels three variables in a K-map simplification.

In Example 3.14, we derived a sum-of-products Boolean expression from a truth table and simplified the expression by grouping minterms that differed by one variable. We made this task easier by breaking up the truth table into groups of four lines. (It is difficult for the eye to grasp an overall pattern in a group of 16 lines.) We chose groups of four because variables A and B are the same in any one group and variables C and D repeat the same binary sequence in each group. This allows us to see more easily when we have terms differing by only one variable.

The **Karnaugh map,** or K-map, is a graphical tool for simplifying Boolean expressions that uses a similar idea. A K-map is a square or rectangle divided into smaller squares called **cells,** each of which represents a line in the truth table of the Boolean expression to be mapped. Thus, the number of cells in a K-map is always a power of 2, usually 4, 8, or 16. The coordinates of each cell are the input variables of the truth table. The cell content is the value of the output variable on that line of the truth table. Figure 3.39 shows the formats of Karnaugh maps for Boolean expressions having two, three, and four variables, respectively.

There are two equivalent ways of labeling the cell coordinates: numerically or by true and complement variables. We will use the numerical labeling since it is always the same, regardless of the chosen variables.

The cells in the Karnaugh maps are set up so that the coordinates of any two **adjacent cells** differ by only one variable. By grouping adjacent cells according to specified rules, we can simplify a Boolean expression by canceling variables in their true and complement forms, much as we did algebraically in the previous section.

Two-Variable Map

Table 3.12 shows the truth table of a two-variable Boolean expression.

The Karnaugh map shown in Figure 3.40 is another way of showing the same information as the truth table. Every line in the truth table corresponds to a cell, or square, in the Karnaugh map.

The coordinates of each cell correspond to a unique combination of input variables (A, B). The content of the cell is the output value for that input combination. If the truth table output is 1 for a particular line, the content of the corresponding cell is also 1. If the output is 0, the cell content is 0.

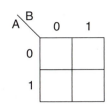

a. Two-variable forms

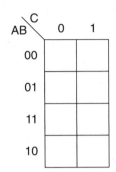

b. Three-variable forms

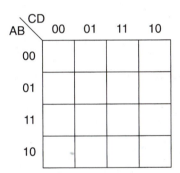

c. Four-variable forms

FIGURE 3.39
Karnaugh Map Formats

Table 3.12 Truth Table for a Two-Variable Boolean Expression

A	B	Y
0	0	1
0	1	1
1	0	0
1	1	0

FIGURE 3.40
Karnaugh Map for Table 3.12

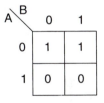

The SOP expression of the truth table is

$$Y = \overline{A}\,\overline{B} + \overline{A}\,B$$

which can be simplified as follows:

$$Y = \overline{A}\,(\overline{B} + B)$$
$$= \overline{A}$$

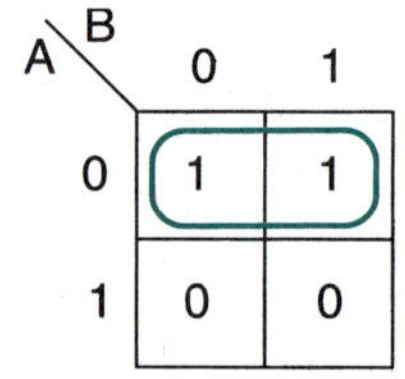

We can perform the same simplification by grouping the adjacent **pair** of 1s in the Karnaugh map, as shown in Figure 3.41.

> **NOTE**
>
> When we circle a pair of 1s in a K-map, we are grouping the common variable in two minterms, then factoring out and canceling the complements.

FIGURE 3.41
Grouping a Pair of Adjacent Cells

To find the simplified form of the Boolean expression represented in the K-map, we examine the coordinates of all the cells in the circled group. We retain coordinate variables that are the same in all cells and eliminate coordinate variables that are different in different cells.

In this case:

$\overline{A}$ is a coordinate of both cells of the circled pair. (Keep $\overline{A}$.)

$\overline{B}$ is a coordinate of one cell of the circled pair, and B is a coordinate of the other. (Discard $B/\overline{B}$.)

$$Y = \overline{A}$$

Three- and Four-Variable Maps

Refer to the forms of three- and four-variable Karnaugh maps shown in Figure 3.39. Each cell is specified by a unique combination of binary variables. This implies that the three-variable map has 8 cells (since $2^3 = 8$) and the four-variable map has 16 cells (since $2^4 = 16$).

The variables specifying the row (both maps) or the column (the four-variable map) do not progress in binary order; they advance such that there is only *one change of variable per row or column*. For example, the numbering of the rows is 00, 01, 11, 10, rather than the binary order 00, 01, 10, 11. If we were to use binary order, adjacent cells in rows 2 and 3 or 3 and 4 would differ by two variables, meaning we could not factor out and cancel a pair of complements by grouping these cells. For instance, we cannot cancel complementary variables from the pair $\overline{A} B C + A \overline{B} C$, which differs by two variables.

> **NOTE**
>
> The number of cells in a group must be a power of 2, such as 1, 2, 4, 8, or 16.

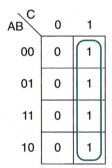

FIGURE 3.42
Quad

A group of four adjacent cells is called a **quad.** Figure 3.42 shows a Karnaugh map for a Boolean function whose terms can be grouped in a quad. The Boolean expression displayed in the K-map is:

$$Y = \overline{A}\,\overline{B}\,C + \overline{A}\,B\,C + A\,B\,C + A\,\overline{B}\,C$$

A and B are both part of the quad coordinates in true and complement form. (Discard A and B.)

C is a coordinate of *each cell* in the quad. (Keep C.)

$$Y = C$$

Grouping cells in a quad is equivalent to factoring two complementary pairs of variables and canceling them.

$$Y = (A + \overline{A})(B + \overline{B})C = C$$

You can verify that this is the same as the original expression by multiplying out the terms.

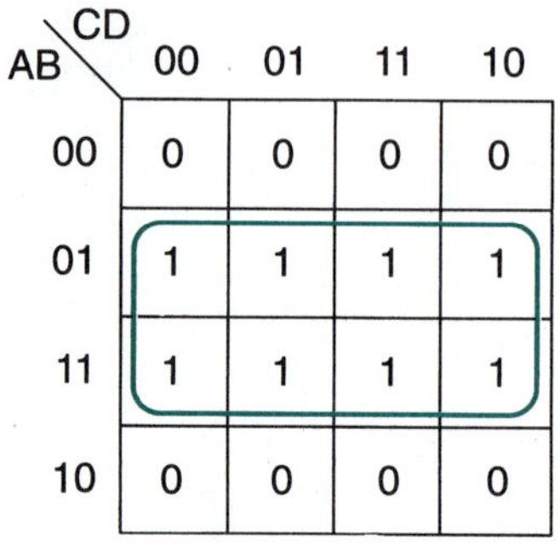

An **octet** is a group of eight adjacent cells. Figure 3.43 shows the Karnaugh map for the following Boolean expression:

FIGURE 3.43
Octet

$$Y = \overline{A}\,B\,\overline{C}\,\overline{D} + \overline{A}\,B\,\overline{C}\,D + \overline{A}\,B\,C\,D + \overline{A}\,B\,C\,\overline{D}$$
$$+ A\,B\,\overline{C}\,\overline{D} + A\,B\,\overline{C}\,D + A\,B\,C\,D + A\,B\,C\,\overline{D}$$

Variables A, C, and D are all coordinates of the octet cells in true and complement form. (Discard A, C, and D.)

B is a coordinate of *each* cell. (Keep B.)

$$Y = B$$

The algebraic equivalent of this octet is an expression where three complementary variables are factored out and canceled.

$$Y = (A + \overline{A})B(C + \overline{C})(D + \overline{D}) = B$$

> **N O T E**
>
> A Karnaugh map completely filled with 1s implies that all input conditions yield an output of 1. For a Boolean expression Y, $Y = 1$.

Grouping Cells Along Outside Edges

The cells along an outside edge of a three- or four-variable map are adjacent to cells along the opposite edge (only one change of variable). Thus we can group cells "around the outside" of the map to cancel variables. In the case of the four-variable map, we can also group the four corner cells as a quad, since they are all adjacent to one another.

▌▌ EXAMPLE 3.15

Use Karnaugh maps to simplify the following Boolean expressions:

a. $Y = \overline{A}\,\overline{B}\,\overline{C} + \overline{A}\,\overline{B}\,C + A\,\overline{B}\,\overline{C} + A\,\overline{B}\,C$

b. $Y = \overline{A}\,\overline{B}\,\overline{C}\,\overline{D} + \overline{A}\,\overline{B}\,C\,\overline{D} + A\,\overline{B}\,\overline{C}\,\overline{D} + A\,\overline{B}\,C\,\overline{D}$

Solutions   Figure 3.44 shows the Karnaugh maps for the Boolean expressions labeled **a** and **b**. Cells in each map are grouped in a quad.

FIGURE 3.44
Example 3.15
K-Maps

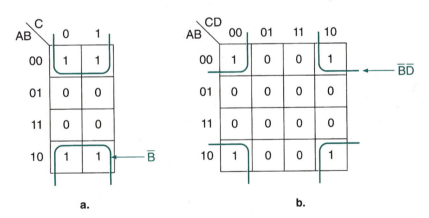

a.

b.

a. A and C are both coordinates of two cells in true form and two cells in complement form. (Discard A and C.)
$\overline{B}$ is a coordinate of each cell. (Keep $\overline{B}$.)

$$Y = \overline{B}$$

b. A and C are both coordinates of two cells in true form and two cells in complement form. (Discard A and C.)
$\overline{B}$ and $\overline{D}$ are coordinates of each cell. (Keep $\overline{B}$ and $\overline{D}$.)

$$Y = \overline{B}\,\overline{D}$$

▌▌

Loading a K-Map From a Truth Table

> **NOTE**
>
> We don't need a Boolean expression to fill a Karnaugh map if we have the function's truth table.

Figures 3.45 and 3.46 show truth table and Karnaugh map forms for three- and four-variable Boolean expressions. The numbers in parentheses show the order of terms in binary sequence for both forms.

The Karnaugh map is not laid out in the same order as the truth table. That is, it is not laid out in a binary sequence. This is due to the criterion for cell adjacency: no more than one variable change between rows or columns is permitted.

FIGURE 3.45

Order of Terms (Three-Variable Function)

A	B	C	Y
0	0	0	(0)
0	0	1	(1)
0	1	0	(2)
0	1	1	(3)
1	0	0	(4)
1	0	1	(5)
1	1	0	(6)
1	1	1	(7)

a. Truth table

AB \ C	0	1
00	(0)	(1)
01	(2)	(3)
11	(6)	(7)
10	(4)	(5)

b. K-map

FIGURE 3.46

Order of Terms (Four-Variable Function)

A	B	C	D	Y
0	0	0	0	(0)
0	0	0	1	(1)
0	0	1	0	(2)
0	0	1	1	(3)
0	1	0	0	(4)
0	1	0	1	(5)
0	1	1	0	(6)
0	1	1	1	(7)
1	0	0	0	(8)
1	0	0	1	(9)
1	0	1	0	(10)
1	0	1	1	(11)
1	1	0	0	(12)
1	1	0	1	(13)
1	1	1	0	(14)
1	1	1	1	(15)

a. Truth table

AB \ CD	00	01	11	10
00	(0)	(1)	(3)	(2)
01	(4)	(5)	(7)	(6)
11	(12)	(13)	(15)	(14)
10	(8)	(9)	(11)	(10)

b. K-map

Filling in a Karnaugh map from a truth table is easy when you understand a system for doing it quickly. For the three-variable map, fill row 1, then row 2, skip to row 4, then go back to row 3. By doing this, you trace through the cells in binary order. Use the mnemonic phrase "1, 2, skip, back" to help you remember this.

The system for the four-variable map is similar but must account for the columns as well. The rows get filled in the same order as the three-variable map, but within each row, fill column 1, then column 2, skip to column 4, then go back to column 3. Again, "1, 2, skip, back."

The four-variable map is easier to fill from the truth table if we break up the truth table into groups of four lines, as we have done in Figure 3.46. Each group is one row in the Karnaugh map. Following this system will quickly fill the cells in binary order.

Go back and follow the order of terms on the four-variable map in Figure 3.46, using this system. (Remember, for both rows and columns, "1, 2, skip, back.")

Multiple Groups

> **NOTE**
>
> If there is more than one group of 1s in a K-map simplification, each group is a term in the maximum SOP simplification of the mapped Boolean expression. The resulting terms are ORed together.

▌▌ EXAMPLE 3.17

Use the Karnaugh map method to simplify the Boolean function represented by Table 3.13.

Table 3.13 Truth Table for Example 4.3

A	B	C	D	Y
0	0	0	0	1
0	0	0	1	0
0	0	1	0	1
0	0	1	1	0
0	1	0	0	0
0	1	0	1	1
0	1	1	0	0
0	1	1	1	1
1	0	0	0	0
1	0	0	1	0
1	0	1	0	0
1	0	1	1	0
1	1	0	0	0
1	1	0	1	1
1	1	1	0	0
1	1	1	1	1

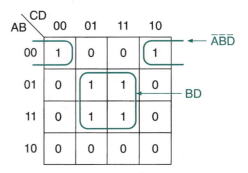

FIGURE 3.47

Example 3.17
K-Map

Solution   Figure 3.47 shows the Karnaugh map for the truth table in Table 3.14. There are two groups of 1s—a pair and a quad.

Pair:

Variables $\overline{A}$, $\overline{B}$, and $\overline{D}$ are coordinates of both cells. (Keep $\overline{A}\,\overline{B}\,\overline{D}$.) C is a coordinate of one cell and $\overline{C}$ is a coordinate of the other. (Discard C.)

Term: $\overline{A}\,\overline{B}\,\overline{D}$

Quad:

Both A and C are coordinates of two cells in true form and two cells in complement form. (Discard A and C.)

B and D are coordinates of all four cells. (Keep $B\,D$.)

Term: $B\,D$

Combine the terms in an OR function:

$$Y = \overline{A}\,\overline{B}\,\overline{D} + B\,D$$

▌▌

Overlapping Groups

> **NOTE**
>
> A cell may be grouped more than once. The only condition is that every group must have at least one cell that does not belong to any other group. Otherwise, redundant terms will result.

█ EXAMPLE 3.18

Simplify the function represented by Table 3.14.

Solution The Karnaugh map for the function in Table 3.14 is shown in Figure 3.48, with two different groupings of terms.

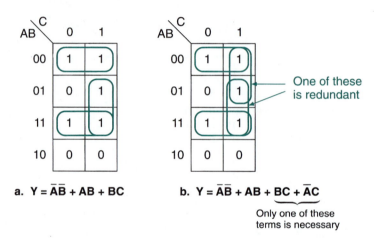

a. $Y = \overline{AB} + AB + BC$

b. $Y = \overline{AB} + AB + BC + \overline{A}C$

One of these is redundant

Only one of these terms is necessary

FIGURE 3.48
Example 3.18
K-Maps

Table 3.14 Truth Table for Example 3.18

A	B	C	Y
0	0	0	1
0	0	1	1
0	1	0	0
0	1	1	1
1	0	0	0
1	0	1	0
1	1	0	1
1	1	1	1

a. The simplified Boolean expression drawn from the first map has three terms.

$$Y = \overline{A}\,\overline{B} + A B + B C$$

b. The second map yields an expression with four terms.

$$Y = \overline{A}\,\overline{B} + A B + B C + \overline{A}\,C$$

One of the last two terms is redundant, since neither of the pairs corresponding to these terms has a cell belonging only to that pair. We could retain either pair of cells and its corresponding term, but not both.

We can show algebraically that the last term is redundant and thus make the expression the same as that in part a.

$$
\begin{aligned}
Y &= \overline{A}\,\overline{B} + A B + B C + \overline{A}\,C \\
&= \overline{A}\,\overline{B} + A B + B C + \overline{A}\,(B + \overline{B})\,C \\
&= \overline{A}\,\overline{B} + A B + B C + \overline{A}\,B\,C + \overline{A}\,\overline{B}\,C \\
&= \overline{A}\,\overline{B}\,(1 + C) + A B + B C\,(1 + \overline{A}) \\
&= \overline{A}\,\overline{B} + A B + B C
\end{aligned}
$$

Conditions for Maximum Simplification

> **NOTE**
>
> The maximum simplification of a Boolean expression is achieved only if the circled groups of cells in its K-map are as large as possible and there are as few groups as possible.

III EXAMPLE 3.19

Find the maximum SOP simplification of the Boolean function represented by Table 3.15.

Solution   The values of Table 3.15 are loaded into the three K-maps shown in Figure 3.49. Three different ways of grouping adjacent cells are shown. One results in maximum simplification; the other two do not.

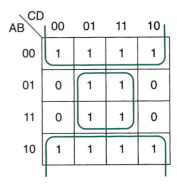

a. **Maximum simplification**
   $Y = \bar{B} + D$

b. **Less than maximum simplification**
   $Y = \bar{A}B + A\bar{B} + D$

c. **Less than maximum simplification**
   $Y = \bar{B} + BD$

FIGURE 3.49
Example 3.19
K-Maps

Table 3.15   Truth Table for
Example 3.19

A	B	C	D	Y
0	0	0	0	1
0	0	0	1	1
0	0	1	0	1
0	0	1	1	1
0	1	0	0	0
0	1	0	1	1
0	1	1	0	0
0	1	1	1	1
1	0	0	0	1
1	0	0	1	1
1	0	1	0	1
1	0	1	1	1
1	1	0	0	0
1	1	0	1	1
1	1	1	0	0
1	1	1	1	1

We get the maximum SOP simplification by grouping the two octets shown in Figure 3.49a. The resulting expression is

a. $Y = \bar{B} + D$

Figures 3.49b and c show two simplifications that are less than the maximum because the chosen cell groups are smaller than they could be. The resulting expressions are:

b. $Y = \bar{A}\,\bar{B} + A\,\bar{B} + D$

c. $Y = \bar{B} + B\,D$

Neither of these expressions is the simplest possible, since both can be reduced by Boolean algebra to the form in Figure 3.49a.

Using K-Maps for Partially Simplified Circuits

Figure 3.50 shows a logic diagram that can be further simplified. If we want to use a Karnaugh map for this process, we must do one of two things:

1. Fill in the K-map from the existing product terms. Each product term that is not a minterm will represent more than one cell in the Karnaugh map. When the map is filled, regroup the cells for maximum simplification.

2. Expand the sum-of-products expression of the circuit to get a sum-of-minterms form. Each minterm represents one cell in the K-map. Group the cells for maximum simplification.

FIGURE 3.50

Logic Diagram That Can Be Further Simplified

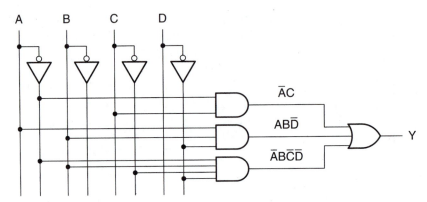

FIGURE 3.51

Further Simplification of Logic Diagram (Figure 3.50)

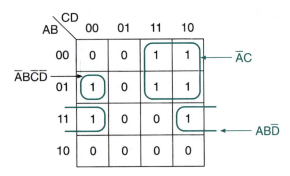

a. K-map from logic diagram (Figure 4.16)

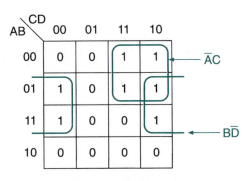

b. Maximum simplification

Figure 3.51 shows the K-map derived from the existing circuit and the regrouped cells that yield the maximum simplification.

The algebraic method requires us to expand the existing Boolean expression to get a sum of minterms. The original expression is:

$$Y = \overline{A} B \overline{C} \overline{D} + A B \overline{D} + \overline{A} C$$

The theorem $(x + \overline{x}) = 1$ implies that we can AND a variable with a term in true and complement form without changing the term. The expanded expression is:

$$Y = \overline{A} B \overline{C} \overline{D} + A B (C + \overline{C}) \overline{D} + \overline{A} (B + \overline{B}) C (D + \overline{D})$$
$$= \overline{A} B \overline{C} \overline{D} + A B C \overline{D} + A B \overline{C} \overline{D}$$
$$+ \overline{A} B C D + \overline{A} B C \overline{D} + \overline{A} \overline{B} C D + \overline{A} \overline{B} C \overline{D}$$

The terms of this expression can be loaded into a K-map and simplified, as shown in Figure 3.51b. Figure 3.52 shows the logic diagram for the simplified expression.

FIGURE 3.52
Simplified Circuit

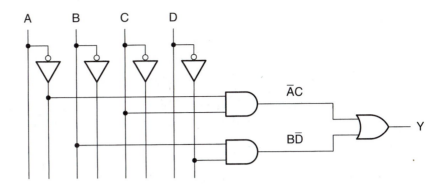

EXAMPLE 3.20

Use a Karnaugh map to find the maximum SOP simplification of the circuit shown in Figure 3.53.

FIGURE 3.53
Example 3.20
Circuit to Be Simplified

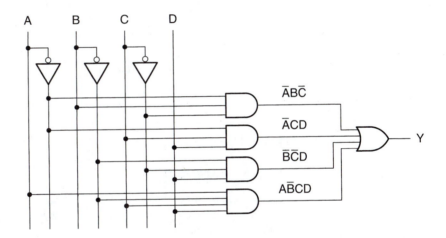

Solution   Figure 3.54a shows the Karnaugh map of Figure 3.53 with terms grouped as shown in the original circuit. Figure 3.54b shows the terms regrouped for the maximum simplification, which is given by:

$$Y = \overline{A}\,D + \overline{B}\,D + \overline{A}\,B\,\overline{C}$$

Alternate method: The Boolean expression for the circuit in Figure 3.53 is:

$$Y = \overline{A}\,B\,\overline{C} + \overline{A}\,C\,D + \overline{B}\,\overline{C}\,D + A\,\overline{B}\,C\,D$$

FIGURE 3.54
Example 3.20
Maximum Simplification of
Figure 3.53

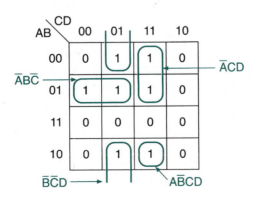

a. K-map from Figure 4.19

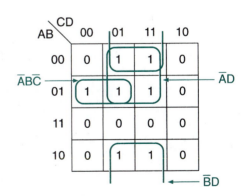

b. Maximum simplification

This expands to the following expression:

$$Y = \overline{A}\,B\,\overline{C}\,(D + \overline{D}) + \overline{A}\,(B + \overline{B})\,C\,D + (A + \overline{A})\,\overline{B}\,\overline{C}\,D + A\,\overline{B}\,C\,D$$
$$= \overline{A}\,B\,\overline{C}\,D + \overline{A}\,B\,\overline{C}\,\overline{D} + \overline{A}\,B\,C\,D + \overline{A}\,\overline{B}\,C\,D + A\,\overline{B}\,\overline{C}\,D$$
$$+ \overline{A}\,\overline{B}\,\overline{C}\,D + A\,\overline{B}\,C\,D$$

This expression can be loaded directly into the K-map and simplified, as shown in Figure 3.54b. The logic diagram for the simplified expression is shown in Figure 3.55.

FIGURE 3.55

Example 3.20
Simplified Circuit

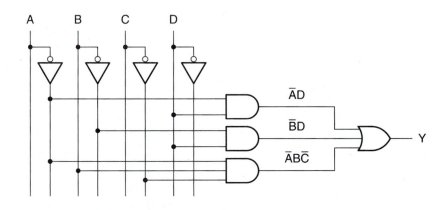

Don't Care States

Sometimes a digital circuit will be intended to work only for certain combinations of inputs; any other input values will never be applied to the circuit.

In such a case, it may be to our advantage to use so-called **don't care states** to simplify the circuit. A don't care state is shown in a K-map cell as an "X" and can be either a 0 or a 1, depending on which case will yield the maximum simplification.

A common application of the don't care state is a digital circuit designed for binary-coded decimal (BCD) inputs. In BCD, a decimal digit (0–9) is encoded as a 4-bit binary number (0000–1001). This leaves six binary states that are never used (1010, 1011, 1100, 1101, 1110, 1111). In any circuit designed for BCD inputs, these states are don't care states.

All cells containing 1s must be grouped if we are looking for a maximum SOP simplification. (If necessary, a group can contain one cell.) The don't care states can be used to maximize the size of these groups. We need not group all don't care states, only those that actually contribute to a maximum simplification.

■■ EXAMPLE 3.21

The circuit in Figure 3.56 is designed to accept binary-coded decimal inputs. The output is HIGH when the input is the BCD equivalent of 5, 7, or 9. If the BCD equivalent of the input is not 5, 7 or 9, the output is LOW. The output is not defined for input values greater than 9.

Find the maximum SOP simplification of the circuit.

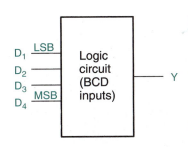

FIGURE 3.56
Example 3.21
Circuit to be Simplified

Solution The Karnaugh map for the circuit is shown in Figure 3.57a.

We can designate three of the don't care cells as 1s—those corresponding to input states 1011, 1101, and 1111. This allows us to group the 1s into two overlapping quads, which yield the following simplification.

$$Y = D_4 D_1 + D_3 D_1$$

The ungrouped don't care states are treated as 0s. The corresponding circuit is shown in Figure 3.57b.

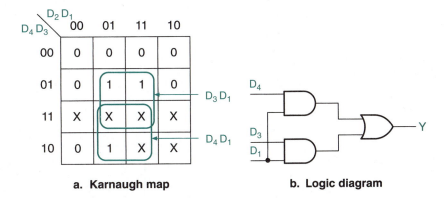

a. Karnaugh map   b. Logic diagram

FIGURE 3.57
Example 3.21
Karnaugh Map and Logic
Diagram

||| EXAMPLE 3.22

Applications

One type of decimal code is called 2421 code, so called because of the positional weights of its bits. (For example, 1011 in 2421 code is equivalent to $2 + 2 + 1 = 5$ in decimal. 1100 is equivalent to decimal $2 + 4 = 6$.) Table 3.16 shows how this code compares to its equivalent decimal digits and to the BCD code used in Example 3.21.

2421 code is sometimes used because it is "self-complementing," a property that BCD code does not have, but that is useful in digital decimal arithmetic circuits.

The bits of the BCD code are designated $D_4 D_3 D_2 D_1$. The bits of the 2421 code are designated $Y_4 Y_3 Y_2 Y_1$.

Use the Karnaugh map method to design a logic circuit that accepts any BCD input and generates an output in 2421 code, as specified by Table 3.16.

Solution The required circuit is called a code converter. Each 4-bit BCD input corresponds to a 4-bit 2421 output. Thus, we must find four Boolean expressions, one for each

Table 3.16 BCD and 2421 Code

Decimal Equivalent	BCD Code				2421 Code			
	D_4	D_3	D_2	D_1	Y_4	Y_3	Y_2	Y_1
0	0	0	0	0	0	0	0	0
1	0	0	0	1	0	0	0	1
2	0	0	1	0	0	0	1	0
3	0	0	1	1	0	0	1	1
4	0	1	0	0	0	1	0	0
5	0	1	0	1	1	0	1	1
6	0	1	1	0	1	1	0	0
7	0	1	1	1	1	1	0	1
8	1	0	0	0	1	1	1	0
9	1	0	0	1	1	1	1	1

FIGURE 3.58
Example 3.22
K-Maps: BCD to 2421

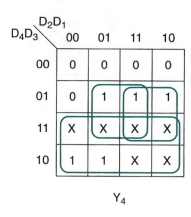

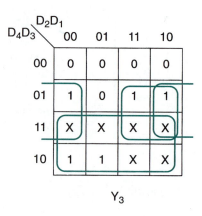

$Y_4 = D_4 + D_3 D_2 + D_3 D_1$
$Y_3 = D_4 + D_3 D_2 + D_3 \overline{D}_1$
$Y_2 = D_4 + \overline{D}_3 D_2 + D_3 \overline{D}_2 D_1$
$Y_1 = D_1$

FIGURE 3.59
Example 3.22
BCD-to-2421 Code Converter

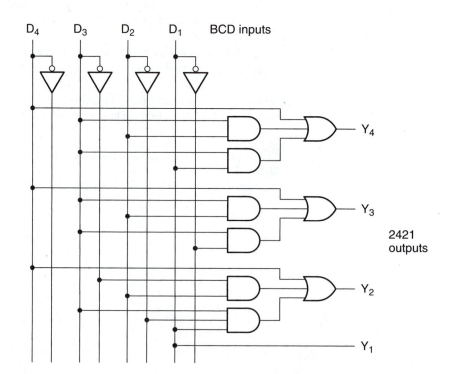

bit of the 2421 code. We can derive each Boolean expression from a truth table represented by the corresponding output column in Table 3.16.

We can load the 2421 values into four different Karnaugh maps, as shown in Figure 3.58. The cells corresponding to the unused input BCD codes 1010, 1011, 1100, 1101, 1110, and 1111 are don't care states in each map.

The K-maps yield the following simplifications:

$$Y_4 = D_4 + D_3 D_2 + D_3 D_1$$
$$Y_3 = D_4 + D_3 D_2 + D_3 \overline{D_1}$$
$$Y_2 = D_4 + \overline{D_3} D_2 + D_3 \overline{D_2} D_1$$
$$Y_1 = D_1$$

Figure 3.59 shows the logic diagram for these equations.

POS Simplification

Until now, we have looked only at obtaining the maximum SOP simplification from a Karnaugh map. It is also possible to find the maximum POS simplification from the same map.

Figure 3.60 shows a Karnaugh map with the cells grouped for an SOP simplification and a POS simplification. The SOP simplification is shown in Figure 3.60a and the POS simplification in Figure 3.60b.

FIGURE 3.60

SOP and POS Forms on a K-Map

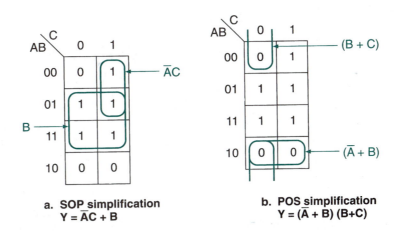

a. **SOP simplification**
   $Y = \overline{A}C + B$

b. **POS simplification**
   $Y = (\overline{A} + B)(B+C)$

When we derive the POS form of an expression from a truth table, we use the lines where the output is 0 and we use the complements of the input variables on these lines as the elements of the selected maxterms. The same principle applies here.

The maxterms are:

$(A + B + C)$	Top left cell
$(\overline{A} + B + C)$	Bottom left cell
$(\overline{A} + B + \overline{C})$	Bottom right cell

The variables are canceled in much the same way as in the SOP form. Remember, however, that the POS variables are the complements of the variables written beside the Karnaugh map.

If there is more than one simplified term, the terms are ANDed together, as in a full POS form.

Cancellations:

Outside pair: A is present in both true and complement form in the pair. (Discard A.)

B and C are present in both cells of the pair. (Keep B and C.)

Term: $B + C$

Bottom pair: $\overline{A}$ and B are present in both cells of the pair. (Keep $\overline{A}$ and B.)

C is present in both true and complement form in the pair. (Discard C.)

Term: $\overline{A} + B$

Maximum POS simplification:

$$Y = (\overline{A} + B)(B + C)$$

Compare this with the maximum SOP simplification:

$$Y = \overline{A}\, C + B$$

By the Boolean theorem $(x + y)(x + z) = x + yz$, we see that the SOP and POS forms are equivalent.

▐▌ EXAMPLE 3.23

Find the maximum POS simplification of the logic function represented by Table 3.17.

Solution Figure 3.61 shows the Karnaugh map from the truth table in Table 3.17. The cells containing 0s are grouped in two quads and there is a single 0 cell left over.

Table 3.17 Truth Table for Example 3.23

A	B	C	D	Y
0	0	0	0	0
0	0	0	1	0
0	0	1	0	0
0	0	1	1	0
0	1	0	0	1
0	1	0	1	1
0	1	1	0	1
0	1	1	1	1
1	0	0	0	0
1	0	0	1	1
1	0	1	0	0
1	0	1	1	1
1	1	0	0	1
1	1	0	1	0
1	1	1	0	1
1	1	1	1	1

Simplification:

Corner quad:	$(B + D)$
Horizontal quad:	$(A + B)$
Single cell:	$(\overline{A} + \overline{B} + C + \overline{D})$

$$Y = (A + B)(B + D)(\overline{A} + \overline{B} + C + \overline{D})$$

FIGURE 3.61
Example 3.23
POS Simplification of Table 3.17

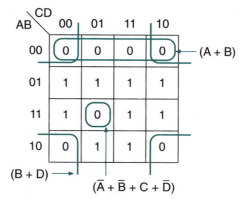

SUMMARY

1. Two or more gates connected together form a logic gate network or combinational logic circuit, which can be described by a truth table, a logic diagram, or a Boolean expression.

2. The output of a combinational logic circuit is always the same with the same combination of inputs, regardless of the order in which they are applied.

3. The order of precedence in a logic gate network is AND, then OR, unless otherwise indicated by parentheses.

4. DeMorgan's theorems: $\overline{x \cdot y} = \overline{x} + \overline{y}$
   $$\overline{x + y} = \overline{x} \cdot \overline{y}$$

5. Inequalities: $\overline{x \cdot y} \neq \overline{x} \cdot \overline{y}$
   $$\overline{x + y} \neq \overline{x} + \overline{y}$$

6. A logic gate network can be drawn to simplify its Boolean expression by ensuring that bubbled (active-LOW) outputs drive bubbled inputs and outputs with no bubble (active-HIGH) drive inputs with no bubble. Some gates might need

to be drawn in their DeMorgan equivalent form to achieve this.

7. In Boolean expressions, logic inversion bars of equal lengths cancel; bars of unequal lengths do not. Bars of equal length represent bubble-to-bubble connections.

8. A logic diagram can be derived from a Boolean expression by order of precedence rules: synthesize ANDs before ORs, unless parentheses indicate otherwise. Inversion bars act as parentheses for a group of variables.

9. A truth table can be derived from a logic gate network either by finding truth tables for intermediate points in the network and combining them by the laws of Boolean algebra, or by simplifying the Boolean expression into a form that can be directly written into a truth table.

10. A sum-of-products (SOP) network combines inputs in AND gates to yield a group of product terms that are combined in an OR gate (logical sum) output.

11. A product-of-sums (POS) network combines inputs in OR gates to yield a group of sum terms that are combined in an AND gate (logical product) output.

12. An SOP Boolean expression can be derived from the lines in a truth table where the output is at logic 1. Each product term contains all inputs in true or complement form, where inputs at logic 0 have a bar and inputs at logic 1 do not.

13. A POS expression is derived from the lines where the output is at logic 0. Each sum term contains all inputs in true or complement form, where inputs at logic 1 have a bar and inputs at logic 0 do not.

14. Theorems of Boolean algebra, summarized in Table 3.8, allow us to simplify logic gate networks.

15. SOP networks can be simplified by grouping pairs of product terms and applying the Boolean identity $xyz + xy\bar{z} = xy$.

16. POS networks can be simplified by grouping pairs of sum terms and applying the Boolean identity $(x + y + z)(x + y + \bar{z}) = (x + y)$.

17. To achieve maximum simplification of an SOP or POS network, each product or sum term should be grouped with another if possible. A product or sum term can be grouped more than once, as long as each group has a term that is only in that group.

18. A Karnaugh map can be used to graphically reduce a Boolean expression to its simplest form by grouping adjacent cells containing 1s. One cell is equivalent to one line of a truth table. A group of adjacent cells that contain 1s represents a simplified product term.

19. Adjacent cells in a K-map differ by only one variable. Cells around the outside of the map are considered adjacent.

20. A group in a K-map must be a power of two in size: 1, 2, 4, 8, or 16. A group of two is called a pair, a group of four is a quad, and a group of eight is an octet.

21. A pair cancels one variable. A quad cancels two variables. An octet cancels three variables.

22. A K-map can have multiple groups. Each group represents one simplified product term in a sum-of-products expression.

23. Groups in K-maps can overlap as long as each group has one or more cells that appear only in that group.

24. Groups in a K-map should be as large as possible for maximum SOP simplification.

25. Don't care states represent output states of input combinations that will never occur in a circuit. They are represented by Xs in a truth table or K-map and can be used as 0s or 1s, whichever is most advantageous for the simplification of the circuit.

GLOSSARY

Adjacent cell   Two cells are adjacent if there is only one variable that is different between the coordinates of the two cells.

Associative property   A mathematical function is associative if its operands can be grouped in any order without affecting the result. For example, addition is associative $((a + b) + c = a + (b + c))$, but subtraction is not $((a - b) - c \neq a - (b - c))$.

Bubble-to-bubble convention   The practice of drawing gates in a logic diagram so that inverting outputs connect to inverting inputs and noninverting outputs connect to noninverting inputs.

Bus form   A way of drawing a logic diagram so that each true and complement input variable is available along a conductor called a bus.

Cell   The smallest unit of a Karnaugh map, corresponding to one line of a truth table. The input variables are the cell's coordinates and the output variable is the cell's contents.

Combinational logic   Digital circuitry in which an output is derived from the combination of inputs, independent of the order in which they are applied.

Combinatorial logic   Another name for combinational logic.

Commutative property   A mathematical operation is commutative if it can be applied to its operands in any order without

affecting the result. For example, addition is commutative $(a + b = b + a)$, but subtraction is not $(a - b \neq b - a)$.

Distributive property   Full name: distributive property of multiplication over addition. The property that allows us to distribute ("multiply through") an AND across several OR functions. For example, $a(b + c) = ac + bc$.

Don't care state   An output state that can be regarded either as HIGH or LOW, as is most convenient. A don't care state is the output state of a circuit for a combination of inputs that will never occur.

Karnaugh map   A graphical tool for finding the maximum SOP or POS simplification of a Boolean expression. A Karnaugh map works by arranging the terms of an expression in such a way that variables can be cancelled by grouping minterms or maxterms.

Levels of gating   The number of gates through which a signal must pass from input to output of a logic gate network.

Logic diagram   A diagram, similar to a schematic, showing the connection of logic gates.

Logic gate network   Two or more logic gates connected together.

Maximum POS simplification   The form of a POS Boolean expression which cannot be further simplified by cancelling variables in the sum terms. It may be possible to get an SOP form with fewer terms or variables.

Maximum SOP simplification   The form of an SOP Boolean expression which cannot be further simplified by cancelling variables in the product terms. It may be possible to get a POS form with fewer terms or variables.

Maxterm   A sum term in a Boolean expression where all possible variables appear once in true or complement form.

Minterm   A product term in a Boolean expression where all possible variables appear once in true or complement form.

Octet   A group of eight cells in a Karnaugh map. An octet cancels three variables in a K-map simplification.

Order of precedence   The sequence in which Boolean functions are performed, unless otherwise specified by parentheses.

Pair   A group of two cells in a Karnaugh map. A pair cancels one variable in a K-map simplification.

Product term   A term in a Boolean expression where one or more true or complement variables are ANDed.

Product-of-sums (POS)   A type of Boolean expression where several sum terms are multiplied (ANDed) together.

Quad   A group of four cells in a Karnaugh map. A quad cancels two variables in a K-map simplification.

Sum term   A term in a Boolean expression where one or more true or complement variables are ORed.

Sum-of-products (SOP)   A type of Boolean expression where several product terms are summed (ORed) together.

Synthesis   The process of creating a logic circuit from a description such as a Boolean equation or truth table.

PROBLEMS

Problem numbers set in color indicate more difficult problems: those with underlines indicate most difficult problems.

Section 3.1   Boolean Expressions, Logic Diagrams, and Truth Tables

3.1   Write the unsimplified Boolean expression for each of the logic gate networks shown in Figure 3.62.

FIGURE 3.62
Problem 3.1
Logic Circuits

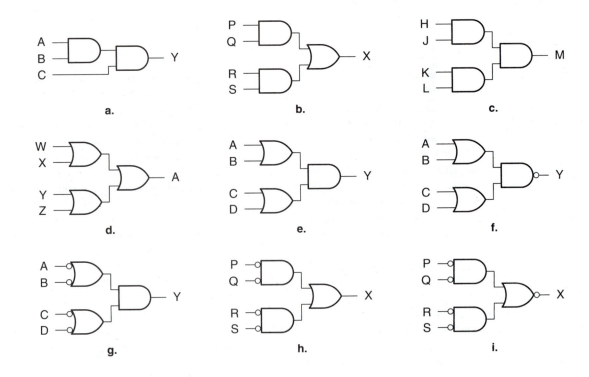

3.2 Write the unsimplified Boolean expression for each of the logic gate networks shown in Figure 3.63.

3.3 Redraw the logic diagrams of the gate networks shown in Figure 3.63 a, e, f, h, i, and j so that they conform to the bubble-to-bubble convention. Rewrite the Boolean expression of each of the redrawn circuits.

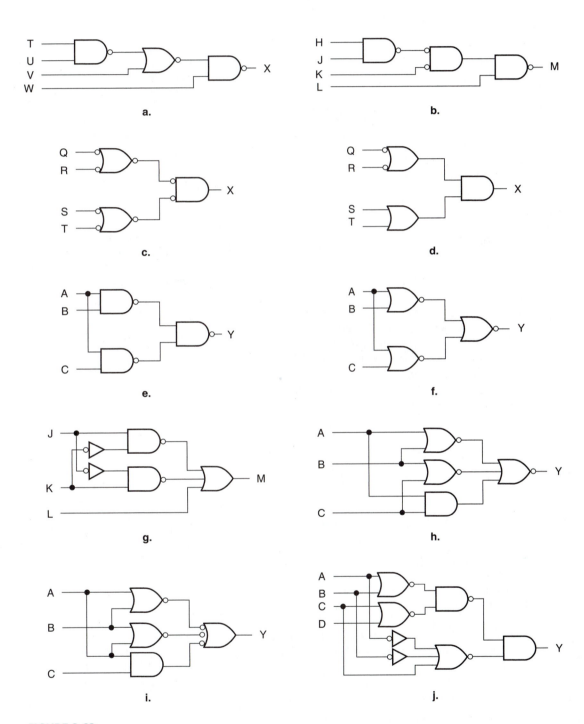

FIGURE 3.63
Problem 3.2
Logic Circuits

3.4   The circuit in Figure 3.64 is called a majority vote circuit. It will turn on an active-HIGH indicator lamp only if a majority of inputs (two out of three) are HIGH. Write the Boolean expression for the circuit.

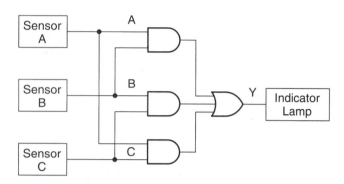

FIGURE 3.64
Problem 3.4
Majority Vote Circuit

3.5   Suppose you wish to design a circuit that indicates when three out of four inputs are HIGH. The circuit has four inputs, D_3, D_2, D_1, and D_0 and an active-HIGH output, Y. Write the Boolean expression for the circuit and draw the logic circuit.

3.6   Draw the logic circuit for each of the following Boolean expressions:

a. $Y = AB + BC$

b. $Y = ACD + BCD$

c. $Y = (A + B)(C + D)$

d. $Y = A + BC + D$

e. $Y = \overline{AC + B + C}$

f. $Y = \overline{\overline{AC} + B} + C$

g. $Y = \overline{\overline{\overline{ABD} + \overline{BC}} + A} + C$

h. $Y = \overline{AB} + \overline{\overline{AC}} + \overline{BC}$

i. $Y = \overline{\overline{AB} + \overline{AC} + BC}$

3.7   Use DeMorgan's theorems to modify the Boolean equations in Problem 3.6, parts e, f, g, h, and i so that there is no bar over any group of variables. Redraw the logic diagrams of the circuits to reflect the changes. (The final circuit versions should conform to the bubble-to-bubble convention.)

3.8   Write the truth tables for the logic diagrams in Figure 3.62, parts b, e, f, and g.

3.9   Write the truth tables for the logic diagrams in Figure 3.63, parts a, h, i, and j.

3.10 Write the truth tables for the Boolean expression in Problem 3.6, parts c, d, e, f, h, and i.

Section 3.2   Sum-of-Products (SOP) and Product-of-Sums (POS) Forms

3.11 Find the Boolean expression, in both sum-of-products (SOP) and product-of-sums (POS) forms, for the logic function represented by the following truth table. Draw the logic diagram for each form.

A	B	C	Y
0	0	0	1
0	0	1	1
0	1	0	1
0	1	1	1
1	0	0	0
1	0	1	0
1	1	0	0
1	1	1	0

3.12 Find the Boolean expression, in both sum-of-products (SOP) and product-of-sums (POS) forms, for the logic function represented by the following truth table. Draw the logic diagram for the SOP form only.

A	B	C	Y
0	0	0	0
0	0	1	1
0	1	0	0
0	1	1	0
1	0	0	1
1	0	1	0
1	1	0	1
1	1	1	0

3.13 Find the Boolean expression, in both sum-of-products (SOP) and product-of-sums (POS) forms, for the logic function represented by the following truth table. Draw the logic diagram for the POS form only.

A	B	C	Y
0	0	0	0
0	0	1	1
0	1	0	1
0	1	1	0
1	0	0	0
1	0	1	1
1	1	0	1
1	1	1	1

3.14 Find the Boolean expression, in both sum-of-products (SOP) and product-of-sums (POS) forms, for the logic function represented by the following truth table. Draw the logic diagram for the SOP form only.

A	B	C	D	Y
0	0	0	0	0
0	0	0	1	0
0	0	1	0	0
0	0	1	1	0
0	1	0	0	1
0	1	0	1	0
0	1	1	0	1
0	1	1	1	0
1	0	0	0	1
1	0	0	1	0
1	0	1	0	0
1	0	1	1	0
1	1	0	0	0
1	1	0	1	1
1	1	1	0	0
1	1	1	1	0

3.15 Write the POS form of the 2-input XOR function. Draw the logic diagram of the POS form of the XOR function.

3.16 Write the POS form of the 2-input XNOR function. Draw the logic diagram of the POS form of the XNOR function.

Section 3.3   Theorems of Boolean Algebra

3.17 Write the Boolean expression for the circuit shown in Figure 3.65 Use the distributive property to transform the circuit into a sum-of-products (SOP) circuit.

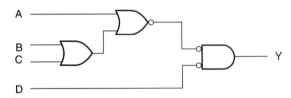

FIGURE 3.65
Problem 3.17
Logic Circuit

3.18 Write the Boolean expression for the circuit shown in Figure 3.66 Use the distributive property to transform the circuit into a sum-of-products (SOP) circuit.

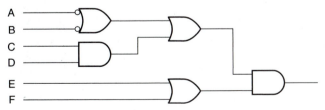

FIGURE 3.66
Problem 3.18
Logic Circuit

3.19 Use the rules of Boolean algebra to simplify the following expressions as much as possible.

a. $Y = A A B + C$

b. $Y = A \overline{A} B + C$

c. $J = K + L \overline{L}$

d. $S = (T + U) V \overline{V}$

e. $S = T + V \overline{V}$

f. $Y = (A \overline{B} + \overline{C})(B \overline{D} + F)$

3.20 Use the rules of Boolean algebra to simplify the following expressions as much as possible.

a. $M = P Q + \overline{P Q} R$

b. $M = P Q + P \overline{Q} R$

c. $S = (\overline{T + U}) V + (T + U)$

d. $Y = (\overline{A} + B + \overline{D}) A C + A \overline{B} D$

e. $Y = (\overline{A} + B + \overline{D}) A C + \overline{A} \overline{B} D$

f. $P = (\overline{Q} R + S T)(\overline{Q} R + Q)$

g. $U = (X + \overline{Y} + \overline{W} Z)(W Y + Y + \overline{W} Z)$

3.21 Use the rules of Boolean algebra to simplify the following expressions as much as possible.

a. $Y = \overline{A B} C D + \overline{(A + B)} \overline{C + D} + \overline{A} + \overline{B}$

b. $Y = \overline{A B} C D + \overline{(A + B)} \overline{C + D} + \overline{\overline{A} + \overline{B}}$

c. $K = (\overline{L} M + L \overline{M})(M \overline{N} + L M N) + M(\overline{N} + L)$

Section 3.4   Simplifying SOP and POS Expressions

3.22 Use the rules of Boolean algebra to find the maximum SOP and POS simplifications of the function represented by the following truth table.

A	B	C	Y
0	0	0	0
0	0	1	1
0	1	0	0
0	1	1	1
1	0	0	0
1	0	1	1
1	1	0	0
1	1	1	1

3.23 Use the rules of Boolean algebra to find the maximum SOP and POS simplifications of the function represented by the following truth table.

A	B	C	Y
0	0	0	1
0	0	1	0
0	1	0	1
0	1	1	0
1	0	0	0
1	0	1	0
1	1	0	1
1	1	1	0

3.24 Use the rules of Boolean algebra to find the maximum SOP and POS simplifications of the function represented by the following truth table.

A	B	C	Y
0	0	0	0
0	0	1	1
0	1	0	0
0	1	1	1
1	0	0	0
1	0	1	1
1	1	0	0
1	1	1	0

3.25 Use the rules of Boolean algebra to find the maximum SOP simplification of the function represented by the following truth table.

A	B	C	D	Y
0	0	0	0	0
0	0	0	1	0
0	0	1	0	0
0	0	1	1	0
0	1	0	0	1
0	1	0	1	1
0	1	1	0	0
0	1	1	1	0
1	0	0	0	0
1	0	0	1	1
1	0	1	0	0
1	0	1	1	1
1	1	0	0	1
1	1	0	1	1
1	1	1	0	0
1	1	1	1	1

3.26 Use the rules of Boolean algebra to find the maximum SOP simplification of the function represented by the following truth table.

A	B	C	D	Y
0	0	0	0	1
0	0	0	1	0
0	0	1	0	1
0	0	1	1	0
0	1	0	0	1
0	1	0	1	0
0	1	1	0	0
0	1	1	1	0
1	0	0	0	0
1	0	0	1	0
1	0	1	0	0
1	0	1	1	0
1	1	0	0	1
1	1	0	1	1
1	1	1	0	0
1	1	1	1	0

3.27 Use the rules of Boolean algebra to find the maximum SOP simplification of the function represented by the following truth table.

A	B	C	D	Y
0	0	0	0	0
0	0	0	1	1
0	0	1	0	0
0	0	1	1	1
0	1	0	0	0
0	1	0	1	1
0	1	1	0	1
0	1	1	1	1
1	0	0	0	0
1	0	0	1	1
1	0	1	0	0
1	0	1	1	0
1	1	0	0	0
1	1	0	1	1
1	1	1	0	1
1	1	1	1	0

3.28 Use the rules of Boolean algebra to find the maximum SOP simplification of the function represented by the following truth table.

A	B	C	D	Y
0	0	0	0	1
0	0	0	1	0
0	0	1	0	0
0	0	1	1	1
0	1	0	0	1
0	1	0	1	0
0	1	1	0	0
0	1	1	1	0
1	0	0	0	1
1	0	0	1	0
1	0	1	0	1
1	0	1	1	1
1	1	0	0	1
1	1	0	1	0
1	1	1	0	1
1	1	1	1	1

3.29 Use the rules of Boolean algebra to find the maximum SOP simplification of the function represented by the following truth table.

A	B	C	D	Y
0	0	0	0	0
0	0	0	1	1
0	0	1	0	0
0	0	1	1	0
0	1	0	0	0
0	1	0	1	0
0	1	1	0	1
0	1	1	1	1
1	0	0	0	1
1	0	0	1	0
1	0	1	0	0
1	0	1	1	0
1	1	0	0	0
1	1	0	1	0
1	1	1	0	1
1	1	1	1	1

3.30 Use the rules of Boolean algebra to find the maximum SOP simplification of the function represented by the following truth table.

A	B	C	D	Y
0	0	0	0	0
0	0	0	1	0
0	0	1	0	0
0	0	1	1	1
0	1	0	0	0
0	1	0	1	1
0	1	1	0	0
0	1	1	1	1
1	0	0	0	0
1	0	0	1	0
1	0	1	0	0
1	0	1	1	1
1	1	0	0	0
1	1	0	1	1
1	1	1	0	0
1	1	1	1	1

Section 3.4 Simplification by the Karnaugh Map Method

3.31 Use the Karnaugh map method to find the maximum SOP simplification of the logic diagram in Figure 3.21.

3.32 Use the Karnaugh map method to reduce the following Boolean expressions to their maximum SOP simplifications:

a. $Y = \overline{A}\,\overline{B}\,C + \overline{A}\,B\,C + A\,B\,C$

b. $Y = \overline{A}\,\overline{B}\,C + \overline{A}\,B\,C + A\,B\,\overline{C} + A\,B\,C + A\,\overline{B}\,C$

c. $Y = \overline{A}\,\overline{B}\,\overline{C} + \overline{A}\,B\,C + A\,B\,C + A\,\overline{B}\,C$

d. $Y = \overline{A}\,\overline{B}\,\overline{C}\,\overline{D} + \overline{A}\,\overline{B}\,\overline{C}\,D + \overline{A}\,\overline{B}\,C\,D + \overline{A}\,B\,C\,\overline{D} + \overline{A}\,B\,C\,\overline{D} + A\,B\,\overline{C}\,D + A\,B\,C\,\overline{D} + A\,\overline{B}\,\overline{C}\,\overline{D} + A\,\overline{B}\,C\,\overline{D}$

3.33 Use the Karnaugh map method to reduce the Boolean expression represented by the following truth table to simplest SOP form.

A	B	C	D	Y
0	0	0	0	0
0	0	0	1	0
0	0	1	0	0
0	0	1	1	1
0	1	0	0	1
0	1	0	1	1
0	1	1	0	1
0	1	1	1	1
1	0	0	0	0
1	0	0	1	0
1	0	1	0	0
1	0	1	1	1
1	1	0	0	0
1	1	0	1	0
1	1	1	0	0
1	1	1	1	1

3.34 Use the Karnaugh map method to reduce the Boolean expression represented by the following truth table to simplest SOP form.

A	B	C	D	Y
0	0	0	0	0
0	0	0	1	1
0	0	1	0	0
0	0	1	1	1
0	1	0	0	0
0	1	0	1	1
0	1	1	0	0
0	1	1	1	1
1	0	0	0	1
1	0	0	1	0
1	0	1	0	0
1	0	1	1	0
1	1	0	0	0
1	1	0	1	1
1	1	1	0	0
1	1	1	1	1

3.35 Use the Karnaugh map method to reduce the Boolean expression represented by the following truth table to simplest SOP form.

A	B	C	D	Y
0	0	0	0	0
0	0	0	1	0
0	0	1	0	1
0	0	1	1	1
0	0	1	0	0
0	1	0	1	0
0	1	1	0	0
0	1	1	1	0
1	0	0	0	0
1	0	0	1	1
1	0	1	0	X
1	0	1	1	X
1	1	0	0	X
1	1	0	1	X
1	1	1	0	X
1	1	1	1	X

3.36 Use the Karnaugh map method to reduce the Boolean expression represented by the following truth table to simplest SOP form.

A	B	C	D	Y
0	0	0	0	1
0	0	0	1	1
0	0	1	0	1
0	0	1	1	1
0	1	0	0	0
0	1	0	1	0
0	1	1	0	1
0	1	1	1	0
1	0	0	0	1
1	0	0	1	0
1	0	1	0	1
1	0	1	1	0
1	1	0	0	0
1	1	0	1	1
1	1	1	0	1
1	1	1	1	0

3.37 Use the Karnaugh map method to reduce the Boolean expression represented by the following truth table to simplest SOP form.

A	B	C	D	Y
0	0	0	0	0
0	0	0	1	1
0	0	1	0	0
0	0	1	1	1
0	1	0	0	0
0	1	0	1	1
0	1	1	0	0
0	1	1	1	1
1	0	0	0	1
1	0	0	1	0
1	0	1	0	0
1	0	1	1	1
1	1	0	0	0
1	1	0	1	0
1	1	1	0	0
1	1	1	1	1

3.38 Use the Karnaugh map method to reduce the Boolean expression represented by the following truth table to simplest SOP form.

A	B	C	D	Y
0	0	0	0	1
0	0	0	1	1
0	0	1	0	0
0	0	1	1	0
0	1	0	0	1
0	1	0	1	1
0	1	1	0	0
0	1	1	1	1
1	0	0	0	0
1	0	0	1	0
1	0	1	0	1
1	0	1	1	1
1	1	0	0	0
1	1	0	1	1
1	1	1	0	1
1	1	1	1	1

3.39 Use the Karnaugh map method to reduce the Boolean expression represented by the following truth table to simplest SOP form.

A	B	C	D	Y
0	0	0	0	0
0	0	0	1	0
0	0	1	0	0
0	0	1	1	0
0	1	0	0	1
0	1	0	1	1
0	1	1	0	1
0	1	1	1	1
1	0	0	0	0
1	0	0	1	1
1	0	1	0	0
1	0	1	1	0
1	1	0	0	1
1	1	0	1	0
1	1	1	0	1
1	1	1	1	0

3.40 Use the Karnaugh map method to reduce the Boolean expression represented by the following truth table to simplest SOP form.

A	B	C	D	Y
0	0	0	0	1
0	0	0	1	0
0	0	1	0	1
0	0	1	1	1
0	1	0	0	0
0	1	0	1	0
0	1	1	0	0
0	1	1	1	1
1	0	0	0	0
1	0	0	1	0
1	0	1	0	1
1	0	1	1	1
1	1	0	0	0
1	1	0	1	0
1	1	1	0	0
1	1	1	1	1

3.41 Use the Karnaugh map method to reduce the Boolean expression represented by the following truth table to simplest SOP form.

A	B	C	D	Y
0	0	0	0	0
0	0	0	1	1
0	0	1	0	0
0	0	1	1	1
0	1	0	0	0
0	1	0	1	1
0	1	1	0	0
0	1	1	1	1
1	0	0	0	0
1	0	0	1	1
1	0	1	0	0
1	0	1	1	1
1	1	0	0	0
1	1	0	1	1
1	1	1	0	0
1	1	1	1	1

3.42 Use the Karnaugh map method to reduce the Boolean expression represented by the following truth table to simplest SOP form.

A	B	C	D	Y
0	0	0	0	1
0	0	0	1	1
0	0	1	0	1
0	0	1	1	1
0	1	0	0	0
0	1	0	1	0
0	1	1	0	1
0	1	1	1	1
1	0	0	0	1
1	0	0	1	1
1	0	1	0	1
1	0	1	1	1
1	1	0	0	0
1	1	0	1	0
1	1	1	0	0
1	1	1	1	0

3.43 Use the Karnaugh map method to reduce the Boolean expression represented by the following truth table to simplest SOP form.

A	B	C	D	Y
0	0	0	0	0
0	0	0	1	0
0	0	1	0	0
0	0	1	1	0
0	1	0	0	0
0	1	0	1	0
0	1	1	0	0
0	1	1	1	1
1	0	0	0	1
1	0	0	1	1
1	0	1	0	1
1	0	1	1	1
1	1	0	0	1
1	1	0	1	1
1	1	1	0	0
1	1	1	1	1

3.44 The circuit in Figure 3.67 represents the maximum SOP simplification of a Boolean function.

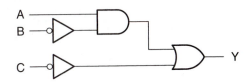

FIGURE 3.67
Problem 3.44:
Logic Circuit

Use a Karnaugh map to derive the circuit for the maximum POS simplification.

3.45 Repeat Problem 3.44 for the circuit in Figure 3.68.

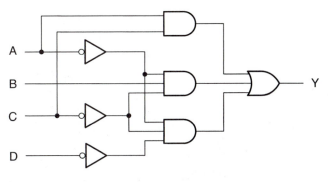

FIGURE 3.68
Problem 3.45:
Logic Circuit

3.46 Refer to the BCD-to-2421 code converter developed in Example 3.22. Use a similar design procedure to develop the circuit of a 2421-to-BCD code converter.

3.47 Excess-3 code is a decimal code that is generated by adding 0011 (= 3_{10}) to a BCD code. Table 3.18 shows the relationship between a decimal digital code, natural BCD code, and Excess-3 code. Draw the circuit of a BCD-to-Excess-3 code converter, using the Karnaugh map method to simplify all Boolean expressions.

3.48 Repeat Problem 3.47 for an Excess-3-to-BCD code converter.

Table 3.18 BCD and Excess-3 Code

Decimal Equivalent	BCD Code				Excess-3			
	D_4	D_3	D_2	D_1	E_4	E_3	E_2	E_1
0	0	0	0	0	0	0	1	1
1	0	0	0	1	0	1	0	0
2	0	0	1	0	0	1	0	1
3	0	0	1	1	0	1	1	0
4	0	1	0	0	0	1	1	1
5	0	1	0	1	1	0	0	0
6	0	1	1	0	1	0	0	1
7	0	1	1	1	1	0	1	0
8	1	0	0	0	1	0	1	1
9	1	0	0	1	1	1	0	0

ANSWERS TO SECTION REVIEW PROBLEMS

Section 3.1

3.1a $Y = \overline{ABC + D}$ **b** $OUT = (\overline{W} + \overline{X} + Y)\overline{Z}$

Section 3.2

A	B	C	Y
0	0	0	0
0	0	1	1
0	1	0	1
0	1	1	1
1	0	0	0
1	0	1	1
1	1	0	0
1	1	1	0

Section 3.3

3.3a SOP: $Y = A\,\overline{B}\,\overline{C} + A\,\overline{B}\,C$

POS: $Y = (A + B + C)(A + B + \overline{C})(A + \overline{B} + C)$
$(A + \overline{B} + \overline{C})(\overline{A} + \overline{B} + C)(\overline{A} + \overline{B} + \overline{C})$

b SOP: $Y = \overline{A}\,\overline{B}\,\overline{C} + A\,\overline{B}\,\overline{C} + A\,\overline{B}\,C + A\,B\,\overline{C}$

POS: $Y = (A + B + \overline{C})(A + \overline{B} + C)(A + \overline{B} + \overline{C})$
$(\overline{A} + \overline{B} + \overline{C})$

Section 3.4

3.4a $Y = \overline{AC}$ or $Y = \overline{A} + \overline{C}$ **3.4b** $Y = \overline{AC} + D$ or
$Y = \overline{A} + \overline{C} + D$

3.4c $Y = A\overline{B}$

Section 3.5

3.5 SOP: $Y = \overline{A}C + B\overline{C}$ POS: $Y = (\overline{A} + \overline{C})(B + C)$

Introduction to PLDs and MAX+PLUS II

CHAPTER OBJECTIVES

Upon successful completion of this chapter you will be able to:

• Describe some advantages of programmable logic over fixed-function logic.

• Name some types of programmable logic devices (PLDs).

• Use Altera's MAX+PLUS II PLD Design Software to enter simple combinational circuits using schematic capture.

• Use VHDL entity declarations, architecture bodies, and concurrent signal assignments to enter simple combinational circuits.

• Create circuit symbols from schematic or VHDL designs and use them in hierarchical designs for PLDs.

• Assign device and pin numbers to schematic or VHDL designs and compile them for programming Altera MAX7000S or FLEX10K20 devices.

• Program Altera PLDs via a JTAG interface and a ByteBlaster Parallel Port Download Cable.

In the first three chapters of this book, we examined logic gates and Boolean algebra. These basic foundations of combinational circuitry, as well as the sequential logic circuits we will study in a later chapter, form the fundamental building blocks of many digital integrated circuits (ICs).

In the past, such digital ICs were fixed in their logic functions; it was not possible to change designs without changing the chips in a circuit. Programmable logic offers the digital circuit designer the possibility of changing design function even after it has been built. A programmable logic device (PLD) can be programmed, erased, and reprogrammed many times, allowing easier prototyping and design modification. (The industry marketing buzz often refers to "rapid prototyping" and "reduced time to market.") The number of IC packages required to implement a design with one or more PLDs is often reduced, compared to a design fabricated using standard fixed-function ICs.

PLDs can be programmed from a personal computer (PC) or workstation running special software. This software is often associated with a set of programs that allow us to design circuits for various PLDs. MAX+PLUS II, owned by Altera Corporation, is such a software package. MAX+PLUS II allows us to enter PLD designs, either as schematics or in several hardware description languages (specialized computer languages for modeling and synthesizing digital hardware). A design can contain components that are in themselves complete digital circuits. MAX+PLUS II converts the design information into a binary form that can be transferred into a PLD via a special interface connected to the parallel port of a PC.   ■

4.1   What Is a PLD?

One of the most far-reaching developments in digital electronics has been the introduction of **programmable logic devices (PLDs).** Prior to the development of PLDs, digital circuits were constructed in various scales of integrated circuit logic, such as small scale integration (SSI) and medium scale integration (MSI) devices. These devices contained logic gates and other digital circuits. The functions were determined at the time of manufacture and could not be changed. This necessitated the manufacture of a large number of device types, requiring shelves full of data books just to describe them. Also, if a designer wanted a device with a particular function that was not in a manufacturer's list of offerings, he or she was forced to make a circuit that used multiple devices, some of which might contain functions neither wanted nor needed, thus wasting circuit board space and design time.

Programmable logic provides a solution to these problems. A PLD is supplied to the user with no logic function programmed in at all. It is up to the designer to make the PLD perform in whatever way a design requires; only those functions required by the design need be programmed. Since several functions can usually be combined in the design and programmed onto a single chip, the package count and required board space can be reduced as well. Also, if a design needs to be changed, a PLD can be reprogrammed with the new design information, often without removing it from the circuit.

PLD is a generic term. There is a wide variety of PLD types, including PAL (programmable array logic), GAL (generic array logic), EPLD (erasable PLD), **CPLD (complex PLD),** FPGA (field-programmable gate array), as well as several others. We will be focussing on CPLDs as a representative type of PLD. Although terminology varies somewhat throughout the industry, we will use the term CPLD to mean a device with several programmable sections that are connected internally. In effect, a CPLD is several interconnected PLDs on a single chip. This structure is not apparent to the user and doesn't really concern us at this time, except as background information. We will look at the structure of PALs, GALs, and CPLDs in Chapter 8. We will use the term "PLD" when we are referring to a generic device and "CPLD" as a more specific type of PLD.

A complication in the use of programmable logic is that we must use specialized computer software to design and program our circuit. Initially, this might seem as though we are adding another level of work to the design, but when these computer techniques are mastered, it shortens the design process greatly and yields a level of flexibility not otherwise available.

Let's look at two examples, comparing the use of SSI logic versus programmable logic.

EXAMPLE 4.1

Figure 4.1 shows a majority vote circuit, as described in Problem 3.4 of Chapter 3. This circuit will produce a HIGH output when two out of three inputs are HIGH. Write the Boolean equation for the circuit and state the minimum number and type of 74HC devices required to build the circuit. How many packages would be required to build two such circuits?

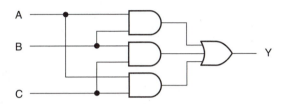

FIGURE 4.1
Majority Vote Circuit

Solution

Boolean equation: $Y = AB + BC + AC$

Figure 4.2 shows the 74HC devices required to build the majority vote circuit: one 74HC08A quad 2-input AND gate and one 74HC4075 triple 3-input OR gate. Figure 4.2 also shows connections between the devices. Note that unused gate inputs are grounded and unused outputs are left open.

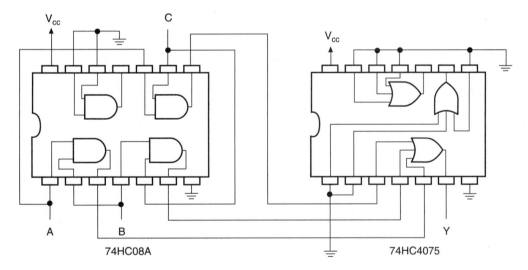

FIGURE 4.2
74HC Devices Required to Build a Majority Vote Circuit

Two majority vote circuits would require 6 ANDs and two ORs. This requires one more 74HC08A package.

EXAMPLE 4.2

Show how a CPLD can be programmed with a majority vote function, using a **schematic capture** tool. State how many CPLDs would be required to build two majority vote circuits.

Solution A CPLD can be programmed by entering the schematic directly, using PLD programming software, such as Altera Corporation's **MAX+PLUS II.** Figure 4.3 shows the circuit as entered in a MAX+PLUS II Graphic Design File.

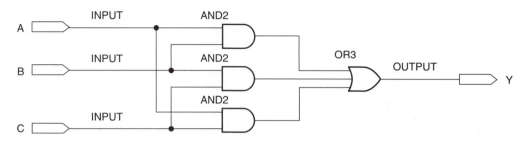

FIGURE 4.3
MAX+PLUS II Graphic Design File of a Majority Vote Circuit

The design can be **compiled** by MAX+PLUS II to create the information required to program the CPLD with the majority vote circuit. If a second copy of the circuit is required, the first circuit can easily be duplicated by a Copy and Paste procedure. The two circuits can than be compiled together and used to program a single CPLD.

4.2   Programming PLDs using MAX+PLUS II

KEY TERMS

Design entry   The process of using software tools to describe the design requirements of a PLD. Design entry can be done by entering a schematic or a text file that describes the required digital function.

Fitting   Assigning internal PLD circuitry, as well as input and output pins, for a PLD design.

Simulation   Verifying design function by specifying a set of inputs and observing the resultant outputs. Simulation is generally shown as a series of input and output waveforms.

Programming   Transferring design information from the computer running PLD design software to the actual PLD chip.

Download   Program a PLD from a computer running PLD design and programming software.

Software tools   Specialized computer programs used to perform specific functions such as design entry, compiling, fitting, and so on. (Sometimes just called "tools.")

Suite (of software tools)   A related collection of tools for performing specific tasks. MAX+PLUS II is a suite of tools for designing and programming digital functions in a PLD.

Target device   The specific PLD for which a digital design is intended.

Altera UP-1 board   A circuit board, part of Altera's University Program Design Laboratory Package, containing two CPLDs and a number of input and output devices.

In order to take a digital design from the idea stage to the programmed silicon chip, we must go through a series of steps known as the PLD Design Cycle. These include **design entry, simulation,** compiling, **fitting,** and **programming.** All steps require the use of PLD software, such as Altera's MAX+PLUS II, a **suite** of **software tools,** to perform the various tasks of the design cycle. Some tasks, such as design entry, require a great deal of attention; others, such as fitting a design to a specified CPLD, are done automatically during the compiling process.

We will be using MAX+PLUS II as a vehicle for learning the concepts that relate to PLD design and programming. The **target devices** for our designs will be two Altera CPLDs, both installed on a circuit board available from Altera called the University Program Design Laboratory Package. We will generally refer to this board, shown in Figure 4.4, as the **Altera UP-1 board.**

FIGURE 4.4
Altera UP-1 Board

Figure 4.5 shows photos of the two CPLDs used in the Altera UP-1 Board. Figure 4.5a shows the CPLD from the MAX7000S family, part number EPM7128SLC84-7. Figure 4.5b shows the CPLD from Altera's FLEX10K series, part number EPF10K20RC240-4. These part numbers are meaningful and will be discussed in detail in Chapter 8.

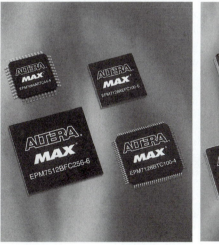

FIGURE 4.5
Altera MAX7000S and FLEX10K CPLDs

In the remaining part of this chapter, we will learn how to enter a design in MAX+PLUS II in both graphical and text format, how to compile the design, and how to **download** it into either one of the CPLDs on the Altera UP-1 circuit board.

Treat this design example as a tutorial in MAX+PLUS II. Follow along with all the steps on your own computer to get the maximum benefit from the chapter. If you do not have access to the Altera UP-1 board or an equivalent, you can still follow through most of the steps.

> **NOTE**
>
> Although the examples in this book are created with the Altera UP-1 board in mind, they will easily adapt to other circuit boards carrying an Altera EPM7128S or other similar CPLD. One such board is available from Intectra Inc. For further information, contact Intectra at:
>
> Intectra, Inc
> 2629 Terminal Blvd
> Mountain View, CA 94043 U.S.A.
> Ph 650-967-8818 Fx 650-967-8836
> intectra@best.com
> www.intectra.com *(Web site in Spanish only)*

4.3 Graphic Design File

KEY TERMS

Graphic Design File (gdf)   A PLD design file in which the digital design is entered as a schematic.

Project   A set of MAX+PLUS II files associated with a particular PLD design.

One way of entering PLD designs is to create a **Graphic Design File.** This type of file contains a representation of a digital circuit, such as in Figure 4.3, showing components and their interconnections, as well as specifying the inputs and output names of the circuit.

MAX+PLUS II automatically generates a number of other files to keep track of the PLD programming information represented by the Graphic Design File. These files, taken together, represent a **project** in MAX+PLUS II. All operations required to create a programming file for a CPLD are performed on a project, not a file. Thus, it is important during the design process to keep track of what the current project is. The MAX+PLUS II toolbar, shown in Figure 4.6, makes this fairly easy.

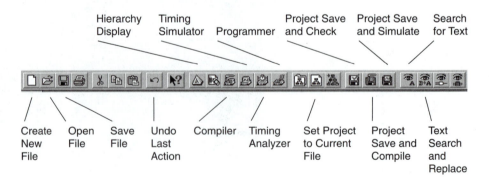

FIGURE 4.6
MAX+PLUS II Toolbar

The toolbar has a number of buttons that pertain to the current project of a PLD design. The operations performed by these buttons can all be done through the regular menus of MAX+PLUS II, but the toolbar offers a quick way to access many available functions. Not all buttons on the toolbar in Figure 4.6 are labeled, just the ones that you will find particularly convenient at this time. You can find out the function of any button by placing the cursor on the button and reading a description at the bottom of the window.

In particular, notice the buttons that create, open, and save files (standard Windows icons) and the button that sets the project to the current file. When creating a new file, make it standard practice to first **Save** the file, then **Set Project to Current File.** If you do this as a habit, you (and MAX+PLUS II) will always know what the current project is. If you don't, you will find that you are saving or compiling some other project and wondering why your last set of changes didn't work.

Another good practice is to create a new Windows folder for each new design that you enter. Since MAX+PLUS II creates many files in the design process, the folders would become unmanageable if designs were not kept in separate folders.

MAX+PLUS II installs a folder for working with design files called **max2work.** The examples in this text will be created in a subfolder of **max2work.** If you are working in a situation where many people share a computer and you have access to a network drive of your own, you may wish to keep your working files in a **max2work** folder on the network drive. Avoid storing your working files on a local hard drive unless you are the only one with regular access to the computer. Examples in this book will not specify a drive letter, but will indicate *drive:***max2work***folder.*

Most of these examples are also available on the accompanying CD in the folder called **Student Files.** A special icon, shown in the margin, will indicate the example filename.

In the following sections, we will go through the process of creating a file in detail, using the majority vote circuit of Figure 4.3 as an example. The example assumes that MAX+PLUS II is properly installed on your computer and running. For installation instructions, see the file **SE_READ** on the accompanying CD or the MAX+PLUS II Installation section of *MAX+PLUS II Getting Started,* available from Altera.

Entering Components

To create a Graphic Design File, click the **New File** icon on the tool bar or choose **New** on the MAX+PLUS II **File** menu. The dialog box, shown in Figure 4.7 appears. Select **Graphic Editor file** and choose **OK.**

FIGURE 4.7
New Dialog Box

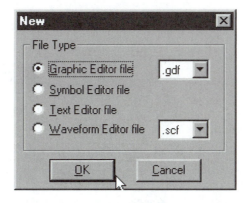

Maximize the window and click the **Save** icon or choose **Save As** or **Save** from the **File** menu. In the dialog box shown in Figure 4.8, save the file in a new folder (e.g., *drive:***max2work****maj_vote****maj_vote.gdf**) and choose **OK.** (If you have not created the new folder, just type the complete path name in the **File Name** box. MAX+PLUS II will

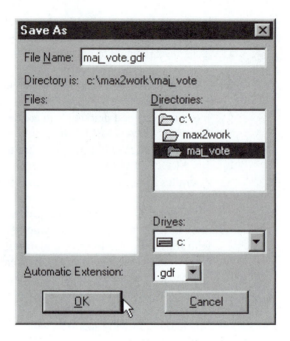

FIGURE 4.8
Save As Dialog Box

create a new folder.) Click the icon to **Set Project to Current File** or choose this action from the **File, Project** menu.

The first design step is to lay out and align the required components. We require three 2-input AND gates, a 3-input OR gate, three input pins, and one output pin. These basic components are referred to as **primitives.** Let us start by entering three copies of the AND gate primitive, called **and2.**

Click the left mouse button to place the cursor (a flashing square) somewhere in the middle of the active window. Right-click to get a pop-up menu, shown in Figure 4.9, and choose **Enter Symbol.** The dialog box in Figure 4.10 appears. Type **and2** in the **Symbol Name** box and choose **OK.** A copy or **instance** of the and2 primitive appears in the active window.

FIGURE 4.9
Enter Symbol Pop-up Menu

You can repeat the above procedure to get two more instances of the and2 primitive, or you can use the **Copy** and **Paste** commands. These are the same icons and **File** commands as for other Windows programs. Highlight the **and2** symbol by clicking it. Right-click the symbol to get the pop-up menu shown in Figure 4.11 and choose **Copy.** You can also click the **Copy** icon on the toolbar or use the **Copy** command in the **File** menu.

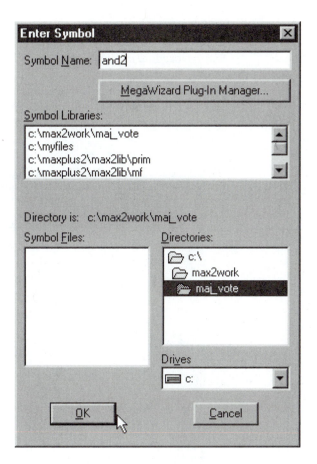

FIGURE 4.10
Enter Symbol Dialog Box

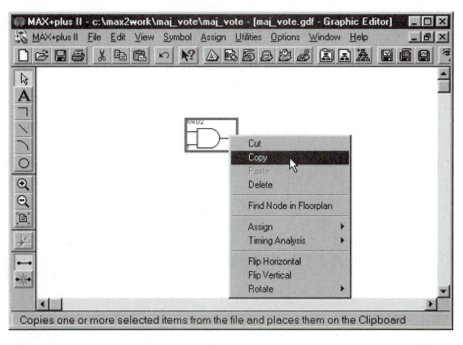

FIGURE 4.11
Copying a Component

Paste an instance of the primitive by clicking to place the cursor, then right-clicking to bring up the menu shown in Figure 4.12. Choose **Paste.** The component will appear at the cursor location, marked in Figure 4.12 by the square at the top left corner of the pop-up menu.

Enter the remaining components by following the **Enter Symbol** procedure outlined above. The primitives are called **or3, input,** and **output.** When all components are entered we can align them, as in Figure 4.13 by highlighting, then dragging each one to a desired location.

FIGURE 4.12
Pasting a Component

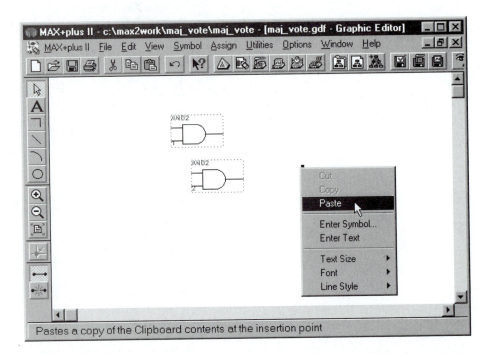

FIGURE 4.13
Aligned Components

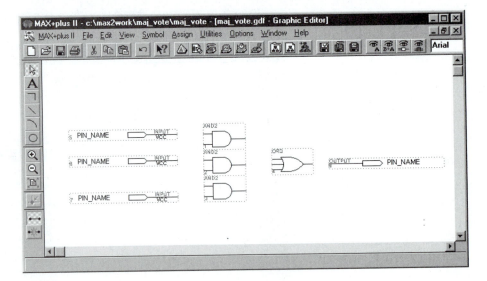

Connecting Components

To connect components, click over one end of one component and drag a line to one end of a second component. When you drag the line, a horizontal and a vertical broken line mark the cursor position, as shown in Figure 4.14. These lines help you align connections properly.

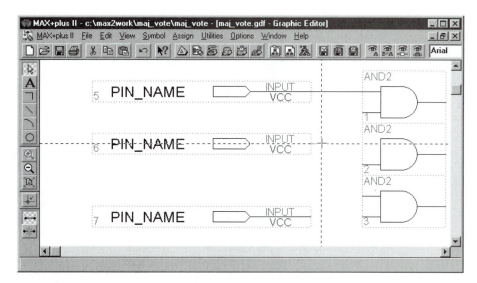

FIGURE 4.14
Dragging a Line to Connect Components

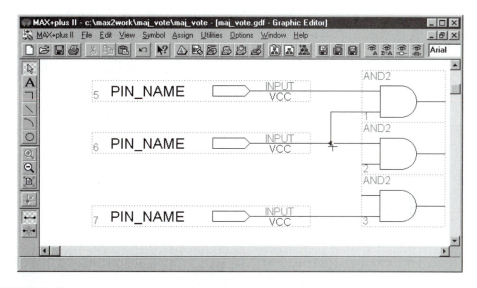

FIGURE 4.15
Making a 90-degree Bend and a Connection

A line will automatically make a connection to a perpendicular line, as shown in Figure 4.15.

A line can have one 90-degree bend, as at the inputs of the AND gates. If a line requires two bends, such as shown at the AND outputs in Figure 4.16, you must draw two separate lines.

Assigning Pin Names

Before a design can be compiled, its inputs and outputs must be assigned names. We could also specify pin numbers, if we wished to make the design conform to a particular CPLD, but it is not necessary to do so at this stage. It may not even be desirable to assign pin numbers, since the design we enter can be used as a component or subdesign of a larger circuit. We may also wish MAX+PLUS II to assign pins to make the best use of the CPLD's internal resources. At any rate, we will leave this step out for now.

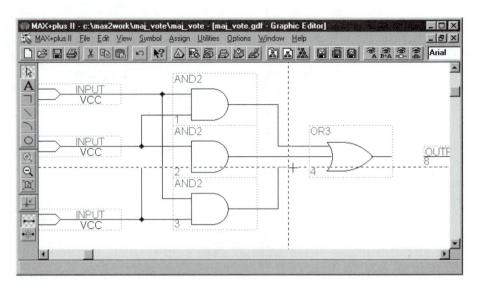

FIGURE 4.16
Line with Two 90-degree Bends

Figure 4.17 shows the naming procedure. Pins A and B have already been assigned names. Highlight a pin by clicking on it. Right-click the highlighted pin and choose **Edit Pin Name** from the pop-up menu. You could also double-click the pin name to highlight it. Type in the new name.

If there are several pins that are spaced one above the other, you can highlight the top pin name, as described above, then highlight successive pin names by using the **Enter** key.

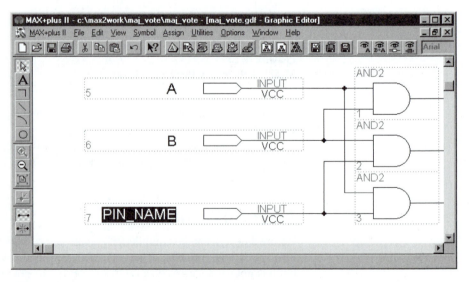

FIGURE 4.17
Assigning Pin Names

4.4 Compiling MAX+PLUS II Files

The MAX+PLUS II compiler converts design entry information into binary files that can be used to program a PLD. Before compiling, we should assign a target device to the design.

From the **Assign** menu, shown in Figure 4.18, select **Device.** From the dialog box in Figure 4.19, select the target device. For the Altera UP-1 board, this would be either the EPM7128SLC84-7 (shown) or the FLEX10K20RC240-4. The device family for the EPM7128S device is MAX7000S.

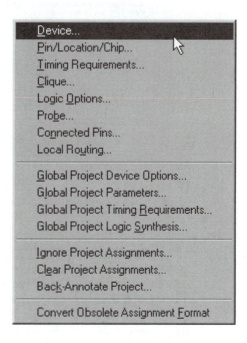

FIGURE 4.18
Assign Menu

NOTE

To see the EPM7128SLC84-7 device, the box that says **Show Only Fastest Speed Grades** must be unchecked.

The compiler has a number of settings that can be chosen prior to the actual compile process. Figure 4.20 shows some of the settings that should be selected from the **Processing** menu of the **Compiler** window. You can open the **Compiler** window from the **MAX+PLUS II** menu or by clicking the **Compiler** button on the toolbar at the top of the screen.

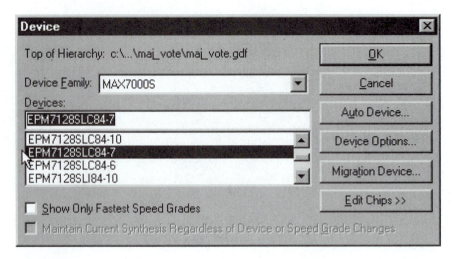

FIGURE 4.19
Device Dialog Box

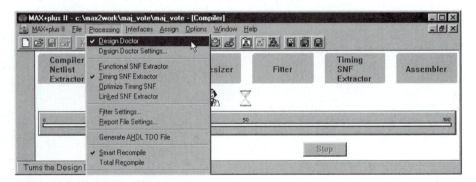

FIGURE 4.20
MAX+PLUS II Compiler Settings

Design Doctor is a utility that checks for adherence to good design practice and will warn you of any bad design choices. (Design Doctor will not stop the design from compiling, but will suggest potential problems that could result from a particular design.) The **Timing SNF Extractor** creates a Simulation Netlist File, which is required to perform a timing simulation of the design. We will perform this step in later MAX+PLUS II designs. (If you are not able to select the **Timing SNF Extractor,** then uncheck the **Functional SNF Extractor** option.) **Smart Recompile** allows the compiler to use previously compiled portions of the design to which no changes have been made. This allows the compiler to avoid having to compile the entire design each time a change is made to one part of the design.

To start the compile process, click **Start** in the **Compiler** window. While in progress, the window will look something like Figure 4.21. Message of three types may appear during the compile process. **Info** messages (green text) are for information only. **Warning** messages (blue text) tell you of potential, but nonfatal, problems with the design. **Error** messages (red text) inform you of design flaws that render the design unusable. A PLD can still be programmed if the compiler generates **info** or **warning** messages, but not if it generates an **error.**

Depending on the device chosen, the compiler generates either a **Programmer Object File (pof)** or **SRAM Object File (sof).** The **pof** is used to *program* a MAX-series PLD. The **sof** is used to *configure* a FLEX-series PLD. The difference is that the MAX device is **nonvolatile,** that is, it retains its programming information after the power has been

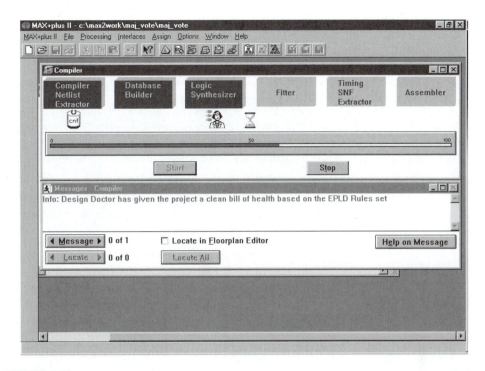

FIGURE 4.21
MAX+PLUS II Compiler Operation

removed. The FLEX-series device is **volatile,** meaning that its programming information must be loaded each time the device powers up.

4.5 Hierarchical Design

> **KEY TERMS**
>
> **Hierarchical design**   A PLD design that is ordered in layers or levels. The highest level of design contains components that are themselves complete designs. These components may, in turn, have lower level designs embedded within them.

A MAX+PLUS II Graphical Design File can be used as part of a **hierarchical design.** That is, it can be represented as a component in a higher-level design. Figure 4.22 shows a gdf that is constructed as a hierarchical design. It contains two majority vote circuits whose outputs are combined in an AND gate. Thus, the output would be HIGH if two out of three

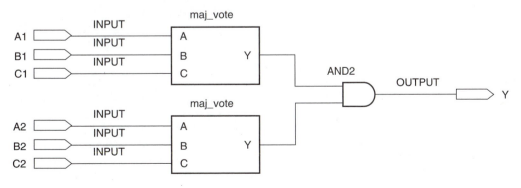

FIGURE 4.22
Two-level Majority Vote Circuit (2votes.gdf)

inputs were HIGH on *both* blocks labeled **maj_vote.** These blocks are complete designs in their own right, and thus form a lower level of the design hierarchy.

Default Symbols and User Libraries

We can create a **default symbol** for the majority vote circuit of Figure 4.3 from the MAX+PLUS II **File** menu, as shown in Figure 4.23. This action will create a symbol file with the same name as the Graphic Design File and the extension **sym.** Before creating the symbol, make sure that the gdf is saved and that the project is set to the current file. The symbol can be embedded into a gdf, as in Figure 4.22.

FIGURE 4.23
Creating a Default Symbol

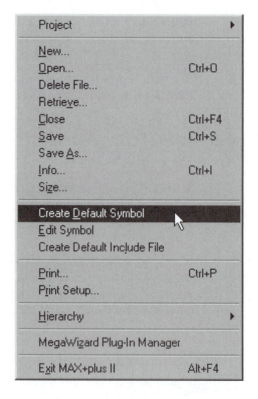

Before we can use the new symbol, we must make sure that MAX+PLUS II knows where to find it. MAX+PLUS II looks for a component first in the present working directory, then in the **user library** folders in the order of priority listed in the User Libraries dialog box.

To create a path to a user library, select **User Libraries** from the **Options** menu (Figure 4.24) in MAX+PLUS II. In the resultant dialog box, shown in Figure 4.25, select the appropriate drive and directories by double-clicking on the name in the **Directories** box. When the desired directory appears in the **Directory Name** box, click **Add,** then **OK.**

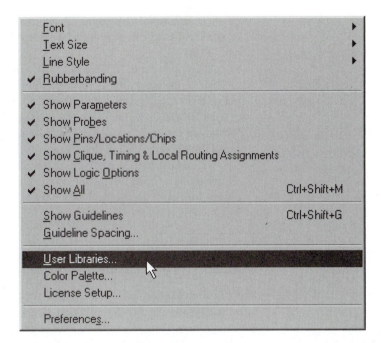

FIGURE 4.24
Options Menu

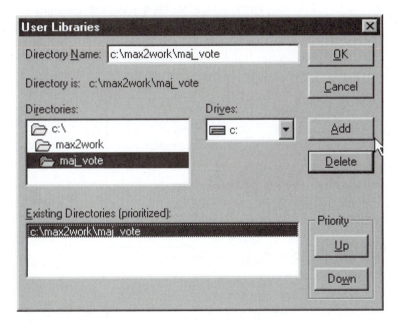

FIGURE 4.25
User Libraries Dialog Box

> **NOTE**
>
> If you are using MAX+PLUS II on a shared computer (e.g., in a computer lab), you should be aware that a library path that points to another user's directory can cause MAX+PLUS II to look there before (or instead of) looking in your directory, resulting in the apparent inability of MAX+PLUS II to find your file.
>
> For example, suppose you have a file called **g:\max2work\my_file.gdf,** where g:\ is a network drive mapped exclusively to your user account. (i.e., everyone has a g:\ drive mapping, unique to their user account.) Further suppose that

another user, against standard lab protocol, has created a file with the same name on the local hard drive: **c:\max2work\my_file.gdf.** (Don't think this doesn't happen. It does.)

At compile time, MAX+PLUS II will look for **my_file.gdf** first in the directory where the active project resides, then in the folders specified in the user library paths. If the user library path **c:\max2work** has a higher priority than **g:\max2work\,** it will compile the version of **myfile.gdf** found on the c: \drive. When you make changes to the copy on the g: \drive, they will not take effect because the file on g:\ is not being compiled.

To remedy this, delete the user libraries that point to local drives, such as a:\ or c:\. If you have no assigned network drive on your system, delete all user libraries except for your own. Since a user library is just the name of a folder where MAX+PLUS II should look for files, this won't do any great harm.

Creating a Design Hierarchy

2votes.gdf
maj_vote.gdf

4votes.gdf

The circuit in Figure 4.22 is saved as **2votes.gdf.** If we double-click on either symbol labeled **maj vote,** the MAX+PLUS II Graphic Editor will bring the file **maj_vote.gdf** to the foreground. Thus, we say that **2votes.gdf** is at the **top level** of the current hierarchy.

We can extend the hierarchy further by making a symbol for **2votes.gdf** and embedding it in a higher-level file called **4votes.gdf,** shown in Figure 4.26. This circuit generates a HIGH output if (two out of three of (A11, B11, C11) are HIGH AND two out of three of (A21, B21, C21) are HIGH) OR the same is true for (A12, B12, C12) AND (A22, B22, C22). If we double-click on either symbol for **2votes,** the Graphic Editor will bring the file **2votes.gdf** to the foreground.

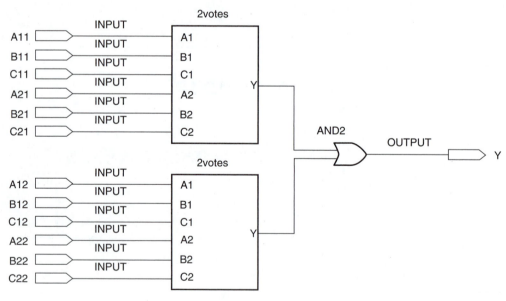

FIGURE 4.26
Further Levels of Hierarchy (4votes.gdf)

MAX+PLUS II can display the hierarchy of a design. To see the hierarchy structure, click the **Hierarchy** icon on the MAX+PLUS II toolbar (the yellow pyramid) or choose **Hierarchy Display** from the **MAX+PLUS II** menu. Figure 4.27 shows the hierarchy for the project **4votes.** Note that the highest level has two subdesigns, each of which breaks down further into two subdesigns. Thus, using hierarchical design and symbols for gdf or

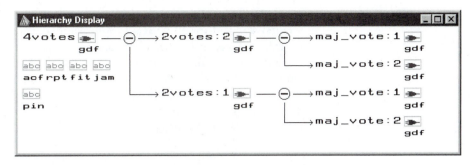

FIGURE 4.27
Hierarchy Display for Project "4votes"

other design files allows us to create multiple instances of a basic design (**maj_vote.gdf**) and use it in many places.

> **N O T E**
>
> In order to correctly show the hierarchy display, the top-level file of the project (in this case **4votes.gdf**) must be compiled first.

4.6   Text Design File (VHDL)

KEY TERMS

Hardware description language   A computer language used to design digital circuits by entering text-based descriptions of the circuits.

AHDL (Altera Hardware Description Language)   Altera's proprietary text-entry design tool for PLDs.

VHDL (VHSIC Hardware Description Language)   An industry-standard computer language used to model digital circuits and produce programming data for PLDs.

VHSIC   Very high speed integrated circuit

Syntax   The "grammar" of a computer language. (i.e., the rules of construction of language statements)

ASICs (application specific integrated circuits)   Integrated circuits that are constructed for a specific design purpose. The term could refer to a PLD, although it usually means a custom-designed fixed function device.

An alternative to schematic entry, and ultimately a more powerful PLD design technique is the use of a text-based design tool, or **hardware description language,** such as Altera's **AHDL (Altera Hardware Description Language)** or the industry-standard **VHDL (VHSIC Hardware Description Language).** A designer creates a text file, framed within a certain set of rules known as the **syntax** of the language and uses a compiler to create programming data much as he or she would with a Graphic Design File. Hardware description languages can be used to generate hardware for hierarchical designs, either as components in graphic or text files or as higher level design entities containing other designs.

AHDL, while very easy to use, has a much narrower application than VHDL because it is one of many proprietary tools on the market aimed at the programming requirements of a particular manufacturer's line of CPLDs. Since VHDL is an industry-standard language and the MAX+PLUS II compiler supports both languages, we will concentrate on VHDL.

VHDL was originally developed by defense contractors in the U.S. and is now the required standard for all **ASICs (application specific integrated circuits)** designed for the U.S. military. It has been standardized by the Institute of Electrical and Electronics Engineers (IEEE) and has been enjoying increasing popularity in the electronics design community. The original VHDL standard was written in 1987 and updated in 1993 (IEEE Std. 1076-1993). This standard and other related ones continue to undergo revision. The current status of Std. 1076 can be determined from the IEEE Standards web site at http://www.standards.ieee.org.

Entity and Architecture

KEY TERMS

Entity   A VHDL structure that defines the inputs and outputs of a design.

Architecture   A VHDL structure than defines the relationship between input, output, and internal signals or variables in a design.

Port   A name assigned to an input or output of a VHDL design entity.

Mode (of a port)   The kind of port, such as input or output.

Signal   A name given to an internal connection in a VHDL architecture.

Variable   A block of working memory used for internal calculation or storage in a VHDL architecture.

Type   A set of characteristics associated with a VHDL port name, signal, or variable that determines the allowable values of the port, signal, or variable.

Library   A collection of VHDL design units that have been previously compiled.

Package   A group of VHDL design elements that can be used by more than one VHDL file.

IEEE Standard 1164   The standard which defines a variety of VHDL types and operations, including the STD_LOGIC and STD_LOGIC_VECTOR types.

Concurrent   Simultaneous.

Concurrent signal assignment   A relationship between an input and output port or signal in which the output is changed as soon as there is a change in input. If the file has more than one concurrent signal assignment, they are all evaluated simultaneously.

Selected signal assignment statement   A concurrent signal assignment in VHDL in which a value is assigned to a signal, depending on the alternative values of another signal or variable.

Comment   Explanatory text in a VHDL (or other computer language) file that is ignored by the computer at compile time.

Vector   A group of digital signals or variables, usually related numerically, that can be treated as a single multibit variable.

Bit string literal   A group of bits assigned to the elements of a vector, enclosed in double quotes (e.g., "001011").

Every VHDL file requires at least two structures: an **entity** declaration and an **architecture** body. The entity declaration defines the *external* aspects of the VHDL function; that is, the input and output names and the name of the function. The architecture body defines the *internal* aspects; that is, how the inputs and outputs behave with respect to one another and with respect to other signals or functions that are internal only.

Let us examine the structure of a VHDL design for the majority vote circuit defined in Figure 4.1. The complete VHDL file for the majority vote circuit is shown next. The double dashes before the first two lines are to indicate that these lines are **comments.** There are also a few other comments to illustrate the use of VHDL.

maj_vot2.vhd

```
-- maj_vot2.vhd
-- VHDL implementation of a majority vote circuit

-- Library contains standard VHDL logic types
LIBRARY ieee;
USE ieee.std_logic_1164.ALL;

-- Entity defines inputs and outputs
ENTITY maj_vot2 IS
PORT (
        a, b, c    : IN STD  LOGIC;
        y          : OUT STD  LOGIC);
END maj_vot2;

-- Architecture describes input/output relationship
ARCHITECTURE majority OF maj_vot2 IS
BEGIN
        y <=    (a and b) or (b and c) or (a and c);
END majority;
```

> **NOTE**
>
> VHDL is not case-sensitive, so statements written in lowercase and uppercase are equivalent. For example, (Y <= A AND B;) is equivalent to (y <= a and b;). However, Altera's style guidelines for VHDL suggest that all keywords, devices, constants, and primitives be capitalized and everything else be written in lowercase letters. The VHDL style guideline can be referred to in the MAX+PLUS II **Help** menu.

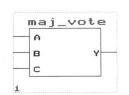

FIGURE 4.28
Graphical Representation of a VHDL Design Entity

The name of the entity, **maj_vot2,** is given in the first and last lines of the entity declaration. The VHDL file that contains this entity must be named **maj_vot2.vhd.** Figure 4.28 shows how the design entity looks if it is converted to a symbol for use in a Graphic Design File.

The Boolean equation for a 3-input majority vote circuit is $Y = AB + BC + AC$. In the architecture body, we can write this operation as:

```
y <= (a and b) or (b and c) or (a and c);
```

The operator <= assigns the value of the right hand side of the equation to the left hand side. Whenever there is a change in a, b, or c, the statement is re-evaluated and the new value is assigned to y. Note that VHDL logical operators (such as **and** and **or**) have equal precedence, so we must make the order of precedence explicit with parentheses.

The Boolean equation above is an example of a **concurrent signal assignment** statement. Concurrent means "simultaneous." The implication is that any number of concurrent signal assignments can be listed in a VHDL architecture body and the order in which they are evaluated does not depend on the order in which they are written, since all statements are concurrent. In this way, a concurrent structure imitates combinational hardware, where a change in one input that is common to several circuits makes all circuits change at the same time.

Enclosed in the entity declaration is a **port** definition. A port is a connection from the PLD to the outside world. Figure 4.29 shows the possible **modes** of a port. Mode IN refers to a port that is only for input. Mode OUT is output only. Mode INOUT is a bidirectional port, in which data can flow in either direction, based on the status of a control input. Mode BUFFER is a special case of OUT that has a feedback connection back into the CPLD logic that can be used as part of another Boolean expression.

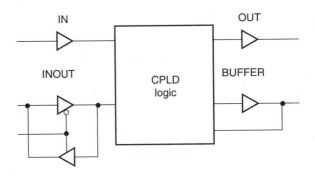

FIGURE 4.29
VHDL Port Modes

Figure 4.30 shows the difference between BUFFER and OUT modes. Port x (defined by x <= a and b;) must be of mode BUFFER because it is fed back and used as part of the expression for port y (defined by y <= x or c;). Port y can be of mode OUT since it has no feedback, only an output.

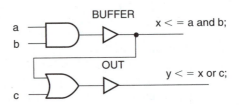

FIGURE 4.30
BUFFER and OUT Modes

In addition to defining the port modes, the entity declaration also defines what **type** each port is. The type of a port, signal, or variable defines what values it is allowed to have. Three common types in VHDL are BIT, STD_LOGIC, and INTEGER. Multibit extensions of these types include BIT_VECTOR and STD_LOGIC_VECTOR.

Ports, signals and variables of type BIT can have a value of '0' or '1'. When using these values, they must be enclosed in single quotes.

The STD_LOGIC (standard logic) type, also called **IEEE Std.1164 Multi-Valued Logic,** has been defined to give a broader range of output values than just '0' and '1'. Any port, signal, or variable of type STD_LOGIC or STD_LOGIC_VECTOR can have any of the values listed below.

```
'U',  -- Uninitialized
'X',  -- Forcing   Unknown
'0',  -- Forcing   0
'1',  -- Forcing   1
'Z',  -- High Impedance
'W',  -- Weak Unknown
'L',  -- Weak 0
'H',  -- Weak 1
'-'   -- Don't care
```

"Forcing" levels are deemed to be the equivalent of a gate output. "Weak" levels are specified by a pull-up or pull-down resistor. ("Weak" levels are usually used in circuit modeling, where it is important to distinguish between gate outputs and pull-up/down.

These levels will not be of importance to us.) The 'Z' state is used as the high-impedance state of a tristate buffer.

The majority of applications can be handled by 'X', '0', '1', and 'Z' values.

To use STD_LOGIC in a VHDL file, you must include the following reference to the VHDL **library** called **ieee** and the **std_logic_1164** package before the entity declaration:

```
LIBRARY ieee;
USE ieee.std_logic_1164.ALL;
```

Why use STD_LOGIC rather than BIT, if we only use '0' and '1' values? The usual reason is for compatibility with existing VHDL components that might be used in our design entities. For example, the Altera Library of Parameterized Modules (LPM) contains predesigned components that are written using STD_LOGIC types. To include these components in a VHDL design, the design must be written with STD_LOGIC types, as well.

The INTEGER type can take on whole-number values. When used in a VHDL file, an integer is written without quotes. Table 4.1 summarizes the BIT, STD_LOGIC, and INTEGER types, as well as the BIT_VECTOR and STD_LOGIC_VECTOR types.

Table 4.1 Some Common VHDL Types

Type	Values	How Written	Examples
BIT	0 or 1	Single quotes	'0', '1'
STD_LOGIC	U, X, 0, 1, Z, W, L, H, -	Single quotes	'X', '0', '1', 'Z'
INTEGER	Whole numbers	No quotes	4095, 7, -120, -1
BIT_VECTOR	Multiple instances of 0 or 1	Double quotes	"100110"
STD_LOGIC_VECTOR	Multiple instances of U, X, 0, 1, Z, W, L, H, -	Double quotes	"1001100", "00ZZ11", "ZZZZZZZZ"

▌▌ EXAMPLE 4.3

Figure 4.31 shows the logic diagram of a 2-line-to-4-line decoder. The circuit detects the presence of a particular binary code and makes one and only one output HIGH, depending on the value of the 2-bit number D_1D_0. Write a VHDL file that describes the decoder.

FIGURE 4.31
2-line-to-4-line Decoder

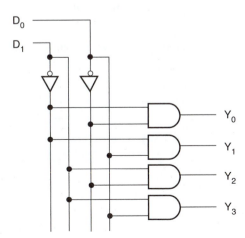

Solution   The circuit has two inputs and four outputs, which are numerically related. We could describe the two inputs as separate names, as we could the four outputs. Or, we could show the inputs and outputs as two groups of related ports, called **vectors.** The elements of the vector can be treated separately or as a group.

Case 1: separate variables

decode1.vhd

```
LIBRARY ieee;
USE ieee.std_logic_1164.ALL;

ENTITY decode1 IS
     PORT(
             d1, d0            : IN  STD_LOGIC;
             y0, y1, y2, y3    : OUT STD_LOGIC);
END decode1;

ARCHITECTURE decoder1 OF decode1 IS
BEGIN
     y0    <=    (not d1) and (not d0);
     y1    <=    (not d1) and (    d0);
     y2    <=    (    d1) and (not d0);
     y3    <=    (    d1) and (    d0);
END decoder1;
```

Case 2: vectors (elements treated separately)

decode2.vhd

```
LIBRARY ieee;
USE ieee.std_logic_1164.ALL;

ENTITY decode2 IS
     PORT (
             d    : IN  STD_LOGIC_VECTOR (1 downto 0);
             y    : OUT STD_LOGIC_VECTOR (3 downto 0));
END decode2;

ARCHITECTURE decoder2 OF decode2 IS
BEGIN
     y(0)    <=    (not d(1)) and (not d(0));
     y(1)    <=    (not d(1)) and (    d(0));
     y(2)    <=    (    d(1)) and (not d(0));
     y(3)    <=    (    d(1)) and (    d(0));
END decoder2;
```

In Case 2, we specify the length of the vector by the construct **(3 downto 0),** indicating that Y3 is the leftmost bit in the vector. We could also use the constructs **(0 to 3),** **(4 downto 1),** or **(1 to 4),** depending on our requirements. Each individual element of the vector is specified by a number in parentheses.

Case 3: vectors (elements treated as a group)

decode2a.vhd

```
-- decode2a.vhd
-- 4-channel decoder
-- Makes one and only one output HIGH for each
-- binary combination of (d1, d0).

LIBRARY ieee;
USE ieee.std_logic_1164.ALL;

ENTITY decode2a IS
     PORT (
             d    : IN  STD_LOGIC_VECTOR (1 downto 0);
             y    : OUT STD_LOGIC_VECTOR (3 downto 0));
END decode2a;
```

```
ARCHITECTURE decoder OF decode2a IS
BEGIN
     -- Choose a signal assignment for y
     -- based on binary value of d
     -- Default case: all outputs deactivated
     WITH d SELECT
          y <=   "0001" WHEN "00",
                 "0010" WHEN "01",
                 "0100" WHEN "10",
                 "1000" WHEN "11",
                 "0000" WHEN others;
END decoder;
```

In Case 3, we use a **selected signal assignment statement** to assign a value to all bits of vector **y** for each combined value of vector **d.** For example, when **d(1) = 0** and **d(0) = 0,** the values assigned to y are: **y(3) = 1, y(2) = 0, y(1) = 0, y(0) = 0.** Similar assignments are made for other values of **d.** The result is a construct that acts much like a truth table of the decoder circuit. The **others** clause is necessary to define a default case since the STD_LOGIC_VECTOR type contains values other than '0' and '1'.

The multibit values assigned to the vectors, called **bit string literals,** must be enclosed in double quotes.

VHDL Templates in MAX+PLUS II

FIGURE 4.32
MAX+PLUS II Template Menu

MAX+PLUS II offers a shortcut to creating VHDL structure in a **Template Menu.** Figure 4.32 shows this menu, which is available in the MAX+PLUS II Text Editor window. To choose a template, select the one desired from the **VHDL Template** dialog box, shown in Figure 4.33.

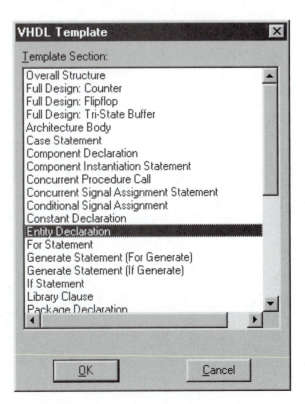

FIGURE 4.33
VHDL Template Dialog Box

Choosing the Entity Declaration template results in the following text:

```
ENTITY __entity_name IS
     GENERIC (__parameter_name : string  := __default_value;
                  __parameter_name : integer:= __default_value);
     PORT (
              __input_name, __input_name   : IN         STD_LOGIC;
              __input_vector_name : IN      STD_LOGIC_VECTOR (__high
downto __low);
              __bidir_name, __bidir_name   : INOUT       STD_LOGIC;
              __output_name, __output_name : OUT     STD_LOGIC);
END __entity_name;
```

To convert this into a valid entity for our use, we delete the lines we do not need and substitute input and output names into the template. For our majority vote circuit, we had inputs called *A, B,* and *C* and an output called *Y*. Thus, we can modify the template to yield the entity declaration:

```
ENTITY maj_vot2 IS
     PORT (
              a, b, c       : IN STD_LOGIC;
              y             : OUT STD_LOGIC);
END maj_vot2;
```

Integrating VHDL and Graphical Design Components

We can create a default symbol for the VHDL majority vote function, much as we did for the same function in the Graphic Design File. In the Text Editor **File** menu, select **Create Default Symbol.** We can integrate this new symbol into a two-level majority vote circuit, as shown in Figure 4.34. This circuit contains primitives (AND gate, input pins, and output pin), a **gdf** symbol (maj_vote), and a symbol created from a VHDL file (MAJ_VOT2). Double-clicking on either symbol will bring forward its original design file.

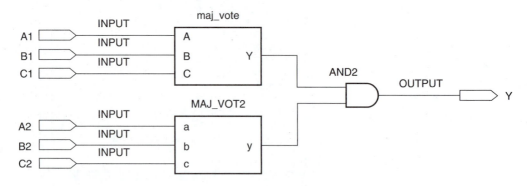

FIGURE 4.34
GDF Containing Symbols from Other GDF and VHDL Files

4.7 Creating a Physical Design

KEY TERMS

Assignment and Configuration File (acf)   A MAX+PLUS II file that contains information about the configuration options for a project, including assigned device and pin numbers.

The previous sections have concentrated on the design aspects of a project. Of course, the ultimate goal of this procedure is to create a physical version of the design. Before we can program our majority vote circuit into hardware, we must assign the input and output pin numbers on the target CPLD. At that point we can recompile the design file and program the CPLD.

Assigning Pin Numbers

Before proceeding with this step, make sure that you have assigned a device part number to the design. Save the file and set the project to the current file.

To assign a pin number, click on the pin to highlight it, then right-click to see the pop-up menu in Figure 4.35. Choose **Assign,** then **Pin/Location/Chip.** You could also do this from the **Assign** menu at the top of the screen.

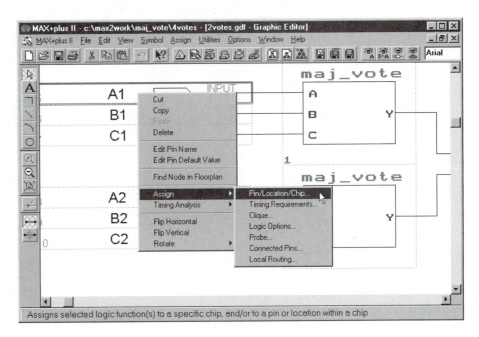

FIGURE 4.35
Pop-up Menu for Pin Assignments

We can assign pin numbers in the dialog box in Figure 4.36.

Type **A1** in the **Node Name** box, **12** in the **Pin** box and click **Add.** Type **B1** in the **Node Name** box, assign this name to pin 16, and click **Add.** Repeat this procedure until all names are assigned, as in Table 4.2. When all assignments are complete, click **OK.**

We can also assign pin numbers by editing the **Assignment and Configuration File (acf),** as shown in Figures 4.37 and 4.38. This technique works especially well if you need to assign pin numbers to a sequence of numerically related inputs and outputs.

Figure 4.37 shows the **acf** with four pin assignments made. We can add the others easily by using a copy-and-paste procedure. Highlight the line you wish to copy and copy it to the Windows clipboard (use **Copy** in the **File** menu or the **Copy** icon on the toolbar or **Ctrl-C**). Paste three copies into the **acf** and modify them so that they represent the remaining required pin assignments, as shown in Figure 4.38.

Figure 4.39 shows the input pin assignments as they appear in the gdf file.

FIGURE 4.36
Pin/Location/Chip Assignment
Dialog Box

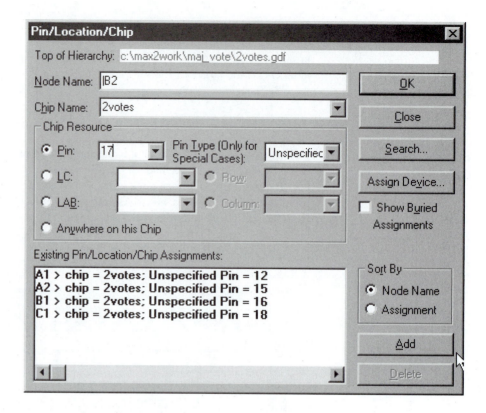

Table 4.2 Pin Assignment
for a Majority Vote Circuit

Pin Name	Pin Number
A1	12
B1	16
C1	18
A2	15
B2	17
C2	21
Y	4

FIGURE 4.37
Pin Assignments in ACF (Before
Copying)

```
2votes.acf - Text Editor                          _ □ ×
--
CHIP 2votes
BEGIN
    |A2 :    PIN = 15;
    |C1 :    PIN = 18;
    |B1 :    PIN = 16;
    |A1 :    PIN = 12;
    DEVICE = EPM7128SLC84-7;
END;


DEFAULT_DEVICES
BEGIN
    ASK BEFORE ADDING EXTRA DEVICES = ON;
Line  24    Col   1    INS
```

```
2votes.acf - Text Editor                          _ □ ✕
--
  CHIP 2votes
  BEGIN
      |Y  :    PIN = 4;
      |C2 :    PIN = 21;
      |B2 :    PIN = 17;
      |A2 :    PIN = 15;
      |C1 :    PIN = 18;
      |B1 :    PIN = 16;
      |A1 :    PIN = 12;
      DEVICE = EPM7128SLC84-7;
  END;
Line   1    Col   1    INS
```

FIGURE 4.38
Pin Assignments in ACF (After Copying)

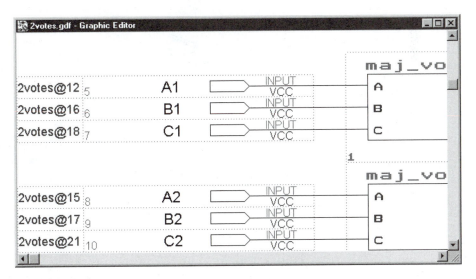

FIGURE 4.39
Pin Assignments as Seen in gdf File

Programming CPLDs on the Altera UP-1 Circuit Board

KEY TERMS

ByteBlaster An Altera ribbon cable and connector used to program or configure Altera CPLDs via the parallel port (LPT port) of an IBM PC or compatible.

JTAG Joint Test Action Group. A standards body that developed the format (called IEEE Std. 1149.1) for testing and programming devices while they are installed in a system.

ISP In-system programmability. The ability of a PLD (such as a MAX7000S) to be programmed without removing it from a circuit board.

ICR In-circuit reconfigurability. The ability of a PLD (such as a FLEX10K) to be configured without removing it from a circuit board.

TDI Test Data In. In a JTAG port, the serial input data to a device.

TDO Test Data Out. The JTAG signal, the serial output data from a device.

TMS Test Mode Select. The JTAG signal that controls the downloading of test or programming data.

TCK Test Clock. The JTAG signal that drives the JTAG downloading process from one state to the next.

JTAG Chain Multiple JTAG-compliant devices whose TDI and TDO ports form a continuous chain connection. Such a chain allows multi-device programming.

The CPLDs on the Altera UP-1 circuit board are programmed via the programming software in MAX+PLUS II and a ribbon cable called the **ByteBlaster.** The ByteBlaster, shown in Figure 4.40, connects to the parallel port of a PC running MAX+PLUS II to a 10-pin male socket that complies with the **JTAG** standard. This standard specifies a four-wire interface, originally developed for testing chips without removing them from a circuit board, but can also be used to program or configure PLDs.

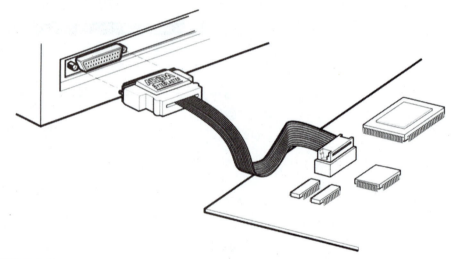

FIGURE 4.40

ByteBlaster Parallel Port Download Cable (By Permission of Altera Corporation)

PLDs that can be programmed or configured while installed on a circuit board are called **in-system programmable (ISP)** or **in-circuit reconfigurable (ICR).** ISP is used to refer to nonvolatile devices, such as MAX7000S; ICR refers to volatile devices, such as FLEX10K.

The JTAG interface has four wires, as well as power and ground connections, as shown in Figure 4.41. Data are sent to a device from a JTAG controller (i.e., the PC) via the **TDI (Test Data In)** line. Data return from the device via **TDO (Test Data Out).** The data transfer is controlled by **TMS (Test Mode Select).** The process is driven from one step to the next by **TCK (Test Clock).**

Multiple devices can be programmed in a **JTAG Chain,** as shown in Figure 4.42. This connection allows both CPLDs on the Altera UP-1 Board to be programmed at the same time. The UP-1 board also has a female 10-pin socket labeled JTAG out, which allows two or more boards to be chained together. The choice of programming one or more CPLDs, or the CPLDs on one or more UP-1 boards, is determined by the placement of four on-board jumpers. These jumper positions are explained in the *Altera University Program Design Laboratory Package User Guide.* A copy of the *User Guide* is included in Appendix A for reference and is available at Altera's Web site.

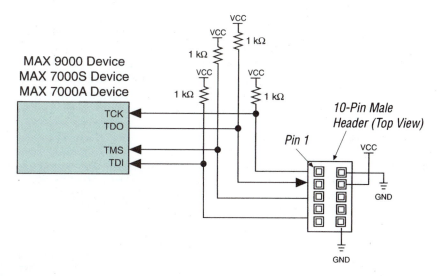

FIGURE 4.41

MAX9000, MAX7000S, and MAX7000A Programming with the ByteBlaster Cable
(By Permission of Altera Corporation)

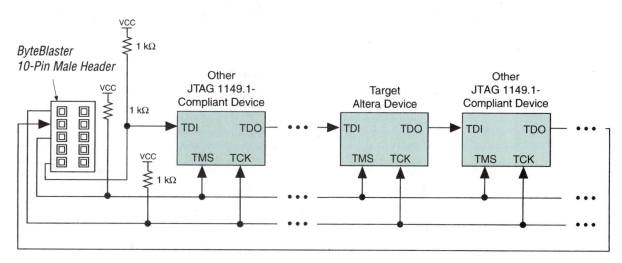

FIGURE 4.42

JTAG Chain Device Programming and Configuration with the ByteBlaster Cable (By Permission
of Altera Corporation)

The operation of the JTAG port is controlled automatically by MAX+PLUS II, so further details are not necessary at this time. For further information on the JTAG interface, refer to *Altera Application Note 39, JTAG Boundary-Scan Testing in Altera Devices,* included in the Altera Documentation folder on the accompanying CD.

MAX+PLUS II Programmer

To program a device on the Altera UP-1 board, set the jumpers to program the EPM7128S or configure the EPF10K20, as shown in the *Altera University Program Design Laboratory Package User Guide.* Connect the ByteBlaster cable from the parallel port of the PC running MAX+PLUS II to the 10-pin JTAG header. (You may have to run a 25-wire cable (male-D-connector-to-female-D-connector) to make it reach.) Plug an AC adapter (9-volt dc output) into the power jack of the UP-1 board.

Open the top-level file of the project you wish to download to the UP-1 board (e.g., **maj_vote.gdf**). Set the project to the current file. Invoke the MAX+PLUS II Programmer from the **MAX+PLUS II** menu or click the Programmer button (the icon showing the blue ribbon cable) on the MAX+PLUS II toolbar.

If you have never programmed a device with your copy of MAX+PLUS II, you will need to set up the hardware configuration. Click **Hardware Setup** in the **Options** menu to get the dialog box in Figure 4.43.

FIGURE 4.43
Hardware Setup Dialog Box

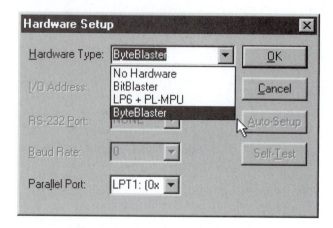

Select **ByteBlaster** in the **Hardware Type** box. Ensure that **Parallel Port** is the same as the port the ByteBlaster is plugged into (usually LPT1:). Click **OK.** (If you have a choice, configure your parallel port as an Enhanced Communications Port (ECP) in your computer's CMOS setup. For most users this step is not necessary, as the port is already configured this way.)

If the current project was compiled with the MAX7000S device selected, the pof file for the project will automatically be available. The **Programmer** dialog box will appear as in Figure 4.44. To download, click **Program.**

FIGURE 4.44
Programmer Dialog Box
(MAX7000S Device)

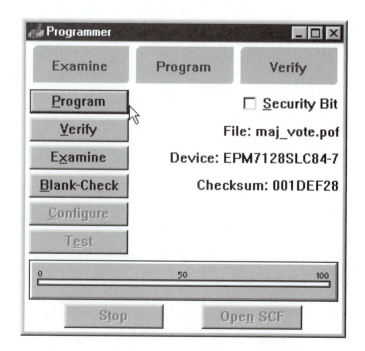

If the project was compiled for the FLEX10K device and the device is to be configured via a ByteBlaster, it must be configured via the **Multi-Device JTAG Chain** available in the **JTAG** menu. Select the **JTAG** menu, shown in Figure 4.45, and choose **Multi-Device JTAG Chain Setup.**

In the **Multi-Device JTAG Chain Setup** window, shown in Figure 4.46, select the pull-down menu for the device name. Select **EPF10K20.** Choose **Delete All** to clear the box of any previous programming file names. Choose the **Select Programming File** button.

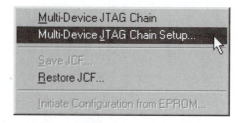

FIGURE 4.45
JTAG Menu

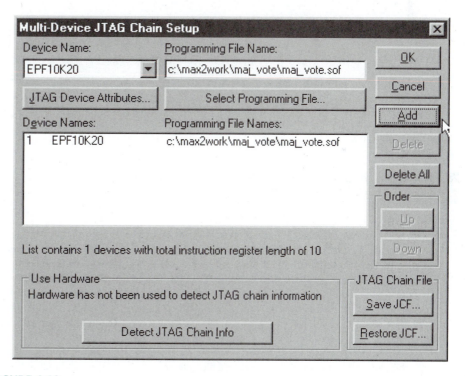

FIGURE 4.46
Multi-Device JTAG Chain Setup

The **Select Programming File** dialog box will appear, as in Figure 4.47. Find and select the file *drive:*\max2work\maj_vote\maj_vote.sof. Click **OK**. Choose the **Add** button in the JTAG setup box to add the SRAM Object File (**sof**) to the list. Choose the **Detect JTAG Chain Info** button to set up the hardware for programming. Choose **OK**. Click the **Configure** button in the **Programmer** dialog box to download the binary information to the FLEX10K CPLD on the UP-1 board.

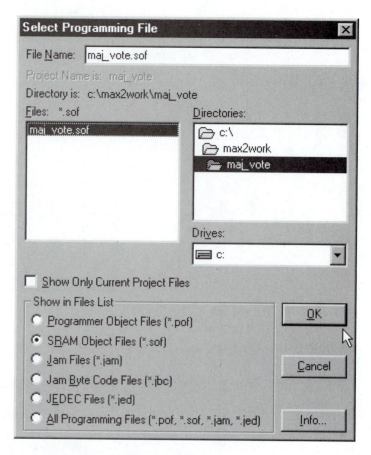

FIGURE 4.47
Select Programming File Dialog Box

SUMMARY

1. A programmable logic device (PLD) is a digital device that is shipped blank and whose function is determined by the end user.

2. PLDs offer design flexibility, reduce board space and package count, and can be used to develop digital designs more quickly than fixed-function logic.

3. Some types of PLDs include PAL (programmable array logic), GAL (generic array logic), EPLD (erasable PLD), CPLD (complex PLD), FPGA (field-programmable gate array).

4. Complex PLDs (CPLDs) are devices with several programmable sections that are interconnected inside the chip.

5. PLD design and programming requires special software, such as Altera's MAX+PLUS II.

6. PLD designs can be entered by schematic capture (Graphic Design Files) or text-based languages, such as Altera Hardware Description Language (AHDL) and VHSIC Hardware Description Language (VHDL).

7. MAX+PLUS II organizes PLD design files in a project. Since many operations in MAX+PLUS II are performed on a project, you should set the project to the current file (File menu) whenever you change windows and make a modification to a design file.

8. *Save your work every time you pause for thought.*

9. A MAX+PLUS II Graphic Design File (gdf) consists of graphical symbols of components that are interconnected by lines drawn between inputs and outputs of the components.

10. Circuit inputs and outputs in a gdf have special symbols. The input and output pins must be named, but need not be numbered in the first stages of a design.

11. The MAX+PLUS II compiler translates the design information from a gdf or text file into binary data that can be downloaded into a PLD. For a MAX7000S, the compiler generates a Programmer Object File (pof) to *program* the device. For a FLEX10K, an SRAM Object File (sof) is generated to *configure* the device.

12. MAX7000S devices are nonvolatile; they stay programmed when the power is removed from the chip. FLEX10K devices are volatile; they lose their programming data when power is removed.

13. If a CPLD part number is not specified, the MAX+PLUS II compiler will automatically select one. It is good practice to assign the part number of the device before compiling, as this can affect the accuracy of certain parts of the design process, such as simulation. The CPLDs on the Altera UP-1 board are EPM7128SLC84-7 and EPF10K20RC240-4.

14. Some useful compiler options are: Design Doctor (checks for good design practice), Timing SNF Extractor (compiles data required for timing simulations), and Smart Recompile (allows part of the compile process to be skipped if only part of a design has changed).

15. Compiler messages can be in green text (Info), blue text (Warning; possible problems, but not fatal), or red text (Error; fatal, compiling stops).

16. MAX+PLUS II files can be arranged in a design hierarchy. That is, a MAX+PLUS II file can contain components that are complete MAX+PLUS II designs in and of themselves.

17. A file that contains other designs, but is not part of a higher-level design, is called the top level of a hierarchy.

18. If the top level of a hierarchy is a gdf, lower-level designs are embedded in the gdf as default symbols that are created from the original design files of the components.

19. MAX+PLUS II looks for default symbols in the present working directory, then in the directories specified as user libraries.

20. VHDL (VHSIC Hardware Description Language) is a text-based programming language used to model and program digital circuits.

21. Every VHDL file requires an entity declaration, which describes the external aspects of the design (inputs and outputs), and an architecture body, which describes the relationship between the inputs and outputs.

22. The entity declaration defines ports (inputs and outputs) and the type of each port (the range of values each port can have).

23. Some common types are BIT (0 or 1), STD_LOGIC (nine-valued standard logic), and INTEGER (whole numbers).

24. The STD_LOGIC type can take on any of the following values:

```
'U', -- Uninitialized
'X', -- Forcing  Unknown
'0', -- Forcing  0
'1', -- Forcing  1
'Z', -- High Impedance
'W', -- Weak Unknown
'L', -- Weak 0
'H', -- Weak 1
'-' -- Don't care
```

25. STD_LOGIC is defined in a library called ieee. To use STD LOGIC, include the following two statements at the beginning of a file

```
LIBRARY ieee;
USE ieee.std_logic_1164.ALL;
```

26. A port in VHDL is an input or output. A signal is an internal connection, like a wire. A variable is a piece of working memory reserved by the VHDL file.

27. The simplest way to relate inputs and outputs in a VHDL design is with a concurrent signal assignment statement, which has the form: `x <= (a and b) or c;` The port or signal on the left side is assigned the value of the logic expression on the right side. (Variables are assigned with a different operator.)

28. A port, signal, or variable can have a multiple-bit construction of type BIT_VECTOR or STD_LOGIC_VECTOR. These structures are called vectors and can be referred to a separate elements (e.g., `y(3) <= d(1) and d(0);`) or as a group (e.g., `y <= "1000"`).

29. A selected signal assignment statement can act as a truth table in VHDL. It assigns alternative values to one or more outputs, depending on the alternative values on one or more inputs.

30. VHDL constructs and statements can be selected in generic form from a template menu in MAX+PLUS II.

31. VHDL designs can be embedded in a gdf as default symbols.

32. Pin numbers must be assigned to a design before it can be downloaded to a CPLD. Pins can be assigned in the Pin/Location/Chip dialog box (accessed by highlighting a pin symbol and right-clicking) or by editing the project's Assignment and Configuration File (acf).

33. An Altera CPLD can be programmed directly from a PC parallel port via a ByteBlaster cable.

34. The ByteBlaster cable implements a programming interface specified by a standard (IEEE Std. 1149.1) of the Joint Test Action Group (JTAG).

35. A JTAG port is a 4-wire interface for loading test and programming information into one or more JTAG-compliant devices. It consists of an input (TDI), output (TDO), mode select (TMS), and clock (TCK).

GLOSSARY

AHDL (Altera Hardware Description Language)   Altera's proprietary text-entry design tool for PLDs.

Altera UP-1 Board   A circuit board, part of Altera's University Program Design Laboratory Package, containing two CPLDs and a number of input and output devices.

Architecture   A VHDL structure than defines the relationship between input, output, and internal signals or variables in a design.

ASICs (application specific integrated circuits)   Integrated circuits that are constructed for a specific design purpose. The term could refer to a PLD, although it usually means a custom-designed fixed function device.

Assignment and Configuration File (acf)   A MAX+PLUS II file that contains information about the configuration options for a project, including assigned device and pin numbers.

Bit string literal   A group of bits assigned to the elements of a vector, enclosed in double quotes (e.g., "001011").

ByteBlaster   An Altera ribbon cable and connector used to program or configure Altera CPLDs via the parallel port (LPT port) of an IBM PC or compatible.

Comment   Explanatory text in a VHDL (or other computer language) file that is ignored by the computer at compile time.

Compile   The process used by CPLD design software to interpret design information (such as a drawing or text file) and create required programming information for a CPLD.

Complex PLD (CPLD)   A digital device consisting of several programmable sections with internal interconnections between the sections.

Concurrent   Simultaneous.

Concurrent signal assignment   A relationship between an input and output port or signal in which the output is changed as soon as there is a change in input. If the file has more than one concurrent signal assignment, they are all evaluated simultaneously.

Default symbol   A graphical symbol that represents a PLD design as a block, showing only the design's inputs and outputs. The symbol can be used as a component in any Graphic Design File.

Design entry   The process of using software tools to describe the design requirements of a PLD. Design entry can be done by entering a schematic or a text file that describes the required digital function.

Download   Program a PLD from a computer running PLD design and programming software.

Entity   A VHDL structure that defines the inputs and outputs of a design.

Fitting   Assigning internal PLD circuitry, as well as input and output pins, for a PLD design.

Graphic Design File (gdf)   A PLD design file in which the digital design is entered as a schematic.

Hardware description language   A computer language used to design digital circuits by entering text-based descriptions of the circuits.

Hierarchical design   A PLD design that is ordered in layers or levels. The highest level of design contains components that are themselves complete designs. These components may, in turn, have lower-level designs embedded within them.

ICR   In-circuit reconfigurability. The ability of a PLD (such as a FLEX10K) to be configured without removing it from a circuit board.

IEEE Standard 1164   The standard which defines a variety of VHDL types and operations, including the STD_LOGIC and STD_LOGIC_VECTOR types.

ISP   In-system programmability. The ability of a PLD (such as a MAX7000S) to be programmed without removing it from a circuit board.

JTAG   Joint Test Action Group. A standards body that developed the format (called IEEE Std. 1149.1) for testing and programming devices while they are installed in a system.

JTAG Chain   Multiple JTAG-compliant devices whose TDI and TDO ports form a continuous chain connection. Such a chain allows multi-device programming.

Library   A collection of VHDL design units that have been previously compiled.

MAX+PLUS II   CPLD design and programming software owned by Altera Corporation.

Mode (of a port)   The kind of port, such as input or output.

Nonvolatile   Able to retain stored information after power is removed.

Package   A group of VHDL design elements that can be used by more than one VHDL file.

Port   A name assigned to an input or output of a VHDL design entity.

Programmable logic device (PLD)   A digital integrated circuit that can be programmed by the user to implement any digital logic function.

Programmer Object File (pof)   Binary file used to program a PLD of the Altera MAX series.

Programming   Transferring design information from the computer running PLD design software to the actual PLD chip.

Project   A set of MAX+PLUS II files associated with a particular PLD design.

Schematic capture   A technique of entering CPLD design information by using a CAD (computer aided design) tool to draw a logic circuit as a schematic. The schematic can then be interpreted by design software to generate programming information for the CPLD.

Selected signal assignment statement   A concurrent signal assignment in VHDL in which a value is assigned to a signal, depending on the alternative values of another signal or variable.

Signal   A name given to an internal connection in a VHDL architecture.

Simulation   Verifying design function by specifying a set of inputs and observing the resultant outputs. Simulation is generally shown as a series of input and output waveforms.

Software tools   Specialized computer programs used to perform specific functions such as design entry, compiling, fitting, and so on. (Sometimes just called "tools".)

SRAM Object File (sof)   Binary file used to configure a PLD of the Altera FLEX series.

Suite (of software tools)   A related collection of tools for performing specific tasks. MAX+PLUS II is a suite of tools for designing and programming digital functions in a PLD.

Syntax   The "grammar" of a computer language (i.e., the rules of construction of language statements).

Target device   The specific PLD for which a digital design is intended.

TCK   Test Clock. The JTAG signal that drives the JTAG downloading process from one state to the next.

TDI   Test Data In. In a JTAG port, the serial input data to a device.

TDO   Test Data Out. The JTAG signal, the serial output data from a device.

TMS   Test Mode Select. The JTAG signal that controls the downloading of test or programming data.

Top level (of a hierarchy)   The file in a hierarchy that contains components specified in other design files and is not itself a component of a higher-level file.

Type   A set of characteristics associated with a VHDL port name, signal, or variable that determines the allowable values of the port, signal, or variable.

User library   A folder containing symbols that can be used in a gdf file.

Variable   A block of working memory used for internal calculation or storage in a VHDL architecture.

Vector   A group of digital signals or variables, usually related numerically, that can be treated as a single multi-bit variable.

VHDL (VHSIC Hardware Description Language)   An industry-standard computer language used to model digital circuits and produce programming data for PLDs.

VHSIC   Very high speed integrated circuit

Volatile   A device is volatile if it does not retain its stored information after the power to the device is removed.

PROBLEMS

Section 4.1   What Is a PLD?

4.1   List some of the advantages of programmable logic over fixed-function logic.

4.2   What does CPLD stand for? How is it different from the term PLD?

4.3   List some types of PLDs other than CPLDs.

4.4   Figure 4.48 shows a 4-to-1 multiplexer circuit. (The circuit switches one of four digital inputs to a single output, depending on the states of two "select inputs.") State the number of 74HC type devices required to make this circuit. You may use the following devices: 74HC04 hex inverter; 74HC11 triple 3-input AND gate; 74HC4002 dual 4-input NOR gate (there are no 4-input OR devices available in the 74HC family). State how many devices are required to make two multiplexers.

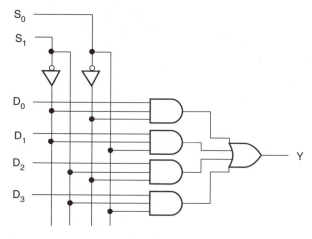

FIGURE 4.48
Problem 4.4
4-to-1 Multiplexer

Section 4.3   Graphic Design File

Section 4.4   Compiling MAX+**PLUS II** Files

4.5   Briefly describe the difference between a design file and a project in MAX+PLUS II.

4.6   State two ways to set the MAX+PLUS II project to the current file.

4.7   State the definitions of the following terms:

a. primitives

b. instance

4.8   Use MAX+PLUS II to create a Graphic Design File for the multiplexer circuit shown in Figure 4.48. Save the file as *drive:***\max2work\chapt4\problems\4to1mux.gdf**. Assign pins as in Table 4.3. Set the project to the current file and compile.

Table 4.3   Pin Assignments for Multiplexer Circuit

Function	Pin
S1	12
S0	16
D0	15
D1	17
D2	21
D3	25
Y	4

4.9   Figure 4.49 shows the circuit for a 4-channel demultiplexer, which switches a digital input to one of four outputs, depending in the states of two "select inputs." Figure 4.49

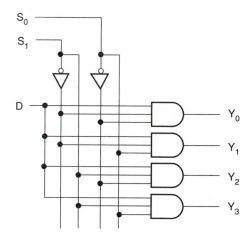

FIGURE 4.49
Problem 4.9
4-channel Demultiplexer

Use MAX+PLUS II to create a Graphic Design File for the demultiplexer circuit. Save the file as *drive:*\max2work\chapt4\problems\4ch_dmux.gdf. Assign pins as in Table 4.4. Set the project to the current file and compile.

Table 4.4 Pin Assignments for Demultiplexer Circuit

Function	Pin
S1	12
S0	16
D	15
Y0	4
Y1	6
Y2	8
Y3	10

4.10 Repeat Problem 4.9 for the 2-bit equality comparator in Figure 4.50. This circuit generates a HIGH output when the two 2-bit numbers A_2A_1 and B_2B_1 are equal. Save the file as *drive:*\max2work\chapt4\problems\eq_comp.gdf. Use the pin assignments in Table 4.5.

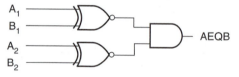

FIGURE 4.50
Problem 4.10
2-bit Equality Comparator

Table 4.5 Pin Assignments for Equality Comparator

Function	Pin
A1	12
A2	16
B1	15
B2	17
AEQB	4

4.11 Use MAX+PLUS II to create a Graphic Design File for the half-adder circuit shown in Figure 4.51. The half-adder adds 2 bits to generate a sum and a carry output. Save the file as *drive:*\max2work\chapt4\problems\halfadd.gdf. Create a default symbol for the file and compile, after setting the project to the current file. Do not assign pin numbers at this time.

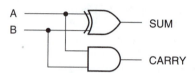

FIGURE 4.51
Problem 4.11
Half Adder

4.12 Use MAX+PLUS II to create a Graphic Design File for the full adder circuit shown in Figure 4.52. The full adder combines two bits A and B, plus an input carry from a previous stage to generate a sum and a carry output.

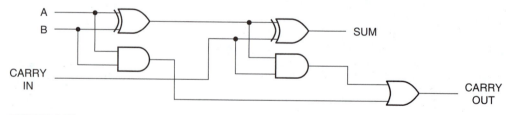

FIGURE 4.52
Problem 4.12
Full Adder

Save the file as *drive:*\max2work\chapt4\problems\fulladd.gdf. Assign pin numbers as shown in Table 4.6. Set the project to the current file and compile.

Table 4.6 Pin Assignments for Full Adder

Function	Pin
A	12
B	15
CARRY IN	33
SUM	6
CARRY OUT	4

4.13 Examine the half adder circuit in Figure 4.51 and the full adder circuit in Figure 4.52. You should find two half adders in the full adder circuit. Use the half adder symbol you created in Problem 4.11 to create a full adder as a hierarchical design, consisting of two half adders and other logic. Save the file as *drive:*\max2work\chapt4\problems\fulladd2.gdf. Assign the pin numbers as in Table 4.6.

Section 4.5 Hierarchical Design

4.14 Use the MAX+PLUS II hierarchy display function for the full adder you created in Problem 4.13. Sketch the resultant hierarchy display.

Section 4.6 Text Design File (VHDL)

4.15 State the full names for which AHDL, VHDL and VHSIC are abbreviations.

4.16 Write one sentence describing the difference between AHDL and VHDL.

4.17 State the two minimum structures required to write a VHDL file and briefly describe each one.

4.18 List four possible modes for a VHDL port. Briefly describe each one.

4.19 Briefly state the difference between a port of mode OUT and a port of mode BUFFER.

4.20 State the difference between type BIT and type STD_LOGIC. Why does STD_LOGIC have so many values?

4.21 Create and compile a VHDL file for the 4-to-1 multiplexer circuit of Figure 4.48. Save the file as *drive:*\max2work\chapt4\problems\mux4.vhd. (Note that VHDL does not permit an identifier (such as an entity name) to start with a digit. Recall that a VHDL file must be saved with the same name as the entity.)

4.22 Create and compile a VHDL file for the 2-line-to-4-line decoder of Figure 4.31. Save the file as *drive:*\max2work\chapt4\problems\decode4.vhd.

4.23 Create and compile a VHDL file for the 4-channel demultiplexer of Figure 4.49. Save the file as *drive:*\max2work\chapt4\problems\dmux4.vhd.

4.24 Create and compile a VHDL file for the equality comparator of Figure 4.50. Save the file as *drive:*\max2work\chapt4\problems\eq_comp.vhd.

(The **xnor** function required by this problem is not supported in VHDL 1987, the default version for the MAX+PLUS II compiler. To change the version to VHDL 1993, which does support the **xnor** function, select **VHDL Netlist Reader Settings** from the **Interfaces** menu in the MAX+PLUS II **Compiler** window. In the box labelled **VHDL Version,** select **VHDL 1993** and click **OK.** Alternatively, use the construct **not(a xor b)** and leave the settings for VHDL 1987.)

4.25 Create and compile a VHDL file for the half adder circuit of Figure 4.51. Save the file as *drive:*\max2work\chapt4\problems\half_add.vhd. Create a default symbol for this file.

4.26 Create and compile a VHDL file for the full adder circuit in Figure 4.52. Save the file as *drive:*\max2work\chapt4\problems\full_add.vhd.

4.27 Create a Graphic Design File that combines two half adder symbols created from VHDL files, as in Problem 4.25, to make a full adder.

Section 4.7 Creating a Physical Design

4.28 State the purpose of a JTAG port and briefly describe its operation.

Combinational Logic Functions

C H A P T E R   O B J E C T I V E S

Upon successful completion of this chapter you will be able to:

• Design binary decoders using logic gates.

• Create decoder designs in MAX+PLUS II, using Graphic Design Files or VHDL.

• Create MAX+PLUS II simulation files to verify the operation of combinational circuits.

• Design BCD-to-seven-segment and hexadecimal-to-seven-segment decoders, including special features such as ripple blanking, using VHDL and Graphic Design Files in MAX+PLUS II.

• Use MAX+PLUS II Graphic Design Files and VHDL to generate the design for a 3-bit binary and a BCD priority encoder.

• Describe the circuit and operation of a simple multiplexer and program these functions in VHDL.

• Draw logic circuits for multiplexer applications, such as single-channel data selection, multibit data selection, waveform generation, and time-division multiplexing (TDM).

• Describe demultiplexer circuits and program them using VHDL.

• Define the operation of a CMOS analog switch and its use in multiplexers and demultiplexers.

• Define the operation of a magnitude comparator and program its function in VHDL.

• Explain the use of parity as an error-checking system and draw simple parity-generation and checking circuits..

A number of standard combinational logic functions have been developed for digital circuits that represent many of the useful tasks that can be performed with digital circuits.

Decoders detect the presence of particular binary states and can activate other circuits based on their input values or can convert an input code to a different output code. Encoders generate a binary or binary coded decimal (BCD) code corresponding to an active input.

Multiplexers and demultiplexers are used for data routing. They select a transmission path for incoming or outgoing data, based on a selection made by a set of binary-related inputs.

Magnitude comparators determine whether one binary number is less than, greater than, or equal to another binary number.

Parity generators and checkers are used to implement a system of checking for errors in groups of data. ■

5.1 Decoders

KEY TERMS

Decoder   A digital circuit designed to detect the presence of a particular digital state.

The general function of a **decoder** is to activate one or more circuit outputs upon detection of a particular digital state. The simplest decoder is a single logic gate, such as a NAND or AND, whose output activates when *all* its inputs are HIGH. When combined with one or more inverters, a NAND or AND can detect any unique combination of binary input values.

An extension of this type of decoder is a device containing several such gates, each of which responds to a different input state. Usually, for an n-bit input, there are 2^n logic gates, each of which decodes a different combination of input variables. A variation is a BCD device with 4 input variables and 10 outputs, each of which activates for a different BCD input.

Some types of decoders translate binary inputs to other forms, such as the decoders that drive seven-segment numerical displays, those familiar figure-8 arrangements of LED or LCD outputs ("segments"). The decoder has one output for every segment in the display. These segments illuminate in unique combinations for each input code.

Single-Gate Decoders

The simplest decoder is a single gate, sometimes in combination with one or more inverters, used to detect the presence of one particular binary value. Figure 5.1 shows two such decoders, both of which detect an input $D_3D_2D_1D_0 = 1111$.

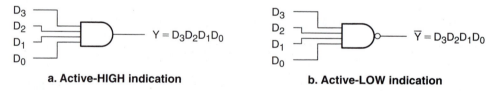

a. Active-HIGH indication   b. Active-LOW indication

FIGURE 5.1
Single-Gate Decoders

The decoder in Figure 5.1a generates a logic HIGH when its input is 1111. The decoder in Figure 5.1b responds to the same input, but makes the output LOW instead.

In Figure 5.1, we designate D_3 as the most significant bit of the input and D_0 the least significant bit. We will continue this convention for multi-bit inputs.

In Boolean expressions, we will indicate the active levels of inputs and outputs separately. For example, in Figure 5.1, the inputs to both gates are the same, so we write $D_3D_2D_1D_0$ for the inputs of both gates. The gates in Figures 5.1a and b have outputs with opposite active levels, so we write the output variables as complements (Y and $\overline{Y}$).

■ EXAMPLE 5.1

Figure 5.2 shows three single-gate decoders. For each one, state the output active level and the input code that activates the decoder. Also write the Boolean expression of each output.

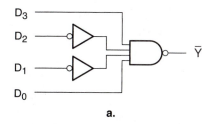

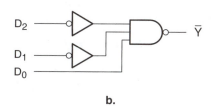

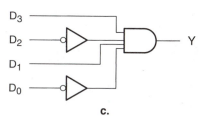

a.   b.   c.

FIGURE 5.2
Example 5.1
Single-Gate Decoders

Solution Each decoder is a NAND or AND gate. For each of these gates, the output is *active* when *all inputs are HIGH*. Because of the inverters, each circuit has a different code that fulfils this requirement.

Figure 5.2a: Output: Active LOW

Input code: $D_3D_2D_1D_0 = 1001$

$$\overline{Y} = D_3\overline{D}_2\overline{D}_1D_0$$

Figure 5.2b: Output: Active LOW

Input code: $D_2D_1D_0 = 001$

$$\overline{Y} = \overline{D}_2\overline{D}_1D_0$$

Figure 5.2c: Output: Active HIGH

Input code: $D_3D_2D_1D_0 = 1010$

$$Y = D_3\overline{D}_2D_1\overline{D}_0$$

Single-gate decoders are often used to activate other digital circuits under various operating conditions, particularly if there is a choice of circuits to activate. For example, single-gate decoders are used to enable peripheral devices in a personal computer (PC). A combination of binary values, called the address, specifies a unique set of conditions to enable a particular peripheral device.

EXAMPLE 5.2

Application

A PC has two serial port cards called COM1 and COM2. Each card is activated when either one of two control inputs called $\overline{IOR}$ (*Input/Output Read*) and $\overline{IOW}$ (*Input/Output Write*) are active and a unique 10-bit address is present. $\overline{IOR}$ and $\overline{IOW}$ are active-LOW. The address is specified by bits $A_9A_8A_7A_6A_5A_4A_3A_2A_1A_0$, which can be represented by three hexadecimal digits. The decoder outputs, $\overline{COM1\_Enable}$ and $\overline{COM2\_Enable}$ are both active-LOW.

The card for COM1 activates when ($\overline{IOR}$ OR $\overline{IOW}$ is LOW) AND the address is between 3F8H and 3FFH.

The card for COM2 activates when ($\overline{IOR}$ OR $\overline{IOW}$ is LOW) AND the address is between 2F8H and 2FFH.

Create a Graphic Design File in MAX+PLUS II that implements the specified decoder.

Solution The lowest address that activates COM1 is

$$A_9A_8A_7A_6A_5A_4A_3A_2A_1A_0 = 3F8H = 11\ 1111\ 1000$$

The highest COM1 address is

$$A_9A_8A_7A_6A_5A_4A_3A_2A_1A_0 = 3FFH = 11\ 1111\ 1111$$

Since *any* address in this range is valid, we can represent the last three bits, $A_2A_1A_0$, as don't care states. Thus, for COM1, we should decode the address:

$$A_9A_8A_7A_6A_5A_4A_3A_2A_1A_0 = 11\ 1111\ 1XXX$$

Similarly, for COM2:

Low address: $A_9A_8A_7A_6A_5A_4A_3A_2A_1A_0 = 2F8H = 10\ 1111\ 1000$

High address: $A_9A_8A_7A_6A_5A_4A_3A_2A_1A_0 = 2FFH = 10\ 1111\ 1111$

Decode: $\quad A_9A_8A_7A_6A_5A_4A_3A_2A_1A_0 = 10\ 1111\ 1XXX$

Figure 5.3 shows the **gdf** representation of the decoder circuit, including inputs for the control signals $\overline{IOR}$ and $\overline{IOW}$.

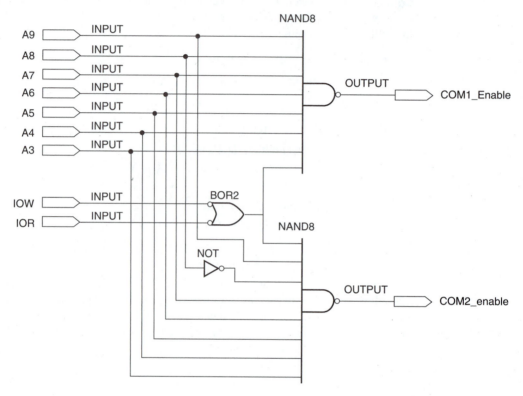

FIGURE 5.3
Example 5.2
COM Port Decoders

III SECTION 5.1A REVIEW PROBLEM

5.1 Draw a single-gate decoder that detects the input state $D_3D_2D_1D_0 = 1100$

    a. with active-HIGH indication

    b. with active-LOW indication

Multiple-Output Decoders

Decoder circuits often are constructed with multiple outputs. In effect, such a device is a collection of decoding gates controlled by the same inputs. A decoder circuit with n inputs can activate up to $m = 2^n$ load circuits. Such a decoder is usually described an n-line-to-m-line decoder.

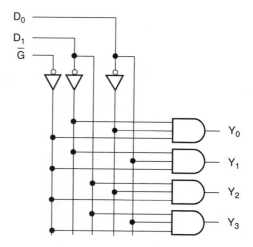

FIGURE 5.4
2-line-to-4-line Decoder with Enable

Figure 5.4 shows the logic circuit of a 2-line-to-4-line decoder. The circuit detects the presence of a particular state of the 2-bit input D_1D_0, as shown by the truth table in Table 5.1. One and only one output is HIGH for any input combination, provided the enable input $\overline{G}$ is LOW. The active input of each line is shown in **boldface**. The subscript of the active output is the same as the value of the 2-bit input. For example, if $D_1D_0 = 10$, output Y_2 is active since 10 (binary) = 2 (decimal).

Table 5.1 Truth Table of a 2-to-4 Decoder with Enable

$\overline{G}$	D_1	D_0	Y_0	Y_1	Y_2	Y_3
0	0	0	**1**	0	0	0
0	0	1	0	**1**	0	0
0	1	0	0	0	**1**	0
0	1	1	0	0	0	**1**
1	X	X	0	0	0	0

If we are using the decoder to activate one of four output loads, it is possible that there are situations where we want no output to be active. In such a case, we can deactivate all outputs (make them all LOW) by setting $\overline{G}$ HIGH.

We can create the 2-line-to-4-line decoder of Figure 5.4 as a graphic or text file in MAX+PLUS II and create a symbol for it that can be used in higher-level graphic files. Figure 5.5 shows the symbol for the decoder.

FIGURE 5.5
MAX+PLUS II Graphic
Symbol for a 2-to-4 Decoder
with Enable

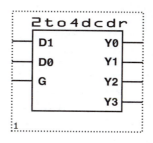

FIGURE 5.6

3-line-to-8-line Decoder with
Enable

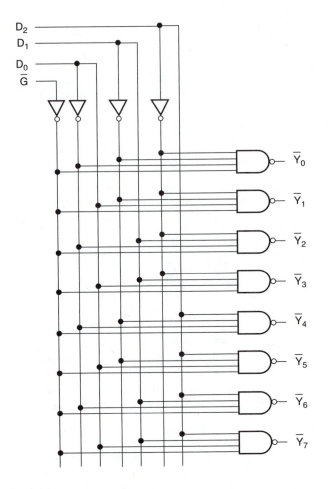

Figure 5.6 shows the circuit for a 3-line-to-8-line decoder, again with an active-LOW enable, $\overline{G}$. In this case, the decoder outputs are active LOW. One and only one output is active for any given combination of $D_2D_1D_0$. Table 5.2 shows the truth table for this decoder. Again if the enable line is HIGH, no output is active.

Table 5.2 Truth Table of a 3-to-8 Decoder with Enable

$\overline{G}$	D_2	D_1	D_0	$\overline{Y}_0$	$\overline{Y}_1$	$\overline{Y}_2$	$\overline{Y}_3$	$\overline{Y}_4$	$\overline{Y}_5$	$\overline{Y}_6$	$\overline{Y}_7$
0	0	0	0	**0**	1	1	1	1	1	1	1
0	0	0	1	1	**0**	1	1	1	1	1	1
0	0	1	0	1	1	**0**	1	1	1	1	1
0	0	1	1	1	1	1	**0**	1	1	1	1
0	1	0	0	1	1	1	1	**0**	1	1	1
0	1	0	1	1	1	1	1	1	**0**	1	1
0	1	1	0	1	1	1	1	1	1	**0**	1
0	1	1	1	1	1	1	1	1	1	1	**0**
1	X	X	X	1	1	1	1	1	1	1	1

▌▌ **EXAMPLE 5.3**

Application

Figure 5.7 shows a partial Graphic Design File, created in MAX+PLUS II, that shows how a 3-line-to-8-line decoder, such as the one shown in Figure 5.6, can be used in a microcomputer memory system as an **address decoder.** Each block labeled **8k_sram** is a memory chip capable of holding 8192 (8K) bytes of data. Since there are eight such de-

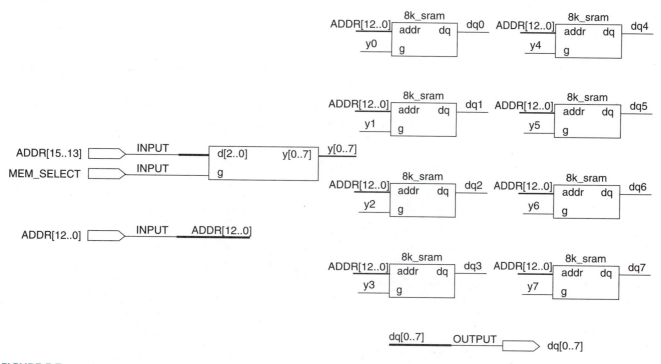

FIGURE 5.7
Example 5.3
Address Decoder for a Memory System

vices, the whole system can hold $8 \times 8192 = 65,536$ (64K) bytes. (Although this amount of memory may seem small by the standards of a desktop computer, it may be typical of a small stand-alone computer system (called an embedded system or a microcontroller) that is used in control applications.)

Each 8K block is enabled by a LOW at its G input. Briefly explain the function of the decoder in the system.

Solution Since only one decoder output is LOW at any one time, the decoder allows only one memory block to be active at any one time. The active block is chosen by inputs $ADDR_{15}ADDR_{14}ADDR_{13}$, which are connected to $D_2D_1D_0$ on the decoder. The active memory block is the one connected to the **y** output whose subscript matches the binary value of these inputs. For example, when $ADDR_{15}ADDR_{14}ADDR_{13} = 110$, the block connected to **y6** is active.

If the decoder is the same as the one in Figure 5.6, no outputs will be active, and therefore no memory block will be enabled, when $\overline{G} = 1$. (Note that the MAX+PLUS II Graphic Editor cannot represent an input or output with an inversion bar. Some conventions would represent an active-LOW terminal with an "n" prefix, indicating "NOT" (e.g., **nG**). This is a matter of personal choice, but without such an indication it is not possible to tell the active level of an input or output from the MAX+PLUS II Graphic Design File.)

The decoders in Figure 5.6 and 5.7 have identical functions, but the symbol in Figure 5.7 shows the D inputs and Y outputs as multibit vectors or busses. Figure 5.7 also shows how the individual signals in a bus can be connected to separate parts of the circuit in a MAX+PLUS II Graphic Design File.

To make the connections, draw and label a line extending from each terminal. To label a line, highlight the line by clicking on it with the left mouse button, then right-click. Select **Enter Node/Bus Name** from the pop-up menu and enter the text. Lines that have the same names are automatically connected by their text references. If a line is a multiple line, it must have signal designators in brackets (e.g., **y[0..7]**). Individual signals from a bus must be numbered in a way that corresponds to the multiple-bit line (e.g., **y0, y1, y2,** and so on).

▮▮▮ SECTION 5.1B REVIEW PROBLEM

5.2 How many inputs are required for a binary decoder with 16 outputs? How many inputs are required for a decoder with 32 outputs?

Simulation of a 2-Line-to-4-Line Decoder

> ### KEY TERMS
>
> **Timing diagram**   A diagram showing how two or more digital waveforms in a system relate to each other over time.
>
> **Simulation**   The verification, using timing diagrams, of the logic of a digital design before programming it into a PLD.
>
> **Stimulus waveforms**   A set of user-defined input waveforms in a simulator file designed to imitate input conditions of a digital circuit.
>
> **Response waveforms**   A set of output waveforms generated by a simulator for a particular digital design in response to a set of stimulus waveforms.
>
> **Propagation delay**   Time difference between a change on a digital circuit input and a change on an output in response to the input change.

An important part of the CPLD design process is **simulation** of the design. A simulation tool allows us to see whether the output responses to a set of circuit inputs are what we expected in our initial design idea. The simulator works by creating a **timing diagram.** We specify a set of input (**stimulus**) waveforms. The simulator looks at the relationship between inputs and outputs, as defined by the design file, and generates a set of **response** outputs.

Figure 5.8 shows a set of simulation waveforms created for the 2-line-to-4-line decoder in Figure 5.4. The inputs D1 and D0 are combined as a single 2-bit value, to which an increasing binary count is applied as a stimulus. The decoder output waveforms are observed individually to determine the decoder's response. Once we have entered the design in the MAX+PLUS II Graphic Editor and compiled it, we can create the waveforms as follows.

FIGURE 5.8
Simulation Waveforms for a 2-to-4 Decoder with Enable

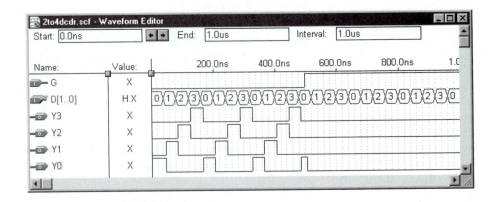

From the **File** menu, select **New.** On the resultant dialog box, select **Waveform Editor File,** with a default file extension **scf.** From the **File** menu, choose **Save As,** then enter *drive:*\max2work\chapt05\decoders\2to4dcdr.scf.

2to4dcdr.gdf
2to4dcdr.scf

We specify the inputs and outputs we want to view by selecting **Enter Nodes from SNF** on the **Node** menu, shown in Figure 5.9. In the dialog box that pops up (Figure 5.10), there are two boxes labelled **Available Nodes & Groups** and **Selected Nodes & Groups,** with an arrow (=>) pointing from one to the other. Select the **List** button to show the "available" signals and click the arrow to transfer them all to the "selected" box. Click **OK** to close the box.

FIGURE 5.9
Node Menu

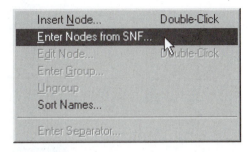

FIGURE 5.10
Selecting Nodes for Waveform Editor

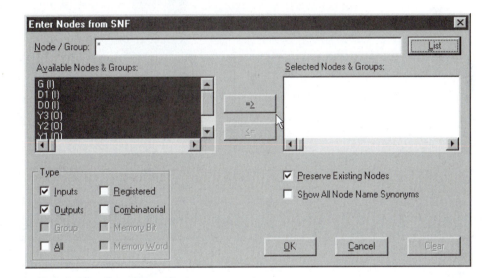

Figure 5.11 shows the simulation waveforms in their uninitialized (default) states. Inputs and outputs are shown by symbols in front of the signal names. Inputs are at logic 0 and outputs are indicated as X or unknown values.

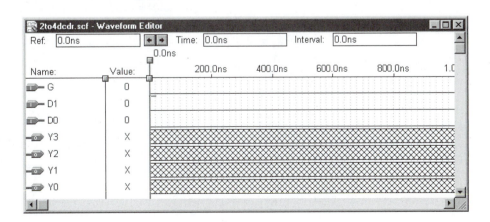

FIGURE 5.11
Default Values of Simulation Waveforms

We now set the timing length of the simulation. The default value is 1 μs, written **1.0us.** For this example, we will leave the end time at the default value. However, if we want to change it, we select **End Time** (**File** menu, Figure 5.12) and enter the new time for the end of simulation in the dialog box of Figure 5.13. Click **OK.**

FIGURE 5.12

Setting the End Time of a Simulation (File Menu)

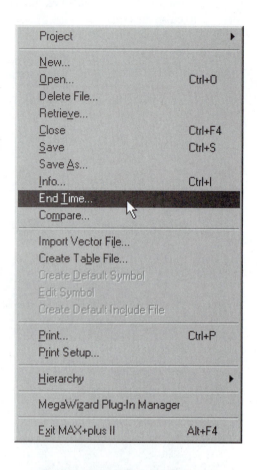

FIGURE 5.13

End Time Dialog Box

The **End Time** dialog sets the end of the simulation. We should also set the **Grid Size,** which determines the size of the smallest time division in the simulation. To do so, select **Grid Size** from the **Options** menu, shown in Figure 5.14. In the dialog box of Figure 5.15, enter the value **20ns** and click **OK.** (We will use this value for many of our simulations

FIGURE 5.14

Setting Simulation Grid Size (Options Menu)

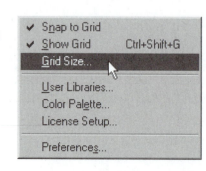

FIGURE 5.15

Grid Size Dialog Box

because it corresponds to one half period of the oscillator on the Altera UP-1 board. In the simulator, one full period requires two grid spaces.)

When we created the simulation file, the D inputs were entered as separate waveforms. We can join these waveforms to make a **Group.** Highlight both D1 and D0 by clicking on one name and dragging the mouse to the next name, as in Figure 5.16. From the **Node** menu or the pop-up menu in Figure 5.17, select **Enter Group.** The dialog box shown in Figure 5.18 appears, containing the most likely name derived from the highlighted group. Either type a new group name or accept the original name by clicking **OK.**

FIGURE 5.16
Highlighting a Group

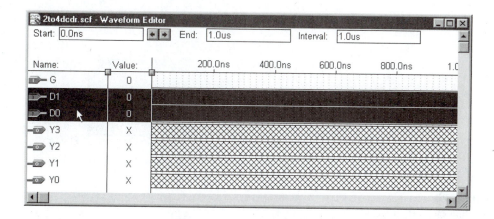

FIGURE 5.17
Pop-up Menu (Enter Group)

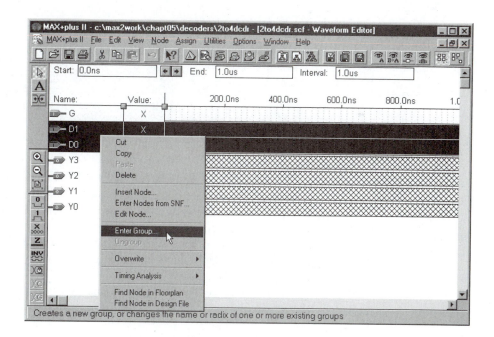

FIGURE 5.18
Enter Group Dialog Box

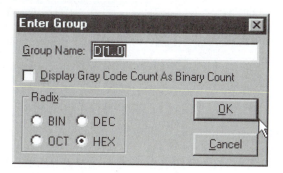

Overwrite Count
Button

As a decoder stimulus, we will define an increasing binary count on the D inputs. Highlight the input group by clicking in the **Value** column. Use the **Overwrite Count** toolbar button to create an increasing binary count on the group, D[1..0]. Fill in the dialog box as shown in Figure 5.19 and click **OK.** The count is increased every 40 ns (2 × 20 ns), as shown in Figure 5.20.

FIGURE 5.19

Overwrite Count Value Dialog
Box

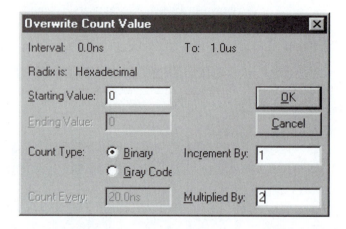

FIGURE 5.20

Group Input with Binary Count

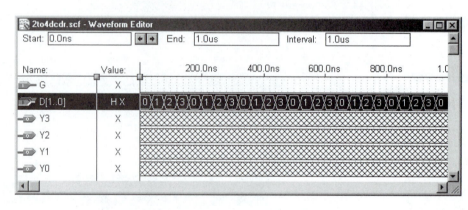

Fit in Window
Button

Save the file. From the **MAX+PLUS II** menu, bring the **Simulator** to the front and click **Start.** When the simulation is finished (almost immediately), click **Open SCF** and maximize the window. From the **View** menu, select **Fit in Window** or select the toolbar button for this function.

The simulator output, shown in Figure 5.21, shows the result of a repeating binary count at the decoder input when the outputs are always enabled. The outputs activate in a repeating sequence, from Y0 to Y3.

You will notice that the D inputs change exactly on the grid lines, but the Y outputs change slightly after. This is due to **propagation delay,** defined as the time between an

FIGURE 5.21

Decoder Simulation with Enable
Always Active

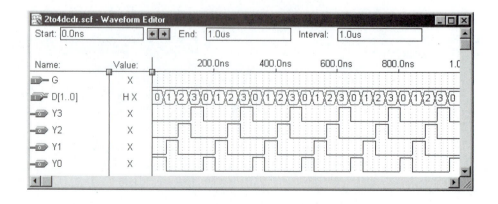

input change and the time an output changes in response to that input. In the EPM7128SLC84-7 CPLD, for which this simulation is created, propagation delay is about 7 nanoseconds. (The MAX+PLUS II simulator accounts for the propagation delay in different CPLDs.) Later simulations in this chapter will not necessarily show the delay, as the timing chosen may be very long compared to delay times.

Overwrite with
HIGH Button

To see the result of the enable input, highlight the G waveform from approximately 500 ns to 1 µs by dragging the mouse along this part of the waveform. Overwrite the highlighted part by clicking the **Overwrite with HIGH** button. When we run the simulation again, we get the waveforms shown in Figure 5.8.

VHDL Binary Decoder

KEY TERMS

Selected signal assignment statement   A concurrent signal assignment in VHDL in which a value is assigned to a signal, depending on the alternative values of another signal or variable.

Conditional signal assignment statement   A concurrent VHDL construct that assigns a value to a signal, depending on a sequence of conditions being true or false.

In Chapter 4, we saw an example of how we can use VHDL to define the function of a 2-line-to-4-line decoder. For reference the description is replicated below, with the difference that the input and output ports are defined as BIT rather than STD_LOGIC types. (This is sufficient for a combinational circuit like a decoder, as the only I/O (input/output) values required are '0' and '1'. If we use BIT types, we do not require a reference to the IEEE library, as we do to define STD_LOGIC types.)

decode1.vhd

```
ENTITY decode1 IS
 PORT(
      d1, d0       : IN  BIT;
      y0, y1, y2, y3 : OUT BIT);
END decode1;

ARCHITECTURE decoder1 OF decode1 IS
BEGIN
 y0  <= (not d1) and (not d0);
 y1  <= (not d1) and (    d0);
 y2  <= (    d1) and (not d0);
 y3  <= (    d1) and (    d0);
END decoder1;
```

The above formulation has no enable input. If we wish to include the enable function, we must modify the entity declaration to include that input and change the signal assignment statements, as well. The new VHDL code is as follows.

decode2.vhd

```
ENTITY decode2 IS

 PORT(
      d1, d0, g      : IN BIT;
      y0, y1, y2, y3 : OUT BIT);
END decode2;

ARCHITECTURE decoder2 OF decode2 IS
BEGIN
 y0  <=  (not d1) and (not d0) and (not g);
 y1  <=  (not d1) and (    d0) and (not g);
 y2  <=  (    d1) and (not d0) and (not g);
 y3  <=  (    d1) and (    d0) and (not g);
END decoder2;
```

In addition to coding the Boolean expressions directly, we can use two types of concurrent signal assignments to create decoder circuits: the **selected signal assignment statement** and the **conditional signal assignment statement.** Both the Altera VHDL manual and the Help menu in MAX+PLUS II have a section on "Golden Rules" for VHDL. The VHDL Golden Rules suggest that you should use a selected signal assignment rather than a conditional signal assignment, if possible. This is because, in certain cases, the selected signal assignment uses the internal circuitry of the CPLD more efficiently.

The selected signal assignment has the form:

```
label: WITH __expression SELECT
 __signal <=__expression WHEN __constant_value,
             __expression WHEN __constant_value,
             __expression WHEN __constant_value,
             __expression WHEN __constant_value;
```

The signal indicated in the second line of the statement template is assigned one of several expressions, depending on the constant value of the expression in the first line. The label is optional. Examine the selected signal statement below:

```
circuit: WITH mode SELECT
    y   <=  q       WHEN "00"
            not q  WHEN "01",
            p       WHEN "11",
            '1'     WHEN others;
```

Signal **y** is assigned one of three values, **p, q,** or **not q,** depending on the status of a two-bit variable called **mode.** Note that the value of **y** for the case when **mode** = "10" is not explicitly stated. This is covered by the last clause **(WHEN others),** which defines a default value for signal **y** of logic 1.

The following VHDL code implements a 2-line-to-4-line decoder using a selected signal assignment statement.

decode3.vhd

```
LIBRARY ieee;
USE ieee.std_logic_1164.ALL;

ENTITY decode3 IS
  PORT(
      d   : IN   STD_LOGIC_VECTOR (1 downto 0);
      y   : OUT  STD_LOGIC_VECTOR (3 downto 0));
END decode3;

ARCHITECTURE decoder OF decode3 IS
BEGIN
  WITH d SELECT
      y <=   "0001" WHEN "00",
             "0010" WHEN "01",
             "0100" WHEN "10",
             "1000" WHEN "11",
             "0000" WHEN others;
END decoder;
```

The selected signal assignment statement evaluates input **d.** For every possible combination of the 2-bit input vector, **d,** a particular value is assigned to the 4-bit vector, **y.** (For example, for the case $d_1 d_0 = 10 (= 2_{10})$, the output y_2 is HIGH: $y_3 y_2 y_1 y_0 = 0100$.)

The default case **("WHEN others")** is required because of the multivalued logic type STD_LOGIC_VECTOR. Since a STD_LOGIC_VECTOR can have values other than '0' and '1', the values listed for **d** don't cover all possible cases. The default output (which will never occur if we only use '0' and '1' inputs) is chosen such that no output is active in the default case. The default case would not be required if we chose to use BIT_VECTOR, rather than STD_LOGIC_VECTOR, since the listed combinations of **d** cover all possible combinations of a BIT_VECTOR. However, it is a good practice to include the default case, in order to account for all possible contingencies.

In order to include an enable input (**g**) in a decoder, we can increase the input vector size to include the **g** input, as shown in the following code.

decode3a.vhd

```
LIBRARY ieee;
USE ieee.std_logic_1164.ALL;

ENTITY decode3a IS
    PORT(
        d  : IN    STD_LOGIC_VECTOR (1 downto 0);
        g  : IN    STD_LOGIC;
        y  : OUT   STD_LOGIC_VECTOR (3 downto 0));
END decode3a;

ARCHITECTURE decoder OF decode3a IS
    SIGNAL inputs : STD_LOGIC_VECTOR (2 downto 0);
BEGIN
    inputs(2)              <= g;
    inputs(1 downto 0)     <= d;
    WITH inputs SELECT
        y  <=     "0001" WHEN "000",
                  "0010" WHEN "001",
                  "0100" WHEN "010",
                  "1000" WHEN "011",
                  "0000" WHEN others;
END decoder;
```

To include **g** and **d** in a single vector, we create a signal called **inputs,** a vector with three elements in the sequence **g, d(1), d(0).** When assigning the **d** to the last two elements of **inputs,** we must be explicit about which elements of **inputs** we want to use. Since **d** only contains two elements and we are assigning them to two elements of **inputs,** we don't need to list the elements of **d** explicitly.

We can use a selected signal assignment statement to evaluate all inputs, including **g** , and assign outputs accordingly. When **g** = '0', the decoder outputs are assigned the same as they were in the example without the enable input. The cases where **g** = '1' are covered by the **others** clause. In this default case, all decoder outputs are LOW (inactive).

Another way to include an enable input is to use a conditional signal assignment statement, which makes an assignment based on a Boolean expression. This template for the conditional signal assignment statement is:

```
__signal  <= __expression WHEN __boolean_expression ELSE
             __expression WHEN __boolean_expression ELSE
             __expression;
```

The first Boolean expression in the statement is evaluated. If it is true, the corresponding expression is assigned to the signal. If false, the next Boolean expression is evaluated, and so on until a true Boolean expression is found. If none are true, the signal is assigned a default expression, listed last in the statement.

The VHDL code below implements the decoder with an active-LOW enable. If **g** is LOW, one decoder output activates, depending on the value of **d.** Note that the **d** inputs are defined as type INTEGER, rather than BIT_VECTOR or STD_LOGIC_VECTOR. In this situation, we don't need to specify the number of inputs; the compiler automatically defines the required inputs **d1** and **d0** when fitting the design to the selected CPLD. Also, since **d** is of type INTEGER, we write its value in the selected signal assignment statement directly, without quotes.

decode4g.vhd

```
LIBRARY ieee;
USE ieee.std_logic_1164.all;

ENTITY decode4g IS
 PORT(
     d  : IN   INTEGER RANGE 0 to 3;
     g  : IN   STD_LOGIC;
     y  : OUT  STD_LOGIC_VECTOR (0 to 3));
END decode4g;

ARCHITECTURE a OF decode4g IS
BEGIN
 y <=    "1000" WHEN (d=0 and g='0') ELSE
         "0100" WHEN (d=1 and g='0') ELSE
         "0010" WHEN (d=2 and g='0') ELSE
         "0001" WHEN (d=3 and g='0') ELSE
         "0000";
END a;
```

MAX+PLUS II Report File

In the Altera Golden Rules, we are told to choose a selected signal assignment over a conditional signal assignment because it uses the CPLD resources more efficiently. How do we check this assertion? Is it always true? This information is stored in a MAX+PLUS II **report file (rpt),** which is created at compile time.

The compile process of MAX+PLUS II goes on behind the scenes; until now we have not enquired about the result of this process. One of many functions of the compiler is to reduce the design information in a graphic or text file to a series of Boolean equations that can be programmed into a PLD.

For example, the report file **decode3a.rpt,** for the file that uses the selected signal assignment, gives us the following information under the EQUATIONS heading.

decode3a.rpt

```
** EQUATIONS **

d0        : INPUT;
d1        : INPUT;
g         : INPUT;

-- Node name is 'y0'
-- Equation name is 'y0', location is LC117, type is output.
 y0       = LCELL( _EQ001 $  GND);
   _EQ001 = !d0 & !d1 & !g;

-- Node name is 'y1'
-- Equation name is 'y1', location is LC115, type is output.
 y1       = LCELL( _EQ002 $  GND);
   _EQ002 =  d0 & !d1 & !g;

-- Node name is 'y2'
-- Equation name is 'y2', location is LC118, type is output.
 y2       = LCELL( _EQ003 $  GND);
   _EQ003 = !d0 &  d1 & !g;

-- Node name is 'y3'
-- Equation name is 'y3', location is LC120, type is output.
 y3       = LCELL( _EQ004 $  GND);
   _EQ004 =  d0 &  d1 & !g;
```

Each output is designated as a node. Let us examine the equation of one node in detail so that we will know how to interpret the others.

The Boolean format in the report file uses different operators than VHDL. They are as follows:

```
! = NOT
& = AND
# = OR
$ = XOR
```

Thus, the equation given as `_EQ001 = !d0 & !d1 & !g` is equivalent to the Boolean expression $\_EQ001 = \overline{d_0} \cdot \overline{d_1} \cdot \overline{g}$.

In the expression (`y0 = LCELL ( _EQ001 $ GND);`), equation _EQ001 is XORed with GND (logic 0) and applied to an LCELL (logic cell) primitive to yield **y0.** The LCELL represents one output of the CPLD. The XOR function is a way to either invert or not invert a logic function by setting one XOR input to GND (noninverting) or VCC (inverting). Thus _EQ001 is applied to a CPLD output without inversion.

A comment in the report file indicates that **y0** is assigned to logic cell LC117 (out of 128), which corresponds to pin 75 (out of 84) on the CPLD. Other equations are assigned to other LCELLs with other Boolean functions, as appropriate. Every pin number on the CPLD package is permanently connected to a specific LCELL. The compiler chooses the LCELL/pin assignments automatically; if we desire specific pin number assignments, we must assign them explicitly before compiling.

How does this compare with the report file for the design with the conditional signal assignment? If you examine **decode4g.rpt,** you will find that the Boolean equations are exactly the same. Thus, we can conclude that for a simple function, such as a 2-line-to-4-line decoder with enable, the two statement forms are easy enough for the compiler to interpret both in the most efficient way.

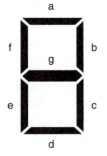
decode4g.rpt

Seven-Segment Decoders

KEY TERMS

Seven-segment display   An array of seven independently controlled light-emitting diode (LED) or liquid crystal display (LCD) elements, shaped like a figure-8, which can be used to display decimal digits and other characters by turning on the appropriate elements.

Common anode display   A seven-segment LED display where the anodes of all the LEDs are connected to the circuit supply voltage. Each segment is illuminated by a logic LOW at its cathode.

Common cathode display   A seven-segment display in which the cathodes of all LEDs are connected together and grounded. A logic HIGH illuminates a segment when applied to its anode.

FIGURE 5.22
Seven-segment Numerical
Display

Display

The **seven-segment display,** shown in Figure 5.22, is a numerical display device used to show digital circuit outputs as decimal digits (and sometimes hexadecimal digits or other alphabetic characters). It is called a seven-segment display because it consists of seven luminous segments, usually LEDs or liquid crystals, arranged in a figure-8. We can display any decimal digit by turning on the appropriate elements, designated by lowercase letters, *a* through *g*. It is conventional to designate the top segment as *a* and progress clockwise around the display, ending with *g* as the center element.

Figure 5.23 shows the usual convention for decimal digit display. Some variation from this convention is possible. For example, we could have drawn the digits 6 and 9 with "tails" (i.e., with segment *a* illuminated for 6 or segment *d* for 9). By convention,

FIGURE 5.23
Convention for Displaying Decimal Digits

we display digit 1 by illuminating segments *b* and *c,* although segments *e* and *f* would also work.

The electrical requirements for an LED circuit are simple. Since an LED is a diode, it conducts when its anode is positive with respect to its cathode, as shown in Figure 5.24a. A decoder/driver for an LED display will illuminate an element by completing this circuit, either by supplying V_{CC} or ground. A series resistor limits the current to prevent the diode from burning out and to regulate its brightness. If the anode is +5 volts with respect to cathode, the resistor value should be in the range of 220 Ω to 470 Ω.

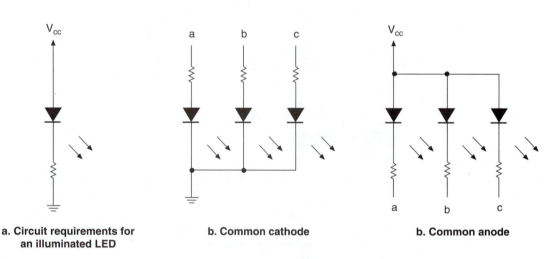

a. Circuit requirements for an illuminated LED   **b. Common cathode**   **b. Common anode**

FIGURE 5.24
Electrical Requirements for LED Displays

Seven-segment displays are configured as **common anode** or **common cathode,** as shown in Figures 5.24b and c. In a common cathode display, the cathodes of all LEDs are connected together and brought out to one or more pin connections on the display package. The cathode pins are wired externally to the circuit ground. We illuminate the segments by applying logic HIGHs to individual anodes.

Similarly, the common anode display has the anodes of the segments brought out to one or more common pins. These pins must be tied to the circuit power supply (V_{CC}). The segments illuminate when a decoder/driver makes their individual cathodes LOW. Figure 5.25 shows how the diodes could be physically laid out in a common anode display.

The two types of displays allow the use of either active HIGH or active LOW circuits to drive the LEDs, thus giving the designer some flexibility. However, it should be noted that the majority of seven-segment decoders are for common-anode displays.

FIGURE 5.25
Physical Placement of LEDs in a
Common Anode Display

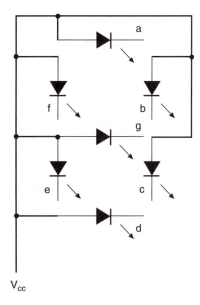

EXAMPLE 5.4

Sketch the segment patterns required to display all 16 hexadecimal digits on a seven-segment display. What changes from the patterns in Figure 5.23 need to be made?

Solution   The segment patterns are shown in Figure 5.26.

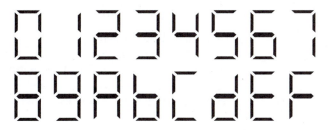

FIGURE 5.26
Hexadecimal Digit Display Format

Hex digits B and D must be displayed as lowercase letters, b and d, to avoid confusion between B and 8 and between D and 0. To make 6 distinct from b, 6 must be given a tail (segment *a*) and to make 6 and 9 symmetrical, 9 should also have a tail (segment *d*).

Decoder

> **KEY TERMS**
>
> **BCD**   Binary coded decimal. A code in which each individual digit of a decimal number is represented by a 4-bit binary number (e.g., 905 (decimal) = 1001 0000 0101 (BCD)).

A BCD-to-seven-segment decoder is a circuit with a 4-bit input for a **BCD** digit and seven outputs for segment selection. To display a number, the decoder must translate the input bits to a combination of active outputs. For example, the input digit $D_3D_2D_1D_0 =$ 0000 must illuminate segments *a, b, c, d, e,* and *f* to display the digit 0. We can make a truth table for each of the outputs, showing which must be active for every digit we wish to display. The truth table for a common-anode decoder (active LOW outputs) is given in Table 5.3.

Table 5.3   Truth Table for Common Anode BCD-to-Seven-Segment Decoder

Digit	D_3	D_2	D_1	D_0	a	b	c	d	e	f	g
0	0	0	0	0	0	0	0	0	0	0	1
1	0	0	0	1	1	0	0	1	1	1	1
2	0	0	1	0	0	0	1	0	0	1	0
3	0	0	1	1	0	0	0	0	1	1	0
4	0	1	0	0	1	0	0	1	1	0	0
5	0	1	0	1	0	1	0	0	1	0	0
6	0	1	1	0	1	1	0	0	0	0	0
7	0	1	1	1	0	0	0	1	1	1	1
8	1	0	0	0	0	0	0	0	0	0	0
9	1	0	0	1	0	0	0	1	1	0	0
Invalid Range	1	0	1	0	X	X	X	X	X	X	X
	1	0	1	1	X	X	X	X	X	X	X
	1	1	0	0	X	X	X	X	X	X	X
	1	1	0	1	X	X	X	X	X	X	X
	1	1	1	0	X	X	X	X	X	X	X
	1	1	1	1	X	X	X	X	X	X	X

The illumination of each segment is determined by a Boolean function of the input variables, $D_3D_2D_1D_0$. From the truth table, the function for segment a is

$$a = \bar{D}_3\bar{D}_2\bar{D}_1D_0 + \bar{D}_3D_2\bar{D}_1\bar{D}_0 + \bar{D}_3D_2D_1\bar{D}_0$$

(Since the display is active-LOW, this means segment a is OFF for digits 1, 4, and 6.)

If we assume that inputs 1010 to 1111 are never going to be used ("don't care states", symbolized by X), we can make any of these states produce HIGH or LOW outputs, depending on which is most convenient for simplifying the segment functions. Figure 5.27a shows a Karnaugh map simplification for segment a. The resultant function is

$$a = \bar{D}_3\bar{D}_2\bar{D}_1D_0 + D_2\bar{D}_0$$

The corresponding partial decoder is shown in Figure 5.27b.

We could do a similar analysis for each of the other segments, but if we are programming the decoder function into a CPLD, it is just as simple to write the truth table directly into a selected signal assignment statement, as shown in the VHDL code that follows.

bcd_7seg.vhd

```
-- bcd_7seg.vhd
-- BCD-to-seven-segment decoder

ENTITY bcd_7seg IS
 PORT(
      d3, d2, d1, d0      : IN  BIT;
      a, b, c, d, e, f, g : OUT BIT);
END bcd_7seg;

ARCHITECTURE seven_segment OF bcd_7seg IS
 SIGNAL input : BIT_VECTOR (3 downto 0);
 SIGNAL output: BIT_VECTOR (6 DOWNTO 0);
BEGIN
 input <= d3 & d2 & d1 & d0;
 WITH input SELECT
      output <=  "0000001" WHEN "0000",
                 "1001111" WHEN "0001",
```

```
            "0010010"  WHEN "0010",
            "0000110"  WHEN "0011",

            "1001100"  WHEN "0100",
            "0100100"  WHEN "0101",
            "1100000"  WHEN "0110",
            "0001111"  WHEN "0111",

            "0000000"  WHEN "1000",
            "0001100"  WHEN "1001",
            "1111111"  WHEN others;

-- Separate the output vector to make individual pin outputs.
    a    <=  output(6);
    b    <=  output(5);
    c    <=  output(4);
    d    <=  output(3);
    e    <=  output(2);
    f    <=  output(1);
    g    <=  output(0);

END seven_segment;
```

FIGURE 5.27

Decoding Segment *a*

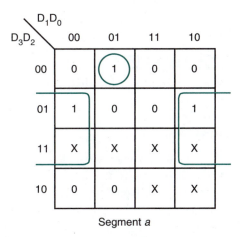

Segment *a*

a. K – map

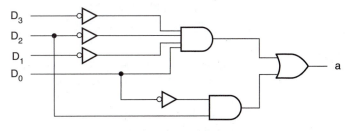

b. Decoder for segment *a* (common anode)

The inputs $D_3D_2D_1D_0$ are defined separately, then concatenated (linked in sequence) by the **&** operator to make a BIT_VECTOR called **input.** This is equivalent to the following four concurrent signal assignments:

```
input (3) <= d3;
input (2) <= d2;
input (1) <= d1;
input (0) <= d0;
```

Why not simply define **d** as a vector? If we wish to create a graphic symbol for the seven-segment decoder, the above method creates a symbol shown with four separate inputs, rather than a single thick line for a 4-bit bus input. The design will work either way.

For each value of **input,** a signal assignment defines the output vector, each bit of which represents the value of one segment. For example, the first clause (`"0000001"` WHEN `"0000"`) sets all segments ON except segment *g*, thus displaying the digit "0".

As a variation, we could define a signal called **d_inputs** of type INTEGER with RANGE 0 to 9. The WHEN clauses would evaluate the integer values 0 to 9, as follows.

```
WITH d_inputs SELECT
     output  <=  "0000001"  WHEN 0,
                 "1001111"  WHEN 1,
                 "0010010"  WHEN 2,
                 "0000110"  WHEN 3,

                 "1001100"  WHEN 4,
                 "0100100"  WHEN 5,
                 "0100000"  WHEN 6,
                 "0001111"  WHEN 7,

                 "0000000"  WHEN 8,
                 "0000100"  WHEN 9,
                 "1111111"  WHEN others; -- blank
```

Ripple Blanking

KEY TERMS

Ripple blanking   A technique used in a multiple-digit numerical display that suppresses leading or trailing zeros in the display, but allows internal zeros to be displayed.

$\overline{\textbf{RBI}}$ Ripple blanking input

$\overline{\textbf{RBO}}$ Ripple blanking output

PROCESS   A VHDL construct that contains statements that are executed if there is a change in a signal in its sensitivity list.

Sensitivity list   A list of signals in a PROCESS statement that are monitored to determine whether the PROCESS should be executed.

CASE statement   A VHDL construct in which there is a choice of statements to be executed, depending on the value of a signal or variable.

IF statement   A VHDL construct within a process that executes a series of statements, if a Boolean test condition is true.

A feature often included in seven-segment decoders is **ripple blanking.** The ripple blanking feature allows for suppression of leading or trailing zeros in a multiple digit display, while allowing zeros to be displayed in the middle of a number.

Each display decoder has a ripple blanking input ($\overline{RBI}$) and a ripple blanking output ($\overline{RBO}$), which are connected in cascade, as shown in Figure 5.28. If the decoder input $D_3D_2D_1D_0$ is 0000, it displays digit 0 if $\overline{RBI} = 1$ and shows a blank if $\overline{RBI} = 0$.

If $\overline{RBI} = 1$ OR $D_3D_2D_1D_0$ is (NOT 0000), then $\overline{RBO} = 1$. When we cascade two or more displays, these conditions suppress leading or trailing zeros (but not both) and still display internal zeros.

To suppress leading zeros in a display, ground the $\overline{RBI}$ of the most significant digit decoder and connect the $\overline{RBO}$ of each decoder to the $\overline{RBI}$ of the next least significant digit. Any zeros preceding the first nonzero digit (9 in this case) will be blanked, as $\overline{RBI} = 0$ AND $D_3D_2D_1D_0 = 0000$ for each of these decoders. The 0 inside the number 904 is displayed since its $\overline{RBI} = 1$.

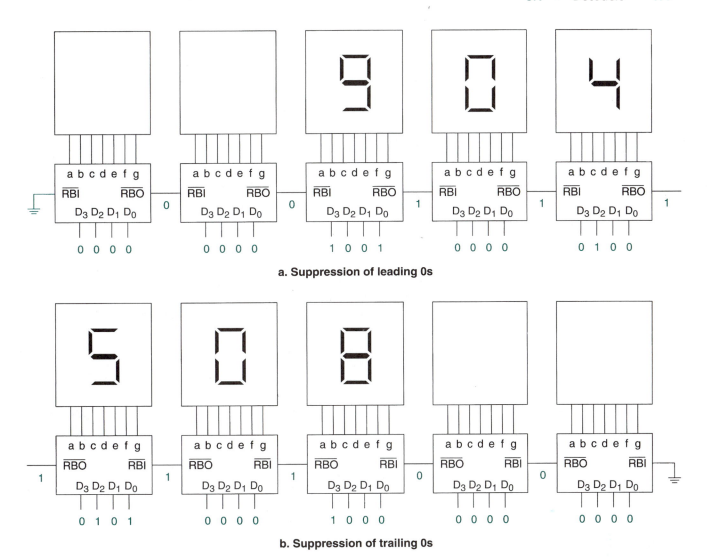

FIGURE 5.28
Zero Suppression in Seven-segment Displays

Trailing zeros are suppressed by reversing the order of $\overline{RBI}$ and $\overline{RBO}$ from the above example. $\overline{RBI}$ is grounded for the least significant digit and the $\overline{RBO}$ for each decoder cascades to the $\overline{RBI}$ of the next most significant digit.

We can implement the ripple blanking feature in a VHDL file by modifying the file for a standard BCD- or hexadecimal-to-seven-segment decoder to include a **CASE statement** within a **PROCESS**. A PROCESS is a construct containing statements that are executed if a signal in the **sensitivity list** of the PROCESS changes. The general form of a PROCESS is:

```
PROCESS (sensitivity list)
BEGIN
    statements;
END PROCESS;
```

A CASE statement can be one of the constructs used inside a process if we want to select among several alternatives. It takes the following form:

```
-- CASE statement within a PROCESS
PROCESS (__signal_name, __signal_name, __signal_name)
BEGIN
 CASE __expression IS
    WHEN __constant_value =>
       __statement;
       __statement;
    WHEN __constant_value =>
       __statement;
       __statement;
    WHEN OTHERS =>
       __statement;
       __statement;
 END CASE;
END PROCESS;
```

Whether the digit "0" is displayed or suppressed is conditional upon the value of $\overline{RBI}$. This can be tested by an **IF statement** within the PROCESS. An IF statement executes one or more VHDL statements, depending on the state of a test condition. It has the following syntax.

```
IF __expression THEN
    __statement;
    __statement;
ELSIF __expression THEN
    __statement;
    __statement;
ELSE
    __statement;
    __statement;
END IF;
```

The following VHDL code demonstrates the ripple blanking function.

```
-- sevsegrb.vhd

ENTITY sevsegrb IS
 PORT(
      nRBI, d3, d2, d1, d0      : IN  BIT;
      a, b, c, d, e, f, g, nRBO : OUT BIT);
END sevsegrb;

ARCHITECTURE seven_segment OF sevsegrb IS
 SIGNAL input: BIT_VECTOR (3 DOWNTO 0);
 SIGNAL output: BIT_VECTOR (6 DOWNTO 0);
BEGIN
 input  <=  d3 & d2 & d1 & d0;
 --  Process Statement
PROCESS (input, nRBI)
BEGIN
    IF (input = "0000" and nRBI ='0') THEN
       -- 0 suppressed
       output  <= "1111111";
       nRBO    <= '0';
    ELSIF (input =  "0000" and nRBI = '1') THEN
       -- 0 displayed
       output  <= "0000001";
       nRBO    <= '1';
    ELSE
       CASE input IS
          WHEN "0001"   => output <= "1001111"; -- 1
```

sevsegrb.vhd

```
            WHEN "0010"    => output <= "0010010"; -- 2
            WHEN "0011"    => output <= "0000110"; -- 3

            WHEN "0100"    => output <= "1001100"; -- 4
            WHEN "0101"    => output <= "0100100"; -- 5
            WHEN "0110"    => output <= "0100000"; -- 6
            WHEN "0111"    => output <= "0001111"; -- 7
            WHEN "1000"    => output <= "0000000"; -- 8
            WHEN "1001"    => output <= "0000100"; -- 9
            WHEN others    => output <= "1111111"; -- blank
        END  CASE;
        nRBO    <= '1';
    END IF;

  -- Separate the output vector to make individual pin outputs.
        a   <=  output(6);
        b   <=  output(5);
        c   <=  output(4);
        d   <=  output(3);
        e   <=  output(2);
        f   <=  output(1);
        g   <=  output(0);

    END PROCESS;
END seven_segment;
```

▪▪ SECTION 5.1C REVIEW PROBLEM

5.3 When would it be logical to suppress trailing zeros in a multiple-digit display and when should trailing zeros be displayed?

5.2 Encoders

KEY TERMS

Encoder   A circuit that generates a binary code at its outputs in response to one or more active input lines.

Priority encoder   An encoder that generates a binary or BCD output corresponding to the subscript of the active input having the highest priority. This is usually defined as the input with the largest subscript value.

The function of a digital **encoder** is complementary to that of a digital decoder. A decoder activates a specified output for a unique digital input code. An encoder operates in the reverse direction, producing a particular digital code (e.g., a binary or BCD number) at its outputs when a specific input is activated.

Figure 5.29 shows an 3-bit binary encoder. The circuit generates a unique 3-bit binary output for every active input provided *only one input* is active at a time.

The encoder has only 8 permitted input states out of a possible 256. Table 5.4 shows the allowable input states, which yield the Boolean equations used to design the encoder. These Boolean equations are:

$$Q_2 = D_7 + D_6 + D_5 + D_4$$
$$Q_1 = D_7 + D_6 + D_3 + D_2$$
$$Q_0 = D_7 + D_5 + D_3 + D_1$$

The D_0 input is not connected to any of the encoding gates, since all outputs are in their LOW (inactive) state when the 000 code is selected.

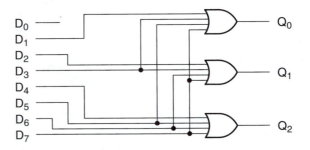

FIGURE 5.29
3-bit Encoder (No Input Priority)

Table 5.4   Partial Truth Table for a 3-bit Encoder

D_7	D_6	D_5	D_4	D_3	D_2	D_1	Q_2	Q_1	Q_0
0	0	0	0	0	0	0	0	0	0
0	0	0	0	0	0	1	0	0	1
0	0	0	0	0	1	0	0	1	0
0	0	0	0	1	0	0	0	1	1
0	0	0	1	0	0	0	1	0	0
0	0	1	0	0	0	0	1	0	1
0	1	0	0	0	0	0	1	1	0
1	0	0	0	0	0	0	1	1	1

Priority Encoder

The shortcoming of the encoder circuit shown in Figure 5.29 is that it can generate wrong codes if more than one input is active at the same time. For example, if we make D_3 and D_5 HIGH at the same time, the output is neither 011 or 101, but 111; the output code does not correspond to either active input.

One solution to this problem is to assign a priority level to each input and, if two or more are active, make the output code correspond to the highest-priority input. This is called a **priority encoder.** Highest priority is assigned to the input whose subscript has the largest numerical value.

◼◼ **EXAMPLE 5.5**

Figures 5.30a through c show a priority encoder with three different combinations of inputs. Determine the resultant output code for each figure. Inputs and outputs are active HIGH.

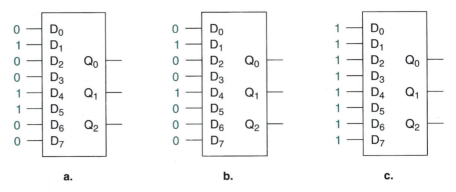

a.   b.   c.

FIGURE 5.30
Example 5.5
Priority Encoder Inputs

Solution

Figure 5.30a: The highest-priority active input is D_5. D_4 and D_1 are ignored. $Q_2Q_1Q_0$ = 101.

Figure 5.30b: The highest-priority active input is D_4. D_1 is ignored. $Q_2Q_1Q_0 = 100$.

Figure 5.30c: The highest-priority active input is D_7. All other inputs are ignored. $Q_2Q_1Q_0 = 111$.

NOTE

The encoding principle of a priority encoder is that a low-priority input must not change the code resulting from a higher-priority input.

For example, if inputs D_3 and D_5 are both active, the correct output code is $Q_2Q_1Q_0 = 101$. The code for D_3 would be $Q_2Q_1Q_0 = 011$. Thus, D_3 must not make $Q_1 = 1$. The Boolean expressions for Q_2, Q_1, and Q_0 covering only these two codes are:

$$Q_2 = D_5 \quad \text{(HIGH if } D_5 \text{ is active.)}$$
$$Q_1 = D_3\overline{D_5} \quad \text{(HIGH if } D_3 \text{ is active AND } D_5 \text{ is NOT active.)}$$
$$Q_0 = D_3 + D_5 \quad \text{(HIGH if } D_3 \text{ OR } D_5 \text{ is active.)}$$

The truth table of an 3-bit priority encoder is shown in Table 5.5.

Table 5.5 Truth Table for an 3-bit Priority Encoder

D_7	D_6	D_5	D_4	D_3	D_2	D_1	Q_2	Q_1	Q_0
0	0	0	0	0	0	0	0	0	0
0	0	0	0	0	0	1	0	0	1
0	0	0	0	0	1	X	0	1	0
0	0	0	0	1	X	X	0	1	1
0	0	0	1	X	X	X	1	0	0
0	0	1	X	X	X	X	1	0	1
0	1	X	X	X	X	X	1	1	0
1	X	X	X	X	X	X	1	1	1

Restating the encoding principle, a bit goes HIGH if it is part of the code for an active input AND it is NOT kept LOW by an input with a higher priority. We can use this principle to develop a mechanical method for generating the Boolean equations of the outputs.

1. Write the codes in order from highest to lowest priority, as in Table 5.6.

Table 5.6 Binary Outputs and Corresponding Decimal Values

Q_2	Q_1	Q_0	Code Value
1	1	1	7
1	1	0	6
1	0	1	5
1	0	0	4
0	1	1	3
0	1	0	2
0	0	1	1
0	0	0	0

2. Examine each code. For a code with value n, add a D_n term to each Q equation where there is a 1. For example, for code 111, add the term D_7 to the equations for Q_2, Q_1, and Q_0. For code 110, add the term D_6 to the equations for Q_2 and Q_1. (Steps 1 and 2 generate the nonpriority encoder equations listed earlier.)

3. Modify any D_n terms to ensure correct priority. Every time you write a D_n term, look at the previous lines in the table. For each previous code with a 0 in the same column as the 1 that generates D_n, use an AND function to combine D_n with a corresponding $\overline{D}$. For example, code 101 generates a D_5 term in the equations for Q_2 and Q_0. The term in the Q_2 equation need not be modified because there are no previous codes with a 0 in the same column. The term in the Q_0 equation must be modified since there is a 0 in the Q_0 column for code 110. This generates the term $\overline{D}_6 D_5$.

The equations from the 3-bit encoder of Figure 5.29 are modified by the priority encoding principle as follows:

$$Q_2 = D_7 + D_6 + D_5 + D_4$$
$$Q_1 = D_7 + D_6 + \overline{D}_5\overline{D}_4D_3 + \overline{D}_5\overline{D}_4D_2$$
$$Q_0 = D_7 + \overline{D}_6D_5 + \overline{D}_6\overline{D}_4D_3 + \overline{D}_6\overline{D}_4\overline{D}_2D_1$$

VHDL Priority Encoder

The most obvious way to program a priority encoder in VHDL is to use the equations derived in the previous section in a set of concurrent signal assignment statements, as follows.

hi_pri8a.vhd

```
-- hi_pri8a.vhd
ENTITY hi_pri8a IS

PORT(
     d  :  IN    BIT_VECTOR(7 downto 0);
     q  :  OUT   BIT_VECTOR (2 downto 0));
END hi_pri8a;

ARCHITECTURE a OF hi_pri8a IS
BEGIN
--   Concurrent Signal Assignments
  q(2)    <=  d(7) or d(6) or d(5) or d(4);

  q(1)    <=  d(7) or d(6)
          or ((not d(5)) and (not d(4)) and d(3))
          or ((not d(5)) and (not d(4)) and d(2));

  q(0)    <=  d(7) or ((not d(6)) and d(5))
          or ((not d(6)) and (not d(4)) and d(3))
          or ((not d(6)) and (not d(4)) and (not d(2)) and d(1));
END a;
```

Although this code works, it is not terribly elegant, nor does it give any insight into the operation of the encoder circuit. Also, if we expand our encoder output by one or more bits, the equations become more cumbersome with each new bit and soon become impractically large and susceptible to typing errors. A VHDL conditional signal assignment statement is an ideal alternative for use in a priority encoder circuit. A section of VHDL code using this format is shown below.

hi_pri8b.vhd
hi_pri8b.scf

```
-- hi_pri8b.vhd
ENTITY hi_pri8b IS
  PORT(
     d  :  IN    BIT_VECTOR (7 downto 0);
     q  :  OUT   INTEGER RANGE 0 to 7);
END hi_pri8b;
```

```
ARCHITECTURE a OF hi_pri8b IS
BEGIN
 --  Conditional Signal Assignment
encoder:
  q   <=  7 WHEN d(7)='1' ELSE
          6 WHEN d(6)='1' ELSE
          5 WHEN d(5)='1' ELSE
          4 WHEN d(4)='1' ELSE
          3 WHEN d(3)='1' ELSE
          2 WHEN d(2)='1' ELSE
          1 WHEN d(1)='1' ELSE
          0;
END   a;
```

Output q is defined as type INTEGER. Since it ranges from 0 to 7, the MAX+PLUS II VHDL compiler will automatically assign three outputs: Q_2, Q_1, and Q_0. The conditional signal assignment statement evaluates the first WHEN clause to determine if its condition ($d(7) = '1'$) is true. If so, it assigns q the value of 7 ($Q_2Q_1Q_0 = 111$). If the first condition is false, the next WHEN clause is evaluated, assigning q the value 6 ($Q_2Q_1Q_0 = 110$) if true, and so on until all WHEN clauses have been evaluated. If no clause is true, then the default value (0: $Q_2Q_1Q_0 = 000$) is assigned to the output.

In the conditional signal assignment, the highest-priority condition is examined first. If it is true, the output is assigned according to that condition and no further conditions are evaluated. If the first condition is false, the condition of next priority is evaluated, and so on until the end. Thus, a low-priority input cannot alter the code resulting from an input of higher priority, as required by the priority encoding principle.

The effect is similar to that of an IF statement, where a sequence of conditions is evaluated, but only one output assignment is made. However, an IF statement must be used within a PROCESS statement, if we choose to use it. The IF statement for a priority encoder is as shown below.

```
PROCESS (d)
BEGIN
 IF (d(7) = '1') THEN
        q <= 7;
    ELSIF  (d(6) = '1') THEN
        q <= 6;
  .
  .
  .
    ELSIF  (d(1) = '1' THEN
        q <= 1;
    ELSE
        q <= 0;
    END IF;
END PROCESS;
```

Figure 5.31 shows the simulation of an 3-bit priority encoder. The **d** inputs are shown separately, so that we can easily determine which inputs are active. The **q** outputs are grouped so as to show the encoded output value as a hexadecimal number.

BCD Priority Encoder

A BCD priority encoder, illustrated in Figure 5.32, accepts ten inputs and generates a BCD code (0000 to 1001), corresponding to the highest-priority active input. The truth table for this circuit is shown in Table 5.7, with a simulation of the circuit shown in Figure 5.33.

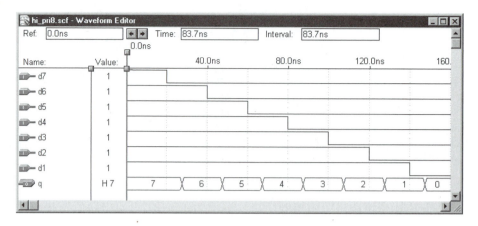

FIGURE 5.31
Simulation File for a 3-bit Priority Encoder

HIPR/BCD

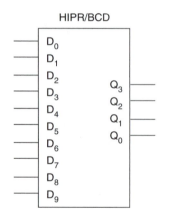

FIGURE 5.32
BCD Priority Encoder

Table 5.7   Truth Table of a BCD Priority Encoder

D_9	D_8	D_7	D_6	D_5	D_4	D_3	D_2	D_1	Q_3	Q_2	Q_1	Q_0
0	0	0	0	0	0	0	0	0	0	0	0	0
0	0	0	0	0	0	0	0	1	0	0	0	1
0	0	0	0	0	0	0	1	X	0	0	1	0
0	0	0	0	0	0	1	X	X	0	0	1	1
0	0	0	0	0	1	X	X	X	0	1	0	0
0	0	0	0	1	X	X	X	X	0	1	0	1
0	0	0	1	X	X	X	X	X	0	1	1	0
0	0	1	X	X	X	X	X	X	0	1	1	1
0	1	X	X	X	X	X	X	X	1	0	0	0
1	X	X	X	X	X	X	X	X	1	0	0	1

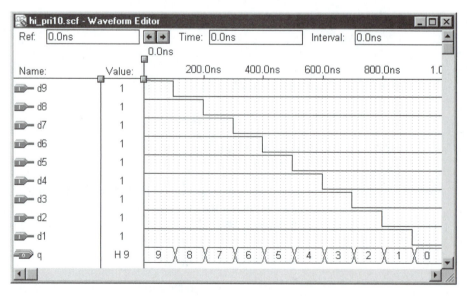

FIGURE 5.33
Simulation File for a BCD Priority Encoder

CD: hi_pri10.scf

Derivation of the BCD priority encoder equations and development of a VHDL description of the circuit are left as exercises in the end-of-chapter problems.

▌▌▌ SECTION 5.2 REVIEW PROBLEM

5.4 State the main limitation of the 3-bit binary encoder shown in Figure 5.29. How can the encoder be modified to overcome this limitation?

5.3 Multiplexers

A multiplexer (abbreviated MUX) is a device for switching one of several digital signals to an output, under the control of another set of binary inputs. The inputs to be switched are called the **data inputs;** those that determine which signal is directed to the output are called the **select inputs.**

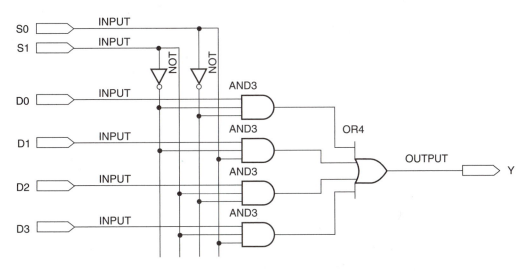

FIGURE 5.34
4-to-1 Multiplexer

Figure 5.34 shows the logic circuit for a 4-to-1 multiplexer, with data inputs labelled D_0 to D_3 and the select inputs labelled S_0 and S_1. By examining the circuit, we can see that the 4-to-1 MUX is described by the following Boolean equation:

$$Y = D_0\bar{S}_1\bar{S}_0 + D_1\bar{S}_1S_0 + D_2S_1\bar{S}_0 + D_3S_1S_0$$

Table 5.8 4-to-1 MUX Truth Table

S_1	S_0	Y
0	0	D_0
0	1	D_1
1	0	D_2
1	1	D_3

For any given combination of S_1S_0, only one of the above four product terms will be enabled. For example, when $S_1S_0 = 10$, the equation evaluates to:

$$Y = (D_0 \cdot 0) + (D_1 \cdot 0) + (D_2 \cdot 1) + (D_3 \cdot 0) = D_2$$

The MUX equation can be described by a truth table as in Table 5.8. The subscript of the selected data input is the decimal equivalent of the binary combination S_1S_0.

Figure 5.35 shows two symbols used for a 4-to-1 multiplexer. The first symbol shows the data and select inputs as individual lines. The second symbol shows the data inputs as a single 4-bit bus line and the select inputs as a 2-bit bus.

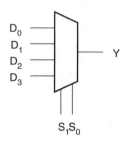 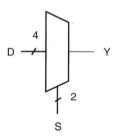

a. 4-to-1 MUX symbol showing individual lines **b. 4-to-1 MUX symbol showing bus lines**

FIGURE 5.35
Multiplexer Symbols

In general, a multiplexer with n select inputs will have $m = 2^n$ data inputs. Thus, other common multiplexer sizes are 8-to-1 (for 3 select inputs) and 16-to-1 (for 4 select inputs). Data inputs can also be multiple-bit busses, as in Figure 5.36. The slash through a thick data line and the number 4 above the line indicate that it represents four related data signals. In this device, the select inputs switch groups of data inputs, as shown in the truth table in Table 5.9.

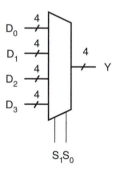

FIGURE 5.36
4-to-1 4-bit Bus Multiplexer

Table 5.9 Truth Table for a 4-to-1 4-bit Bus MUX

S_1	S_0	$Y_3\ Y_2\ Y_1\ Y_0$
0	0	$D_{03}D_{02}D_{01}D_{00}$
0	1	$D_{13}D_{12}D_{11}D_{10}$
1	0	$D_{23}D_{22}D_{21}D_{20}$
1	1	$D_{33}D_{32}D_{31}D_{30}$

The naming convention shown in Table 5.9, known as **double-subscript notation,** is used frequently for identifying variables that are bundled in numerically related groups, the elements of which are themselves numbered. The first subscript identifies the group that a variable belongs to; the second subscript indicates which element of the group a variable represents.

Multiplexing of Time-Varying Signals

We can observe the function of a multiplexer by using time-varying waveforms, such as a series of digital pulses. If we apply a different digital signal to each data input, and step the select inputs through an increasing binary sequence, we can see the different input waveforms appear at the output in a predictable sequence, as shown by the simulation waveforms in Figure 5.37. The frequencies shown in the simulation were chosen to make as great a contrast as possible between adjacent inputs so that the different selected inputs could easily be seen.

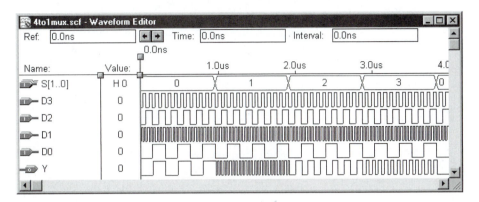

FIGURE 5.37
Simulation Waveforms for a 4-to-1 MUX

In Figure 5.37, we initially see the D_0 waveform appearing at the Y output when $S_1S_0 = 00$, followed in sequence by the D_1, D_2, and D_3 waveforms when $S_1S_0 = 01$, 10, and 11, respectively. (The S_1S_0 input combination is shown as a single hexadecimal value between 0 and 3, labelled S[1..0].)

This simulation can be created in the MAX+PLUS II simulator by defining a base clock pulse length (e.g., 40 ns) and assigning that to one of the inputs (D_1 in this case). Other input waveforms are set to periods of 2, 4, and 8 times the base waveform period (for D_3, D_2, and D_0, respectively). The select input count waveforms are set to allow three cycles of the longest waveform (D_0) to appear at Y when selected.

VHDL Implementation of Multiplexers

A multiplexer can be represented in MAX+PLUS II as a Graphic Design File, similar to the diagram of Figure 5.34, or in a hardware description language such as VHDL.

Several different VHDL constructs can be used to define a multiplexer. We can use a concurrent signal assignment statement, a selected signal assignment statement, or a CASE statement within a PROCESS. We will briefly look at each form for a 4-to-1 multiplexer. Later, you will be required to extend these constructs to larger multiplexer circuits.

Concurrent Signal Assignment

Recall that the concurrent signal assignment statement takes the form:

```
__signal <= __expression;
```

We can use this to encode the Boolean expression that describes a 4-to-1 MUX. The VHDL file that incorporates this statement is as follows.

```
-- mux4.vhd
-- 4-to-1 multiplexer
-- Directs one of four input signals (d0 to d3) to output,
--    depending on status of select bits (s1, s0).
```

mux4.vhd
mux4.scf

```
ENTITY mux4 IS
PORT(
      d0, d1, d2, d3 : IN   BIT;
      s              : IN   BIT_VECTOR (1 downto 0);
      y              : OUT  BIT);
END mux4;

ARCHITECTURE mux4to1 OF mux4 IS
BEGIN
--  Concurrent Signal Assignment
y<=  ((not s(1)) and (not s(0)) and d0)
  or ((not s(1)) and (     s(0)) and d1)
  or ((     s(1)) and (not s(0)) and d2)
  or ((     s(1)) and (     s(0)) and d3);
END mux4to1;
```

While the concurrent signal assignment is fairly easy to use, it becomes cumbersome for larger multiplexers, such as 8-to-1 or greater.

The entity declaration will be identical for the other VHDL examples. The only change we will make will be to replace the concurrent signal assignment in the architecture body with some other VHDL construct.

Selected Signal Assignment Statement

This construct has the following form (the label is optional):

```
__label:
WITH __expression SELECT
  __signal <=    __expression WHEN __constant_value,
                 __expression WHEN __constant_value,
                 __expression WHEN __constant_value,
                 __expression WHEN __constant_value;
```

The 4-to-1 MUX can be described in VHDL as follows, using a selected signal assignment:

mux4sel.vhd

```
ENTITY mux4sel IS
 PORT(
      d0, d1, d2, d3 : IN   BIT;
      s              : IN   BIT_VECTOR (1 downto 0);
      y              : OUT  BIT);
END mux4sel;

ARCHITECTURE mux4to1 OF mux4sel IS
BEGIN
M:   WITH s SELECT
     y  <=  d0 WHEN "00",
            d1 WHEN "01",
            d2 WHEN "10",
            d3 WHEN "11";
END mux4to1;
```

The selected signal assignment evaluates the expression in the WITH clause (in this case, the 2-bit vector, s) and, depending on its value, selects an expression to assign to y. Thus, if $s_1 s_0 = 00$, $y = d_0$. If $s_1 s_0 = 01$, then $y = d_1$, and so on for the remaining values of $s_1 s_0$.

CASE Statement within a PROCESS

In our MUX example, we could use a CASE statement as follows:

mux4case.vhd

```
ENTITY mux4case IS
 PORT(
     d0, d1, d2, d3 : IN   BIT;
     s              : IN   BIT_VECTOR (1 downto 0);
     y              : OUT  BIT);
END mux4case;

ARCHITECTURE mux4to1 OF mux4case IS
BEGIN
-- CASE statement within a PROCESS
-- Monitor select inputs and execute if they change
     PROCESS (s)
     BEGIN
       CASE s IS
            WHEN "00"    =>  y  <=  d0;
            WHEN "01"    =>  y  <=  d1;
            WHEN "10"    =>  y  <=  d2;
            WHEN "11"    =>  y  <=  d3;
            WHEN others  =>  y  <=  '0';
       END CASE;
     END PROCESS;
END mux4to1;
```

If the select inputs change, the PROCESS statements are executed. The CASE statement evaluates the select input vector, *s*, and chooses a signal assignment based on its value. It is good design practice to include a default case (the "others" clause) even when there are no obvious other cases. A default case is essential when using STD_LOGIC types rather than BIT types, as '0' and '1' values do not cover all possible cases for STD LOGIC signals. (Recall from Chapter 4 that STD_LOGIC is a nine-valued logic type, incorporating such things as "Don't Care" ('-'), "Unknown" ('X'), and "High Impedance" ('Z'), as well as '0' and '1'.)

Multiplexer Applications

Multiplexers are used for a variety of applications, including selection of one data stream out of several choices, switching multiple-bit data from several channels to one multiple-bit output, sharing data on one output over time, and generating bit patterns or waveforms.

Single-Channel Data Selection

The simplest way to use a multiplexer is to switch the select inputs manually in order to direct one data source to the MUX output. Example 5.6 shows a pair of single-pole single-throw (SPST) switches supplying the select input logic for this type of application.

EXAMPLE 5.6

Figure 5.38 shows a digital audio switching system. The system shown can select a signal from one of four sources (compact disc (CD) players, labelled CD_0 to CD_3) and direct it to a digital signal processor (DSP) at its output. We assume we have direct access to the audio signals in digital form.

Make a table listing which digital audio source in Figure 5.38 is routed to the DSP for each combination of the multiplexer select inputs, S_1 and S_0.

FIGURE 5.38
Example 5.6
Single-Channel Data Selection

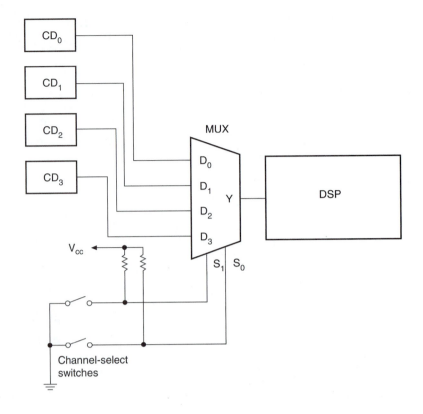

Solution

Table 5.10   Sources Selected by a 4-to-1 MUX in Figure 5.38

S_1	S_0	Selected Input	Selected Source
0	0	D_0	CD_0
0	1	D_1	CD_1
1	0	D_2	CD_2
1	1	D_3	CD_3

Multi-Channel Data Selection

Example 5.6 assumes that the output of a multiplexer is a single bit or stream of bits. Some applications require several bits to be selected in parallel, such as when data would be represented on a numerical display.

Figure 5.39 shows a circuit, based on a quadruple (4-channel) 2-to-1 multiplexer, that will direct one of two BCD digits to a seven-segment display. The bits $D_{03}D_{02}D_{01}D_{00}$ act as a 4-bit group input, since the first digit of all four subscripts is 0. When the MUX select input (S) is 0, these inputs are all connected to the outputs $Y_3Y_2Y_1Y_0$. Similarly, when the select input is 1, inputs $D_{13}D_{12}D_{11}D_{10}$ are connected to the Y outputs.

The seven-segment display in Figure 5.39 will display "4" if $S = 0$ (D_0 inputs selected) and "9" if $S = 1$ (D_1 inputs selected).

FIGURE 5.39
Quadruple 2-to-1 MUX as a
Digital Output Selector

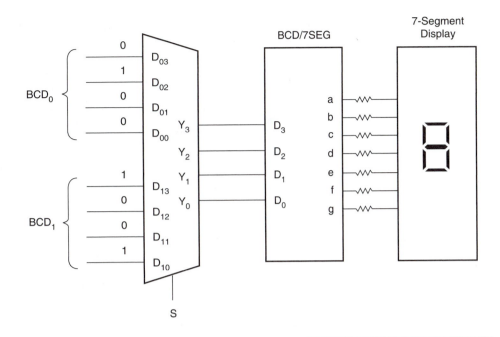

EXAMPLE 5.7

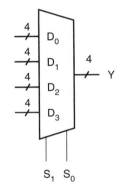

FIGURE 5.40
Example 5.7
4-channel 4-bit MUX

quad4to1.vhd
quad4to1.scf

Draw the symbol for a multiplexer that will select one of four 4-bit channels and direct it to a 4-bit output. Create a VHDL file that implements this function and a simulation showing the operation of the device.

Solution Figure 5.40 shows the symbol for the 4-channel, 4-bit multiplexer. This symbol is shown with the data inputs and outputs in bus form. The data inputs are labelled in groups D_0 to D_3, which contain the individual inputs $[D_{03}..D_{00}]$ to $[D_{33}..D_{30}]$.

A VHDL file describing this function is listed below.

```
-- quad4to1.vhd

ENTITY quad4to1 IS
  PORT(
        s    : IN     INTEGER RANGE 0 to 3;
        d0   : IN     BIT_VECTOR (3 downto 0);
        d1   : IN     BIT_VECTOR (3 downto 0);
        d2   : IN     BIT_VECTOR (3 downto 0);
        d3   : IN     BIT_VECTOR (3 downto 0);
        y    : OUT    BIT_VECTOR (3 downto 0));
END quad4to1;

ARCHITECTURE mux4 OF quad4to1 IS
BEGIN
  --  Selected Signal Assignment
MUX4:   WITH s SELECT
        y   <=  d0 WHEN 0,
                d1 WHEN 1,
                d2 WHEN 2,
                d3 WHEN 3;
END mux4;
```

Figure 5.41 shows a set of simulation waveforms for the multiplexer. The D inputs are shown in groups of four, the value of each shown as a steady hexadecimal value. The select inputs are grouped, showing an increasing 2-bit binary count as a hexadecimal value (0 to 3, then repeating). As the S inputs select each group of D inputs, their combined value is directed to the Y output group.

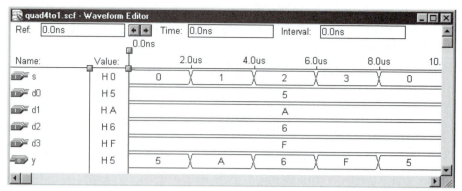

FIGURE 5.41
Example 5.7
Simulation for a 4-channel 4-bit MUX

Time-Dependent Multiplexer Applications

> **KEY TERMS**
>
> **Counter**   A digital circuit whose output produces a fixed sequence of binary states when an input called the clock receives a series of pulses. The output advances by one for each clock pulse (e.g., the output state of a 4-bit binary counter progresses in order from 0000 to 1111, then repeats).
>
> **Clock**   A signal that controls the operation of a sequential digital circuit, such as a counter, by advancing its outputs to the next state when it receives a pulse.
>
> **Positive edge**   The point on a digital waveform where the logic level of the waveform makes a LOW-to-HIGH transition.

A time-dependent multiplexer application is one that uses the MUX input channels one after the other in a repeating time sequence. We can create such an application by applying a set of changing binary signals to the MUX select inputs. For this function, we can use a 3-bit binary **counter** to generate a binary sequence that goes from 000 to 111 (8 states) and repeats indefinitely, the outputs advancing by one with every pulse applied to the **clock** input of the counter.

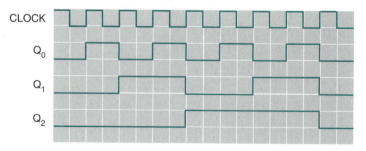

FIGURE 5.42
Timing Diagram of a 3-bit Counter

Figure 5.42 shows the timing diagram of a 3-bit counter. The outputs $Q_2Q_1Q_0$ change every time the clock signal makes a transition from LOW to HIGH. If you read the Q

waveforms from bottom to top, you will see that they generate a repeating binary sequence (000, 001, 010, 011, 100, 101, 110, 111, 000 . . .).

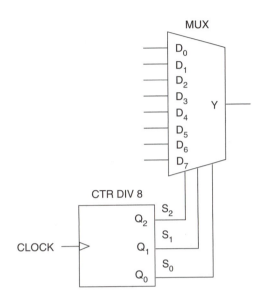

FIGURE 5.43
Time-Dependent Selection of Eight Multiplexer Channels

If we connect the counter outputs $Q_2Q_1Q_0$ to the select inputs of an 8-to-1 MUX, as in Figure 5.43, we will select the channels in sequence, one after the other. The counter is labelled CTR DIV 8 because its most significant bit output has a frequency equal to the clock frequency divided by eight. The triangle on the clock input indicates that it is active when the clock waveform makes a transition from one logic level to another. Since there is no inverting bubble on the clock input, we know that the active clock transition is from LOW to HIGH (i.e., a **positive edge**).

Waveform Generation.   A multiplexer and counter can be used as a programmable waveform generator. The output waveform can be programmed to any pattern by switching the logic levels on the data inputs. This is an easy way to generate an asymmetrical waveform, a task which is more complicated using other digital circuits. The circuit can also generate symmetrical waveforms by alternating the logic levels of consecutive groups of inputs.

▌▌ EXAMPLE 5.8

Draw a circuit that uses an 8-to-1 multiplexer to generate a programmable 8-bit repeating pattern. Draw the timing diagram of the select inputs and the output waveform for the following pattern of data inputs.

D_7	D_6	D_5	D_4	D_3	D_2	D_1	D_0
0	1	1	0	0	1	0	1

Solution   Figure 5.44a shows the waveform generator circuit. The output waveform with respect to the counter inputs is shown in Figure 5.44b. This pattern is relatively difficult to generate by other means since it has several unequal HIGH and LOW sequences in one period.

FIGURE 5.44
Example 5.8
Programmable Waveform
Generator

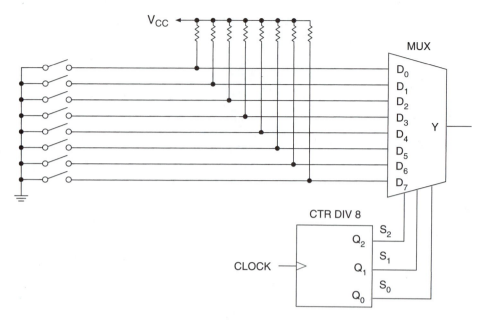

a. Circuit

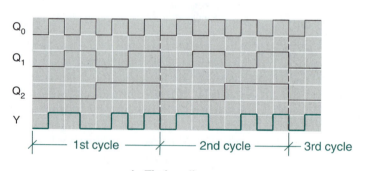

b. Timing diagram

▌▌ **EXAMPLE 5.9**

The programmable waveform generator in Figure 5.44 generates a symmetrical pulse waveform having a frequency of 1 kHz when the data inputs are set as follows.

D_7	D_6	D_5	D_4	D_3	D_2	D_1	D_0
0	0	0	0	1	1	1	1

How should the switches be set to generate a symmetrical 2 kHz waveform? A symmetrical 4 kHz waveform?

Solution

Pattern for 2 kHz:

D_7	D_6	D_5	D_4	D_3	D_2	D_1	D_0
0	0	1	1	0	0	1	1

Pattern for 4 kHz:

D_7	D_6	D_5	D_4	D_3	D_2	D_1	D_0
0	1	0	1	0	1	0	1

Time Division Multiplexing

Time division multiplexing is a method of improving the efficiency of a transmission system by sharing one transmission path among many signals. For example, if we wish to send four 4-bit numbers over a single transmission line, we can transmit the bits one after the other, as shown in Figure 5.45.

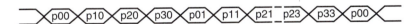

FIGURE 5.45
4 × 4 Data Stream (Bit Multiplexing)

In Figure 5.45, we see the least significant bit of the 4-bit word $p0$ transmitted, followed by the LSB of $p1$, $p2$, then $p3$. After that, the second bit of each word is transmitted in sequence, then all the third bits, and finally, all MSBs in sequence. Each bit is assigned a **time slot** in the sequence. During that time, the bit has sole access to the transmission line. When its time elapses, the next bit is sent and so on in sequence, until the channel assignment returns to the original location. This technique, known as **bit multiplexing,** can be implemented by a circuit similar to the waveform generator shown in Figure 5.44. Rather than fixed switch inputs, the data inputs would be some data source, such as a digitized audio signal.

We can also arrange our circuit so that one byte (8 bits) or one word (a group of bits) is sent through a selected channel. In this case, we must keep the channel selected for enough clock pulses to transmit the byte or word, then move to the next one. This technique is called **byte (or word) multiplexing.** Figure 5.46 shows a data stream of four 4-bit words that are word-multiplexed down a data transmission path.

FIGURE 5.46
4 × 4 Data Stream (Word Multiplexing)

Telephone companies use TDM to maximize the use of their phone lines. Speech or data is digitally encoded for transmission. Each speech or data channel becomes a multiplexer data input which shares time with all other channels on a single phone line. A counter on the MUX selects the speech channels one after the other in a continuous sequence. The counter must switch the channels fast enough so that there is no apparent interruption of the transmitted conversation or data stream.

▉▉ EXAMPLE 5.10

Draw a diagram of a circuit that uses an 8-to-1 multiplexer to share one telephone line among eight digitized speech channels.

Write a VHDL file for the multiplexer and create a simulation to show its operation.

Solution   Figure 5.47 shows the required multiplexer circuit. Each channel is connected to a data input and a 3-bit binary counter is connected to the select inputs.

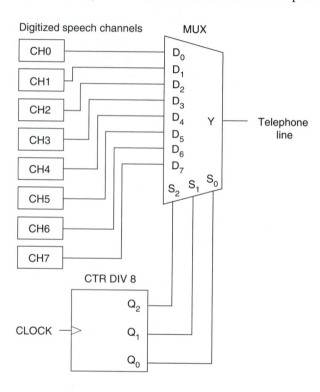

FIGURE 5.47
Example 5.10
Time-Division Multiplexing of Telephone Channels

The VHDL code for the multiplexer is:

mux_8ch.vhd
mux_8ch.scf

```
ENTITY mux_8ch IS
 PORT(
    sel : IN   BIT_VECTOR (2 downto 0);
    d   : IN   BIT_VECTOR (7 downto 0);
    y   : OUT  BIT);
END mux_8ch;

ARCHITECTURE a OF mux_8ch IS
BEGIN
 --  Selected Signal Assignment
MUX8:   WITH sel SELECT
     y   <=  d(0) WHEN "000",
```

```
                   d(1)  WHEN  "001",
                   d(2)  WHEN  "010",
                   d(3)  WHEN  "011",
                   d(4)  WHEN  "100",
                   d(5)  WHEN  "101",
                   d(6)  WHEN  "110",
                   d(7)  WHEN  "111";

END a;
```

The simulation is shown in Figure 5.48. For clarity, digital data are present on the MUX inputs just before and after they are switched to the *Y* output. The output shows the channel data in sequence, starting with channel 0.

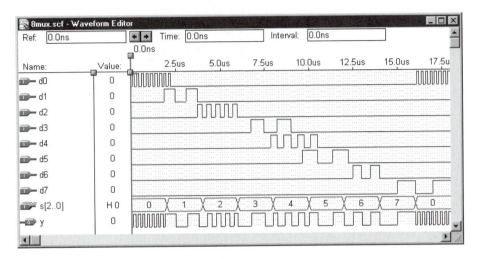

FIGURE 5.48
Simulation for an 8-bit Time-Division Multiplexer

III **SECTION 5.3 REVIEW PROBLEM**

5.5 What defines whether a multiplexer application is time-dependent or not? What additional component can be added to make a MUX application time-dependent?

5.4 Demultiplexers

KEY TERMS

Demultiplexer   A circuit that uses a binary decoder to direct a digital signal from a single source to one of several destinations.

A **demultiplexer** performs the reverse function of a multiplexer. A multiplexer (MUX) directs one of several input signals to a single output; a demultiplexer (DMUX) directs a single input signal to one of several outputs. In both cases, the selected input or output is chosen by the state of an internal decoder.

Figure 5.49 shows the logic circuit for a 1-to-4 demultiplexer. Compare this to Figure 5.4, a 4-output decoder. This circuits are the same except that the active-LOW enable input has been changed to an active-HIGH data input. The circuit in Figure 5.49 could still be used as a decoder, except that its enable input would be active-HIGH.

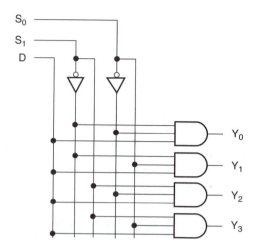

FIGURE 5.49
4-bit Decoder/Demultiplexer

Each AND gate in the demultiplexer enables or inhibits the signal output according to the state of the select inputs, thus directing the data to one of the output lines. For instance, $S_1 S_0 = 10$ directs incoming digital data to output Y_2.

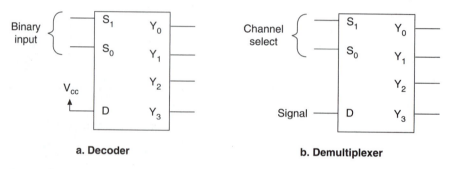

a. Decoder   **b. Demultiplexer**

FIGURE 5.50
Same Device Used as a Decoder or Demultiplexer

Figure 5.50 illustrates the use of a single device as either a decoder or a demultiplexer. In Figure 5.50a, input D is tied HIGH. When an output is selected by S_1 and S_0, it goes HIGH, acting as a decoder with active-HIGH outputs. In Figure 5.50b, D acts as a demultiplexer data input. The data are directed to the output selected by S_1 and S_0.

> **N O T E**
>
> Since a single device can be used either way, this implies that any of the VHDL binary decoder designs used in this chapter can also be used as demultiplexers.

A decoder/demultiplexer can have active-LOW outputs, but only if the D input is also active-LOW. This is important because the demultiplexer data must be inverted twice to retain its original logic values.

Demultiplexing a TDM Signal

In Example 5.10, we saw how a multiplexer could be used to send 8 digital channels across a single line, multiplexed over time. Obviously, such a system is not of much value if the signals cannot be sorted out at the receiving end. The received digital data must be demultiplexed and sent to their appropriate destinations.

The process is the reverse of multiplexing; data are sent to an output selected by a counter at the DMUX select inputs. (We assume that the counters at the MUX and DMUX select inputs are somehow synchronized or possibly, if located close together, are the same counter.)

III EXAMPLE 5.11

Draw a demultiplexing circuit that will take the multiplexed output of the circuit in Figure 5.47 and distribute it to 8 different local telephone circuits. Write a VHDL file for the demultiplexer and create a simulation file that shows its operation. Use active-LOW outputs for the demultiplexer. How does this affect the outputs when they are not transmitting data?

Solution Figure 5.51 shows the original multiplexing circuit connecting to the new demultiplexing circuit. The diagram indicates that the two sides of the circuit are separated by some distance. The clock is shared between both sides of the circuit, but is generated on the MUX side. Both sides share a common ground. Each side of the circuit has its own 3-bit counter.

FIGURE 5.51
Example 5.11
Time-Division Multiplexing and
Demultiplexing

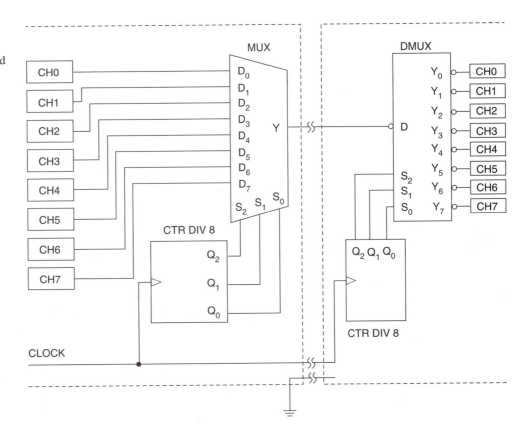

The VHDL code for the demultiplexer is as follows. (This is the same implementation as a 3-line-to-8-line decoder with an enable input.)

dmux8.vhd
dmux8.scf

```
-- dmux8.vhd
-- 1-to-8 demultiplexer/decoder
-- Decoder: set d to '0'; outputs are activated by
--    binary combination of s.
-- Demultiplexer: apply data stream to d; data directed to
--    y output with subscript same as value of s.
-- Outputs and d are active-LOW. DMUX data are inverted twice
--    to keep them true.
```

```
ENTITY dmux8 IS
  PORT(
       s  : IN    INTEGER Range 0 to 7;
       d  : IN    BIT;
       y  : OUT   BIT_VECTOR (0 to 7));
END dmux8;

ARCHITECTURE a OF dmux8 IS
  SIGNAL output : BIT_VECTOR (0 to 7);
BEGIN
  PROCESS (d, s)
  BEGIN
    IF  (d = '1')  THEN
        output <=  "11111111";
      ELSE
        CASE s IS
            WHEN 0 => output      <=  "01111111";
            WHEN 1 => output      <=  "10111111";
            WHEN 2 => output      <=  "11011111";
            WHEN 3 => output      <=  "11101111";
            WHEN 4 => output      <=  "11110111";
            WHEN 5 => output      <=  "11111011";
            WHEN 6 => output      <=  "11111101";
            WHEN 7 => output      <=  "11111110";
            WHEN OTHERS => output <=  "11111111";
        END CASE;
      END IF;
      y    <=  output;
  END PROCESS;
END a;
```

www.electronictech.com

The simulation, shown in Figure 5.52, has as its input data the output of the original MUX simulation in Figure 5.48. Data are distributed to the outputs in sequence. Compare the DMUX output data to the MUX input data in Figure 5.48.

FIGURE 5.52
Example 5.11
Demultiplexer Simulation

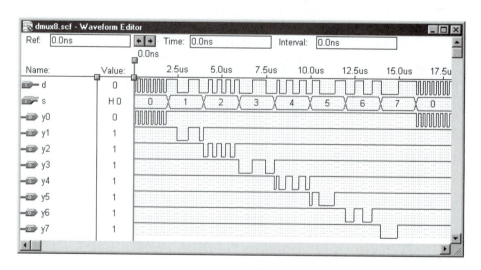

Note that idle channels sit HIGH. This is opposite from the status of the idle MUX lines and may affect circuit operation. If so, a DMUX with active-HIGH outputs and active-HIGH enable should be used.

CMOS Analog Multiplexer/Demultiplexer

KEY TERMS

CMOS analog switch   A CMOS device that will pass an analog or digital signal in either direction, when enabled. Also called a transmission gate. There is no TTL equivalent.

An interesting device used in some CMOS medium-scale integration multiplexers and demultiplexers, as well as other applications, is the **CMOS analog switch,** or transmission gate. This device has the property of allowing signals to pass in two directions, instead of only one, thus allowing both positive and negative voltages and currents to pass. It also has no requirement that the voltages be of a specific value such as $+5$ volts. These properties make the device suitable for passing analog signals.

FIGURE 5.53

Line Drivers

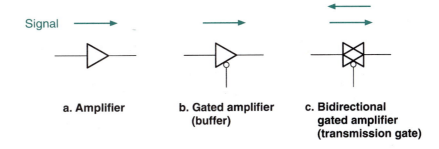

a. Amplifier   b. Gated amplifier   c. Bidirectional
                     (buffer)   gated amplifier
                                               (transmission gate)

Figure 5.53 shows several symbols, indicating the development of the transmission gate concept. Figures 5.53a and b show amplifiers whose output and input are clearly defined by the direction of the triangular amplifier symbol. A signal has one possible direction of flow. Figure 5.53b includes an active-LOW gating input, which can turn the signal on and off.

Figure 5.53c shows two opposite-direction overlapping amplifier symbols, with a gating input to enable or inhibit the bidirectional signal flow. The signal through the transmission gate may be either analog or digital.

Analog switches are available in packages of four switches with part numbers such as 4066B (standard CMOS) or 74HC4066 (high-speed CMOS).

Several available CMOS MUX/DMUX chips use analog switches to send signals in either direction. Figure 5.54 illustrates the design principle as applied to a 4-channel MUX/DMUX.

FIGURE 5.54

4-Channel CMOS MUX/DMUX

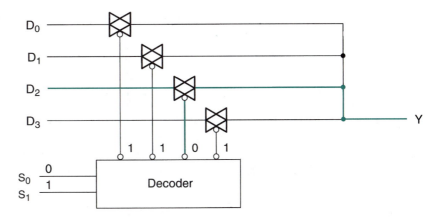

If four signals are to be multiplexed, they are connected to inputs D_0 to D_3. The decoder, activated by S_1 and S_0, selects which one of the four switches is enabled. Figure 5.54 shows Channel 2 active ($S_1S_0 = 10$).

Since all analog switch outputs are connected together, any selected channel connects to Y, resulting in a multiplexed output. To use the circuit in Figure 5.54 as a demultiplexer, the inputs and outputs are merely reversed.

■■ EXAMPLE 5.12   A CMOS 4097B dual 8-channel MUX/DMUX can be used simultaneously as a multiplexer on one half of the device and as a demultiplexer on the other side.

FIGURE 5.55
Example 5.12
4097B MUX/DMUX as a Time
Division MUX/DMUX

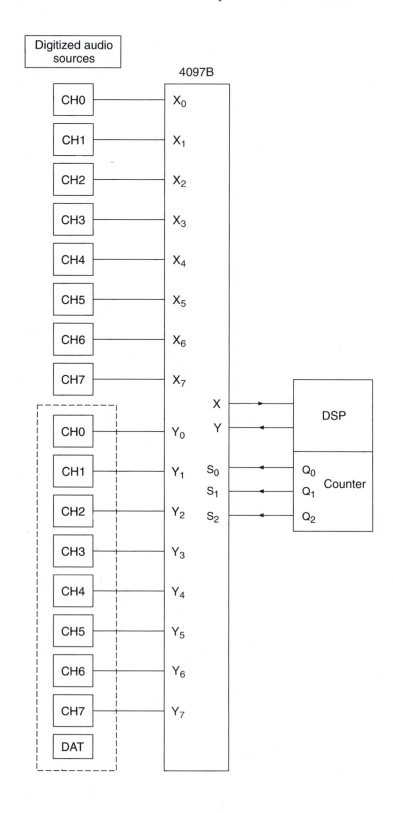

A circuit in a recording studio uses one side of a 4097B MUX/DMUX to multi-plex 8 digital audio channels into a digital signal processor (DSP), using time division multiplexing. The other half of the 4097B takes the processed signals from a DSP output and distributes them to 8 channels on a digital audio tape (DAT) unit. Draw the circuit.

Solution   Figure 5.55 shows a possible circuit. The counter can be part of the DSP. An audio source channel is selected by the counter inputs, data are sent to the DSP, where they are processed and sent to the same channel of the DAT. The counter advances by one, selecting a new channel and repeating the process.

Some analog MUX/DMUX devices in high-speed CMOS include: 74HC4051 8-channel MUX/DMUX, 74HC4052 dual 4-channel MUX/DMUX, and 74HC4053 triple 2-channel MUX/DMUX.

5.5 Magnitude Comparators

FIGURE 5.56
Exclusive NOR Gate

KEY TERMS

Magnitude comparator   A circuit that compares two *n*-bit binary numbers, indicates whether or not the numbers are equal, and, if not, which one is larger.

If we are interested in finding out whether or not two binary numbers are the same, we can use a **magnitude comparator.** The simplest comparison circuit is the Exclusive NOR gate, whose circuit symbol is shown in Figure 5.56 and whose truth table is given in Table 5.11.

The output of the XNOR gate is 1 if its inputs are the same ($A = B$, symbolized *AEQB*) and 0 if they are different. For this reason, the XNOR gate is sometimes called a coincidence gate.

We can use several XNORs to compare each bit of two multi-bit binary numbers. Figure 5.57 shows a 2-bit comparator with one output that goes HIGH if all bits of *A* and *B* are identical.

Table 5.11   XNOR Truth Table

A	B	Y
0	0	1
0	1	0
1	0	0
1	1	1

FIGURE 5.57
2-bit Magnitude Comparator

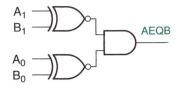

If the most significant bit (MSB) of *A* equals the MSB of *B*, the output of the upper XNOR is HIGH. If the least significant bits (LSBs) are the same, the output of the lower XNOR is HIGH. If both these conditions are satisfied, then $A = B$, which is indicated by a HIGH at the AND output. This general principle applies to any number of bits:

$$AEQB = \overline{(A_{n-1} \oplus B_{n-1})} \cdot \overline{(A_{n-2} \oplus B_{n-2})} \ldots \overline{(A_1 \oplus B_1)} \cdot \overline{(A_0 \oplus B_0)}$$

for two *n*-bit numbers, *A* and *B*.

Some magnitude comparators also include an output that activates if *A* is greater than *B* (symbolized $A > B$ or *AGTB*) and another that is active when *A* is less than *B* (symbolized $A < B$ or *ALTB*). Figure 5.58 shows the comparator of Figure 5.57 expanded to include the "greater than" and "less than" functions.

Let us analyze the *AGTB* circuit. The *AGTB* function has two AND-shaped gates that compare *A* and *B* bit-by-bit to see which is larger.

FIGURE 5.58

2-bit Comparator With AEQB, AGTB, and ALTB Outputs

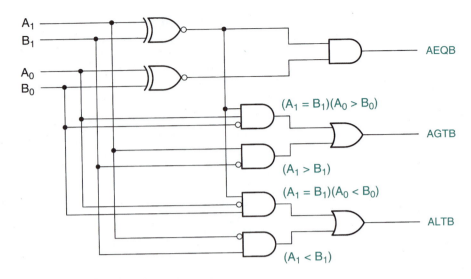

1. The 2-input gate examines the MSBs of A and B. If $A_1 = 1$ AND $B_1 = 0$, then we know that A > B. (This implies one of the following inequalities: 10 > 00; 10 > 01; 11 > 00; or 11 > 01.)

2. If $A_1 = B_1$, then we don't know whether or not $A > B$ until we compare the next most significant bits, A_0 and B_0. The 3-input gate makes this comparison. Since this gate is enabled by the XNOR, which compares the two MSBs, it is only active when $A_1 = B_1$. This yields the term $\overline{(A_1 \oplus B_1)}A_0\overline{B_0}$ in the Boolean expression for the $AGTB$ function.

3. If $A_1 = B_1$ AND $A_0 = 1$ AND $B_0 = 0$, then the 3-input gate has a HIGH output, telling us, via the OR gate, that $A > B$. (The only possibilities are (01 > 00) and (11 > 10).)

Similar logic works in the $ALTB$ circuit, except that inversion is on the A, rather than the B bits. Alternatively, we can simplify either the $AGTB$ or the $ALTB$ function by using a NOR function. For instance, if we have developed a circuit to indicate $AEQB$ and $ALTB$, we can make the $AGTB$ function from the other two, as follows:

$$AGTB = \overline{AEQB + ALTB}$$

This Boolean expression implies that if A is not equal to or less than B, then it must be greater than B.

Figure 5.59 shows a 4-bit comparator with $AEQB$, $ALTB$, and $AGTB$ outputs.

The Boolean expressions for the outputs are:

$$AEQB = \overline{(A_3 \oplus B_3)}\,\overline{(A_2 \oplus B_2)}\,\overline{(A_1 \oplus B_1)}\,\overline{(A_0 \oplus B_0)}$$

$$ALTB = \overline{A_3}B_3 + \overline{(A_3 \oplus B_3)}\overline{A_2}\,B_2 + \overline{(A_3 \oplus B_3)}\overline{(A_2 \oplus B_2)}\overline{A_1}B_1 + \overline{(A_3 \oplus B_3)}$$
$$\overline{(A_2 \oplus B_2)}\,\overline{(A_1 \oplus B_1)}\overline{A_0}B_0$$

$$AGTB = \overline{AEQB + ALTB}$$

This comparison technique can be expanded to as many bits as necessary. A 4-bit comparator requires four AND-shaped gates for its $ALTB$ function. We can interpret the Boolean expression for this function as follows.

$A < B$ if:

1. The MSB of A is less than the MSB of B, OR

2. The MSBs are equal, but the second bit of A is less than the second bit of B, OR

3. The first two bits are equal, but the third bit of A is less than the third bit of B, OR

4. The first three bits are equal, but the LSB of A is less than the LSB of B

Expansion to more bits would use the same principle of comparing bits one at a time, beginning with the MSBs.

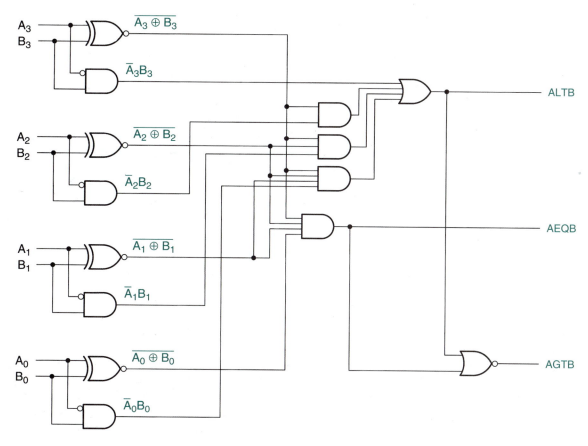

FIGURE 5.59
4-bit Magnitude Comparator

ⅠⅠⅠ EXAMPLE 5.13

Application

A digital thermometer has two input probes. A circuit in the thermometer converts the measured temperature at each probe to an 8-bit number, as shown by the block in Figure 5.60.

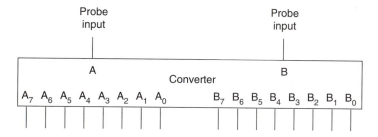

FIGURE 5.60
Example 5.13
Two-channel Digital Thermometer

In addition to measuring the temperature at each input, the thermometer has a comparison function that indicates whether the temperature at one input is greater than, equal to, or less than the temperature at the other input.

Draw a logic diagram showing how a magnitude comparator could be connected to light a green LED for *AGTB,* an amber LED for *AEQB,* and a red LED for *ALTB.*

Solution   Figure 5.61 shows the logic diagram of the magnitude comparator connected to the thermometer's digital output.

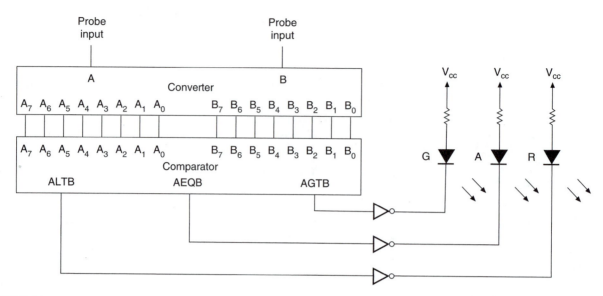

FIGURE 5.61
Example 5.13
Temperature Comparator Block Diagram

When one of the comparator outputs goes HIGH, it sets the output of the corresponding inverter LOW. This provides a current path to ground for the indicator LED for that output, causing it to illuminate.

VHDL Magnitude Comparators

The most obvious way to create a VHDL representation of a magnitude comparator is to use a concurrent signal assignment statement for each comparing function. For example, the following VHDL code can represent the 2-bit magnitude comparator of Figure 5.57:

```
-- compare2.vhd

ENTITY compare2 IS
  PORT(
      a, b              : IN   BIT_VECTOR (1 downto 0);
      agtb, aeqb, altb  : OUT  BIT);
END compare2;
```

compare2.vhd
compare2.scf

```
ARCHITECTURE a OF compare2 IS
BEGIN
      altb   <=  (not (a(1)) and b(1))
             or  ((not (a(1) xor b(1))) and (not (a(0)) and b(0)));
      aeqb   <=  (not (a(1) xor b(1))) and (not (a(0) xor b(0)));
      agtb   <=  (a(1) and not (b(1)))
             or  ((not (a(1) xor b(1))) and (a(0) and not (b(0))));
END a;
```

A simulation for this file is shown in Figure 5.62. The comparison outputs go HIGH to indicate A = B, A < B, or A > B.

Although this approach works, it is not a very good one. Due to the complexity of the Boolean equations for *ALTB* and *AGTB,* it is difficult to type them without making errors. (Try it!) The difficulty increases greatly with the number of required inputs.

The following code for a 4-bit comparator illustrates a much more efficient method. Since VHDL allows inputs to be represented as integers, we can define the required size of

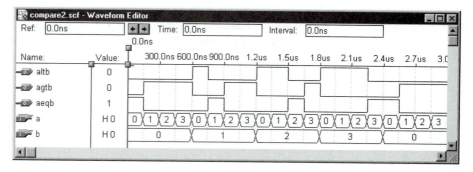

FIGURE 5.62

Simulation for a 2-bit Magnitude Comparator

inputs *A* and *B* and compare them using IF statements. For every comparison, we assign an output vector consisting of bits for *ALTB, AEQB,* and *AGTB* one of the values 110, 101, or 011, for active-LOW outputs. For example, if *A* = 12 and *B* = 9, then the output vector would be 011 (i.e., A > B). An active-LOW output will illuminate a LOW-sense LED, such as those on the Altera UP-1 board.

compare4.vhd
compare4.scf

```
-- compare4.vhd
LIBRARY ieee;
USE ieee.std_logic_1164.ALL;

ENTITY compare4 IS
    PORT(
        a, b               : IN   INTEGER RANGE 0 TO 15;
        agtb, aeqb, altb : OUT  STD_LOGIC);
END compare4;

ARCHITECTURE a OF compare4 IS
    SIGNAL compare : STD_LOGIC_VECTOR (2 downto 0);
BEGIN
    PROCESS (a,b)
    BEGIN
        IF a<b THEN
            compare    <=  "110";
        ELSIF a=b THEN
            compare    <=  "101";
        ELSIF a>b THEN
            compare    <=  "011";
        ELSE
            compare    <=  "111";
        END IF;
        agtb   <=  compare(2);
        aeqb   <=  compare(1);
        altb   <=  compare(0);
    END PROCESS;
END a;
```

The beauty of this method is that the number of input bits can be changed by modifying one number: the range of the INTEGER-type input. For example, a 12-bit comparator is identical to the 4-bit comparator in the previous VHDL code, except that the inputs have a range of 0 to 4095 ($= 2^{12}-1$). Using this method, we can program an EPM7128S CPLD with a comparator up to 28 bits wide (range of 0 to 268,435,455). If we do, however, there is no room for anything else.

EXAMPLE 5.14

Write a VHDL file that uses IF statements to compare two 8-bit numbers *A* and *B*. The design should have outputs for *AEQB*, *ALTB*, and *AGTB*.

Solution

```
-- compare8.vhd
LIBRARY ieee;
USE ieee.std_logic_1164.ALL;

ENTITY compare8 IS
    PORT(
        a, b              : IN    INTEGER RANGE 0 TO 255;
        agtb, aeqb, altb  : OUT   STD_LOGIC);
END compare8;

ARCHITECTURE a OF compare8 IS
    SIGNAL compare : STD_LOGIC_VECTOR (2 downto 0);
BEGIN
    PROCESS (a,b)
    BEGIN
        IF a<b THEN
            compare   <=  "110";
        ELSIF a=b THEN
            compare   <=  "101";
        ELSIF a>b THEN
            compare   <=  "011";
        ELSE
            compare   <=  "111";
        END IF;
        agtb  <= compare(2);
        aeqb  <= compare(1);
        altb  <= compare(0);
    END PROCESS;
END a;
```

compare8.vhd

5.6 Parity Generators and Checkers

KEY TERMS

Parity   A system that checks for errors in a multi-bit binary number by counting the number of 1s.

Even parity   An error-checking system that requires a binary number to have an even number of 1s.

Odd parity   An error-checking system that requires a binary number to have an odd number of 1s.

Parity bit   A bit appended to a binary number to make the number of 1s even or odd, depending on the type of parity.

When data are transmitted from one device to another, it is necessary to have a system of checking for errors in transmission. These errors, which appear as incorrect bits, occur as a result of electrical limitations such as line capacitance or induced noise.

   Parity error checking is a way of encoding information about the correctness of data before they are transmitted. The data can then be verified at the system's receiving end. Figure 5.63 shows a block diagram of a parity error-checking system.

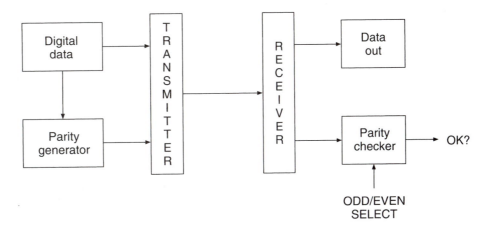

FIGURE 5.63
Parity Error Checking

The parity generator in Figure 5.63 examines the outgoing data and adds a bit called the **parity bit** that makes the number of 1s in the transmitted data odd or even, depending on the type of parity. Data with **EVEN parity** have an even number of 1s, including the parity bit, and data with **ODD parity** have an odd number of 1s.

The data receiver "knows" whether to expect EVEN or ODD parity. If the incoming number of 1s matches the expected parity, the parity checker responds by indicating that correct data have been received. Otherwise, the parity checker indicates an error.

III EXAMPLE 5.16

Data are transmitted from a PC serial port to a modem in groups of 7 data bits plus a parity bit. What should the parity bit, P, be for each of the following data if the parity is EVEN? If the parity is ODD?

a. 0110110

b. 1000000

c. 0010101

Solution

a. 0110110   Four 1s in data. (4 is an even number.)
             EVEN parity: $P = 0$
             ODD parity: $P = 1$

b. 1000000   One 1 in data. (1 is an odd number.)
             EVEN parity: $P = 1$
             ODD parity: $P = 0$

c. 0010101   Three 1s in data. (3 is an odd number.)
             EVEN parity: $P = 1$
             ODD parity: $P = 0$

III

A ⎓⎓⎓⎓ $A \oplus B$
B ⎓⎓⎓⎓

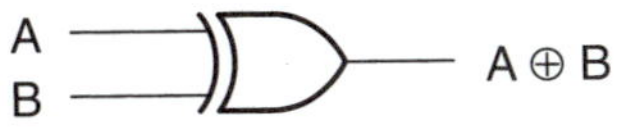

FIGURE 5.64
Exclusive OR Gate

An Exclusive OR gate can be used as a parity generator or a parity checker. Figure 5.64 shows the gate, and Table 5.12 is the XOR truth table. Notice that each line of the XOR truth table has an even number of 1s if we include the output column.

Figure 5.65 shows the block diagram of a circuit that will generate an EVEN parity bit from 2 data bits, A and B, and transmit the three bits one after the other, that is, serially, to a data receiver.

Figure 5.66 shows a parity checker for the parity generator in Figure 5.65. Data are received serially, but read in parallel. The parity bit is re-created from the received values of A and B, and then compared to the received value of P to give an error indication, P'. If P

Table 5.12 Exclusive OR Truth Table

A	B	$A \oplus B$
0	0	0
0	1	1
1	0	1
1	1	0

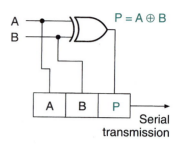

FIGURE 5.65
Even Parity Generation

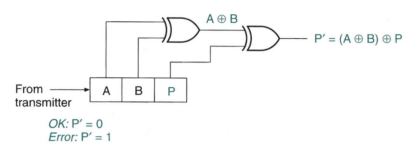

OK: P' = 0
Error: P' = 1

FIGURE 5.66
Even Parity Checking

and $A \oplus B$ are the same, then $P' = 0$ and the transmission is correct. If P and $A \oplus B$ are different, then $P' = 1$ and there has been an error in transmission.

▌▌ EXAMPLE 5.17

The following data and parity bits are transmitted four times: $ABP = 101$.

1. State the type of parity used.
2. The transmission line over which the data are transmitted is particularly noisy and the data arrive differently each time as follows:
   a. $ABP = 101$
   b. $ABP = 100$
   c. $ABP = 111$
   d. $ABP = 110$

   Indicate the output P' of the parity checker in Figure 5.66 for each case and state what the output means.

Solution

1. The system is using EVEN parity.
2. The parity checker produces the following responses:

   a. $ABP = 101$
       $A \oplus B = 1 \oplus 0 = 1$
       $P' = (A \oplus B) \oplus P = 1 \oplus 1 = 0$ Data received correctly.

   b. $ABP = 100$
       $A \oplus B = 1 \oplus 0 = 1$
       $P' = (A \oplus B) \oplus P = 1 \oplus 0 = 1$ Transmission error. (Parity bit incorrect.)

   c. $ABP = 111$
       $A \oplus B = 1 \oplus 1 = 0$
       $P' = (A \oplus B) \oplus P = 0 \oplus 1 = 1$ Transmission error. (Data bit B incorrect.)

d. $ABP = 110$
$$A \oplus B = 1 \oplus 1 = 0$$
$$P' = (A \oplus B) \oplus P = 0 \oplus 0 = 0 \qquad \text{Transmission error undetected. } (B \text{ and } P \text{ incorrectly received.})$$

The second and third cases in Example 5.17 show that parity error-detection cannot tell which bit is incorrect.

The fourth case points out the major flaw of parity error detection: An even number of errors cannot be detected. This is true whether the parity is EVEN or ODD. If a group of bits has an even number of 1s, a single error will change that to an odd number of 1s, but a double error will change it back to even. (Try a few examples to convince yourself this is true.)

An ODD parity generator and checker can be made using an Exclusive NOR, rather than an Exclusive OR, gate. If a set of transmitted data bits require a 1 for EVEN parity, it follows that they require a 0 for ODD parity. This implies that EVEN and ODD parity generators must have opposite-sense outputs.

▥ EXAMPLE 5.18

Modify the circuits in Figures 5.65 and 5.66 to operate with ODD parity. Verify their operation with the data bits $AB = 11$ transmitted twice and received once as $AB = 11$ and once as $AB = 01$.

Solution   Figure 5.67a shows an ODD parity generator and Figure 5.67b shows an ODD parity checker. The checker circuit still has an Exclusive OR output since it presents the same error codes as an EVEN parity checker. The parity bit is re-created at the receive end of the transmission path and compared with the received parity bit. If they are the same, $P' = 0$ (correct transmission). If they are different, $P' = 1$ (transmission error).

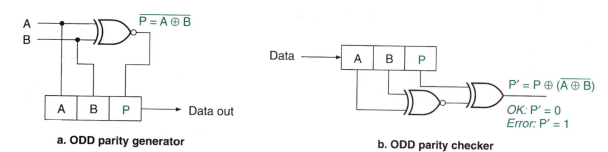

a. ODD parity generator   b. ODD parity checker

FIGURE 5.67
Example 5.18
ODD Parity Generator and Checker

Verification:

Generator:

Data: $AB = 11$ Parity: $P = \overline{A \oplus B} = \overline{1 \oplus 1} = 1$

Checker:

Received data: $AB = 11$

$P' = (A \oplus B) \oplus P = (1 \oplus 1) \oplus 1 = 1 \oplus 1 = 0$ (Correct transmission)

Generator:

Data: $AB = 11$ Parity: $P = \overline{A \oplus B} = \overline{1 \oplus 1} = 1$

Checker:

Received data: $AB = 01$

$P' = (A \oplus B) \oplus P = (0 \oplus 1) \oplus 1 = 0 \oplus 1 = 1$ (Incorrect transmission)

Parity generators and checkers can be expanded to any number of bits by using an XOR gate for each pair of bits and combining the gate outputs in further stages of 2-input XOR gates. The true form of the generated parity bit is P_E, the EVEN parity bit. The complement form of the bit is P_O, the ODD parity bit.

Table 5.13 shows the XOR truth table for 4 data bits and the ODD and EVEN parity bits. The EVEN parity bit P_E is given by $\overline{(A \oplus B) \oplus (C \oplus D)}$. The ODD parity bit P_O is given by $\overline{P_E} = \overline{(A \oplus B) \oplus (C \oplus D)}$. For every line in Table 5.13, the bit combination $ABCDP_E$ has an even number of 1s and the group $ABCDP_O$ has an odd number of 1s.

Table 5.13   Even and Odd Parity Bits for 4-bit Data

A	B	C	D	$A \oplus B$	$C \oplus D$	P_E	P_O
0	0	0	0	0	0	0	1
0	0	0	1	0	1	1	0
0	0	1	0	0	1	1	0
0	0	1	1	0	0	0	1
0	1	0	0	1	0	1	0
0	1	0	1	1	1	0	1
0	1	1	0	1	1	0	1
0	1	1	1	1	0	1	0
1	0	0	0	1	0	1	0
1	0	0	1	1	1	0	1
1	0	1	0	1	1	0	1
1	0	1	1	1	0	1	0
1	1	0	0	0	0	0	1
1	1	0	1	0	1	1	0
1	1	1	0	0	1	1	0
1	1	1	1	0	0	0	1

III EXAMPLE 5.19

Use Table 5.13 to draw a 4-bit parity generator and a 4-bit parity checker that can generate and check either EVEN or ODD parity, depending on the state of one select input.

Solution   Figure 5.68 shows the circuit for a 4-bit parity generator. The XOR gate at the output is configured as a programmable inverter to give P_E or P_O. When $\overline{\text{EVEN}}/\text{ODD} = 0$, the parity output is not inverted and the circuit generates P_E. When $\overline{\text{EVEN}}/\text{ODD} = 1$, the XOR inverts the parity bit, giving P_O.

FIGURE 5.68
Example 5.19
4-bit Parity Generator

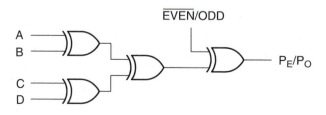

The 4-bit parity checker, shown in Figure 5.69, is the same circuit, with an additional XOR gate to compare the parity bit re-created from data and the previously encoded parity bit.

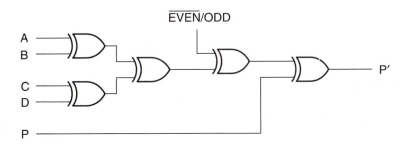

FIGURE 5.69
Example 5.19
4-bit Parity Checker

III EXAMPLE 5.20

Draw the circuit for an 8-bit *EVEN/ODD* parity generator.

Solution An 8-bit parity generator is an expanded version of the 4-bit generator in the previous example. The circuit is shown in Figure 5.70.

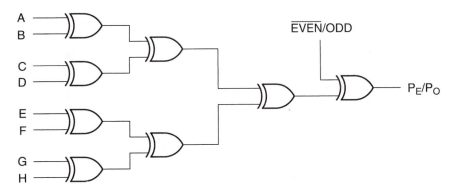

FIGURE 5.70
Example 5.20
8-bit Parity Generator

III SECTION 5.6 REVIEW PROBLEM

5.6 Data (including a parity bit) are detected at a receiver configured for checking ODD parity. Which of the following data do we know are incorrect? Could there be errors in the remaining data? Explain.

a. 010010

b. 011010

c. 1110111

d. 1010111

e. 1000101

SUMMARY

1. A decoder detects the presence of a particular binary code. The simplest decoder is an AND or NAND gate, which can detect a binary code when combined with the right combination of input inverters.

2. Multiple-output decoders are implemented by a series of single-gate decoders, each of which responds to a different input code.

3. For an n-input decoder, there can be as many as 2^n unique outputs.

4. MAX+PLUS II can simulate the function of a digital circuit by generating a set of output waveforms in response to a defined set of input waveforms.

5. VHDL constructs such as selected signal assignment statements and conditional signal assignments can describe decoders. Both statement types assign alternative values to a VHDL port or signal, based on the state of another port or signal.

6. A selected signal assignment statement has the form:

```
label: WITH __expression SELECT
__signal <=__expression WHEN __constant_value,
          __expression WHEN __constant_value,
          __expression WHEN __constant_value,
          __expression WHEN __constant_value;
```

7. A conditional signal assignment statement has the form:

```
__signal <= __expression WHEN __boolean_expression ELSE
            __expression WHEN __boolean_expression ELSE
            __expression;
```

8. SIGNALs act as internal connections in a VHDL design entity. They can be single lines or vectors and are declared before the BEGIN clause of an ARCHITECTURE body.

9. The report file of a MAX+PLUS II project contains design and configuration information, including the Boolean equations that the compiler derives from the design entry file(s) of the project.

10. A seven-segment display is an array of seven luminous segments (usually LED or LCD), arranged in a figure-8 pattern, used to display numerical digits.

11. The segments in a seven-segment display are designated by lowercase letters a through g. The sequence of labels goes clockwise, starting with segment a at the top and ending with g in the center.

12. Seven-segment displays are configured as common anode (active-LOW inputs) or common cathode (active-HIGH segments).

13. A seven-segment decoder can be described with a truth table or Boolean equation for each segment function. Since the segment functions do not simplify very much, it is often easier to program a CPLD with a VHDL truth table, in the form of a selected signal assignment statement, rather than with the Boolean equations of the decoder.

14. A multiplexer (MUX) is a circuit that directs a signal or group of signals (called the data inputs) to an output, based on the status of a set of select inputs.

15. Generally, for n select inputs in a multiplexer, there are $m = 2^n$ data inputs. Such a multiplexer is referred to as an m-to-1 multiplexer.

16. The selected data input in a MUX is usually denoted by a subscript that is the decimal equivalent of the combined binary value of the select inputs. For example, if the select inputs in an 8-to-1 MUX are set to $S_2 S_1 S_0 = 100$, data input D_4 is selected since 100 (binary) = 4 (decimal).

17. A MUX can be designed to switch groups of signals to a multi-bit output. The inputs can be denoted by double subscript notation, where the first subscript indicates the number of the signal group and the second subscript the element in the group. For example, a MUX can have a 4-bit set of inputs called $D_{03} D_{02} D_{01} D_{00}$ and another 4-bit input group called $D_{13} D_{12} D_{11} D_{10}$, each of which can be switched to a 4-bit output called $Y_3 Y_2 Y_1 Y_0$ by the state of one select input.

18. A multiplexer can be used in time-dependent applications if a binary counter is applied to its select inputs.

19. Some examples of time-dependent MUX applications are waveform or bit pattern generation and time-division multiplexing (TDM).

20. In time division multiplexing, several digital signals share a single transmission path by allotting a time slot for every signal, during which that signal has sole access to the transmission path.

21. TDM can be configured for bit multiplexing, in which a channel transmits one bit each time it is selected, or byte (or word) multiplexing, in which a channel transmits and entire byte or word each time it is selected.

22. A demultiplexer (DMUX) receives data from a single source and directs the data to one of several outputs, which is selected by the status of a set of select inputs.

23. A decoder with an enable input can also act as a demultiplexer if the enable input of the decoder is used as a data input for a demultiplexer.

24. A TDM signal can be demultiplexed by applying a binary count to the DMUX's select inputs at the same rate as the count is applied to the select input of the multiplexer that originally sent the data.

25. A CMOS analog multiplexer or demultiplexer works by using a decoder to enable a set of analog data transmission switches. It can be used in either direction.

26. A magnitude comparator determines whether two binary numbers are equal and, if not, which one is greater.

27. The simplest equality comparator is an XNOR gate, whose output is HIGH if both inputs are the same.

28. A pair of multiple-bit numbers can be compared by a set of XNOR gates whose outputs are ANDed. The circuit compares the two numbers bit-by-bit.

29. Given two numbers A and B, the Boolean function $\overline{A}_n B_n$, if true, indicates that the nth bit of A is less than the nth bit of B.

30. Given two numbers A and B, the Boolean function $A_n \overline{B}_n$, if true, indicates that the nth bit of A is greater than the nth bit of B.

31. The less-than and greater-than functions can be combined with an equality comparator to determine, bit-by-bit, how two numbers compare in magnitude to one another.
32. A magnitude comparator can be best implemented in VHDL by using INTEGER types for the inputs and using IF statements to compare their respective magnitudes.
33. Parity checking is a system of error detection that works by counting the number of 1s in a group of bits.
34. Even parity requires a group of bits to have an even number of 1s. Odd parity requires a group of bits to have an odd number of 1s. This is achieved by appending a parity bit to the data whose value depends on the number of 1s in the data bits.
35. An XOR gate is the simplest even parity generator. Each line in its truth table has an even number of 1s, if the output column is included.
36. An XNOR gate can be used to generate an odd parity bit from two data bits.
37. A parity checker consists of a parity generator on the receive end of a transmission system and a comparator to determine if the locally generated parity bit is the same as the transmitted parity bit.
38. Parity generators and checkers can be expanded to any number of bits by using an XOR gate for each pair of bits and combining the gate outputs in further stages of 2-input XOR gates.

GLOSSARY

BCD   Binary coded decimal. A code in which each individual digit of a decimal number is represented by a 4-bit binary number. (e.g., 905 (decimal) = 1001 0000 0101 (BCD)).

Bit multiplexing   A TDM technique in which one bit is sent from each channel during its assigned time slot.

Byte (or word) multiplexing   A TDM technique in which a byte (or word) is sent from each channel during its assigned time slot. (A byte is eight bits; a word is a group of bits whose size varies with the particular system.)

CASE statement   A VHDL construct in which there is a choice of statements to be executed, depending on the value of a signal or variable.

Clock   A signal that controls the operation of a sequential digital circuit, such as a counter, by advancing its outputs to the next state when it receives a pulse.

CMOS analog switch   A CMOS device that will pass an analog or digital signal in either direction, when enabled. Also called a transmission gate. There is no TTL equivalent.

Common anode display   A seven-segment LED display where the anodes of all the LEDs are connected to the circuit supply voltage. Each segment is illuminated by a logic LOW at its cathode.

Common cathode display   A seven-segment display in which the cathodes of all LEDs are connected together and grounded. A logic HIGH illuminates a segment when applied to its anode.

Conditional signal assignment statement   A concurrent VHDL construct that assigns a value to a signal, depending on a sequence of conditions being true or false.

Counter   A digital circuit whose output produces a fixed sequence of binary states when an input called the clock receives a series of pulses. The output advances by one for each clock pulse (e.g., the output state of a 4-bit binary counter progresses in order from 0000 to 1111, then repeats).

Data inputs   The multiplexer inputs that feed a digital signal to the output when selected.

Decoder   A digital circuit designed to detect the presence of a particular digital state.

Demultiplexer   A circuit that uses a binary decoder to direct a digital signal from a single source to one of several destinations.

Double-subscript notation   A naming convention where two or more numerically related groups of signals are named using two

subscript numerals. Generally, the first digit refers to a group of signals and the second to an element of a group. (e.g., X_{03} represents element 3 of group 0 for a set of signal groups, X.)

Encoder   A circuit that generates a digital code at its outputs in response to one or more active input lines.

Even parity   An error-checking system that requires a binary number to have an even number of 1s.

IF statement   A VHDL construct within a process that executes a series of statements, if a Boolean test condition is true.

Magnitude comparator   A circuit that compares two n-bit binary numbers, indicates whether or not the numbers are equal, and, if not, which one is larger.

Multiplexer   A circuit that directs one of several digital signals to a single output, depending on the states of several select inputs.

Odd parity   An error-checking system that requires a binary number to have an odd number of 1s.

Parity   A system that checks for errors in a multi-bit binary number by counting the number of 1s.

Parity bit   A bit appended to a binary number to make the number of 1s even or odd, depending on the type of parity.

Positive edge   The point on a digital waveform where the logic level of the waveform makes a LOW-to-HIGH transition.

Priority encoder   An encoder that generates a binary or BCD output corresponding to the subscript of the active input having the highest priority. This is usually defined as the input with the largest subscript value.

PROCESS   A VHDL construct that contains statements that are executed if there is a change in a signal in its sensitivity list.

Propagation delay   Time difference between a change on a digital circuit input and a change on an output in response to the input change.

$\overline{RBI}$ Ripple blanking input.

$\overline{RBO}$ Ripple blanking output.

Response waveforms   A set of output waveforms generated by a simulator tool for a particular digital design in response to a set of stimulus waveforms.

Ripple blanking   A technique used in a multiple-digit numerical display that suppresses leading or trailing zeros in the display, but allows internal zeros to be displayed.

Select inputs   The multiplexer inputs which select a digital input channel.

Selected signal assignment statement   A concurrent signal assignment in VHDL in which a value is assigned to a signal, depending on the alternative values of another signal or variable.

Sensitivity list   A list of signals in a PROCESS statement that are monitored to determine whether the PROCESS should be executed.

Seven-segment display   An array of seven independently controlled light-emitting diode (LED) or liquid crystal display (LCD) elements, shaped like a figure-8, which can be used to display decimal digits and other characters by turning on the appropriate elements.

Simulation   The verification of the logic of a digital design before programming it into a PLD.

Stimulus waveforms   A set of user-defined input waveforms on a simulator file designed to imitate input conditions of a digital circuit.

Time division multiplexing (TDM)   A technique of using one transmission line to send many signals simultaneously by making them share the line for equal fractions of time.

Time slot   A period of time during which a transmitted data element has sole access to a transmission path.

Timing diagram   A diagram showing how two or more digital waveforms in a system relate to each other over time.

PROBLEMS

Section 5.1   Decoders

5.1   When a HIGH is on the outputs of each of the decoding circuits shown in Figure 5.71, what is the binary code appearing at the inputs? Write the Boolean expression for each decoder output.

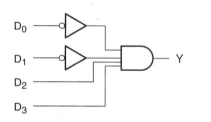 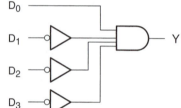 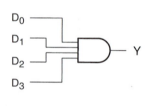

FIGURE 5.71
Problem 5.1
Decoding Circuits

5.2   Draw the decoding circuit for each of the following Boolean expressions:

a. $\overline{Y} = \overline{D_3} D_2 \overline{D_1} D_0$
b. $\overline{Y} = \overline{D_3} \overline{D_2} D_1 D_0$
c. $Y = \overline{D_3} \overline{D_2} D_1 D_0$
d. $\overline{Y} = D_3 D_2 \overline{D_1} \overline{D_0}$
e. $Y = D_3 D_2 \overline{D_1} D_0$

5.3   Use a Graphic Design File in MAX+PLUS II to draw the logic diagram of a 2-line-to-4-line decoder with active-HIGH outputs and an active-LOW enable input. Create a simulation file to show the operation of the circuit.

5.4   Use a Graphic Design File in MAX+PLUS II to draw the logic diagram of a 3-line-to-8-line decoder with active-HIGH outputs and an active-LOW enable input. Create a simulation file to show the operation of the circuit.

5.5   For a generalized n-line-to-m-line decoder, state the value of m if n is:

a. 5
b. 6
c. 8

Write the equation giving the general relation between n and m.

5.6   A microcomputer system has a RAM capacity of 128 megabytes (MB), split into 16 MB portions. Each RAM device is enabled by a low at a $\overline{G}$ input. Draw a logic diagram showing how a binary decoder can select one particular RAM device.

5.7   Briefly describe the difference between a selected signal assignment statement and a conditional signal assignment statement in VHDL. State which one is the preferred statement in VHDL files and why.

5.8a.   Write a VHDL file for a 3-line-to-8-line decoder with active-LOW outputs and no enable. Use a selected signal assignment statement. Assign the device as an EPM 7128SLC84.

b. Write the Boolean equations for the decoder in part a, as reported in the decoder's MAX+PLUS II report file. (Use the form $(\overline{x} \cdot \overline{y} \cdot \overline{z} + x \cdot y \cdot z)$ rather than (!x & !y & !z #x & y & z).)

c. Change the decoder in part a so that its outputs are active-HIGH. Compile the design and examine the resulting report file to find the Boolean equations of the modified design. Write the equations and state how the compiler deals the change in output active level.

5.9 Create a MAX+PLUS II simulation file for the decoder in Problem 5.8.

5.10 Write a VHDL file for a 3-line-to-8-line decoder with active-LOW outputs and an active-LOW enable input.

5.11 Create a MAX+PLUS II simulation file for the decoder in Problem 5.10.

5.12 Write a truth table for a hexadecimal-to-seven-segment decoder for a common anode display. Use the digit patterns of Figure 5.26 as a model.

5.13 Use the truth table derived in Problem 5.12 to derive the Boolean equations for each segment driver. Simplify the equations as much as possible, using any convenient method.

5.14 Write a VHDL file for the hexadecimal-to-seven-segment decoder described in Problem 5.12.

5.15 Modify the VHDL file for the hexadecimal-to-seven-segment decoder from Problem 5.14 to add a ripple-blanking feature.

5.16 Draw a diagram consisting of four seven-segment displays, each driven by a BCD-to-seven-segment decoder with ripple blanking. The circuit should be configured to suppress all leading zeros. Show the displayed digits and $\overline{RBO}/\overline{RBI}$ logic levels for each of the following displayed values: 100, 217, 1024.

Section 5.2 Encoders

5.17 Figure 5.72 shows a BCD priority encoder with three different sets of inputs. Determine the resulting output code for each input combination. Inputs and outputs are active HIGH.

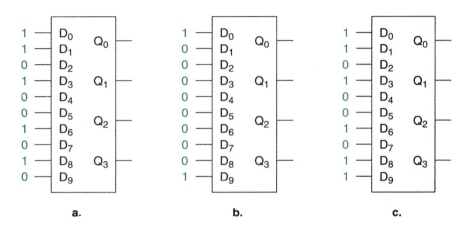

FIGURE 5.72
Problem 5.17
BCD Priority Encoder

5.18 Derive the Boolean equations for the outputs of a BCD priority encoder, based on the encoding principle stated in Section 5.2. Show all work.

5.19 Create a Graphic Design File in MAX+PLUS II for a BCD priority encoder, based on the equations in Problem 5.18. Also generate a simulation for this function.

5.20 Write a VHDL file that implements the function of a BCD priority encoder. Create a simulation file for this function. Write the Boolean equations of the encoder, as shown in the encoder's report file. State how the equations from the report file compare to the equations you derived in Problem 5.18.

5.21 Write a VHDL file that implements the function of a 4-bit binary priority encoder. Create a simulation file for this function.

Section 5.3 Multiplexers

5.22 Make a table listing which digital audio source in Figure 5.73 is routed to output Y for each combination of

the multiplexer select inputs. (CD = compact disc; DAT = digital audio tape.)

5.23 Draw symbols for an 8-to-1 and a 16-to-1 multiplexer. Write the truth table for each multiplexer, showing which data input is selected for every binary combination of the select inputs.

5.24 Make a Graphic Design File in MAX+PLUS II for an 8-to-1 multiplexer circuit. Also create a simulation that shows the operation of the device.

5.25 Write the Boolean expression describing an 8-to-1 multiplexer. Evaluate the equation for the case where input D_5 is selected.

5.26 Draw the symbol for a quadruple 8-to-1 multiplexer (i.e., a MUX with eight switched groups of 4 bits each). Write the truth table for this device, showing which data inputs are selected for every binary combination of the select inputs. Use double-subscript notation.

FIGURE 5.73
Problem 5.22
Digital Audio Multiplexer

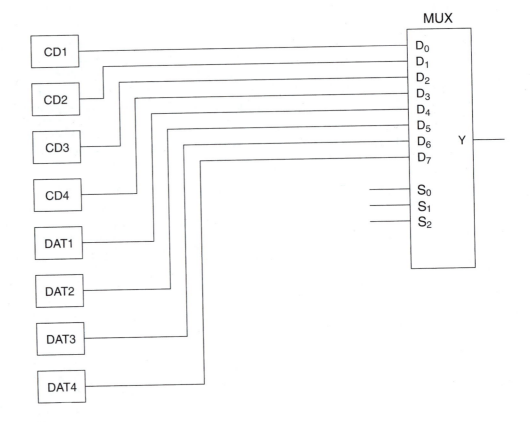

5.27 Write a VHDL file for the quadruple 8-to-1 multiplexer in Problem 5.26. Create a MAX+PLUS II simulation for the design to verify its operation.

5.28 Draw the symbol for an octal 4-to-1 multiplexer (i.e., a MUX with four switched groups of 8 bits each). Write the truth table for this device, showing which data inputs are selected for every binary combination of the select inputs. Use double-subscript notation.

5.29 Write a VHDL file for the octal 4-to-1 multiplexer in the Problem 5.28. Create a MAX+PLUS II simulation for the design to verify its operation. Write its Boolean equations from the project report file.

5.30 Write a VHDL file for an 8-to-1 multiplexer using a concurrent signal assignment statement to encode the multiplexer's Boolean equation directly. Would this be a good method for encoding a larger device, such as a 16-to-1 multiplexer? Explain your answer.

5.31 Write a VHDL file for an 8-to-1 multiplexer using a selected signal assignment statement. Would this be a good method for encoding a larger device, such as a 16-to-1 multiplexer? Explain your answer.

5.32 Write a VHDL file for a 16-to-1 multiplexer using the method you believe to be most efficient.

5.33 Draw the circuit of a programmable waveform generator based on an 8-to-1 multiplexer. Draw a timing diagram of this circuit for the following input data:

  a. $D_7D_6D_5D_4D_3D_2D_1D_0 = 01100101$

  b. $D_7D_6D_5D_4D_3D_2D_1D_0 = 01010101$

5.34 The data pattern in Problem 5.34b generates a symmetrical 12 kHz waveform. Write the data patterns required to produce a 6 kHz waveform and a 3 kHz waveform at the output of a MUX-based programmable waveform generator.

Section 5.4 Demultiplexers

5.35 Make a Graphic Design File in MAX+PLUS II for a 1-to-4 demultiplexer circuit with active-LOW outputs and an active-LOW enable input. Create a simulation that shows how this device can be used as a demultiplexer or decoder.

5.36 Make a Graphic Design File in MAX+PLUS II for a 1-to-8 demultiplexer circuit with active-HIGH outputs. Create a simulation that shows the operation of the device.

5.37 Write a VHDL file that implements the function of a 1-to-16 demultiplexer.

5.38 Briefly state what characteristics of an analog switch make it suitable for transmitting analog signals.

5.39 Draw a diagram showing how eight analog switches can be connected to a decoder to form an 8-channel MUX/DMUX circuit. Briefly explain why the same circuit can be used as a multiplexer or as a demultiplexer.

5.40 Draw a circuit showing how a 74HC4052 dual 4-channel analog MUX/DMUX can be used to multiplex four transmitted digital audio channels onto a phone line and demultiplex four received audio channels from another phone line.

Section 5.5   Magnitude Comparators

5.41   Briefly explain the operation of the *ALTB* portion of the 2-bit magnitude comparator shown in Figure 5.58.

5.42   Draw the *ALTB* portion of a 4-bit magnitude comparator as a Graphic Design File in MAX+PLUS II. Create a simulation for the circuit and briefly explain its operation.

5.43   Use MAX+PLUS II to create a 3-bit magnitude comparator that has outputs for *AEQB, AGTB,* and *ALTB* functions. Create a simulation that shows the operation of this circuit.

5.44   Write the Boolean expressions for the *AEQB, ALTB,* and *AGTB* outputs of a 6-bit magnitude comparator.

5.45   Write a VHDL file that implements the functions $A < B$, $A = B$ and $A > B$ for two 16-bit numbers.

5.46   Write a VHDL file that implements the following six comparison functions in a single device for two 4-bit inputs *A* and *B*: $A < B$, $A \leq B$, $A = B$, $A \neq B$, $A \geq B$, and $A > B$. Make the outputs indicate active-LOW.

5.47   Create a simulation that verifies the operation of the six-function comparator in Problem 5.46.

Section 5.6   Parity Generators and Checkers

5.48   What parity bit, *P,* should be added to the following data if the parity is EVEN? If the parity is ODD?

   a. 1111100

   b. 1010110

   c. 0001101

5.49   The following data are transmitted in a serial communication system (*P* is the parity bit). What parity is being used in each case?

   a. *ABCDEFGHP* = 010000101

   b. *ABCDEFGHP* = 011000101

   c. *ABCDP* = 01101

   d. *ABCDEP* = 101011

   e. *ABCDEP* = 111011

5.50   The data *ABCDEFGHP* = 110001100 are transmitted in a serial communication system. Give the output *P′* of a receiver parity checker for the following received data. State the meaning of the output *P′* for each case.

   a. *ABCDEFGHP* = 110101100

   b. *ABCDEFGHP* = 110001101

   c. *ABCDEFGHP* = 110001100

   d. *ABCDEFGHP* = 110010100

5.51   Use MAX+PLUS II to create a Graphic Design File for a 5-bit parity generator with a switchable *EVEN/ODD* output. Create a simulation file to show the operation of the device.

5.52   Use MAX+PLUS II to create a Graphic Design File for a 5-bit parity checker corresponding to the parity generator in Problem 5.51. Create a simulation file to show the operation of the device.

ANSWERS TO SECTION REVIEW PROBLEMS

Section 5.1a
5.1 The decoders are shown in Figure 5.74.

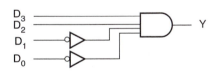

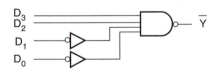

FIGURE 5.74
Decoders

Section 5.1b
5.2 A decoder with 16 outputs requires 4 inputs. A decoder with 32 outputs requires 5 inputs.

Section 5.1c
5.3 Trailing zeros could logically be suppressed after a decimal point or if there are digits displaying a power-of-ten exponent (e.g., 455. or 4.55 02), that is, if the zeros are nonsignificant. The zeros should be displayed if they set the location of the decimal point (e.g., 450).

Section 5.2
5.4 The encoder in Figure 5.29 can have only one input active at any time. If more than one input is active, it may generate incorrect output codes. The circuit can be modified according to the priority encoding principle, as expressed by the Boolean equations for the 3-bit priority encoder, to ensure that a low-priority input is not able to modify the code generated by a higher-priority input.

Section 5.3
5.5 A multiplexer application is time-dependent if its channels are selected in a repeating sequence. This can be accomplished by connecting a binary counter to the select inputs of the multiplexer.

Section 5.6
5.6 Parts a and c are certainly incorrect because each has an even number of 1s. Items b, d, and e could have an even number of errors, which is undetectable by parity checking.

Digital Arithmetic and Arithmetic Circuits

CHAPTER OBJECTIVES

Upon successful completion of this chapter, you will be able to:

- Add or subtract two unsigned binary numbers.
- Write a signed binary number in true-magnitude, 1's complement, or 2's complement form.
- Add or subtract two signed binary numbers.
- Explain the concept of overflow.
- Calculate the maximum sum or difference of two signed binary numbers that will not result in an overflow.
- Add or subtract two hexadecimal numbers.
- Write decimal numbers in BCD codes, such as 8421 (Natural BCD) and Excess-3 code.
- Construct a Gray code sequence.
- Use the ASCII table to convert alphanumeric characters to hexadecimal or binary numbers and vice versa.
- Derive the logic gate circuits for full and half adders, given their truth tables.
- Demonstrate the use of full and half adder circuits in arithmetic and other applications.
- Add and subtract n-bit binary numbers, using parallel binary adders and logic gates.
- Explain the difference between ripple carry and parallel carry.
- Design a circuit to detect sign-bit overflow in a parallel adder.
- Draw circuits to perform BCD arithmetic and explain their operation.
- Use VHDL to program CPLD devices to perform various arithmetic functions, such as parallel adders, overflow detectors, and 1's complementers.

There are two ways of performing binary arithmetic: with unsigned binary numbers or with signed binary numbers. Signed binary numbers incorporate a bit defining the sign of a number; unsigned binary numbers do not. Several ways of writing signed binary numbers are true-magnitude form, which maintains the magnitude of the number in binary value, and 1's complement and 2's complement forms, which modify the magnitude but are more suited to digital circuitry.

Hexadecimal arithmetic is used for calculations that would be awkward in binary due to the large number of bits involved. Important applications of hexadecimal arithmetic are found in microcomputer systems.

In addition to positional number systems, binary numbers can be used in a variety of nonpositional number codes, which can represent numbers, letters, and computer control codes. Binary coded decimal (BCD) codes represent decimal digits as individually encoded groups of bits. Gray code is a binary code used in special applications. American Standard Code for Information Interchange (ASCII) represents alphanumeric and control code characters in a 7- or 8-bit format.

There are a number of different digital circuits for performing digital arithmetic, most of which are based on the parallel binary adder, which in turn is based on the full adder and half adder circuits. The half adder adds two bits and produces a sum and a carry. The full adder also allows for an input carry from a previous adder stage. Parallel adders have many full adders in cascade, with carry bits connected between the stages.

Specialized adder circuits are used for adding and subtracting binary numbers, generating logic functions, and adding numbers in binary-coded decimal (BCD) form. ■

6.1 Digital Arithmetic

> **KEY TERMS**
>
> **Signed binary number**   A binary number of fixed length whose sign is represented by one bit, usually the most significant bit, and whose magnitude is represented by the remaining bits.
>
> **Unsigned binary number**   A binary number whose sign is not specified by a sign bit. A positive sign is assumed unless explicitly stated otherwise.

Digital arithmetic usually means binary arithmetic, or perhaps BCD arithmetic. Binary arithmetic can be performed using **signed binary numbers,** in which the MSB of each number indicates a positive or negative sign, or **unsigned binary numbers,** in which the sign is presumed to be positive.

The usual arithmetic operations of addition and subtraction can be performed using signed or unsigned binary numbers. Signed binary arithmetic is often used in digital circuits for two reasons:

1. Calculations involving real-world quantities require us to use both positive and negative numbers.

2. It is easier to build circuits to perform some arithmetic operations, such as subtraction, with certain types of signed numbers than with unsigned numbers.

Unsigned Binary Arithmetic

> **KEY TERMS**
>
> **Operand**   A number upon which an arithmetic function operates (e.g., in the expression $x + y = z$, x and y are the operands).
>
> **Augend**   The number in an addition operation to which another number is added.
>
> **Addend**   The number in an addition operation that is added to another.
>
> **Sum**   The result of an addition operation.
>
> **Carry**   A digit that is "carried over" to the next most significant position when the sum of two single digits is too large to be expressed as a single digit.

Sum bit (single-bit addition)   The least significant bit of the sum of two 1-bit binary numbers.

Carry bit   A bit that holds the value of a carry (0 or 1) resulting from the sum of two binary numbers.

Addition

When we add two numbers, they combine to yield a result called the **sum.** If the sum is larger than can be contained in one digit, the operation generates a second digit, called the **carry.** The two numbers being added are called the **augend** and the **addend,** or more generally, the **operands.**

For example, in the decimal addition $9 + 6 = 15$, 9 is the augend, 6 is the addend, and 15 is the sum. Since the sum cannot fit into a single digit, a carry is generated into a second digit place.

Four binary sums give us all of the possibilities for adding two n-bit binary numbers:

$$0 + 0 = 00$$
$$1 + 0 = 01$$
$$1 + 1 = 10 \qquad (1_{10} + 1_{10} = 2_{10})$$
$$1 + 1 + 1 = 11 \qquad (1_{10} + 1_{10} + 1_{10} = 3_{10})$$

Each of these results consists of a **sum bit** and a **carry bit.** For the first two results above, the carry bit is 0. The final sum in the table is the result of adding a carry bit from a sum in a less significant position.

When we add two 1-bit binary numbers in a logic circuit, the result *always* consists of a sum bit and a carry bit, even when the carry is 0, since each bit corresponds to a measurable voltage at a specific circuit location. Just because the value of the carry is 0 does not mean it has ceased to exist.

▌▌ EXAMPLE 6.1

Calculate the sum $10010 + 1010$.

SOLUTION

```
                ┌──── (Carry from sum of 2nd LSBs)
            1
        10010
      +  1010
        11100
```

▌▌ EXAMPLE 6.2

Calculate the sum $10111 + 10010$.

SOLUTION

```
              ┌─┬┬──── (Carry bits)
         1  11
          10111
      +   10010
         101001
```

▌▌

▌▌ SECTION 6.1A REVIEW PROBLEMS

6.1 Add $11111 + 1001$.

6.2 Add $10011 + 1101$.

Subtraction

> **KEY TERMS**
>
> **Difference**   The result of a subtraction operation.
>
> **Minuend**   The number in a subtraction operation from which another number is subtracted.
>
> **Subtrahend**   The number in a subtraction operation that is subtracted from another number.
>
> **Borrow**   A digit brought back from a more significant position when the subtrahend digit is larger than the minuend digit.

In unsigned binary subtraction, two operands, called the **subtrahend** and the **minuend,** are subtracted to yield a result called the **difference.** In the operation $x = a - b$, x is the difference, a is the minuend, and b is the subtrahend. To remember which comes first, think of the minuend as the number that is di*mini*shed (i.e., something is taken away from it).

Unsigned binary subtraction is based on the following four operations:

$$0 - 0 = 0$$
$$1 - 0 = 1$$
$$1 - 1 = 0$$
$$10 - 1 = 1 \quad (2_{10} - 1_{10} = 1_{10})$$

The last operation shows how to obtain a positive result when subtracting a 1 from a 0: **borrow** 1 from the next most significant bit.

Borrowing Rules:

1. If you are borrowing from a position that contains a 1, leave behind a 0 in the borrowed-from position.

2. If you are borrowing from a position that already contains a 0, you must borrow from a more significant digit that contains a 1. All 0s up to that point become 1s, and the last borrowed-from digit becomes a 0.

▮▮ EXAMPLE 6.3

Subtract $1110 - 1001$.

SOLUTION

```
(New 2nd LSB) ┐ ┌ (Bit borrowed from 2nd LSB)
             01
           1110
         − 1001
           ─────
           0101
```

▮▮ EXAMPLE 6.4

Subtract $10000 - 101$.

SOLUTION

```
                         1111
  10000  (original     10000  (After borrowing
−   101   problem)    −  101  from higher-order bits)
                       ─────
                        1011
```

■■ SECTION 6.1B REVIEW PROBLEMS

6.3 Subtract $10101 - 10010$.

6.4 Subtract $10000 - 1111$.

6.2 Representing Signed Binary Numbers

KEY TERMS

Sign bit   A bit, usually the MSB, that indicates whether a signed binary number is positive or negative.

Magnitude bits   The bits of a signed binary number that tell us how large the number is (i.e., its magnitude).

True-magnitude form   A form of signed binary number whose magnitude is represented in true binary.

1's complement   A form of signed binary notation in which negative numbers are created by complementing all bits of a number, including the sign bit.

2's complement   A form of signed binary notation in which negative numbers are created by adding 1 to the 1's complement form of the number.

NOTE

Positive numbers are the same in all three notations.

Binary arithmetic operations are performed by digital circuits that are designed for a fixed number of bits, since each bit has a physical location within a circuit. It is useful to have a way of representing binary numbers within this framework that accounts not only for the magnitude of the number, but for the sign as well.

This can be accomplished by designating one bit of a binary number, usually the most significant bit, as the **sign bit** and the rest as **magnitude bits.** When the number is negative, the sign bit is 1, and when the number is positive, the sign bit is 0.

There are several ways of writing the magnitude bits, each having its particular advantages. **True-magnitude** form represents the magnitude in straight binary form, which is relatively easy for a human operator to read. Complement forms, such as **1's complement** and **2's complement,** modify the magnitude so that it is more suited to digital circuitry.

True-Magnitude Form

In true-magnitude form, the magnitude of a number is translated into its true binary value. The sign is represented by the MSB, 0 for positive and 1 for negative.

■■ EXAMPLE 6.5

Write the following numbers in 6-bit true-magnitude form:

a. 25_{10} b. -25_{10} c. 12_{10} d. -12_{10}

SOLUTION   Translate the magnitudes of each number into 5-bit binary, padding with leading zeros as required, and set the sign bit to 0 for a positive number and 1 for a negative number.

a. 011001   b. 111001   c. 001100   d. 101100

1's Complement Form

True-magnitude and 1's complement forms of binary numbers are the same for positive numbers—the magnitude is represented by the true binary value and the sign bit is 0. We can generate a negative number in one of two ways:

1. Write the positive number of the same magnitude as the desired negative number. Complement each bit, including the sign bit; or

2. Subtract the n-bit positive number from a binary number consisting of n 1s.

⫶⫶ EXAMPLE 6.6

Convert the following numbers to 8-bit 1's complement form:

a. 57_{10} b. -57_{10} c. 72_{10} d. -72_{10}

SOLUTION   Positive numbers are the same as numbers in true-magnitude form. Negative numbers are the bitwise complements of the corresponding positive number.

a. $57_{10} = 00111001$
b. $-57_{10} = 11000110$
c. $72_{10} = 01001000$
d. $-72_{10} = 10110111$

We can also generate an 8-bit 1's complement negative number by subtracting its positive magnitude from 11111111 (eight 1s). For example, for part b:

$$\begin{array}{r} 11111111 \\ -\underline{00111001} \ (\ \ 57_{10}) \\ 11000110 \ (-57_{10}) \end{array}$$

2's Complement Form

Positive numbers in 2's complement form are the same as in true-magnitude and 1's complement forms. We create a negative number by adding 1 to the 1's complement form of the number.

⫶⫶ EXAMPLE 6.7

Convert the following numbers to 8-bit 2's complement form:

a. 57_{10} b. -57_{10} c. 72_{10} d. -72_{10}

SOLUTION

a. $57 = 00111001$
b. $-57 = 11000110$ (1's complement)
$$\begin{array}{r} \underline{1} \\ 11000111 \end{array}$$ (2's complement)
c. $72 = 01001000$
d. $-72 = 10110111$ (1's complement)
$$\begin{array}{r} \underline{1} \\ 10111000 \end{array}$$ (2's complement)

A negative number in 2's complement form can be made positive by 2's complementing it again. Try it with the negative numbers in Example 6.7.

6.3 Signed Binary Arithmetic

> **KEY TERM**
>
> **Signed binary arithmetic**   Arithmetic operations performed using signed binary numbers.

Signed Addition

Signed addition is done in the same way as unsigned addition. The only difference is that both operands *must* have the same number of magnitude bits, and each has a sign bit.

■ EXAMPLE 6.8

Add $+30_{10}$ and $+75_{10}$. Write the operands and the sum as 8-bit signed binary numbers.

SOLUTION

$$
\begin{array}{r r}
+30 & 00011110 \\
+75 & +01001011 \\
\hline
+105 & 01101001
\end{array}
$$

     (Magnitude bits)
     (Sign bit)

Subtraction

The real advantage of complement notation becomes evident when we subtract signed binary numbers. In complement notation, we add a negative number instead of subtracting a positive number. We thus have only one kind of operation—addition—and can use the same circuitry for both addition and subtraction.

This idea does not work for true-magnitude numbers. In the complement forms, the magnitude bits change depending on the sign of the number. In true-magnitude form, the magnitude bits are the same regardless of the sign of the number.

Let us subtract $80_{10} - 65_{10} = 15_{10}$ using 1's complement and 2's complement addition. We will also show that the method of adding a negative number to perform subtraction is not valid for true-magnitude signed numbers.

1's Complement Method

> **KEY TERM**
>
> **End-around carry**   An operation in 1's complement subtraction where the carry bit resulting from a sum of two 1's complement numbers is added to that sum.

Add the 1's complement values of 80 and -65. If the sum results in a carry beyond the sign bit, perform an **end-around carry.** That is, add the carry to the sum.

$$
\begin{aligned}
80_{10} &= 01010000 \\
65_{10} &= 01000001 \\
-65_{10} &= 10111110 \quad \text{(1's complement)}
\end{aligned}
$$

$$
\begin{array}{r l}
80 & 01010000 \\
-65 & + \ 10111110 \\
\hline
& 1\ 00001110 \\
& \quad \longrightarrow 1 \quad \text{(End-around carry)} \\
\hline
+15 & \quad 00001111
\end{array}
$$

2's Complement Method

Add the 2's complement values of 80 and -65. If the sum results in a carry beyond the sign bit, discard it.

$$80_{10} = 01010000$$

$$
\begin{array}{rl}
65_{10} = & 01000001 \\
-65_{10} = & 10111110 \quad \text{(1's complement)} \\
+ & \underline{\qquad 1} \\
& 10111111 \quad \text{(2's complement)}
\end{array}
$$

$$
\begin{array}{rl}
80 & 01010000 \\
-65 & + \underline{10111111} \\
+15 & 1\ 00001111
\end{array}
$$

|_____ (Discard carry)

True-Magnitude Method

$$80_{10} = 01010000$$

$$
\begin{array}{rl}
65_{10} = & 01000001 \\
-65_{10} = & 11000001
\end{array}
$$

$$
\begin{array}{rl}
80 & 01010000 \\
-65 & + \underline{11000001} \\
? & 1\ 00010001
\end{array}
$$

If we perform an end-around carry, the result is $00010010 = 18_{10}$. If we discard the carry, the result is $00010001 = 17_{10}$. Neither answer is correct. Thus, adding a negative true-magnitude number is not equivalent to subtraction.

Negative Sum or Difference

All examples to this point have given positive-valued results. When a 2's complement addition or subtraction yields a negative sum or difference, we can't just read the magnitude from the result, since a 2's complement operation modifies the bits of a negative number. We must calculate the 2's complement of the sum or difference, which will give us the positive number that has the same magnitude. That is, $-(-x) = +x$.

III EXAMPLE 6.9

Subtract $65_{10} - 80_{10}$ in 2's complement form.

SOLUTION

$$
\begin{array}{rl}
65_{10} = & 01000001 \\
\\
80_{10} = & 01010000 \\
-80_{10} = & 10101111 \quad \text{(1's complement)} \\
+ & \underline{\qquad 1} \\
& 10110000 \quad \text{(2's complement)}
\end{array}
$$

$$
\begin{array}{rl}
65 & 01000001 \\
-80 & + \underline{10110000} \\
& 11110001
\end{array}
$$

Take the 2's complement of the difference to find the positive number with the same magnitude.

$$
\begin{array}{rl}
& 11110001 \\
& 00001110 \quad \text{(1's complement)} \\
+ & \underline{\qquad 1} \\
& 00001111 \quad \text{(2's complement)}
\end{array}
$$

$(-15) \leftarrow$

$(+15) \longrightarrow$

$00001111 = +15_{10}$. We generated this number by complementing 11110001. Thus, $11110001 = -15_{10}$.

Range of Signed Numbers

Table 6.1 4-bit 2's Complement Numbers

Decimal	2's Complement
+7	0111
+6	0110
+5	0101
+4	0100
+3	0011
+2	0010
+1	0001
0	0000
−1	1111
−2	1110
−3	1101
−4	1100
−5	1011
−6	1010
−7	1001
−8	1000

The largest positive number in 2's complement notation is a 0 followed by n 1s for a number with n magnitude bits. For instance, the largest positive 4-bit number is $0111 = +7_{10}$. The negative number with the largest magnitude is *not* the 2's complement of the largest positive number. We can find the largest negative number by extension of a sequence of 2's complement numbers.

The 2's complement form of -7_{10} is $1000 + 1 = 1001$. The positive and negative numbers with the next largest magnitudes are $0110 (= +6_{10})$ and $1010 (= -6_{10})$. If we continue this process, we will get the list of numbers in Table 6.1.

We have generated the 4-bit negative numbers from -1_{10} (1111) through -7_{10} (1001) by writing the 2's complement forms of the positive numbers 1 through 7. Notice that these numbers count down in binary sequence. The next 4-bit number in the sequence (which is the only binary number we have left) is 1000. By extension, $1000 = -8_{10}$. This number is its own 2's complement. (Try it.) It exemplifies a general rule for the n-bit negative number with the largest magnitude.

> **NOTE**
>
> A 2's complement number consisting of a 1 followed by n 0s is equal to -2^n. Therefore, the range of a signed number, x, is $-2^n \leq x \leq 2^n - 1$ for a number with n magnitude bits.

▌▌ EXAMPLE 6.10

Write the largest positive and negative numbers for an 8-bit signed number in decimal and 2's complement notation.

SOLUTION

$$01111111 = +127 \quad (\text{7 magnitude bits: } 2^7 - 1 = 127)$$
$$10000000 = -128 \quad (\text{1 followed by seven 0s: } -2^7 = -128)$$

▌▌ EXAMPLE 6.11

Write -16_{10}

a. As an 8-bit 2's complement number

b. As a 5-bit 2's complement number

(8-bit numbers are more common than 5-bit numbers in digital systems, but it is useful to see how we must write the same number differently with different numbers of bits.)

SOLUTION

a. An 8-bit number has 7 magnitude bits and 1 sign bit.

$$+16 = 00010000$$

$$
\begin{array}{rl}
-16 = & 11101111 \quad (\text{1's complement}) \\
+ & \underline{1} \\
& 11110000 \quad (\text{2's complement})
\end{array}
$$

b. A 5-bit number has 4 magnitude bits and 1 sign bit. Four magnitude bits are not enough to represent 16. However, a 1 followed by n 0s is equal to -2^n. For a 1 and four 0s, $-2^n = -2^4 = -16$. Thus, $10000 = -16_{10}$.

The last five bits of the binary equivalent of -16 are the same in both the 5-bit and 8-bit numbers.

NOTE

The 8-bit number is padded with leading 1s. This same general pattern applies for any negative number with a power-of-2 magnitude. ($-2^n = n$ 0s preceded by all 1s within the defined number size.)

▌▌ SECTION 6.3 REVIEW PROBLEM

6.5 Write -32 as an 8-bit 2's complement number.

6.6 Write -32 as a 6-bit 2's complement number.

Sign Bit Overflow

KEY TERM

Overflow   An erroneous carry into the sign bit of a signed binary number that results from a sum or difference larger than can be represented by the number of magnitude bits.

Signed addition of positive numbers is performed in the same way as unsigned addition. The only problem occurs when the number of bits in the sum of two numbers exceeds the number of magnitude bits and **overflows** into the sign bit. This causes the number to appear to be negative when it is not. For example, the sum $75 + 96 = 171$ causes an overflow in 8-bit signed addition. In unsigned addition the binary equivalent is:

$$\begin{array}{r} 1001011 \\ +\ 1100000 \\ \hline 10101011 \end{array}$$

In signed addition, the sum is the same, but has a different meaning.

$$\begin{array}{r} 0\ \ 1001011 \\ +\ 0\ \ 1100000 \\ \hline 1\ \ 0101011 \end{array}$$

(Sign bit) ⌐┘ └──────┘ (Magnitude bits)

The sign bit is 1, indicating a negative number, which cannot be true, since the sum of two positive numbers is always positive.

NOTE

A sum of positive signed binary numbers must not exceed $2^n - 1$ for numbers having n magnitude bits. Otherwise, there will be an overflow into the sign bit.

Overflow in Negative Sums

Overflow can also occur with large negative numbers. For example, the addition of -80_{10} and -65_{10} should produce the result:

$$-80_{10} + (-65_{10}) = -145_{10}$$

In 2's complement notation, we get:

$$+80_{10} = 01010000$$
$$-80_{10} = 10101111 \quad \text{(1's complement)}$$
$$+ \quad\quad\quad\quad 1$$
$$\overline{10110000} \quad \text{(2's complement)}$$

$$+65_{10} = 01000001$$
$$-65_{10} = 10111110 \quad \text{(1's complement)}$$
$$+ \quad\quad\quad\quad 1$$
$$\overline{10111111} \quad \text{(2's complement)}$$

$$
\begin{array}{rl}
-80 & 10110000 \\
+ (-65) & + \underline{10111111} \\
\hline
? & 1\ 01101111
\end{array}
$$

(Incorrect magnitude $= 111_{10}$)
(Erroneous sign bit $= 0$)
(Discard carry)

This result shows a positive sum of two negative numbers—clearly incorrect. We can extend the statement we made earlier about permissible magnitudes of sums to include negative as well as positive numbers.

> **NOTE**
>
> A sum of signed binary numbers must be within the range of $-2^n \leq \text{sum} \leq 2^n - 1$ for numbers having n magnitude bits. Otherwise, there will be an overflow into the sign bit.

For an 8-bit signed number in 2's complement form, the permissible range of sums is $10000000 \leq \text{sum} \leq 01111111$. In decimal, this range is $-128 \leq \text{sum} \leq +127$.

> **NOTE**
>
> A sum of two positive numbers is always positive. A sum of two negative numbers is always negative. Any 2's complement addition or subtraction operation that appears to contradict these rules has produced an overflow into the sign bit.

▐▌ EXAMPLE 6.12

Which of the following sums will produce a sign bit overflow in 8-bit 2's complement notation? How can you tell?

a. $67_{10} + 33_{10}$

b. $67_{10} + 63_{10}$

c. $-96_{10} - 22_{10}$

d. $-96_{10} - 42_{10}$

SOLUTION A sign bit overflow is generated if the sum of two positive numbers appears to produce a negative result or the sum of two negative numbers appears to produce a positive result. In other words, overflow occurs if the operand sign bits are both 1 and the sum sign bit is 0 or vice versa. We know this will happen if an 8-bit sum is outside the range ($-128 \leq \text{sum} \leq +127$).

a.
$$
\begin{array}{rl}
+67_{10} & 01000011 \quad \text{(no overflow;} \\
+33_{10} & \underline{00100001} \quad \text{sum of positive numbers} \\
\hline
100_{10} & 01100100 \quad \text{is positive.)}
\end{array}
$$

b.
$$\begin{array}{rl} +67_{10} & 01000011 \\ +63_{10} & 00111111 \\ \hline 130_{10} & 10000010 \end{array}$$
(Overflow; sum of positive numbers is negative. Sum $> +127$; out of range.)

c.
$$\begin{array}{rl} +96 = & 01100000 \\ -96 = & 10011111 \\ & +\underline{\qquad 1} \\ & 10100000 \end{array}$$
(1's complement)

(2's complement)

$$\begin{array}{rl} +22 = & 00010110 \\ -22 & 11101001 \\ & +\underline{\qquad 1} \\ & 11101010 \end{array}$$
(1's complement)

(2's complement)

$$\begin{array}{rl} -96 & 10100000 \\ -22 & 11101010 \\ \hline -118 & 1 \;\; 10001010 \end{array}$$

(Magnitude bits)
(Sign bit)
(Discard carry)

(No overflow; sum of two negative numbers is negative.)

d.
$$\begin{array}{rl} +96 = & 01100000 \\ -96 = & 10011111 \\ & +\underline{\qquad 1} \\ & 10100000 \end{array}$$
(1's complement)

(2's complement)

$$\begin{array}{rl} +42 = & 00101010 \\ -42 & 11010101 \\ & +\underline{\qquad 1} \\ & 11010110 \end{array}$$
(1's complement)

(2's complement)

$$\begin{array}{rl} -96 & 10100000 \\ -42 & 11010110 \\ \hline -138 & 1 \;\; 01110110 \end{array}$$

(Magnitude bits)
(Sign bit)
(Discard carry)

(Overflow; sum of two negative numbers is positive. Sum < -128; out of range.)

> **NOTE**
>
> The carry bit generated in 1's and 2's complement operations is not the same as an overflow bit. (See Example 6.12, parts c and d.) An overflow is a change in the sign bit, which leads us to believe that the number is opposite in sign from its true value. A carry is the result of an operation carrying beyond the physical limits of an n-bit number. It is similar to the idea of an odometer rolling over from 999999.9 to 1 000000.0. There are not enough places to hold the new number, so it goes back to the beginning and starts over.

6.4 Hexadecimal Arithmetic

(This section may be omitted without loss of continuity.)
The main reason to be familiar with addition and subtraction in the hexadecimal system is that it is useful for calculations related to microcomputer and memory systems.

Microcomputer systems often use binary numbers of 8, 16, 20, or 32 bits. Rather than write out all these bits, we use hex numbers as shorthand. Binary numbers having 8, 16, 20, or 32 bits can be represented by 2, 4, 5, or 8 hex digits, respectively.

Hex Addition

Hex addition is very much like decimal addition, except that we must remember how to deal with the hex digits A to F. A few sums are helpful:

$$F + 1 = 10$$
$$F + F = 1E$$
$$F + F + 1 = 1F$$

The positional multipliers for the hexadecimal system are powers of 16. Thus, the most significant bit of the first sum is the 16's column. The equivalent sum in decimal is:

$$15_{10} + 1_{10} = 16_{10} = 10H$$

The second sum is the largest possible sum of two hex digits; the carry to the next position is 1. This shows that the sum of two hex digits will never produce a carry larger than 1. The second sum can be calculated as follows:

$$FH + FH = 15_{10} + 15_{10}$$
$$= 30_{10}$$
$$= 16_{10} + 14_{10}$$
$$= 10H + EH$$
$$= 1EH$$

The third sum shows that if there is a carry from a previous sum, the carry to the next bit will still be 1.

NOTE

It is useful to think of any digits larger than 9 as their decimal equivalents. For any digit greater than 15_{10} (FH), subtract 16_{10}, convert the difference to its hex equivalent, and carry 1 to the next digit position.

▌▌ EXAMPLE 6.13

Add 6B3H + A9CH.

SOLUTION

Hex	Decimal Equivalents
6B3	(6) (11) (3)
+A9C	+ (10) (9) (12)
	(16) (20) (15)

For sums greater than 15, subtract 16 and carry 1 to the next position:

	Hex	Decimal Equivalents
(Carry) ——	11	(1) (1)
	6B3	(6) (11) (3)
+	A9C	+ (10) (9) (12)
	114F	(1) (1) (4) (15)

Sum: 6B3H + A9CH = 114FH.

▌▌

Hex Subtraction

There are two ways to subtract hex numbers. The first reverses the addition process in the previous section. The second is a complement form of subtraction.

III EXAMPLE 6.14

Subtract 6B3H − 49CH.

SOLUTION

Hex	Decimal Equivalent
6B3	(6) (11) (3)
− 49C	− (4) (9) (12)

To subtract the least significant digits, we must borrow 10H (16_{10}) from the previous position. This leaves the subtraction looking like this:

Hex	Decimal Equivalent
(Borrow)———— 1	
6A3	(6) (10) (16 + 3)
− 49C	− (4) (9) (12)
217	(2) (1) (7)

The second subtraction method is a complement method, where, as in 2's complement subtractions, we add a negative number to subtract a positive number.

Calculate the 15's complement of a hex number by subtracting it from a number having the same number of digits, all Fs. Calculate the 16's complement by adding 1 to this number. This is the negated value of the number.

III EXAMPLE 6.15

Negate the hex number 15AC by calculating its 16's complement.

SOLUTION

```
    FFFF
 −  15AC
    EA53    (15's complement)
 +     1
    EA54    (16's complement)
```

The original value, 15AC, can be restored by calculating the 16's complement of EA54. Try it.

III EXAMPLE 6.16

Subtract 8B63 − 55D7 using the complement method.

SOLUTION   Find the 16's complement of 55D7.

```
    FFFF
 −  55D7
    AA28    (15's complement)
 +     1
    AA29    (16's complement)
```

Therefore, $-55D7 = AA29$.

$$
\begin{array}{r}
1 \\
8B63 \\
+\ AA29 \\
\hline
1\ \ 358C
\end{array}
$$

(Discard ⌐‾‾‾⌐
carry)

Difference: $8B63 - 55D7 = 358C$.

▌▌ SECTION 6.4 REVIEW PROBLEM

6.7 Perform the following hexadecimal calculations:

    a. $A25F + 74A2$

    b. $7380 - 5FFF$

6.5 Numeric and Alphanumeric Codes

BCD Codes

> **KEY TERM**
>
> **Binary-coded decimal (BCD).**   A code that represents each digit of a decimal number by a binary value.

BCD stands for **binary-coded decimal.** As the name implies, BCD is a system of writing decimal numbers with binary digits. There is more than one way to do this, as BCD is a *code,* not a positional number system. That is, the various positions of the bits do not necessarily represent increasing powers of a specified number base.

Two commonly used BCD codes are 8421 code, where the bits for *each decimal digit* are weighted, and Excess-3 code, where each decimal digit is represented by a binary number that is 3 larger than the true binary value of the digit.

Table 6.2 Decimal Digits and Their 8421 BCD Equivalents

Decimal Digit	BCD (8421)
0	0000
1	0001
2	0010
3	0011
4	0100
5	0101
6	0110
7	0111
8	1000
9	1001

8421 Code

> **KEY TERM**
>
> **8421 code**   A BCD code that represents each digit of a decimal number by its 4-bittrue binary value.

The most straightforward BCD code is the **8421 code,** also called Natural BCD. Each decimal digit is represented by its 4-bit true binary value. When we talk about BCD code, this is usually what we mean.

This code is called 8421 because these are the positional weights of each digit. Table 6.2 shows the decimal digits and their BCD equivalents.

8421 BCD is not a positional number system, because each decimal digit is encoded separately as a 4-bit number.

▌▌ EXAMPLE 6.17

Write 4987_{10} in both binary and 8421 BCD.

SOLUTION   The binary value of 4987_{10} can be calculated by repeated division by 2:

$$4987_{10} = 1\ 0011\ 0111\ 1011_2$$

The BCD digits are the binary values of each decimal digit, encoded separately. We can break bits into groups of 4 for easier reading. Note that the first and last BCD digits each have a leading zero to make them 4 bits long.

$$4987_{10} = 0100\ 1001\ 1000\ 0111_{BCD}$$

Excess-3 Code

> **KEY TERMS**
>
> **Excess-3 Code** A BCD code that represents each digit of a decimal number by a binary number derived by adding 3 to its 4-bit true binary value.
>
> **9's complement** A way of writing decimal numbers where a number is made negative by subtracting each of its digits from 9 (e.g., $-726 = 999 - 726 = 273$ in 9's complement).
>
> **Self-complementing** A code that automatically generates a negative equivalent (e.g., 9's complement for a decimal code) when all its bits are inverted.

Table 6.3 Decimal Digits and Their 8421 and Excess-3 Equivalents

Decimal Digit	8421	Excess-3
0	0000	0011
1	0001	0100
2	0010	0101
3	0011	0110
4	0100	0111
5	0101	1000
6	0110	1001
7	0111	1010
8	1000	1011
9	1001	1100

Excess-3 code is a type of BCD code that is generated by adding 11_2 (3_{10}) to the 8421 BCD codes. Table 6.3 shows the Excess-3 codes and their 8421 and decimal equivalents.

The advantage of this code is that it is **self-complementing.** If the bits of the Excess-3 digit are inverted, they yield the **9's complement** of the decimal equivalent.

We can generate the 9's complement of an n-digit number by subtracting it from a number made up of n 9s. Thus, the 9's complement of 632 is $999 - 632 = 367$.

The Excess-3 equivalent of 632 is 1001 0110 0101. If we invert all the bits, we get 0110 1001 1010. The decimal equivalent of this Excess-3 number is 367, the 9's complement of 632.

This property is useful for performing decimal arithmetic digitally.

Gray Code

> **KEY TERM**
>
> **Gray code** A binary code that progresses such that only one bit changes between two successive codes.

Table 6.4 4-Bit Gray Code

Decimal	True Binary	Gray Code
0	0000	0000
1	0001	0001
2	0010	0011
3	0011	0010
4	0100	0110
5	0101	0111
6	0110	0101
7	0111	0100
8	1000	1100
9	1001	1101
10	1010	1111
11	1011	1110
12	1100	1010
13	1101	1011
14	1110	1001
15	1111	1000

Table 6.4 shows a 4-bit **Gray code** compared to decimal and binary values. Any two adjacent Gray codes differ by exactly one bit.

Gray code can be extended indefinitely if you understand the relationship between the binary and Gray digits. Let us name the binary digits $b_3 b_2 b_1 b_0$, with b_3 as the most significant bit, and the Gray code digits $g_3 g_2 g_1 g_0$ for a 4-bit code. For a 4-bit code:

$$g_3 = b_3$$
$$g_2 = b_3 \oplus b_2$$
$$g_1 = b_2 \oplus b_1$$
$$g_0 = b_1 \oplus b_0$$

For an n-bit code, the MSBs are the same in Gray and binary ($g_n = b_n$). The other Gray digits are generated by the Exclusive OR function of the binary digits in the same position and the next most significant position.

Another way to generate a Gray code sequence is to recognize the inherent symmetry in the code. For example, a 2-bit Gray code sequence is given by:

00
01
11
10

To generate a 3-bit Gray code, write the 2-bit sequence, then write it again in reverse order.

00
01
11
10
10
11
01
00

Add an MSB of 0 to the first four codes and an MSB of 1 to the last four codes. The sequence followed by the last two bits of all codes is symmetrical about the center of the sequence.

000
001
011
010
110
111
101
100

We can apply a similar process to generate a 4-bit Gray code. Write the 3-bit sequence, then again in reverse order. Add an MSB of 0 to the first half of the table and an MSB of 1 to the second half. This procedure yields the code in Table 6.4.

ASCII Code

KEY TERMS

Alphanumeric code A code used to represent letters of the alphabet and numerical characters.

ASCII American Standard Code for Information Interchange. A 7-bit code for representing alphanumeric and control characters.

Case shift Changing letters from capitals (uppercase) to small letters (lowercase) or vice versa.

Digital systems and computers could operate perfectly well using only binary numbers. However, if there is any need for a human operator to understand the input and output data of a digital system, it is necessary to have a system of communication that is understandable to both a human operator and the digital circuit.

A code that represents letters (alphabetic characters) and numbers (numeric characters) as binary numbers is called an **alphanumeric code.** The most commonly used alphanumeric code is **ASCII** ("askey"), which stands for American Standard Code for Information Interchange. ASCII code represents letters, numbers, and other "typewriter characters" in 7 bits. In addition, ASCII has a repertoire of "control characters," codes that are used to send control instructions to and from devices such as video display terminals, printers, and modems.

Table 6.5 shows the ASCII code in both binary and hexadecimal forms. The code for any character consists of the bits in the column heading, then those in the row heading. For example, the ASCII code for "A" is 1000001_2 or 41H. The code for "a" is 1100001_2 or 61H. The codes for capital (uppercase) and lower case letters differ only by the second most significant bit, for all letters. Thus, we can make an alphabetic **case shift,** like using the Shift key on a typewriter or computer keyboard, by switching just one bit.

Numeric characters are listed in column 3, with the least significant digit of the ASCII code being the same as the represented number value. For example, the numeric character "0" is equivalent to 30H in ASCII. The character "9" is represented as 39H.

The codes in columns 0 and 1 are control characters. They cannot be displayed on any kind of output device, such as a printer or video monitor, although they may be used to control the device. For instance, if the codes 0AH (Line Feed) and ODH (Carriage Return)

Table 6.5 ASCII Code

	MSBs							
	000 **(0)**	**001** **(1)**	**010** **(2)**	**011** **(3)**	**100** **(4)**	**101** **(5)**	**110** **(6)**	**111** **(7)**
LSBs								
0000 (0)	NUL	DLE	SP	0	@	P	`	p
0001 (1)	SOH	DC1	!	1	A	Q	a	q
0010 (2)	STX	DC2	"	2	B	R	b	r
0011 (3)	ETX	DC3	#	3	C	S	c	s
0100 (4)	EOT	DC4	$	4	D	T	d	t
0101 (5)	ENQ	NAK	%	5	E	U	e	u
0110 (6)	ACK	SYN	&	6	F	V	f	v
0111 (7)	BEL	ETB	'	7	G	W	g	w
1000 (8)	BS	CAN	(	8	H	X	h	x
1001 (9)	HT	EM	)	9	I	Y	i	y
1010 (A)	LF	SUB	*	:	J	Z	j	z
1011 (B)	VT	ESC	+	;	K	[	k	{
1100 (C)	FF	FS	,	<	L	\	l	|
1101 (D)	CR	GS	-	=	M	]	m	}
1110 (E)	SO	RS	.	>	N	^	n	~
1111 (F)	SI	US	/	?	O	—	o	DEL

Control Characters:

NUL–NUll	DLE–Data Link Escape
SOH–Start of Header	DC1–Device Control 1
STX–Start Text	DC2–Device Control 2
ETX–End Text	DC3–Device Control 3
EOT–End of Transmission	DC4–Device Control 4
ENQ–Enquiry	NAK–No Acknowledgment
ACK–Acknowledge	SYN–Synchronous Idle
BEL–Bell	ETB–End of Transmission Block
BS–Backspace	CAN–Cancel
HT–Horizontal Tabulation	EM–End of Medium
LF–Line Feed	SUB–Substitute
VT–Vertical Tabulation	ESC–Escape
FF–Form Feed	FS–Form Separator
CR–Carriage Return	GS–Group Separator
SO–Shift Out	RS–Record Separator
SI–Shift In	US–Unit Separator
SP–Space	DEL–Delete

are sent to a printer, the paper will advance by one line and the print head will return to the beginning of the line.

The displayable characters begin at 20H ("space") and continue to 7EH ("tilde"). Spaces are considered ASCII characters.

▐▐ EXAMPLE 6.17

Encode the following string of characters into ASCII (hexadecimal form). Do not include quotation marks.

"Total system cost: $4,000,000. @ 10%"

SOLUTION   Each character, including spaces, is represented by two hex digits as follows:

54	6F	74	61	6C	20	73	79	73	74	65	6D	20	63	6F	73	74	3A	20
T	o	t	a	l	SP	s	y	s	t	e	m	SP	c	o	s	t	:	SP

24	34	2C	30	30	30	2C	30	30	30	2E	20	40	20	31	30	25
$	4	,	0	0	0	,	0	0	0	.	SP	@	SP	1	0	%

▐▐

▐▐ SECTION 6.5 REVIEW PROBLEM

6.8 Decode the following sequence of hexadecimal ASCII codes.

54	72	75	65	20	6F	72	20	46	61	6C	73	65	3A	20	31

2F	34	20	3C	20	31	2F	32

6.6   Binary Adders and Subtractors

Half and Full Adders

KEY TERMS

Half adder   A circuit that will add two bits and produce a sum bit and a carry bit.

Full adder   A circuit that will add a carry bit from another full or half adder and two operand bits to produce a sum bit and a carry bit.

There are only three possible sums of two 1-bit binary numbers:

$$0 + 0 = 00$$
$$0 + 1 = 01$$
$$1 + 1 = 10$$

We can build a simple combinational logic circuit to produce the above sums. Let us designate the bits on the left side of the above equalities as inputs to the circuit and the bits on the right side as outputs. Let us call the LSB of the output the sum bit, symbolized by Σ, and the MSB of the output the carry bit, designated C_{OUT}.

Figure 6.1 shows the logic symbol of the circuit, which is called a **half adder.** Its truth table is given in Table 6.6. Since addition is subject to the commutative property, ($A + B = B + A$), the second and third lines of the truth table are the same.

The Boolean functions of the two outputs, derived from the truth table, are:

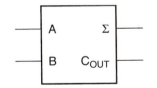

FIGURE 6.1
Half Adder

Table 6.6   Half Adder Truth Table

A	B	C_{OUT}	Σ
0	0	0	0
0	1	0	1
1	0	0	1
1	1	1	0

NOTE

$$C_{OUT} = AB$$
$$\Sigma = \overline{A}B + A\overline{B} = A \oplus B$$

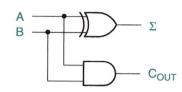

FIGURE 6.2
Half Adder Circuit

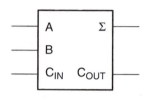

FIGURE 6.3
Full Adder

Table 6.7   Full Adder Truth Table

A	B	C_{IN}	C_{OUT}	Σ
0	0	0	0	0
0	0	1	0	1
0	1	0	0	1
0	1	1	1	0
1	0	0	0	1
1	0	1	1	0
1	1	0	1	0
1	1	1	1	1

The corresponding logic circuit is shown in Figure 6.2.

The half adder circuit cannot account for an *input* carry, that is, a carry from a lower-order 1-bit addition. A **full adder,** shown in Figure 6.3, can add two 1-bit numbers *and* accept a carry bit from a previous adder stage. Operation of the full adder is based on the following sums:

$$0 + 0 + 0 = 00$$
$$0 + 0 + 1 = 01$$
$$0 + 1 + 1 = 10$$
$$1 + 1 + 1 = 11$$

Designating the left side of the above equalities as circuit inputs A, B, and C_{IN} and the right side as outputs C_{OUT} and Σ, we can make the truth table in Table 6.7. (The second and third of the above sums each account for three lines in the full adder truth table.)

The unsimplified Boolean expressions for the outputs are:

$$C_{OUT} = \overline{A} B C_{IN} + A \overline{B} C_{IN} + A B \overline{C}_{IN} + A B C_{IN}$$
$$\Sigma = \overline{A} \overline{B} C_{IN} + \overline{A} B \overline{C}_{IN} + A \overline{B} \overline{C}_{IN} + A B C_{IN}$$

There are a couple of ways to simplify these expressions.

Karnaugh Map Method

Since we have expressions for Σ and C_{OUT} in sum-of-products form, let us try to use the Karnaugh maps in Figure 6.4 to simplify them. The expression for Σ doesn't reduce at all. The simplified expression for C_{OUT} is:

$$C_{OUT} = A B + A C_{IN} + B C_{IN}$$

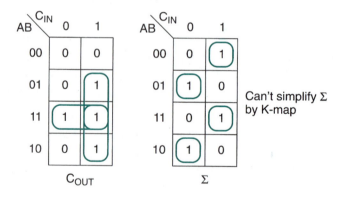

FIGURE 6.4
K-Maps for a Full Adder

The corresponding logic circuits for Σ and C_{OUT}, shown in Figure 6.5, don't give us much of a simplification.

Boolean Algebra Method

The simplest circuit for C_{OUT} and Σ involves the Exclusive OR function, which we cannot derive from K-map groupings. This can be shown by Boolean algebra, as follows:

$$C_{OUT} = \overline{A} B C_{IN} + A \overline{B} C_{IN} + A B \overline{C}_{IN} + A B C_{IN}$$
$$= (\overline{A} B + A \overline{B})C_{IN} + A B (\overline{C}_{IN} + C_{IN})$$
$$= (A \oplus B) C_{IN} + A B$$

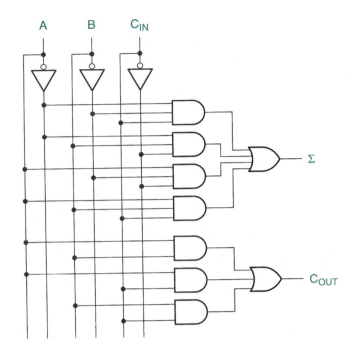

FIGURE 6.5
Full Adder from K-Map Simplification

$$
\begin{aligned}
\Sigma &= (\overline{A}\ \overline{B} + AB)\ C_{IN} + (\overline{A}\ B + A\ \overline{B})\ \overline{C}_{IN} \\
&= (\overline{A \oplus B})\ C_{IN} + (A \oplus B)\ \overline{C}_{IN} \qquad Let\ x = A \oplus B \\
&= \overline{x}\ C_{IN} + x\ \overline{C}_{IN} \\
&= x \oplus C_{IN} \\
&= (A \oplus B) \oplus C_{IN}
\end{aligned}
$$

The simplified expressions are as follows:

N O T E

$$
\begin{aligned}
C_{OUT} &= (A \oplus B)\ C_{IN} + A\ B \\
\Sigma &= (A \oplus B) \oplus C_{IN}
\end{aligned}
$$

Figure 6.6 shows the logic circuit derived from these equations. If you refer back to the half adder circuit in Figure 6.2, you will see that the full adder can be constructed from two half adders and an OR gate, as shown in Figure 6.7.

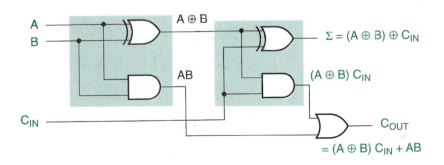

FIGURE 6.6
Full Adder from Logic Gates

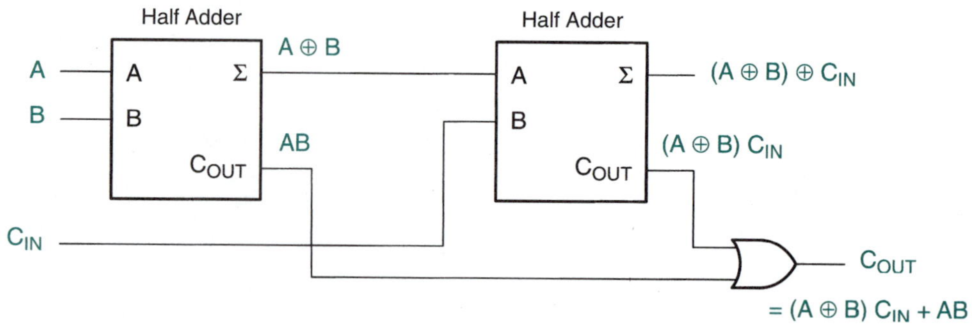

Half Adder   Half Adder

A ——[A   Σ]—— A ⊕ B ——[A   Σ]—— (A ⊕ B) ⊕ C_{IN}

B ——[B   ]   ——[B   ]

          C_{OUT} —— AB C_{OUT} —— (A ⊕ B) C_{IN}

C_{IN} ————————————————————

                                                    C_{OUT}
                                            = (A ⊕ B) C_{IN} + AB

FIGURE 6.7

Full Adder From Two Half Adders

▌▌ **EXAMPLE 6.18**

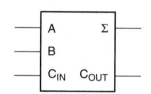

FIGURE 6.8

Example 6.18

Full Adder

Evaluate the Boolean expression for Σ and C_{OUT} of the full adder in Figure 6.8 for the following input values. What is the binary value of the outputs in each case?

a. $A = 0, B = 0, C_{IN} = 1$

b. $A = 1, B = 0, C_{IN} = 0$

c. $A = 1, B = 0, C_{IN} = 1$

d. $A = 1, B = 1, C_{IN} = 0$

SOLUTION   The output of a full adder for any set of inputs is simply given by $C_{OUT}\,\Sigma = A + B + C_{IN}$. For each of the stated sets of inputs:

a. $C_{OUT}\,\Sigma = A + B + C_{IN} = 0 + 0 + 1 = 01$

b. $C_{OUT}\,\Sigma = A + B + C_{IN} = 1 + 0 + 0 = 01$

c. $C_{OUT}\,\Sigma = A + B + C_{IN} = 1 + 0 + 1 = 10$

d. $C_{OUT}\,\Sigma = A + B + C_{IN} = 1 + 1 + 0 = 10$

We can verify each of these sums algebraically by plugging the specified inputs into the full adder Boolean equations:

$$C_{OUT} = (A \oplus B)\, C_{IN} + A\,B$$
$$\Sigma = (A \oplus B) \oplus C_{IN}$$

a. $C_{OUT} = (0 \oplus 0) \cdot 1 + 0 \cdot 0$

   $\qquad = 0 \cdot 1 + 0$

   $\qquad = 0 + 0 = 0$

   $\Sigma = (0 \oplus 0) \oplus 1$

   $\qquad = 0 \oplus 1 = 1$ (Binary equivalent: $C_{OUT}\,\Sigma = 01$)

b. $C_{OUT} = (1 \oplus 0) \cdot 0 + 1 \cdot 0$

   $\qquad = 1 \cdot 0 + 0$

   $\qquad = 0 + 0 = 0$

   $\Sigma = (1 \oplus 0) \oplus 0$

   $\qquad = 1 \oplus 0 = 1$ (Binary equivalent: $C_{OUT}\,\Sigma = 01$)

c. $C_{OUT} = (1 \oplus 0) \cdot 1 + 1 \cdot 0$

   $\qquad = 1 \cdot 1 + 0$

   $\qquad = 1 + 0 = 1$

   $\Sigma = (1 \oplus 0) \oplus 1$

   $\qquad = 1 \oplus 1 = 0$ (Binary equivalent: $C_{OUT}\,\Sigma = 10$)

d. $C_{OUT} = (1 \oplus 1) \cdot 0 + 1 \cdot 1$
$$= 0 \cdot 0 + 1$$
$$= 0 + 1 = 1$$

$\Sigma = (1 \oplus 1) \oplus 0$
$$= 0 \oplus 0 = 0 \qquad \text{(Binary equivalent: } C_{OUT}\Sigma = 10)$$

In each case, the binary equivalent is the same as the number of HIGH inputs, regardless of which inputs they are.

▌▎ EXAMPLE 6.19

Combine a half adder and a full adder to make a circuit that will add two 2-bit numbers. Check that the circuit will work by adding the following numbers and writing the binary equivalents of the inputs and outputs:

a. $A_2 A_1 = 01$, $B_2 B_1 = 01$
b. $A_2 A_1 = 11$, $B_2 B_1 = 10$

SOLUTION The 2-bit adder is shown in Figure 6.9. The half adder combines A_1 and B_1; A_2, B_2, and C_1 are added in the full adder. The carry output, C_1, of the half adder is connected to the carry input of the full adder. (A half adder can be used only in the LSB of a multiple-bit addition.)

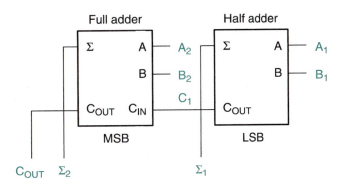

FIGURE 6.9
Example 6.19
2-Bit Adder

Sums:

a. $01 + 01 = 010$

$\quad A_1 = 1, B_1 = 1 \qquad\qquad C_1 = 1, \Sigma_1 = 0$

$\quad A_2 = 0, B_2 = 0, C_1 = 1 \quad C_2 = 0, \Sigma_2 = 1$

(Binary equivalent: $A_2 A_1 + B_2 B_1 = C_2 \Sigma_2 \Sigma_1 = 010$)

b. $11 + 10 = 101$

$\quad A_1 = 1, B_1 = 0 \qquad\qquad C_1 = 0, \Sigma_1 = 1$

$\quad A_2 = 1, B_2 = 1, C_1 = 0 \quad C_2 = 1, \Sigma_2 = 0$

(Binary equivalent: $A_2 A_1 + B_2 B_1 = C_2 \Sigma_2 \Sigma_1 = 101$)

▌▎

Parallel Binary Adder/Subtractor

KEY TERMS

Parallel binary adder A circuit, consisting of *n* full adders, that will add two *n*-bit binary numbers. The output consists of *n* sum bits and a carry bit.

Ripple carry A method of passing carry bits from one stage of a parallel adder to the next by connecting C_{OUT} of one full adder to C_{IN} of the following stage.

Cascade To connect an output of one device to an input of another, often for the purpose of expanding the number of bits available for a particular function.

As Example 6.19 implies, a binary adder can be expanded to any number of bits by using a full adder for each bit addition and connecting their carry inputs and outputs in **cascade.** Figure 6.10 shows four full adders connected as a 4-bit **parallel binary adder.**

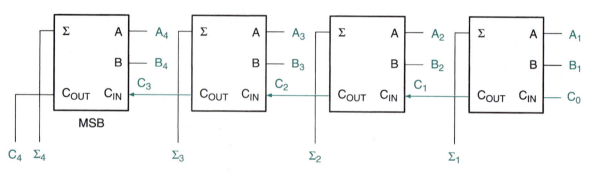

FIGURE 6.10
4-Bit Parallel Binary Adder

The first stage (LSB) can be either a full adder with its carry input forced to logic 0 or a half adder, since there is no previous stage to provide a carry. The addition is done one bit at a time, with the carry from each adder propagating to the next stage.

EXAMPLE 6.20

Verify the summing operation of the circuit in Figure 6.10 by calculating the output for the following sets of inputs:

a. $A_4 A_3 A_2 A_1 = 0101$, $B_4 B_3 B_2 B_1 = 1001$
b. $A_4 A_3 A_2 A_1 = 1111$, $B_4 B_3 B_2 B_1 = 0001$

SOLUTION At each stage, $A + B + C_{IN} = C_{OUT} \Sigma$.

a. $0101 + 1001 = 1110$
$(5_{10} + 9_{10} = 14_{10})$
$A_1 = 1, B_1 = 1, C_0 = 0; C_1 = 1, \Sigma_1 = 0$
$A_2 = 0, B_2 = 0, C_1 = 1; C_2 = 0, \Sigma_2 = 1$
$A_3 = 1, B_3 = 0, C_2 = 0; C_3 = 0, \Sigma_3 = 1$
$A_4 = 0, B_4 = 1, C_3 = 0; C_4 = 0, \Sigma_4 = 1$

(Binary equivalent: $C_4 \Sigma_4 \Sigma_3 \Sigma_2 \Sigma_1 = 01110$)

b. $1111 + 0001 = 10000$
   $(15_{10} + 1_{10} = 16_{10})$
   $A_1 = 1, B_1 = 1, C_0 = 0; C_1 = 1, \Sigma_1 = 0$
   $A_2 = 1, B_2 = 0, C_1 = 1; C_2 = 1, \Sigma_2 = 0$
   $A_3 = 1, B_3 = 0, C_2 = 1; C_3 = 1, \Sigma_3 = 0$
   $A_4 = 1, B_4 = 0, C_3 = 1; C_4 = 1, \Sigma_4 = 0$

   (Binary equivalent: $C_4 \Sigma_4 \Sigma_3 \Sigma_2 \Sigma_1 = 10000$)

The internal carries in the parallel binary adder in Figure 6.10 are achieved by a system called **ripple carry.** The carry output of one full adder cascades directly to the carry input of the next. Every time a carry bit changes, it "ripples" through some or all of the following stages. A sum is not complete until the carry from another stage has arrived. The equivalent circuit of a 4-bit ripple carry is shown in Figure 6.11.

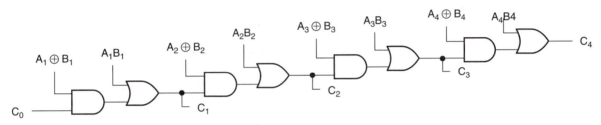

FIGURE 6.11
4-bit Ripple Carry Chain

A potential problem with this design is that the adder circuitry does not switch instantaneously. A carry propagating through a ripple adder adds delays to the summation time and, more importantly, can introduce unwanted intermediate states.

Examine the sum ($1111 + 0001 = 10000$). For a parallel adder having a ripple carry, the output goes through the following series of changes as the carry bit propagates through the circuit:

$$C_4 \Sigma_4 \Sigma_3 \Sigma_2 \Sigma_1 = 01111$$
$$01110$$
$$01100$$
$$01000$$
$$10000$$

If the output of the full adder is being used to drive another circuit, these unwanted intermediate states may cause erroneous operation of the load circuit.

Fast Carry

> ### KEY TERM
>
> **Fast carry (or look-ahead carry)**   A gate network that generates a carry bit directly from all incoming operand bits, independent of the operation of each full adder stage.

An alternative carry circuit is called **fast carry** or **look-ahead carry.** The idea behind fast carry is that the circuit will examine all the A and B bits simultaneously and produce an output carry that uses fewer levels of gating than a ripple carry circuit. Also, since there is

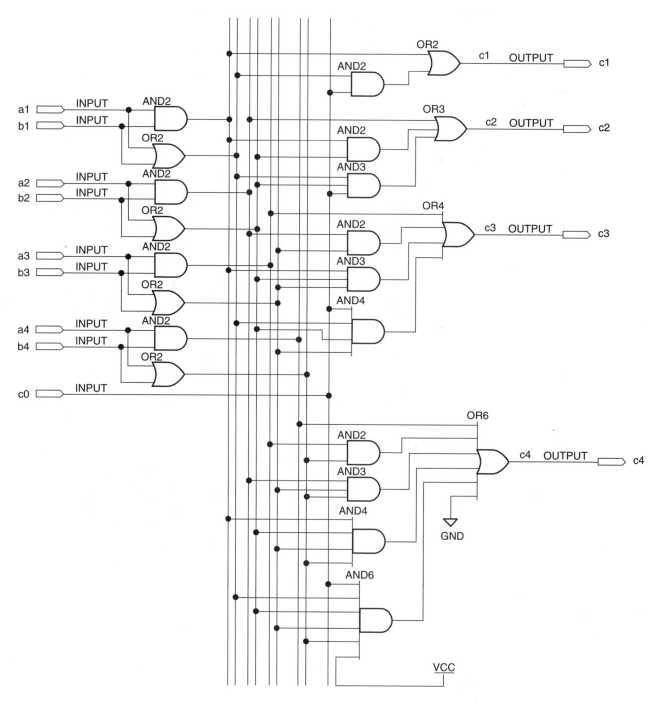

FIGURE 6.12
4-bit Fast Carry Circuit

a carry bit gate network for each internal stage, the propagation delay is the same for each full adder, regardless of the input operands.

The algebraic relation between operand bits and fast carry output is presented below, without proof. It can be developed from the fast carry circuit of Figure 6.12 by tracing the logic of the gates in the circuit.

$$C_4 = A_4 B_4 + A_3 B_3 (A_4 + B_4) + A_2 B_2 (A_4 + B_4)(A_3 + B_3)$$
$$+ A_1 B_1 (A_4 + B_4)(A_3 + B_3)(A_2 + B_2)$$
$$+ C_0 (A_4 + B_4)(A_3 + B_3)(A_2 + B_2)(A_1 + B_1)$$

We can make some intuitive sense of the above expression by examining it a term at a time. The first term says if the MSBs of both operands are 1, there will be a carry (e.g., 1000 + 1000 = 10000; carry generated).

The second term says if both second bits are 1 AND at least one MSB is 1, there will be a carry (e.g., 0100 + 1100 = 10000, or 1100 + 1100 = 11000; carry generated in either case). This pattern can be followed logically through all the terms.

The internal carry bits are generated by similar circuits that drive the carry input of each full adder stage in the parallel adder. In general, we can generate each internal carry by expanding the following expression:

$$C_n = A_n B_n + C_{n-1} (A_n + B_n)$$

The algebraic expressions for the remaining carry bits are:

$$C_1 = A_1 B_1 + C_0 (A_1 + B_1)$$
$$C_2 = A_2 B_2 + A_1 B_1 (A_2 + B_2) + C_0 (A_2 + B_2)(A_1 + B_1)$$
$$C_3 = A_3 B_3 + A_2 B_2 (A_3 + B_3) + A_1 B_1 (A_3 + B_3)(A_2 + B_2)$$
$$+ C_0 (A_3 + B_3)(A_2 + B_2)(A_1 + B_1)$$

▌▌▌ SECTION 6.6A REVIEW PROBLEM

6.9 Refer to the logic diagrams for the ripple carry and fast carry circuits (Figures 6.11 and 6.12). How many gates must a carry bit propagate through in each device if the effect of the carry input ripples through to the Σ_4 bit? (See Figure 6.32 on page 273 and Figure 6.33 on page 273.)

Using VHDL Components to Implement a Parallel Adder

KEY TERMS

Hierarchy   A group of design entities associated in a series of levels or layers in which complete designs form portions of another, more general design entity. The more general design is considered to be the higher level of the hierarchy.

Component   A complete VHDL design entity that can be used as a part of a higher-level file in a hierarchical design.

Port   An input or output of a VHDL design entity or component.

Component declaration statement   A statement that defines the input and output port names of a component used in a VHDL design entity.

Instantiate   To use an instance of a component.

Component instantiation statement   A statement that maps port names of a VHDL component to the port names, internal signals, or variables of a higher-level VHDL design entity.

VHDL designs can be created using a **hierarchy** of design entities. Certain functions, such as full adders, decoders, and so on, can be created once and used in many designs or multiple times in a single design.

We can create a parallel adder in VHDL by using multiple instances of a full adder **component** in the top-level file of a VHDL design hierarchy. Figure 6.13 shows a graphical illustration of this concept. Each full adder shown is an instance of a component written in VHDL, as shown in the following.

```
-- full_add.vhd
-- Full adder: adds two bits, a and b, plus input carry
-- to yield sum bit and output carry.
```

Full_add.vhd

```
ENTITY full_add IS
PORT (
      a, b, c_in : IN  BIT;
      c_out, sum : OUT  BIT);
END full_add;

ARCHITECTURE adder OF full_add IS
BEGIN
      c_out <= ((a xor b) and c_in) or (a and b) ;
      sum   <= (a xor b) xor c_in;
END adder;
```

FIGURE 6.13

4-bit Parallel Adder with Ripple Carry

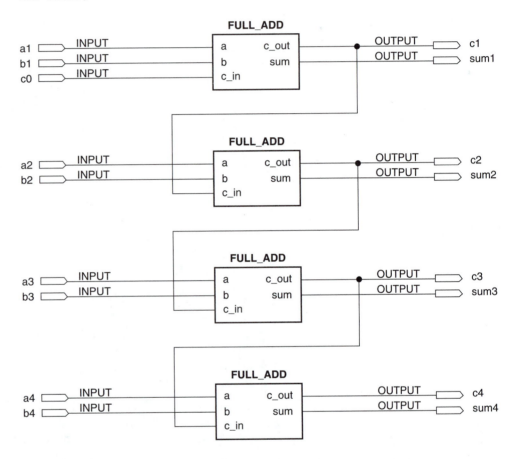

We can create the same design as in Figure 6.13 using VHDL only. To make this hierarchical design we require:

1. A separate component file for a full adder **(full_add.vhd),** saved in a folder where the compiler can find it (i.e., on a library path)

2. A **component declaration statement** in the top-level file of the design hierarchy

3. A **component instantiation statement** for each instance of the full adder component

The general form of a design entity using components is:

```
ENTITY entity_name IS
    PORT ( input and output definitions);
END entity_name;

ARCHITECTURE arch_name OF entity_name IS
    component declaration(s);
    signal declaration(s);
```

```
BEGIN
    Component instantiation(s);
    Other statements;
END arch_name;
```

The VHDL file for a 4-bit parallel adder using full adder components is shown next.

add4par.vhd

```
-- add4par.vhd
-- 4-bit parallel adder, using 4 instances
-- of the component full_add

ENTITY add4par IS
    PORT(
        c0      : IN   BIT;
        a, b    : IN   BIT_VECTOR (4 downto 1);
        c4      : OUT  BIT;
        sum     : OUT  BIT_VECTOR (4 downto 1));
END add4par;

ARCHITECTURE adder OF add4par IS

    -- Component declaration
    COMPONENT full_add
        PORT (
            a, b, c_in : IN  BIT;
            c_out, sum : OUT BIT);
    END COMPONENT;

    -- Define a signal for internal carry bits
    SIGNAL c : BIT_VECTOR (3 downto 1);

BEGIN
    -- Four Component Instantiation Statements
    adder1: full_add
        PORT MAP ( a     => a(1),
                   b     => b(1),
                   c_in  => c0,
                   c_out => c(1),
                   sum   => sum (1));
    adder2: full_add
        PORT MAP ( a     => a(2),
                   b     => b(2),
                   c_in  => c(1),
                   c_out => c(2),
                   sum   => sum (2));
    adder3: full_add
        PORT MAP ( a     => a(3),
                   b     => b(3),
                   c_in  => c(2),
                   c_out => c(3),
                   sum   => sum (3));
    adder4: full_add
        PORT MAP ( a     => a(4),
                   b     => b(4),
                   c_in  => c(3),
                   c_out => c4,
                   sum   => sum (4));
END adder;
```

The component declaration statement defines the ports of the component with the same names as in the **full_add.vhd.** Note that the form of the component declaration statement is almost the same as that of the component's entity declaration. In effect, we are redefining the component entity in the top-level file of the design hierarchy.

The component instantiation statement is of the following form:

```
__instance_name: __component_name
   GENERIC MAP (__parameter_name => __parameter_value ,
                __parameter_name => __parameter_value)
   PORT MAP (__component_port => __connect_port,
             __component_port => __connect_port);
```

In the generic map, a generalized parameter name can be mapped to a specific value when the component is instantiated. For example, a parameter name can be given a value that specifies the number of component output bits. We will not use this feature in our present examples.

In the port map, component ports are the names of the ports used in the component file and connect ports are the names of the ports, variables, or signals used in the higher-level design entity. For example, the component ports of the full adder component are **a, b, c in, c_out,** and **sum.** The connect ports for the instance **adder1** are **a(1), b(1), c0, c(1),** and **sum(1).** The ripple carry from **adder1** to **adder2** is achieved by mapping the port **c_in** of **adder2** to **c(1),** which is also mapped to the port **c_out** of **adder1.**

We can write the component instantiation statements more efficiently if we decide to use all ports of the component in the order they are defined. In this case, we can simply list the connect ports in the port map in the correct order, as follows:

```
adder1: full_add PORT MAP (a(1),b(1),c0,  c(1),sum(1));
adder2: full_add PORT MAP (a(2),b(2),c(1),c(2),sum(2));
adder3: full_add PORT MAP (a(3),b(3),c(2),c(3),sum(3));
adder4: full_add PORT MAP (a(4),b(4),c(3),c4,  sum(4));
```

If we only wish to use some of the component ports or use them in a different order than the order in which theywere originally defined, we must use the previous form of port map (i.e., a => a(1), etc.).

GENERATE Statements

> **KEY TERM**
>
> **GENERATE statement**   A VHDL construct that is used to create repetitive portions of hardware.

The four component instantiation statements shown previously can be written in a more general form:

```
adder(i): full_add PORT MAP (a(i), b(i), c(i-1), c(i), sum(i));
```

A statement that can be written in this indexed form can be implemented using a **GENERATE statement,** which has the form:

```
label:
FOR index IN range GENERATE
   statements;
END GENERATE;
```

The VHDL code that follows shows how to use the statement to create a 4-bit adder.

```
-- add4gen.vhd
-- 4-bit parallel adder, using a generate statement
-- and components
```

add4gen.vhd

```
ENTITY add4gen IS
   PORT (
        c0      : IN   BIT;
        a, b    : IN   BIT_VECTOR (4 downto 1);
        c4      : OUT  BIT;
        sum     : OUT  BIT_VECTOR (4 downto 1));
END add4gen;

ARCHITECTURE adder OF add4gen IS

   --Component declaration
   COMPONENT full_add
      PORT (
           a, b, c_in : IN   BIT;
           c_out, sum : OUT  BIT);
   END COMPONENT;

   -- Define a signal for internal carry bits
   SIGNAL c : BIT_VECTOR (4 downto 0);

BEGIN
   c(0)    <=  c0;
   adders:
   FOR i IN 1 to 4 GENERATE
      adder: full_add PORT MAP (a(i),b(i),c(i-1),c(i),sum(i));
   END GENERATE;

   c4  <= c(4);
END adder;
```

The GENERATE statement will create hardware that corresponds to the range of the index variable, i. In this case i goes from 1 to 4, so the statement instantiates four instances of the full adder. Since we have an input carry, an output carry and three internal carries, we must use a 5-bit signal (BIT_VECTOR (4 downto 0)) if we are to include all carry bits in indexed form. The input carry, c0, defined in the entity declaration, is assigned to the vector element c(0). Similarly, the output, c4, is assigned the value of the element c(4).

It is easy to expand the adder width by changing the range of the FOR GENERATE statement. For example, to make an 8-bit adder, we change the vectors to have a width of eight bits. The required VHDL code, shown next, requires the same number of lines of code as the 4-bit adder.

```
-- add8gen.vhd
-- 8-bit parallel adder, using a generate statement
-- and components

ENTITY add8gen IS
   PORT (
        C0      : IN   BIT;
        a, b    : IN   BIT_VECTOR (8 downto 1);
        c8      : OUT  BIT;
        sum     : OUT  BIT_VECTOR (8 downto 1));
END add8gen;

ARCHITECTURE adder OF add8gen IS
   -- Component declaration
   COMPONENT full_add
      PORT (
           a, b, c_in : IN   BIT;
           c_out, sum : OUT  BIT);
```

add8gen.vhd

```
END COMPONENT;
-- Define a signal for internal carry bits
SIGNAL c : BIT_VECTOR (8 downto 0);
BEGIN
    c(0)   <=  c0;
    adders:
    FOR i IN 1 to 8 GENERATE
        adder: full_add PORT MAP (a(i), b(i), c(i-1), c(i),
sum(i));
    END GENERATE;
    c8  <=  c(8);
END adder;
```

2's Complement Subtractor

Recall the technique for subtracting binary numbers in 2's complement notation. For example, to find the difference $0101 - 0011$ by 2's complement subtraction:

1. Find the 2's complement of 0011:

$$
\begin{array}{ll}
0011 & \\
1100 & \text{(1's complement)} \\
\underline{+1} & \\
1101 & \text{(2's complement)}
\end{array}
$$

2. Add the 2's complement of the subtrahend to the minuend:

$$
\begin{array}{ll}
0101 & (+5) \\
+\ \underline{1101} & (-3) \\
1\ 0010 & (+2)
\end{array}
$$

(Discard carry) ⌐⌡

We can easily build a circuit to perform 2's complement subtraction, using a parallel binary adder and an inverter for each bit of one of the operands. The circuit shown in Figure 6.14 performs the operation $(A - B)$.

FIGURE 6.14

2's Complement Subtractor

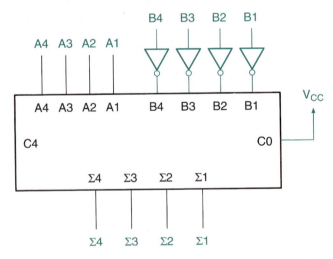

The four inverters generate the 1's complement of B. The parallel adder generates the 2's complement by adding the carry bit (held at logic 1) to the 1's complement at the B inputs. Algebraically, this is expressed as:

$$A - B = A + (-B) = A + \overline{B} + 1$$

where $\overline{B}$ is the 1's complement of B, and $(\overline{B} + 1)$ is the 2's complement of B.

EXAMPLE 6.21   Verify the operation of the 2's complement subtractor in Figure 6.14 by subtracting:

a. $1001 - 0011$ (unsigned)

b. $0100 - 0111$ (signed)

SOLUTION   Let $\overline{B}$ be the 1's complement of B.

a. Inverter inputs (B):   0011

    Inverter outputs ($\overline{B}$):   1100

    Sum ($A + \overline{B} + 1$):   1001   (9)

                                         1100   + (−3)

                           +   1

                          1 0110   (6)

    (Discard carry) ——————⌡

b. Inverter inputs (B):   0111

    Inverter outputs ($\overline{B}$):   1000

    Sum ($A + \overline{B} + 1$):   0100   (+4)

                                        1000   + (−7)

                           +   1

    Negative result: 1101   (−3) ←

    1's complement of 1101:   0010

                              +   1

    2's complement of 1101:   0011   (+3)

Parallel Binary Adder/Subtractor

Figure 6.15 shows a parallel binary adder configured as a programmable adder/subtractor. The Exclusive OR gates work as programmable inverters to pass B to the parallel adder in either true or complement form, as shown in Figure 6.16.

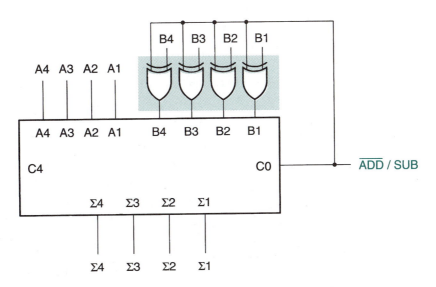

FIGURE 6.15
2's Complement Adder/Subtractor

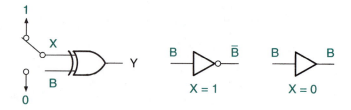

FIGURE 6.16
XOR as a Programmable Inverter

The $\overline{ADD/SUB}$ input is tied to the XOR inverter/buffers and to the carry input of the parallel adder. When $\overline{ADD/SUB} = 1$, B is complemented and the 1 from the carry input is added to the complement sum. The effect is to subtract $(A - B)$. When $\overline{ADD/SUB} = 0$, the B inputs are presented to the adder in true form and the carry input is 0. This produces an output equivalent to $(A + B)$.

This circuit can add or subtract 4-bit signed or unsigned binary numbers.

▌▌ EXAMPLE 6.22

Write a VHDL file to implement the 4-bit adder/subtractor shown in Figure 6.15. Also create a simulation file to test a representative selection of addition and subtraction operations.

SOLUTION The VHDL file is as follows:

addsub4g.vhd

```
-- addsub4g.vhd
ENTITY addsub4g IS
    PORT (
          sub    : IN    BIT;
          a, b   : IN    BIT_VECTOR (4 downto 1);
          c4     : OUT   BIT;
          sum    : OUT   BIT_VECTOR (4 downto 1));
END addsub4g;

ARCHITECTURE adder OF addsub4g IS
    -- Component declaration
    COMPONENT full_add
        PORT (
            a, b, c_in : IN    BIT;
            c_out, sum : OUT   BIT);
    END COMPONENT;
    -- Define a signal for internal carry bits
    SIGNAL c      : BIT_VECTOR (4 downto 0);
    SIGNAL b_comp : BIT_VECTOR (4 downto 1);
BEGIN
    -- add/subtract select to carry input (sub=1 for subtract)
    c(0)    <= sub;
    adders:
    FOR i IN 1 to 4 GENERATE
        -- invert b for subtract (b(i) xor 1),
        -- do not invert for add (b(i) xor 0)
        b_comp(i) <= b(i) xor sub;
        adder: full_add PORT MAP (a(i), b_comp(i), c(i-1), c(i),
sum(i));
    END GENERATE;
    c4   <= c(4);
END adder;
```

The VHDL code for the adder/subtractor is the same as that for the 4-bit adder created using a GENERATE statement, except that there is an input to select the add or subtract function. This input (**sub**) is tied to **c(0)** and to a set of XOR functions that invert **b** for subtraction. Input **b** is transferred through the XOR functions without inversion for the add function.

FIGURE 6.17

Example 6.21 Simulation of a 4-bit Adder/Subtractor

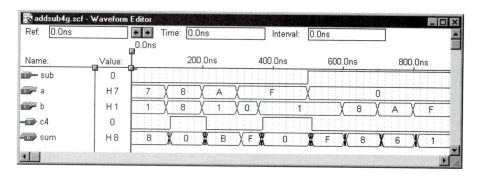

addSub4g.scf

Figure 6.17 shows the simulation for the adder/subtractor. Table 6.8 shows the operations included in the simulation in both hexadecimal and binary form. Note that the sums are interpreted as unsigned operations and the differences are interpreted as signed operations. Any sum or difference can be interpreted either way, but this will sometimes result in a sign bit overflow. (e.g., the sums $8 + 8 = 10$ and $F + 1 = 10$ both indicate an overflow if they are interpreted as signed additions.)

Table 6.8 Add/Subtract Results

Hexadecimal Sum/Difference	Binary Equivalent
$7 + 1 = 8$	$0111 + 0001 = 0\ 1000$ (Unsigned)
$8 + 8 = 10$	$1000 + 1000 = 1\ 0000$ (Unsigned)
$A + 1 = B$	$1010 + 0001 = 0\ 1011$ (Unsigned)
$F + 0 = F$	$1111 + 0000 = 0\ 1111$ (Unsigned)
$F + 1 = 10$	$1111 + 0001 = 1\ 0000$ (Unsigned)
$0 - 1 = F$	$0000 - 0001 = 1111$ (Signed: -1)
$0 - 8 = 8$	$0000 - 1000 = 1000$ (Signed: -8)
$0 - A = 6$	$0000 - 1010 = 0110$ (Signed: $0 - (-6) = +6$)
$0 - F = 1$	$0000 - 1111 = 0001$ (Signed: $0 - (-1) = +1$)

▌▌ EXAMPLE 6.23

Note that the simulation in Figure 6.17 shows some intermediate states on the **sum** waveform in between steady state values. Examine the transition from the sum $F + 0 = F$ to the sum $F + 1 = 10$ by using the Zoom In function in the Simulator window. Briefly explain how the intermediate states arise in this transition.

SOLUTION Figures 6.18 and 6.19 show the transition from $F + 0 = F$ to $F + 1 = 10$. The transition on the **sum** waveform is from F to E to 0 or in binary from 1111 to 1110 to

FIGURE 6.18

Example 6.22 Interval from F to E

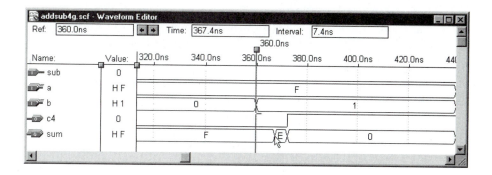

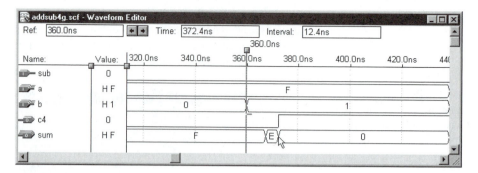

FIGURE 6.19
Example 6.22
Interval from F to 0

0000. This transition is the result of a change from 0 to 1 on the **b1** input of the adder/subtractor.

Figure 6.18 shows the interval from F to E (the time difference between the vertical line marking 36 ns and the arrow cursor, shown in the box labeled **Interval**) as 7.4 ns. This is the delay from **b1** to **sum1.**

Figure 6.19 shows the interval from F to 0 on the **sum** waveform, given as 12.6 ns. This interval represents the time required for **sum2, sum3,** and **sum4** to change after a change on **b1.**

Overflow Detection

We will examine two methods for detecting overflow in a binary adder/subtractor: one that requires access to the sign bits of the operands and result and another that requires access to the internal carry bits of the circuit.

Recall from Example 6.12 the condition for detecting a sign bit overflow in a sum of two binary numbers.

> **NOTE**
>
> If the sign bits of both operands are the same and the sign bit of the sum is different from the operand sign bits, an overflow has occurred.

This implies that overflow is not possible if the sign bits of the operands are different from each other. This is true because the sum of two opposite-sign numbers will always be smaller in magnitude than the larger of the two operands.

Here are two examples:

1. $(+15) + (-7) = (+8)$; $+8$ has a smaller magnitude than $+15$.
2. $(-13) + (+9) = (-4)$; -4 has a smaller magnitude than -13.

No carry into the sign bit will be generated in either case.

An 8-bit parallel binary adder will add two signed binary numbers as follows:

$$\begin{array}{ll} S_A\, A_7\, A_6\, A_5\, A_4\, A_3\, A_2\, A_1 & (S_A = \text{Sign bit of } A) \\ \underline{S_B\, B_7\, B_6\, B_5\, B_4\, B_3\, B_2\, B_1} & (S_B = \text{Sign bit of } B) \\ S_\Sigma\, \Sigma_7\, \Sigma_6\, \Sigma_5\, \Sigma_4\, \Sigma_3\, \Sigma_2\, \Sigma_1 & (S_\Sigma = \text{Sign bit of sum}) \end{array}$$

From our condition for overflow detection, we can make a truth table for an overflow variable, V, in terms of S_A, S_B, and S_Σ. Let us specify that $V = 1$ when there is an overflow condition. This condition occurs when $(S_A = S_B) \neq S_\Sigma$. Table 6.9 shows the truth table for the overflow detector function.

Table 6.9 Overflow Detector Truth Table

S_A	S_B	S_Σ	V
0	0	0	0
0	0	1	1
0	1	0	0
0	1	1	0
1	0	0	0
1	0	1	0
1	1	0	1
1	1	1	0

The SOP Boolean expression for the overflow detector is:

$$V = S_A S_B \overline{S_\Sigma} + \overline{S_A} \overline{S_B} S_\Sigma$$

Figure 6.20 shows a logic circuit that will detect a sign bit overflow in a parallel binary adder. The inputs S_A, S_B, and S_Σ are the MSBs of the adder A and B inputs and Σ outputs, respectively.

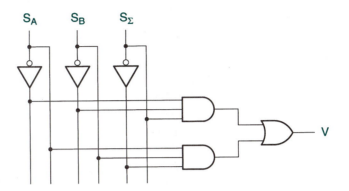

FIGURE 6.20
Overflow Detector

▌▌ EXAMPLE 6.24

Combine two instances of the 4-bit counter shown in Figure 6.15 and other logic to make an 8-bit adder/subtractor that includes a circuit to detect sign bit overflow.

SOLUTION Figure 6.21 represents the 8-bit adder/subtractor with an overflow detector of the type shown in Figure 6.20.

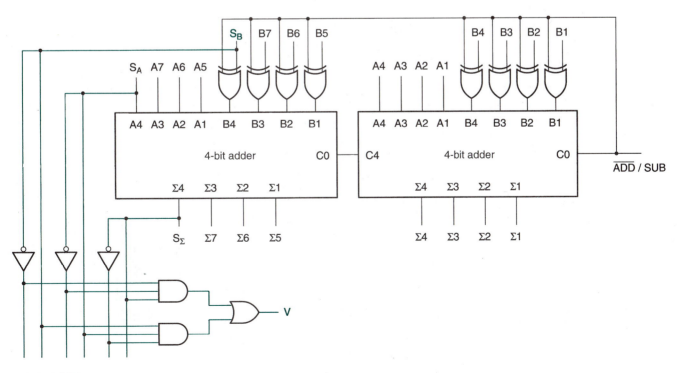

FIGURE 6.21
Example 6.24 8-Bit Adder With Overflow Detector

A second method of overflow detection generates an overflow indication by examining the carry bits into and out of the MSB of a 2's complement adder/subtractor.

Consider the following 8-bit 2's complement sums. We will use our previous knowledge of overflow to see whether overflow occurs and then compare the carry bits into and out of the MSB.

a. 80H + 80H

b. 7FH + 01H

c. 7FH + 80H

d. 7FH + C0H   ·

a. 80H = 10000000

$$\begin{array}{r} 10000000 \\ + \ 10000000 \\ \hline 1\ 00000000 \end{array}$$ (Sign bit overflow; V = 1)

Carry into MSB = 0
Carry out of MSB = 1

b. 7FH = 01111111
    01H = 00000001

$$\begin{array}{r} 01111111 \\ + \ 00000001 \\ \hline 0\ 10000000 \end{array}$$ (Sign bit overflow; V = 1)

Carry into MSB = 1
Carry out of MSB = 0

c. 7FH = 01111111
    80H = 10000000

$$\begin{array}{r} 01111111 \\ + \ 10000000 \\ \hline 0\ 11111111 \end{array}$$ (No sign bit overflow; V = 0)

Carry into MSB = 0
Carry out of MSB = 0

d. 7FH = 01111111
    C0H = 11000000

$$\begin{array}{r} 01111111 \\ + \ 11000000 \\ \hline 1\ 00111111 \end{array}$$ (No sign bit overflow; V = 0)

Carry into MSB = 1
Carry out of MSB = 1

The above examples suggest that a 2's complement sum has overflowed if there is a carry into or out of the MSB, but not both. For an 8-bit adder/subtractor, we can write the Boolean equation for this condition as $V = C_8 \oplus C_7$. More generally, for an n-bit adder/subtractor, $V = C_n \oplus C_{n-1}$.

Figure 6.22 shows a circuit that can implement the overflow detection from the carry into and out of the MSB of an 8-bit adder.

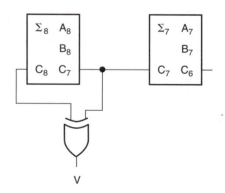

FIGURE 6.22

||. SECTION 6.6B REVIEW PROBLEM

6.10 What is the permissible range of values of a sum or difference, x, in a 12-bit parallel binary adder if it is written as:

a. A signed binary number?

b. An unsigned binary number?

6.7 BCD Adders

(This section may be omitted without loss of continuity.)

> **KEY TERM**
>
> **BCD adder** A parallel adder whose output is in groups of 4 bits, each group representing a BCD digit.

It is sometimes convenient to have the output of an adder circuit available as a BCD number, particularly if the result is to be displayed numerically. The problem is that most parallel adders have binary outputs, and 6 of the 16 possible 4-bit binary sums—1010 to 1111—are not within the range of the BCD code.

BCD numbers range from 0000 to 1001, or 0 to 9 in decimal. The unsigned binary sum of any two BCD numbers plus an input carry can range from 00000 ($= 0000 + 0000 + 0$) to 10011 ($= 1001 + 1001 + 1 = 19_{10}$).

For any sum up to 1001, the BCD and binary values are the same. Any sum greater than 1001 must be modified, since it requires a second BCD digit. For example, the binary value of 19_{10} is 10011_2. The BCD value of 19_{10} is $0001\ 1001_{BCD}$. (The most significant digit of a sum of two BCD digits and a carry will never be larger than 1, since the largest such sum is 19_{10}.)

Table 6.10 shows the complete list of possible binary sums of two BCD digits (A and B) and a carry (C), their decimal equivalents, and their corrected BCD values. The MSD of the BCD sum is shown only as a carry bit, with leading zeros suppressed.

Table 6.10 Binary Sums of Two BCD Digits and a Carry Bit

BinarySum ($A + B + C$)	Decimal	Corrected BCD (Carry + BCD)
00000	0	0 + 0000
00001	1	0 + 0001
00010	2	0 + 0010
00011	3	0 + 0011
00100	4	0 + 0100
00101	5	0 + 0101
00110	6	0 + 0110
00111	7	0 + 0111
01000	8	0 + 1000
01001	9	0 + 1001
01010	10	1 + 0000
01011	11	1 + 0001
01100	12	1 + 0010
01101	13	1 + 0011
01110	14	1 + 0100
01111	15	1 + 0101
10000	16	1 + 0110
10001	17	1 + 0111
10010	18	1 + 1000
10011	19	1 + 1001

FIGURE 6.23

BCD Adder (1½ Digit Output)

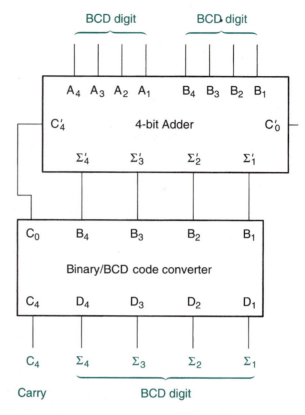

Figure 6.23 shows how we can add two BCD digits and get a corrected output. The **BCD adder** circuit consists of a standard 4-bit parallel adder to get the binary sum and a code converter to translate it into BCD.

The Binary-to-BCD code converter operates on the binary inputs as follows:

1. A carry output is generated if the binary sum is in the range $01010 \leq$ sum ≤ 10011 (BCD equivalent: $1\ 0000 \leq$ sum $\leq 1\ 1001$).

2. If the binary sum is less than 01001, the output is the same as the input.

3. If the sum is in the range $01010 \leq$ sum ≤ 10011, the four LSBs of the input must be corrected to a BCD value. This can be done by adding 0110 to the four LSBs of the input and discarding any resulting carry. We add 0110_2 (6_{10}) because we must account for six unused codes.

Let's look at how each of these requirements can be implemented by a digital circuit.

Carry Output

The carry output will be automatically 0 for any uncorrected sum from 00000 to 01001 and automatically 1 for any sum from 10000 to 10011. Thus, if the binary adder's carry output, which we will call C_4', is 1, the BCD adder's carry output, C_4, will also be 1.

Any sum falling between these ranges, that is, between 01010 and 01111, must have its MSB modified. This modifying condition is a function, designated C_4'', of the binary adder's sum outputs when its carry output is 0. This function can be simplified by a Karnaugh map, as shown in Figure 6.24, resulting in the following Boolean expression.

$$C_4'' = \Sigma_4'\ \Sigma_3' + \Sigma_4'\ \Sigma_2'$$

The BCD carry output C_4 is given by:

$$C_4 = C_4' + C_4''$$
$$= C_4' + \Sigma_4'\ \Sigma_3' + \Sigma_4'\ \Sigma_2'$$

The BCD carry circuit is shown in Figure 6.25.

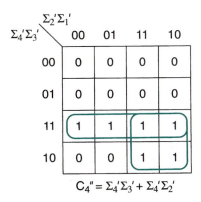

FIGURE 6.24

Carry as a Function of Sum
Bits When $C_4' = 0$

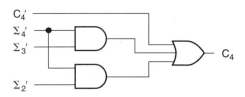

FIGURE 6.25

BCD Carry Circuit

Sum Correction

The four LSBs of the binary adder output need to be corrected if the sum is 01010 or greater and need not be corrected if the binary sum is 01001 or less. This condition is indicated by the BCD carry. Let us designate the binary sum outputs as $\Sigma_4' \Sigma_3' \Sigma_2' \Sigma_1'$ and the BCD sum outputs as $\Sigma_4 \Sigma_3 \Sigma_2 \Sigma_1$.

$$\text{If } C_4 = 0, \Sigma_4 \Sigma_3 \Sigma_2 \Sigma_1 = \Sigma_4' \Sigma_3' \Sigma_2' \Sigma_1' + 0000;$$
$$\text{If } C_4 = 1, \Sigma_4 \Sigma_3 \Sigma_2 \Sigma_1 = \Sigma_4' \Sigma_3' \Sigma_2' \Sigma_1' + 0110.$$

Figure 6.26 shows a BCD adder, complete with a binary adder, BCD carry, and sum correction. A second parallel adder is used for sum correction. The B inputs are the uncorrected binary sum inputs. The A inputs are either 0000 or 0110, depending on the value of the BCD carry.

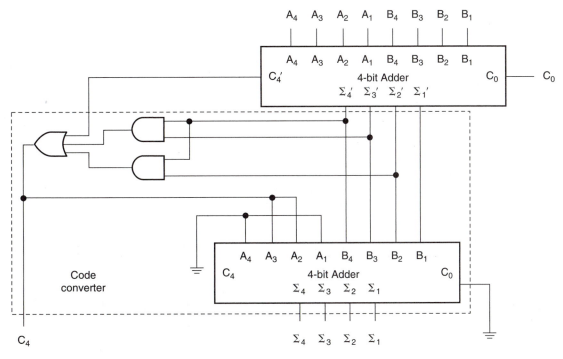

FIGURE 6.26

BCD Adder

■ **EXAMPLE 6.25**

bcd_add.vhd

Write a VHDL file for a BCD adder, using two parallel adder components, such as in the logic diagram in Figure 6.26.

SOLUTION

```
-- bcd_add.vhd
-- BCD adder, using 2 instances of the component add4par

ENTITY bcd_add IS
    PORT (
        c0     : IN    BIT;
        a, b   : IN    BIT_VECTOR (4 downto 1);
        c4     : OUT   BIT;
        sum    : OUT   BIT_VECTOR (4 downto 1));
END bcd_add;

ARCHITECTURE adder OF bcd  add IS
    -- Component declaration
    COMPONENT add4par
    PORT (
        c0     : IN    BIT;
        a, b   : IN    BIT_VECTOR (4 downto 1);
        c4     : OUT   BIT;
        sum    : OUT   BIT_VECTOR (4 downto 1));
    END COMPONENT;
    SIGNAL c4_bin : BIT;
    SIGNAL sum_bin : BIT_VECTOR (4 downto 1);
    SIGNAL a_bcd : BIT_VECTOR (4 downto 1);
    SIGNAL b_bcd : BIT_VECTOR (4 downto 1);
    SIGNAL c0_bcd: BIT;
BEGIN
    -- Instantiate 4-bit adder (binary sum)
    add_bin: add4par
        PORT MAP ( c0  =>  c0,
                   a   =>  a,
                   b   =>  b,
                   c4  =>  c4_bin,
                   sum =>  sum_bin);
    --Instantiate 4-bit adder (binary-BCD converter)
    converter: add4par
        PORT MAP ( c0  => c0_bcd,
                   a   => a_bcd,
                   b   => b_bcd,
                   sum => sum);
    --Connect components
    c0_bcd     <= '0';
    b_bcd      <= sum_bin;
    a_bcd(4)   <= '0';
    a_bcd(3)   <= c4  bin or (sum_bin(4) and sum_bin(3))
                  or (sum_bin(4) and sum_bin(2));
    a_bcd(2)   <= c4  bin or (sum_bin(4) and sum_bin(3))
                  or (sum_bin(4) and sum_bin(2));
    a_bcd(1)   <= '0';
    c4         <= c4_bin or (sum_bin(4) and sum_bin(3))
                  or (sum_bin(4) and sum_bin(2));
END adder;
```

Multiple-Digit BCD Adders

Several BCD adders can be cascaded to add multidigit BCD numbers. Figure 6.27 shows a $4\frac{1}{2}$-digit BCD adder. The carry output of the most significant digit is considered to be a half-digit since it can only be 0 or 1. The output range of the $4\frac{1}{2}$-digit BCD adder is 00000 to 19999.

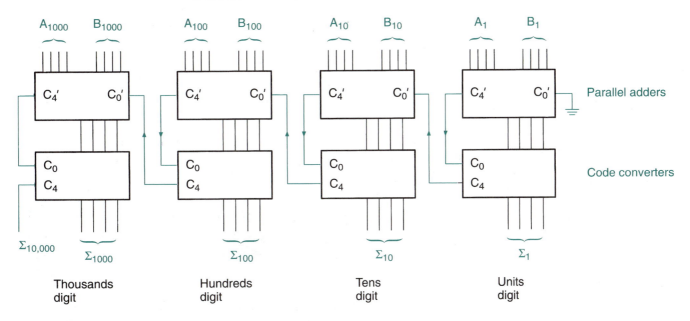

FIGURE 6.27
$4^1/_2$-Digit BCD Adder

BCD adders are cascaded by connecting the code converter carry output of one stage to the binary adder carry input of the next most significant stage. Each BCD output digit represents a decade, designated as the units, tens, hundreds, thousands, and ten thousands digits.

▌▌ SECTION 6.7 REVIEW PROBLEM

6.11 What is the maximum BCD sum of two 3-digit numbers with no carry input? How many digits are required to display this result on a numerical output?

6.8 Carry Generation in MAX+PLUS II

> ### KEY TERMS
>
> **Speed grade** A specification that indicates the internal delay time that can be expected of a CPLD.
>
> **Expander buffer** A MAX+PLUS II primitive that supplies an inverted product term for general use within a CPLD.

The VHDL adder circuits implemented in Section 6.6 were all defined using a ripple carry format. Is it necessary to design a fast carry circuit when compiling one of these adder designs in MAX+PLUS II? Probably not.

Recall that the design strategy behind the fast carry circuit was to flatten the gate network, that is, to replace a long network (many gates for the carry bit to pass through) with a wide one (fewer levels of gating). Also recall that any combinational logic function can be implemented as a sum-of-products (SOP) network, which inherently is a very flat network form.

The internal circuit of a MAX7000S CPLD is a programmable SOP network. In order to program such a device, the MAX+PLUS II compiler must analyze the design entity, break it into product terms and reassemble it as an SOP network. (This is an oversimplification. Sometimes SOP outputs are fed back into the circuit to be reused by other parts of the circuit, thus lengthening the logic path.)

MAX+PLUS II allows us to choose a style of logic synthesis that balances circuit speed and chip area occupied by the programmed circuit. The styles can be user-defined or we can use one of three predefined synthesis styles called Normal, Fast, and WYSIWYG (What You See Is What You Get). Each one of these styles is optimized for speed, area, or a compromise. The Normal and Fast styles disassemble the design entity and reassemble it after optimizing the logic according to the style rules. The WYSIWYG style allows us (rather than the compiler) to largely define the logic synthesis without altering our design format by very much.

To choose a synthesis style, select **Global Project Logic Synthesis,** from the MAX+PLUS II **Assign** menu, as shown in Figure 6.28. A drop-down menu in the resulting dialog box, shown in Figure 6.29, allows us to select one of the three Altera-defined synthesis styles.

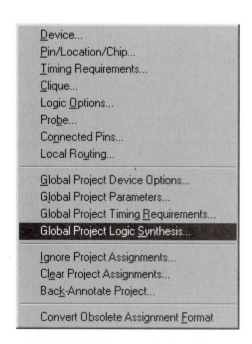

FIGURE 6.28
Assigning a Synthesis Style (Assign Menu)

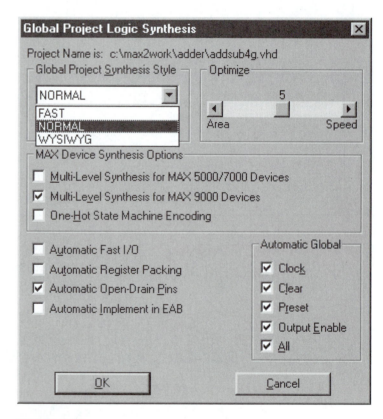

FIGURE 6.29
Assigning a Synthesis Style

N O T E

To use the WYSIWYG style, you must also check the box that says **Multi-Level Synthesis for MAX 5000/7000 Devices.**

We can calculate the circuit delays for an adder by running the compiled design through the MAX+PLUS II Timing Analyzer. Figure 6.30 shows an example of such an analysis for the parallel adder **add4par.vhd** with a Normal synthesis style and an EPM7128SLC84-7 as the selected device.

FIGURE 6.30

Delay matrix for a 4-bit adder (Normal Synthesis)

The values in the Destination columns are the delays from logic level changes on the inputs specified by the Source rows. For example, a change on input **a1** reaches the output **sum1** in 7.5 ns and **sum4** in 12.5 ns. Most of the entries in the **c4** column have two values (7.5 ns and 11.5 ns), indicating that the delay to output carry is the same from all input bits (i.e., a fast carry). The actual delay to **c4** will depend on the logic level change that takes place on the source line and thus which logic path is taken. The delay time of 7.5 ns is about the lowest value possible in the EPM7128SLC84-7 device. (The "−7" tells us that the chip has a **speed grade** of minus 7, meaning an internal delay of about 7 ns.)

Figure 6.31 shows the timing analysis of the same adder with a WYSIWYG synthesis style. (The synthesis style is the only design change.) In this analysis, the delay from an input bit to **c4** varies from 7.5 ns (from **a4** or **b4**) to as much as 16.5 ns (for **a1, b1, a2,** or **b2**). Since the lower-order bits result in a longer delay to the carry bit, we can infer that the compiler has not synthesized a fast carry circuit.

FIGURE 6.31

Delay matrix for a 4-bit adder (WYSIWYG Synthesis)

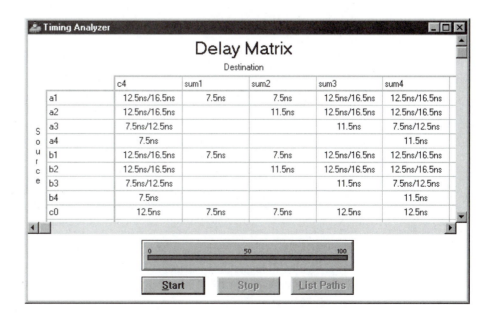

We can examine the actual equations from the MAX+PLUS II report file to confirm our assessment. The synthesized equations for **c4** are given below.

WYSIWYG Synthesis:

```
-- Node name is 'c4' = '|full_add:adder4| :12'
-- Equation name is 'c4', type is output
 c4       = LCELL( _EQ001 $  GND);
 _EQ001 =    a3 &  b3 & b4
        #    b4 &  _LC113 & _LC114
        #    a3 &  a4 & b3
        #    a4 &  _LC113 & _LC114
        #    a4 &  b4;

-- Node name is '|full_add:adder2|:12'
-- Equation name is '_LC113', type is buried
_LC113   = LCELL( _EQ008 $  GND);
 _EQ008 =    a2 &  b2
        #    a2 &  c0 &  _X007
        #    b2 &  c0 &  _X007
        #    a1 &  b1 &  _X002;
 _X007  = EXP (!a1 & !b1);
 _X002  = EXP (!a2 & !b2);

-- Node name is '|full_add:adder3|:9'
-- Equation name is '_LC114', type is buried
_LC114   = LCELL( b3 $ a3);
```

Normal synthesis:

```
-- Node name is 'c4'
-- Equation name is 'c4', location is LC123, type is output.
 c4       = LCELL ( EQ001 $ VCC);
 _EQ001 =  !a1 &  _X001 & _X002 & _X003  & _X004
        #  !b1 & !c0 &  _X001 & _X003 & _X004
        #  !a2 & !b2 &  _X003 & _X004
        #  !a3 & !b3 &  _X004
        #  !a4 & !b4;
 _X001  = EXP ( a2 & b2);
 _X002  = EXP ( b1 & c0);
 _X003  = EXP ( a3 & b3);
 _X004  = EXP ( a4 & b4);
```

The function EXP(signal) is for a MAX+PLUS II primitive called an **expander buffer,** which represents a **shared logic expander** in the CPLD. There will be more detail about this type of buffer in Chapter 8, but for now, just be aware that this type of buffer supplies inverted product terms for general use within the CPLD.

The Normal synthesis mode generates a sum-of-products equation that uses a number of expanders, but only one logic cell (i.e., one SOP output), indicated as LC123. The WYSIWYG synthesis uses an unnumbered output logic cell, which in turn uses two other logic cell outputs (LC113 and LC114) as inputs. Thus, in the Normal synthesis mode, the input signals propagate through one logic cell, and in the WYSIWYG mode, the input signals go through two layers of logic cells, increasing the path length, and thus the delay.

What can we conclude? If MAX+PLUS II is allowed to synthesize a design for a full adder in the defined Normal style, it will optimize the design equations to produce as flat a network as possible. Thus we do not need to explicitly design an adder circuit to have a fast carry function.

SUMMARY

1. Addition combines an addend (x) and an augend (y) to get a sum $(z = x + y)$.

2. Binary addition is based on four sums:

$$0 + 0 = 0$$
$$0 + 1 = 1$$
$$1 + 1 = 10$$
$$1 + 1 + 1 = 11$$

3. A sum of two bits generates a sum bit and a carry bit. (For the first two sums above, the carry bit is 0; the last two sums have a carry of 1. The last sum includes a carry from a lower-order bit.)

4. Subtraction combines a minuend (x) and a subtrahend (y) to get a difference $(z = x - y)$.

5. Binary subtraction is based on the following four differences:

$$0 - 0 = 0$$
$$1 - 0 = 1$$
$$1 - 1 = 0$$
$$10 - 1 = 1$$

6. If the subtrahend bit is larger than the minuend bit, as in the fourth difference above, a 1 must be borrowed from the next higher-order bit.

7. Binary addition or subtraction can be unsigned, where the magnitudes of the operands and result are presumed to be positive, or signed, where the operands and result can be positive or negative. The sign is indicated by a sign bit.

8. The sign bit (usually MSB) of a binary number indicates that the number is positive if it is 0 and negative if it is 1.

9. Signed binary numbers can be written in true-magnitude, 1's complement, or 2's complement form. True magnitude has the same binary value for positive and negative numbers, with only the sign bit changed. A 1's complement negative number is generated by inverting all bits of the positive number of the same magnitude. A 2's complement negative number is generated by adding 1 to the equivalent 1's complement number. Positive numbers are the same in all three forms.

10. 1's complement or 2's complement binary numbers are used in signed addition or subtraction. Subtraction is performed by adding a negative number in complement form to another number in complement form (i.e., $x - y = x + (-y)$). This technique does not work for true-magnitude form.

11. A negative sum or difference in 2's complement subtraction must be converted to a positive form to read its magnitude (i.e., $-(-x) = +x$).

12. A signed binary number, x, with n bits has a valid range of $-2^n \leq x \leq +(2^n - 1)$.

13. A negative number with a power-of-2 magnitude (i.e., -2^n) is written in 2's complement form as n 0s preceded by all 1s to fill the defined size of the number (e.g., in 8-bit 2's complement form, $-128 = 10000000$ (1 followed by seven 0s; $128 = 2^7$); in 8-bit 2's complement form, $-8 = 11111000$ (all 1s, followed by three 0s; $8 = 2^3$).

14. If a sum or difference falls outside the permissible range of magnitudes for a 2's complement number, it generates an overflow into the sign bit of the number. The result is that the sum of two positive numbers appears to be negative (e.g., $01111111 + 00000010 = 10000001$; $127 + 2 = 129$) or the sum of two negative numbers appears to be positive (e.g., $11111111 + 10000000 = 01111111$, where the carry beyond the 8th place is discarded: $-1 + (-128) = -129$).

15. When adding two hexadecimal digits, any digit sum greater than 15 (F) can be converted to a hexadecimal value by subtracting 16 and carrying a 1 to the next digit position.

16. Hexadecimal numbers can be subtracted conventionally or by a complement method. To get the 16's complement of a number, obtain the 15's complement by subtracting the number from all Fs and adding 1 to the result.

17. Binary numbers can be used in nonpositional codes to represent numbers or alphanumeric characters.

18. Binary coded decimal (BCD) codes represent decimal numbers as a series of 4-bit groups of numbers. Natural BCD or 8421 code does this as a positionally weighted code for each digit (e.g., $158 = 0001\ 0101\ 1000$ (NBCD)). Other codes, such as Excess-3, are not positionally weighted.

19. Gray code is a binary code that has a difference of one bit between adjacent codes. It can be generated by a set of XOR functions or by recognizing the symmetry inherent in the code. In any Gray code sequence, the MSB is 0 for the first half of the sequence and 1 for the second half. The remaining bits are symmetrical about the halfway point of the sequence.

20. ASCII code represents alphanumeric characters and computer control codes as a 7-bit group of binary numbers. Alpha characters are listed in uppercase in columns 4 and 5 of the ASCII table. Lowercase alpha characters are in columns 6 and 7. Numeric characters are in column 3.

21. A half adder combines two bits to generate a sum and a carry. It can be represented by the following truth table:

A	B	C_{OUT}	Σ
0	0	0	0
0	1	0	1
1	0	0	1
1	1	1	0

22. From the half adder truth table, we can derive two equations:

$$C_{OUT} = AB$$
$$\Sigma = A \oplus B$$

23. A full adder can accept an input carry from a lower-order adder and combine the input carry with two operands to generate

a sum and output carry. Its operation can be summarized in the following truth table:

A	B	C_{IN}	C_{OUT}	Σ
0	0	0	0	0
0	0	1	0	1
0	1	0	0	1
0	1	1	1	0
1	0	0	0	1
1	0	1	1	0
1	1	0	1	0
1	1	1	1	1

24. The following Boolean equations for a full adder can be derived from the truth table and Boolean algebra:

$$C_{OUT} = (A \oplus B)\, C_{IN} + AB$$
$$\Sigma = (A \oplus B) \oplus C_{IN}$$

25. Two half adders can be combined to make a full adder. Operands A and B go to the first half adder. The sum output of the first half adder and the carry input go to the inputs of the second half adder. The carry outputs of both half adders are combined in an OR gate.

26. Multiple full adders can be cascaded to make a parallel binary adder. Operands A_1 and B_1 are applied to the first full adder. Carry bit C_0 is grounded. A_2 and B_2 go to the second adder stage, and so on. The carry output of one stage is cascaded to the carry input of the following stage. This connection is called ripple carry.

27. Ripple carry has the disadvantage of increasing the time required to generate an output result as more stages are added. Fast carry, or look-ahead carry, examines all adder inputs simultaneously and generates each internal and output carry with a separate circuit. This makes the carry circuit wider, but flatter, thus reducing the delay time of the circuit.

28. A parallel adder can be implemented in VHDL by creating a design entity for a full adder, then using multiple instances of the full adder as components in the parallel adder.

29. To use a component in a VHDL design hierarchy, we require a design entity that defines the component, a component declaration in the design entity that uses the component, and a component instantiation statement for every instance of the component in the higher-level design entity.

30. The general form of a design entity using components is:

```
ENTITY entity_name IS
    PORT ( input and output definitions);
END entity_name;
ARCHITECTURE arch_name OF entity_name IS
    component declaration (s);
    signal declaration(s);
BEGIN
    Component instantiation(s);
    Other statements;
END arch_name;
```

31. The port map of a component maps the port names defined in a component to the port, signal, or variable names defined in the design entity that uses the component.

32. If all ports of a component are to be used in the same order as in the component definition in the original component design entity, the port map can simply contain the user names in the same order. For example:

```
adder1: full_add PORT MAP (a(1), b(1),
c0, c(1), sum(1));
```

33. If only a portion of the component ports are to be used or they are not used in the same sequence as they are declared, the port map must be more explicit. For example:

```
adder1: full_add
    PORT MAP ( b      => b(1),
               a      => a(1),
               c_in   => c0,
               sum    => sum(1)),
               c_out  => c(1);
```

34. A GENERATE statement can be used to instantiate multiple instances of a component. The GENERATE statement has the form:

```
label:
FOR index IN range GENERATE
    statements;
END GENERATE;
```

35. MAX+PLUS II will synthesize an adder to minimize carry delays without much intervention.

36. A parallel binary adder can be made into a 2's complement subtractor by inverting one set of inputs and tying the input carry to a logic HIGH.

37. A parallel binary adder can be made into a 2's complement adder/subtractor by using a set of XOR gates as programmable inverters and connecting the XOR control line to the carry input of the adder.

38. One method of detecting a sign bit overflow in a 2's complement adder/subtractor is to compare the sign bits of the operands to the sign bit of the result. If the sign bits of the operands are the same as each other, but different from the sign bit of the result, there has been an overflow. The Boolean equation for this detector is given by $V = \overline{S_\Sigma} \cdot S_A \cdot S_B + S_\Sigma \cdot \overline{S_A} \cdot \overline{S_B}$.

39. Another method of overflow detection compares the carry out of the MSB of the adder/subtractor to the carry into the MSB. An overflow occurs if there is a carry out of or into the MSB, but not both. The Boolean equation for this detector is given by $V = C_n \oplus C_{n-1}$, for an n-bit adder/subtractor.

40. A BCD adder adds two binary coded decimal (BCD) digits and generates a BCD digit and a carry bit.

41. Since BCD is a 4-bit code, BCD addition can be done with a 4-bit binary adder and a code converter. The code converter can be synthesized from another 4-bit binary adder and a circuit to generate a carry.

GLOSSARY

1's complement   A form of signed binary notation in which negative numbers are created by complementing all bits of a number, including the sign bit.

10's complement   A way of writing decimal numbers where a negative number is generated by adding 1 to its 9's complement.

2's complement   A form of signed binary notation in which negative numbers are created by adding 1 to the 1's complement form of the number.

8421 Code (or NBCD; natural binary coded decimal)   A BCD code that represents each digit of a decimal number by its 4-bit true binary value.

9's complement   A way of writing decimal numbers where a number is made negative by subtracting each of its digits from 9 (e.g., $-726 = 999 - 726 = 273$ in 9's complement).

Addend   The number in an addition operation that is added to another.

Alphanumeric code   A code used to represent letters of the alphabet and numerical characters.

ASCII   American Standard Code for Information Interchange. A 7-bit code for representing alphanumeric and control characters.

Augend   The number in an addition operation to which another number is added.

BCD   Binary coded decimal. A code that represents each digit of a decimal number by a 4-bit binary value.

BCD adder   A parallel adder whose output is in groups of 4 bits, each group representing a BCD digit.

Borrow   A digit brought back from a more significant position when the subtrahend digit is larger than the minuend digit.

Carry   A digit which is "carried over" to the next most significant position when the sum of two single digits is too large to be expressed as a single digit.

Carry bit   A bit that holds the value of a carry (0 or 1) resulting from the sum of two binary numbers.

Cascade   To connect an output of one device to a input of another, often for the purpose of expanding the number of bits available for a particular function.

Case shift   Changing letters from capitals (UPPERCASE) to small letters (lowercase) or vice versa.

Component   A complete VHDL design entity that can be used as a part of a higher-level file in a hierarchical design.

Component declaration statement   A statement that defines the input and output port names of a component used in a VHDL design entity.

Component instantiation statement   A statement that maps port names of a VHDL component to the port names, internal signals, or variables of a higher-level VHDL design entity.

Difference   The result of a subtraction operation.

End-around carry   An operation in 1's complement subtraction where the carry bit resulting from a sum of two 1's complement numbers is added to that sum.

Excess-3 code   A BCD code that represents each digit of a decimal number by a binary number derived by adding 3 to its 4-bit true binary value. Excess-3 code has the advantage of being "self-complementing."

Expander buffer   A MAX+PLUS II primitive that supplies an inverted product term for general use within a CPLD.

Fast carry (or look-ahead carry)   A gate network which generates a carry bit directly from *all* incoming operand bits, independent of the operation of each full adder stage.

Full adder   A circuit that will add a carry bit from another full or half adder and two operand bits to produce a sum bit and a carry bit.

GENERATE statement   A VHDL construct that is used to create repetitive portions of hardware.

Gray code   A binary code which progresses such that only one bit changes between two successive codes.

Half adder   A circuit that will add two bits and produce a sum bit and a carry bit.

Hierarchy   A group of design entities associated in a series of levels or layers in which complete designs form portions of another, more general design entity. The more general design is considered to be the higher level of the hierarchy.

Instantiate   To use an instance of a component.

Magnitude bits   The part of a signed binary number that tell us how large the number is (i.e., its magnitude).

Minuend   The number in a subtraction operation from which another number is subtracted.

Operand   A number upon which an arithmetic function operates (e.g., in the expression $x + y = z$, x and y are the operands).

Overflow   An erroneous carry into the sign bit of a signed binary number which results from a sum larger than can be represented by the number of magnitude bits.

Parallel binary adder   A circuit, consisting of n full adders, which will add two n-bit binary numbers. The output consists of n sum bits and a carry bit.

Port   An input or output of a VHDL design entity or component.

Ripple carry   A method of passing carry bits from one stage of a parallel adder to the next by connecting C_{OUT} of one full adder to C_{IN} of the following stage.

Self-complementing   A code that automatically generates a negative-equivalent (e.g., 9's complement for a decimal code) when all its bits are inverted.

Sign bit   A bit, usually the MSB, that indicates whether a signed binary number is positive or negative.

Signed binary arithmetic   Arithmetic operations performed using signed binary numbers.

Signed binary number   A binary number of fixed length whose sign is represented by one bit, usually the most significant bit, and whose magnitude is represented by the remaining bits.

Speed grade   A specification that indicates the internal delay time that can be expected of a CPLD.

Subtrahend   The number in a subtraction operation that is subtracted from another number.

Sum   The result of an addition operation.

Sum bit (single-bit addition)   The least significant bit of the sum of two 1-bit binary numbers.

True-magnitude form   A form of signed binary number whose magnitude is represented in true binary.

Unsigned binary arithmetic   Arithmetic operations performed using unsigned binary numbers.

Unsigned binary number   A binary number whose sign is not indicated by a sign bit. A positive sign is assumed unless explicitly stated otherwise.

PROBLEMS

Section 6.1 Digital Arithmetic

6.1   Add the following unsigned binary numbers.

   a. 10101 + 1010
   b. 10101 + 1011
   c. 1111 + 1111
   d. 11100 + 1110
   e. 11001 + 10011
   f. 111011 + 101001

6.2   Subtract the following unsigned binary numbers.

   a. 1100 − 100
   b. 10001 − 1001
   c. 10101 − 1100
   d. 10110 − 1010
   e. 10110 − 1001
   f. 10001 − 1111
   g. 100010 − 10111
   h. 1100011 − 100111

Section 6.2   Representing Signed Binary Numbers

6.3   Write the following decimal numbers in 8-bit true-magnitude, 1's complement, and 2's complement forms.

   a. −110
   b. 67
   c. −54
   d. −93
   e. 0
   f. −1
   g. 127
   h. −127

Section 6.3   Signed Binary Arithmetic

6.4   Perform the following arithmetic operations in the true-magnitude (addition only), 1's complement, and 2's complement systems. Use 8-bit numbers consisting of a sign bit and 7 magnitude bits. (The numbers shown are in the decimal system.)

   Convert the results back to decimal to prove the correctness of each operation. Also demonstrate that the idea of adding a negative number to perform subtraction is not valid for the true-magnitude form.

   a. 37 + 25
   b. 85 + 40
   c. 95 − 63
   d. 63 − 95
   e. −23 − 50
   f. 120 − 73
   g. 73 − 120

6.5   What are the largest positive and negative numbers, expressed in 2's complement notation, that can be represented by an 8-bit signed binary number?

6.6   Perform the following *signed* binary operations, using 2's complement notation where required. State whether or not sign bit overflow occurs. Give the signed decimal equivalent values of the sums in which overflow does *not* occur.

   a. 01101 + 00110
   b. 01101 + 10110
   c. 01110 − 01001
   d. 11110 + 00010
   e. 11110 − 00010

6.7   Without doing any binary complement arithmetic, indicate which of the following operations will result in 2's complement overflow. (Assume 8-bit representation consisting of a sign bit and 7 magnitude bits.) Explain the reasons for each choice.

   a. −109 + 36
   b. 109 + 36
   c. 65 + 72
   d. −110 − 29
   e. 117 + 11
   f. 117 − 11

6.8   Explain how you can know, by examining sign or magnitude bits of the numbers involved, when overflow has occurred in 2's complement addition or subtraction.

Section 6.4   Hexadecimal Arithmetic

6.9   Add the following hexadecimal numbers.

   a. 27H + 16H
   b. 87H + 99H
   c. A55H + C5H
   d. C7FH + 380H
   e. 1FFFH + A80H

6.10   Subtract the following hexadecimal numbers.

    a. F86H − 614H

    b. E72H − 229H

    c. 37FFH − 137FH

    d. 5764H − ACBH

    e. 7D30H − 5D33H

    f. 5D33H − 7D30H

    g. 813AH − A318H

Section 6.5   Numeric and Alphanumeric Codes

6.11   Convert the following decimal numbers to true binary, 8421 BCD code, and Excess-3 code.

    a. 709_{10}

    b. 1889_{10}

    c. 2395_{10}

    d. 1259_{10}

    e. 3972_{10}

    f. 77300_{10}

6.12   Make a table showing the equivalent Gray codes corresponding to the range from 0_{10} to 31_{10}.

6.13   Write your name in ASCII code.

6.14   Encode the following text into ASCII code: "10% off purchases over $50. (Monday only)"

6.15   Decode the following string of ASCII code.

    57 41 52 4E 49 4E 47 21 20 54 68 69 73 20 63 6F 6D

    6D 61 6E 64 20 65 72 61 73 65 73 20 36 34 30 4D 20

    6F 66 20 6D 65 6D 6F 72 79 2E

Section 6.6   Binary Adders and Subtractors

6.16   Write the truth table for a half adder, and from the table derive the Boolean expressions for both C_o (carry output) and Σ (sum output) in terms of inputs A and B. Draw the half adder circuit.

6.17   Write the truth table for a full adder, and from the table derive the simplest possible Boolean expressions for C_{OUT} and Σ in terms of A, B, and C_{IN}.

6.18   From the equations in Problems 6.16 and 6.17, draw a circuit showing a full adder constructed from two half adders.

6.19   Evaluate the Boolean expression for Σ and C_{OUT} of the full adder in Figure 6.7 for the following input values. What is the binary value of the outputs in each case?

    a. $A = 0, B = 0, C_{IN} = 0$

    b. $A = 0, B = 1, C_{IN} = 0$

    c. $A = 0, B = 1, C_{IN} = 1$

    d. $A = 1, B = 1, C_{IN} = 1$

6.20   Verify the summing operation of the circuit in Figure 6.10, as follows. Determine the output of each full adder based on the inputs shown below. Calculate each sum manually and compare it to the 5-bit output $(C_4\,\Sigma_4\,\Sigma_3\,\Sigma_2\,\Sigma_1)$ of the parallel adder circuit.

    a. $A_4 A_3 A_2 A_1 = 0100, B_4 B_3 B_2 B_1 = 1001$

    b. $A_4 A_3 A_2 A_1 = 1010, B_4 B_3 B_2 B_1 = 0110$

    c. $A_4 A_3 A_2 A_1 = 0101, B_4 B_3 B_2 B_1 = 1101$

    d. $A_4 A_3 A_2 A_1 = 1111, B_4 B_3 B_2 B_1 = 0111$

6.21   Briefly describe the differences in the underlying design strategies of the ripple carry adder and the fast carry adder (i.e., what makes the fast carry faster than the ripple carry?). What is the main limitation for the fast carry circuit?

6.22   Write the general form of the fast carry equation. Use it to generate Boolean expression for C_1, C_2, and C_3 for a fast carry adder.

6.23   The following equation describes the carry output function for a parallel binary adder:

$$
\begin{aligned}
C_{OUT} &= A_4 B_4 + A_3 B_3 (A_4 + B_4) \\
&\quad + A_2 B_2 (A_4 + B_4)(A_3 + B_3) \\
&\quad + A_1 B_1 (A_4 + B_4)(A_3 + B_3)(A_2 + B_2) \\
&\quad + C_{IN} (A_4 + B_4)(A_3 + B_3)(A_2 + B_2) \\
&\qquad (A_1 + B_1)
\end{aligned}
$$

Briefly explain how to interpret the third term of this equation.

6.24   Write a VHDL file for an 8-bit parallel adder, using eight instances of a full adder component.

6.25   Create a simulation for the 8-bit adder of Problem 6.24, showing a representative sample of sums. How many different sums would be required to show all possible combinations of inputs?

6.26   Write a VHDL file that creates a 12-bit adder using a GENERATE statement.

6.27   Create a simulation file for the 12-bit adder of Problem 6.26 showing only one transition, as follows. Set input **a** to 000 from 0 to 500 ns, then 001 from 500 ns to 1 μs. Set input **b** to FFF from 0 to 1 μs. From the simulation, determine the internal delays from **a1** to each of the **sum** bits. Confirm your observations with the delay matrix from a timing analysis.

6.28   Use MAX+PLUS II to create a Graphic Design File for a 2's complement subtractor based on a 4-bit parallel binary adder. Explain how the circuit generates the 2's complement of B for the subtraction $A − B$.

6.29   Use MAX+PLUS II to create a Graphic Design File for a 2's complement adder/subtractor based on a 4-bit parallel binary adder. Explain how the circuit is programmed to add or subtract and how it produces the 2's complement of B for the subtraction $A − B$.

6.30   Use MAX+PLUS II to draw a circuit that will detect an overflow condition in a 4-bit 2's complement adder/subtractor. The detector output should go HIGH upon overflow detection. Draw the circuit truth table, explain what all input and output variables are, and show any Boolean equations you need to complete the circuit design.

6.31   Modify the 4-bit adder/subtractor drawn in Figure 6.15 to include an overflow detection circuit.

6.32 Create a simulation for the 4-bit adder/subtractor with overflow detection (Problem 6.31), using the following representative hexadecimal input values: $F + 1 = 10$ (carry, but no overflow); $7 + 1 = 8$ (overflow, but no carry); $8 + 8 = 10$ (carry and overflow); $0 - 1 = F$ (result $= -1$).

6.33 Modify the VHDL file for the 4-bit parallel binary adder/subtractor (**addsub4g.vhd**) to include an overflow detection circuit. Use two different methods.

6.34 What is the permissible range of values that a sum or difference, x, can have in a 16-bit parallel binary adder if it is written as:

a. A signed binary number

b. An unsigned binary number

Section 6.7   BCD Adders

6.35 What is the maximum BCD sum of two 3-digit BCD numbers plus an input carry? How many digits are needed to display the result?

6.36 What is the maximum BCD sum of two 4-digit BCD numbers plus an input carry? How many digits are needed to display the result?

6.37 Based on the answers to Problems 6.35 and 6.36, formulate a general rule to calculate the maximum BCD sum of two n-digit BCD numbers plus a carry bit.

6.38 Derive the Boolean expression for a BCD carry output as a function of the sum of two BCD digits.

6.39 Draw the circuit for a binary-to-BCD code converter.

6.40 Write a VHDL file to implement a binary-to-BCD code converter for a BCD adder. Use a selected signal assignment or CASE statement.

6.41 Write a VHDL file that uses the binary-to-BCD code converter of Problem 6.40 and a 4-bit parallel binary adder as components in a BCD adder.

6.42 Write a VHDL file that uses a code converter and parallel adder as components in a design that will add two 2-digit BCD numbers and produce a 2½ digit result.

6.43 Draw the block diagram of a circuit that will add two 3-digit BCD numbers and display the result as a series of decimal digits. How many digits will the output display?

ANSWERS TO SECTION REVIEW PROBLEMS

Section 6.1a

6.1 101000;   **6.2** 100000.

Section 6.1b

6.3 11;   **6.4** 1

Section 6.3

6.5 11100000;   6.6 100000.

Section 6.4

6.7a 11701H   6.7b 1281H

Section 6.5

6.8 "True or False: 1/4 < 1/2"

Section 6.6a

6.9 Figures 6.32 and 6.33 show the propagation paths for the carry bits.

Fast carry: 3 gates

Ripple carry: 8 gates

Section 6.6b

6.10a Signed: $-2048 \le x \le +2047$ (11 magnitude bits, 1 sign bit)

6.10b Unsigned: $0 \le x \le +4095$ (12 magnitude bits, no sign bit: positive implied)

Section 6.7

6.11 Maximum BCD sum $= 1001\ 1001\ 1001 + 1001\ 1001\ 1001 = 1\ 1001\ 1001\ 1000_{BCD} = 1998_{10}$. This sum requires a $3\frac{1}{2}$-digit numerical display.

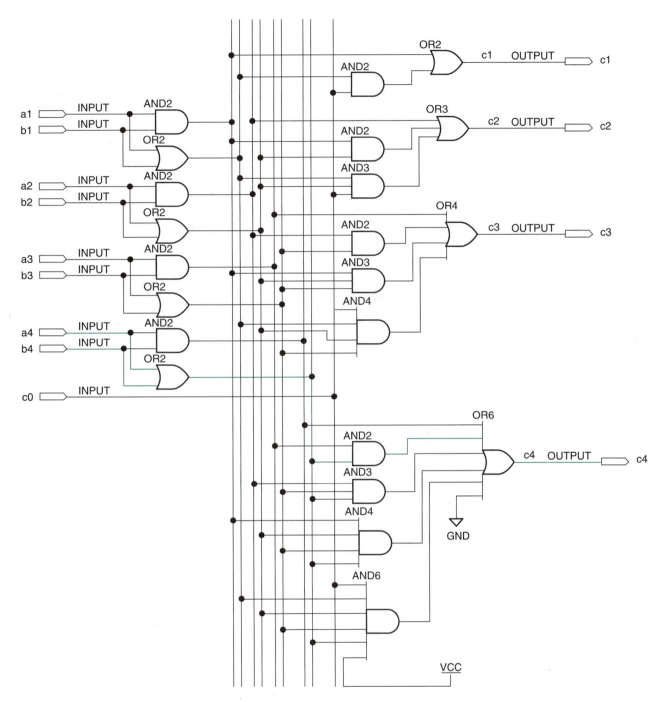

FIGURE 6.32
Fast Carry from A4/B4 to C4.

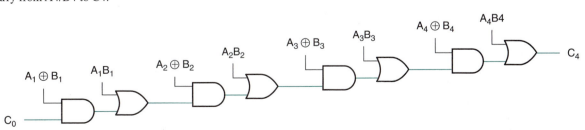

FIGURE 6.33
Ripple Carry from C_0 to C_4

Introduction to Sequential Logic

C H A P T E R   O B J E C T I V E S

Upon successful completion of this chapter, you will be able to:

• Explain the difference between combinational and sequential circuits.

• Define the set and reset functions of an SR latch.

• Draw circuits, function tables, and timing diagrams of NAND and NOR latches.

• Explain the effect of each possible input combination to a NAND and a NOR latch, including set, reset, and no change functions, as well as the ambiguous or forbidden input condition.

• Design circuit applications that employ NAND and NOR latches.

• Describe the use of the *ENABLE* input of a gated SR or D latch as an enable/inhibit function and as a synchronizing function.

• Outline the problems involved with using a level-sensitive *ENABLE* input on a gated SR or D latch.

• Explain the concept of edge-triggering and why it is an improvement over level-sensitive enabling.

• Draw circuits, function tables, and timing diagrams of edge-triggered D, JK, and T flip-flops.

• Describe the toggle function of a JK flip-flop and a T flip-flop.

• Describe the operation of the asynchronous preset and clear functions of D, JK, and T flip-flops and be able to draw timing diagrams showing their functions.

• Use MAX+PLUS II to create simple circuits and simulations with D latches and D, JK, and T flip-flops.

• Create simple flip-flop designs using VHDL.

The digital circuits studied to this point have all been combinational circuits, that is, circuits whose outputs are functions only of their present inputs. A particular set of input states will always produce the same output state in a combinational circuit.

This chapter will introduce a new category of digital circuitry: the sequential circuit. The output of a sequential circuit is a function both of the present input conditions and the previous conditions of the inputs and/or outputs. The output depends on the sequence in which the inputs are applied.

We will begin our study of sequential circuits by examining the two most basic sequential circuit elements: the latch and the flip-flop, both of which are part of the general class of circuits called bistable multivibrators. These are similar devices, each being used to store a single bit of information indefinitely. The difference between a latch and a flip-flop is the condition under which the stored bit is allowed to change.

Latches and flip-flops are also used as integral parts of more complex devices, such as programmable logic devices (PLDs), usually when an input or output state must be stored.

7.1 Latches

KEY TERMS

Sequential circuit   A digital circuit whose output depends not only on the present combination of inputs, but also on the history of the circuit.

Latch   A sequential circuit with two inputs called *SET* and *RESET*, which make the latch store a logic 0 (reset) or 1 (set) until actively changed.

SET   1. The stored HIGH state of a latch circuit.
2. A latch input that makes the latch store a logic 1.

RESET   1. The stored LOW state of a latch circuit.
2. A latch input that makes the latch store a logic 0.

All the circuits we have seen up to this point have been combinational circuits. That is, their present outputs depend only on their present inputs. The output state of a combinational circuit results only from a combination of input logic states.

The other major class of digital circuits is the **sequential circuit.** The present outputs of a sequential circuit depend not only on its present inputs, but also on its past input states.

The simplest sequential circuit is the SR **latch,** whose logic symbol is shown in Figure 7.1a. The latch has two inputs, *SET (S)* and *RESET (R),* and two complementary outputs, Q and $\overline{Q}$. If the latch is operating normally, the outputs are always in opposite logic states.

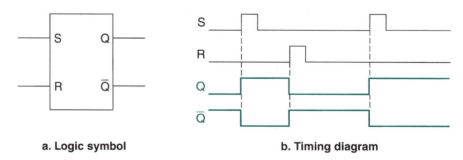

a. Logic symbol   **b. Timing diagram**

FIGURE 7.1
SR Latch (Active HIGH Inputs)

The latch operates like a momentary-contact pushbutton with START and STOP functions, shown in Figure 7.2. A momentary-contact switch operates only when it is held down. When released, a spring returns the switch to its rest position.

Suppose the switch in Figure 7.2 is used to control a motor starter. When you push the START button, the motor begins to run. Releasing the START switch does not turn the motor off; that can be done only by pressing the STOP button. If the motor is running,

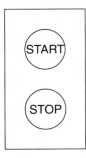

FIGURE 7.2
Industrial Pushbutton (e.g., Motor Starter)

pressing the START button again has no effect, except continuing to let the motor run. If the motor is not running, pressing the STOP switch has no effect, since the motor is already stopped.

There is a conflict if we press both switches simultaneously. In such a case we are trying to start and stop the motor at the same time. We will come back to this point later.

The latch *SET* input is like the START button in Figure 7.2. The *RESET* input is like the STOP button.

NOTE

By definition:

A latch is set when $Q = 1$ and $\bar{Q} = 0$.
A latch is reset when $Q = 0$ and $\bar{Q} = 1$.

The latch in Figure 7.1 has active-HIGH *SET* and *RESET* inputs. To set the latch, make $R = 0$ and make $S = 1$. This makes $Q = 1$ until the latch is actively reset, as shown in the timing diagram in Figure 7.1b. To activate the reset function, make $S = 0$ and make $R = 1$. The latch is now reset ($Q = 0$) until the set function is next activated.

Combinational circuits produce an output by *combining inputs*. In sequential circuits, it is more accurate to think in terms of *activating functions*. In the latch described, S and R are not *combined* by a Boolean function to produce a particular result at the output. Rather, the set function is *activated* by making $S = 1$, and the reset function is *activated* by making $R = 1$, much as we would activate the START or STOP function of a motor starter by pressing the appropriate pushbutton.

The timing diagram in Figure 7.1b shows that the inputs need not remain active after the set or reset functions have been selected. In fact, the S or R input *must* be inactive before the opposite function can be applied, in order to avoid conflict between the two functions.

▌▌ EXAMPLE 7.1

Latches can have active-HIGH or active-LOW inputs, but in each case $Q = 1$ after the set function is applied and $Q = 0$ after reset. For each latch shown in Figure 7.3, complete the timing diagram shown. Q is initially LOW in both cases. (The state of Q before the first active *SET* or *RESET* is unknown unless specified, since the present state depends on previous history of the circuit.)

FIGURE 7.3
Example 7.1
SR Latch

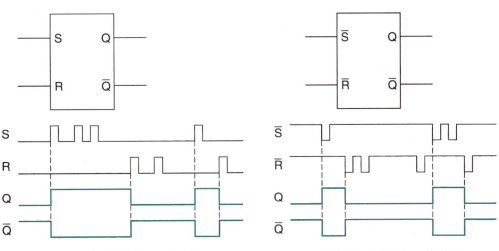

a. Active-HIGH input latch   b. Active-LOW input latch

SOLUTION   The Q and $\overline{Q}$ waveforms are shown in Figure 7.3. Note that the outputs respond only to the first set or reset command in a sequence of several pulses.

▌▌ EXAMPLE 7.2

Figure 7.4 shows a latching HOLD circuit for an electronic telephone. When HIGH, the *HOLD* output allows you to replace the handset without disconnecting a call in progress.

FIGURE 7.4
Example 7.2
Latching HOLD Button

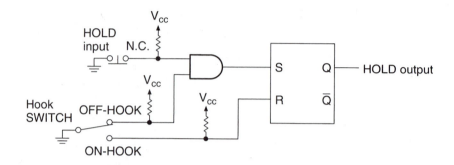

The two-position switch is the telephone's hook switch (the switch the handset pushes down when you hang up), shown in the off-hook (in-use) position. The normally closed pushbutton is a momentary-contact switch used as a *HOLD* button. The circuit is such that the *HOLD* button does not need to be held down to keep the *HOLD* active. The latch "remembers" that the switch was pressed, until told to "forget" by the reset function.

Describe the sequence of events that will place a caller on hold and return the call from hold. Also draw timing diagrams showing the waveforms at the *HOLD* input, hook switch inputs, S input, and *HOLD* output for one hold-and-return sequence. (*HOLD* output = 1 means the call is on hold.)

SOLUTION   To place a call on hold, we must set the latch. We can do so if we press and hold the *HOLD* switch, then the hook switch. This combines two HIGHs—one from the *HOLD* switch and one from the on-hook position of the hook switch—into the AND gate, making $S = 1$ and $R = 0$. Note the sequence of events: press *HOLD*, hang up, release *HOLD*. The S input is HIGH only as long as the *HOLD* button is pressed. The handset can be kept on-hook and the *HOLD* button released. The latch stays set, as $S = R = 0$ (neither *SET* not *RESET* active) as long as the handset is on-hook.

To restore a call, lift the handset. This places the hook switch into the off-hook position and now $S = 0$ and $R = 1$, which resets the latch and turns off the *HOLD* condition. Figure 7.5 shows the timing diagram for the sequence described.

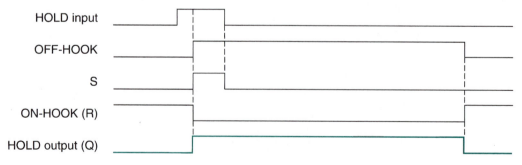

FIGURE 7.5
Example 7.2
HOLD Timing Diagram

■■ SECTION 7.1 REVIEW PROBLEM

7.1 A latch with active-HIGH S and R inputs is initially set. R is pulsed HIGH three times, with $S = 0$. Describe how the latch responds.

7.2 NAND/NOR Latches

An SR latch is easy to build with logic gates. Figure 7.6 shows two such circuits, one made from NOR gates and one from NANDs. The NAND gates in the second circuit are drawn in DeMorgan equivalent form.

FIGURE 7.6
SR Latch Circuits

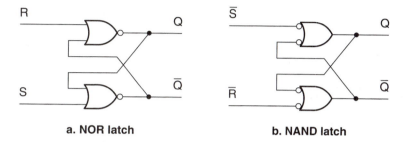

a. NOR latch   **b. NAND latch**

The two circuits both have the following three features:

1. OR-shaped gates
2. Logic level inversion between the gate input and output
3. Feedback from the output of one gate to an input of the opposite gate

During our examination of the NAND and NOR latches, we will discover why these features are important.

A significant difference between the NAND and NOR latches is the placement of *SET* and *RESET* inputs with respect to the Q and $\overline{Q}$ outputs. Once we define which output is Q and which is $\overline{Q}$, the locations of the *SET* and *RESET* inputs are automatically defined.

In a NOR latch, the gates have active-HIGH inputs and active-LOW outputs. When the input to the Q gate is HIGH, $Q = 0$, since either input HIGH makes the output LOW. Therefore, this input must be the *RESET* input. By default, the other is the *SET* input.

In a NAND latch, the gate inputs are active LOW (in DeMorgan equivalent form) and the outputs are active HIGH. A LOW input on the Q gate makes $Q = 1$. This, therefore, is the *SET* input, and the other gate input is *RESET*.

Since the NAND and NOR latch circuits have two binary inputs, there are four possible input states. Table 7.1 summarizes the action of each latch for each input combination. The functions are the same for each circuit, but they are activated by opposite logic levels.

Table 7.1   NOR and NAND Latch Functions

S	R	Action (NOR Latch)	$\overline{S}$	$\overline{R}$	Action (NAND Latch)
0	0	Neither *SET* nor *RESET* active; output does not change from previous state	0	0	Both *SET* and *RESET* active; forbidden condition
0	1	*RESET* input active	0	1	*SET* input active
1	0	*SET* input active	1	0	*RESET* input active
1	1	Both *SET* and *RESET* active; forbidden condition	1	1	Neither *SET* nor *RESET* active; output does not change from previous state

We will examine the NAND latch circuit for each of the input conditions in Table 7.1. The analysis of a NOR latch is similar and will be left as an exercise.

NAND Latch Operation

Figure 7.7 shows a NAND latch in its two possible stable states. In each case the inputs $\overline{S}$ and $\overline{R}$ are both HIGH (inactive).

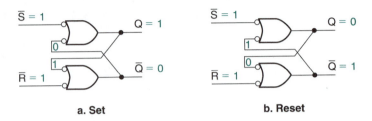

a. Set   **b. Reset**

FIGURE 7.7
NAND Latch Stable States

Figure 7.7a shows the latch in its *SET* condition ($Q = 1$). The feedback connections from each gate output to the input of the opposite gate keep the latch in a stable condition. The upper gate has a LOW on the "inner" input. Since, for a NAND gate, either input LOW makes the output HIGH, this makes $Q = 1$. This HIGH value is fed to the gate on the other side of the latch. The lower gate has both inputs HIGH, thus keeping its output LOW. The LOW at $\overline{Q}$ feeds back to the upper gate, forming a closed loop of consistent logic levels. There is no tendency for the outputs to change under these conditions.

Figure 7.7b shows a similar state for the latch in a *RESET* condition ($Q = 0$). As with the *SET* state, the stability of the latch depends on the feedback connections. The logic values of the latch gate inputs are the same as before, except that the LOW input is on the lower gate, not the upper gate as in the SET condition.

Figure 7.8 shows a NAND latch as a Graphic Design File created with MAX+PLUS II. The inputs are labeled nS and nR and one output as nQ as we cannot enter input names with bars over them. (BOR2 = "Bubbled OR, 2-inputs".)

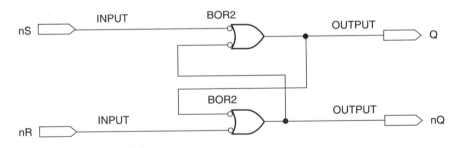

FIGURE 7.8
Graphic Design File representation of a NAND Latch.

> **NOTE**
>
> The documentation for MAX+PLUS II recommends that you do not create your own latch circuits or similar cross-coupled structures. Rather, you should use primitives such as **LATCH,** or components such as **lpm_latch,** which can be used in **gdf** or **vhd** files. We will use the design in Figure 7.8 only to illustrate the function of a NAND latch and to generate some timing data with the MAX+PLUS II simulator.

In order to make MAX+PLUS II synthesize this circuit as we have drawn it in Figure 7.8, we must select **Global Project Logic Synthesis** from the **Assign Menu** (Figure 7.9).

In the resulting dialog box (Figure 7.10), we must choose the WYSIWYG (What You See Is What You Get) synthesis style and check the box that says **Multi-Level Synthesis for MAX5000/7000 Devices.**

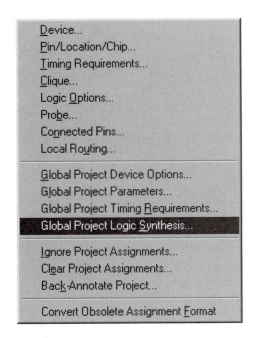

FIGURE 7.9
Assign Menu

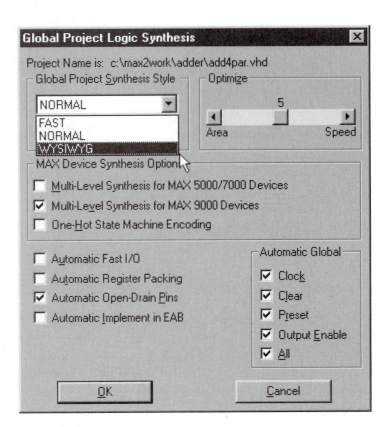

FIGURE 7.10
Choosing WYSIWYG Synthesis Style

When we compile the graphic file, MAX+PLUS II synthesizes the following equations, which we can read in the project report file:

```
** EQUATIONS **

nR      : INPUT;
nS      : INPUT;

-- Node name is 'nQ' = ':3'
-- Equation name is 'nQ', type is output
 nQ      = LCELL( _EQ001 $ GND);
 _EQ001 = !nR
         # !Q;

-- Node name is 'Q' = ':2'
-- Equation name is 'Q', type is output
 Q       = LCELL ( _EQ002 $ GND);
 _EQ002 = !nS
         # !nQ;
```

We can rewrite the synthesized latch equations as:

$$Q = \overline{nS} + \overline{nQ}$$
$$nQ = \overline{nR} + \overline{Q}$$

When we run the MAX+PLUS II Timing Analyzer, we get the delay matrix shown in Figure 7.11. The delays are symmetrical for this circuit. The delay from nS to Q (7.5 ns) is through one gate; from nS to nQ (12.5 ns) is through two gates. These values are the same for the path from nR to nQ (7.5 ns; one gate) and from nR to Q (12.5 ns; two gates). We can see these changes on simulation waveforms for the *SET* and *RESET* functions.

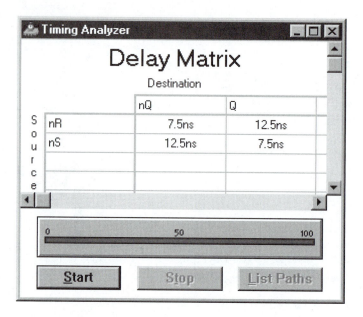

FIGURE 7.11
NAND Latch Delay Matrix (WYSIWYG Synthesis)

www.electronictech.com

Figures 7.12 and 7.13 show the transition of a NAND latch from the *RESET* to the *SET* condition. In Figure 7.12a, the latch is stable in the *RESET* condition ($Q = 0$) at time $t < 0$ (i.e., before a SET pulse is applied to the latch). At time $t = 0$, the $\overline{S}$ input goes LOW (Figure 7.12b) and 7.5 ns later, the output Q goes HIGH (Figure 7.12c). This applies a HIGH to the lower gate in the latch and at $t = 12.5$ ns (Figure 7.12d), the $\overline{Q}$ output goes LOW, closing the loop. The latch is now in a new stable configuration and the $\overline{S}$ input can go back HIGH, as shown in Figure 7.12e.

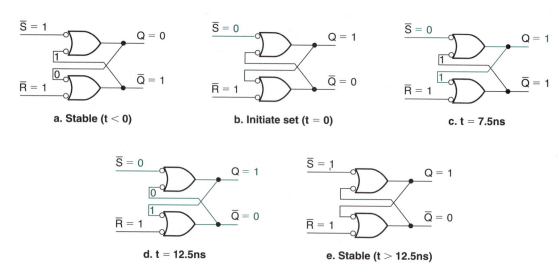

FIGURE 7.12
RESET-to-SET transition

FIGURE 7.13
NAND Latch SET function
simulation

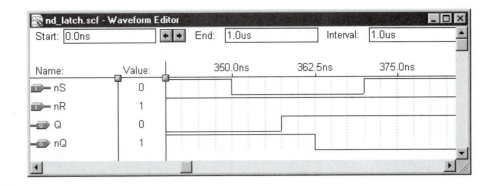

The waveforms in Figure 7.13 also show this transition. The simulation window has a 2.5 ns grid, so three grid spaces are equivalent to 7.5 ns and five grid spaces to 12.5 ns. The waveforms show Q going HIGH 7.5 ns after nS goes LOW, followed by nQ going LOW at 12.5 ns after nS.

Figures 7.14 and 7.15 show the same thing for the RESET function. The latch is in a stable *SET* condition at time $t < 0$ (Figure 7.14a). Input $\overline{R}$ goes LOW at $t = 0$ (Figure 7.14b). At time $t = 7.5$ ns, $\overline{Q}$ goes HIGH, which is transferred to the upper gate in the latch circuit Figure 7.14c). Since both inputs of the upper gate are now HIGH, Q goes LOW at time $t = 12.5$ ns (Figure 7.14d). At this point the latch is stable in the *RESET* condition and the input $\overline{R}$ can return to the HIGH (inactive) state, as shown in Figure 7.14e. Figure 7.15 shows the simulation waveforms for this transition.

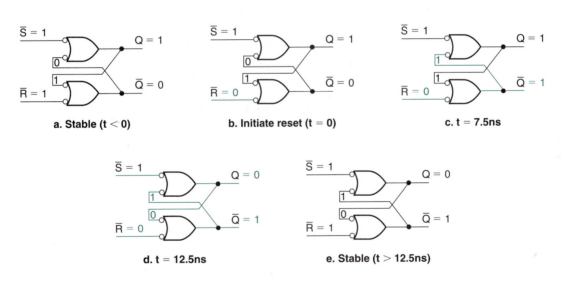

a. Stable (t < 0)   b. Initiate reset (t = 0)   c. t = 7.5ns

d. t = 12.5ns   e. Stable (t > 12.5ns)

FIGURE 7.14
SET-to-RESET Transition

FIGURE 7.15
NAND latch RESET function
simulation

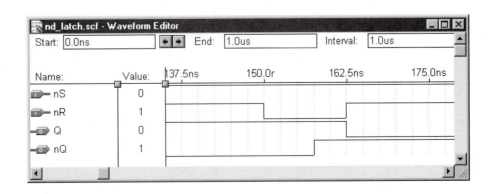

Note that the latch is not stable in its new condition until the new logic levels have propagated through both gates. Figure 7.16 shows the result of a *RESET* pulse that only lasts for 7.5 ns. This pulse is too short to allow both gates to change states. The outputs both oscillate, since the changing logic levels never "catch up" as they move around the latch. This is due to the fact that both paths (nS-Q-nQ and nR-nQ-Q) are the same length. If one path were slightly longer, the logic level controlled by the longer path would dominate and the latch would stabilize in one state or the other.

FIGURE 7.16
NAND latch oscillation due to a RESET pulse that is too short

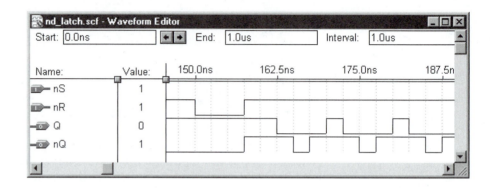

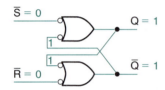

FIGURE 7.17
NAND Latch Forbidden State

Figure 7.17 shows a NAND latch with $\overline{S} = \overline{R} = 0$. This implies that both *SET* and *RESET* functions are active. Since a NAND gate requires at least one input LOW to make the output HIGH, both outputs respond by going HIGH. This condition is not unstable in and of itself, but instability can result when the inputs change.

There are three possible results when the outputs go back to the HIGH state.

1. The *SET* input goes HIGH before the *RESET* input. In this case the latch resets, as *RESET* is the last input active. This is shown in the simulation in Figure 7.18.

FIGURE 7.18
SET goes HIGH before RESET

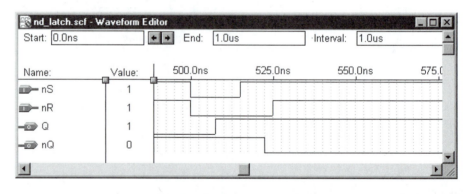

FIGURE 7.19
RESET goes HIGH before SET

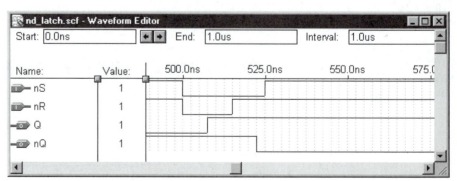

2. The *RESET* input goes HIGH before *SET*. In this case, the latch sets, as shown in Figure 7.19.

3. The *SET* and *RESET* inputs go HIGH at the same time. This is an unstable case. Figure 7.20 shows how the latch will oscillate under this condition. When the inputs $\overline{S}$ and $\overline{R}$ are both LOW (Figure 7.20a), both latch outputs are HIGH. When $\overline{S}$ and $\overline{R}$ go HIGH (Figure 7.20b), all gate inputs are HIGH. This makes both outputs LOW (Figure 7.20c). The LOWs transfer across the latch to the opposite gates and, after a delay, make both outputs HIGH (Figure 7.20d). At this point, oscillations will be sustained until the latch is *SET* or *RESET*. The simulation waveforms in Figure 7.21 show the oscillatory condition of the latch outputs.

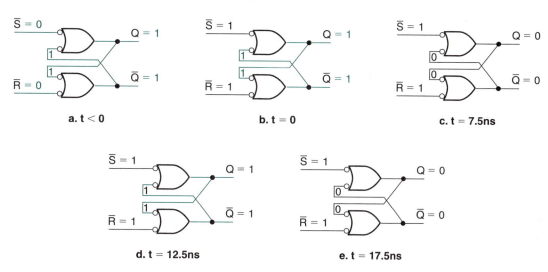

FIGURE 7.20
NAND Latch Forbidden State Transition

In practice, the oscillatory condition of Figure 7.21 is unlikely to be sustained for very long. One of the two gates is likely to be slightly faster than the other, which will allow one state or the other to dominate.

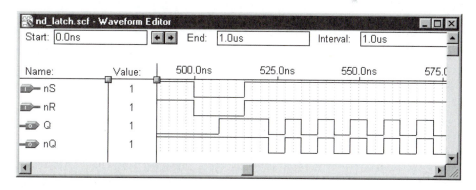

FIGURE 7.21
SET and RESET go HIGH simultaneously

The operation of the NAND latch can be summarized in a function table, shown in Table 7.2. The notation Q_{t+1} indicates that the column shows the value of Q *after* the specified input is applied. Q_t indicates the present state of the Q input.* Thus, the entry for the no change state indicates that after the inputs $\overline{S} = 0, \overline{R} = 0$ are applied, the next state of the output is the same as its present state.

*Many sources (such as data sheets) use the notation Q_0 to refer to the previous state of Q. We will use the notation indicated (Q_t for present state and Q_{t+1} for next state) so as to be able to reserve Q_0 for the least significant bit of a circuit requiring multiple Q outputs.

Table 7.2 NAND Latch Function Table

$\overline{S}$	$\overline{R}$	Q_{t+1}	$\overline{Q}_{t+1}$	Function
0	0	1	1	Forbidden
0	1	1	0	Set
1	0	0	1	Reset
1	1	Q_t	$\overline{Q}_t$	No Change

Table 7.3 NOR Latch Function Table

S	R	Q_{t+1}	$\overline{Q}_{t+1}$	Function
0	0	Q_t	$\overline{Q}_t$	No Change
0	1	0	1	Reset
1	0	1	0	Set
1	1	0	0	Forbidden

Table 7.3 shows the function table for the NOR latch.

Practical Synthesis in MAX+PLUS II

The NAND latch shown previously (Figure 7.8) was synthesized in MAX+PLUS II, using the WYSIWYG synthesis style. We did this so as to be able to use the MAX+PLUS II simulation tool to get waveforms for a standard NAND latch. However, if we allow MAX+PLUS II to synthesize the latch circuit in the Normal synthesis style, the software will choose a more stable configuration, shown in Figure 7.22.

FIGURE 7.22

NAND Latch as synthesized by MAX+PLUS II (NORMAL synthesis)

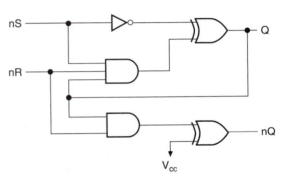

The equations for the circuit in Figure 7.22 from the MAX+PLUS II report file are given as:

```
** EQUATIONS **

nR        : INPUT;
nS        : INPUT;

-- Node name is 'nQ'
-- Equation name is 'nQ', location is LC117, type is output.
  nQ       = LCELL( _EQ001 $ VCC);
  _EQ001 =   nR &  Q;

-- Node name is 'Q' = '~2~1'
-- Equation name is 'Q', location is LC115, type is output.
  Q        = LCELL( _EQ002 $ !nS);
  _EQ002 = nR & nS & Q;
```

We can rewrite the above equations as:

$$Q = (nR \cdot nS \cdot Q) \oplus \overline{nS}$$
$$nQ = (nR \cdot Q) \oplus 1 = \overline{nR \cdot Q}$$

Without going into a detailed analysis, we will just note that the latching occurs through a combination of the XOR gate at Q and the feedback from the Q output to the 3-input AND. The lower AND/XOR structure simply serves to invert the Q output to provide a complementary value at nQ.

This configuration is more stable because both *SET* and *RESET* functions go through the *same* path (the 3-input AND gate). Delay is the same from nS to Q and from nR to Q. In the WYSIWYG version, the path is equal from nS to nQ and from nR to Q, but not from nS to Q and nR to Q. The *SET* and *RESET* pulses thus go through *different* paths in the

WYSIWYG synthesis, resulting in unequal delays from the inputs to the Q output, which can lead to instability.

Latch as a Switch Debouncer

Pushbutton or toggle switches are sometimes used to generate pulses for digital circuit inputs, as illustrated in Figure 7.23. However, when a switch is operated and contact is made on a new terminal, the contact, being mechanical, will bounce a few times before settling into the new position. Figure 7.23d shows the effect of contact bounce on the waveform for a pushbutton switch. The contact bounce is shown only on the terminal where contact is being made, not broken.

FIGURE 7.23

Switches as Pulse Generators

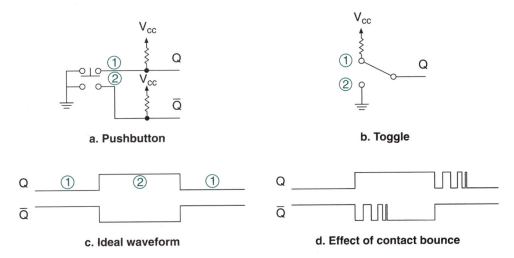

a. Pushbutton   b. Toggle

c. Ideal waveform   d. Effect of contact bounce

Contact bounce can be a serious problem, particularly when a switch is used as an input to a digital circuit that responds to individual pulses. If the circuit expects to receive one pulse, but gets several from a bouncy switch, it will behave unpredictably.

A latch can be used as a switch debouncer, as shown in Figure 7.24a. When the pushbutton is in the position shown, the latch is set, since $\overline{S} = 0$ and $\overline{R} = 1$. (Recall that the NAND latch inputs are active LOW.) When the pushbutton is pressed, the $\overline{R}$ contact

FIGURE 7.24

NAND Latch as a Switch Debouncer

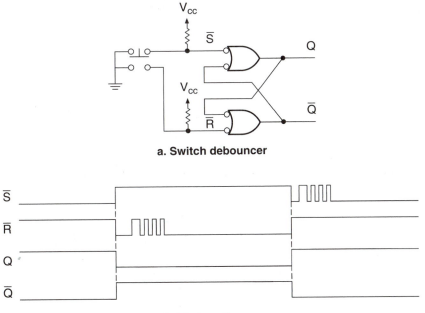

a. Switch debouncer

b. Timing diagram

bounces a few times, as shown in Figure 7.24b. However, on the first bounce, the latch is reset. Any further bounces are ignored, since the resulting input state is either $\overline{S} = \overline{R} = 1$ (no change) or $\overline{S} = 1, \overline{R} = 0$ (reset).

Similarly, when the pushbutton is released, the $\overline{S}$ input bounces a few times, setting the latch on the first bounce. The latch ignores any further bounces, since they either do not change the latch output ($\overline{S} = \overline{R} = 1$) or set it again ($\overline{S} = 0, \overline{R} = 1$). The resulting waveforms at Q and $\overline{Q}$ are free of contact bounce and can be used reliably as inputs to digital sequential circuits.

▌▌ EXAMPLE 7.3

A NOR latch can be used as a switch debouncer, but not in the same way as a NAND latch. Figure 7.25 shows two NOR latch circuits, only one of which works as a switch debouncer. Draw a timing diagram for each circuit, showing R, S, Q, and $\overline{Q}$, to prove that the circuit in Figure 7.25b eliminates switch contact bounce but the circuit in Figure 7.25a does not.

FIGURE 7.25
Example 7.3
NOR Latch Circuits

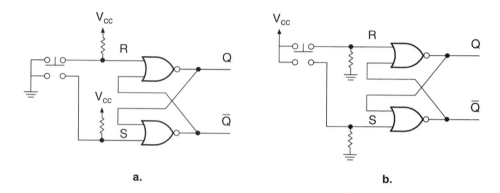

a.   b.

SOLUTION   Figure 7.26 shows the timing diagrams of the two NOR latch circuits. In the circuit in Figure 7.25a, contact bounce causes the latch to oscillate in and out of the forbidden state of the latch ($S = R = 1$). This causes one of the two outputs to bounce for each contact closure. (Use the function table of the NOR latch to examine each part of the timing diagram to see that this is so.)

By making the resistors pull down rather than pull up, as in Figure 7.25b, the latch oscillates in and out of the no change state ($S = R = 0$) as a result of contact bounce. The first

FIGURE 7.26
Example 7.3
NOR Latch Circuits

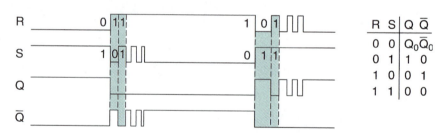

R	S	Q	$\overline{Q}$
0	0	Q_0	$\overline{Q}_0$
0	1	1	0
1	0	0	1
1	1	0	0

a. Timing diagram for circuit of Figure 6.13a

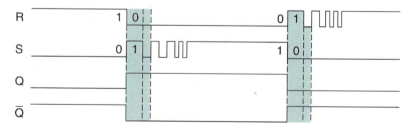

b. Timing diagram for circuit of Figure 6.13b

bounce on the *SET* terminal sets the latch, and other oscillations are disregarded. The first bounce on the *RESET* input resets the latch, and further pulses on this input are ignored.

The principle illustrated here is that a closed switch must present the active input level to the latch, since switch bounce is only a problem on contact closure. Thus, a closed switch must make the input of a NOR latch HIGH or the input of a NAND latch LOW to debounce the switch waveform.

NOTE

The NOR latch is seldom used in practice as a switch debouncer. The pull-down resistors need to be about 500 Ω or less to guarantee a logic LOW at the input of a TTL NOR gate. In such a case, a constant current of about 10 mA flows through the resistor connected to the normally closed portion of the switch. This value is unacceptably high in most circuits, as it draws too much idle current from the power supply. For this reason, the NAND latch, which uses higher-value pull-up resistors (about 1 kΩ or larger) and therefore draws less idle current, is preferred for a switch debouncer.

SECTION 7.2 REVIEW PROBLEM

7.2 Why is the input state $S = R = 1$ considered forbidden in the NOR latch? Why is the same state in the NAND latch the no change condition?

7.3 Gated Latches

KEY TERMS

Gated SR latch   An SR latch whose ability to change states is controlled by an extra input called the *ENABLE* input.

Steering gates   Logic gates, controlled by the *ENABLE* input of a gated latch, that steer a *SET* or *RESET* pulse to the correct input of an SR latch circuit.

Transparent latch (gated D latch)   A latch whose output follows its data input when its *ENABLE* input is active.

Gated SR Latch

It is not always desirable to allow a latch to change states at random times. The circuit shown in Figure 7.27, calleda **gated SR latch,** regulates the times when a latch is allowed to change state.

The gated SR latch has two distinct subcircuits. One pair of gates is connected as an SR latch. A second pair, called the **steering gates,** can be enabled or inhibited by a control signal, called *ENABLE,* allowing one or the other of these gates to pass a *SET* or *RESET* signal to the latch gates.

The *ENABLE* input can be used in two principal ways: (1) as an ON/OFF signal, and (2) as a synchronizing signal.

Figure 7.27b shows the *ENABLE* input functioning as an ON/OFF signal. When *ENABLE* = 1, the circuit acts as an active-HIGH latch. The upper gate converts a HIGH at S to a LOW at $\bar{S}$, setting the latch. The lower gate converts a HIGH at R to a LOW at $\bar{R}$, thus resetting the latch.

When *ENABLE* = 0, the steering gates are inhibited and do not allow *SET* or *RESET* signals to reach the latch gate inputs. In this condition, the latch outputs cannot change.

FIGURE 7.27
Gated SR Latch

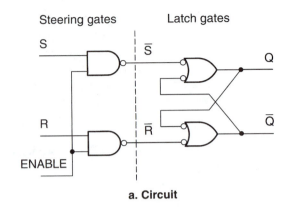

a. Circuit

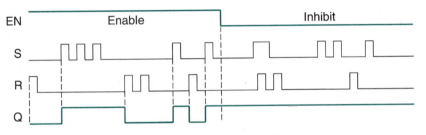

b. ENABLE used as an ON/OFF signal

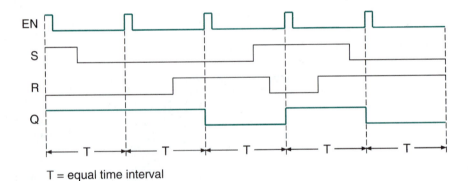

T = equal time interval

c. ENABLE used as a synchronizing signal

Figure 7.27c shows the *ENABLE* input as a synchronizing signal. A periodic pulse waveform is present on the *ENABLE* line. The *S* and *R* inputs are free to change at random, but the latch outputs will change only when the *ENABLE* input is active. Since the *ENABLE* pulses are equally spaced in time, changes to the latch output can occur only at fixed intervals. The outputs can change out of synchronization if *S* or *R* change when *ENABLE* is HIGH. We can minimize this possibility by making the *ENABLE* pulses as short as possible.

Table 7.4 represents the function table for a gated SR latch.

Table 7.4   Gated SR Latch Function Table

EN	*S*	*R*	Q_{t+1}	$\bar{Q}_{t+1}$	**Function**
1	0	0	Q_t	$\bar{Q}_t$	No change
1	0	1	0	1	Reset
1	1	0	1	0	Set
1	1	1	0	0	Forbidden
0	X	X	Q_t	$\bar{Q}_t$	Inhibited

▋▋ EXAMPLE 7.4

Figure 7.28 shows two gated latches with the same S and R input waveforms but different *ENABLE* waveforms. EN_1 has a 50% duty cycle. EN_2 has a duty cycle of 16.67%.

Draw the output waveforms, Q_1 and Q_2. Describe how the length of the *ENABLE* pulse affects the output of each latch, assuming that the intent of each circuit is to synchronize the output changes to the beginning of the *ENABLE* pulse.

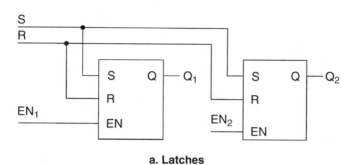

a. Latches

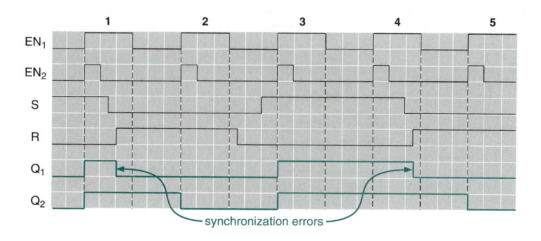

FIGURE 7.28
Example 7.4
Effect of ENABLE Pulse Width

SOLUTION   Figure 7.28b shows the completed timing diagram. The longer *ENABLE* pulse at latch 1 allows the output to switch too soon during pulses 1 and 4. ("Too soon" means before the beginning of the next *ENABLE* pulse.) In each of these cases, the S and R inputs change while the *ENABLE* input is HIGH. This premature switching is eliminated in latch 2 because the S and R inputs change after the shorter *ENABLE* pulse is finished. A shorter pulse gives less chance for synchronization error, since the time for possible output changes is minimized.   ▋▋

Transparent Latch (Gated D Latch)

Figure 7.29 shows the equivalent circuit of a gated D ("data") latch, or **transparent latch.** This circuit has two modes. When the *ENABLE* input is HIGH, the latch is *transparent* because the output Q goes to the level of the data input, D. (We say, "Q follows D.") When the *ENABLE* input is LOW, the latch *stores* the data that was present at D when *ENABLE* was last HIGH. In this way, the latch acts as a simple memory circuit.

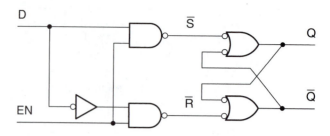

FIGURE 7.29
Transparent Latch

The latch in Figure 7.29 is a modification of the gated SR latch, configured so that the S and R inputs are always opposite. Under these conditions, the states $S = R = 0$ (no change) and $S = R = 1$ (forbidden) can never occur. However, the equivalent of the no change state happens when the *ENABLE* input is LOW, when the latch steering gates are inhibited.

Figure 7.30 shows the operation of the transparent latch in the inhibit (no change), set, and reset states. When the latch is inhibited, the steering gates block any LOW pulses to the latch gates; the latch does not change states, regardless of the logic level at D.

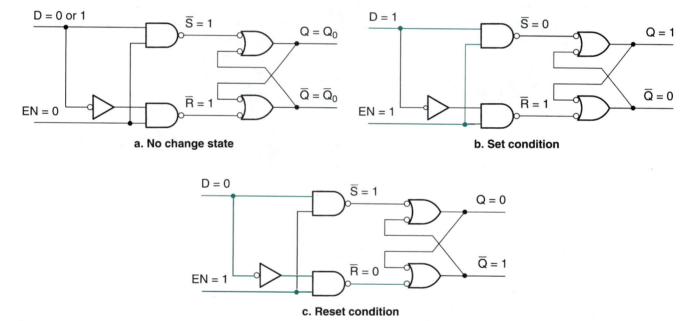

a. No change state

b. Set condition

c. Reset condition

FIGURE 7.30
Operation of Transparent Latch

If $EN = 1$, Q follows D. When $D = 1$, the upper steering gate transmits a LOW to the *SET* input of the latch and $Q = 1$. When $D = 0$, the lower steering gate transmits a LOW to the *RESET* input of the output latch and $Q = 0$.

Table 7.5 shows the function table for a transparent latch.

Table 7.5 Function Table of a Transparent Latch

EN	*D*	Q_{t+1}	$\bar{Q}_{t+1}$	**Function**	**Comment**
0	X	Q_t	$\bar{Q}_t$	No Change	Store
1	0	0	1	Reset	Transparent
1	1	1	0	Set	

Implementing D Latches in MAX+PLUS II

A D latch can be implemented in MAX+PLUS II as a primitive in a Graphic Design File or in a VHDL design entity. It can also be created with a behavioral or structural description in a VHDL file.

Figure 7.31 shows a D latch primitive in a MAX+PLUS II Graphic Design File. Figure 7.32 shows a simulation of the latch. From 0 to 500 ns, *ENABLE* is HIGH and the latch is in the transparent mode (*Q* follows *D*). When *ENABLE* goes LOW, the last value of *D* (0) is stored until *ENABLE* goes high again, just before 800 ns. When *ENABLE* goes LOW again, a new value of *D* (1) is stored until the end of the simulation.

d_latch.gdf
d_latch.scf

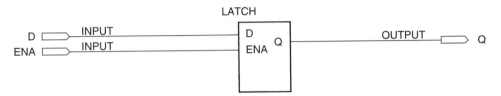

FIGURE 7.31
D-Latch in a MAX+PLUS II Graphic Design File

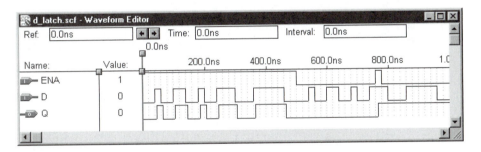

FIGURE 7.32
Simulation for a D Latch

In VHDL, a PROCESS statement is *concurrent,* but the statements inside the PROCESS are *sequential.* In other words, anything described by a PROCESS acts like a separate component in a design entity. However, the interior of the component so described acts as a sequential circuit. Since the behavior of a D latch is sequential, its description can be created inside a PROCESS. (You can pull a latch out of a bin of parts and connect it in a circuit, but the inside of the part is sequential.) The following VHDL code describes a D latch.

d_lch.vhd
d_lch.scf

```
-- d_lch.vhd
-- D latch with active-HIGH level-sensitive enable

ENTITY d_lch IS
    PORT(
        d, ena : IN    BIT;
        q      : OUT   BIT);
END d_lch;

ARCHITECTURE a OF d_lch IS
BEGIN
    PROCESS  (d, ena)
    BEGIN
        IF (ena = '1') THEN
            q  <= d;
        END IF;
    END PROCESS;
END a;
```

Another method, recommended by the MAX+PLUS II documentation, is to instantiate a LATCH primitive in a VHDL file. The primitive is contained in the **altera** library, in a package called **maxplus2.** The component declaration for this primitive is:

```
COMPONENT LATCH
    PORT (d    : IN STD_LOGIC;
            ena : IN STD_LOGIC;
            q    : OUT STD_LOGIC);
END COMPONENT;
```

Since the component declaration is in the **maxplus2** package, you do not have to declare it in the file in which you are using it. A VHDL file that uses the latch primitive is listed next. The component declaration uses STD   LOGIC types, so we must include the type definitions in the **ieee** library (**std_logic_1164** package).

lch_prim.vhd

```
-- lch_prim.vhd
-- D latch with active-HIGH level-sensitive enable

LIBRARY ieee;
USE ieee.std_logic_1164.ALL;
LIBRARY altera;
USE altera.maxplus2.ALL;

ENTITY lch_prim IS
    PORT(
        d_in, enable  : IN   STD_LOGIC;
        q_out            : OUT  STD_LOGIC);
END lch_prim;

ARCHITECTURE a OF lch_prim IS
BEGIN
    -- Instantiate a latch from a MAX+PLUS II primitive
    latch_primitive: latch
        PORT MAP (d     => d_in,
                    ena  => enable,
                    q     => q_out);
END a;
```

More information about MAX+PLUS II primitives can be found in MAX+PLUS II Help. In the **Help** menu, select **Primitives.** By clicking on the name of a particular primitive, you can determine whether it can be instantiated in a VHDL file and what its component declaration is, if available.

❚❚❚ EXAMPLE 7.5

A system for monitoring automobile traffic is set up at an intersection, with four sensors, placed as shown in Figure 7.33. Each sensor monitors traffic for a particular direction. When a car travels over a sensor, it produces a logic HIGH. The status of the sensor system

FIGURE 7.33
Example 7.5
Sensor Placement in a Traffic
Intersection

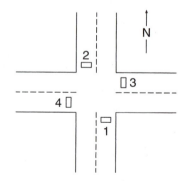

FIGURE 7.34

Example 7.5
D Latch Collection of Data

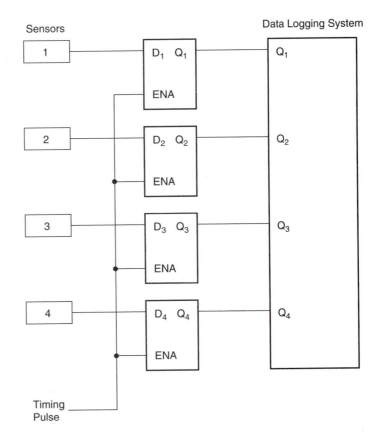

is captured for later analysis by a set of D latches, as shown in Figure 7.34. A timing pulse enables the latches once every five seconds and thus stores the system status as a "snapshot" of the traffic pattern.

Figure 7.35 shows the timing diagram of a typical traffic pattern at the intersection. The D inputs show the cars passing through the intersection in the various lanes. Complete this timing diagram by drawing the Q outputs of the latches.

How should we interpret the Q output waveforms?

FIGURE 7.35

Example 7.5
Latch Configuration and Timing Diagram

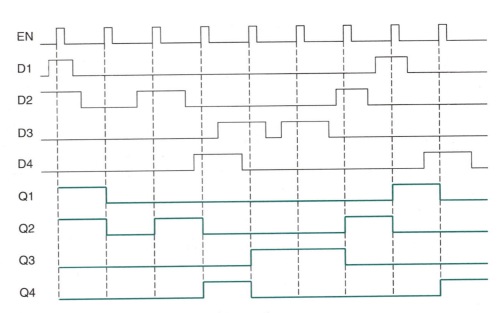

b. Timing diagram

SOLUTION   Figure 7.35 shows the completed timing diagram. The *ENABLE* input synchronizes the random sensor pattern to a 5-second standard interval. A HIGH on any Q output indicates a car over a sensor at the beginning of the interval. For example, at the beginning of the first interval, there is a car in the northbound lane ($Q1$) and one in the southbound lane ($Q2$). Similar interpretations can be made for each interval.   ▐▌

Multi-bit Latches in VHDL

> **KEY TERMS**
>
> **Library of Parameterized Modules (LPM)**   A standardized set of components for which certain properties can be specified when the component is instantiated.
>
> **Parameter (in an LPM component)**   A property of a component that can be specified when the component is instantiated.
>
> **Generic map**   A VHDL construct that maps one or more parameters of a component to a value for that instance of the component.
>
> **Port map**   A VHDL construct that maps the name of a port in a component to the name of a port, variable, or signal in a design entity that uses the component.

We can easily use VHDL to implement latches with multiple D inputs and Q outputs, but with a common *ENABLE* line, as in Figure 7.34. Three approaches are:

1. Use a behavioral description, as we did earlier for a single latch (**d_lch.vhd**). Use STD_LOGIC_VECTOR types for D and Q, rather than STD_LOGIC.

2. Altera recommends using a latch primitive or predefined component, rather than creating your own latch structures. We can use multiple LATCH primitives, instantiated by a GENERATE statement, as we did for multiple instances of a full adder in Chapter 6.

3. Use a latch component from the **Library of Parameterized Modules (LPM)**. These components are specified in the **lpm_components** package in the **lpm** library.

Certain properties of an LPM component, such as the number of inputs or outputs, can be specified when the component is instantiated. These properties are referred to as **parameters,** and are listed in a **generic map**. For example, to make the latch output and input four bits wide, we set the parameter called LPM_WIDTH to a value of 4. The various parameters of an LPM component can be found in the LPM Quick Reference on the CD that accompanies this book or in the MAX+PLUS II **Help** menu under **Megafunctions/LPM.**

An input or output of an LPM component is called a **port.** A port map is used to make a correspondence between the port names in the component declaration and the port names used in the file containing the component. Since LPM components are declared in a separate package, we must refer to the MAX+PLUS II Help or the LPM Quick Reference to determine the port names for a component. LPM components are instantiated the same as any other component.

The three VHDL files that follow each specify a 4-bit latch with common enable, each using one of the above methods.

Behavioral Description:

```
-- ltch4bhv.vhd
-- D latch with active-HIGH level-sensitive enable

LIBRARY ieee;
USE ieee.std_logic_1164.ALL;
```

```
ENTITY ltch4bhv IS
    PORT(d      : IN   STD_LOGIC_VECTOR (3 downto 0);
         enable : IN   STD_LOGIC;
         q      : OUT  STD_LOGIC_VECTOR (3 downto 0));
END ltch4bhv;

ARCHITECTURE a OF ltch4bhv IS
BEGIN
    PROCESS (enable, d)
    BEGIN
        IF (enable = ´1´) THEN
            q <= d;
        END IF;
    END PROCESS;
END a;
```

ltch4bhv.vhd

4 LATCH Primitives and a GENERATE Statement:

```
-- ltch4prm.vhd
-- D latch with active-HIGH level-sensitive enable

LIBRARY ieee;
USE ieee.std_logic_1164.ALL;
LIBRARY altera;
USE altera.maxplus2.ALL;

ENTITY ltch4prm IS
    PORT(d_in  : IN   STD_LOGIC_VECTOR (3 downto 0);
         enable : IN   STD_LOGIC;
         q_out  : OUT  STD_LOGIC_VECTOR (3 downto 0));
END ltch4prm;

ARCHITECTURE a OF ltch4prm IS
BEGIN
    -- Instantiate a latch from a MAX+PLUS II primitive
    latch4:
    FOR i IN 3 downto 0 GENERATE
    latch_primitive: latch
        PORT MAP (d => d_in (i), ena => enable, q => q_out (i));
    END GENERATE;
END a;
```

ltch4prm.vhd

LPM Latch:

```
-- ltch4lpm.vhd
-- 4-BIT D latch with active-HIGH level-sensitive enable
-- Uses a latch component from the Library of Parameterized
-- Modules (LPM)

LIBRARY ieee;
USE ieee.std_logic_1164.ALL;
LIBRARY lpm;
USE lpm.lpm_components.ALL;

ENTITY ltch4lpm IS
    PORT(d_in  : IN   STD_LOGIC_VECTOR (3 downto 0);
         enable : IN   STD_LOGIC;
         q_out  : OUT  STD_LOGIC_VECTOR (3 downto 0) );
END ltch4lpm;
```

ltch4lpm.vhd
ltch4lpm.scf

```
ARCHITECTURE a OF ltch4lpm IS
BEGIN
    -- Instantiate latch from an LPM component
    latch4: lpm_latch
        GENERIC MAP (LPM_WIDTH => 4)
        PORT MAP (data => d_in,
                  gate => enable,
                  q    => q_out);
END a;
```

All three files can be tested with the same simulation, shown in Figure 7.36. The inputs, **d_in,** represent a 4-bit group of signals, as do the outputs, **q_out.** An increasing count, from 5 to C (0101 to 1100) is applied to **d_in.** This count contains both states (0 and 1) for each input bit. For each applied input state, the output bus, **q_out,** does not change until the enable line goes HIGH.

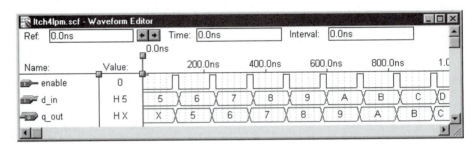

FIGURE 7.36
Simulation of a 4-bit D Latch

▐▐ SECTION 7.3 REVIEW PROBLEM

7.3 Write the VHDL code for a 16-bit latch with common active-HIGH enable, using MAX+PLUS II **latch** primitives.

7.4 Edge-Triggered D Flip-Flops

> **KEY TERMS**
>
> **Edge**   The HIGH-to-LOW (negative edge) or LOW-to-HIGH (positive edge) transition of a pulse waveform.
>
> **CLOCK**   An enabling input to a sequential circuit that is sensitive to the positive- or negative-going edge of a waveform.
>
> **Edge-triggered**   Enabled by the positive or negative edge of a digital waveform.
>
> **Edge-sensitive**   Edge-triggered.
>
> **Level-sensitive**   Enabled by a logic HIGH or LOW level.
>
> **Flip-flop**   A sequential circuit based on a latch whose output changes when its *CLOCK* input receives an edge.

In Example 7.4, we saw how a shorter pulse width at the *ENABLE* input of a gated latch increased the chance of the output being synchronized to the *ENABLE* pulse waveform. This is because a shorter *ENABLE* pulse gives less chance for the *SET* and *RESET* inputs to change during the time the latch is enabled.

A logical extension of this idea is to enable the latch for such a small time that the width of the *ENABLE* pulse is almost zero. The best approximation we can make to this is to allow changes to the circuit output only when an enabling, or **CLOCK,** input receives the **edge** of an input waveform. An edge is the part of a waveform that is in transition from

LOW to HIGH (positive edge) or HIGH to LOW (negative edge), as shown in Figure 7.37. We can say that a device enabled by an edge is **edge-triggered** or **edge-sensitive.**

FIGURE 7.37
Edges of a CLOCK Waveform

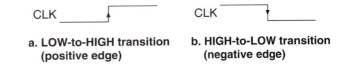

**a. LOW-to-HIGH transition
(positive edge)**   **b. HIGH-to-LOW transition
(negative edge)**

Since the *CLOCK* input enables a circuit only while in transition, we can refer to it as a "dynamic" input. This is in contrast to the *ENABLE* input of a gated latch, which is **level-sensitive** or "static," and will enable a circuit for the entire time it is at its active level.

Latches vs. Flip-Flops

> **KEY TERM**
>
> **Edge detector** A circuit in an edge-triggered flip-flop that converts the active edge of a CLOCK input to an active-level pulse at the internal latch's SET and RE-SET inputs.

A gated latch with a clock input is called a **flip-flop.** Although the distinction is not always understood, we will define a *latch* as a circuit with a *level-sensitive enable* (e.g., gated D latch) or *no enable* (e.g., NAND latch) and a *flip-flop* as a circuit with an *edge-triggered clock* (e.g., D flip-flop). A NAND or NOR latch is sometimes called an SR flip-flop. By our definition this is not correct, since neither of these circuits has a clock input. (An SR flip-flop would be like the gated SR latch of Figure 7.27 with a clock instead of an enable input.)

The symbol for the D, or data, flip-flop is shown in Figure 7.38. The D flip-flop has the same behavior as a gated D latch, except that the outputs change only on the positive edge of the clock waveform, as opposed to the HIGH state of the enable input. The triangle on the *CLK* (clock) input of the flip-flop indicates that the device is edge-triggered.

Table 7.6 shows the function table of a positive edge-triggered D flip-flop.

Figure 7.39 shows the equivalent circuit of a positive edge-triggered D flip-flop. The circuit is the same as the transparent latch of Figure 7.29, except that the enable input (called *CLK* in the flip-flop) passes through an **edge detector,** a circuit that converts a positive edge to a brief positive-going pulse. (A negative edge detector converts a negative edge to a positive-going pulse.)

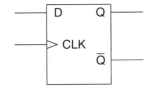

FIGURE 7.38
D Flip-Flop Logic Symbol

Table 7.6 Function Table for a Positive Edge-Triggered D Flip-Flop

CLK	D	Q_{t+1}	$\bar{Q}_{t+1}$	Function
↑	0	0	1	Reset
↑	1	1	0	Set
0	X	Q_t	$\bar{Q}_t$	Inhibited
1	X	Q_t	$\bar{Q}_t$	Inhibited
↓	X	Q_t	$\bar{Q}_t$	Inhibited

FIGURE 7.39
D Flip-Flop Equivalent Circuit

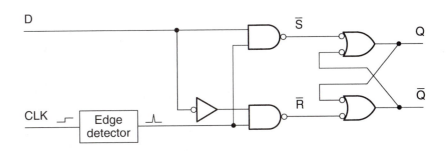

Figure 7.40 shows a circuit that acts as a simplified positive edge detector. Edge detection depends on the fact that a gate output does not switch immediately when its input switches. There is a delay of about 3 to 10 ns from input change to output change, called propagation delay.

FIGURE 7.40

Positive Edge Detector

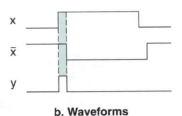

a. Simplified circuit

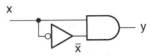

b. Waveforms

When input x, shown in the timing diagram of Figure 7.40, goes from LOW to HIGH, the inverter output, $\bar{x}$, goes from HIGH to LOW after a short delay. This delay causes both x and $\bar{x}$ to be HIGH for a short time, producing a high-going pulse at the circuit output immediately following the positive edge at x.

When x returns to LOW, $\bar{x}$ goes HIGH after a delay. However, there is no time in this sequence when both AND inputs are HIGH. Therefore, the circuit output stays LOW after the negative edge of the input waveform.

Figure 7.41 shows how the D flip-flop circuit operates. When $D = 0$ and the edge detector senses a positive edge at the *CLK* input, the output of the lower NAND gate steers a low-going pulse to the *RESET* input of the latch, thus storing a 0 at Q. When $D = 1$, the upper NAND gate is enabled. The edge detector sends a high-going pulse to the upper steering gate, which transmits a low-going *SET* pulse to the output latch. This action stores a 1 at Q.

FIGURE 7.41

Operation of a D Flip-Flop

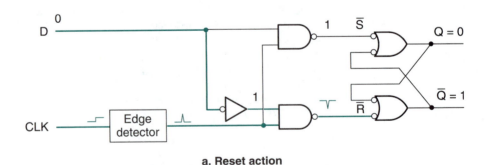

a. Reset action

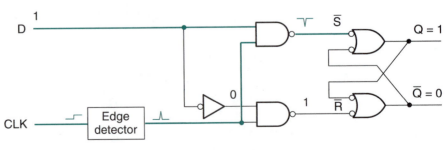

b. Set action

■■ EXAMPLE 7.6

Figure 7.42 shows a MAX+PLUS II Graphic Design File with a D latch and a D flip-flop connected to the same data input and clock. Create a MAX+PLUS II simulation that illustrates the difference between the latch (level-sensitive enable) and the flip-flop (edge-triggered clock).

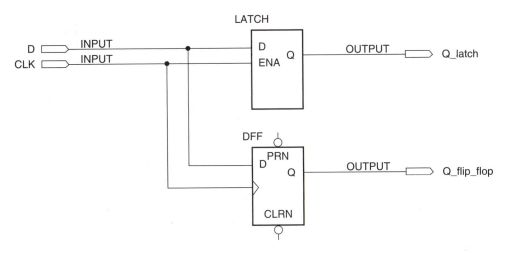

FIGURE 7.42
D Latch and D Flip-Flop

SOLUTION   The simulation, shown in Figure 7.43, has a 200 ns grid. Several points on the waveform indicate the similarities and differences between the latch and flip-flop operation.

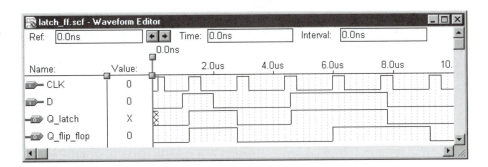

FIGURE 7.43
Simulation showing the Difference between D Latch and D Flip Flop

latch_ff.gdf
latch_ff.scf

1. At 1.2 μs, D goes HIGH. The latch output *(Q_latch)* and the flip-flop output *(Q_flip_flop)* both go HIGH at 1.4 μs, since the beginning of the enable HIGH state and the positive edge of the *CLK* both correspond to this time.

2. D goes LOW at 2 μs. Both Q outputs go LOW at 2.8 μs since the positive edge of the *CLK* and its HIGH level occur at the same time.

3. The D input goes HIGH at 4.4 μs, in the middle of a *CLK* pulse. Since the *CLK* line is HIGH, *Q_ latch* changes immediately. *Q_flip_flop* does not change until the next positive edge, at 6 μs.

4. D goes LOW at 7.8 μs. *Q_latch* also changes at this time, since *CLK* is HIGH. *Q_flip_flop* changes on the next positive edge, at 9.2 μs.

■■

Note that the latch output is in an unknown state until the first *CLK* pulse, whereas the flip-flop output is LOW, even before the first *CLK* pulse. This is because Altera CPLDs have power-on reset circuitry that ensures that flip-flop outputs in a CPLD are LOW immediately after power is applied to the device. The MAX+PLUS II simulator accounts for this condition.

▌▌ EXAMPLE 7.7

Two positive edge-triggered D flip-flops are connected as shown in Figure 7.44a. Inputs D_0 and *CLK* are shown in the timing diagram. Complete the timing diagram by drawing the waveforms for Q_0 and Q_1, assuming that both flip-flops are initially reset.

FIGURE 7.44
Example 7.7
Circuit and Timing Diagram

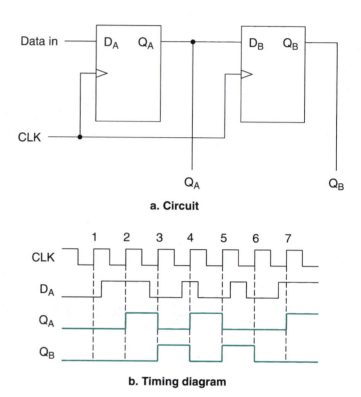

a. Circuit

b. Timing diagram

SOLUTION   Figure 7.44b shows the output waveforms. Q_0 follows D_0 at each point where the clock input has a positive edge. One result of this is that the HIGH pulse on D_0 between clock pulses 5 and 6 is ignored, since $D_0 = 0$ on positive edges 5 and 6.

Since $D_1 = Q_0$ and Q_1 follows D_1, the waveform at Q_1 is the same as at Q_0, but delayed by one clock cycle. If Q_0 changes due to *CLK,* we assume that the value of D_1 is the same as Q_0 just *before* the clock pulse. This is because delays within the circuitry of the flip-flops ensure that their outputs will not change for several nanoseconds after an applied clock pulse. Therefore, the level at D_1 remains constant long enough for it to be clocked into the second flip-flop.

The data entering the circuit at D_0 are moved, or shifted, from one flip-flop to the next. This type of data movement, called "serial shifting," is frequently used in data communication and digital arithmetic circuits.

▌▌ SECTION 7.4 REVIEW PROBLEM

7.4 Which part of a D flip-flop accounts for the difference in operation between a D flip-flop and a D latch? How does it work?

7.5 Edge-Triggered JK Flip-Flops

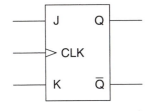

a. Positive edge-triggered

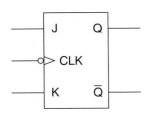

b. Negative edge-triggered

FIGURE 7.45
Edge-Triggered JK Flip-Flops

A versatile and widely used sequential circuit is the JK flip-flop.

Figure 7.45 shows the logic symbols of a positive- and a negative-edge triggered JK flip-flop. *J* acts as a *SET* input and *K* acts as a *RESET* input, with the output changing on the active clock edge in response to *J* and *K*. When *J* and *K* are both HIGH, the flip-flop will **toggle** between opposite logic states with each applied clock pulse. The function tables of the devices in Figure 7.45 are shown in Table 7.7.

Figure 7.46 shows the simplified circuit of a negative-edge triggered JK flip-flop. The circuit is like that of a gated SR latch with an edge detector (an SR flip-flop), except that there are two extra feedback lines from the latch outputs to the steering gate inputs. This extra feedback is responsible for the flip-flop's toggling action.

Figure 7.47 illustrates how the additional two lines cause the flip-flop to toggle. The cross-feedback from *Q* to *K* and from $\bar{Q}$ to *J* enables one, but not both, of the steering gates. The edge detector just after the *CLK* input produces a short positive-going pulse upon detecting a negative edge on the *CLK* waveform. The enabled steering gate complements and transmits this pulse to the latch, activating either the set or reset function. This in turn changes the latch state and enables the opposite steering gate.

Since all inputs of the steering gates must be HIGH to enable one of the latch functions, *J* and *K* must both be HIGH to sustain a repeated toggling action. Under these conditions, $\bar{Q}$ and *Q* alternately enable one of the steering gates.

Table 7.7   Function Tables for Edge-Triggered JK Flip-Flops

CLK	J	K	Q_{t+1}	$\bar{Q}_{t+1}$	Function	CLK	J	K	Q_{t+1}	$\bar{Q}_{t+1}$	Function
↑	0	0	Q_t	$\bar{Q}_t$	No change	↓	0	0	Q_t	$\bar{Q}_t$	No change
↑	0	1	0	1	Reset	↓	0	1	0	1	Reset
↑	1	0	1	0	Set	↓	1	0	1	0	Set
↑	1	1	$\bar{Q}_t$	Q_t	Toggle	↓	1	1	$\bar{Q}_t$	Q_t	Toggle
0	X	X	Q_t	$\bar{Q}_t$	Inhibited	0	X	X	Q_t	$\bar{Q}_t$	Inhibited
1	X	X	Q_t	$\bar{Q}_t$	Inhibited	1	X	X	Q_t	$\bar{Q}_t$	Inhibited
↓	X	X	Q_t	$\bar{Q}_t$	Inhibited	↑	X	X	Q_t	$\bar{Q}_t$	Inhibited
			Positive Edge-Triggered						Negative Edge-Triggered		

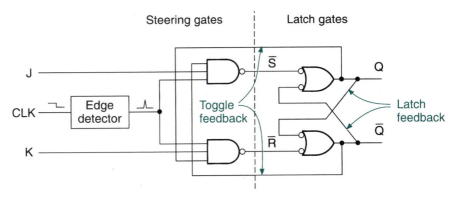

FIGURE 7.46
JK Flip-Flop Circuit (Simplified)

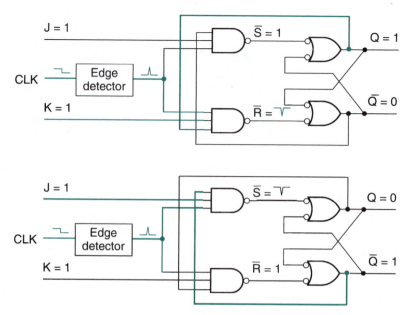

FIGURE 7.47
Toggle Action of a JK Flip-Flop

▌▌ EXAMPLE 7.8

The *J*, *K*, and *CLK* inputs of a negative edge-triggered JK flip-flop are as shown in the timing diagram in Figure 7.48. Complete the timing diagram by drawing the waveforms for Q and $\overline{Q}$. Indicate which function (no change, set, reset, or toggle) is performed at each clock pulse. The flip-flop is initially reset.

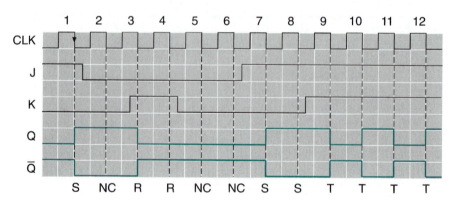

FIGURE 7.48
Example 7.8
Timing Diagram (Negative-Edge-Triggered JK Flip-Flop)

SOLUTION   The completed timing diagram is shown in Figure 7.48. The outputs change only on the negative edges of the *CLK* waveform. Note that the same output sometimes results from different inputs. For example, the function at clock pulse 4 is reset and the function at pulses 5 and 6 is no change, but the Q waveform is LOW in each case.

▌▌ EXAMPLE 7.9

The toggle function of a JK flip-flop is often used to generate a desired output sequence from a series of flip-flops. The circuit shown in Figure 7.49 is configured so that all flip-flops are permanently in toggle mode.

Assume that all flip-flops are initially reset. Draw a timing diagram showing the *CLK*, Q_0, Q_1, and Q_2 waveforms when eight clock pulses are applied. Make a table showing each

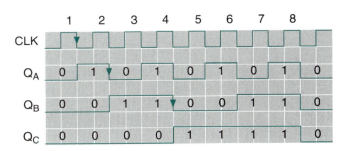

FIGURE 7.49
Example 7.9
Circuit

combination of Q_2, Q_1, and Q_0. What pattern do the outputs form over the period shown on the timing diagram?

SOLUTION   The circuit timing diagram is shown in Figure 7.50. All flip-flops are in toggle mode. Each time a negative clock edge is applied to the flip-flop *CLK* input, the Q output will change to the opposite state.

Table 7.8 Sequence of Outputs for Circuit in Figure 7.49

Clock Pulse	Q_2	Q_1	Q_0
0	0	0	0
1	0	0	1
2	0	1	0
3	0	1	1
4	1	0	0
5	1	0	1
6	1	1	0
7	1	1	1
8	0	0	0

FIGURE 7.50
Example 7.9
Timing Diagram

For flip-flop 0, this happens with every clock pulse, since it is clocked directly by the CLK waveform. Each of the other flip-flops is clocked by the Q output waveform of the previous stage. Flip-flop 1 is clocked by the negative edge of the Q_0 waveform. Flip-flop 2 toggles when Q_1 goes from HIGH to LOW.

Table 7.8 shows the flip-flop outputs after each clock pulse. The outputs form a 3-bit number that counts from 000 to 111 in binary sequence, then returns to 000 and repeats.

This flip-flop circuit is called a 3-bit asynchronous counter. ∎

Synchronous versus Asynchronous Circuits

KEY TERMS

Synchronous   Synchronized to the system clock.

Asynchronous   Not synchronized to the system clock

The **asynchronous** counter in Figure 7.49 has the advantage of being simple to construct and analyze. However, because it is asynchronous (that is, not synchronized to a single clock), it is seldom used in modern digital designs. The main problem with this and other asynchronous circuits is that their outputs do not change at the same time, due to delays in the flip-flops. This yields intermediate states that are not part of the desired output sequence.

Figure 7.51 shows a simulation of a circuit similar to that in Figure 7.49. The outputs are shown separately, and also as a group labeled Q[2..0] that shows the combined binary value of the outputs.

asynch3.gdf
asynch3.scf

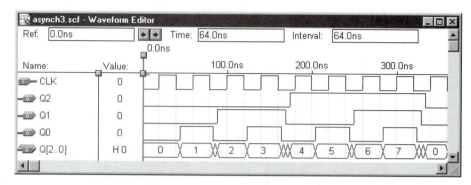

FIGURE 7.51
Simulation of a 3-bit Asynchronous Counter

Figure 7.52 shows a detail of the simulation at the point where the output goes from 7 to 0 (111 to 000). At 300 ns, the circuit output is 111. A negative clock edge, applied to flip-flop 0, makes Q_0 toggle after a short delay. The output is now 110 (=6_{10}). The resulting negative edge on Q_0 clocks flip-flop 1, making it toggle, and yields a new output of 100 (=4_{10}). The negative edge on Q_1 clocks flip-flop 2, making the output equal to 000 after a short delay.

FIGURE 7.52

Detail of simulation for a 3-bit Asynchronous Counter

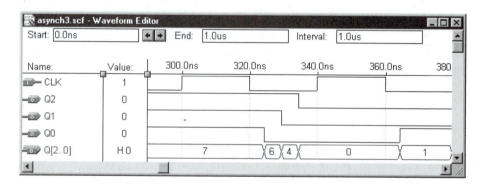

Thus, the output goes through two short intermediate states that are not in the desired output sequence. Instead of going directly from 111 to 000, as in Figure 7.50, the output goes in the sequence 111–110–100–000. We see in Figure 7.51 that the counter output goes through one or more intermediate transitions after each negative edge of the Q_0 waveform. In other words, intermediate states arise whenever a change propagates through more than one flip-flop. This happens because the flip-flops are clocked from different sources.

sync3.gdf
sync3.scf

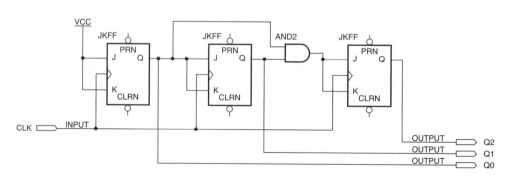

FIGURE 7.53
3-bit Synchronous Counter

Figure 7.53 shows the circuit of a 3-bit **synchronous** counter. Unlike the circuit in Figure 7.49, the flip-flops in this circuit are clocked from a common source. Therefore, flip-flop delays do not add up through the circuit, and all the outputs change at the same time. Figure 7.54 shows a simulation of the circuit of Figure 7.53. Note that the outputs progress in a binary sequence, and there are no intermediate states.

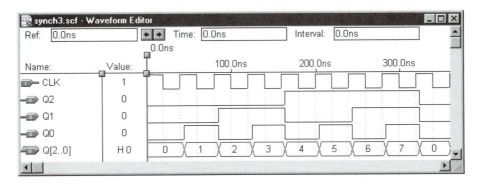

FIGURE 7.54
Simulation of a 3-bit Synchronous Counter

The circuit works as follows:

1. Flip-flop 0 is configured for toggle mode ($J_0 K_0 = 11$). Since the flip-flops in Figure 7.53 are positive edge-triggered, Q_0 toggles on each positive clock edge.

2. Q_0 is connected to inputs J_1 and K_1. Since these inputs are tied together, only two states are possible: no change ($JK = 00$) or toggle ($JK = 11$). If $Q_0 = 1$, Q_1 toggles. Otherwise, it does not change. This results in a Q_1 waveform that toggles at half the rate of Q_0.

3. J_2 and K_2 are both tied to the output of an AND gate. The AND gate output is HIGH if *both* Q_1 and Q_0 are HIGH. This makes Q_2 toggle, since $J_2 K_2 = 11$. In all other cases, there is no change on Q_2. The result of this is that Q_2 toggles every fourth clock pulse, the only times when Q_1 and Q_0 are both HIGH.

Asynchronous Inputs (Preset and Clear)

> ### KEY TERMS
>
> **Synchronous inputs**   The inputs of a flip-flop that do not affect the flip-flop's Q outputs unless a clock pulse is applied. Examples include D, J, and K inputs.
>
> **Asynchronous inputs**   The inputs of a flip-flop that change the flip-flop's Q outputs immediately, without waiting for a pulse at the CLK input. Examples include preset and clear inputs.
>
> **Preset**   An asynchronous set function.
>
> **Clear**   An asynchronous reset function.

The *D, J,* and *K* inputs of the flip-flops examined so far are called **synchronous inputs.** This is because any effect they have on the flip-flop outputs is synchronized to the *CLK* input.

Another class of input is also provided on many flip-flops. These inputs, called **asynchronous inputs,** do not need to wait for a clock pulse to make a change at the output. The two functions usually provided are **preset,** an asynchronous set function, and **clear,** an asynchronous reset function. These functions are generally active LOW, and are abbreviated $\overline{PRE}$ and $\overline{CLR}$.

Figure 7.55 shows a modification to the JK flip-flop of Figure 7.46. The $\overline{PRE}$ and $\overline{CLR}$ inputs have direct access to the latch gates of the flip-flop and thus are not affected by the

FIGURE 7.55
$\overline{PRE}$ and $\overline{CLR}$ Inputs

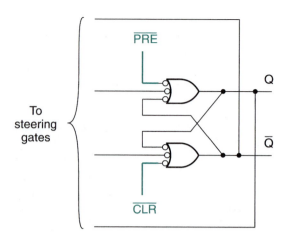

CLK input. They act exactly the same as the *SET* and *RESET* inputs of an SR latch and will override any synchronous input functions currently active.

█ EXAMPLE 7.10

The waveforms for the *CLK, J, K, $\overline{PRE}$*, and *$\overline{CLR}$* inputs of a negative edge-triggered JK flip-flop are shown in the timing diagram of Figure 7.56. Complete the diagram by drawing the waveform for output *Q*.

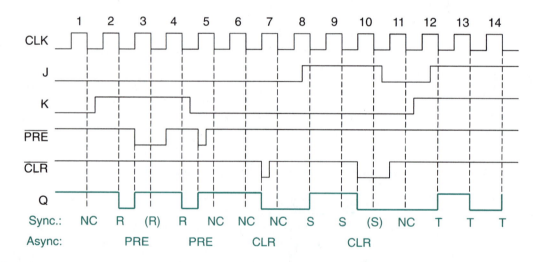

FIGURE 7.56
Example 7.10
Waveforms

SOLUTION The *Q* waveform is shown in Figure 7.56. The asynchronous inputs cause an immediate change in *Q*, whereas the synchronous inputs must wait for the next negative clock edge. If asynchronous and synchronous inputs are simultaneously active, the asynchronous inputs have priority. This occurs in two places: pulse 3 *(K, $\overline{PRE}$)* and pulse 10 *(J, $\overline{CLR}$)*.

The diagram shows the synchronous functions (no change, reset, set, and toggle) at each clock pulse and the asynchronous functions (preset and clear) at the corresponding transition points.

█

The function table of a negative edge-triggered JK flip-flop with preset and clear functions is shown in Table 7.9.

Table 7.9 Function Table of a Negative Edge-Triggered JK Flip-Flop with Preset and Clear Functions

	$\overline{PRE}$	$\overline{CLR}$	CLK	J	K	Q_{t+1}	$\overline{Q}_{t+1}$	Function
Synchronous Functions	1	1	↓	0	0	Q_t	$\overline{Q}_t$	No change
	1	1	↓	0	1	0	1	Reset
	1	1	↓	1	0	1	0	Set
	1	1	↓	1	1	$\overline{Q}_t$	Q_t	Toggle
Asynchronous Functions	0	1	X	X	X	1	0	Preset
	1	0	X	X	X	0	1	Clear
	0	0	X	X	X	1	1	Forbidden
	1	1	0	X	X	Q_t	$\overline{Q}_t$	Inhibited
	1	1	1	X	X	Q_t	$\overline{Q}_t$	Inhibited
	1	1	↑	X	X	Q_t	$\overline{Q}_t$	Inhibited

X = Don't care   ↓ = HIGH-to-LOW transition
Q_t = Present state of Q   ↑ = LOW-to-HIGH transition
Q_{t+1} = Next state of Q

> **NOTE**
>
> If preset and clear functions are not used, they should be disabled by connecting them to logic HIGH (for active-LOW inputs). This prevents them from being activated inadvertently by circuit noise. The synchronous functions of some flip-flops will not operate properly unless $\overline{PRE}$ and $\overline{CLR}$ are HIGH. In MAX+PLUS II, the asynchronous inputs of all flip-flop primitives are set to a default level of HIGH.

Using Asynchronous Reset in a Synchronous Circuit

> **KEY TERM**
>
> **Master Reset**   An asynchronous reset input used to set a sequential circuit to a known initial state.

Figure 7.57 shows an application of asynchronous clear inputs in a 3-bit synchronous counter. An input called *RESET* is tied to the asynchronous $\overline{CLR}$ inputs of all flip-flops. The counter output is set to 000 when the *RESET* line goes LOW.

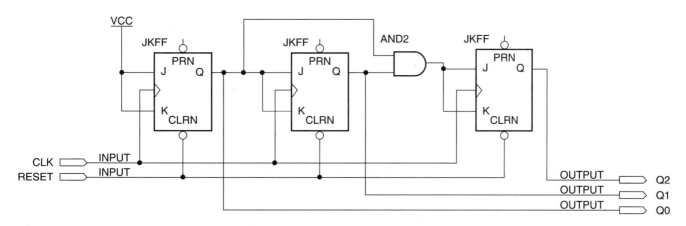

FIGURE 7.57
Synchronous Counter with Asynchronous Reset

Figure 7.58 shows a set of simulation waveforms that illustrate the asynchronous clear function. When *RESET* is HIGH, the count proceeds normally. The positive clock edge at 440 ns drives the counter to state 011. The reset pulse at 460 ns sets the counter to 000 as soon as it goes LOW. On the next clock edge, the count proceeds from 000.

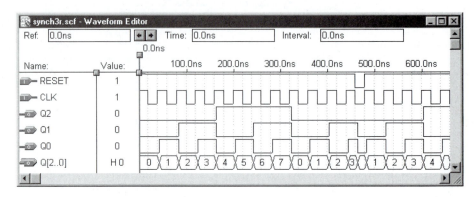

FIGURE 7.58
Simulation of Synchronous Counter with Asynchronous Reset

The function that sets all flip-flops in a circuit to a known initial state is sometimes called **Master Reset.**

▮▮ SECTION 7.5 REVIEW PROBLEM

7.5 What is the main difference between synchronous and asynchronous circuits, such as the two counters in Figures 7.49 and 7.53? What disadvantage is there to an asynchronous circuit?

7.6 Edge-Triggered T Flip-Flops

KEY TERM

T (toggle) flip-flop A flip-flop whose output toggles between HIGH and LOW states on each applied clock pulse when a synchronous input, called *T*, is active.

In the section on the JK flip-flop, we saw how that device can be set to toggle between HIGH and LOW output states. Other types of flip-flops can perform this function, as well. For example, Figure 7.59 shows a D flip-flop configured for toggle operation. Since Q follows D and $D = \overline{Q}$ in this circuit, then the flip-flop output must change to its opposite state with each clock pulse. Figure 7.60 shows a MAX+PLUS II simulation of this circuit.

notg2d.gdf
notg2d.scf

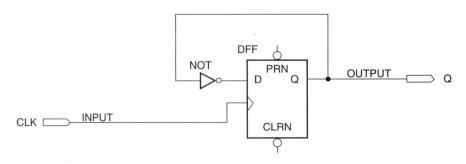

FIGURE 7.59
D Flip-Flop Configured for Toggle Function

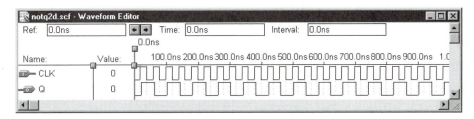

FIGURE 7.60
Simulation of D Flip-Flop in Toggle Mode

It is seldom useful for flip-flops in synchronous circuits to be permanently configured in toggle mode. What made the JK flip-flops suitable elements for the synchronous counter in Figure 7.53 was the fact that sometimes they toggled and sometimes they didn't, depending on the current point in the output sequence of the counter. Figure 7.61 shows a D flip-flop configured for a switchable toggle function.

FIGURE 7.61
Switchable Toggle Function for a D Flip-Flop

d_toggle.gdf
d_toggle.scf

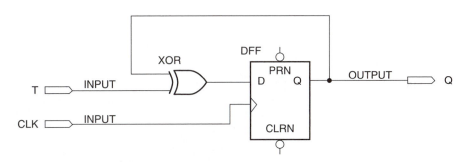

The XOR gate acts as an inverter when the T input is HIGH and as a noninverting buffer when T is LOW. Thus, when T is LOW, the Q output is circulated back to the D input of the flip-flop and the current value of Q is reloaded on the next clock pulse. When T is HIGH, the circuit acts like that of Figure 7.59 and toggles.

A **T flip-flop** has this equivalent function. Figure 7.62 shows the symbol of a T flip-flop in a MAX+PLUS II Graphic Design File. A MAX+PLUS II simulation in Figure 7.63 shows the operation of this device. The Q output toggles on each clock pulse when

FIGURE 7.62
T Flip-Flop

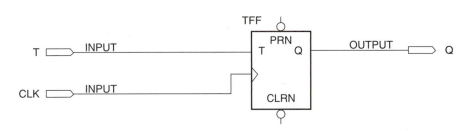

t_flipflop.gdf
t_flipflop.scf

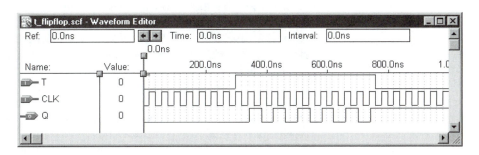

FIGURE 7.63
Simulation of T Flip-Flop

T is HIGH; otherwise Q retains its last value. A function table for the T flip-flop is shown in Table 7.10.

Table 7.10 Function Table for a T Flip-Flop

CLK	T	Q_{t+1}	Function
↑	0	Q_t	No Change
↑	1	$\overline{Q}_t$	Toggle
0	X	Q_t	Inhibited
1	X	Q_t	Inhibited
↓	X	Q_t	Inhibited

▐▐ SECTION 7.6 REVIEW PROBLEM

7.6 Draw a circuit showing how the JK flip-flops in Figure 7.53 can be replaced by T flip-flops.

7.7 Timing Parameters

> **KEY TERMS**
>
> **Setup time (t_{su})**   The time required for the synchronous inputs of a flip-flop to be stable before a *CLK* pulse is applied.
>
> **Hold time (t_h)**   The time that the synchronous inputs of a flip-flop must remain stable after the active *CLK* transition is finished.
>
> **Pulse width (t_w)**   Minimum time required for an active-level pulse applied to a *CLK*, $\overline{CLR}$, or $\overline{PRE}$ input, as measured from the midpoint of the leading edge of the pulse to the midpoint of the trailing edge.
>
> **Recovery time (t_{rec})**   Minimum time from the midpoint of the trailing edge of a $\overline{CLR}$ or $\overline{PRE}$ pulse to the midpoint of an active *CLK* edge.
>
> **Propagation delay**   The time required for the output of a digital circuit to change states after a change at one or more of its inputs.

Flip-flops are electrical devices with inherent internal switching delays. As such, they have specific requirements for the timing of the input and output waveforms in order for them to operate reliably. We will examine the basic timing requirements for two small scale integration (SSI) devices: the 74LS107A JK flip-flop (LSTTL family) and the 74HC107 JK flip-flop (high-speed CMOS family). Figure 7.64 shows some of the basic timing requirements of a JK flip-flop.

Figure 7.64a illustrates the definitions of **setup time** (t_{su}), **hold time** (t_h), *and* **pulse width** (t_W). The notation used for the "J or K" waveform indicates that the J or K input could be at either logic level and makes a transition to the opposite level at some point. The setup time is measured from the midpoint of the J or K transition to the midpoint of the active *CLK* edge. The logic level on the J or K input must be steady for at least this time for the flip-flop to operate correctly. Setup time for both LSTTL and high-speed CMOS flip-flops is about 20 ns.

Similarly, the hold time is measured from the midpoint of the *CLK* transition to the midpoint of the next J or K transition. The J or K level must be held steady for at least this time to ensure dependable operation. Hold time is 0 for LSTTL and 3 ns for a high-speed CMOS flip-flop.

The pulse width, t_w, shows how long the *CLK* needs to be held LOW after an active *CLK* edge. Although the LOW level does not itself latch data into the flip-flop, internal logic levels must reach a steady state before the device can accept a new clock pulse. This

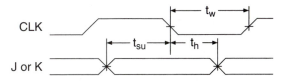

a. Setup, hold, and CLK pulse width

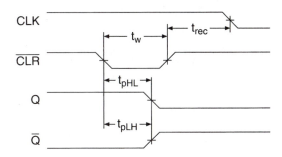

b. CLR pulse width, propagation delay, and recovery time

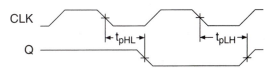

c. Propagation delay from CLK

FIGURE 7.64
Timing Parameters of a JK Flip-Flop

minimum pulse width allows the necessary time for these internal transitions. The data sheet for a 74HC107 flip-flop (high-speed CMOS) gives the clock pulse width as 16 ns; a data sheet for a 74LS107A device gives the value as 20 ns.

Figure 7.64b shows the pulse width required at the $\overline{CLR}$ input, the **propagation delay** from $\overline{CLR}$ to Q and $\overline{Q}$, and the **recovery time** that must be allowed from the end of a $\overline{CLR}$ pulse to the beginning of a CLK pulse. These times also apply to a pulse on the $\overline{PRE}$ input of a flip-flop.

Propagation delay is the result of internal electrical delays, primarily the charging and discharging of internal capacitances of the gate transistor junctions. The practical result of this is that a pulse at the $\overline{CLR}$ input makes Q go LOW, but not immediately; there is a delay of several nanoseconds between input pulse and output response.

Propagation delay is defined by the direction of the *output* transition. The delay at Q, which goes from HIGH to LOW, is called t_{pHL}. The delay at $\overline{Q}$, which goes from LOW to HIGH when cleared, is called t_{pLH}. Values for propagation delay from $\overline{CLR}$ to Q or $\overline{Q}$ are about 20 ns for LSTTL and 31 ns for high-speed CMOS.

The recovery time, t_{rec}, allows the internal logic levels of the flip-flop to reach a steady state after a $\overline{CLR}$ pulse. When the internal levels are stable, the device is ready to accept an active CLK edge. The recovery time for high-speed CMOS is 20 ns and 25 ns for an LSTTL device. (The LSTTL data sheet treats this parameter as a species of setup time; it is shown as setup time after the $\overline{CLR}$ is inactive. Same thing.)

Finally, Figure 7.64c shows the propagation delay from CLK to Q. This is the time from the midpoint of an active CLK edge to the midpoint of a transition at Q caused by that CLK edge. The parameters are defined, as before, by the direction of the output transition. Propagation delays t_{pLH} and t_{pHL} are 20 ns, maximum, for a 74LS107A device and 25 ns for a 74HC107 flip-flop.

The timing restrictions of a flip-flop imply that there is a maximum *CLK* frequency beyond which the device will not operate reliably. Data sheets give these values as about 30 MHz for both LSTTL and high-speed CMOS devices.

Table 7.11 summarizes the timing parameters of a 74LS107A flip-flop and a 74HC107 device. The values for the latter device are for V_{cc} = 4.5 V and a temperature range of −55°C to 25°C; they increase with a higher temperature range or a lower supply voltage.

Table 7.11   Timing Parameters of an LSTTL and a High-Speed CMOS Flip-Flop

Symbol	Parameter	74LS107A	74HC107
t_{su}	Setup time	20 ns	20 ns
t_h	Hold time	0 ns	3 ns
t_w	$\overline{CLR}$ pulse width	25 ns	16 ns
	CLK pulse width	20 ns	16 ns
t_{rec}	Recovery time	25 ns	20 ns
t_{pHL}	Propagation delay		
t_{pLH}	(from $\overline{CLR}$)	20 ns	31 ns
	(from *CLK*)	20 ns	25 ns
f_{max}	Maximum frequency	30 MHz	30 MHz

▌▌ EXAMPLE 7.11

The timing diagrams in Figure 7.65 represent some of the timing parameters of a JK flip-flop. From these diagrams, determine the setup and hold times and the propagation delays from *CLK* and $\overline{CLR}$ to Q and $\overline{Q}$.

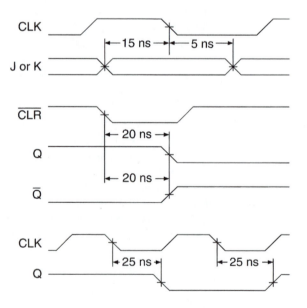

FIGURE 7.65
Example 7.11
Timing Parameters

SOLUTION   The values are as follows:

Setup time = 15 ns

Hold time = 5 ns

Propagation delays (from $\underline{CLK}$): 25 ns (t_{pLH} and t_{pHL})
(from $\overline{CLR}$): 20 ns (t_{pLH} and t_{pHL})

■■ SECTION 7.7 REVIEW PROBLEM

7.7 An active edge on the clock input of a JK flip-flop makes Q go from HIGH-to-LOW. Name the timing parameter that measures the delay between the input and output change. Write the symbol for the parameter.

SUMMARY

1. A combinational circuit combines inputs to generate a particular output logic level that is always the same, regardless of the order in which the inputs are applied. A sequential circuit might generate different outputs for the same inputs, depending on the sequence in which the inputs were applied.
2. An SR latch is a sequential circuit with *SET (S)* and *RESET (R)* inputs and complementary outputs (Q and $\overline{Q}$). By definition, a latch is set when $Q = 1$ and reset when $Q = 0$.
3. A latch sets when its S input activates. When S returns to the inactive state, the latch remains in the set condition until explicitly reset by activating its R input.
4. A latch can have active-HIGH inputs (designated S and R) or active-LOW inputs (designated $\overline{S}$ and $\overline{R}$).
5. Two basic SR latch circuits are the NAND latch and the NOR latch, each consisting of two gates with cross-coupled feedback. In the NAND form, we draw the gates in their DeMorgan equivalent form so that each circuit has OR-shaped gates, inversion from input to output, and feedback to the opposite gate.
6. A NOR latch has active-HIGH inputs. It is described by the following function table:

S	R	Q_{t+1}	$\overline{Q}_{t+1}$	Function
0	0	Q_t	$\overline{Q}_t$	No change
0	1	0	1	Reset
1	0	1	0	Set
1	1	0	0	Forbidden

7. A NAND latch has active-LOW inputs and is described by the following function table:

$\overline{S}$	$\overline{R}$	Q_{t+1}	$\overline{Q}_{t+1}$	Function
0	0	1	1	Forbidden
0	1	1	0	Set
1	0	0	1	Reset
1	1	Q_t	$\overline{Q}_t$	No change

8. A NAND latch can be used as a switch debouncer for a switch with a grounded common terminal, a normally open, and a normally closed contact. When the switch operates, one contact closes, resetting the latch on the first bounce. Further bounces are ignored. When the switch returns to its normal position, it sets the latch on the first bounce and further bounces are ignored.

9. A gated SR latch controls the times when a latch can switch. The circuit consists of a pair of latch gates and a pair of steering gates. The steering gates are enabled or inhibited by a control signal called *ENABLE*. When the steering gates are enabled, they can direct a set or reset pulse to the latch gates. When inhibited, the steering gates block any set or reset pulses to the latch gates so the latch output cannot change.
10. A gated D ("data") latch can be constructed by connecting opposite logic levels to the S and R inputs of an SR latch. Since S and R are always opposite, the D latch has no forbidden state. The no change state is provided by the inhibit property of the *ENABLE* input.
11. In a gated D latch (or transparent latch), Q follows D when *ENABLE* is active. This is the transparent mode of the latch. When *ENABLE* is inactive, the latch stores the last value of D.
12. A D latch can be described in VHDL by an IF statement within a PROCESS. The PROCESS statement in VHDL is concurrent, but the statements inside the PROCESS are sequential.
13. A D latch can also be implemented in VHDL by instantiating a LATCH primitive as a component in a VHDL design entity or by instantiating a component called **lpm_latch** from the Library of Parameterized Modules (LPM).
14. An LPM component is a standard component with certain properties, called parameters, that can be specified when the component is instantiated. The inputs and outputs of an LPM component are called ports. Parameter values are assigned in the generic map of a component instantiation statement. Component port names are associated with user port names in the port map of a component instantiation statement.
15. A flip-flop is like a gated latch that responds to the edge of a pulse applied to an enable input called *CLOCK*. A flip-flop output will change only when the input makes a transition from LOW to HIGH (for a positive edge-triggered device) or HIGH to LOW (for a negative edge-triggered device).
16. In a positive edge-triggered D flip-flop, Q follows D when there is a positive edge on the clock input.
17. D flip-flops are used primarily for data storage and transfer.
18. A JK flip-flop has two synchronous inputs, called J and K. J acts as an active-HIGH set input. K acts as an active-HIGH reset function. When both inputs are asserted, the flip-flop toggles between 0 and 1 with each applied clock pulse.
19. The toggle function in a JK flip-flop is implemented with additional cross-coupled feedback from the latch gate outputs to the steering gate inputs.
20. A chain of JK flip-flops can implement an asynchronous binary counter if the Q of each flip-flop is connected to the

clock input of the next. Although this is an easy way to create a counter, it is seldom used because internal flip-flop delays result in unwanted intermediate states in the count sequence.

21. JK flip-flops can be combined with a network of logic gates to make a synchronous binary counter. The gates are connected in such a way that each flip-flop toggles when all previous bits are HIGH; otherwise the flip-flops are in a no change state. Although more complex than an asynchronous counter, a synchronous counter is free of unwanted intermediate states.

22. Many flip-flops are provided with asynchronous preset (set) and clear (reset) functions. Since these functions are connected directly to the latch gates of a flip-flop, they act immediately, without waiting for the clock. In most cases, these functions are active-LOW.

23. Asynchronous inputs, such as preset and clear, are usually designed so that they will override the synchronous inputs, such as *D* or *JK*.

24. Unused asynchronous inputs should be disabled by tying them to a logic HIGH (for an active-LOW input). Flip-flop primitives in MAX+PLUS II automatically have their asynchronous inputs connected to HIGH unless otherwise specified by a design entry file.

25. The outputs of a T (toggle) flip-flop toggle with each clock pulse when the *T* input is HIGH and do not change when *T* is LOW.

26. Several important timing parameters for a flip-flop include: setup and hold time, propagation delay, minimum pulse width, and recovery time.

27. Setup time is the time before a clock edge that a synchronous input must be held steady. Hold time is the time after an applied clock edge that an input level must be held constant.

28. Propagation delay is the time for an input change, such as on *CLK* or $\overline{CLR}$, to have an effect on an output, such as *Q*. Propagation time is always indicated with respect to the change in output level: t_{pLH} for a LOW-to-HIGH output transition and t_{pHL} for a HIGH-to-LOW output change.

29. Minimum pulse width, t_w, indicates how long a *CLK* or $\overline{CLR}$ input must be held after an active edge or level is applied before returning to the original level.

30. Recovery time is the minimum time required from the end of an active level on one input (such as $\overline{CLR}$) to an active *CLK* edge.

GLOSSARY

Asynchronous   Not synchronized to the system clock.

Asynchronous inputs   The inputs of a flip-flop that change the flip-flop's *Q* outputs immediately, without waiting for a pulse at the *CLK* input. Examples include preset and clear inputs.

Clear   An asynchronous reset function.

CLOCK   An enabling input to a sequential circuit that is sensitive to the positive- or negative-going edge of a waveform.

Edge   The HIGH-to-LOW (negative edge) or LOW-to-HIGH (positive edge) transition of a pulse waveform.

Edge detector   A circuit in an edge-triggered flip-flop that converts the active edge of a *CLOCK* input to an active-level pulse at the internal latch's *SET* and *RESET* inputs.

Edge-sensitive   Edge-triggered.

Edge-triggered   Enabled by the positive or negative edge of a digital waveform.

Flip-flop   A sequential circuit based on a latch whose output changes when its *CLOCK* input receives either an edge or a pulse, depending on the device.

Gated SR latch   An SR latch whose ability to change states is controlled by an extra input called the *ENABLE* input.

Generic map   A VHDL construct that maps one or more parameters of a component to a value for that instance of the component.

Hold time (t_h) The time that the synchronous inputs of a flip-flop must remain stable after the active *CLK* transition is finished.

Latch   A sequential circuit with two inputs called *SET* and *RESET*, which make the latch store a logic 0 (reset) or 1 (set) until actively changed.

Level-sensitive   Enabled by a logic HIGH or LOW level.

Library of Parameterized Modules (LPM)   A standardized set of components for which certain properties can be specified when the component is instantiated.

Master Reset   An asynchronous reset input used to set a sequential circuit to a known initial state.

Parameter (in an LPM component)   A property of a component that can be specified when the component is instantiated.

Preset   An asynchronous set function.

Port map   A VHDL construct that maps the name of a port in a component to the name of a port, variable, or signal in a design entity that uses the component.

Propagation delay   The time required for the output of a digital circuit to change states after a change at one or more of its inputs.

Pulse width (t_w) Minimum time required for an active-level pulse applied to a *CLK*, $\overline{CLR}$, or $\overline{PRE}$ input, as measured from the midpoint of the leading edge of the pulse to the midpoint of the trailing edge.

Recovery time (t_{rec}) Minimum time from the midpoint of the trailing edge of a $\overline{CLR}$ or $\overline{PRE}$ pulse to the midpoint of an active *CLK* edge.

Reset   1. The stored LOW state of a latch circuit. 2. A latch input that makes the latch store a logic 0.

Sequential circuit   A digital circuit whose output depends not only on the present combination of inputs, but also on the history of the circuit.

Set   1. The stored HIGH state of a latch circuit. 2. A latch input that makes the latch store a logic 1.

Setup time (t_{su}) The time required for the synchronous inputs of a flip-flop to be stable before a *CLK* pulse is applied.

Steering gates   Logic gates, controlled by the *ENABLE* input of a gated latch, that steer a *SET* or *RESET* pulse to the correct input of an SR latch circuit.

Synchronous   Synchronized to the system clock.

Synchronous inputs   The inputs of a flip-flop that do not affect the flip-flop's Q outputs unless a clock pulse is applied. Examples include *D, J,* and *K* inputs.

Toggle   Alternate between binary states with each applied clock pulse.

T (toggle) flip-flop   A flip-flop whose output toggles between HIGH and LOW states on each applied clock pulse when a synchronous input, called *T,* is active.

Transparent latch (gated D latch)   A latch whose output follows its data input when its *ENABLE* input is active.

PROBLEMS

Section 7.1   Latches

7.1   Complete the timing diagram in Figure 7.66 for the active-HIGH latch shown. The latch is initially set.

7.2   Repeat Problem 7.1 for the timing diagram shown in Figure 7.67.

7.3   Complete the timing diagram in Figure 7.68 for the active-LOW latch shown.

7.4   Figure 7.69 shows an active-LOW latch used to control a motor starter. The motor runs when $Q = 1$ and stops when $Q = 0$. (Problem continues . . .)

FIGURE 7.66
Problem 7.1
Timing Diagram

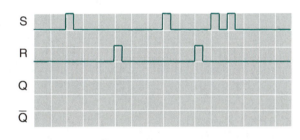
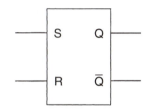

FIGURE 7.67
Problem 7.2
Timing Diagram

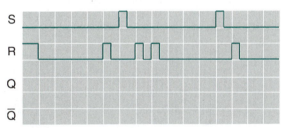

FIGURE 7.68
Problem 7.3
Timing Diagram

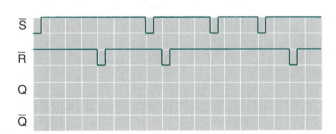
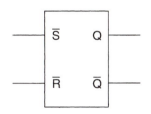

FIGURE 7.69
Problem 7.4
Latch for Motor Starter

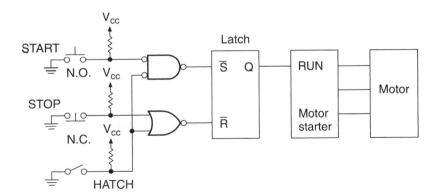

The motor is housed in a safety enclosure that has an access hatch for service. A safety interlock prevents the motor from running when the hatch is open. The *HATCH* switch opens when the hatch opens, supplying a logic HIGH to the circuit. The *START* switch is a normally open momentary-contact pushbutton (LOW when pressed). The *STOP* switch is a normally closed momentary-contact pushbutton (HIGH when pressed).

Draw the timing diagram of the circuit, showing *START, STOP, HATCH, S̄, R̄, and Q* for the following sequence of events:

a. *START* is pressed and released.

b. The hatch cover is opened.

c. *START* is pressed and released.

d. The hatch cover is closed.

e. *START* is pressed and released.

f. *STOP* is pressed and released.

Briefly describe the functions of the three switches and how they affect the motor operation.

Section 7.2 NAND/NOR Latches

7.5 Draw a NAND latch, correctly labeling the inputs and outputs. Describe the operation of a NAND latch for all four possible combinations of $\overline{S}$ and $\overline{R}$.

7.6 Draw a NOR latch, correctly labeling the inputs and outputs. Describe the operation of a NOR latch for all four possible combinations of S and R.

7.7 The timing diagram in Figure 7.70 shows the input waveforms of a NAND latch. Complete the diagram by showing the output waveforms.

7.8 Figure 7.71 shows the input waveforms to a NOR latch. Draw the corresponding output waveforms.

7.9 Figure 7.72 represents two input waveforms to a latch circuit.

a. Draw the outputs Q and $\overline{Q}$ if the latch is a NAND latch.

b. Draw the output waveforms if the latch is a NOR latch.

(Note that in each case, the waveforms will produce the forbidden state at some point. Even under this condition, it is still possible to produce unambiguous output waveforms. Refer to Figures 7.18 and 7.19 for guidance.)

7.10 a. Draw a timing diagram for a NAND latch showing each of the following sequences of events:

   i. $\overline{S}$ and $\overline{R}$ are both LOW; $\overline{S}$ goes HIGH before $\overline{R}$.

   ii. $\overline{S}$ and $\overline{R}$ are both LOW; $\overline{R}$ goes HIGH before $\overline{S}$.

   iii. $\overline{S}$ and $\overline{R}$ are both LOW; $\overline{S}$ and $\overline{R}$ go HIGH simultaneously.

b. State why $\overline{S} = \overline{R} = 0$ is a forbidden state for the NAND latch.

c. Briefly explain what the final result is for each of the above transitions.

7.11 a Draw a timing diagram for a NOR latch showing each of the following sequences of events:

   i. S and R are both HIGH; S goes LOW before R.

   ii. S and R are both HIGH, R goes LOW before S.

   iii. S and R are both HIGH, S and R go LOW simultaneously.

FIGURE 7.70
Problem 7.7
Timing Diagram

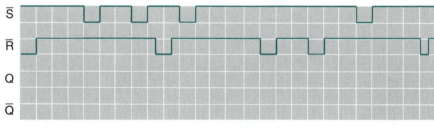

FIGURE 7.71
Problem 7.8
Input Waveforms to a NOR Latch

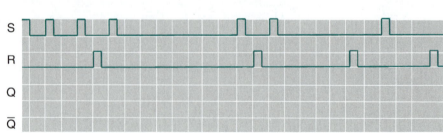

FIGURE 7.72
Problem 7.9
Input Waveforms to a Latch

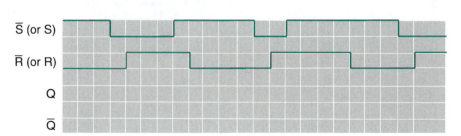

b. Briefly explain what the final result is for each of the transitions listed in part a of this question.

c. State why $S = R = 1$ is a forbidden state for the NOR latch.

7.12 Figure 7.73 shows the effect of mechanical bounce on the switching waveforms of a single-pole double-throw (SPDT) switch.

a. Briefly explain how this effect arises.

b. Draw a NAND latch circuit that can be used to eliminate this mechanical bounce, and briefly explain how it does so.

Section 7.3 Gated Latches

7.13 Complete the timing diagram for the gated latch shown in Figure 7.74.

7.14 Complete the timing diagram for the gated latch shown in Figure 7.75.

FIGURE 7.73
Problem 7.12
Effect of Mechanical Bounce on
a SPDT Switch

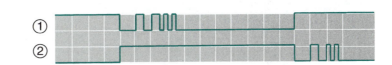

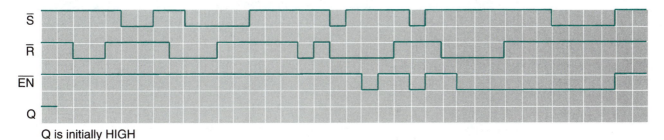

FIGURE 7.74
Problem 7.13
Gated Latch

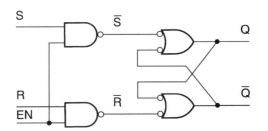

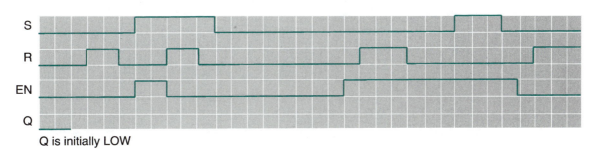

Q is initially LOW

FIGURE 7.75
Problem 7.14
Gated Latch

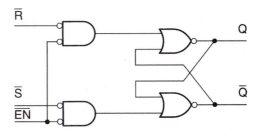

Q is initially HIGH

7.15 A pump motor can be started at two different locations with momentary-contact pushbuttons S_1 and S_2. It can be stopped by momentary-contact pushbuttons ST_1 and ST_2. As in Problem 7.4, a *RUN* input on the motor controller must be kept HIGH to keep the motor running. After the motor is stopped, a timer prevents the motor from starting for 5 minutes.

Draw a circuit block diagram showing how an SR latch and some additional gating logic can be used in such an application. The timer can be shown as a block activated by the *STOP* function. Assume that the timer output goes HIGH for 5 minutes when activated.

7.16 The *S* and *R* waveforms in Figure 7.76 are applied to two different gated latches. The *ENABLE* waveforms for the latches are shown as EN_1 and EN_2. Draw the output waveforms Q_1 and Q_2, assuming that *S*, *R*, and *EN* are all active HIGH. Which output is least prone to synchronization errors? Why?

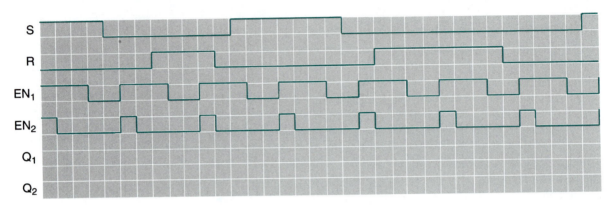

FIGURE 7.76
Problem 7.16
Waveforms

FIGURE 7.77
Problem 7.17
Waveforms

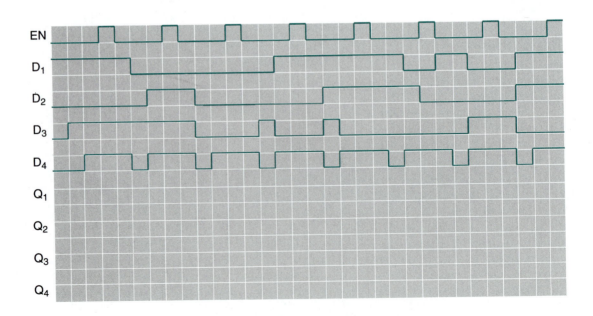

7.17 Figure 7.77 represents the waveforms of the *EN* and *D* inputs of a 4-bit transparent latch. Complete the timing diagram by drawing the waveforms for Q_1 to Q_4.

7.18 An electronic direction finder aboard an aircraft uses a 4-bit number to distinguish 16 different compass points as follows:

Direction	Degrees	Gray Code
N	0/360	0000
NNE	22.5	0001
NE	45	0011
ENE	67.5	0010
E	90	0110
ESE	112.5	0111
SE	135	0101
SSE	157.5	0100
S	180	1100
SSW	202.5	1101
SW	225	1111
WSW	247.5	1110
W	270	1010
WNW	295.5	1011
NW	315	1001
NNW	337.5	1000

The output of the direction finder is stored in a 4-bit latch so that the aircraft flight path can be logged by a computer. The latch is periodically updated by a continuous pulse on the latch enable line.

Figure 7.78 shows a sample reading of the direction finder's output as presented to the latch. (Problem continues . . .)

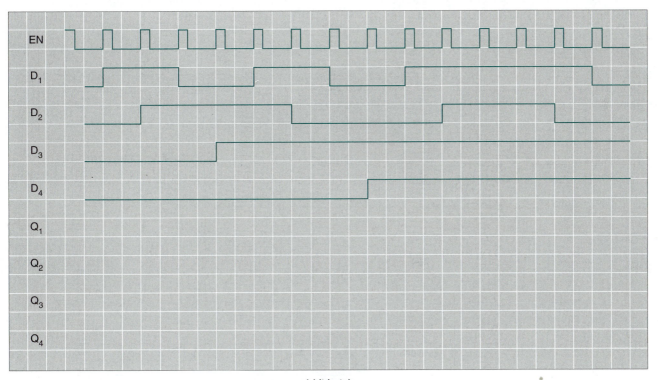

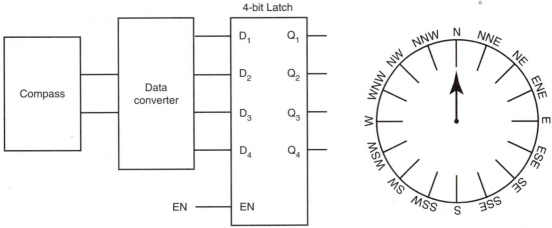

FIGURE 7.78
Problem 7.18
Direction Finder and Sample Output

a. Complete the timing diagram by filling in the data for the Q outputs.

b. Based on the completed timing diagram of Figure 7.78, make a rough sketch of the aircraft's flight path for the monitored time.

7.19 Write a VHDL file for an 8-bit latch, using LATCH primitives. Create a simulation file that demonstrates the operation of all eight bits.

7.20 Write a VHDL file for an 8-bit latch, using a component from the Library of Parameterized Modules. Create a simulation file that tests the latch for all eight bits.

Section 7.4 Edge-Triggered D Flip-Flops

7.21 The waveforms in Figure 7.79 are applied to the inputs of a positive edge-triggered D flip-flop and a gated D latch. Complete the timing diagram where Q_1 is the output of the flip-flop and Q_2 is the output of the gated latch. Ac-

count for any differences between the Q_1 and Q_2 waveforms.

7.22 Complete the timing diagram for a positive edge-triggered D flip-flop if the waveforms shown in Figure 7.80 are applied to the flip-flop inputs.

7.23 Repeat Problem 7.22 for the waveforms shown in Figure 7.81.

7.24 Repeat Problem 7.22 for the waveforms shown in Figure 7.82.

7.25 Draw a logic diagram of a D flip-flop configured for toggle mode. (Hint: The D input must always be the opposite of the Q output.)

7.26 Write a VHDL file that defines a 12-bit D flip-flop with a clock common to all flip-flops, using MAX+PLUS II primitives. The component declaration for the DFF component is as follows:

FIGURE 7.79
Problem 7.21
Waveforms

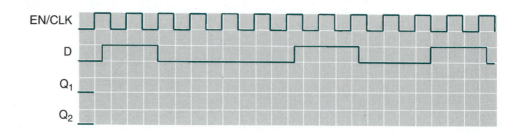

FIGURE 7.80
Problem 7.22
Waveforms

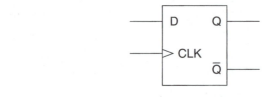

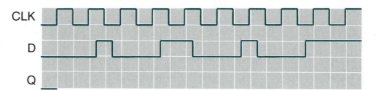

FIGURE 7.81
Problem 7.23
Waveforms

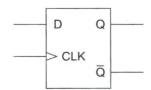

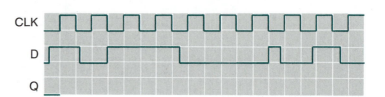

FIGURE 7.82
Problem 7.24
Waveforms

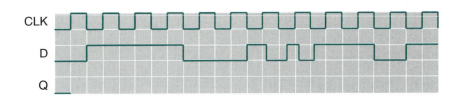

```
COMPONENT DFF
    PORT (d : IN STD_LOGIC;
       clk  : IN STD_LOGIC;
       clrn : IN STD_LOGIC;
       prn  : IN STD_LOGIC;
       q    : OUT STD_LOGIC;
END COMPONENT;
```

    Disregard the **clrn** (active-LOW clear) and **prn** (active-LOW preset) ports for this problem. (Hint: you may have to use a component declaration in your file that only declares the ports **d, clk,** and **q.**)

7.27 Write a VHDL file that creates a 12-bit D flip-flop, using the LPM component lpm_ff. (This component is instantiated as a D flip-flop by default. The required LPM component port names are: **data, clock,** and **q.**)

Section 7.5   Edge-Triggered JK Flip-Flops

7.28 The waveforms in Figure 7.83 are applied to a negative edge-triggered JK flip-flop. Complete the timing diagram by drawing the Q waveform.

7.29 Repeat Problem 7.28 for the waveforms in Figure 7.84.

7.30 Given the inputs x, y, and z to the circuit in Figure 7.85, draw the waveform for output Q.

FIGURE 7.83
Problem 7.28
Waveforms

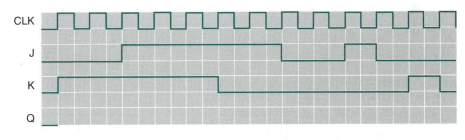

FIGURE 7.84
Problem 7.29
Waveforms

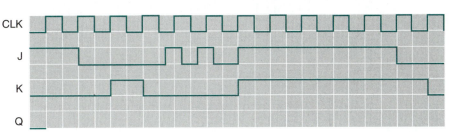

FIGURE 7.85
Problem 7.30
Inputs to Circuit

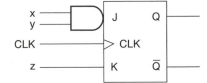

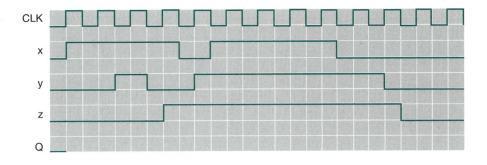

FIGURE 7.86
Problem 7.31
Flip-Flops

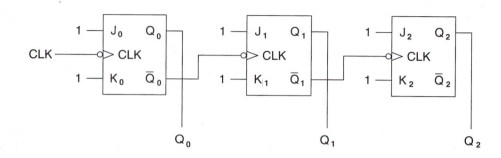

7.31 Assume that all flip-flops in Figure 7.86 are initially set. Draw a timing diagram showing the *CLK*, Q_0, Q_1, and Q_2 waveforms when eight clock pulses are applied. Make a table showing each combination of Q_2, Q_1, and Q_0. What pattern do the outputs form over the period shown on the timing diagram?

7.32 Refer to the JK flip-flop circuit in Figure 7.87. Is the circuit synchronous or asynchronous? Explain your answer.

7.33 Assume all flip-flops in the circuit in Figure 7.87 are reset. Analyze the operation of the circuit when six-teen clock pulses are applied by making a table showing the sequence of states of $Q_3Q_2Q_1Q_0$, beginning at 0000.

7.34 Draw a timing diagram showing the sequence of states from the table derived in Problem 7.33.

7.35 The waveforms shown in Figure 7.88 are applied to a negative edge-triggered JK flip-flop. The flip-flop's Preset and Clear inputs are active LOW. Complete the timing diagram by drawing the output waveforms.

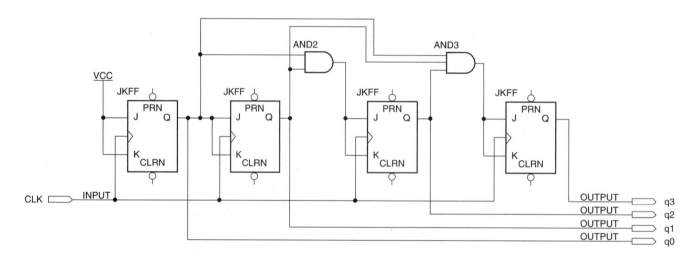

FIGURE 7.87
Problem 7.32
Flip-Flop Circuit

FIGURE 7.88
Problem 7.35
Waveforms

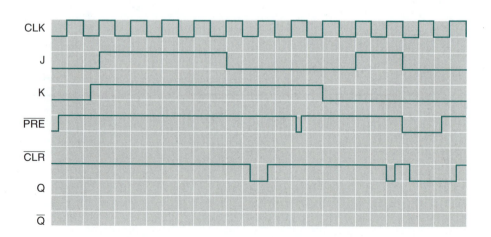

7.36 Repeat Problem 7.35 for the waveforms in Figure 7.89.

7.37 Create a MAX+PLUS II Graphic Design File for the synchronous circuit in Figure 7.87. Modify the circuit to add an asynchronous Master Reset function. Create a simulation file to verify the circuit operation.

7.38 Modify the **gdf** created in Problem 7.37 to include a Master Reset function *and* an asynchronous preset function that will set the state of the circuit to $Q_3Q_2Q_1Q_0 = 1010$ when activated. Create a simulation file to verify the circuit operation.

7.39 The term *asynchronous* is sometimes used to refer to the configuration of a circuit (e.g., a 3-bit asynchronous counter) and sometimes to a type of input to a device (e.g., an asynchronous clear input). Briefly explain how these two usages are similar and how they are different.

7.40 Write a VHDL file for a 12-bit D flip-flop that uses MAX+PLUS II DFF primitives, similar to that in Problem 7.26. Include active-LOW asynchronous clear (CLRN)

and preset (PRN) inputs. Create a simulation file to verify the operation of your design.

7.41 Write a VHDL file for a 12-bit D flip-flop with asynchronous preset and clear, using the LPM component lpm_ff, similar to that in Problem 7.27. Required ports: **data, clock, aclr** (asynchronous clear), **aset** (asynchronous set), and **q.** Ports **aset** and **aclr** are active-HIGH. Add two signals to the VHDL design to make them active-LOW. Create a simulation file to verify the operation of your design.

Section 7.6   Edge-Triggered T Flip-Flops

7.42 The *T* and *CLK* waveforms for a positive-edge triggered T flip-flop is shown in Figure 7.90. Complete the timing diagram.

7.43 The *T* and *CLK* waveforms for a positive-edge triggered T flip-flop is shown in Figure 7.91. Complete the timing diagram.

FIGURE 7.89
Problem 7.36
Waveforms

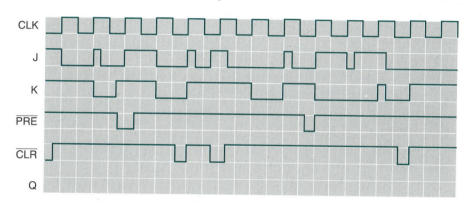

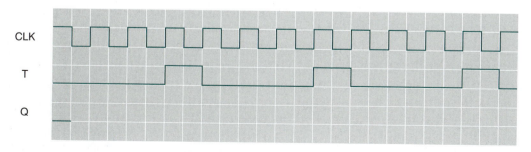

FIGURE 7.90
Problem 7.42
Timing Diagram

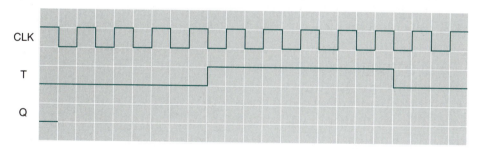

FIGURE 7.91
Problem 7.43
Timing Diagram

7.44 Refer to the synchronous circuit in Figure 7.87. Create a MAX+PLUS II Graphic Design File for a circuit with the same function, using T flip-flops rather than JK flip-flops. Include an asynchronous reset input in the circuit. Create a simulation file to test the operation of the circuit.

7.45 Write a VHDL file that implements the circuit you drew in Problem 7.44. Use TFF primitives in the design.

Section 7.7 Timing Parameters

7.46 Use a TTL or high-speed CMOS data sheet, as appropriate, to look up the setup and hold times of the following devices:

   a. 74LS74A

   b. 74HC76

   c. 74LS76A

   d. 74LS107A

   e. 74ALS112A

   f. 74HC112

7.47 Draw a timing diagram showing the setup and hold times for a 74LS76A flip-flop.

7.48 Draw timing diagrams (to scale) showing setup and hold times, minimum *CLK* and $\overline{CLR}$ pulse widths, recovery time, and propagation delay times from *CLK* and $\overline{CLR}$ for both 74LS107A and 74HC107 flip-flops.

7.49 Write names and values of the JK flip-flop timing parameters illustrated in Figure 7.92.

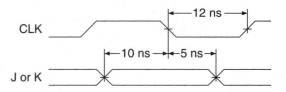

FIGURE 7.92
Problem 7.49
Timing Parameters

7.50 Repeat Problem 7.49 for the timing diagram in Figure 7.93.

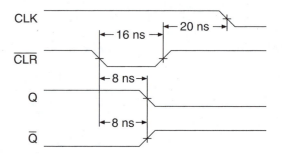

FIGURE 7.93
Problem 7.50
Timing Diagram

ANSWERS TO SECTION REVIEW PROBLEMS

Section 7.1

7.1 The latch resets (i.e., *Q* goes LOW) upon receiving the first reset pulse. At that point, the latch is already reset, so further pulses are ignored.

Section 7.2

7.2 The NOR latch has active-HIGH inputs. If you make both inputs HIGH, you are attempting to set and reset the latch at the same time, which is a contradictory action. A NAND latch has active-LOW inputs. Therefore, if both inputs are HIGH, neither the set nor reset function activates and there is no change on the latch output.

Section 7.3

7.3

```
LIBRARY ieee;
USE ieee.std_logic_1164.ALL;
LIBRARY altera;
USE altera.maxplus2.ALL;

ENTITY lch16prm IS
    PORT(d_in    : IN STD_LOGIC_VECTOR (15 downto 0);
         enable  : IN STD_LOGIC;
         q_out   : OUT STD_LOGIC_VECTOR (15 downto 0) );
END lch16prm;
```

```
ARCHITECTURE a OF lch16prm IS
BEGIN
    -- Instantiate a latch from a MAX+PLUS II primitive
    latch4:
    FOR i IN 15 downto 0 GENERATE
    latch_primitive: latch
        PORT MAP (d => d_in (i),
            ena => enable, q => q_out (i) );
    END GENERATE;
END a;
```

Section 7.4

7.4 The edge detector circuit in the clock circuit accounts for the operational difference between a D flip-flop and a D latch. It works by using the difference in internal delay times between the gates that comprise the flip-flop's clock input circuit.

Section 7.5

7.5 The flip-flops in asynchronous circuits are not all clocked at the same time; they are asynchronous with respect to the system clock. The flip-flops in a synchronous circuit have a common clock connection, which makes them synchronous to the system clock. The disadvantage to asynchronous circuits is that the internal delays of flip-flops can lead to unwanted intermediate states, since the flip-flops do not all change at the same time.

Section 7.6

7.6 The circuit is shown in Figure 7.94.

Section 7.7

7.7 The parameter is called propagation delay. For the specified output transition, the symbol is t_{pHL}.

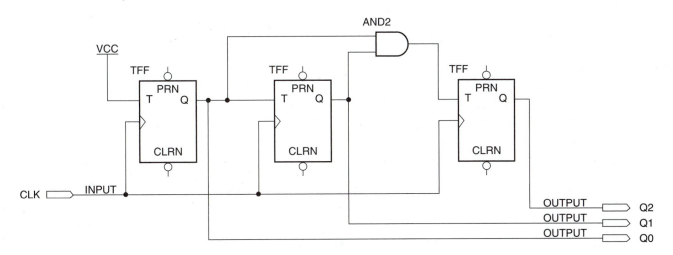

FIGURE 7.94
Solution to Section Review Problem 7.6

CHAPTER 8

Introduction to Programmable Logic Architectures

CHAPTER OBJECTIVES

Upon successful completion of this chapter, you will be able to:

- Draw a diagram showing the basic hardware conventions for a sum-of-products-type programmable logic device.

- Describe the structure of a programmable array logic (PAL) AND matrix.

- Draw fuses on the logic diagram of a PAL to implement simple logic functions.

- Describe the structures of combinational, programmable polarity, and registered PAL outputs.

- Determine the number and type of outputs from a PAL/GAL part number.

- Explain the structure of an output logic macrocell (OLMC).

- State differences between Universal PAL and generic array logic (GAL) and standard PAL.

- Interpret the logic diagrams of Universal PAL and GAL devices to determine the number of outputs and product terms and the type of control signals available in a device.

- Interpret block diagrams to determine the basic structure of an Altera MAX7000S CPLD, including macrocell configuration, Logic Array Blocks (LABs), control signals, and product term expanders.

- State the differences between PLDs based on sum-of-products (SOP) architecture versus look-up table (LUT) architecture.

- Interpret block diagrams to determine the basic structure of a logic element in an Altera FLEX10K CPLD, including look-up tables, cascade chains, carry chains, and control signals.

- Interpret block diagrams to determine how a logic element in a FLEX10K device relates to the overall structure of the device.

- Interpret block diagrams to determine how logic array blocks and embedded array blocks relate to the overall structure of a FLEX10K CPLD.

I n the past several chapters, we have been using Altera's MAX+PLUS II software to make circuit designs for downloading into a complex programmable logic device (CPLD). We have treated this device as a black box—something whose function we design, but whose structure we do not really understand. In this chapter, we will look inside the box. ∎

Before we examine the structure of an Altera MAX7000S CPLD, we will look at the internal structure of several simpler devices that are based on similar technologies, such as the PAL16L8 and PAL16R8 low-density PLDs (largely for an historical overview), the PALCE16V8, and the GAL22V10.

These devices are based on programmable matrices of sum-of-products (SOP) circuits, as is the Altera MAX series of devices. The main programming element is the EEPROM (electrically erasable programmable read-only memory) cell. EEPROM-based devices will retain their programmed data when power is removed from the device.

The Altera FLEX series of CPLDs is based on another technology altogether. It stores logic functions in look-up tables (LUTs) that act as truth tables with four input bits. The main logic element of the FLEX series is the SRAM (static random access memory) cell. SRAM-based CPLDs must have their programming data loaded every time they are powered up. They have the advantage of being faster than EEPROM devices, with a higher bit capacity.

8.1 Programmable Sum-of-Products Arrays

> **KEY TERMS**
>
> **Product line** A single line on a logic diagram used to represent all inputs to an AND gate (i.e., one product term) in a PLD sum-of-products array.
>
> **Input line** A line that applies the true or complement form of an input variable to the AND matrix of a PLD.
>
> **PAL** Programmable array logic. Programmable logic with a fixed OR matrix and a programmable AND matrix.

The original **programmable logic devices (PLDs)** consisted of a number of AND and OR gates organized in sum-of-products (SOP) arrays in which connections were made or broken by a matrix of fuse links. An intact fuse allowed a connection to be made; a blown fuse would break a connection.

Figure 8.1a shows a simple fuse matrix connected to a 4-input AND gate. True and complement forms of two variables, *A* and *B,* can be connected to the AND gate in any combination by blowing selected fuses. In Figure 8.1a, fuses for $\overline{A}$ and B are blown. The output of the AND gate represents the product term $A\overline{B}$, the logical product of the intact fuse lines.

Figure 8.1b shows a more compact notation for the AND-gate fuse matrix. Rather than showing each AND input individually, a single line, called the **product line,** goes into the AND gate, crossing the true and complement **input lines.** An intact connection to an input line is shown by an "X" on the junction between the input line and the product line.

A symbol convention similar to Figure 8.1b has been developed for programmable logic. Figure 8.2 shows an example.

The circuit shown in Figure 8.2 is a sum-of-products network whose Boolean expression is given by:

$$F = \overline{A}\,\overline{B}\,C + A\,\overline{B}\,\overline{C}$$

The product terms are accumulated by the AND gates as in Figure 8.1b. A buffer having true and complement outputs applies each input variable to the AND matrix, thus producing two **input lines.** Each product line can be joined to any input line by leaving the corresponding fuse intact at the junction between the input and product lines.

If a product line, such as for the third AND gate, has all its fuses intact, we do not show the fuses on that product line. Instead, this condition is indicated by an "X" through the gate. The output of the third AND gate is a logic 0, since $(A\,A\,\overline{B}\,B\,\overline{C}\,C) = 0$. This is necessary to enable the OR gate output:

$$\overline{A}\,\overline{B}\,C + A\,\overline{B}\,\overline{C} + 0 = \overline{A}\,\overline{B}\,C + A\,\overline{B}\,\overline{C}$$

FIGURE 8.1
Crosspoint Fuse Matrix

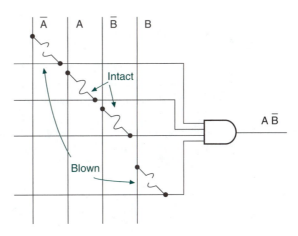

a. Crosspoint fuse matrix (A and $\overline{B}$ intact)

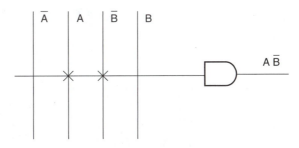

b. PLD notation for fuse matrix

Unconnected inputs are HIGH (e.g., $\overline{A} \cdot 1 \cdot \overline{B} \cdot 1 \cdot 1 \cdot C = \overline{A}\,\overline{B}\,C$ for the the first product line).

If the unused AND output was HIGH, the function F would be:

$$\overline{A}\,\overline{B}\,C + A\,\overline{B}\,\overline{C} + 1 = 1$$

The configuration in Figure 8.2, with a programmable AND matrix and a hardwired OR connection, is called **PAL (programmable array logic)** architecture.[1]

Since any combinational logic function can be written in SOP form, any Boolean function can be programmed into these PLDs by blowing selected fuses. The programming

FIGURE 8.2
PLD Symbology

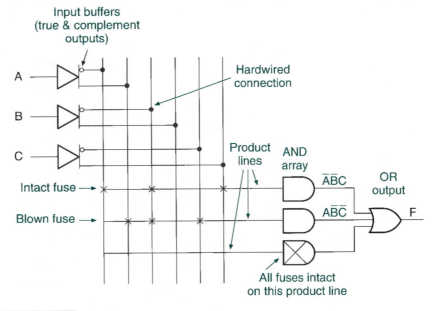

---

[1]PAL is a registered trademark of Vantis Semiconductor.

is done by special equipment and its associated software. The hardware and software selects each fuse individually and applies a momentary high-current pulse if the fuse is to be blown.

The main problem with fuse-programmable PLDs is that they can be programmed one time only; if there is a mistake in the design and/or programming or if the design is updated, we must program a new PLD. More recent technology has produced several types of erasable PLDs, based not on fuses but on floating-gate metal-oxide-semiconductor transistors. These transistors also form the basis of memory technologies such as electrically erasable programmable read-only memory (EEPROM or E²PROM).

8.2 PAL Fuse Matrix and Combinational Outputs

> **KEY TERMS**
>
> **JEDEC** Joint Electron Device Engineering Council
>
> **JEDEC file** An industry-standard form of text file indicating which fuses are blown and which are intact in a programmable logic device.
>
> **Text file** An ASCII-coded document stored on disk.
>
> **Checksum** An error-checking code derived from the accumulated sum of the data being checked.
>
> **Cell** A programmable location in a PLD, specified by the intersection of an input line and a product line.
>
> **Product line first cell number** The lowest cell number on a particular product line in a PAL AND matrix where all cells are consecutively numbered.
>
> **Input line number** A number assigned to a true or complement input line in a PAL AND matrix.
>
> **Multiplexer** A circuit that selects one of several signals to be directed to a single output.

Figure 8.3 shows the logic diagram of a PAL16L8 PAL circuit. This device can produce up to eight different sum-of-products expressions, one for each group of AND and OR gates. The device has active-LOW tristate outputs, as indicated by the "L" in the part number. Each is controlled by a product line from the related AND matrix.

The pins that can be used only as inputs or outputs are marked "I" or "O," respectively. Six of the pins can be used as inputs or outputs and are marked "I/O." The I/O pins can also feed back a derived Boolean expression into the matrix, where it can be employed as part of another function. A detail of an I/O section is shown in Figure 8.4.

The part number of a PAL device gives the designer information about the number of inputs and outputs and their configurations, as follows:

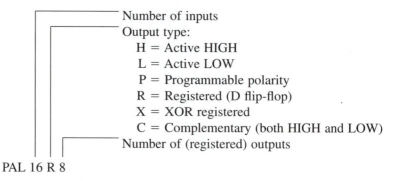

Number of inputs
Output type:
    H = Active HIGH
    L = Active LOW
    P = Programmable polarity
    R = Registered (D flip-flop)
    X = XOR registered
    C = Complementary (both HIGH and LOW)
Number of (registered) outputs

PAL 16 R 8

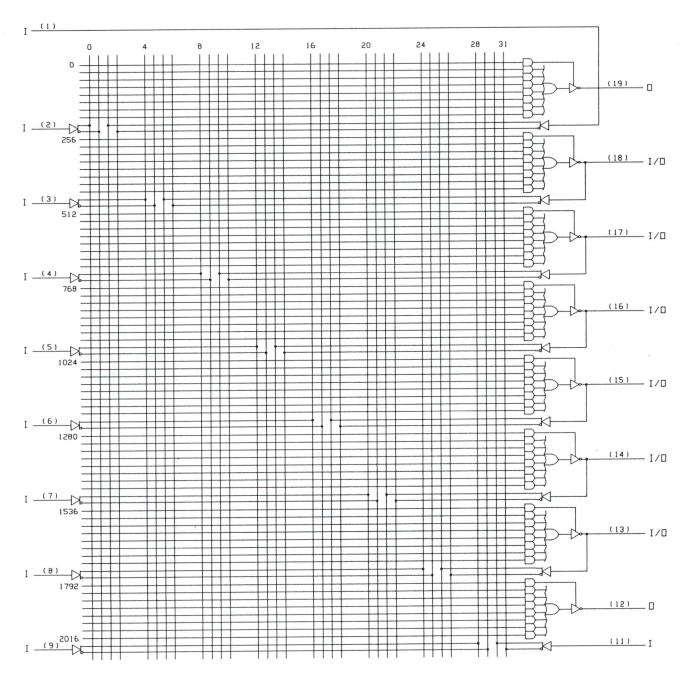

FIGURE 8.3
Unprogrammed PAL16L8

The numbering system has some potential ambiguities. For example, it is not possible to use 16 inputs and 8 outputs in a PAL16L8 device at the same time; 6 of the inputs are actually input/output pins. Some possible configurations are as follows:

16 inputs (10 dedicated + 6 I/O) and 2 dedicated outputs

10 dedicated inputs and 8 outputs (2 dedicated + 6 I/O)

12 inputs (10 dedicated + 2 I/O) and 6 outputs (2 dedicated + 4 I/O)

Each of the outputs of the PAL16L8 is buffered by a tristate inverter, whose *ENABLE* input is controlled by its own product line. When the *ENABLE* line of the tristate inverter

FIGURE 8.4
PAL16L8 I/O Section

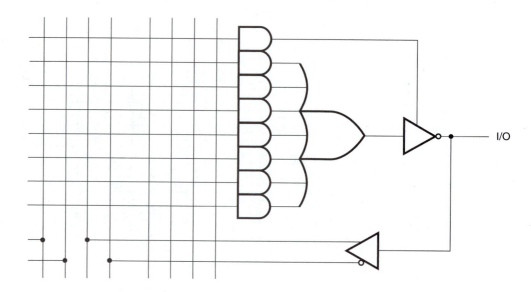

is HIGH, the inverter output is the same as it would normally be—a logic HIGH or LOW, determined by the state of the corresponding OR gate output.

When the *ENABLE* line is LOW, the inverter output is in the high-impedance state. The output acts as an open circuit, neither HIGH nor LOW; it is as though the output was completely disconnected from the circuit. The inverter is permanently enabled if all fuses on the *ENABLE* product line are blown, and permanently disabled if these fuses are all intact.

Published logic diagrams of PAL devices generally do not have fuses drawn on them. This allows us to draw fuses for any application. In practice, PLDs have become too complex to manually draw fuse maps for most applications.

Historically, PLD programming would begin with fuses drawn on a logic diagram, and each fuse would be selected and blown individually by someone operating a hardware device constructed for such a purpose.

Fuse assignment is now done with special software such as ABEL, CUPL, or PALASM. These programs will take inputs such as Boolean equations, truth tables, or other forms and produce the simplest SOP solution to the particular problem. (MAX+PLUS II is not configured to generate programming data for low-density PALs, although it can generate data for similar devices in the Altera Classic PLD series.)

The end result of such software is a **JEDEC file,** an industry-standard way of listing which fuses in the PLD should remain intact and which should be blown. The JEDEC file is stored on disk as an ASCII **text file.** Most PLD programmers will accept the JEDEC file and use it as a template for blowing fuses in the target device.

Fuse locations, called **cells,** are specified by two numbers: the **product line first cell number,** shown along the left side of the diagram, and the **input line number,** shown along the top. The address of any particular fuse is the sum of its product line first cell number and its input line number. The fuses on the PAL16L8 device are numbered from 0000 to 2047 (= 2016 + 31).

Figure 8.5 shows an example of a JEDEC file for a PAL16L8 application. The file starts with an ASCII "Start Text" character (^B). Next is some information required by the PAL programmer about the type of device (PAL16L8), number of fuses (2048), and so forth. The fuse information starts with the line L0000, which is the first product line. The 1s and 0s which follow show the programmed state of each cell in each product line; a 1 is a blown fuse and a 0 is an intact fuse. In other words, each 0 in the JEDEC file represents an X in the same position on the PAL logic diagram.

The product terms for first sum-of-products output are set by the states of fuses 0000 to 0255 (eight product lines). In the file shown, all fuses are blown in the first product

FIGURE 8.5

Sample JEDEC File

```
^B
PAL16L8
*
QF2048*QP20*F0*
L0000
1111 1111 1111 1111 1111 1111 1111 1111
1111 1111 0111 1111 1111 1111 0111 1111
1111 1111 0111 1111 1111 1111 1101 1111
1111 1111 1111 1111 1111 1111 0101 1111*
L0256
1111 1111 1111 1111 1111 1111 1111 1111
1110 1111 1111 1011 1111 1111 1111 0111
1110 1111 1111 0111 1111 1111 1111 1011
1101 1111 1111 1011 1111 1111 1111 1011
1101 1111 1111 0111 1111 1111 1111 0111*
L0512
1111 1111 1111 1111 1111 1111 1111 1111
1101 1111 1111 0111 1111 1111 1111 1111
1101 1111 1111 1111 1111 1111 1111 0111
1111 1111 1111 0111 1111 1111 1111 0111*
L0768
1111 1111 1111 1111 1111 1111 1111 1111
1011 1111 1101 1111 1011 1111 1111 1111
1011 1111 1110 1111 0111 1111 1111 1111
0111 1111 1110 1111 1011 1111 1111 1111
0111 1111 1101 1111 0111 1111 1111 1111*
L1024
1111 1111 1111 1111 1111 1111 1111 1111
0111 1111 1111 1111 0111 1111 1111 1111
0111 1111 1101 1111 1111 1111 1111 1111
1111 1111 1101 1111 0111 1111 1111 1111*
L1280
1111 1111 1111 1111 1111 1111 1111 1111
1111 1011 1111 1111 1101 1011 1111 1111
1111 1011 1111 1111 1110 0111 1111 1111
1111 0111 1111 1111 1110 1011 1111 1111
1111 0111 1111 1111 1101 0111 1111 1111*
L1536
1111 1111 1111 1111 1111 1111 1111 1111
1111 0111 1111 1111 1111 0111 1111 1111
1111 0111 1111 1111 1101 1111 1111 1111
1111 1111 1111 1111 1101 0111 1111 1111*
L1792
1111 1111 1111 1111 1111 1111 1111 1111
1111 1111 1011 1111 1111 1111 1001 1111
1111 1111 1011 1111 1111 1111 0110 1111
1111 1111 0111 1111 1111 1111 1010 1111
1111 1111 0111 1111 1111 1111 0101 1111*
L2048
1111 1111*
C8DCF*
^C0000
```

line, the second product line shows three intact fuses, and so forth. Since all fuses are intact in the last three lines, they need not be shown in the JEDEC file.

Whenever some unprogrammed product lines are omitted from the fuse map, the last fuse line shown ends with an asterisk (*). The next line with programmed fuses is indicated by a new fuse number. For example, the second group of fuses (0256 to 0511) in Figure 8.5 begins after the line marked L0256 in the JEDEC file. The remaining fuse lines are similarly indicated.

The JEDEC file in Figure 8.5 ends with a hexadecimal **checksum** (C8DCF), an error-checking code derived from the programming data, and an ASCII "End Text" code (^C).

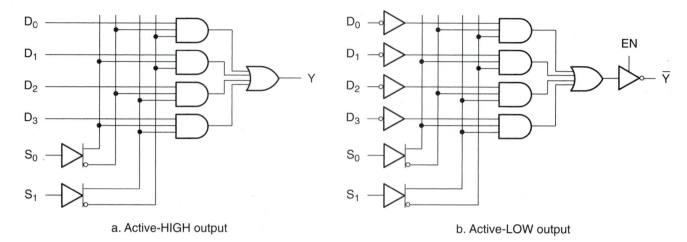

a. Active-HIGH output   b. Active-LOW output

FIGURE 8.6

4-to-1 Multiplexer Circuits

In order to examine the general principle of fuse programming, let us develop the programmed logic diagram for a common combinational circuit: a 4-to-1 **multiplexer.** (After developing the fuse maps for several examples, we will not refer to this technique again.)

This circuit, shown in Figure 8.6a, directs one of four input logic signals, D_0 to D_3, to output Y, depending on the state of two select inputs S_0 and S_1. The circuit works on the enable/inhibit principle; each AND gate is enabled by a different combination of $S_1 S_0$. The binary state of the select inputs is the same as the decimal subscript of the selected data input. For instance, $S_1 S_0 = 10$ selects data input D_2; the AND gate corresponding to D_2 is enabled and the other three ANDs are inhibited.

The logic equation for output Y is given by:

$$Y = D_0 \overline{S_1} \overline{S_0} + D_1 \overline{S_1} S_0 + D_2 S_1 \overline{S_0} + D_3 S_1 S_0$$

Since the outputs of the PAL16L8 are active LOW, as illustrated in Figure 8.6b, we should rewrite the equation as follows:

$$\overline{Y} = \overline{D_0} \overline{S_1} \overline{S_0} + \overline{D_1} \overline{S_1} S_0 + \overline{D_2} S_1 \overline{S_0} + \overline{D_3} S_1 S_0$$

The D inputs must be complemented to reverse the effect of the active-LOW output. The output is enabled when the EN input is HIGH. Figure 8.7 shows the PAL16L8A logic diagram with fuses for the multiplexer application.

8.3 PAL Outputs With Programmable Polarity

The multiplexer application developed in the previous section uses a PAL device whose output is always fixed at the active-LOW polarity. This fixed polarity is suitable for most applications, but Boolean functions that would normally have active-HIGH outputs must be implemented in DeMorgan equivalent form, which is not always very straightforward.

Some applications require both active-HIGH and active-LOW outputs. In such cases, it is useful to have a device whose output polarity is fuse programmable.

Figure 8.8 shows the logic diagram of a PAL20P8 PAL device. This device is the same as a PAL16L8, except that there are four more dedicated inputs, and the polarity of each output is programmable. The Exclusive OR gate on each output is programmed to act as either an inverter or a buffer. When its associated fuse is intact, the XOR input is grounded and passes the output of its related SOP network in true form. When combined with the output inverter, this produces an active-LOW output. When the polarity fuse is blown, the fused XOR input floats to the HIGH state, inverting the SOP output; the output pin becomes active HIGH.

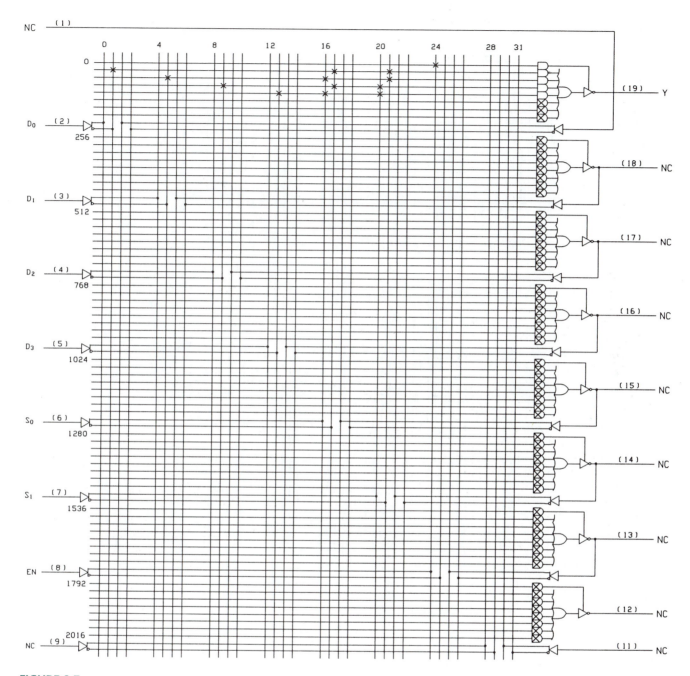

FIGURE 8.7

Programmed Logic Diagram for a 4-to-1 Multiplexer

The polarity fuses are given numbers higher than those of the main fuse array. In this case, the product line fuses are numbered 0000 to 2559 and the output polarity fuses are numbered 2560 to 2567.

Figure 8.9 illustrates the selection of output polarity. Two Boolean functions, *F1* and *F2,* are programmed into the fuse array, with outputs at pins (17) and (15), respectively. The equations are:

$$F1 = A\,B + \overline{A}\,\overline{B}$$
$$F2 = \overline{A\,B + \overline{A}\,\overline{B}}$$

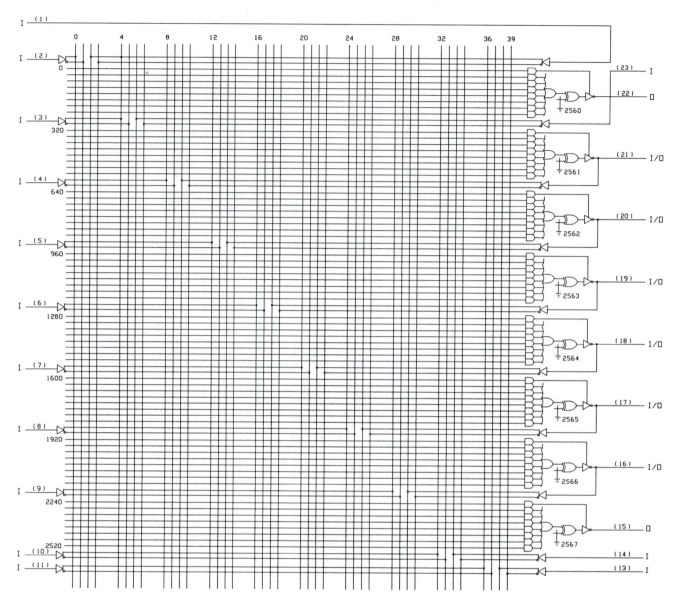

FIGURE 8.8
PAL20P8 Logic Diagram

We could, if we chose, rewrite $F2$ to show the output as active LOW:

$$\overline{F2} = A\,B + \overline{A}\,\overline{B}$$

The portion of the PAL20P8 logic diagram shown in Figure 8.9 represents the fuses required to program $F1$ and $F2$. Pins (14) and (16) supply inputs A and B to the matrix. The *ENABLE* lines of the tristate output buffers float HIGH, since all fuses are blown on the corresponding product lines, thus permanently enabling the output buffers.

The fuses numbered 2565 and 2567 select the polarity at pins (15) and (17). Fuse 2565 is blown. The fused input to the corresponding XOR gate floats HIGH, thus making the gate into an inverter. Combined with the tristate buffer, this makes pin (17) active HIGH.

Fuse 2567 is intact. This grounds the input to the corresponding XOR gate, making the gate into a noninverting buffer. Combined with the tristate output buffer, this makes pin (15) active LOW.

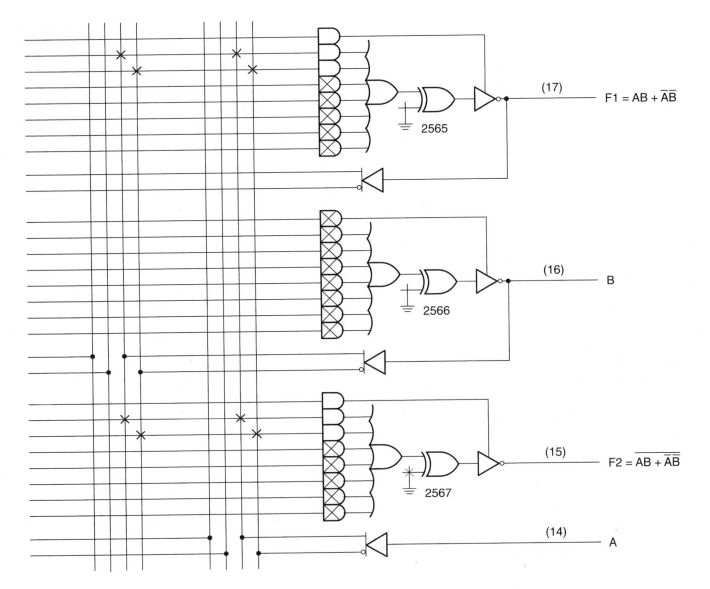

FIGURE 8.9
PAL Outputs With Programmable Polarity

▌ EXAMPLE 8.1

Show how a PAL20P8 device can be used to implement the following logic functions by drawing fuses on the device's logic diagram.

$$
\begin{aligned}
\text{NOT:} \quad & F_1 = \overline{A} \\
\text{AND:} \quad & F_2 = BC \\
\text{OR:} \quad & F_3 = D + E \\
\text{NAND:} \quad & F_4 = \overline{FG} \\
\text{NOR:} \quad & F_5 = \overline{H + J} \\
\text{XOR:} \quad & F_6 = K \oplus L = \overline{K} L + K \overline{L} \\
\text{XNOR:} \quad & F_7 = \overline{M \oplus N} = \overline{M} \, \overline{N} + M N
\end{aligned}
$$

How would the implementation of these logic functions differ if only active-LOW outputs were available, as in a PAL16L8?

SOLUTION The PAL20P8 has 14 dedicated inputs, 2 dedicated outputs, and 6 lines that can be used as inputs or outputs. Our functions need 13 input variables and 7 output variables. We will use six I/O pins (pins (16) through (21)) and one dedicated output (pin (15)) for the output variables.

All functions must be in SOP form. Outputs for NOT, AND, OR, Exclusive OR, and Exclusive NOR are active HIGH. Therefore, polarity fuses on the outputs for *F1, F2, F3, F6,* and *F7* are blown. NAND and NOR outputs are active LOW; the polarity fuses for *F4* and *F5* remain intact.

Figure 8.10 shows the logic diagram of the programmed PAL. If only active-LOW outputs were available, we would need to rewrite some of the equations to make the outputs correspond to their DeMorgan equivalent forms, as follows:

$$\text{AND:} \quad F_2 = \overline{\overline{B} + \overline{C}}$$
$$\text{OR:} \quad F_3 = \overline{\overline{D}\,\overline{E}}$$
$$\text{XOR:} \quad F_6 = K \oplus L = \overline{\overline{K}\,\overline{L} + KL}$$
$$\text{XNOR:} \quad F_7 = \overline{M \oplus N} = \overline{\overline{M}\,N + M\,\overline{N}}$$

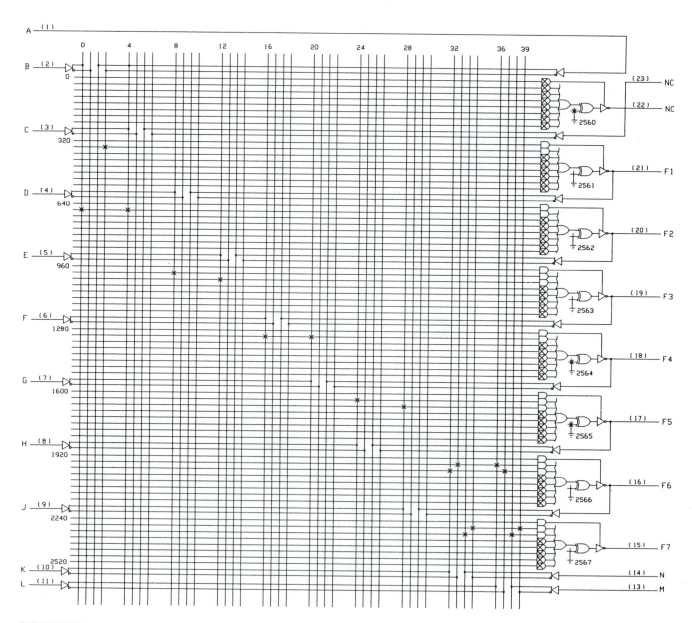

FIGURE 8.10
Programmed Logic Diagram for Seven Logic Functions

8.4 PAL Devices With Registered Outputs

Flip-flops are generally found in programmable logic devices as **registered outputs.** A **register** is one or more flip-flops used to store data. Registered outputs in programmable array logic (PAL) devices can be used for the same functions as individual flip-flops.

Figure 8.11 shows the logic diagram of a PAL device with eight registered outputs: a PAL16R8. The fuse matrix is identical to that of a PAL16L8 device; the differences

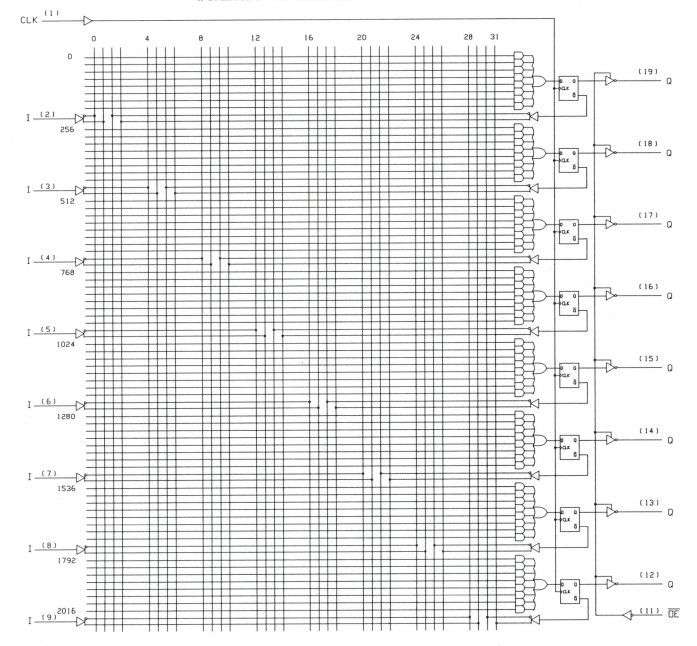

FIGURE 8.11
PAL16R8 Logic Diagram

between the two devices are the registered outputs, a dedicated clock input (pin 1), and a pin for enabling all registered outputs (pin 11).

With Registered PAL, the number of outputs shown in the part number indicates the number of registered outputs. For example, a PAL16R4 device has four registered outputs and four combinational I/O pins, a PAL16R6 device has six registered outputs and two combinational I/O pins, and a PAL16R8 has eight registered outputs.

▮▮ EXAMPLE 8.2

A common data operation is that of "rotation." Figure 8.12 illustrates how a 4-bit number can be rotated to the right by 0, 1, 2, or 3 places by a circuit called a "barrel shifter." To rotate the data, move all bits the required number of places to the right. As data reach the rightmost position, move them to the beginning so that they are transferred in a closed loop.

FIGURE 8.12
Example 8.2
Rotation to the Right (4-bit Data)

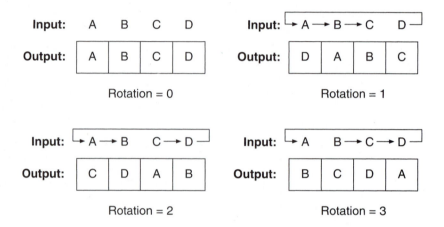

This operation is usually performed by serially shifting the data the required number of places and feeding back the last output to the first input of a serial shift register.

Rotation can also be accomplished by a parallel transfer operation. We can load the bits of the input into four D flip-flops in the order determined by two select inputs, S_1 and S_0. Assume that the binary number $S_1 S_0$ is the same as the rotation number in Figure 8.12. Table 8.1 summarizes the contents of the circuit after one clock pulse is applied.

Table 8.1 Rotation to the Right by a Selectable Number of Bits

S_1	S_0	Q_A	Q_B	Q_C	Q_D	Rotation
0	0	A	B	C	D	0
0	1	D	A	B	C	1
1	0	C	D	A	B	2
1	1	B	C	D	A	3

Sketch a circuit, using gates and flip-flops, that can accomplish this rotation as a parallel transfer function. Briefly explain its operation.

Write the Boolean expression(s) for the circuit.

Show how the circuit can be implemented by a PAL16R4 device by drawing fuses on its logic diagram.

SOLUTION Figure 8.13 shows a parallel transfer circuit (barrel shifter) that will perform the specified rotation. The circuit works by enabling one AND gate in each group of four for each combination of S_1 and S_0. For example, when $S_1 S_0 = 00$, the rotation is 0 and the leftmost AND gate of each group is enabled, transferring the parallel data into the flip-flops so that $D_A = A$, $D_B = B$, $D_C = C$, and $D_D = D$. After one clock pulse, $Q_A Q_B Q_C Q_D = ABCD$.

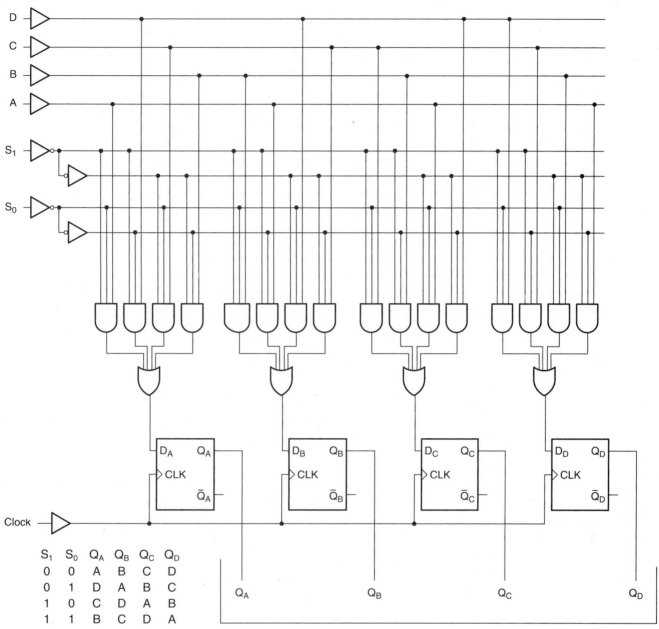

FIGURE 8.13

Example 8.2

Rotation by Parallel Transfer (Barrel Shifter)

Similarly, if $S_1 S_0 = 10$, we select a rotation of 2. The third AND gate from the left is selected in each group of four. This makes the data $D_A = C$, $D_B = D$, $D_C = A$, and $D_D = B$ appear at the flip-flop inputs. After one clock pulse, $Q_A Q_B Q_C Q_D = CDAB$.

The same principle governs the circuit operation for the other two select codes. The Boolean equations for the circuit are:

$$Q_A = \overline{S_1}\, \overline{S_0}\, A + \overline{S_1}\, S_0\, D + S_1\, \overline{S_0}\, C + S_1\, S_0\, B$$
$$Q_B = \overline{S_1}\, \overline{S_0}\, B + \overline{S_1}\, S_0\, A + S_1\, \overline{S_0}\, D + S_1\, S_0\, C$$
$$Q_C = \overline{S_1}\, \overline{S_0}\, C + \overline{S_1}\, S_0\, B + S_1\, \overline{S_0}\, A + S_1\, S_0\, D$$
$$Q_D = \overline{S_1}\, \overline{S_0}\, D + \overline{S_1}\, S_0\, C + S_1\, \overline{S_0}\, B + S_1\, S_0\, A$$

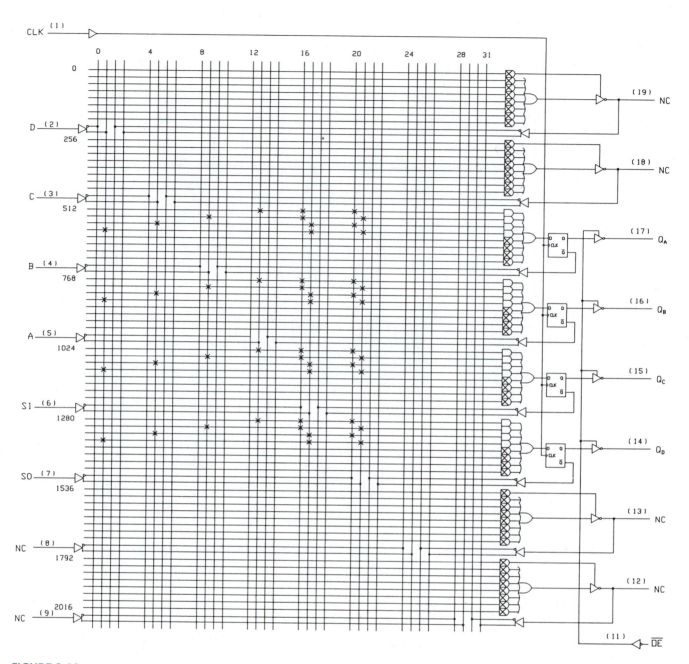

FIGURE 8.14
Example 8.2
Programmed PLD for Selectable Bit Rotation

These equations imply that each registered output requires us to use four product lines, one for each product term. The programmed logic diagram is shown in Figure 8.14.

8.5   Universal PAL and Generic Array Logic (GAL)

KEY TERMS

One-time programmable (OTP)   A property of some PLDs that allows them to be programmed, but not erased.

Generic array logic (GAL)   A type of programmable logic device whose outputs can be configured as combinational or registered and whose programming matrix is based on electrically erasable logic cells.

Universal PAL   A PLD based on erasable cells and configurable outputs, much like GAL, but primarily designed to emulate PAL devices, such as PAL16L8.

Output logic macrocell (OLMC)   An input/output circuit that can be programmed for a variety of input or output configurations, such as active HIGH or active LOW, combinational or registered. Often just called a **macrocell.**

In-system programmability (ISP)   The ability of a PLD to be programmed through a standard four-wire interface while installed in a circuit.

JTAG port   A four-wire interface specified by the Joint Test Action Group (JTAG) used for loading test data or programming data into a PLD installed in a circuit.

Architecture cell   A programmable cell that, in combination with other architecture cells, sets the configuration of a macrocell.

Global architecture cell   An architecture cell that affects the configuration of all macrocells in a device.

Local architecture cell   An architecture cell that affects the configuration of one macrocell only.

Global clock   A clock signal in a PLD that clocks all registered outputs in the device.

There are several limitations of standard low-density PALs. First, these devices are **one-time programmable (OTP).** Since the AND matrix of a PAL is programmable by blowing metal fuse links, programming is permanent; there is no opportunity to correct or update a design. In development of a new design, where many modifications must be made to the original design, this can be particularly wasteful. Second, standard PAL outputs are permanently configured either as combinational or registered. A given PAL has a certain number of each type of output, which may not be optimum for the design. Third, a standard PAL cannot be programmed while it is installed in a circuit.

A number of low-density PLDs have been developed to address these concerns. Devices such as the PALCE16V8 **Universal PAL** (Vantis Corporation), and the GAL16V8 and GAL22V10 **Generic Array Logic** (Lattice Semiconductor)* are based on sum-of-products fuse matrices, just as the earlier-version PALs. However, these devices are based on electrically erasable read only memory (EEPROM or E^2PROM) cells, rather than fuses, which allow them to be erased and reprogrammed about 10,000 times. A programmed device will hold its data for about 20 years.

Universal PALs and GALs also have programmable input/output configurations. An I/O pin can be configured as a registered output, a combinational output, or a dedicated input, as required. Additionally, an output can be specified as active-HIGH or active-LOW.

---

*Vantis has recently been acquired by Lattice, so these devices are really produced by the same company

Devices such as the ispGAL22V10 or the Altera MAX7000S series can be programmed while installed in a circuit via a standard four-wire interface called a **JTAG port.** This property is known as **in-system programmability (ISP).**

PALCE16V8

Figure 8.15 shows one I/O pin and its associated circuitry for a PALCE16V8 Universal PAL. (The "V" stands for "variable" or "versatile" architecture.) It consists of a programmable SOP array with 8 product terms and an **output logic macrocell (OLMC),** or just "macrocell", which determines the I/O configuration for that pin. The various configuration options are selected by a network of four multiplexers that are programmed by a set of **architecture cells** that set the MUX select inputs HIGH or LOW.

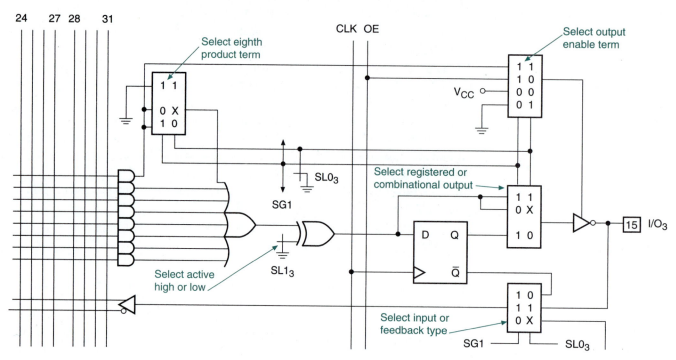

FIGURE 8.15
Output Logic Macrocell for a PALCE16V8 PLD

A **global architecture cell,** SG1, selects configuration options for all macrocells in the device. Two **local configuration cells,** $SL0_n$ and $SL1_n$, select configurations for I/O_n only. (In this case, the cells shown are $SL0_3$ and $SL1_3$ for configuration of I/O_3.)

Figure 8.16 shows the different macrocell configurations for a PALCE16V8 Universal PAL. Most of these configurations are designed to emulate an I/O of a standard PAL, so that an old-style PAL can be replaced by a Universal PAL, and can be programmed by data for the older PAL. The macrocells can also be configured in a pattern that does not conform to an older device.

Figure 8.17 shows the logic diagram of a PALCE16V8 Universal PAL. The device has eight dedicated inputs, eight macrocells, a Clock pin and an Output Enable pin. The latter two signals are shown in the macrocell diagram of Figure 8.15 as the lines labeled *CLK* and *OE*.

If there are registered outputs, the clock input (pin 1) provides a **global clock** function. That is, all registered outputs are clocked simultaneously by this signal. (Some other PLDs provide an option to clock a registered output from a product term in the AND matrix, allowing several clock functions in one chip.) If there are no registered outputs used in the PLD, pin 1 can be used as an input.

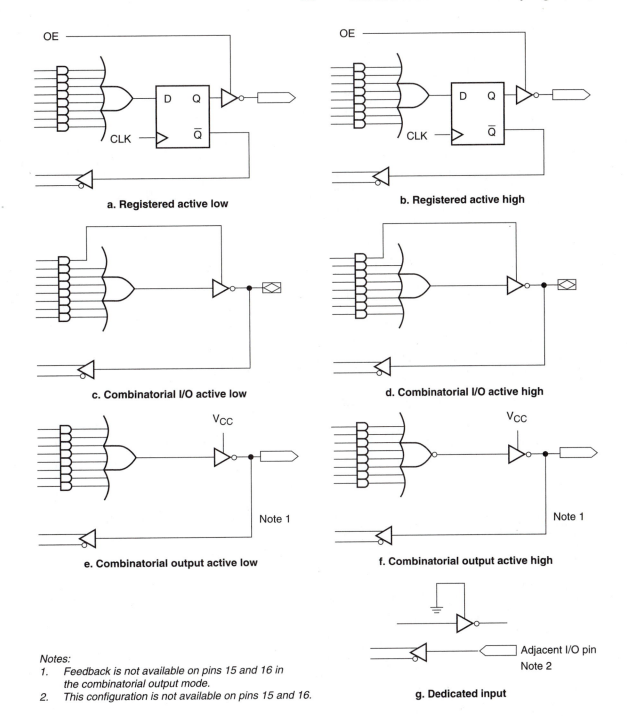

a. Registered active low

b. Registered active high

c. Combinatorial I/O active low

d. Combinatorial I/O active high

e. Combinatorial output active low

f. Combinatorial output active high

g. Dedicated input

Notes:
1. Feedback is not available on pins 15 and 16 in the combinatorial output mode.
2. This configuration is not available on pins 15 and 16.

FIGURE 8.16
Macrocell Configurations for a PALCE16V8 PLD (Courtesy of Lattice Semiconductor Corporation)

Pin 11 provides an active-LOW Output Enable function. This is selected by local architecture cells to provide control of the output tristate buffer, either from the $\overline{OE}$ pin or from a product term in the AND matrix. If the $\overline{OE}$ function is unused, the pin can be used as an input.

GAL22V10

Figure 8.18 shows the logic diagram of a GAL22V10 generic array logic device. This industry-standard device has a number of features that make it superior to the PALCE16V8.

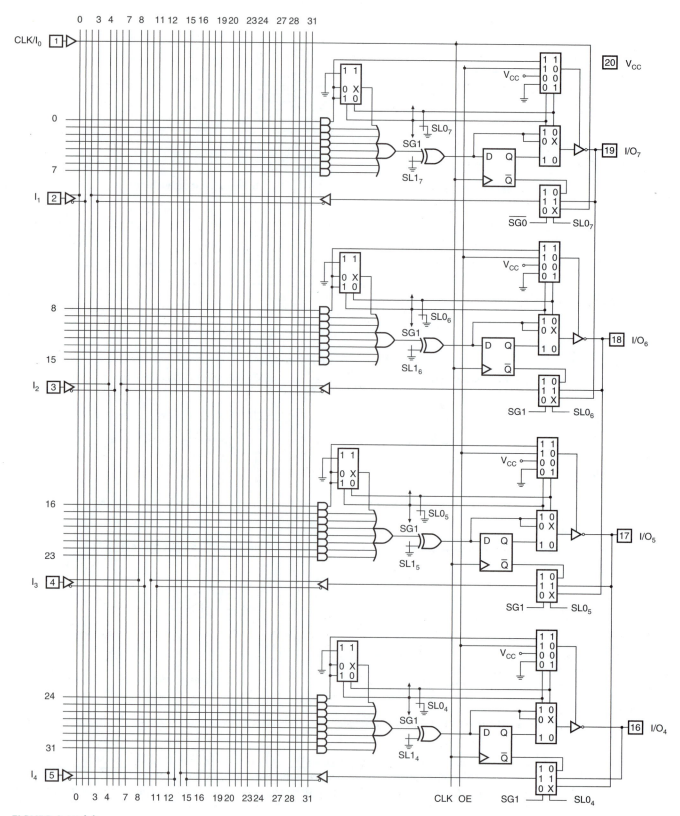

FIGURE 8.17 (a)
PALCE16V8 Logic Diagram (Courtesy of Lattice Semiconductor Corporation)

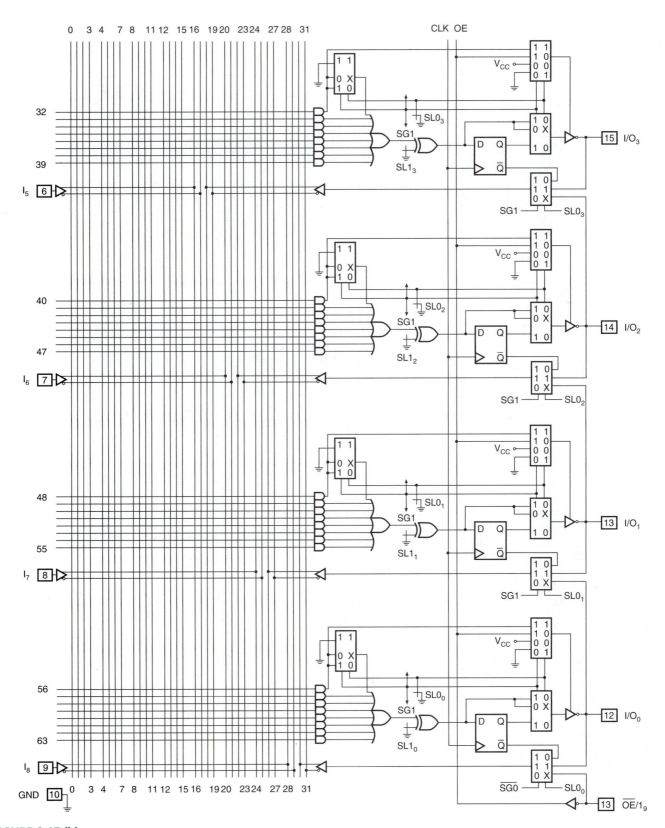

FIGURE 8.17 (b)
(PALCE16V8 Logic Diagram

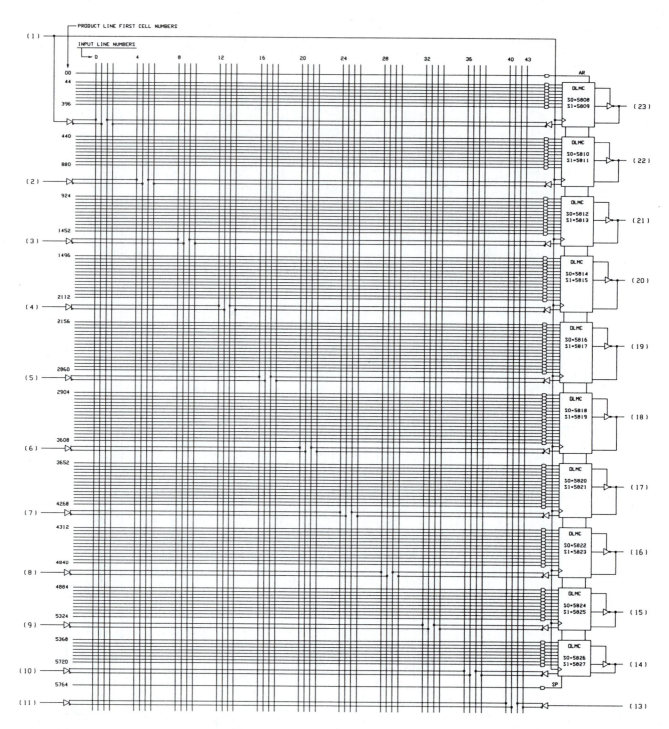

FIGURE 8.18
GAL22V10 Logic Diagram

1. There are more outputs (10 as opposed to 8 for the 16V8).

2. There are more inputs (11 dedicated inputs, plus any I/O lines used as inputs).

3. The output logic macrocells are of different sizes, allowing expressions with larger numbers of product terms in some OLMCs than others. There are two OLMCs with each of the following numbers of product lines: 8, 10, 12, 14, and 16. This allows more flexibility in design, while minimizing the number of product lines.

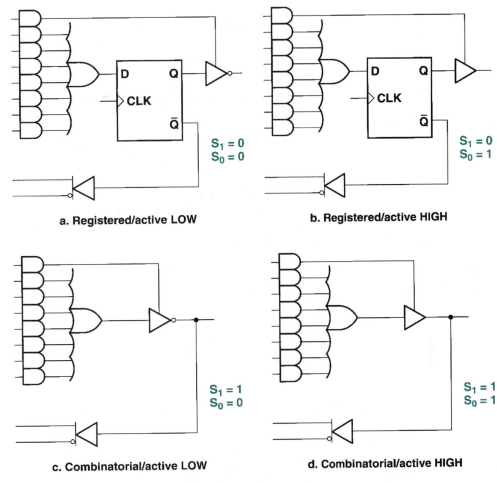

FIGURE 8.19
GAL22V10 OLMC Configurations

4. OLMC configuration is much simpler than that of a PALCE16V8. Two architecture cells per macrocell, S_0 and S_1, select the output type, as shown in Figure 8.19.

5. There are product lines for Synchronous Preset *(SP)* and Asynchronous Reset *(AR)*. The *SP* line sets all flip-flops HIGH on the first clock pulse after it becomes active. The *AR* line sets all flip-flops LOW as soon as it activates, without waiting for the clock pulse. (Note that these lines set or reset the *Q* output of each flip-flop. An active-LOW registered output inverts this state at the output pin.)

8.6 MAX7000S CPLD

> **KEY TERMS**
>
> **CPLD** Complex programmable logic device. A programmable logic device consisting of several interconnected programmable blocks.
>
> **Logic Array Block (LAB)** A group of macrocells that share common resources in a CPLD.
>
> **Programmable Interconnect Array (PIA)** An internal bus with programmable connections that link together the Logic Array Blocks of a CPLD.

> **Buried logic**   Logic circuitry in a PLD that has no connection to the input or output pins of the PLD, but is used solely as internal logic.
>
> **I/O Control Block**   A circuit in an Altera CPLD that controls the type of tristate switching used in a macrocell output.
>
> **Parallel logic expanders**   Product terms that are borrowed from neighboring macrocells in the same LAB.
>
> **Shared logic expanders**   Product terms that are inverted and fed back into the programmable AND matrix of an LAB for use by any other macrocell in the LAB.

Figure 8.20 shows the block diagram of an Altera MAX7000S **Complex PLD (CPLD).** A device of this type—the EPM7128SLC84—is one of the two devices installed on the Altera UP-1 University Program board, so we will use it as a specific example of the MAX7000S family of devices.

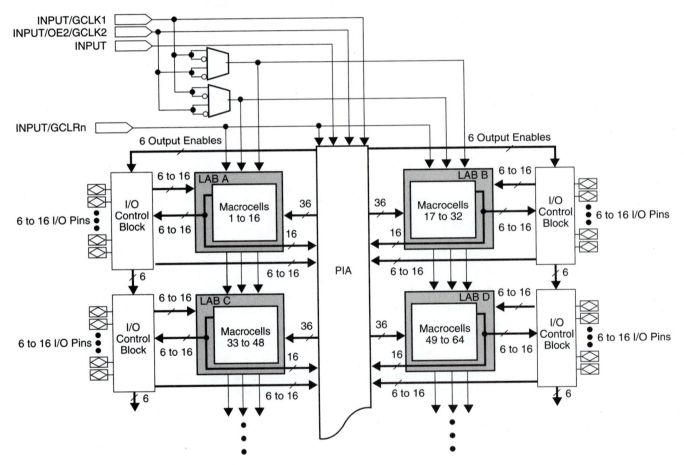

FIGURE 8.20
MAX 7000E and MAX 7000S Device Block Diagram (Courtesy of Altera)

The part number breaks up as follows:

EPM7	MAX7000 family
128	number of macrocells
S	in-system programmable
LC84	84-pin PLCC package

The main structure of the MAX7000S is a series of **Logic Array Blocks (LABs),** linked by a **Programmable Interconnect Array (PIA).** Each LAB is a group of 16 macrocells that can share common product terms and lend or borrow unused product terms among each other. A single LAB has similar I/O and programming capability to a low-density PLD, so a CPLD like the MAX7000S can be thought of as an array of interconnected PALs or GALs on a single chip.

An EPM7128S has 8 LABs, for a total of $8 \times 16 = 128$ macrocells. However, these are not all available to the user as I/Os; the number of available I/O pins depends on the device package. Figure 8.20 indicates that each LAB in a MAX7000S device has from 6 to 16 I/O pins. For an EPM7128S in a 160-pin PQFP package, there are 12 I/Os per LAB, for a total of 96 available pins. For the same device in an 84-pin PLCC package, there are only 8 I/Os per LAB, for a total of 64 pins.

In practice, if an EPM7128SLC84 is to be programmed in-circuit (i.e., while installed on a circuit board), there are only 60 I/Os available, as four pins are required for the programming interface. The macrocells that are not connected to user I/O pins can only be used for **buried logic,** or logic that is internal to the chip only.

As implied in Figure 8.20, all I/O pins connect to and from their associated LAB via an **I/O Control Block** (a circuit that controls the tristate switching of signals at an I/O pin). The I/O pin signals also connect directly to the PIA, where they are available for use in other LABs. Sixteen lines connect the macrocell outputs of each LAB to the PIA, again for use throughout the device. The PIA communicates to each LAB via 36 product lines to provide connections from other LABs.

The MAX7000S family has four pins that can be configured as control signals or inputs. *GCLK1* is a global clock that is common to all macrocells in the device and can be used to synchronously clock all registers. *OE1* is an output enable that can globally activate or disable the tristate outputs of the device macrocells. *GCLRn* is an active-LOW global clear function. The fourth control pin can be configured as an input, as can the other three pins, or as a second global clock *(GCLK2)* or output enable *(OE2)*. If the control functions are not used, these pins add four inputs to the available total. These assignments can be made by the MAX+PLUS II software during the design process.

Figure 8.21 shows a macrocell from a MAX7000S device. The macrocell is similar to that of a GAL or Universal PAL in that it provides a sum-of-products function with active-HIGH or -LOW options and the choice of registered or combinational output. Registered outputs can be clocked with one of two global clocks or by a product term from the AND matrix. The register can be cleared globally or by a product term and preset with a product term.

The macrocell has five dedicated product terms, which is fewer than found in the PAL and GAL matrices we examined earlier. This is generally sufficient to implement most logic functions. If more terms are required, they can be supplied by a set of **shared logic expanders** or **parallel logic expanders.**

Shared logic expanders do not add more product terms to a given macrocell. They do make the programming of the entire LAB more efficient by allowing a product term to be programmed once and used in several macrocells of the same LAB. One product term per macrocell is inverted and fed back into the shared expander pool of product terms. Since there are 16 macrocells per LAB, the shared logic expander pool has up to 16 product terms.

Parallel logic expanders allow a macrocell to borrow up to 15 product terms from its three lower-numbered neighbors (5 product terms per neighboring macrocell). For example, macrocell 4 can borrow up to 5 terms each from macrocells 3, 2, and 1. By using its 5 dedicated product terms and the maximum number of parallel expanders, a macrocell can have up to 20 product terms at its disposal. These borrowed terms are not usable by the macrocell from which they were borrowed. The parallel expanders are set up so that a lower-number cell lends product terms to a higher-number cell, so the number of available terms depends on how close to the end of a chain a macrocell is. Expander assignments are done automatically by MAX+PLUS II at compile time.

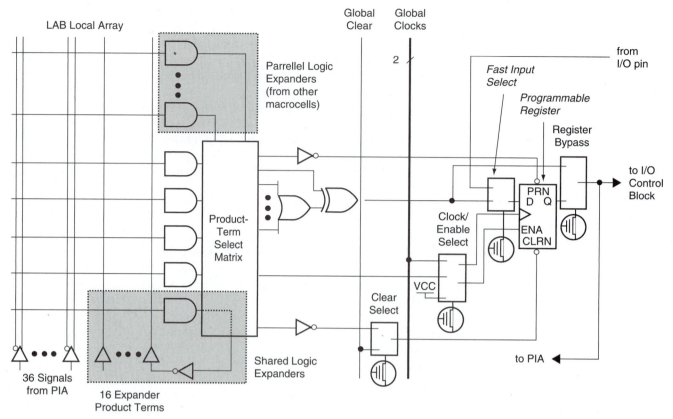

FIGURE 8.21
MAX 7000E and MAX 7000S Device Macrocell (Courtesy of Altera)

8.7 FLEX10K CPLD

KEY TERMS

Look-up table (LUT)   A circuit that implements a combinational logic function by storing a list of output values that correspond to all possible input combinations.

Logic element (LE)   A circuit internal to a CPLD used to implement a logic function as a look-up table.

Cascade chain   A circuit in a CPLD that allows the input width of a Boolean function to expand beyond the width of one logic element.

Carry chain   A circuit in a CPLD that is optimized for efficient operation of carry functions between logic elements.

Embedded array block (EAB)   A relatively large block of storage elements in a CPLD (2048 bits in a FLEX10K device), used for implementing complex logic functions in look-up table format.

All programmable logic devices we have seen until now have been based on sum-of-products arrays. Another major type of PLD is based on **look-up table (LUT)** architecture. In this architecture, a number of storage elements are used to synthesize logic functions by storing each function as a truth table. To illustrate the look-up table concept, let us use the truth table of a 2-bit equality comparator, shown in Table 8.2.

The comparator examines inputs A_1A_0 and B_1B_0 and makes output $AEQB$ equal to logic 1 if $A_1A_0 = B_1B_0$. If we were to implement the circuit as an SOP array, we would first find the Boolean expression by combining the four product terms from the truth table and then program the appropriate cells in a CPLD AND matrix. The look-up table implementation of this function is based on a totally different concept.

Table 8.2 Truth Table for a 2-bit Equality
Comparator

A_1	A_0	B_1	B_0	Decimal	$AEQB$
0	0	0	0	0	1
0	0	0	1	1	0
0	0	1	0	2	0
0	0	1	1	3	0
0	1	0	0	4	0
0	1	0	1	5	1
0	1	1	0	6	0
0	1	1	1	7	0
1	0	0	0	8	0
1	0	0	1	9	0
1	0	1	0	10	1
1	0	1	1	11	0
1	1	0	0	12	0
1	1	0	1	13	0
1	1	1	0	14	0
1	1	1	1	15	1

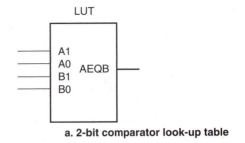

a. 2-bit comparator look-up table

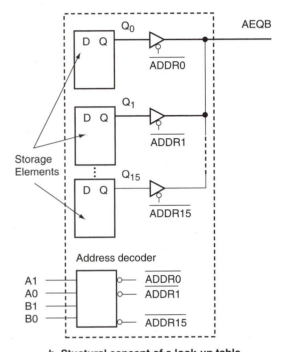

b. Stuctural concept of a look-up table

FIGURE 8.22
Look-up Table

Figure 8.22 shows the structural concept of a 4-bit look-up table circuit. An array of 16 flip-flops (Q_0 through Q_{15}) contain data for all possible combinations of $A_1A_0B_1B_0$, one flip-flop per combination. The LUT inputs $A_1A_0B_1B_0$ are decoded by an internal **address decoder.** Each decoder output activates a tristate buffer that passes or blocks the output of one flip-flop. The active buffer passes the contents of the flip-flop to $AEQB$; all other buffers are in the high-impedance state, blocking the data from the other flip-flops.

The contents of the flip-flops are loaded when the look-up table is configured (programmed) with the required function. After that the flip-flops retain their information until they are reconfigured. For our comparator example, flip-flops 0, 5, 10, and 15 are all set ($Q = 1$). All other flip-flops are reset ($Q = 0$). Examine Table 8.2 to confirm that this is true.

The 16-bit storage element in Figure 8.22, combined with switching to choose a combinational or registered output and to interconnect with other parts of the chip, is called a **logic element (LE).** A logic element performs a function similar to that of a macrocell in SOP-type PLDs.

Figure 8.23 shows the structure of a logic element in an Altera FLEX10K CPLD. In addition to the LUT, the LE has circuitry to select various control functions, such as clock and reset, a flip-flop for registered output, some expansion circuitry (cascade and carry), and interconnections to local and global busses.

The **cascade chain** circuit, shown in Figure 8.24 allows the user to program Boolean functions with more than four inputs, thus requiring more than one LUT. The

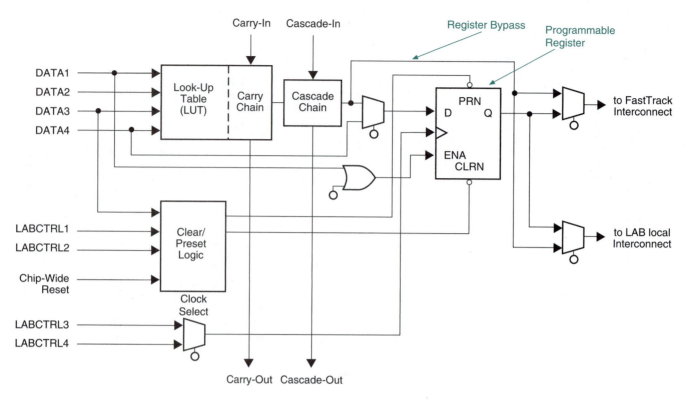

FIGURE 8.23
FLEX10K Logic Element (Courtesy of Altera)

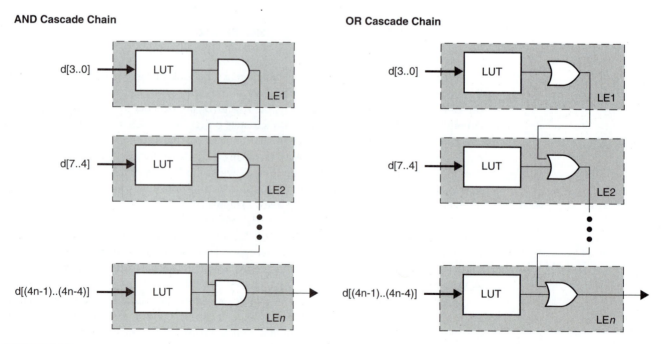

FIGURE 8.24
Cascade Chain Operation (Courtesy of Altera)

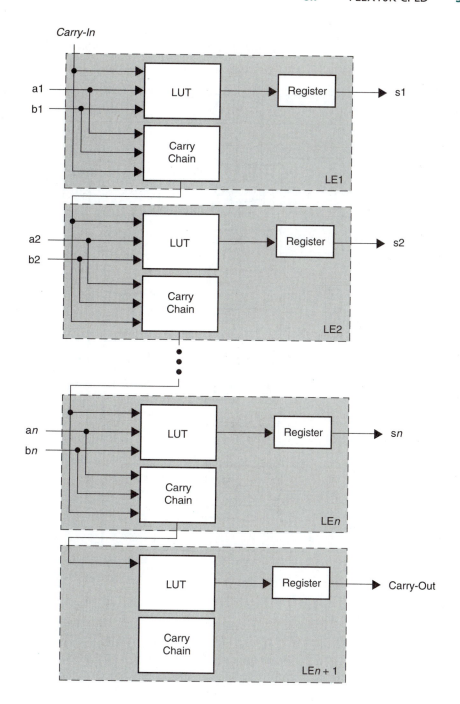

FIGURE 8.25

Carry Chain Operation
(*n*-bit Full Adder)
(Courtesy of Altera)

cascade chain can be AND- or OR-type, depending on what DeMorgan equivalent form is most appropriate.

The **carry chain,** shown in Figure 8.25 allows for efficient fast-carry implementation of adders, comparators, and other circuits that depend on the combination of low-order bits to define high-order functions (i.e., circuits whose inputs become wider with higher-order bits). Figure 8.25 shows the carry chain as implemented by an *n*-bit adder.

A Logic Array Block (LAB), shown in Figure 8.26, consists of eight logic elements and a local interconnect. The LAB is connected to the rest of the device by a series of row and column interconnects, which Altera calls a FastTrack Interconnect. Figure 8.27

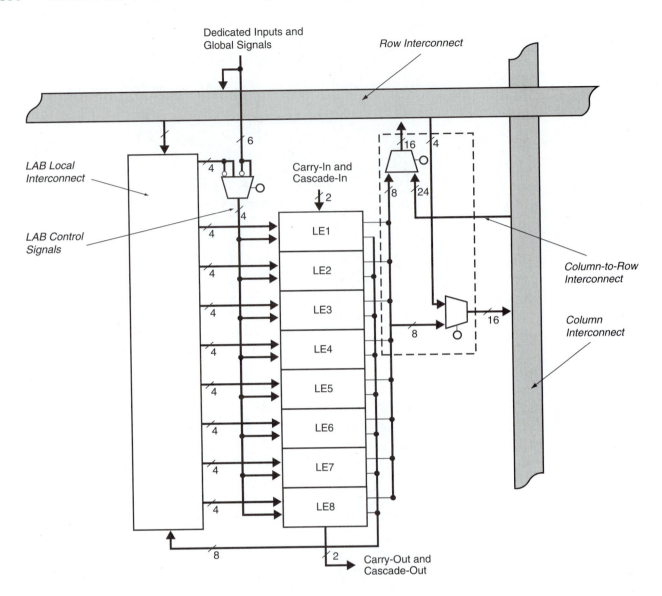

FIGURE 8.26
FLEX10K LAB (Courtesy of Altera)

shows the overall structure of a FLEX10K device, with several LABs and a number of **Embedded Array Blocks (EABs).** An EAB is an array of 2048 storage elements that can be used to efficiently implement complex logic functions.

The FLEX10K device found on the Altera UP-1 board—the EPF10K20RC240-4—has an array of 6 rows by 24 columns of LABs, which gives a total of 144 LABs (= 8 × 144 = 1152 logic elements). The device also has 6 EABs (6 × 2048 = 12288 bits of EAB storage). Note that one EAB has significantly more storage capacity than all LABs combined.

The FLEX10K series of CPLDs (and LUT-based devices generally) are based on static random access memory (SRAM) technology. The advantage of this configuration is that it can be manufactured with a very high density of storage cells and it programs quickly compared to an EEPROM-based SOP device. The disadvantage is that SRAM cells are volatile; that is, they do not retain their data when power is removed from the circuit. An SRAM-based device must be reconfigured every time it is powered up.

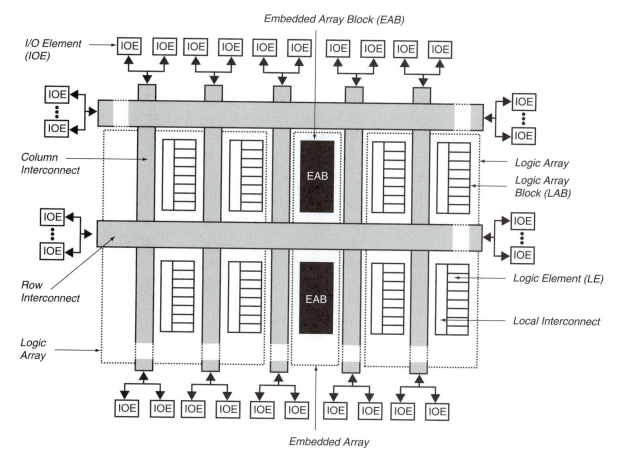

Embedded Array Block (EAB)

I/O Element (IOE)

Column Interconnect

Logic Array

Logic Array Block (LAB)

Row Interconnect

Logic Element (LE)

Logic Array

Local Interconnect

Embedded Array

FIGURE 8.27
FLEX10K Device Block Diagram (Courtesy of Altera)

S U M M A R Y

1. Programmable logic devices (PLDs) are configured in two basic architectures: sum-of-products (SOP), which usually consist of a series of programmable AND/OR circuits, and look-up table (LUT), that stores the truth table of a Boolean function in a small memory.

2. Programmable array logic (PAL) is an SOP-type architecture in which there are a series of programmable AND gates that have a fixed connection to an OR-gate output.

3. Connections from PLD inputs to PAL AND arrays were historically made by leaving intact selected fuses in a crosspoint fuse array. In modern PLDs, these connections are made by programming EEPROM (electrically erasable programmable read only memory) cells.

4. An AND-gate input in a PAL array is called a product line.

5. A PAL16L8 PLD is an SOP device with up to 16 inputs and up to 8 outputs. There are 10 dedicated inputs, 2 dedicated outputs, and 6 pins that can be configured as input or output. All outputs in the PAL16L8 are active-LOW.

6. A PAL is programmed by a computer and programming hardware that uses a JEDEC file as a template for determining which fuses to blow and which to leave intact.

7. Some PAL devices have programmable-polarity outputs. This is achieved with an XOR gate that has a programmable cell or fuse on one input to switch the output between inverting and noninverting levels.

8. A registered PLD output consists of a flip-flop (usually D-type) on the output of an SOP matrix.

9. A PAL part number indicates the number of registered outputs (e.g., a PAL16R8 has eight registered outputs).

10. Early-version standard PALs are limited in that they are one-time programmable (OTP), their outputs are permanently configured as combinational or registered, and they cannot be programmed in-system. Later-version PALs (e.g., PAL16CE16V8 Universal PAL) and GALs (generic array logic such as GAL22V10) overcome these limitations.

11. PALs and GALs with configurable architecture have outputs that can be combinational or registered, with various input or feedback options.

12. Configurable output circuits in a PLD are called output logic macrocells (OLMCs) or just macrocells.

13. Macrocells are configured by programming architecture cells. Global architecture cells affect all macrocells in a device. A local architecture cell affects only the macrocell in which it is found.

14. GALs and Universal PALs have global control signals, such as clock, clear, and output enable, that can be applied to all macrocells in the device.

15. A GAL22V10 has ten macrocells, a global clock that can be used as a combinational input for nonclocked designs, and eleven dedicated inputs.

16. The GAL22V10 macrocells are not all the same size. There are two macrocells with each of the following numbers of product terms: 8, 10, 12, 14, 16.

17. PLDs that can be programmed while installed in a circuit are called in-system programmable (ISP). They are programmed by a 4-wire interface that complies to a standard published by the Joint Test Action Group (JTAG) and the IEEE (Std. 1149.1).

18. An Altera MAX7000S CPLD consists of groups of 16 macrocells, called Logic Array Blocks (LABs), that are interconnected by an internal bus called a Programmable Interconnect Array (PIA).

19. The number of macrocell outputs in an LAB that are connected to I/O pins depends on the CPLD package type. Macrocells that do not have external connections can still be used for buried logic function.

20. MAX7000S devices have four programmable control pins: global clock *(GCLK1)*, Global Output Enable *(OE1)*, Global Clear *(GCLRn)*, and a pin that can be configured as a second global clock *(GCLK2)* or as a second global output enable *(OE2)*. If these functions are not used, the associated pins can be used as standard I/Os.

21. If the ISP capability of a CPLD is to be used, there are four fewer pins available on the CPLD for user I/O.

22. Each MAX7000S macrocell has five dedicated product lines and capability to borrow or share additional product terms with neighboring macrocells in the same LAB.

23. Shared logic expanders allow one product term per macrocell to be shared with other macrocells in the LAB, totaling 16 product terms per LAB. The expander inverts the product term and feeds it back into the LAB AND matrix.

24. Parallel logic expanders allow a macrocell to borrow product lines from neighboring macrocells. These borrowed product lines are only available to one macrocell.

25. Expander assignments are done automatically by MAX+ PLUS II at compile time.

26. MAX7000S devices are based on EEPROM cells and are thus nonvolatile.

27. The Altera FLEX10K series of CPLDs is based on a look-up table (LUT) architecture. A look-up table consists of a 16-bit array of storage elements that are selected by four logic inputs.

28. An LUT combined with switching, configuration, and expansion circuitry comprises a logic element (LE), whose function is equivalent to a macrocell in an SOP-type device.

29. Eight logic elements and a local interconnect make up a Logic Array Block (LAB).

30. LABs in a FLEX10K device are interconnected by global row and column busses.

31. The number of inputs in a logic function can be expanded beyond the capacity of one logic element by using cascade chains.

32. Carry chains can be used to more efficiently implement carry functions in adders, counters, and comparators.

33. FLEX10K devices are based on SRAM technology and are therefore volatile; they must be reconfigured each time power is applied to the circuit.

GLOSSARY

Architecture cell   A programmable cell that, in combination with other architecture cells, sets the configuration of a macrocell.

Buried logic   Logic circuitry in a PLD that has no connection to the input or output pins of the PLD, but is used solely as internal logic.

Carry chain   A circuit in a CPLD that is optimized for efficient operation of carry functions between logic elements.

Cascade chain   A circuit in a CPLD that allows the input width of a Boolean function to expand beyond the width of one logic element.

Cell   A fuse location in a programmable logic device, specified by the intersection of an input line and a product line.

Checksum   An error-checking code derived from the accumulating sum of the data being checked.

CPLD   Complex programmable logic device. A programmable logic device consisting of several interconnected programmable blocks.

Embedded array block (EAB)   A relatively large block of storage elements in a CPLD (2048 bits in a FLEX10K device), used for implementing complex logic functions in look-up table format.

Generic array logic (GAL)   A type of programmable logic device whose outputs can be configured as combinational or registered and whose programming matrix is based on electrically erasable logic cells.

Global architecture cell   An architecture cell that affects the configuration of all macrocells in a device.

Global clock   A clock signal in a PLD that clocks all registered outputs in the device.

I/O Control Block   A circuit in an Altera CPLD that controls the type of tristate switching used in a macrocell output.

Input line   A line which applies the true or complement form of an input variable to the AND matrix of a PLD.

Input line number   A number assigned to a true or complement input line in a PAL AND matrix.

In-system programmability (ISP)   The ability of a PLD to be programmed through a standard four-wire interface while installed in a circuit.

JEDEC   Joint Electron Device Engineering Council

JEDEC file   An industry standard form of text file indicating which fuses are blown and which are intact in a programmable logic device.

JTAG Port   A four-wire interface specified by the Joint Test Action Group (JTAG) used for loading test data or programming data into a PLD installed in a circuit.

Local architecture cell   An architecture cell that affects the configuration of one macrocell only.

Logic Array Block (LAB)   A group of macrocells that share common resources in a CPLD.

Logic element (LE)   A circuit internal to a CPLD used to implement a logic function as a look-up table.

Look-up table (LUT)   A circuit that implements a combinational logic function by storing a list of output values that correspond to all possible input combinations.

Multiplexer   A circuit which selects one of several signals to be directed to a single output.

One-time programmable (OTP)   A property of some PLDs that allows them to be programmed, but not erased.

Output logic macrocell (OLMC)   An input/output circuit that can be programmed for a variety of input or output configurations, such as active HIGH or active LOW, combinational, or registered. Often just called a **macrocell.**

PAL   Programmable array logic. Programmable logic with a fixed OR matrix and a programmable AND matrix.

Parallel logic expanders   Product terms that are borrowed from neighboring macrocells in the same LAB.

Product line   A single line on a logic diagram used to represent all inputs to an AND gate (i.e., one product term) in a PLD sum-of-products array.

Product line first cell number   The lowest cell number on a particular product line in a PAL AND matrix where all cells are consecutively numbered.

Programmable Interconnect Array (PIA)   An internal bus with programmable connections that link together the Logic Array Blocks of a CPLD.

Programmable logic device (PLD)   A logic device whose function can be programmed by the user, usually in sum-of-products form.

Register   A digital circuit such as a flip-flop that stores one or more bits of digital information.

Registered output   An output of a programmable array logic (PAL) device having a flip-flop (usually D-type) which stores the output state.

Shared logic expanders   Product terms that are inverted and fed back into the programmable AND matrix of an LAB for use by any other macrocell in the LAB.

Text file   An ASCII-coded document stored on a magnetic disk.

Universal PAL   A PLD based on erasable cells and configurable outputs, much like GAL, but primarily designed to emulate PAL devices, such as PAL16L8.

PROBLEMS

Problem numbers set in color indicate more difficult problems; those with underlines indicate most difficult problems.

Section 8.1   Introduction to Progammable Logic

Section 8.2   PAL Fuse Matrix and Combinational Outputs

Section 8.3   PAL Outputs With Programmable Polarity

8.1   Draw a diagram showing the basic configuration and symbology for a PLD sum-of-products array.

8.2   Draw a basic PAL circuit having four inputs, eight product terms, and one active-LOW combinational output. Draw fuses on your diagram showing how to make the following Boolean expression:

$$\overline{F} = \overline{A}\,B\,\overline{C} + \overline{B}\,C\,D + \overline{A}\,C\,D + A\,\overline{C}\,D$$

8.3   Modify the PAL circuit drawn in Problem 8.2 to make two outputs having eight product terms and programmable polarity. Draw fuses on the diagram for each of the following functions:

$$F1 = A\,B\,\overline{C} + \overline{B}\,C\,D + \overline{A}\,C\,D + A\,\overline{C}\,D$$
$$\overline{F2} = \overline{A}\,B\,\overline{C} + \overline{B}\,C\,D + \overline{A}\,C\,D + A\,\overline{C}\,D$$

8.4   Make a photocopy of Figure 8.8 (PAL20P8 logic diagram). Draw fuses on the PAL20P8 logic diagram showing how to make a BCD-to-2421 code converter, as developed in Example 3.22.

Table 8.3 shows how the two codes relate to each other. The equations are listed on page 362.

Table 8.3   BCD and 2421 Code

Decimal Equivalent	BCD Code				2421 Code			
	D_4	D_3	D_2	D_1	Y_4	Y_3	Y_2	Y_1
0	0	0	0	0	0	0	0	0
1	0	0	0	1	0	0	0	1
2	0	0	1	0	0	0	1	0
3	0	0	1	1	0	0	1	1
4	0	1	0	0	0	1	0	0
5	0	1	0	1	1	0	1	1
6	0	1	1	0	1	1	0	0
7	0	1	1	1	1	1	0	1
8	1	0	0	0	1	1	1	0
9	1	0	0	1	1	1	1	1

The Boolean equations for the BCD-to-2421 decoder are:

$$Y_4 = D_4 + D_3D_2 + D_3D_1$$
$$Y_3 = D_4 + D_3D_2 + D_3\overline{D_1}$$
$$Y_2 = D_4 + \overline{D_3}D_2 + D_3\overline{D_2}D_1$$
$$Y_1 = D_1$$

8.5   Repeat Problem 8.4 for a 2421-to-BCD code converter.

8.4   PAL Devices with Registered Outputs

8.6   What is a registered output?

8.7   State the number of registered outputs for each of the following PAL devices:

a. PAL16R4
b. PAL16R6
c. PAL16R8

8.5   Universal PAL and Generic Array Logic (GAL)

8.8   Name two features of a PALCE16V8 that make it superior to a PAL16L8.

8.9   State the difference between a global architecture cell and a local architecture cell in a PALCE16V8.

8.10   How many macrocells are there in a GAL22V10? How many product lines do these macrocells have?

8.11   State the four configurations possible with a macrocell in a GAL22V10.

8.12   Is there a global output enable function available for a PALCE16V8? For a GAL22V10?

8.13   Can the registered outputs of a PALCE16V8 be clocked by a product term function from the PAL AND matrix?

8.14   Can the registered outputs of a GAL22V10 be clocked by a product term function from the GAL AND matrix?

8.15   Are the Asynchronous Reset *(AR)* and Synchronous Preset *(SP)* functions in a GAL22V10 global or local? Explain your answer in one sentence.

8.6   MAX7000S CPLD

8.16   State one way in which a Complex PLD, such as an Altera MAX7000S, differs from a low-density PAL or GAL.

8.17   How many macrocells are available in the following CPLDs:

a. EPM7032
b. EPM7064
c. EPM7128S
d. EPM7160S

8.18   Which of the CPLDs listed in Problem 8.17 are in-system programmable? What does it mean when a device is in-system programmable?

8.19   How many logic array blocks (LABs) are there in an Altera MAX7000S CPLD?

8.20   How many user I/O pins are there in an EPM7128SLC84 CPLD? How many pins per LAB does this represent?

8.21   What can be done with the macrocells in an LAB that do not connect to I/O pins?

8.22   State the possible clock configurations of a MAX7000S macrocell.

8.23   State the possible reset configurations of a MAX7000S macrocell.

8.24   State the possible preset configurations of a MAX7000S macrocell.

8.25   How many dedicated product terms are available in a MAX7000S macrocell? How can this number of product terms be supplemented? What is the maximum number of product terms available to a macrocell?

8.26   How many shared logic expanders are available in an LAB?

8.7   FLEX10K CPLD

8.27   Briefly state the difference between CPLDs having sum-of-products architecture and look-up table architecture.

8.28   How many inputs can a look-up table accept in an Altera FLEX10K logic element? How can this be expanded?

8.29   What is the purpose of the carry chain in a FLEX10K CPLD?

8.30   How many logic elements are there in a FLEX10K LAB?

8.31   How many bits of storage are there in an Embedded Array Block in a FLEX10K CPLD?

Counters and Shift Registers

CHAPTER OBJECTIVES

Upon successful completion of this chapter you will be able to:

- Determine the modulus of a counter.

- Determine the number of outputs required by a counter for a given modulus.

- Determine the maximum modulus of a counter, given the number of circuit outputs.

- Draw the count sequence table, state diagram, and timing diagram of a counter.

- Determine the recycle point of a counter's sequence.

- Calculate the frequencies of each counter output, given the input clock frequency.

- Draw a circuit for any full sequence synchronous counter.

- Determine the count sequence, state diagram, timing diagram, and modulus of any synchronous counter.

- Complete the state diagram of a synchronous counter to account for unused states.

- Design the circuit of a truncated sequence synchronous counter, using flip-flops and logic gates.

- Use MAX+PLUS II to create a graphic design file for any synchronous counter circuit.

- Use behavioral descriptions in VHDL to design synchronous counters of any modulus.

- Use a parameterized counter from the Library of Parameterized Modules in a VHDL file.

- Use the MAX+PLUS II simulation tool to verify the operation of synchronous counters.

- Implement various counter control functions, such as parallel load, clear, count enable, and count direction, both in Graphic Design Files and in VHDL.

- Design a circuit to decode the output of the counter, both in a MAX+PLUS II Graphic Design File or in VHDL.

- Draw a logic circuit of a serial shift register and determine its contents over time given any input data.

- Draw a timing diagram showing the operation of a serial shift register.
- Draw the logic circuit of a general parallel-load shift register.
- Draw a timing diagram showing the operation of a parallel-load shift register.
- Draw the general logic circuit of a bidirectional shift register and explain the concepts of right-shift and left-shift.
- Use timing diagrams to explain the operation of a bidirectional shift register.
- Describe the operation of a universal shift register.
- Design shift registers, ring counters, and Johnson counters with the MAX+PLUS II Graphic Editor or VHDL.
- Verify the operation of shift registers, ring counters, and Johnson counters using the MAX+PLUS II simulation tool.
- Design a decoder for a Johnson counter.
- Use a ring counter or a Johnson counter as an event sequencer.
- Compare binary, ring, and Johnson counters in terms of the modulus and the required decoding for each circuit.

Counters and shift registers are two important classes of sequential circuits. In the simplest terms, a counter is a circuit that counts pulses. As such, it is used in many circuit applications, such as event counting and sequencing, timing, frequency division, and control. A basic counter can be enhanced to incorporate functions such as synchronous or asynchronous parallel loading, synchronous or asynchronous clear, count enable, directional control, and output decoding. In this chapter, we will design counters using schematic entry, VHDL, and counters from the Library of Parameterized Modules and verify their operation using the MAX+PLUS II simulator.

Shift registers are circuits that store and move data. They can be used in serial data transfer, serial/parallel conversion, arithmetic functions, and delay elements. As with counters, many shift registers have additional functions such as parallel load, clear, and directional control. We can implement these circuits using schematic entry, VHDL, and LPM components. ∎

9.1 Basic Concepts of Digital Counters

> **KEY TERMS**
>
> **Counter** A sequential digital circuit whose output progresses in a predictable repeating pattern, advancing by one state for each clock pulse.
>
> **Recycle** To make a transition from the last state of the count sequence to the first state.
>
> **Count sequence** The specific series of output states through which a counter progresses.
>
> **State diagram** A diagram showing the progression of states of a sequential circuit.
>
> **Modulus** The number of states through which a counter sequences before repeating.
>
> **Modulo-n (or mod-n) counter** A counter with a modulus of n.
>
> **UP counter** A counter with an ascending sequence.
>
> **DOWN counter** A counter with a descending sequence.

The simplest definition of a **counter** is "a circuit that counts pulses." Knowing only this, let us look at an example of how we might use a counter circuit.

▌▌ EXAMPLE 9.1

Figure 9.1 shows a 10-bit binary counter that can be used to count the number of people passing by an optical sensor. Every time the sensor detects a person passing by, it produces a pulse. Briefly describe the counter's operation. What is the maximum number of people it can count? What happens if this number is exceeded?

FIGURE 9.1
Example 9.1
10-bit Counter

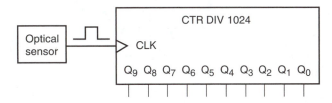

Solution   The counter has a 10-bit output, allowing a binary number from 00 0000 0000 to 11 1111 1111 (0 to 1023) to appear at its output. The sensor causes the counter to advance by one binary number for every pulse applied to the counter's clock *(CLK)* input. If the counter is allowed to register *no people* (i.e., 00 0000 0000), then the circuit can count 1023 people, since there are 1024 unique binary combinations of a 10-bit number, including 0. (This is because $2^{10} = 1024$.) When the 1024th pulse is applied to the clock input, the counter rolls over to 0 (or **recycles**) and starts counting again. (After this point, the counter would not accurately reflect the number of people counted.)

The counter is labeled CTR DIV 1024 to indicate that one full cycle of the counter requires 1024 clock pulses (i.e., the frequency of the MSB output signal (Q_9) is the clock frequency divided by 1024).

A counter is a digital circuit that has a number of binary outputs whose states progress through a fixed sequence. This **count sequence** can be ascending, descending, or nonlinear.

The output sequence of a counter is usually defined by its **modulus,** that is, the number of states through which the counter progresses. An **UP counter** with a modulus of 12 counts through 12 states from 0000 up to 1011 (0 to 11 in decimal), recycles to 0000, and continues. A **DOWN counter** with a modulus of 12 counts from 1011 down to 0000, recycles to 1011, and continues downward. Both types of counter are called **modulo-12,** or just **mod-12** counters, since they both have sequences of 12 states.

State Diagram

The states of a counter can be represented by a **state diagram.** Figure 9.2 compares the state diagram of a mod-12 UP counter to an analog clock face. Each counter state is illustrated in the state diagram by a circle containing its binary value. The progression is shown by a series of directional arrows.

Both the clock face and the state diagram represent a closed system of counting. In each case, when we reach the end of the count sequence, we start over from the beginning of the cycle.

For instance, if it is 10:00 a.m. and we want to meet a friend in four hours, we know we should turn up for the appointment at 2:00 p.m. We arrive at this figure by starting at 10 on the clock face and counting 4 digits forward in a "clockwise" circle. This takes us two digits past 12, the "recycle point" of the clock face.

Similarly, if we want to know the 8th state after 0111 in a mod-12 UP counter, we start at state 0111 and count 8 positions in the direction of the arrows. This brings us to state 0000 (the recycle point) in 5 counts and then on to state 0011 in another 3 counts.

FIGURE 9.2
Mod-12 State Diagram and
Analog Clock Face

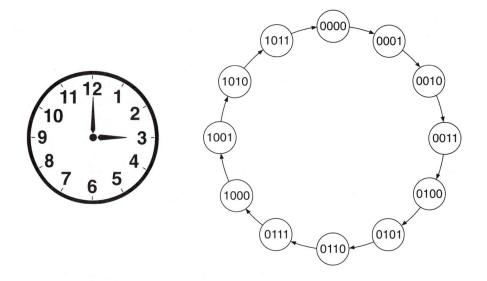

Number of Bits and Maximum Modulus

The state diagram of Figure 9.2 represents the states of a mod-12 counter as a series of 4-bit numbers. Counter states are always written with a fixed number of bits, since each bit represents the logic level of a physical location in the counter circuit. A mod-12 counter requires four bits because its highest count value is a 4-bit number: 1011.

The **maximum modulus** of a 4-bit counter is 16 ($= 2^4$). The count sequence of a mod-16 UP counter is from 0000 to 1111 (0 to 15 in decimal), as illustrated in the state diagram of Figure 9.3.

In general, an n-bit counter has a maximum modulus of 2^n and a count sequence from 0 to $2^n - 1$ (i.e., all 0s to all 1s). Since a mod-16 counter has a modulus of 2^n ($= m_{max}$), we say that it is a **full-sequence counter.** We can also call this a **binary counter** if it generates the sequence in binary order. A counter, such as a mod-12 counter, whose modulus is less than 2^n, is called a **truncated sequence counter.**

Count-Sequence Table and Timing Diagram

Two ways to represent a count sequence other than a state diagram are by a **count sequence table** and by a timing diagram. The count sequence table is simply a list of counter states in the same order as the count sequence. Tables 9.1 and 9.2 show the count sequence tables of a mod-16 UP counter and a mod-12 UP counter, respectively.

FIGURE 9.3
State Diagram of a Mod-16
Counter

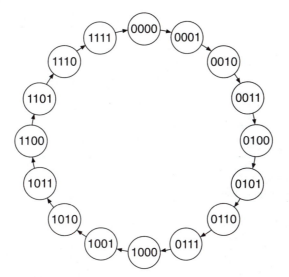

Table 9.1 Mod-16
Count Sequence Table

$Q_3Q_2Q_1Q_0$
0000
0001
0010
0011
0100
0101
0110
0111
1000
1001
1010
1011
1100
1101
1110
1111

Table 9.2 Mod-12
Count-Sequence Table

$Q_3Q_2Q_1Q_0$
0000
0001
0010
0011
0100
0101
0110
0111
1000
1001
1010
1011

We can derive timing diagrams from each of these tables. We know that each counter advances by one state with each applied clock pulse. The mod-16 count sequence shows us that the Q_0 waveform changes state with each clock pulse. Q_1 changes with every two clock pulses, Q_2 with every four, and Q_3 with every eight. Figure 9.4 shows this pattern for the mod-16 UP counter, assuming the counter is a positive edge-triggered device.

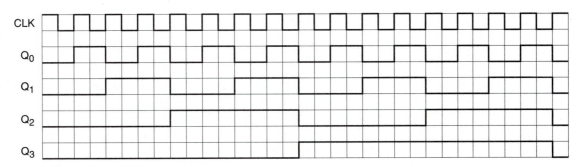

FIGURE 9.4
Mod-16 Timing Diagram

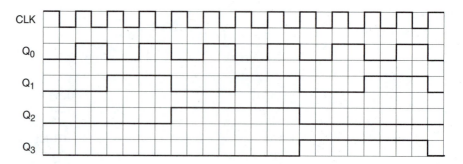

FIGURE 9.5
Mod-12 Timing Diagram

A divide-by-two ratio relates the frequencies of adjacent outputs of a binary counter. For example, if the clock frequency is $f_c = 16$ MHz, the frequencies of the output waveforms are: 8 MHz ($f_0 = f_c/2$); 4 MHz ($f_1 = f_c/4$); 2 MHz ($f_2 = f_c/8$); 1 MHz ($f_3 = f_c/16$).

We can construct a similar timing diagram, illustrated in Figure 9.5, for a mod-12 UP counter. The changes of state can be monitored by noting where Q_0 (the least significant bit) changes. This occurs on each positive edge of the *CLK* waveform. The sequence progresses by 1 with each *CLK* pulse until the outputs all go to 0 on the first *CLK* pulse after state $Q_3Q_2Q_1Q_0 = 1011$.

The output waveform frequencies of a truncated sequence counter do not necessarily have a simple relationship to one another as do binary counters. For the mod-12 counter the relationships between clock frequency, f_c, and output frequencies are: $f_0 = f_c/2$; $f_1 = f_c/4$; $f_2 = f_c/12$; $f_3 = f_c/12$. Note that both Q_2 and Q_3 have the same frequencies (f_2 and f_3), but are out of phase with one another.

▌▌ EXAMPLE 9.2

Draw the state diagram, count sequence table, and timing diagram for a mod-12 DOWN counter.

Solution   Figure 9.6 shows the state diagram for the mod-12 DOWN counter. The states are identical to those of a mod-12 UP counter, but progress in the opposite direction. Table 9.3 shows the count sequence table of this circuit.

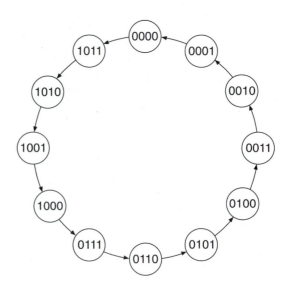

FIGURE 9.6
Example 9.2
State Diagram of a Mod-12 DOWN Counter

Table 9.3 Count-
Sequence Table for a
Mod-12 DOWN Counter

$Q_3Q_2Q_1Q_0$
1011
1010
1001
1000
0111
0110
0101
0100
0011
0010
0001
0000

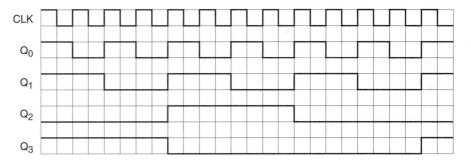

FIGURE 9.7
Example 9.2
Timing Diagram of a Mod-12 DOWN Counter

The timing diagram of this counter is illustrated in Figure 9.7. The output starts in state $Q_3Q_2Q_1Q_0 = 1011$ and counts DOWN until it reaches 0000. On the next pulse, it recycles to 1011 and starts over.

■|■ SECTION 9.1 REVIEW PROBLEM

9.1 How many outputs does a mod-24 counter require? Is this a full-sequence or a truncated sequence counter? Explain your answer.

9.2 Synchronous Counters

KEY TERMS

Synchronous counter   A counter whose flip-flops are all clocked by the same source and thus change in synchronization with each other.

Present state   The current state of flip-flop outputs in a synchronous sequential circuit.

Next state   The desired future state of flip-flop outputs in a synchronous sequential circuit after the next clock pulse is applied.

Memory section   A set of flip-flops in a synchronous circuit that hold its present state.

Control section   The combinational logic portion of a synchronous circuit that determines the next state of the circuit.

Status lines   Signals that communicate the present state of a synchronous circuit from its memory section to its control section.

Command lines   Signals that connect the control section of a synchronous circuit to its memory section and direct the circuit from its present to its next state.

In Chapter 7, we briefly examined the circuits of a 3-bit and a 4-bit **synchronous counter** (Figures 7.53 and 7.87, respectively). A synchronous counter is a circuit consisting of flip-flops and control logic, whose outputs progress through a regular predictable sequence, driven by a clock signal. The counter is synchronous because all flip-flops are clocked at the same time.

Figure 9.8 shows the block diagram of a synchronous counter, which consists of a **memory section** to keep track of the **present state** of the counter and a **control section** to direct the counter to its **next state.** The memory section is a sequential circuit (flip-flops) and the control section is combinational (gates). They communicate through a set of **status lines** that go from the Q outputs of the flip-flops to the control gate inputs and **command lines** that connect the control gate outputs to the synchronous inputs *(J, K, D,* or *T)* of the flip-flops. Outputs can be tied directly to the status lines or can be decoded to give a sequence other than that of the flip-flop output states. The circuit might have inputs to implement one or more control functions, such as changing the count direction, clearing the counter, or presetting the counter to a specific value.

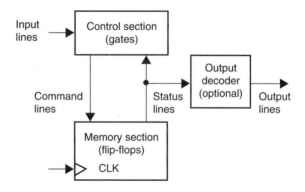

FIGURE 9.8
Synchronous Counter Block Diagram

Analysis of Synchronous Counters

A 3-bit synchronous binary counter based on JK flip-flops is shown in Figure 9.9. Let us analyze its count sequence in detail so that we can see how the *J* and *K* inputs are affected by the *Q* outputs and how transitions between states are made. Later we will look at the function of truncated sequence counter circuits and counters that are made from flip-flops other than JK.

The synchronous input equations are given by:

$$J_2 = K_2 = Q_1 \cdot Q_0$$
$$J_1 = K_1 = Q_0$$
$$J_0 = K_0 = 1$$

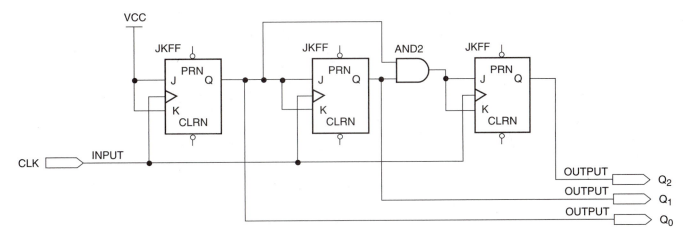

FIGURE 9.9
3-bit Synchronous Binary Counter

For reference, the JK flip-flop function table is shown in Table 9.4:

Table 9.4 Function Table of a JK Flip-Flop

J	K	Q_{t+1}	Function
0	0	Q_t	No change
0	1	0	Reset
1	0	1	Set
1	1	$\overline{Q_t}$	Toggle

Q_t indicates the state of Q before a clock pulse is applied. Q_{t+1} indicates the state of Q after the clock pulse.

Assume the counter output is initially $Q_2Q_1Q_1 = 000$. Before any clock pulses are applied, the J and K inputs are at the following states:

$$J_2 = K_2 = Q_1 \cdot Q_0 = 0 \cdot 0 = 0 \quad \text{(No change)}$$
$$J_1 = K_1 = Q_0 = 0 \quad \text{(No change)}$$
$$J_0 = K_0 = 1 \text{ (Constant)} \quad \text{(Toggle)}$$

The transitions of the outputs after the clock pulse are:

$$Q_2: 0 \rightarrow 0 \quad \text{(No change)}$$
$$Q_1: 0 \rightarrow 0 \quad \text{(No change)}$$
$$Q_0: 0 \rightarrow 1 \quad \text{(Toggle)}$$

The output goes from $Q_2Q_1Q_1 = 000$ to $Q_2Q_1Q_1 = 001$ (see Figure 9.10). The transition is defined by the values of J and K *before* the clock pulse, since the propagation delays of the flip-flops prevent the new output conditions from changing the J and K values until after the transition.

The new conditions of the J and K inputs are:

$$J_2 = K_2 = Q_1 \cdot Q_0 = 0 \cdot 1 = 0 \quad \text{(No change)}$$
$$J_1 = K_1 = Q_0 = 1 \quad \text{(Toggle)}$$
$$J_0 = K_0 = 1 \text{ (Constant)} \quad \text{(Toggle)}$$

The transitions of the outputs generated by the second clock pulse are:

$$Q_2: 0 \rightarrow 0 \quad \text{(No change)}$$
$$Q_1: 0 \rightarrow 1 \quad \text{(Toggle)}$$
$$Q_0: 1 \rightarrow 0 \quad \text{(Toggle)}$$

The new output is $Q_2Q_1Q_0 = 010$, since both Q_0 and Q_1 change and Q_2 stays the same. The J and K conditions are now:

$$J_2 = K_2 = Q_1 \cdot Q_0 = 1 \cdot 0 = 0 \quad \text{(No change)}$$
$$J_1 = K_1 = Q_0 = 0 \qquad\qquad \text{(No change)}$$
$$J_0 = K_0 = 1 \text{ (Constant)} \qquad \text{(Toggle)}$$

The output transitions are:

$$Q_2: 0 \rightarrow 0 \quad \text{(No change)}$$
$$Q_1: 1 \rightarrow 1 \quad \text{(No change)}$$
$$Q_0: 0 \rightarrow 1 \quad \text{(Toggle)}$$

The output is now $Q_2Q_1Q_0 = 011$, which results in the JK conditions:

$$J_2 = K_2 = Q_1 \cdot Q_0 = 1 \cdot 1 = 1 \quad \text{(Toggle)}$$
$$J_1 = K_1 = Q_0 = 1 \qquad\qquad \text{(Toggle)}$$
$$J_0 = K_0 = 1 \text{ (Constant)} \qquad \text{(Toggle)}$$

The above conditions result in output transitions:

$$Q_2: 0 \rightarrow 1 \quad \text{(Toggle)}$$
$$Q_1: 1 \rightarrow 0 \quad \text{(Toggle)}$$
$$Q_0: 1 \rightarrow 0 \quad \text{(Toggle)}$$

All the outputs toggle and the new output state is $Q_2Q_1Q_0 = 100$. The J and K values repeat the above pattern in the second half of the counter cycle (states 100 to 111). Go through the exercise of calculating the J, K, and Q values for the rest of the cycle. Compare the result with the timing diagram in Figure 9.10.

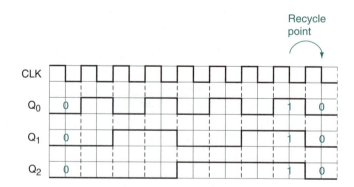

FIGURE 9.10

Timing Diagram for a Synchronous 3-bit Binary Counter

In the counter we have just analyzed, the combinational circuit generates either a toggle ($JK = 11$) or a no change ($JK = 00$) state at each point through the count sequence. We could use any combination of JK modes (no change, reset, set, or toggle) to make the transitions from one state to the next. For instance, instead of using only the no change and toggle modes, the $000 \rightarrow 001$ transition could also be done by making Q_0 set ($J_0 = 1$,

$K_0 = 0$) and Q_1 and Q_2 reset ($J_1 = 0$, $K_1 = 1$ and $J_2 = 0$, $K_2 = 1$). To do so we would need a different set of combinational logic in the circuit.

The simplest synchronous counter design uses only the no change ($JK = 00$) or toggle ($JK = 11$) modes, since the J and K inputs of each flip-flop can be connected together. The no change and toggle modes allow us to make any transition (i.e., not just in a linear sequence), even though for truncated sequence and nonbinary counters this is not usually the most efficient design.

There is a simple progression of algebraic expressions for the J and K inputs of a synchronous binary (full sequence) counter, which uses only the no change and toggle states:

$$J_0 = K_0 = 1$$
$$J_1 = K_1 = Q_0$$
$$J_2 = K_2 = Q_1 \cdot Q_0$$
$$J_3 = K_3 = Q_2 \cdot Q_1 \cdot Q_0$$
$$J_4 = K_4 = Q_3 \cdot Q_2 \cdot Q_1 \cdot Q_0$$
etc.

The J and K inputs of each stage are the ANDed outputs of all previous stages. This implies that a flip-flop toggles only when the outputs of *all* previous stages are HIGH. For example, Q_2 doesn't change unless *both* Q_1 AND Q_0 are HIGH (and therefore $J_2 = K_2 = 1$) before the clock pulse. In a 3-bit counter, this occurs only at states 011 and 111, after which Q_2 will toggle, along with Q_1 and Q_0, giving transitions to states 100 and 000 respectively. Look at the timing diagram of Figure 9.10 to confirm this.

Determining the Modulus of a Synchronous Counter

We can use a more formal technique to analyze any synchronous counter, as follows.

1. Determine the equations for the synchronous inputs *(JK, D,* or *T)* in terms of the Q outputs for all flip-flops. (For counters other than straight binary full sequence types, the equations will *not* be the same as the algebraic progressions previously listed.)

2. Lay out a table with headings for the Present State of the counter (Q outputs before *CLK* pulse), each Synchronous Input before *CLK* pulse, and Next State of the counter (Q outputs after the clock pulse).

3. Choose a starting point for the count sequence, usually 0, and enter the starting point in the Present State column.

4. Substitute the Q values of the initial present state into the synchronous input equations and enter the results under the appropriate columns.

5. Determine the action of each flip-flop on the next *CLK* pulse (e.g., for a JK flip-flop, the output either will not change ($JK = 00$), or will reset ($JK = 01$), set ($JK = 10$), or toggle ($JK = 11$)).

6. Look at the Q values for every flip-flop. Change them according to the function determined in Step 5 and enter them in the column for the counter's next state.

7. Enter the result from Step 6 on the next line of the column for the counter's present state (i.e., this line's next state is the next line's present state).

8. Repeat the above process until the result in the next state column is the same as the initial state.

■ **EXAMPLE 9.3**

Find the count sequence of the synchronous counter shown in Figure 9.11 and, from the count sequence table, draw the timing diagram and state diagram. What is the modulus of the counter?

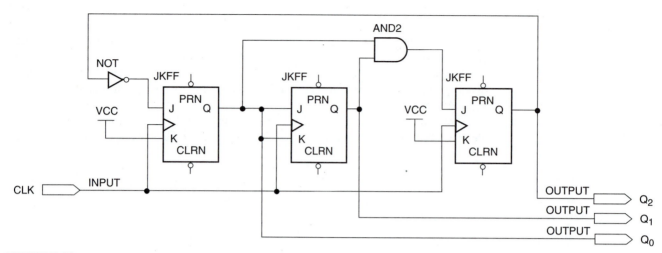

FIGURE 9.11
Synchronous Counter of Unknown Modulus

Solution   The J and K equations are:

$$J_2 = Q_1 \cdot Q_0 \qquad J_1 = Q_0 \qquad J_0 = \overline{Q_2}$$
$$K_2 = 1 \qquad\quad K_1 = Q_0 \qquad K_0 = 1$$

The output transitions can be determined from the values of the J and K functions before each clock pulse, as shown in Table 9.5.

Table 9.5   State Table for Figure 9.11

Present State	Synchronous Inputs						Next State
$Q_2Q_1Q_0$	J_2K_2		J_1K_1		J_0K_0		$Q_2Q_1Q_0$
000	01	(R)	00	(NC)	11	(T)	001
001	01	(R)	11	(T)	11	(T)	010
010	01	(R)	00	(NC)	11	(T)	011
011	11	(T)	11	(T)	11	(T)	100
100	01	(R)	00	(NC)	01	(R)	000

Since there are five unique output states, the counter's modulus is 5.
The timing diagram and state diagram are shown in Figure 9.12. Since this circuit produces one pulse on Q_2 for every 5 clock pulses, we can use it as a divide-by-5 circuit.

FIGURE 9.12
Example 9.3
Timing Diagram and State
Diagram of a Mod-5 Counter

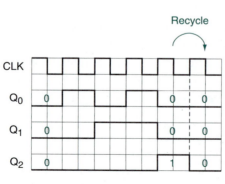

a. Timing diagram

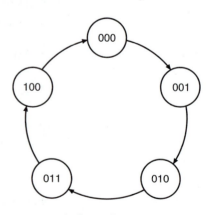

b. State diagram

The analysis in Example 9.3 did not account for the fact that the counter uses only 5 of a possible 8 output states. In any truncated sequence counter, it is good practice to determine the next state for each unused state to ensure that if the counter powers up in one of these unused states, it will eventually enter the main sequence.

EXAMPLE 9.4

Extend the analysis of the counter in Example 9.3 to include its unused states. Redraw the counter's state diagram to show how these unused states enter the main sequence (if they do).

Solution   The synchronous input equations are:

$$J_2 = Q_1 \cdot Q_0 \qquad J_1 = Q_0 \qquad J_0 = \overline{Q_2}$$
$$K_2 = 1 \qquad K_1 = Q_0 \qquad K_0 = 1$$

The unused states are $Q_2Q_1Q_0 = 101$, 110, and 111. Table 9.6 shows the transitions made by the unused states. Figure 9.13 shows the completed state diagram.

Table 9.6   State Table for Mod-5 Counter Including Unused States

Present State	Synchronous Inputs						Next State
$Q_2Q_1Q_0$	J_2K_2		J_1K_1		J_0K_0		$Q_2Q_1Q_0$
000	01	(R)	00	(NC)	11	(T)	001
001	01	(R)	11	(T)	11	(T)	010
010	01	(R)	00	(NC)	11	(T)	011
011	11	(T)	11	(T)	11	(T)	100
100	01	(R)	00	(NC)	01	(R)	000
101	01	(R)	11	(T)	01	(R)	010
110	01	(R)	00	(NC)	01	(R)	010
111	11	(T)	11	(T)	01	(R)	000

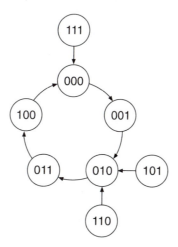

FIGURE 9.13
Example 9.4
Complete State Diagram

SECTION 9.2 REVIEW PROBLEM

9.2   A 4-bit synchronous counter based on JK flip-flops is described by the following set of equations:

$$J_3 = Q_2Q_1Q_0 \qquad J_2 = Q_1Q_0 \qquad J_1 = \overline{Q_3}Q_0 \qquad J_0 = 1$$
$$K_3 = Q_0 \qquad K_2 = Q_1Q_0 \qquad K_1 = Q_0 \qquad K_0 = 1$$

Assume the counter output is at 1000 in the count sequence. What will the output be after one clock pulse? After two clock pulses?

9.3 Design of Synchronous Counters

A synchronous counter can be designed using established techniques that involve the derivation of Boolean equations for the counter's next state logic. Alternatively, several VHDL structures can be used to define counters; we can use a behavioral description of the counter, or we can use a **state machine** definition in VHDL that specifies each present and next state explicitly.

In addition to the classical counter design techniques, we will examine the design of a counter through a behavioral description in VHDL. We will leave the state machine design for the following chapter.

Classical Design Technique

There are several steps involved in the classical design of a synchronous counter.

1. Define the problem. Before you can begin design of a circuit, you have to know what its purpose is and what it should do under all possible conditions.

2. Draw a state diagram showing the progression of states under various input conditions and what outputs the circuit should produce, if any.

3. Make a state table which lists all possible Present States and the Next State for each one. *List the present states in* **binary order.**

4. Use flip-flop **excitation tables** to determine at what states the flip-flop synchronous inputs must be to make the circuit go from each Present State to its Next State.

5. The logic levels of the synchronous inputs are Boolean functions of the flip-flop outputs and the control inputs. Simplify the expression for each input and write the simplified Boolean expression.

6. Use the Boolean expressions found in step 5 to draw the required logic circuit.

Flip-flop Excitation Tables

In the synchronous counter circuits we examined earlier in this chapter, we used JK flip-flops that were configured to operate only in toggle or no change mode. We can use any type of flip-flop for a synchronous sequential circuit. If we choose to use JK flip-flops, we can use any of the modes (no change, reset, set, or toggle) to make transitions from one state to another.

A flip-flop excitation table shows all possible transitions of a flip-flop output and the synchronous input levels needed to effect these transitions. Table 9.7 is the excitation table of a JK flip-flop.

If we want a flip-flop to make a transition from 0 to 1, we can use either the toggle function ($JK = 11$) or the set function ($JK = 10$). It doesn't matter what K is, as long as $J = 1$. This is reflected by the variable pair ($JK = 1X$) beside the $0 \rightarrow 1$ entry in Table 9.7. The X is a don't care state, a 0 or 1 depending on which is more convenient for the simplification of the Boolean function of the J or K input affected.

Table 9.8 shows a condensed version of the JK flip-flop excitation table.

Table 9.7 JK Flip-Flop Excitation Table

Transition	Function	JK	
$0 \rightarrow 0$	No change . or reset	00 01	0X
$0 \rightarrow 1$	Toggle or set	11 10	1X
$1 \rightarrow 0$	Toggle or reset	11 01	X1
$1 \rightarrow 1$	No change or set	00 10	X0

Table 9.8 Condensed Excitation Table for a JK Flip-Flop

Transition	JK
$0 \rightarrow 0$	0X
$0 \rightarrow 1$	1X
$1 \rightarrow 0$	X1
$1 \rightarrow 1$	X0

Design of a Synchronous Mod-12 Counter

We will follow the procedure outlined above to design a synchronous mod-12 counter circuit, using JK flip-flops. The aim is to derive the Boolean equations of all *J* and *K* inputs and to draw the counter circuit.

1. *Define the problem.* The circuit must count in binary sequence from 0000 to 1011 and repeat. The output progresses by 1 for each applied clock pulse. Since the outputs are 4-bit numbers, we require 4 flip-flops.

2. *Draw a state diagram.* The state diagram for this problem is shown in Figure 9.14.

3. *Make a state table showing each present state and the corresponding next state.*

4. *Use flip-flop excitation tables to fill in the J and K entries in the state table.* Table 9.9 shows the combined result of steps 3 and 4. Note that all present states are in binary order.

   We assume for now that states 1100 to 1111 never occur. If we assign their corresponding next states to be don't care states, they can be used to simplify the *J* and *K* expressions we derive from the state table.

FIGURE 9.14

State Diagram for a Mod-12 Counter

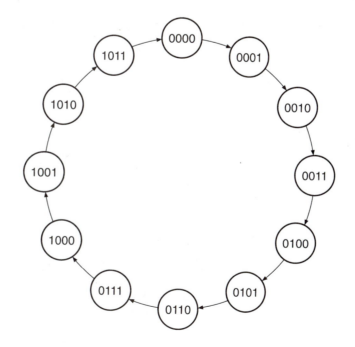

Table 9.9 State Table for a Mod-12 Counter

Present State	Next State	Synchronous Inputs			
$Q_3Q_2Q_1Q_0$	$Q_3Q_2Q_1Q_0$	J_3K_3	J_2K_2	J_1K_1	J_0K_0
0000	0 0 0 1	0 X	0 X	0 X	1 X
0001	0 0 1 0	0 X	0 X	1 X	X 1
0010	0 0 1 1	0 X	0 X	X 0	1 X
0011	0 1 0 0	0 X	1 X	X 1	X 1
0100	0 1 0 1	0 X	X 0	0 X	1 X
0101	0 1 1 0	0 X	X 0	1 X	X 1
0110	0 1 1 1	0 X	X 0	X 0	1 X
0111	1 0 0 0	1 X	X 1	X 1	X 1
1000	1 0 0 1	X 0	0 X	0 X	1 X
1001	1 0 1 0	X 0	0 X	1 X	X 1
1010	1 0 1 1	X 0	0 X	X 0	1 X
1011	0 0 0 0	X 1	0 X	X 1	X 1
1100	XXXX	XX	XX	XX	XX
1101	XXXX	XX	XX	XX	XX
1110	XXXX	XX	XX	XX	XX
1111	XXXX	XX	XX	XX	XX

Let us examine one transition to show how the table is completed. The transition from $Q_3Q_2Q_1Q_0 = 0101$ to $Q_3Q_2Q_1Q_0 = 0110$ consists of the following individual flip-flop transitions.

$Q_3: 0 \rightarrow 0$	(No change or reset;	$J_3K_3 = 0X$)
$Q_2: 1 \rightarrow 1$	(No change or set;	$J_2K_2 = X0$)
$Q_1: 0 \rightarrow 1$	(Toggle or set;	$J_1K_1 = 1X$)
$Q_0: 1 \rightarrow 0$	(Toggle or reset;	$J_0K_0 = X1$)

The other lines of the table are similarly completed.

5. *Simplify the Boolean expression for each input.* Table 9.9 can be treated as eight truth tables, one for each J or K input. We can simplify each function by Boolean algebra or by using a Karnaugh map.

Figure 9.15 shows K-map simplification for all 8 synchronous inputs. These maps yield the following simplified Boolean expressions.

$$J_0 = 1$$
$$K_0 = 1$$

$$J_1 = Q_0$$
$$K_1 = Q_0$$

$$J_2 = \overline{Q_3}Q_1Q_0$$
$$K_2 = Q_1Q_0$$

$$J_3 = Q_2Q_1Q_0$$
$$K_3 = Q_1Q_0$$

6. *Draw the required logic circuit.* Figure 9.16 shows the circuit corresponding to the above Boolean expressions.

We have assumed that states 1100 to 1111 will never occur in the operation of the mod-12 counter. This is normally the case, but when the circuit is powered up, there is no guarantee that the flip-flops will be in any particular state.

If a counter powers up in an unused state, the circuit should enter the main sequence after one or more clock pulses. To test whether or not this happens, let us make a state

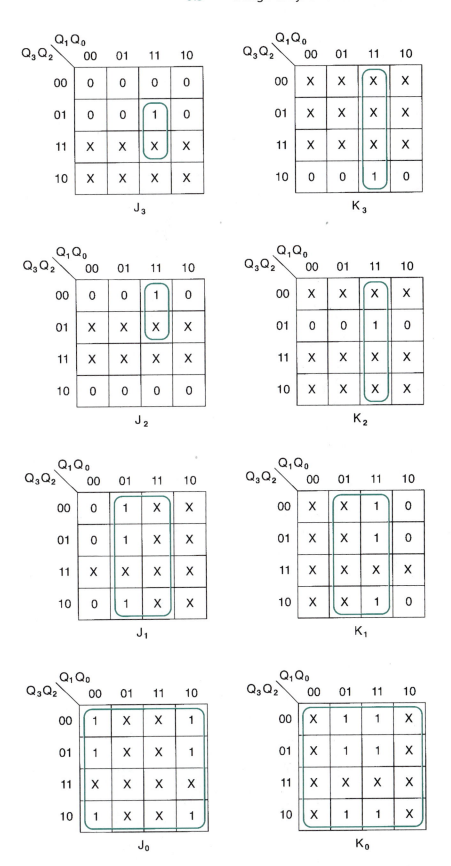

FIGURE 9.15
K-Map Simplification of Table 9.9

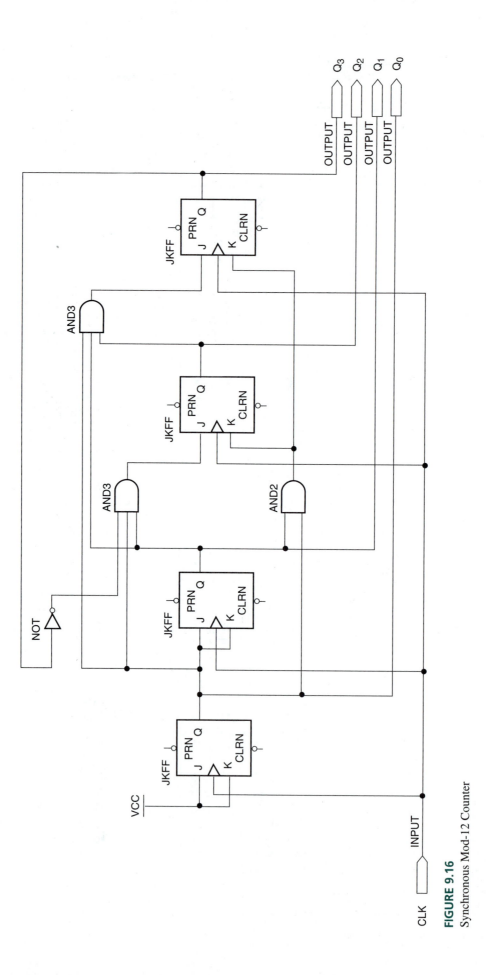

FIGURE 9.16
Synchronous Mod-12 Counter

Table 9.10 Unused States in a Mod-12 Counter

Present State	Synchronous Inputs				Next State
$Q_3Q_2Q_1Q_0$	J_3K_3	J_2K_2	J_1K_1	J_0K_0	$Q_3Q_2Q_1Q_0$
0000	00	00	00	11	1101
1101	00	00	11	11	1110
1110	00	00	00	11	1111
1111	11	01	11	11	0000

table, applying each unused state to the J and K equations as implemented, to see what the Next State is for each case. This analysis is shown in Table 9.10.

Figure 9.17 shows the complete state diagram for the designed mod-12 counter. If the counter powers up in an unused state, it will enter the main sequence in no more than four clock pulses.

If we want an unused state to make a transition directly to 0000 in one clock pulse, we have a couple of options:

1. We could reset the counter asynchronously and otherwise leave the design as is.

2. We could rewrite the state table to specify these transitions, rather than make the unused states don't cares.

Option 1 is the simplest and is considered perfectly acceptable as a design practice. Option 2 would yield a more complicated set of Boolean equations and hence a more complex circuit, but might be worthwhile if a direct synchronous transition to 0000 were required.

FIGURE 9.17

Complete State Diagram of Mod-12 Counter in Figure 9.16

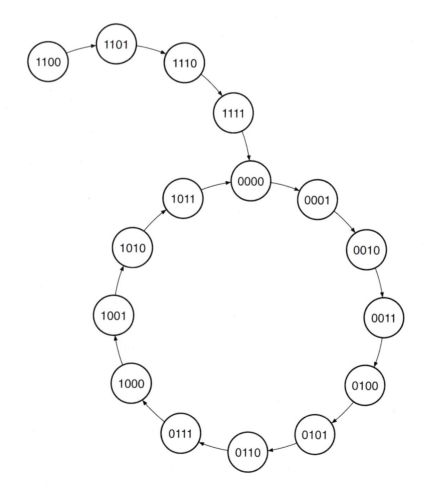

Derive the synchronous input equations of a 4-bit synchronous binary counter based on D flip-flops. Draw the corresponding counter circuit.

Solution   The first step in the counter design is to derive the excitation table of a D flip-flop. Recall that Q follows D when the flip-flop is clocked. Therefore the next state of Q is the same as the input D for any transition. This is illustrated in Table 9.11.

Table 9.11 Excitation Table of a D Flip-Flop

Transition	D
$0 \rightarrow 0$	0
$0 \rightarrow 1$	1
$1 \rightarrow 0$	0
$1 \rightarrow 1$	1

Next, we must construct a state table, shown in Table 9.12, with present and next states for all possible transitions. Note that the binary value of $D_3D_2D_1D_0$ is the same as the next state of the counter.

Table 9.12   State Table for a 4-bit Binary Counter

Present State	Next State	Synchronous Inputs
$Q_3Q_2Q_1Q_0$	$Q_3Q_2Q_1Q_0$	$D_3D_2D_1D_0$
0000	0001	0001
0001	0010	0010
0010	0011	0011
0011	0100	0100
0100	0101	0101
0101	0110	0110
0110	0111	0111
0111	1000	1000
1000	1001	1001
1001	1010	1010
1010	1011	1011
1011	1100	1100
1100	1101	1101
1101	1110	1110
1110	1111	1111
1111	0000	0000

This state table yields four Boolean equations, for D_3 through D_0, in terms of the present state outputs. Figure 9.18 shows four Karnaugh maps used to simplify these functions. The simplified equations are:

$$D_3 = \overline{Q_3}Q_2Q_1Q_0 + Q_3\overline{Q_2} + Q_3\overline{Q_1} + Q_3\overline{Q_0}$$
$$D_2 = \overline{Q_2}Q_1Q_0 + Q_2\overline{Q_1} + Q_1\overline{Q_0}$$
$$D_1 = \overline{Q_1}Q_0 + Q_1\overline{Q_0}$$
$$D_0 = \overline{Q_0}$$

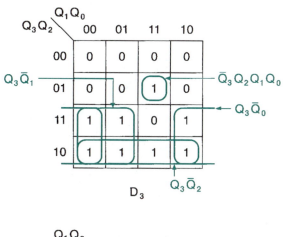

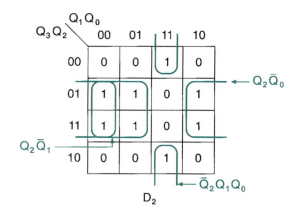

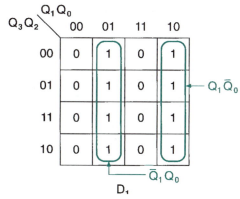

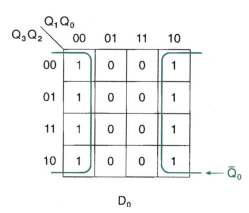

FIGURE 9.18
Example 9.5
K-Maps for a 4-bit Counter Based on D Flip-Flops

These equations represent the maximum SOP simplifications of the input functions. However, we can rewrite them to make them more compact. For example the equation for D_3 can be rewritten, using DeMorgan's theorem ($\overline{x} + \overline{y} + \overline{z} = \overline{xyz}$) and our knowledge of Exclusive OR (XOR) functions ($\overline{x}y + x\overline{y} = x \oplus y$).

$$
\begin{aligned}
D_3 &= \overline{Q_3}Q_2Q_1Q_0 + Q_3\overline{Q_2} + Q_3\overline{Q_1} + Q_3\overline{Q_0} \\
&= \overline{Q_3}(Q_2Q_1Q_0) + Q_3(\overline{Q_2} + \overline{Q_1} + \overline{Q_0}) \\
&= \overline{Q_3}(Q_2Q_1Q_0) + Q_3(\overline{Q_2Q_1Q_0}) \\
&= Q_3 \oplus Q_2Q_1Q_0
\end{aligned}
$$

We can write similar equations for the other D inputs as follows:

$$
\begin{aligned}
D_2 &= Q_2 \oplus Q_1Q_0 \\
D_1 &= Q_1 \oplus Q_0 \\
D_0 &= Q_0 \oplus 1
\end{aligned}
$$

These equations follow a predictable pattern of expansion. Each equation for an input D_n is simply Q_n XORed with the logical product (AND) of all previous Qs.

Figure 9.19 shows the circuit for the 4-bit counter, including an asynchronous reset.

FIGURE 9.19
Example 9.5
4-bit Counter Using D
Flip-Flops

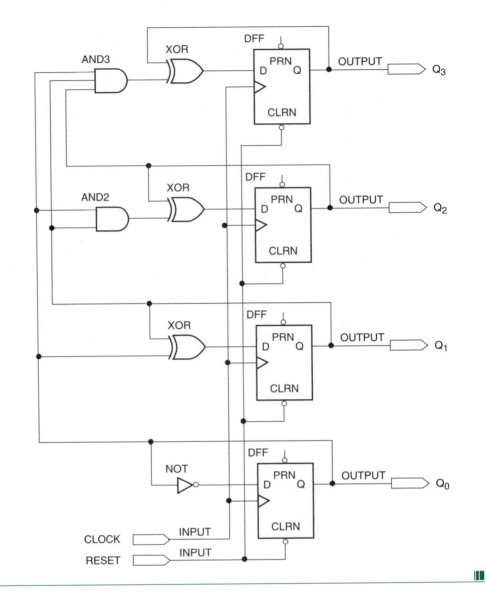

In Section 7.6 (Edge-Triggered T Flip-Flops) of Chapter 7, we saw how a D flip-flop could be configured for a switchable toggle function (refer to Figure 7.59). The flip-flops in Figure 9.19 are similarly configured. Each flip-flop output, except Q_0, is fed back to its input through an Exclusive OR gate. The other input to the XOR controls whether this feedback is inverted (for toggle mode) or not (for no change mode). Recall that $x \oplus 0 = x$ and $x \oplus 1 = \overline{x}$.

For example, Q_3 is fed back to D_3 through an XOR gate. The feedback is inverted only if the 3-input AND gate has a HIGH output. Thus, the Q_3 output toggles only if all previous bits are HIGH ($Q_3Q_2Q_1Q_0 = 0111$ or 1111). The flip-flop toggle mode is therefore controlled by the states of the XOR and AND gates in the circuit.

▌▌ SECTION 9.3 REVIEW PROBLEM

9.3 A 4-bit synchronous counter must make a transition from state $Q_3Q_2Q_1Q_0 = 1011$ to $Q_3Q_2Q_1Q_0 = 1100$. Write the required states of the synchronous inputs for a set of four JK flip-flops used to implement the counter. Write the required states of the synchronous inputs if the counter is made from D flip-flops.

9.4 Programming Binary Counters in VHDL

> **KEY TERMS**
>
> **If statement**   A VHDL construct in which statements within the IF statement are executed only when a specified Boolean condition is satisfied.
>
> **Attribute**   A property associated with a named identifier in VHDL. (For example, the attribute **EVENT,** when associated with the identifier **clk** (written **clk'EVENT**), indicates, when true, that a transition has occurred on the input called **clk.**)

When using VHDL to create a counter, we can take several approaches. We can encode the Boolean equations of the counter directly with concurrent signal assignment statements; we can use VHDL code to describe the behavior of the counter; we can use a CASE statement to implement the state diagram of the counter; or we can use a predefined counter, such as those found in the MAX+PLUS II Library of Parameterized Modules (LPM) and map its ports to the ports of a VHDL design entity.

If we chose to use concurrent signal assignments to encode the Boolean equations of a counter, we could derive the following equations for a 4-bit counter with D flip-flops.

```
d(3)<= q(3)xor(q(2)and q(1)and q(0));,
d(2)<= q(2)xor(q(1)and q(0));
d(1)<= q(1)xor q(1);,
d(0)<= not q(0);,
```

In Chapter 5, we saw that using concurrent signal assignment statements is an inefficient way to code many digital functions. (For one thing, if we use this procedure, we must know what the equations are. Getting to that point requires a lot of work that can be done by the VHDL compiler.) While acknowledging this as a possible option, we will not examine this method any further for the count logic of binary counters.

In this section, we will design a counter using a behavioral description and using an LPM counter. The design of a counter as a state machine will be examined in the next chapter.

Behavioral Description of Counters

The following VHDL code shows the behavioral description of a simple 8-bit counter (**ct_simp.vhd**) with asynchronous clear.

ct_simp.vhd

```
ENTITY ct_simp IS
    PORT(
        clk      : IN     BIT;
        clear    : IN     BIT;
        q        : OUT    INTEGER RANGE 0 TO 255);
END ct_simp;

ARCHITECTURE a OF ct_simp IS
    BEGIN
    PROCESS (clk, clear)
        VARIABLE count : INTEGER RANGE 0 TO 255;
    BEGIN
        If (clear = '0') THEN
            count := 0;
        ELSE
            IF (clk'EVENT AND clk = '1') THEN
                count := count + 1;
            END IF;
```

```
        END IF;
     q <= count;
     END PROCESS;
END a;
```

Recall that the PROCESS statement has the following syntax:

```
PROCESS (sensitivity list)
   [VARIABLE variable name :type [range]; ]
BEGIN
   Process statements
END PROCESS;
```

Square brackets [ ] indicate an optional part of the code.

When there is a change in an item in the sensitivity list, the process statements are executed. For a synchronous counter, the list would often only include **clock,** since any action in a synchronous circuit depends on a clock transition. Since the clear function in this counter is asynchronous, the **clear** input must also be monitored for any changes.

To hold the accumulating output value of the counter, we define a variable called **count,** presumed to have an initial value of 0, but defined for the range of 0 to 255. (This 8-bit value rolls over to 0 when the count exceeds 255.) The variable (*any* variable) is local to the process in which it is defined. We update the value of **count** by an **IF statement,** with the form:

```
IF (condition) THEN
   Statement[s];
[ELSIF (condition) THEN
   statement[s];]
[ELSE
   statement[s];]
END IF;
```

The clause (IF (clear='0') THEN) monitors the asynchronous clear function independently of the clock and executes the variable assignment that sets the output to 0 if the Boolean condition (clear='0') is true. Otherwise, the clock is monitored for a positive edge by the condition (clk'EVENT AND clk = '1'). The clause clk'EVENT (pronounced "clock tick event") is a predefined **attribute** of the clock signal and is true if there has just been a change on clock. The combination of this and the condition clk = '1' indicates that a positive edge has just occurred. If this is true, the count is incremented.

As a final step, the accumulated count must be assigned to an output port. This is done in the concurrent signal assignment q <= count at the end of the process.

Note the difference in types of assignments. A *variable* is assigned by the := operator (e.g., count := count + 1;). A *signal* is assigned by the <= operator (eg., q <= count).

LPM Counters in VHDL

lpm_simp.vhd

We can use a component **(lpm_counter)** from the Library of Parameterized Modules (LPM) to instantiate a counter in VHDL. When using an LPM counter, we don't need to describe the behavior of the counter, as this has been done for us in the module itself. All we need to do is map the ports and parameters of the LPM component to the ports of the VHDL design entity. We do this by using a **generic map** to specify the parameters we need and a **port map** to map the ports of the LPM device either to an external port or an internal signal. The VHDL code below shows the VHDL implementation **(lpm_simp.vhd)** of the same 8-bit counter as in the previous behavioral example.

```
-- lpm_simp.vhd
-- Eight-bit binary counter based on a component
```

```
--    from the Library of Parameterized Modules (LPM)
--  Counter has an active-LOW asynchronous clear.

LIBRARY ieee;
USE ieee.std_logic_1164.ALL;
LIBRARY lpm;
USE lpm.lpm_components.ALL;

ENTITY lpm_simp IS
    PORT(
        clk, clear : IN  STD_LOGIC;
        q          : OUT STD_LOGIC_VECTOR (7 downto 0));
END lpm_simp;

ARCHITECTURE count OF lpm_simp IS
    SIGNAL clrn : STD_LOGIC;
BEGIN
    count8: lpm_counter
      GENERIC MAP (LPM_WIDTH   => 8)
      PORT MAP ( clock  => clk,
                 aclr   => clrn,
                 q      => q(7 downto 0));

    clrn <= not clear;
END count;
```

LPM components require us to use two packages: the **std_logic_1164** package in the **ieee** library to define STD_LOGIC types used in the LPM components and the **lpm_components** package in the **lpm** library to define the components themselves. Since LPM components are defined using STD_LOGIC and STD_LOGIC_VECTOR types, we should use these types for our other identifiers as well.

The entity declaration defines the inputs and outputs of our counter and need not correspond to the port names for the LPM counter. That correspondence is defined in the architecture body, where we instantiate the counter module. The counter is defined in a component instantiation statement, which takes the following form:

```
__instance_name: __component_name
    GENERIC MAP (__parameter_name => __parameter_value,
                 __parameter_name => __parameter_value)
    PORT MAP (__component_port => __connect_port,
              __component_port => __connect_port);
```

The component name is the name of the LPM component. Parameter names are those defined in the LPM component, such as LPM_WIDTH. Parameter values are those values assigned in the instance of the component. Component ports are the LPM port names. Connect ports are the names of identifiers declared in the entity or as signals or variables.

If we want to invert the active level of an LPM input port, we must use a signal assignment statement. (e.g., `clrn <= not clear;`) We need to do this because a VHDL input port cannot be "updated" (modified); only an output can be assigned a new value as a result of a Boolean expression. Thus, we create a signal called **clrn** that maps to the **aclr** (asynchronous clear) port of the LPM counter. This is connected to the **clear** input of the counter circuit via an inverter. Figure 9.20 shows the graphic equivalent of this mapping.

▮▮ SECTION 9.4 REVIEW PROBLEM

9.4 Write a VHDL code segment that increments a variable called **count** upon detection of a negative edge of an input called **clock.**

FIGURE 9.20
Graphic Equivalent of an LPM
Counter with Active-Low Clear

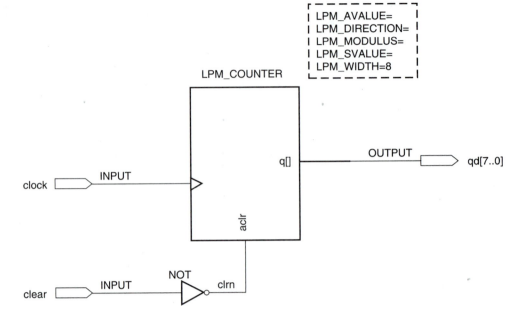

9.5 Control Options for Synchronous Counters

> **KEY TERMS**
>
> **Parallel load** A function that allows simultaneous loading of binary values into all flip-flops of a synchronous circuit. Parallel loading can be synchronous or asynchronous.
>
> **Presettable counter** A counter with a parallel load function.
>
> **Clear** Reset (synchronous or asynchronous).
>
> **Count enable** A control function that allows a counter to progress through its count sequence when active and disables the counter when inactive.
>
> **Bidirectional counter** A counter that can count up or down, depending on the state of a control input.
>
> **Terminal count** The last state in a count sequence before the sequence repeats (e.g., 1111 is the terminal count of a 4-bit binary UP counter; 0000 is the terminal count of a 4-bit binary DOWN counter).
>
> **Ripple carry out or ripple clock out (RCO)** An output that produces one pulse with the same period as the clock upon terminal count.

Synchronous counters can be designed with a number of features other than just straight counting. Some of the most common features include:

- Synchronous or asynchronous **parallel load,** which allows the count to be set to any value whenever a *LOAD* input is asserted
- Synchronous or asynchronous **clear** (reset), which sets all of the counter outputs to zero
- **Count enable,** which allows the count sequence to progress when asserted and inhibits the count when deasserted
- **Bidirectional** control, which determines whether the counter counts up or down
- **Output decoding,** which activates one or more outputs when detecting particular states on the counter outputs
- **Ripple carry out or ripple clock out (RCO),** a special case of output decoding that produces a pulse upon detecting the **terminal count,** or last state, of a count sequence.

We will examine the implementation of these functions, first as Graphic Design Files in MAX+PLUS II, and then, in the next section, in VHDL, both as behavioral descriptions and as functions of LPM counters.

Parallel Loading

Figure 9.21 shows the symbol of a 4-bit **presettable counter** (i.e., a counter with a parallel load function). The parallel inputs, P_3 to P_0, have direct access to the flip-flops of the counter. When the *LOAD* input is asserted, the values at the P inputs are loaded directly into the counter and appear at the Q outputs.

> ### N O T E
>
> Parallel loading requires at least two sets of inputs: the load *data* (P_3 to P_0) and the load *command (LOAD)*. If the load function is synchronous, as described below, it also requires a clock input.

FIGURE 9.21

4-bit Counter with Parallel Load

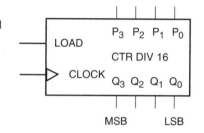

4b_al_sl.scf

Parallel loading can be synchronous or asynchronous. The MAX+PLUS II simulation in Figure 9.22 shows the difference. Two waveforms, QS[3..0] and QA[3..0], represent the outputs of two 4-bit counters with synchronous and asynchronous load, respectively. Both counters have the same clock, load, and P inputs. The count is already in progress at the beginning of the simulation window and shows both counters advancing with each clock pulse: 4, 5, 6.

When *LOAD* goes HIGH at 500 ns, the value of P[3..0] (= AH) is loaded into the asynchronously loading counter (QA[3..0]) immediately after a short propagation delay (12.5 ns). The counter with synchronous load (QS[3..0]) is not loaded until the next positive clock edge, shown at 560 ns.

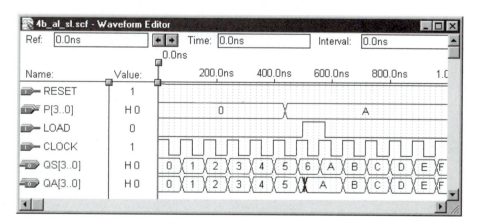

FIGURE 9.22

Synchronous vs. Asynchronous Load

Synchronous Load

The logic diagram of Figure 9.23 shows the concept of synchronous parallel load. Depending on the status of the *LOAD* input, the flip-flop will either count according to its

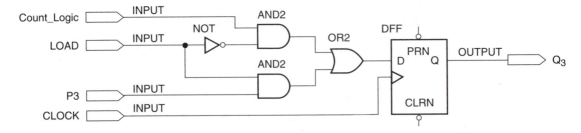

FIGURE 9.23
Count/Load Selection

count logic (the next-state combinational circuit) or load an external value. The flip-flop shown is the most significant bit of a 4-bit binary counter, such as shown in Figure 9.19, but with the count logic represented only by an input pin. (For the fourth bit of a counter, the Boolean equation of the count logic is given by $D_3 = Q_3 \oplus Q_2Q_1Q_0$. It is left out in order to more clearly show the operation of the count/load function select circuit.)

The *LOAD* input selects whether the flip-flop synchronous input will be fed by the count logic or by the parallel input P_3. When $LOAD = 0$, the upper AND gate steers the count logic to the flip-flop, and the count progresses with each clock pulse. When $LOAD = 1$, the lower AND gate loads the logic level at P_3 directly into the flip-flop on the next clock pulse.

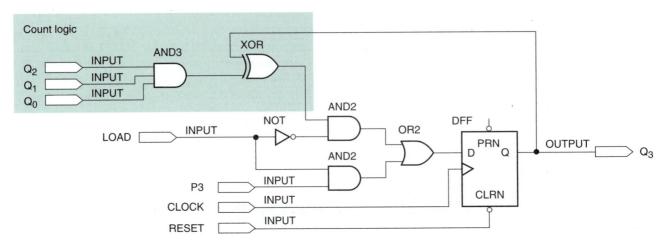

FIGURE 9.24
Counter Element with Synchronous Load and Asynchronous Clear

sl_count.gdf
4bit_sl.gdf
4bit_sl.scf

Figure 9.24 shows the same circuit, but includes the count logic. If we leave out the 3-input AND gate, as in Figure 9.25, we have a circuit that can be used as a general element (called **sl_count**) in a synchronous presettable counter. Figure 9.26 shows the logic diagram of a 4-bit synchronously presettable counter consisting of four instances of the counter element of Figure 9.25 and appropriate AND gates for a synchronous counter. This diagram implements a synchronous counter like that of Figure 9.19, but also incorporates a synchronous load function.

Figure 9.27 shows a simulation of the counter in Figure 9.26. The first 19 clock pulses drive the counter through its normal 4-bit cycle from 0H to FH, then up to 2H. At this point, we set the *LOAD* input HIGH and the value at the *P* inputs (9H) is loaded into the counter on the rising edge of the next clock pulse. An asynchronous *RESET* pulse at 880 ns drives the counter outputs to 0H, after which the count resumes.

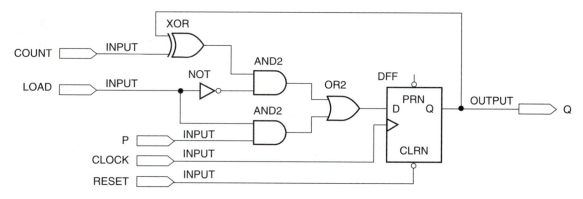

FIGURE 9.25
Counter Element with Synchronous Load and Asychronous Reset (sl_count)

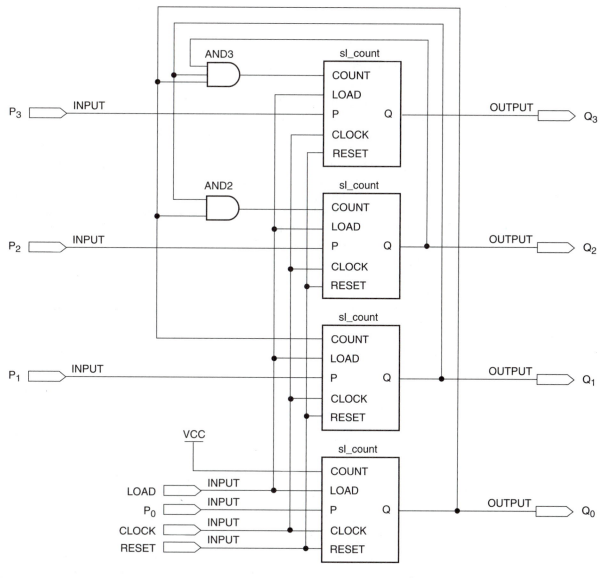

FIGURE 9.26
4-bit Counter with Synchronous Load and Asynchronous Reset

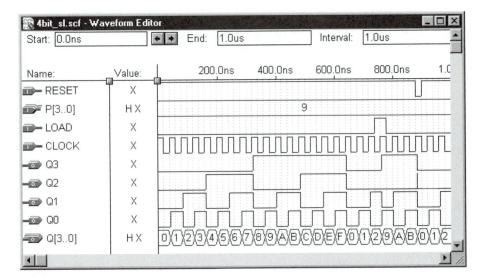

FIGURE 9.27
Simulation of 4-bit Counter with Synchronous Load and Asynchronous Reset

Asynchronous Load

The asynchronous load function of a counter makes use of the asynchronous preset and clear inputs of the counter's flip-flops. Figure 9.28 shows the circuit implementation of the asynchronous load function, without any count logic.

When *ALOAD* (Asynchronous LOAD) is HIGH, both NAND gates in Figure 9.28 are enabled. If the *P* input is HIGH, the output of the upper NAND gate goes LOW, activating the flip-flop's asynchronous *PRESET* input, thus setting $Q = 1$. The lower NAND gate has a HIGH output, thus deactivating the flip-flop's *CLEAR* input.

If *P* is LOW the situation is reversed. The upper NAND output is HIGH and the lower NAND has a LOW output, activating the flip-flop's *CLEAR* input, resetting *Q*. Thus, *Q* will be the same value as *P* when the *ALOAD* input is asserted. When *ALOAD* is not asserted ($= 0$), both NAND outputs are HIGH and thus do not activate either the preset or clear function of the flip-flop.

Figure 9.29 shows the asynchronous load circuit with an asynchronous clear (reset) function added. The flip-flop can be cleared by a logic LOW either from the *P* input (via the lower NAND gate) or the *CLEAR* input pin. The clear function disables the upper NAND gate when it is LOW, preventing the flip-flop from being cleared and preset simultaneously. This extra connection also ensures that the clear function has priority over the load function.

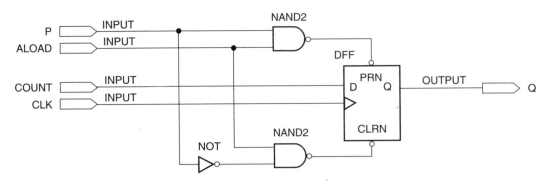

FIGURE 9.28
Asynchronous LOAD Element

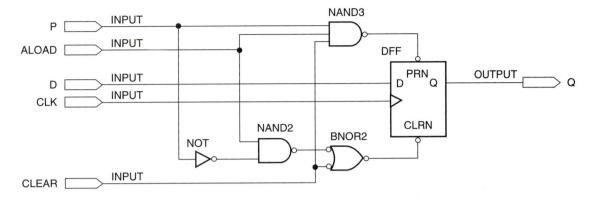

FIGURE 9.29
Asynchronous LOAD Element with Asynchronous Clear

■■ EXAMPLE 9.6

al_count.gdf

Use MAX+PLUS II to redraw the circuit in Figure 9.29 to create a general element called **al_count** that can be used in a synchronous counter with asynchronous load and clear. (Refer to Figure 9.25 for a similar element with *synchronous* load.)

Solution   Figure 9.30 shows the modified circuit, which includes an XOR gate for part of the count logic. The remainder of the count logic must be supplied externally to this element for each bit of the counter.

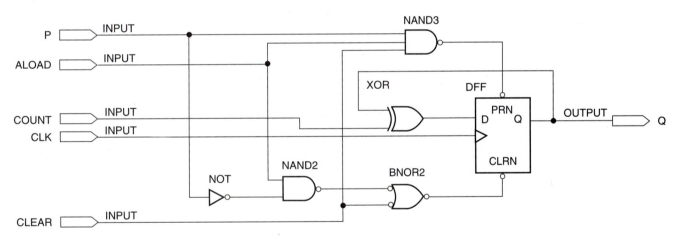

FIGURE 9.30
Example 9.6
Counter Element with Asynchronous Load and Clear (al_count)

■■ EXAMPLE 9.7

4bit_al.gdf
4bit_al.scf

Draw a circuit with four instances of **al_count** (from Example 9.6) to make a 4-bit synchronous counter with asynchronous load and reset. Create a simulation that tests the function of the counter.

Solution   Figure 9.31 shows the circuit. (Compare this circuit to the counter with synchronous load in Figure 9.26. This difference between the two is in the load function, not the count logic.)

The Boolean function applied to the *COUNT* input of each instance of **al_count** consists of the logical product of all previous output bits. ($COUNT_3 = Q_2Q_1Q_0$, $COUNT_2 = Q_1Q_0$, $COUNT_1 = Q_0$, $COUNT_0 = 1$.) When combined with the XOR at the *COUNT* input

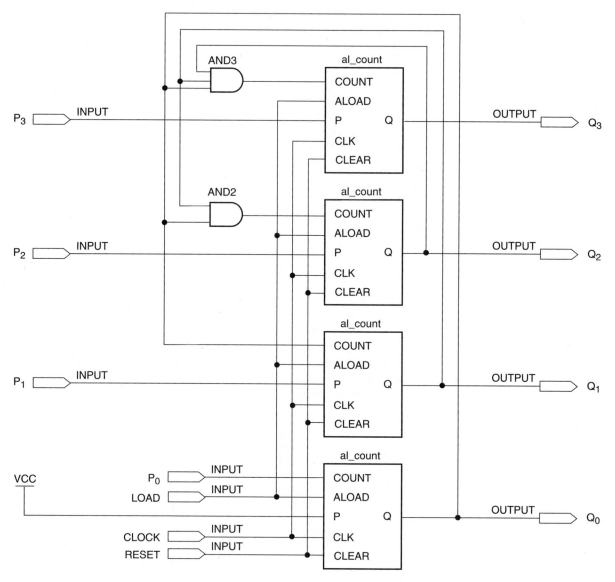

FIGURE 9.31
Example 9.7
4-bit Counter with Asynchronous Load and Reset

of each element, this yields the Boolean equations for a binary counter based on D flip-flops, as derived in Example 9.5. The circuitry inside each instance of **al_count** also generates the asynchronous load and clear functions.

Figure 9.32 shows a MAX+PLUS II simulation of the counter. The counter cycles through its full range and continues. A pulse at 700 ns loads the counter with the value 9H ($= 1001_2$), after which the count continues from that point.

FIGURE 9.32
Example 9.7
Simulation of a 4-bit Counter
with Asynchronous Load and
Reset

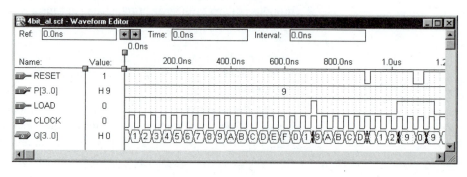

The reset pulse at 900 ns clears the counter. The *LOAD* pulse starting at 1.02 μs shows how the load function has precedence over the count function. When *LOAD* is asserted, 9H is loaded and the count does not increase until *LOAD* is deasserted. The *RESET* pulse at 1.08 μs overrides both load and count functions. When *RESET* is deasserted, 9H is asynchronously reloaded.

Count Enable

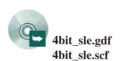

4bit_sle.gdf
4bit_sle.scf

The counter elements in Figures 9.25 (**sl_count**) and 9.30 (**al_count**) are just D flip-flops configured for switchable toggle operation with additional circuitry for load and clear functions. Normally, when these elements are used in synchronous counters, the count progresses when the input to the element's XOR gate goes HIGH. In other words, the count progresses when the counter element is switched from a no change to a toggle mode.

In order to arrest the count sequence, we must disable the count logic of the counter circuit. Figure 9.33 shows a simple modification to the 4-bit counter circuit of Figure 9.26 that can achieve this function. Each AND gate has an extra input which is used to enable or inhibit the count logic function to each flip-flop.

Figure 9.34 shows a simulation of the counter. Note that the count progresses normally when *COUNT_ENA* is HIGH and stops when *COUNT_ENA* is LOW, even though the clock pulses remain constant throughout the simulation.

Also note that the count enable has no effect on the synchronous load and asynchronous reset functions. In the latter part of the simulation, the count stops at AH ($Q_3Q_2Q_1Q_0$ = 1010_2), when *COUNT_ENA* goes LOW. At 760 ns, the synchronous load function loads the value of 9H into the counter. The counter stays at this value, even after LOAD is no longer active, since the count is still disabled. At 880 ns, an asynchronous reset pulse clears the counter. The count resumes on the first clock pulse after *COUNT_ENA* goes HIGH again.

Bidirectional Counters

Figure 9.35 shows the logic diagram of a 4-bit synchronous DOWN counter. Its count sequence starts at 1111 and counts backwards to 0000, then repeats. The Boolean equations for this circuit will not be derived at this time, but will be left for an exercise in an end-of-chapter problem.

We can intuitively analyze the operation of the counter if we understand that the upper three flip-flops will each toggle when their associated XOR gates have a HIGH input from the rest of the count logic.

element.gdf

Q_0 is set to toggle on each clock pulse. Q_1 toggles whenever Q_0 is LOW (every second clock pulse, at states 1110, 1100, 1010, 1000, 0110, 0100, 0010, and 0000). Q_2 toggles when Q_1 AND Q_0 are LOW (1100, 1000, 0100, and 0000). Q_3 toggles when Q_2 AND Q_1 AND Q_0 are LOW (1000 and 0000). The result of this analysis can be represented by a timing diagram, such as the simulation shown in Figure 9.36. As we expect, the counter will count down from 1111 (FH) to 0000 (0H) and repeat.

We can create a bidirectional counter by including a circuit to select count logic for an UP or DOWN sequence. Figure 9.37 shows a basic synchronous counter element that can be used to create a synchronous counter. The element is simply a D flip-flop configured for switchable toggle mode.

Four of these elements can be combined with selectable count logic to make a 4-bit bidirectional counter, as shown in Figure 9.38. Each counter element has a pair of AND-shaped gates and an OR gate to steer the count logic to the XOR in the element. When *DIR* = 1, the upper gate in each pair is enabled and the lower gates disabled,

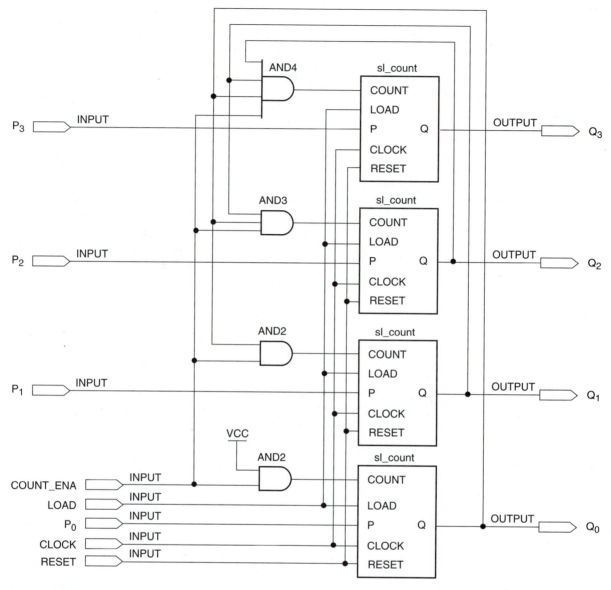

FIGURE 9.33

4-bit Counter with Synchronous Load, Asynchronous Reset, and Count Enable

FIGURE 9.34

Simulation of 4-bit Counter with Synchronous Load, Asynchronous Reset, and Count Enable

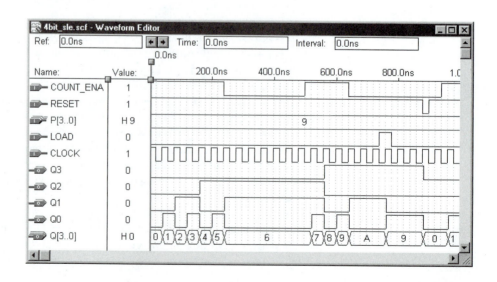

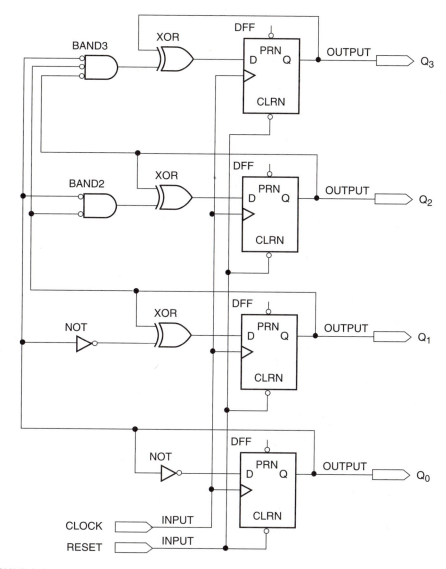

FIGURE 9.35
4-bit Synchronous DOWN Counter

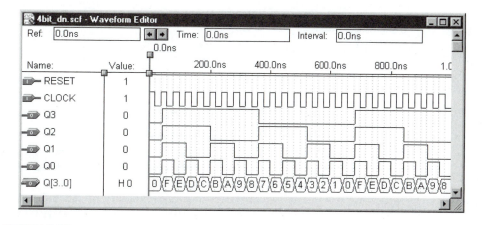

FIGURE 9.36
4-bit DOWN Counter Simulation

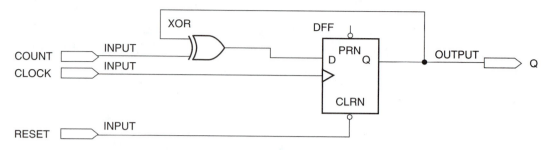

FIGURE 9.37
Synchronous Counter Element (T Flip-Flop)

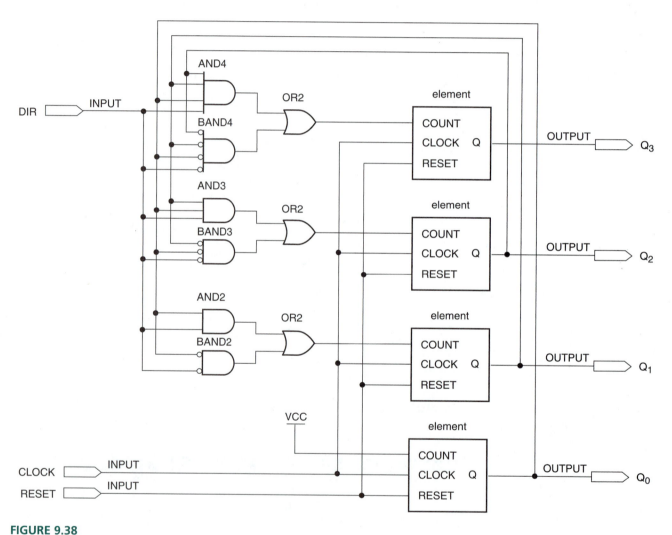

FIGURE 9.38
4-bit Bidirectional Counter

steering the UP count logic to the counter element. When $DIR = 0$, the lower gate in each pair is enabled, steering the DOWN count logic to the counter element. The directional function can also be combined with the load and count enable functions, as was shown for unidirectional UP counters.

Figure 9.39 shows a simulation of the bidirectional counter of Figure 9.38. The waveforms show the UP count when DIR is HIGH and the DOWN count when DIR is LOW.

4bit_dir.gdf
4bit_dir.scf

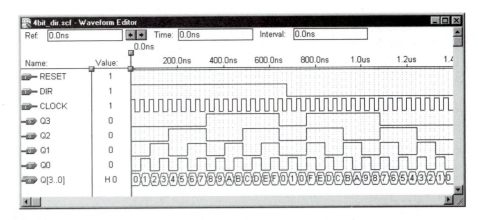

FIGURE 9.39
Simulation of 4-bit Bidirectional Counter

Decoding the Output of a Counter

Figure 9.40 shows a graphic design file of a 4-bit bidirectional counter with an output decoder. The counter is the one shown in Figure 9.38, represented as a logic circuit symbol. The decoder component **decode16** is a module written in VHDL, as listed below.

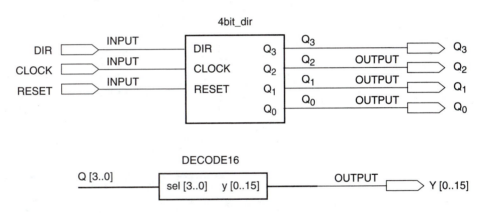

FIGURE 9.40
4-bit Bidirectional Counter with Output Decoder

```
-- decode16.vhd

LIBRARY ieee;
USE ieee.std_logic_1164.ALL;

ENTITY decode16 IS
   PORT(
        sel : IN   INTEGER RANGE 0 to 15;
        y   : OUT  BIT_VECTOR (0 to 15));
END decode16;

ARCHITECTURE a OF decode16 IS
BEGIN
   WITH sel SELECT
        y <=   x"7FFF" WHEN 0,
               x"BFFF" WHEN 1,
               x"DFFF" WHEN 2,
               x"EFFF" WHEN 3,
               x"F7FF" WHEN 4,
               x"FBFF" WHEN 5,
```

CD: decode16.vhd
4bit_dcd.gdf
4bit_dcd.scf

```
                    x"FDFF"  WHEN  6,
                    x"FEFF"  WHEN  7,
                    x"FF7F"  WHEN  8,
                    x"FFBF"  WHEN  9,
                    x"FFDF"  WHEN  10,
                    x"FFEF"  WHEN  11,
                    x"FFF7"  WHEN  12,
                    x"FFFB"  WHEN  13,
                    x"FFFD"  WHEN  14,
                    x"FFFE"  WHEN  15,
                    X"FFFF"  WHEN  others;
END a;
```

The decoder has 16 outputs, one for each state of the counter. For each state, one and only one output will be low. (Refer to the section on binary decoders in Chapter 5 for a more detailed description of n-line-to-m-line binary decoders.)

Figure 9.41 shows a portion of the simulation waveforms (i.e., only the count value and the decoder outputs) for the circuit in Figure 9.40. As the count progresses up or down, as shown by the waveform for Q[3..0], the decoder outputs respond by going LOW in sequence.

Output decoders for binary counters can also be configured to have active HIGH outputs. In this case, one and only one output would be HIGH for each output state of the counter.

Terminal Count and RCO

A special case of output decoding is a circuit that will detect the **terminal count,** or last state, of a count sequence and activate an output to indicate this state. The terminal count depends on the count sequence. A 4-bit binary UP counter has a terminal count of 1111; a 4-bit binary DOWN counter has a terminal count of 0000. A circuit to detect these conditions must detect the *maximum value* of an UP count and the *minimum value* of a DOWN count.

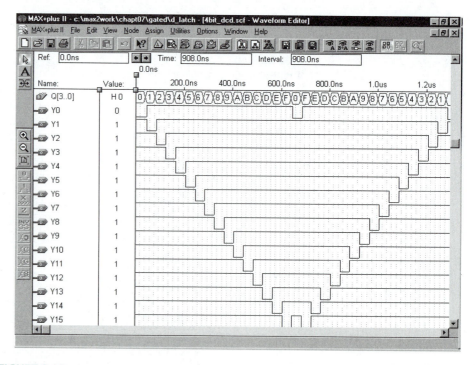

FIGURE 9.41
Simulation of 4-bit Decoder

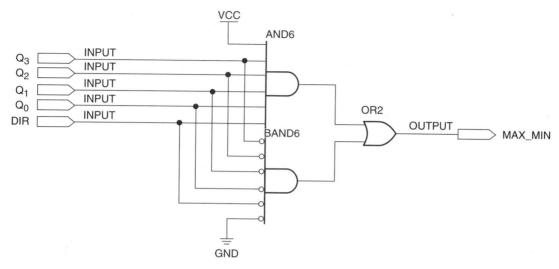

FIGURE 9.42
Terminal Count Decoder for a 4-bit Bidirectional Counter

term_dcd.gdf

4bit_rco.gdf
4bit_rco.scf

The decoder shown in Figure 9.42 fulfills both of these conditions. The directional input *DIR* enables the upper gate when HIGH and the lower gate when LOW. Thus, the upper gate generates a HIGH output when $DIR = 1$ AND $Q_3Q_2Q_1Q_0 = 1111$. The lower gate generates a HIGH when $DIR = 0$ AND $Q_3Q_2Q_1Q_0 = 0000$.

Figure 9.43 shows the terminal count decoder combined with a 4-bit bidirectional counter. The decoder is also used to enable a NAND gate output that generates an *RCO* signal. RCO stands for ripple carry out or ripple clock out. The purpose of *RCO* is to produce exactly one clock pulse upon terminal count and have the positive edge of *RCO* at the end of the counter cycle, for a counter that has a positive edge-triggered clock.

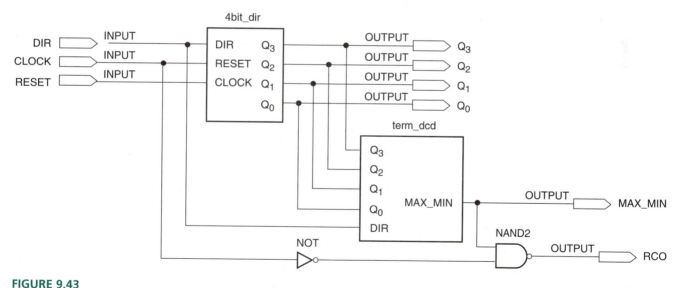

FIGURE 9.43
4-bit Bidirectional Counter with Terminal Count Detection

This function is generally found in counters with a fixed number of bits (i.e., fixed-function counter chips, not PLDs) and is used to asynchronously clock a further counter stage, as in Figure 9.44. This allows us to extend the width of the counter beyond the number of bits available in the fixed-function device. This is not necessary when designing synchronous counters in programmable logic, but is included for the sake of completeness.

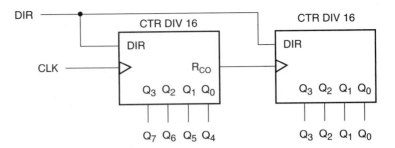

FIGURE 9.44
Counter Expansion Using RCO

The NAND gate in Figure 9.43 is enabled upon terminal count and passes the clock signal through to *RCO*. The NAND output sits HIGH when inhibited. The clock is inverted in the RCO circuit so that when the NAND gate inverts it again, the circuit generates a clock pulse in true form.

Figure 9.45 shows the simulation of the circuit of Figure 9.43. In the first half of the simulation, the counter is counting DOWN. The terminal count decoder output, *MAX_MIN*, goes HIGH when $Q_3Q_2Q_1Q_0 = 0000$. *RCO* generates a pulse at that time. For the second half, the counter is counting UP. *MAX_MIN* is HIGH when $Q_3Q_2Q_1Q_0 = 1111$ and *RCO* generates a pulse at that time.

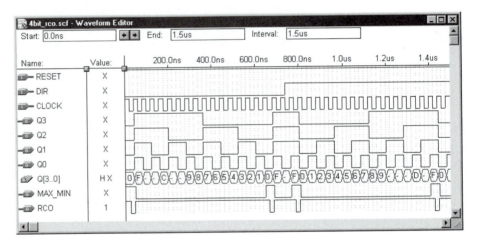

FIGURE 9.45
Simulation of a 4-bit Bidirectional Counter with Terminal Count Detection

Note that the *RCO* pulse appears to be half the width of the *MAX_MIN* pulse. Although the NAND gate that generates *RCO* is enabled for the whole *MAX_MIN* pulse, the clock input is HIGH for the first half-period, which is the same as the *RCO* inhibit level.

The positive edge of *RCO* is at the *end* of the pulse. The idea is to synchronize the positive edge of the clock with the positive edge of *RCO*. However, since the RCO decoder is combinational, a propagation delay of about 7 ns is introduced.

▌▌▌ SECTION 9.5 REVIEW PROBLEM

9.5 Figure 9.46 shows two presettable counters, one with asynchronous load and clear, the other with synchronous load and clear. The counter with asynchronous functions has a 4-bit output labeled *QA*. The synchronously loaded counter has a 4-bit output labeled *QS*. The load *and* reset inputs to both counters are *active LOW*.

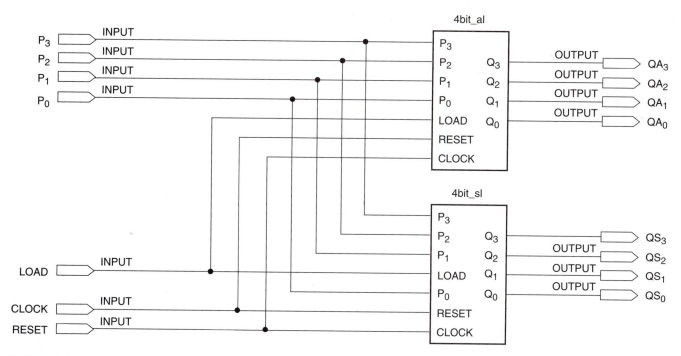

FIGURE 9.46

Section Review Problem 9.5
Two Presettable Counters

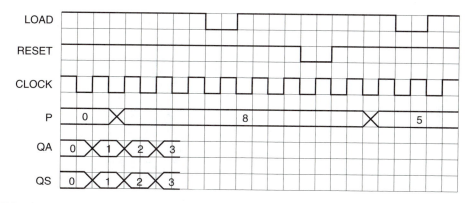

FIGURE 9.47

Timing Diagram for Counters in Figure 9.46

Figure 9.47 shows a partial timing diagram for the counters. Complete the diagram.

9.6 Programming Presettable and Bidirectional Counters in VHDL

The presettable counters and bidirectional counters described in the previous section can be easily implemented in VHDL, either as behavioral descriptions or as LPM components. We will initially examine the behavioral descriptions of two counters, one with asynchronous load and clear and one with synchronous load and clear. We will then examine some options available in the module **lpm_counter.**

Behavioral Description

The following lists the VHDL code for an 8-bit bidirectional counter with count enable, terminal count decoding, and asynchronous load and clear:

```
ENTITY pre_ct8a IS
  PORT (
        clk, count_ena        : IN   BIT;
        clear, load, direction : IN   BIT;
        p                      : IN   INTEGER RANGE 0 TO 255;
        max_min                : OUT BIT;
        qd                     : OUT INTEGER RANGE 0 TO 255);
END pre_ct8a;

ARCHITECTURE a OF pre_ct8a IS
 BEGIN
 PROCESS (clk, clear, load)
     VARIABLE cnt : INTEGER RANGE 0 TO 255;
     BEGIN
         IF (clear = '0') THEN              -- Asynchronous clear
            cnt := 0;
         ELSIF (load = '1' and clear = '1') THEN -- Asynchronous load
            cnt := p;
         ELSE
            IF (clk'EVENT AND clk = '1') THEN
               IF (count_ena = '1' and direction = '0') THEN
                  cnt := cnt - 1;
               ELSIF (count_ena = '1' and direction = '1') THEN
                  cnt := cnt + 1;
               END IF;
            END IF;
         END IF;
         qd <= cnt;

         -- Terminal count decoder
         IF (cnt = 0 and direction = '0') THEN
            max_min <=  '1';
         ELSIF (cnt = 255 and direction = '1') THEN
            max_min <=  '1';
         ELSE
            max_min <=  '0';
         END IF;
    END PROCESS;
END a;
```

pre_ct8a.vhd

The load and clear functions of this counter are asynchronous, so these identifiers are part of the sensitivity list of the PROCESS statement; the statements in the process will execute if there is a change on the clear, load, or clock inputs. Load and clear are checked by IF statements, independently of the clock. Since load and clear are checked first, they have precedence over the clock. Clear has precedence over load since load can only activate if clear is not active.

If clear and load are not asserted, the clock status is checked by a clause in an IF statement: (IF (clk'EVENT and CLK = '1') THEN). If this condition is true, then a count variable is incremented or decremented, depending on the states of a count enable input and a directional control input. If the count enable input is not asserted, the count is neither incremented nor decremented.

The count value is assigned to the counter outputs by the signal assignment statement (qd <= cnt;) after the clear, load, clock, count enable, and direction inputs have been evaluated. Possible results from the signal assignment are:

- **qd = 0** (clear = 0),
- **qd = p** (load = 1 AND clear = 1),
- **increment qd** (count_ena = 1 AND direction = 1),
- **decrement qd** (count_ena = 1 AND direction = 0), or
- **no change on qd** (count_ena = 0).

The terminal count is decoded by determining the count direction and value of the count variable. If the count is UP and the count value is maximum (255_{10} = FFH) or the count is DOWN and the count value is minimum (0 = 00H), a terminal count decoder output called **max_min** goes HIGH.

The code for the same 8-bit counter, but with synchronous clear and load, is shown next.

pre_ct8s.vhd

```
ENTITY pre_ct8s IS
    PORT (
            clk, count_ena        : IN BIT;
            clear, load, direction : IN BIT;
            p                      : IN INTEGER RANGE 0 TO 255;
            max_min                : OUT BIT;
            qd                     : OUT INTEGER RANGE 0 TO 255);
END pre_ct8s;

ARCHITECTURE a OF pre_ct8s IS
    BEGIN
    PROCESS (clk)
        VARIABLE cnt : INTEGER RANGE 0 TO 255;
        BEGIN
            IF (clk'EVENT AND clk = '1') THEN
                IF (clear = '0') THEN -- Synchronous clear
                    cnt := 0;
                ELSIF (load = '1') THEN -- Synchronous load
                    cnt := p;
                ELSIF (count_ena = '1' and direction = '0') THEN
                    cnt := cnt - 1;
                ELSIF (count_ena = '1' and direction = '1') THEN
                    cnt := cnt + 1;
                END IF;
            END IF;
            qd <= cnt;

            -- Terminal count decoder
            IF (cnt = 0 and direction = '0') THEN
                max_min <= '1';
            ELSIF (cnt = 255 and direction = '1') THEN
                max_min <= '1';
            ELSE
                max_min <= '0';
            END IF;
        END PROCESS;
    END a;
```

The PROCESS statement in the synchronous counter has only one identifier in its sensitivity list—that of the clock input. Load and clear status are not evaluated until after the

pre_ct8a.scf

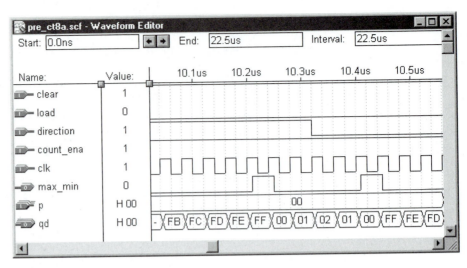

FIGURE 9.48
Simulation Detail of 8-bit VHDL Counter (Bidirectional with Terminal Count Detection)

process checks for a positive clock edge. Otherwise the code is the same as for the asynchronously loading counter.

Figure 9.48 shows a detail of a simulation of the asynchronously loading counter. It shows the point where the count rolls over from FFH to 00H and activates the **max_min** output. The directional output changes shortly after this point and shows the terminal count decoding for a DOWN count, the point where the counter rolls over from 00H to FFH. In the UP count, **max_min** is HIGH when the counter output is FFH, but not 00H. In the DOWN count, **max_min** goes HIGH when the counter output is 00H, but not FFH.

Figure 9.49 shows the operation of the asynchronous load and clear functions. Figure 9.50 show the synchronous load and clear. The inputs are identical for each simulation; each has two pairs of load pulses and a pair of clear pulses. The first pulse of each pair is arranged so that it immediately *follows* a positive clock edge; the second pulse of each pair immediately *precedes* a positive clock edge.

In the counter with asynchronous load and clear, these functions are activated by the first pulse of each pair and again on the second pulse. For the counter with synchronous load and clear, only the second pulse of each pair has an effect, since the load and clear functions must be active *during* or *just prior to* an active clock edge, in order to satisfy

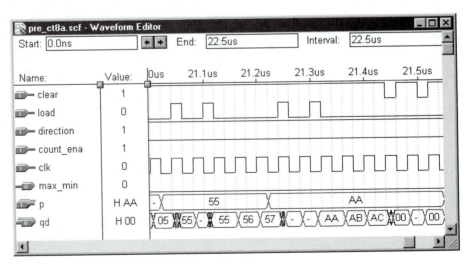

FIGURE 9.49
Simulation Detail of 8-bit VHDL Counter with Asynchronous Load and Clear

FIGURE 9.50

Simulation Detail of 8-bit VHDL Counter with Synchronous Load and Clear

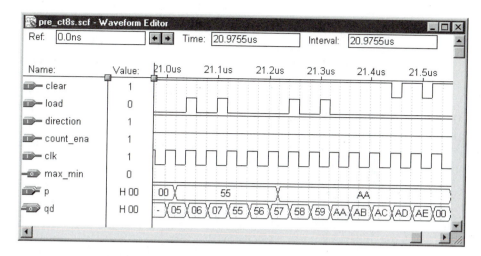

setup-time requirements of the counter flip-flops. The end of the load and clear pulse can correspond to the positive clock edge, as the flip-flop hold time is zero.

Also note that the counter with synchronous load and clear has no intermediate glitch states on its outputs. (The simulation for the asynchronously loading counter shows glitch states on output qd at 21.04 μs, 21.10 μs, 21.22 μs, and 21.44 μs. Refer to the section on Synchronous versus Asynchronous Circuits in Chapter 7 for further discussion of intermediate states in asynchronous circuits.)

Figures 9.51 and 9.52 show further simulation details for our two VHDL counters. Both show the priority of the load, clear, and count enable functions. Both diagrams show that load and clear are independent of count enable and that clear has precedence over load. Again note that the counter with synchronous load and clear is free of intermediate glitch states.

FIGURE 9.51

Simulation Detail of 8-bit VHDL Counter Showing Priority of Count Enable, Asynchronous Load, and Asynchronous Clear

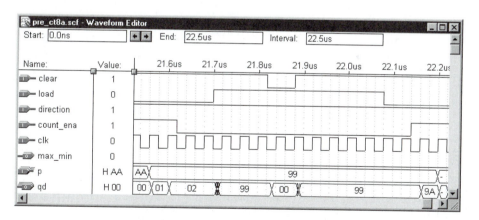

FIGURE 9.52

Simulation Detail of 8-bit VHDL Counter Showing Priority of Count Enable, Synchronous Load, and Synchronous Clear

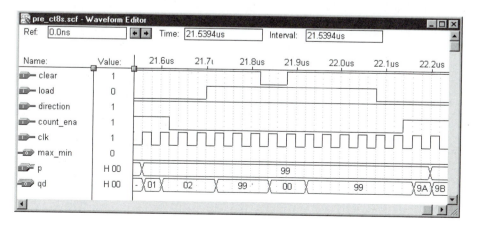

LPM Counters

Earlier in this chapter, we saw how a parameterized counter from the Library of Parameterized Modules (LPM) could be used as a simple 8-bit counter. The component **lpm_counter** has a number of other functions that can be implemented using specific ports and parameters. These functions are indicated in Table 9.13.

Table 9.13 Available Functions of an LPM counter

Function	Ports	Parameters	Description
Basic count operation	clock, q []	LPM_WIDTH	Output **q[]** increases by one with each positive clock edge. **LPM_WIDTH** is the number of output bits.
Synchronous load	sload, data []	none	When **sload** = 1, output **q[]** goes to the value at input **data[]** on the next positive clock edge. **data[]** has the same width as **q[].**
Synchronous clear	sclr	none	When **sclr** = 1, output **q[]** goes to zero on positive clock edge.
Synchronous set	sset	LPM_SVALUE	When **sset** = 1, output goes to value of **LPM_SVALUE** on positive clock edge. If **LPM_SVALUE** is not specified, **q[]** goes to all 1s.
Asynchronous load	aload, data[]	none	Output goes to value at **data[]** when **aload** = 1.
Asynchronous clear	aclr	none	Output goes to zero when **aclr** = 1.
Asynchronous set	aset	LPM_AVALUE	Output goes to value of **LPM_AVALUE** when **aset** = 1. If **LPM_AVALUE** is not specified, outputs all go HIGH when **aset** = 1.
Directional control	updown	LPM_DIRECTION	Optional direction control. Default direction is UP. Only one of **updown** and **LPM_DIRECTION** can be used. **updown** = 1 for UP count, **updown** = 0 for DOWN count. **LPM_DIRECTION** = "UP", "DOWN", or "DEFAULT"
Count enable	cnt_en	none	When **cnt_en** = 1, count proceeds upon positive clock edges. No effect on other synchronous functions **(sload, sclr, sset).** Defaults to "enabled" when not specified.
Clock enable	clk_en	none	All synchronous functions are enabled when **clk_en** = 1. Defaults to "enabled" when not specified.
Modulus control	none	LPM_MODULUS	Modulus of counter is set to value of **LPM_MODULUS**
Output decoding (GDF or AHDL only; not available in VHDL)	eq[15..0]	none	Sixteen active-HIGH decoded outputs, one for each internal counter value from 0 to 15.

The only ports that are required by an LPM counter are **clock,** and one of **q[]** (counter outputs) or **eq[]** (decoder outputs). The only required parameter is **LPM_WIDTH,** which specifies the number of counter output bits. All other ports and parameters are optional, although certain ones must be used together. (For instance, ports **sload** and **data[]** are optional, but both must be used for the synchronous load function.) If unused, a port or parameter will be held at a default logic level.

To use any of the functions of an LPM component in a VHDL file, we use a component instantiation statement and specify the required parameters in a generic map and the ports in a port map.

NOTE

The VHDL component declaration, shown below, indicates that all parameters except LPM_WIDTH are defined as having type STRING, which requires the parameter value to be written in double quotes, even if numeric. (e.g, `LPM_MODULUS => "12"`). Since LPM_WIDTH is defined as type POSITIVE (i.e., any integer > 0) it must be written without quotes (e.g., `LPM_WIDTH => 8`). Default values of all ports and parameters are also included in the component declaration (e.g., `clk_en: IN STD_LOGIC := '1'`; the default value of the clock enable input is '1'). The LPM component declaration can also be found in the MAX+PLUS II Help menu (**Help; Megafunctions/LPM; lpm_counter**).

VHDL Component Declaration for lpm_counter:

```
COMPONENT lpm_counter
    GENERIC (LPM_WIDTH: POSITIVE;
        LPM_MODULUS: STRING := "UNUSED";
        LPM_AVALUE: STRING:= "UNUSED";
        LPM_SVALUE: STRING := "UNUSED";
        LPM_DIRECTION: STRING := "UNUSED";
        LPM_TYPE: STRING := "L_COUNTER";
        LPM_PVALUE: STRING := "UNUSED";
        LPM_HINT : STRING := "UNUSED");
    PORT (data: IN STD_LOGIC_VECTOR (LPM_WIDTH-1 DOWNTO 0) := (OTHERS => '0');
        clock: IN STD_LOGIC;
        cin: IN STD_LOGIC := '0';
        clk_en: IN STD_LOGIC := '1';
        cnt_en: IN STD_LOGIC := '1';
        updown: IN STD_LOGIC := '1'
        sload: IN STD_LOGIC := '0';
        sset: IN STD_LOGIC := '0';
        sclr: IN STD_LOGIC := '0';
        aload: IN STD_LOGIC := '0';
        aset: IN STD_LOGIC := '0';
        aclr: IN STD_LOGIC := '0';
        cout: OUT STD_LOGIC;
        q: OUT STD_LOGIC_VECTOR (LPM_WIDTH-1 DOWNTO 0) );
END COMPONENT;
```

■ EXAMPLE 9.8

Write a VHDL file for an 8-bit LPM counter with ports for the following functions: asynchronous load, asynchronous clear, directional control, and count enable.

Solution The required VHDL file is shown below. Note that no behavioral descriptions are required for the functions, only a mapping from the defined port names to the entity inputs and outputs.

```
-- pre_lpm8
-- 8-bit presettable counter with asynchronous clear and load,
-- count enable, and a directional control port

LIBRARY ieee;
USE ieee.std_logic_1164.ALL;
LIBRARY lpm;
USE lpm.lpm_components.ALL;
```

pre_lpm8.vhd

```
ENTITY pre_lpm8 IS
    PORT (
        clk, count_ena   : IN    STD_LOGIC;
        clear, load, direction   : IN   STD_LOGIC;
        p                : IN    STD_LOGIC_VECTOR(7 downto 0);
        qd               : OUT   STD_LOGIC_VECTOR(7 downto 0));
END pre_lpm8;

ARCHITECTURE a OF pre_lpm8 IS
BEGIN
    counter1: lpm_counter
        GENERIC MAP (LPM_WIDTH => 8)
        PORT MAP ( clock => clk,
                   updown => direction,
                   cnt_en => count_ena,
                   data   => p,
                   aload  => load,
                   aclr   => clear,
                   q      => qd);
END a;
```

▌▌ EXAMPLE 9.9

Write a VHDL file that uses an LPM counter to generate a DOWN counter with a modulus of 500. Create a MAX+PLUS II simulation file to verify the counter's operation.

Solution   A mod-500 counter requires nine bits since $2^8 < 500 < 2^9$. Since the counter always counts DOWN, we can use the parameter LPM_DIRECTION to specify the DOWN counter rather than using an unnecessary port. The required VHDL code is given below.

Note that the value of LPM_WIDTH is written without quotes, since it is defined as type POSITIVE in the component declaration. LPM_MODULUS and LPM_DIRECTION are written in double quotes, since the component declaration defines them as type STRING.

mod5c_lpm.vhd
mod5c_lpm.scf

```
LIBRARY ieee;
USE ieee.std_logic_1164.ALL;
LIBRARY lpm;
Use lpm.lpm_components.ALL;

ENTITY mod5c_lpm IS
    PORT (
        clk : IN     STD_LOGIC;
        q   : OUT    STD_LOGIC_VECTOR (8 downto 0) );
END mod5c_lpm;

ARCHITECTURE a OF mod5c_lpm IS
BEGIN
    counter1: lpm_counter
        GENERIC MAP(LPM_WIDTH     => 9,
                    LPM_DIRECTION => "DOWN",
                    LPM_MODULUS   => "500")
        PORT MAP ( clock => clk,
                   q     => q);
END a;
```

Figure 9.53 shows a partial simulation of the counter, indicating the point at which the output rolls over from 0 to 499 (decimal).

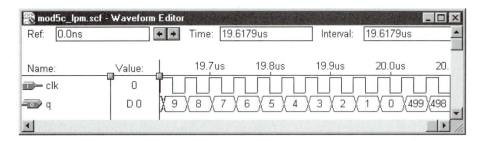

FIGURE 9.53
Example 9.9
Partial Simulation of a Mod-500 LPM DOWN Counter

If we are designing a counter for the Altera UP-1 circuit board, we can simulate the on-board oscillator by choosing a clock period of 40 ns, which corresponds to a clock frequency of 25 MHz. The default simulation period is from 0 to 1 μs, which only gives 1 μs ÷ 40 ns/clock period = 25 clock periods. This is not enough time to show the entire count cycle. The minimum value for the end of the simulation time is:

$$40 \text{ ns/clock period} \times 500 \text{ clock periods} = 20000 \text{ ns} = 20 \text{ μs}.$$

If we wish to see a few clock cycles past the recycle point, we can set the simulation end time to 20.1 μs. (In the MAX+PLUS II Simulator window, select **File menu; End Time.** Enter the value **20.1us** (no spaces) into the **Time** window and click **OK**.)

To view the count waveform, **q**, in decimal rather than hexadecimal, select the waveform by clicking on it. Either right-click to get a pop-up menu or select **Enter Group** from the simulator **Node** menu, as in Figure 9.54. This will bring up the **Enter Group** dialog box shown in Figure 9.55. Select **DEC** (for decimal) and click **OK**.

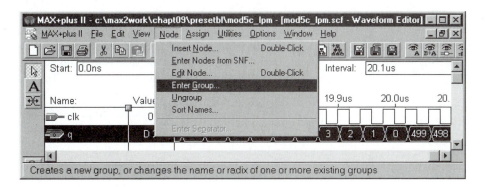

FIGURE 9.54
Selecting a Group in a MAX+PLUS II Simulation

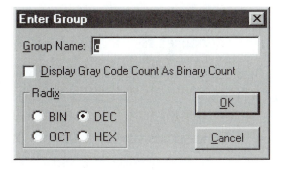

FIGURE 9.55
Changing the Name or Radix of a Group

■ **EXAMPLE 9.10**

Write a VHDL file that instantiates a 12-bit LPM counter with asynchronous clear and synchronous set functions. Design the counter to set to 2047 (decimal). Create a simulation to verify the counter operation.

Solution   The required VHDL file is:

```
-- sset_lpm.vhd
-- 12-bit LPM counter with sset and aclr

LIBRARY ieee;
USE ieee.std_logic_1164.ALL;
LIBRARY lpm;
USE lpm.lpm_components.ALL;

ENTITY sset_lpm IS
    PORT(
            clk        :  IN   STD_LOGIC;
            clear, set :  IN   STD_LOGIC;
            q          :  OUT  STD_LOGIC_VECTOR (11 downto 0) );
END sset_lpm;

ARCHITECTURE a OF sset_lpm IS
BEGIN
    counter1: lpm_counter
        GENERIC MAP    (LPM_WIDTH => 12,
                        LPM_SVALUE => "2047")
        PORT MAP ( clock    => clk,
                   sset     => set,
                   aclr     => clear,
                   q        => q);
END a;
```

Figure 9.56 shows the simulation file of the counter. The full count sequence would take over 160 μs, so we will assume the count portion of the design works properly. Only the set and clear functions are fully simulated. The count waveform is shown in decimal.

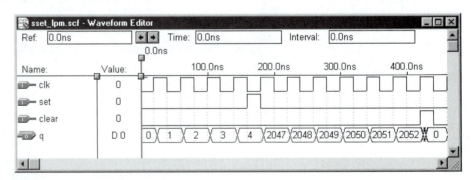

FIGURE 9.56
Example 9.10
Simulation of a 12-bit Counter with Synchronous Set to 2047 and Asynchronous Clear

■ **SECTION 9.6 REVIEW PROBLEM**

9.6 The first part of a VHDL process statement includes a sensitivity list: PROCESS (*sensitivity list*). How should this be written for a counter with asynchronous clear and for a counter with synchronous clear?

9.7 Shift Registers

A **shift register** is a synchronous sequential circuit used to store or move data. It consists of several flip-flops, connected so that data are transferred into and out of the flip-flops in a standard pattern.

Figure 9.57 represents three types of data movement in three 4-bit shift registers. The circuits each contain four flip-flops, configured to move data in one of the ways shown.

Figure 9.57a shows the operation of **serial shifting.** The stored data are taken in one at a time from the input and moved one position toward the output with each applied clock pulse.

Parallel transfer is illustrated in Figure 9.57b. As with the synchronous parallel load function of a presettable counter, data move simultaneously into all flip-flops when a clock pulse is applied. The data are available in parallel at the register outputs.

Rotation, depicted in Figure 9.57c, is similar to serial shifting in that data are shifted one place to the right with each clock pulse. In this operation, however, data are continuously circulated in the shift register by moving the rightmost bit back to the leftmost flip-flop with each clock pulse.

Serial Shift Registers

srg4_sr.gdf
srg4_sr.scf

Figure 9.58 shows the most basic shift register circuit: the serial shift register, so called because data are shifted through the circuit in a linear or serial fashion. The circuit shown consists of four D flip-flops connected in cascade and clocked synchronously.

For a D flip-flop, *Q* follows *D*. The value of a bit stored in any flip-flop *after* a clock pulse is the same as the bit in the flip-flop to its left *before* the pulse. The result is that when a clock pulse is applied to the circuit, the contents of the flip-flops move one position to the

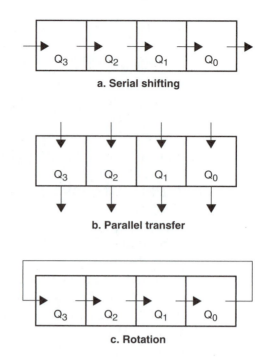

a. Serial shifting

b. Parallel transfer

c. Rotation

FIGURE 9.57
Data Movement in a 4-bit Shift Register

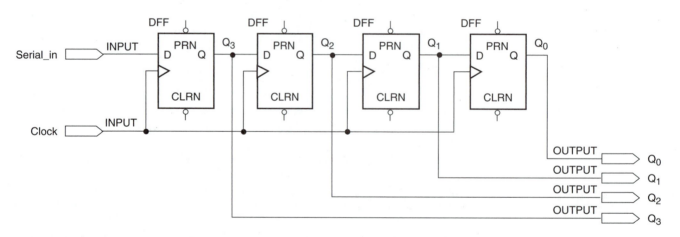

FIGURE 9.58
4-bit Serial Shift Register Configured to Shift Right

right and the bit at the circuit input is shifted into Q_3. The bit stored in Q_0 is overwritten by the former value of Q_1 and is lost. Since the data move from left to right, we say that the shift register implements a **right shift** function. (Data movement in the other direction, requiring a different circuit connection, is called **left shift.**)

Let us track the progress of data through the circuit in two cases. All flip-flops are initially cleared in each case.

Case 1: A 1 is clocked into the shift register, followed by a string of 0s, as shown in Figure 9.59. The flip-flop containing the 1 is shaded.

Before the first clock pulse, all flip-flops are filled with 0s. Data In goes to a 1 and on the first clock pulse, the 1 is clocked into the first flip-flop. After that, the input goes to 0. The 1 moves one position right with each clock pulse, the register filling up with 0s behind it, fed by the 0 at Data In. After four clock pulses, the 1 reaches the Data Out flip-flop. On the fifth pulse, the 0 coming behind overwrites the 1 at Q_0, leaving the register filled with 0s.

FIGURE 9.59

Shifting a "1" Through a Shift Register (Shift Right)

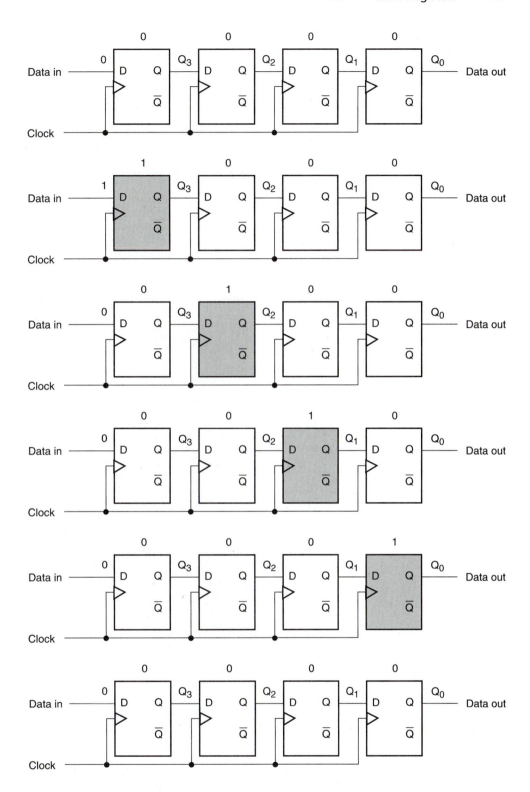

Case 2: Figure 9.60 shows a shift register, initially cleared, being filled with 1s.

As before, the initial 1 is clocked into the shift register and reaches the Data Out line on the fourth clock pulse. This time, the register fills up with 1s, not 0s, because the Data input remains HIGH.

Figure 9.61 shows a MAX+PLUS II simulation of the 4-bit serial shift register in Figure 9.58 through 9.60. The first half of the simulation shows the circuit operation for Case

FIGURE 9.60
Filling a Shift Register with "1"s (Shift Right)

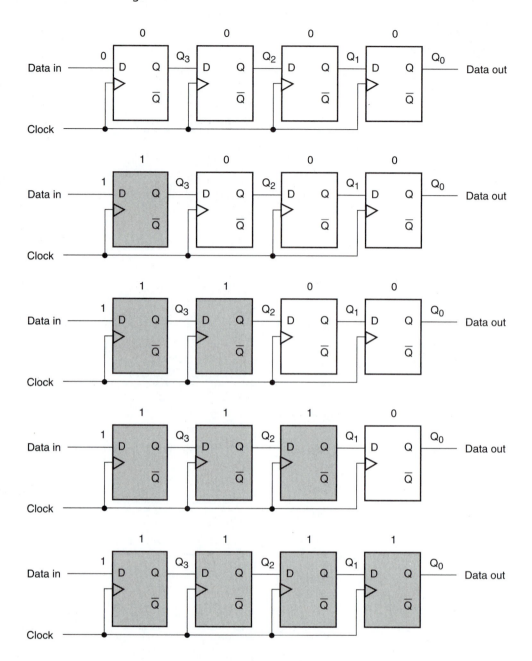

FIGURE 9.61
Simulation of a 4-bit Shift Register (Shift Right)

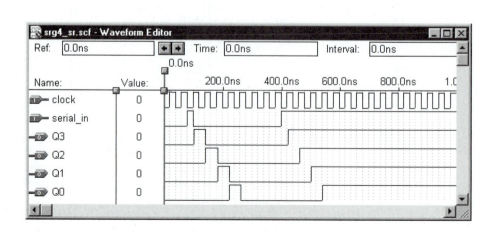

1, above. The 1 enters the register at Q_3 on the first clock pulse after **serial_in** (Data In) goes HIGH. The 1 moves one position for each clock pulse, which is seen in the simulation as a pulse moving through the Q outputs.

Case 2 is shown in the second half of the simulation. Again, a 1 enters the register at Q_3. The 1 continues to be applied to **serial_in,** so all Q outputs stay HIGH after receiving the 1 from the previous flip-flop.

> **NOTE**
>
> Conventions differ about whether the rightmost or leftmost bit in a shift register should be considered the most significant bit. The Altera Library of Parameterized Modules uses the convention of the leftmost bit being the MSB, so this is the convention we will follow. The convention has no physical meaning; the concept of right or left shift only makes sense on a logic diagram. The actual flip-flops may be laid out in any configuration at all in the physical circuit and still implement the right or left shift functions as defined on the logic diagram. (That is to say, wires, circuit board traces, and internal programmable logic connections can run wherever you want; left and right are defined on the logic diagram.)

▌▌ EXAMPLE 9.11

srg4_sl.gdf
srg4_sl.scf

Use the MAX+PLUS II Graphic Editor to create the logic diagram of a 4-bit serial shift register that shifts left, rather than right.

Solution   Figure 9.62 shows the required logic diagram. The flip-flops are laid out the same way as in Figure 9.58, with the MSB (Q_3) on the left. The D input of each flip-flop is connected to the Q output of the flip-flop to its right, resulting in a looped-back connection. A bit at D_0 is clocked into the rightmost flip-flop. Data in the other flip-flops are moved one place to the left. The bit in Q_2 overwrites Q_3. The previous value of Q_3 is lost.

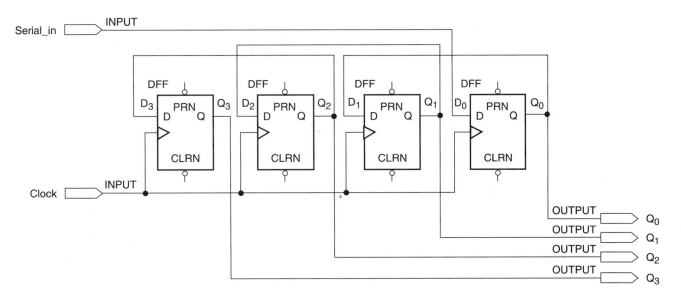

FIGURE 9.62
4-bit Serial Shift Register Configured to Shift Left

▌▌ EXAMPLE 9.12

Draw a diagram showing the movement of a single 1 through the register in Figure 9.62. Also draw a diagram showing how the register can be filled up with 1s.

Solution   Figures 9.63 and 9.64 show the required data movements.

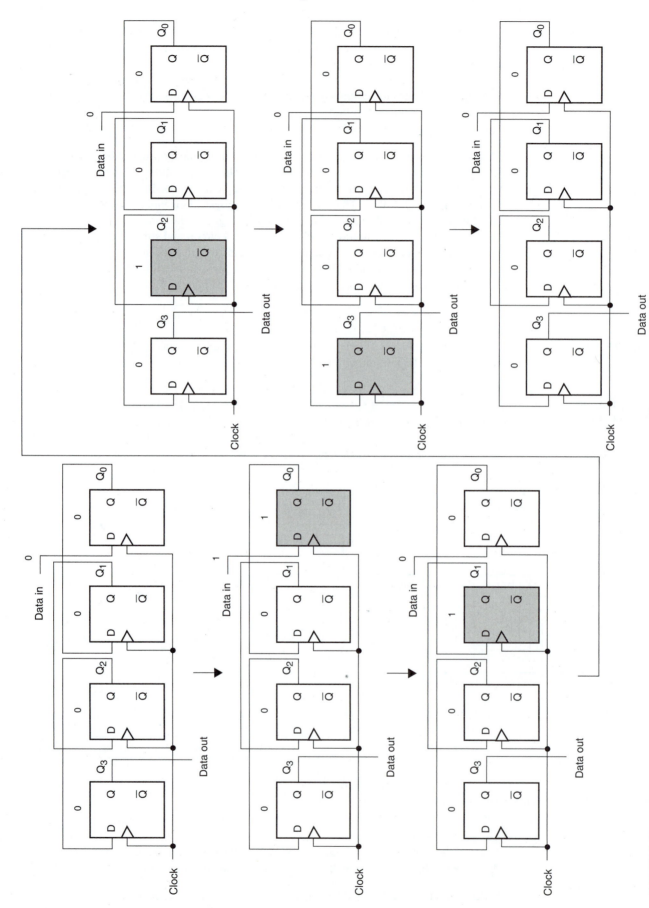

FIGURE 9.63

Shifting a "1" Through a Shift Register (Shift Left)

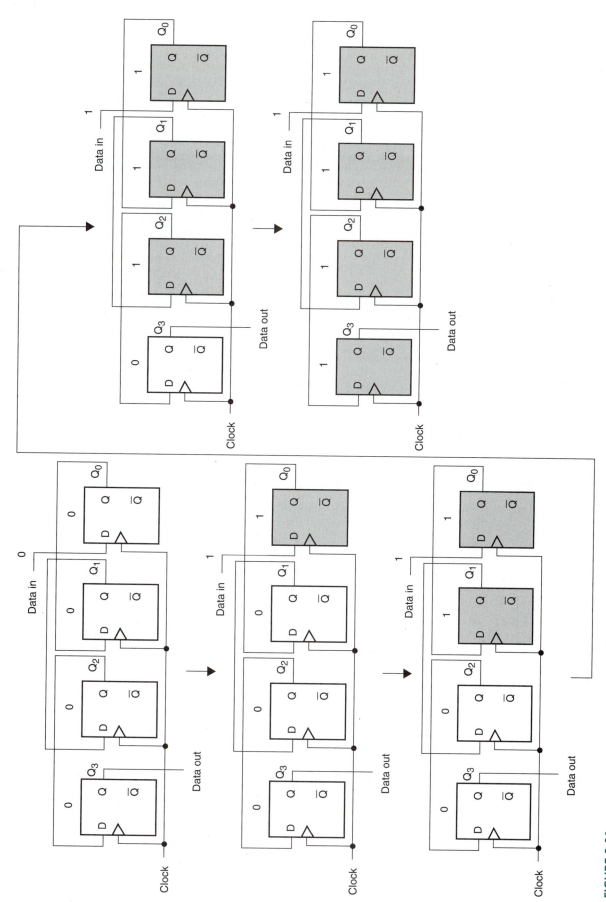

FIGURE 9.64

Filling a Shift Register with "1"s (Shift Left)

▐▌ EXAMPLE 9.13

Use the MAX+PLUS II simulator to verify the operation of the shift-left serial shift register in Figure 9.62.

Solution   Figure 9.65 shows the simulation of the shift operations shown in Example 9.12. Compare this simulation to the one in Figure 9.61 to see how the opposite shift direction appears on a timing diagram.

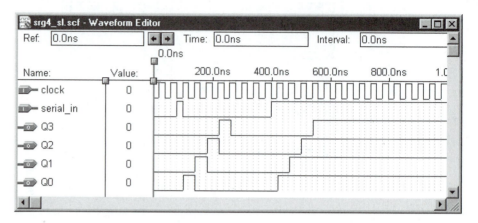

FIGURE 9.65
Simulation of a 4-bit Shift Register (Shift Left)

Bidirectional Shift Registers

srg4_bi.gdf
srg4_bi.scf

Figure 9.66 shows the logic diagram of a **bidirectional shift register.** This circuit combines the properties of the right shift and left shift circuits, seen earlier in Figures 9.58 and 9.62. This circuit can serially move data right or left, depending on the state of a control input, called *DIRECTION*.

The shift direction is controlled by enabling or inhibiting four pairs of AND-OR circuit paths that direct the bits at the flip-flop outputs to other flip-flop inputs. When *DIRECTION* = 0, the right-hand AND gate in each pair is enabled and the flip-flop outputs are directed to the *D* inputs of the flip-flops one position left. Thus the enabled pathway is from *Left_Shift_In* to Q_0, then to Q_1, Q_2, and Q_3.

When *DIRECTION* = 1, the left-hand AND gate of each pair is enabled, directing the data from *Right_Shift_In* to Q_3, then to Q_2, Q_1, and Q_0. Thus, *DIRECTION* = 0 selects left shift and *DIRECTION* = 1 selects right-shift.

Figure 9.67 shows a MAX+PLUS II simulation of the bidirectional shift register in Figure 9.66. The simulation shows the left shift function from 0 to 500 ns and right shift after 500 ns. Both *Right_Shift_In* and *Left_Shift_In* are applied in both parts of the simulation, but the circuit responds only to one for each function.

For the left shift function, a 1 is applied to Q_0 at 140 ns and shifted left. The *Right_Shift_In* pulse is ignored. Similarly, for the right shift function, a 1 is applied to Q_3 at 540 ns and shifted right. *Left_Shift_In* is ignored.

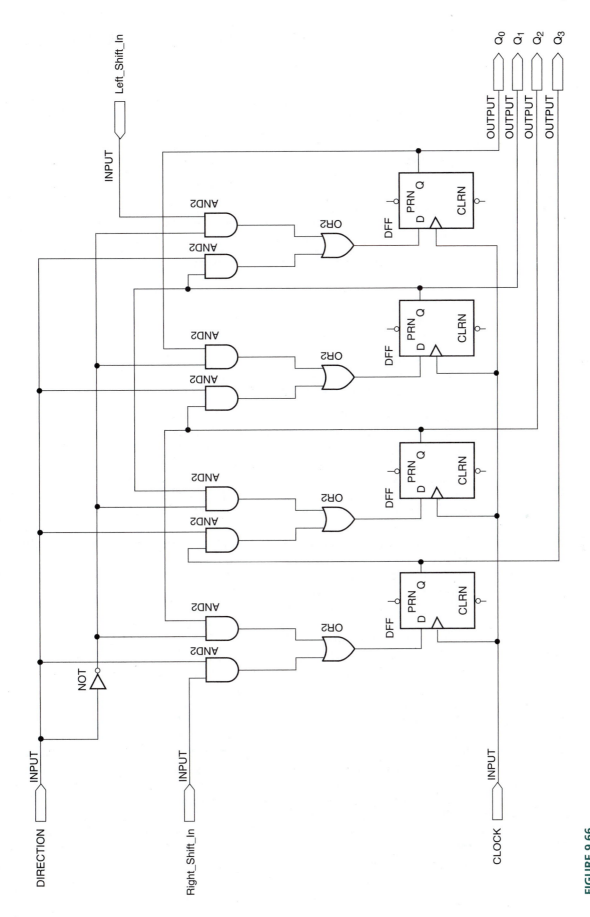

FIGURE 9.66
Bidirectional Shift Register

FIGURE 9.67
Simulation of a 4-bit
Bidirectional Shift Register

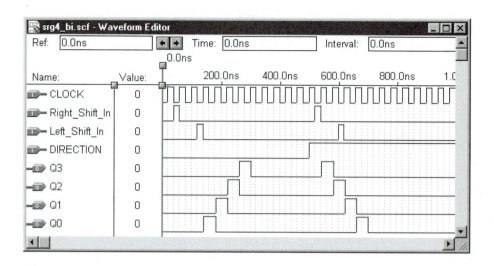

srg4_par.gdf
srg4_par.scf

srg4_uni.gdf
srg4_uni.scf

Shift Register with Parallel Load

Earlier in this chapter, we saw how a counter could be set to any value by synchronously loading a set of external inputs directly into the counter flip-flops. We can implement the same function in a shift register, as shown in Figure 9.68.

The circuit is similar to that of the bidirectional shift register in Figure 9.66. The synchronous input of each flip-flop is fed by an AND-OR circuit that directs one of two signals to the flip-flop: the output of the previous flip-flop (shift function) or a parallel input (load function). The circuit is configured such that the shift function is enabled when $LOAD = 0$ and the load function is enabled when $LOAD = 1$.

Figure 9.69 shows a simulation of the parallel-load shift register circuit of Figure 9.68. In the first part of the simulation, the shift function is selected. This is tested by sending a 1 through the circuit in a right-shift pattern. Next, at 400 ns, $LOAD$ goes HIGH, and the parallel input value AH ($= 1010_2$) is synchronously loaded into the circuit. The $LOAD$ input goes LOW, thus causing the circuit to revert to the shift function. The data in the register are right-shifted out, followed by 0s. At 640 ns, the value FH ($= 1111_2$) is loaded into the circuit, then right-shifted out.

Figure 9.70 shows the logic circuit of a **universal shift register.** This circuit can implement any combination of serial and parallel inputs and outputs. It can also serially shift data left or right or hold data, depending on the states of S_1 and S_0, which form a 2-bit function select input.

Each AND-OR circuit acts as a multiplexer to direct one of several possible data sources to the synchronous inputs of each flip-flop. For instance, if we trace the paths through the corresponding AND-OR circuit, we find that the possible sources of data at D_2, the synchronous input of the second flip-flop, are Q_3 ($S_1 S_0 = 01$), P_2 ($S_1 S_0 = 11$), Q_1 ($S_1 S_0 = 10$), and Q_2($S_1 S_0 = 00$). These are the inputs required for the right-shift, parallel load, left-shift, and hold functions, respectively. All functions are synchronous, including the parallel load and hold functions.

The hold function is a synchronous no change function, implemented by feeding back the Q output of a flip-flop to its synchronous (D) input. It is necessary to have this function, so that the flip-flops will not synchronously clear when none of the other functions is selected.

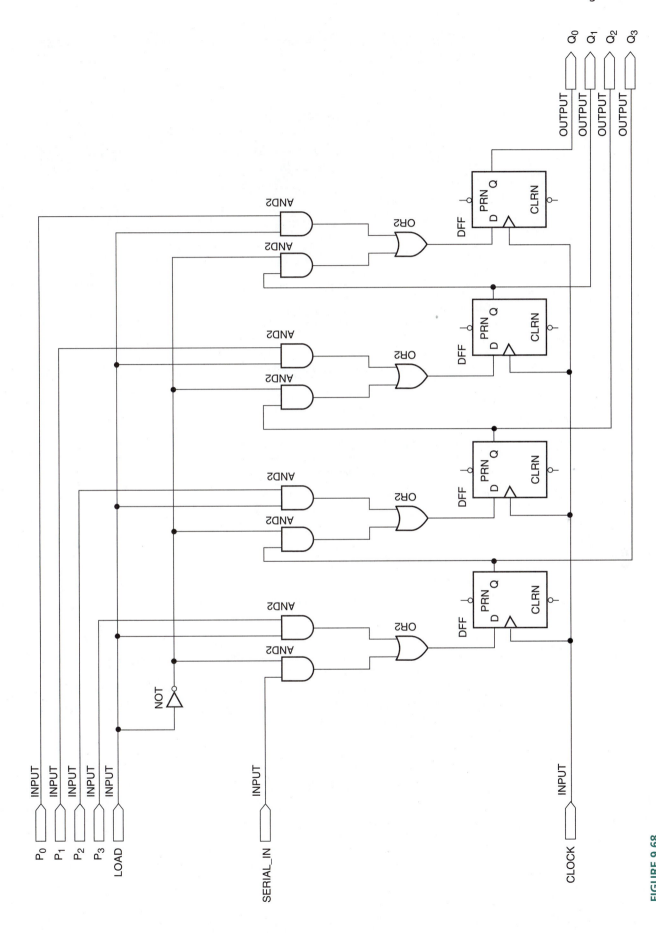

FIGURE 9.68
Serial Shift Register with Parallel Load

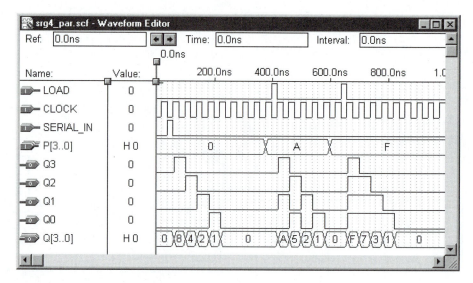

FIGURE 9.69
Simulation of a 4-bit Serial Shift Register with Parallel Load

Table 9.14 summarizes the various possible inputs to each flip-flop as a function of S_1 and S_0.

Table 9.14 Flip-Flop Inputs as a Function of S_1S_0 in a Universal Shift Register

S_1	S_0	Function	D_3	D_2	D_1	D_0
0	0	Hold	Q_3	Q_2	Q_1	Q_0
0	1	Shift Right	RSI^*	Q_3	Q_2	Q_1
1	0	Shift Left	Q_2	Q_1	Q_0	LSI^{**}
1	1	Load	P_3	P_2	P_1	P_0

$*RSI$ = Right-shift input
$**LSI$ = Left-shift input

■■ EXAMPLE 9.14

Create a simulation file to verify the operation of the universal shift register of Figure 9.70.

Solution Figure 9.71 shows a possible solution. The following functions are tested: hold, right shift (*LSI* ignored), hold, left shift (*RSI* ignored), load FH, asynchronous clear, load FH, shift right for two clocks, shift left for three clocks.

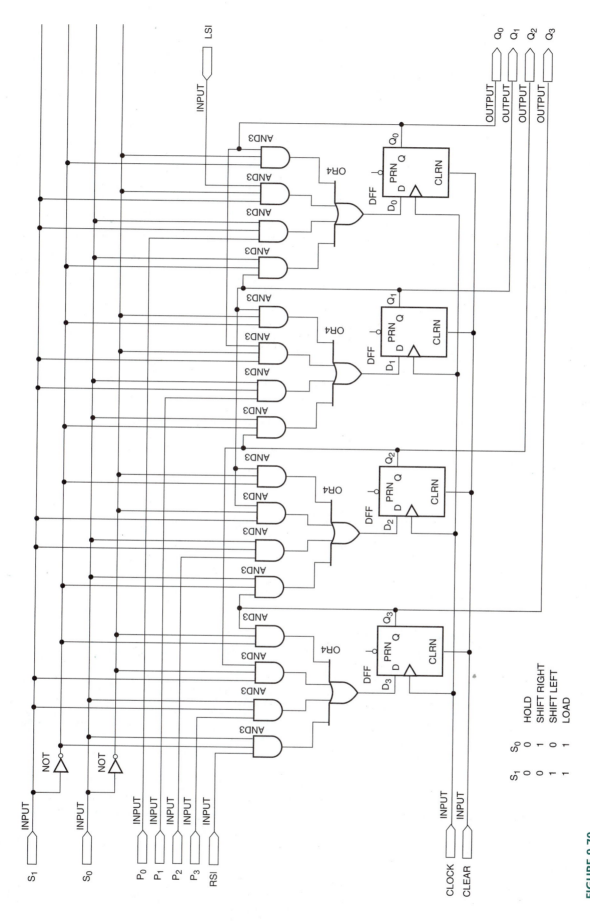

FIGURE 9.70

4-bit Universal Shift Register

425

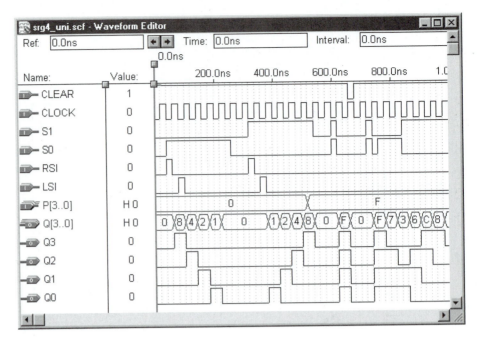

FIGURE 9.71
Example 9.14
Simulation of a 4-bit Universal Shift Register

■■ SECTION 9.7 REVIEW PROBLEM

9.7 Can the D flip-flops in Figure 9.58 be replaced by JK flip-flops? If so, what modifications to the existing circuit are required?

9.8 Programming Shift Registers in VHDL

> **KEY TERMS**
>
> **Structural design**   A VHDL design technique that connects predesigned components using internal signals.
>
> **Dataflow design**   A VHDL design technique that uses Boolean equations to define relationships between inputs and outputs.
>
> **Behavioral design**   A VHDL design technique that uses descriptions of required behavior to describe the design.

As with other circuit applications, we can take several approaches to programming shift registers in VHDL. Three basic design techniques are **structural, dataflow,** and **behavioral** descriptions. We will use each of these techniques to design a 4-bit shift register, such as the one shown in Figure 9.58.

Structural Design

Structural design is like taking components out of a bin and connecting them together to make a circuit. We can use the **DFF** component from the MAX+PLUS II primitives library and instantiate enough components to make a shift register, with connections made by internal signals. The code to make a 4-bit shift register using the structural design technique is shown here in the file **srg4strc.vhd**.

srg4strc.vhd
srgstrc.scf

```
-- srg4strc.vhd
-- Structural description of a 4-bit serial shift register
LIBRARY ieee;
USE ieee.std_logic_1164.ALL;
LIBRARY altera;
USE altera.maxplus2.ALL;

ENTITY srg4strc IS
    PORT(
        serial_in, clk : IN      STD_LOGIC;
        qo                : BUFFER  STD_LOGIC_VECTOR(3 downto 0) );
END srg4strc;

ARCHITECTURE right_shift of srg4strc IS
    COMPONENT DFF
        PORT (d : IN STD_LOGIC;
            clk : IN STD_LOGIC;
            q   : OUT STD_LOGIC);
    END COMPONENT;
BEGIN
    flip_flop_3: dff
        PORT MAP (serial_in, clk, qo(3) );

    dffs:
    FOR i IN 2 downto 0 GENERATE
    flip_flops_2_to_0: dff
        PORT MAP (qo(i + 1), clk, qo(i) );
    END GENERATE;
END right_shift;
```

The design entity **srg4strc.vhd** instantiates four D flip-flops from the **altera. maxplus2** package and connects them by assigning common inputs and outputs to related components. A different way of writing the component instantiations would be as follows.

```
flip_flop_3: dff
    PORT MAP (serial_in, clk, qo(3) );
flip_flop_2: dff
    PORT MAP(qo(3), clk, qo(2) );
flip_flop_1: dff
    PORT MAP(qo(2), clk, qo(1) );
flip_flop_0: dff
    PORT MAP(qo(1), clk, qo(0) );
```

Since the component ports are in the sequence (D, clk, Q), the component instantiations shown above imply that the D input of a flip-flop is fed by the Q of the previous flip-flop.

The port identifier **qo** is defined as mode BUFFER, not as OUT, because it is sometimes used as an input and sometimes as an output. A port of mode OUT can only be used as an output. A port of mode BUFFER has a feedback connection so that the output can be reused in the programmed AND matrix of the CPLD macrocell. Figure 9.72 illustrates the difference between these modes.

Rather than defining connections in the component instantiations, we would also be able to use an internal signal to connect the flip-flops. This method allows us to use a port of mode OUT, rather than BUFFER. The file **srg4str2.vhd** shows this alternative way.

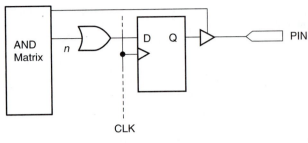

a. Driver of mode OUT

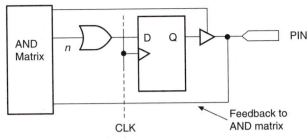

b. Driver of mode BUFFER

FIGURE 9.72
OUT vs. BUFFER

srg4str2.vhd
srg4str2.scf

```
--srg4str2.vhd
-- Structural description of a 4-bit serial shift register
LIBRARY ieee;
USE ieee.std_logic_1164.ALL;
LIBRARY altera;
USE altera.maxplus2.ALL;

ENTITY srg4str2 IS
    PORT (
          serial_in, clk : IN  STD_LOGIC;
          qo             : OUT STD_LOGIC_VECTOR(3 downto 0) );
END srg4str2;

ARCHITECTURE right_shift of srg4str2 IS
    COMPONENT DFF
        PORT (d   : IN STD_LOGIC;
              clk : IN STD_LOGIC;
              q   : OUT STD_LOGIC);
    END COMPONENT;
    SIGNAL connect : STD_LOGIC_VECTOR(3 downto 0);
BEGIN
    flip_flop_3: dff
        PORT MAP (serial_in, clk, connect(3) );
    dffs:
    FOR i IN 2 downto 0 GENERATE
    flip_flops_2_to_0: dff
        PORT MAP (connect(i + 1), clk, connect(i) );
    END GENERATE;

    qo <= connect;
END right_shift;
```

In this case, the internal signal **connect** is used to tie the flip-flops together. The circuit output derives from a signal assignment statement at the end of the file. Since the internal signal **connect** is used to fulfil the flip-flop input/output functions, **qo** can be defined solely as an output.

Dataflow Design

Dataflow design describes a design entity in terms of the Boolean relationships between different parts of the circuit. The Boolean relationships in a 4-bit shift register are defined by the expressions for the flip-flop synchronous inputs:

$$D_3 = serial\_in$$
$$D_2 = Q_3$$
$$D_1 = Q_2$$
$$D_0 = Q_1$$

The design entity **srg4dflw.vhd** illustrates the use of the dataflow design method for a 4-bit serial shift register.

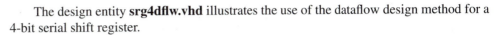
srg4dflw.vhd
srg4dflw.scf

```
-- srg4dflw.vhd
-- Dataflow description of a 4-bit serial shift register
LIBRARY ieee;
USE ieee.std_logic_1164.ALL;

ENTITY srg4dflw IS
    PORT (
        serial_in, clk : IN       STD_LOGIC;
        q              : BUFFER   STD_LOGIC_VECTOR(3 downto 0) );
END srg4dflw;

ARCHITECTURE right_shift of srg4dflw IS
    SIGNAL d : STD_LOGIC_VECTOR(3 downto 0);
BEGIN

    PROCESS (clk)
    BEGIN
        -- Define a 4-bit D flip-flop
        IF clk'EVENT and clk = '1' THEN
            q <= d;
        END IF;
    END PROCESS;

    d <= serial_in & q(3 downto 1);
END right_shift;
```

Before the flip-flops can be connected, they must be defined in a PROCESS statement. The statements inside the process are sequential, as they must be to define a flip-flop, but the process itself is a concurrent statement. Signals are applied concurrently (simultaneously) to the construct implied by the process (the flip-flops) and all other concurrent constructs in the design entity (the connections between **q** and **d** and the serial input).

A signal assignment statement implements the Boolean equations for the shift register. It is written as a single statement for efficiency, but could also be written as four separate assignment statements, as follows:

```
d(3) <= serial_in;
d(2) <= q(3);
d(1) <= q(2);
d(0) <= q(1);
```

We must define **q** as mode BUFFER, since we are using it as both input and output.

Table 9.15 Next States of Flip-Flops in a Serial Shift Register

Q_3	Q_2	Q_1	Q_0
serial_in	Q_3	Q_2	Q_1

Behavioral Design

We can create a VHDL design entity from the description of its desired behavior. In the case of a shift register, we know that after a clock pulse all data move over one position and the first flip-flop in the chain accepts a bit from a serial input, as indicated in Table 9.15. We can use this behavioral description to implement a serial shift register, as shown in the VHDL file **srg4behv.vhd**.

```
-- srg4behv.vhd
-- Behavioral description of a 4-bit serial shift register
LIBRARY ieee;
USE ieee.std_logic_1164.ALL;

ENTITY srg4behv IS
    PORT (
          serial_in, clk : IN      STD_LOGIC;
          q                : BUFFER  STD_LOGIC_VECTOR(3 downto 0) );
END srg4behv;

ARCHITECTURE right_shift of srg4behv IS
BEGIN
    PROCESS (clk)
    BEGIN
        IF (clk'EVENT and clk = '1') THEN
            q <= serial_in & q(3 downto 1);
        END IF;
    END PROCESS;
END right_shift;
```

srg4behv.vhd
srg4behv.scf

In the behavioral design, we are not concerned with the flip-flop inputs or other internal connections; the behavioral description is sufficient for the VHDL compiler to synthesize the required hardware. Compare this to the dataflow description, where we created a set of flip-flops, then assigned Boolean functions to the *D* inputs. In this case, the behavioral design method combines these two steps into one.

▍▍ **EXAMPLE 9.15**

Write the code for a VHDL design entity that implements a 4-bit bidirectional shift register with asynchronous clear. Create a simulation that verifies the design function.

Solution   The VHDL code for the bidirectional shift register, **srg4bidi.vhd**, follows. A CASE statement monitors the directional control of the shift register. We require the **others** clause of the CASE statement since the identifier **direction** is of type STD_LOGIC; the cases '0' and '1' do not cover all possible values of STD_LOGIC. Since we want no action to be taken in the default case, we use the keyword **NULL**.

www.electronictech.com

srg4bidi.vhd
srg4bidi.scf

```
LIBRARY ieee;
USE ieee.std_logic_1164.ALL;

ENTITY srg4bidi IS
    PORT (
          clk, clear : IN STD_LOGIC;
          rsi, lsi   : IN STD_LOGIC;
          direction  : IN STD_LOGIC;
          q          : BUFFER STD_LOGIC_VECTOR(3 downto 0) );
END srg4bidi;

ARCHITECTURE bidirectional_shift of srg4bidi IS
BEGIN
    PROCESS (clk, clear)
    BEGIN
        IF clear = '0' THEN
            q <= "0000"; -- asynchronous clear
        ELSIF (clk'EVENT and clk = '1') THEN
            CASE direction IS
                WHEN '0' =>
                    q <= q(2 downto 0) & lsi; -- left shift
```

```
            WHEN '1' =>
               q <= rsi & q(3 downto 1); -- right shift
            WHEN others =>
               NULL;
         END CASE;
      END IF;
   END PROCESS;
END bidirectional_shift;
```

Figure 9.73 shows the simulation of the shift register, with the left shift function in the first half of the simulation and the right shift function in the second half.

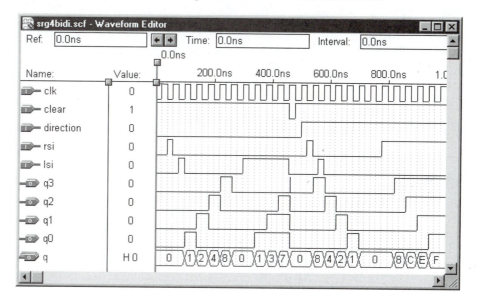

FIGURE 9.73
Example 9.15
4-bit Bidirectional Shift Register

Shift Registers of Generic Width

> **KEY TERM**
>
> **GENERIC** A clause in the entity declaration of a VHDL component that lists the parameters that can be specified when the component is instantiated.

All multibit VHDL components we have examined until now have been of a specified width (e.g., 2-to-4 decoder, 8-bit MUX, 8-bit adder, 4-bit counter). VHDL allows us to create components having a generic, or unspecified, width or other parameter which is specified when the component is instantiated. In the entity declaration of such a component, we indicate an unspecified parameter (such as width) in a **GENERIC** clause. The unspecified parameter must be given a default value in the GENERIC clause, indicated by := *value*.

When we instantiate the component, we specify the parameter value in a generic map, as we have done with components from the Library of Parameterized Modules. The design entity **srt_bhv.vhd** below behaviorally defines an *n*-bit right-shift register, with a default width of four bits given by the statement (`GENERIC (width : POSITIVE := 4);`).

The entity **srt8_bhv.vhd** instantiates the *n*-bit register as an 8-bit circuit by specifying the bit width in a generic map. If no value is specified, the component is presumed to have a default width of four, as defined in the component's entity declaration.

srt_bhv.vhd
srt8_bhv.vhd
srt8_bhv.scf

```
-- srt_bhv.vhd
-- Behavioral description of an n-bit shift register
LIBRARY ieee;
USE ieee.std_logic_1164.ALL;

ENTITY srt_bhv IS
  GENERIC (width : POSITIVE := 4);
  PORT (
      serial_in, clk : IN       STD_LOGIC;
      q              : BUFFER   STD_LOGIC_VECTOR(width-1 downto 0) );
END srt_bhv;

ARCHITECTURE right_shift of srt_bhv IS
BEGIN
  PROCESS (clk)
  BEGIN
      IF (clk'EVENT and clk = '1') THEN
            q(width-1 downto 0) <= serial_in & q(width-1 downto 1);
      END IF;
  END PROCESS;
END right_shift;

-- srt8_bhv.vhd
-- 8-bit shift register that instantiates srt_bhv
LIBRARY ieee;
USE ieee.std_logic_1164.ALL;

ENTITY srt8_bhv IS
 PORT(
      data_in, clock : IN       STD_LOGIC;
      qo             : BUFFER   STD_LOGIC_VECTOR(7 downto 0) );
END srt8_bhv;

ARCHITECTURE right_shift of srt8_bhv IS
COMPONENT srt_bhv
  GENERIC (width : POSITIVE);
  PORT (
      serial_in, clk  : IN     STD_LOGIC;
      q               : OUT    STD_LOGIC_VECTOR(7 downto 0) );
END COMPONENT;
BEGIN
  Shift_right_8: srt_bhv
     GENERIC MAP (width=> 8)
     PORT MAP (serial_in => data_in,
               clk       => clock,
               q         => qo);
END right_shift;
```

▌▌ EXAMPLE 9.16

Write the code for a VHDL design entity that defines a universal shift register with a generic width. (The default width is eight bits.) Instantiate this entity as a component in a file for a 16-bit universal shift register.

Solution

```
-- srg_univ.vhd
-- Universal shift register with generic width
```

```
-- Default width = 8 bits
LIBRARY ieee;
USE ieee.std_logic_1164.ALL;
USE ieee.std_logic_arith.ALL;

ENTITY srg_univ IS
    GENERIC (width : POSITIVE := 8);
    PORT (
        clk, clear      : IN STD_LOGIC;
        rsi, lsi        : IN STD_LOGIC;
        function_select : IN STD_LOGIC_VECTOR(1 downto 0);
        p               : IN STD_LOGIC_VECTOR(width-1 downto 0);
        q               : BUFFER STD_LOGIC_VECTOR(width-1 downto 0) );
END srg_univ;
```

srg_univ.vhd
srg16uni.vhd

```
ARCHITECTURE universal_shift of srg_univ IS
BEGIN
  PROCESS (clk, clear)
  BEGIN
    IF clear = '0' THEN
        -- Conversion function to convert integer 0 to vector
        -- of any width. Requires ieee.std_logic_arith package.
        q <= CONV_STD_LOGIC_VECTOR(0, width);
    ELSIF (clk'EVENT and clk = '1') THEN
        CASE function_select IS
            WHEN "00" =>
                q <= q; -- Hold
            WHEN "01" =>
                q <= rsi & q(width-1 downto 1); -- Shift right
            WHEN "10" =>
                q <= q(width-2 downto 0) & lsi; -- Shift left
            WHEN "11" =>
                q <= p; -- Load
            WHEN OTHERS =>
                NULL;
        END CASE;
    END IF;
  END PROCESS;
END universal_shift;

-- srg16uni.vhd
-- 16-bit universal shift register (instantiates srg_univ)
LIBRARY ieee;
USE ieee.std_logic_1164.ALL;

ENTITY srg16uni IS
    PORT (
        clock, clr  : IN      STD_LOGIC;
        rsi, lsi    : IN      STD_LOGIC;
        s           : IN      STD_LOGIC_VECTOR(1 downto 0);
        parallel_in : IN      STD_LOGIC_VECTOR(15 downto 0);
        qo          : BUFFER  STD_LOGIC_VECTOR(15 downto 0) );
END srg16uni;

ARCHITECTURE universal_shift of srg16uni IS
COMPONENT srg_univ
  GENERIC (width : POSITIVE);
  PORT (
```

```
        clk, clear      : IN STD_LOGIC;
        rsi, lsi        : IN STD_LOGIC;
        function_select    : IN STD_LOGIC_VECTOR(1 downto 0);
        p               : IN STD_LOGIC_VECTOR(width-1 downto 0);
        q               : BUFFER STD_LOGIC_VECTOR(width-1 downto 0));
END COMPONENT;
BEGIN
    Shift_universal_16: srg_univ
        GENERIC MAP (width=> 16)
        PORT MAP (clk              => clock,
                  clear            => clr,
                  rsi              => rsi,
                  lsi              => lsi,
                  function_select  => s,
                  p                => parallel_in,
                  q                => qo);
END universal_shift;
```

When we are designing the clear function in **srg_univ.vhd**, we must account for the fact that we must set all bits of a vector of unknown width to '0'. To get around this problem, we use a conversion function that changes an INTEGER value of 0 to a STD_LOGIC_VECTOR of **width** bits and assigns the value to the output. The required conversion function, CONV_STD_LOGIC_VECTOR(*value, number_of_bits*), is found in the **std_logic_arith** package in the **ieee** library. We could also use the construct

```
q <= (others => '0');
```

which states that the default case is to set all bits of **q** to 0 when **clear** is 0. Since there is no other case specified, all bits of **q** are cleared.

LPM Shift Registers

The Library of Parameterized Modules contains a shift register component, **lpm_shiftreg**, that we can instantiate in a VHDL design entity. The various functions of **lpm_shiftreg** are listed in Table 9.16.

The following VHDL code instantiates **lpm_shiftreg** as an 8-bit shift register with serial input and serial output. In this case, the LPM component is declared explicitly, with the component declaration statement listing only the ports and parameters used by the design entity. The component instantiation statement lists the port names from the design entity in the same order as the corresponding component port names. By default the register direction is LEFT (i.e., toward the MSB).

```
-- srg8_lpm.vhd
-- 8-bit serial shift register (shift left by default)
LIBRARY ieee;
USE ieee.std_logic_1164.ALL;
LIBRARY lpm;
USE lpm.lpm_components.ALL;
```

Table 9.16 Available Functions for *lpm_shiftreg*

Function	Ports	Parameters	Description
Basic serial operation	clock, shiftin, shiftout, q[]	LPM_WIDTH	Data moves serially from **shiftin** to **shiftout.** Parallel outputs appear at **q[].**
Load	sload, data[]	none	When **sload** = 1, **q[]** goes to the value at input **data[]** on the next positive clock edge. **Data[]** has the same width as **LPM_WIDTH.**
Synchronous clear	sclr	none	When **sclr** = 1, **q[]** goes to zero on positive clock edge
Synchronous set	sset	LPM_SVALUE	When **sset** = 1, output goes to value of **LPM_SVALUE** on positive clock edge. If **LPM_SVALUE** is not specified **q[]** goes to all 1s.
Asynchronous clear	aclr	none	Output goes to zero when **aclr** = 1.
Asynchronous set	aset	LPM_AVALUE	Output goes to value of **LPM_AVALUE** when **aset** = 1. If **LPM_AVALUE** is not specified, outputs all go HIGH when **aset** = 1.
Directional control	none	LPM_DIRECTION	Optional direction control. Default direction is LEFT. **LPM_DIRECTION** = "LEFT" or "RIGHT". If **shiftin** and **shiftout** are used, the serial shift always goes through the entire shift register, in the direction given by **LPM_DIRECTION.**
Clock enable	enable	none	All synchronous functions are enabled when **enable** = 1. Defaults to "enabled" when not specified.

srg8_lpm.vhd
srg8_lpm.scf

```
ENTITY srg8_lpm IS
    PORT (
        clk        : IN  STD_LOGIC;
        serial_in  : IN  STD_LOGIC;
        serial_out : OUT STD_LOGIC);
END srg8_lpm;

ARCHITECTURE lpm_shift of srg8_lpm IS
    COMPONENT lpm_shiftreg
    GENERIC(LPM_WIDTH: POSITIVE);
        PORT (
            clock, shiftin : IN  STD_LOGIC;
            shiftout       : OUT STD_LOGIC);
END COMPONENT;
BEGIN
    Shift_8: lpm_shiftreg
        GENERIC MAP (LPM_WIDTH=> 8)
        PORT MAP (clk, serial_in, serial_out);
END lpm_shift;
```

Figure 9.74 shows a simulation of the shift register, with the data shifting from right to left (LSB to MSB). Since there is no parallel output (**q[]**) instantiated in our design, we would not normally be able to monitor the progress of bits from flip-flop to flip-flop; we would only see **shiftin**, and then, eight clock cycles later, **shiftout**. However, we are able to monitor the flip-flop states as buried nodes (**Shift_8|dffs[7..0].Q**). These buried nodes are the last eight lines in the simulation.

FIGURE 9.74
Simulation of an 8-bit LPM Shift
Register (Shift Left)

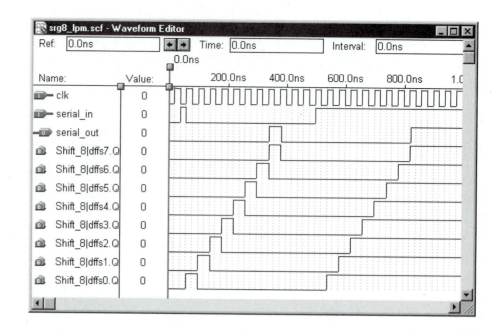

⊪ EXAMPLE 9.17

Modify the VHDL code just shown to make the serial shift register shift right, rather than left. Create a simulation to verify the circuit function. How do the positions of **shiftin** and **shiftout** ports relate to the internal flip-flops for the right-shift and left-shift implementations?

Solution   The modified VHDL code is shown next as design entity **srg8lpm2.** The only difference is the addition of the parameter LPM_DIRECTION to both the component declaration and component instantiation statements.

```
-- srg8_lpm2.vhd
-- 8-bit serial shift register (shift right)
LIBRARY ieee;
USE ieee.std_logic_1164.ALL;
LIBRARY lpm;
USE lpm.lpm_components.ALL;

ENTITY srg8lpm2 IS
    PORT (
        clk         : IN   STD_LOGIC;
        serial_in   : IN   STD_LOGIC;
        serial_out  : OUT STD_LOGIC);
END srg8_lpm2;

ARCHITECTURE lpm_shift of srg8_lpm2 IS
COMPONENT lpm_shiftreg
    GENERIC(LPM_WIDTH: POSITIVE; LPM_DIRECTION: STRING);
    PORT (
        clock, shiftin : IN   STD_LOGIC;
        shiftout       : OUT STD_LOGIC);
END COMPONENT;
BEGIN
    shift_8: lpm)_shiftreg
        GENERIC MAP (LPM_WIDTH=> 8, LPM_DIRECTION => "RIGHT")
        PORT MAP (clk, serial_in, serial_out);
END lpm_shift;
```

srg8_lpm2.vhd
srg8_lpm2.scf

FIGURE 9.75
Example 9.17
Simulation of an 8-bit LPM Shift
Register (Shift Right)

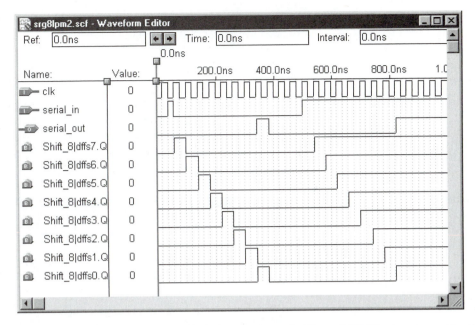

The simulation for the right-shift register is shown in Figure 9.75. The inputs are identical to those of Figure 9.74, but the internal shift direction is opposite. The LPM component configures the serial shift input and output such that they allow data to go through the entire register, regardless of shift direction. For left-shift, **serial_in (shiftin)** is applied to **D0**, and is shifted toward **Q7**. For right-shift, the same **serial_in** is applied to **D7** and shifted toward **Q0**. Thus, there is no right shift input or left shift input in this component, and also no bidirectional shift that can be controlled by an input port. Shift direction can only be set by the value of a parameter and is therefore fixed when a component is instantiated.

▍ EXAMPLE 9.18

Write the VHDL code for an 8-bit LPM shift register with both parallel and serial outputs, parallel load and asynchronous clear. Create a simulation to verify the design operation.

Solution   The VHDL code for the parallel load shift register follows as **srg8lpm3**.

```
LIBRARY ieee;
USE ieee.std_logic_1164.ALL;
LIBRARY lpm;
USE lpm.lpm_components.ALL;

ENTITY srg8lpm3 IS
    PORT (
        clk, ld, clr  : IN    STD_LOGIC;
        p             : IN    STD_LOGIC_VECTOR(7 downto 0);
        q_out         : OUT   STD_LOGIC_VECTOR(7 downto 0);
        serial_out    : OUT   STD_LOGIC);
END srg8lpm3;

ARCHITECTURE lpm_shift of srg8lpm3 IS
COMPONENT lpm_shiftreg
    GENERIC(LPM_WIDTH: POSITIVE);
    PORT (
        clock, load  : IN    STD_LOGIC;
        aclr         : IN    STD_LOGIC;
        data         : IN    STD_LOGIC_VECTOR (7 downto 0);
        q            : OUT   STD_LOGIC_VECTOR (7 downto 0);
        shiftout     : OUT   STD_LOGIC);
END COMPONENT;
```

srg8lpm3.vhd
srg8lpm3.scf

```
BEGIN
    Shift_8: lpm_shiftreg
        GENERIC MAP (LPM_WIDTH=> 8)
        PORT MAP (clk, ld, clr, p, q_out, serial_out);
END lpm_shift;
```

The simulation for **srg8lpm3** is shown in Figure 9.76. The load input is initially HIGH, causing the shift register to load 55H (= 01010101) on the first clock pulse. Since we have not instantiated the serial input **shiftin**, the serial input reverts to a default value of '1', causing the register to be filled with 1s. If we did not want this to be the case, we would have to instantiate the **shiftin** port and set it to '0'.

FIGURE 9.76

Example 9.18
Simulation of an 8-bit LPM Shift
Register with Parallel Load

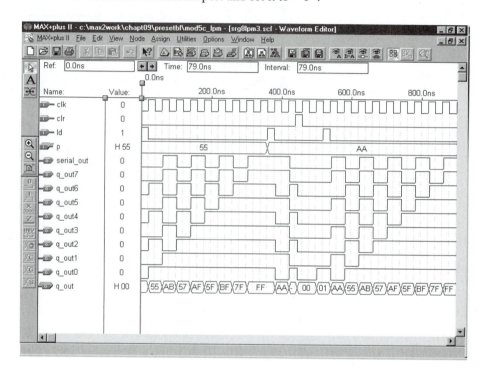

▌▌▌ SECTION 9.8 REVIEW PROBLEM

9.8 When a shift register is encoded in VHDL, why are its outputs defined as BUFFER, not OUT?

9.9 Shift Register Counters

KEY TERMS

Ring counter   A serial shift register with feedback from the output of the last flip-flop to the input of the first.

Johnson counter   A serial shift register with complemented feedback from the output of the last flip-flop to the input of the first. Also called a twisted ring counter

By introducing feedback into a serial shift register, we can create a class of synchronous counters based on continuous circulation, or rotation, of data.

If we feed back the output of a serial shift register to its input without inversion, we create a circuit called a **ring counter**. If we introduce inversion into the feedback loop, we have a circuit called a **Johnson counter**. These circuits can be decoded more easily than binary counters of similar size and are particularly useful for event sequencing.

Ring Counters

Figure 9.77 shows a 4-bit ring counter made from D flip-flops. This circuit could also be constructed from SR or JK flip-flops, as can any serial shift register.

FIGURE 9.77

4-bit Ring Counter

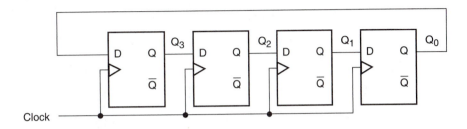

A ring counter circulates the same data in a continuous loop. This assumes that the data have somehow been placed into the circuit upon initialization, usually by synchronous or asynchronous preset and clear inputs, which are not shown.

Figure 9.78 shows the circulation of a logic 1 through a 4-bit ring counter. If we assume that the circuit is initialized to the state $Q_3Q_2Q_1Q_0 = 1000$, it is easy to see that the 1 is shifted one place right with each clock pulse. The feedback connection from Q_0 to D_3 ensures that the input of flip-flop 3 will be filled by the contents of Q_0, thus recirculating the initial data. The final transition in the sequence shows the 1 recirculated to Q_3.

A ring counter is not restricted to circulating a logic 1. We can program the counter to circulate any data pattern we happen to find convenient.

Figure 9.79 shows a ring counter circulating a 0 by starting with an initial state of $Q_3Q_2Q_1Q_0 = 0111$. The circuit is the same as before; only the initial state has changed. Figure 9.80 shows the timing diagrams for the circuit in Figures 9.78 and 9.79.

Ring Counter Modulus and Decoding

The maximum modulus of a ring counter is the maximum number of unique states in its count sequence. In Figures 9.78 and 9.79, the ring counters each had a maximum modulus of 4. We say that 4 is the *maximum* modulus of the ring counters shown, since we can change the modulus of a ring counter by loading different data at initialization.

For example, if we load a 4-bit ring counter with the data $Q_3Q_2Q_1Q_0 = 1000$, the following unique states are possible: 1000, 0100, 0010, and 0001. If we load the same circuit with the data $Q_3Q_2Q_1Q_0 = 1010$, there are only two unique states: 1010 and 0101. Depending on which data are loaded, the modulus is 4 or 2.

Most input data in this circuit will yield a modulus of 4. Try a few combinations.

NOTE

The maximum modulus of a ring counter is the same as the number of bits in its output.

A ring counter requires more flip-flops than a binary counter to produce the same number of unique states. Specifically, for n flip-flops, a binary counter has 2^n unique states and a ring counter has n.

This is offset by the fact that a ring counter requires no decoding. A binary counter used to sequence eight events requires three flip-flops and eight 3-input decoding gates. To perform the same task, a ring counter requires eight flip-flops and no decoding gates.

As the number of output states of an event sequencer increases, the complexity of the decoder for the binary counter also increases. A circuit requiring 16 output states can be implemented with a 4-bit binary counter and sixteen 4-input decoding gates. If you need 18 output states, you must have a 5-bit counter ($2^4 \leq 18 \leq 2^5$) and eighteen 5-input decoding gates.

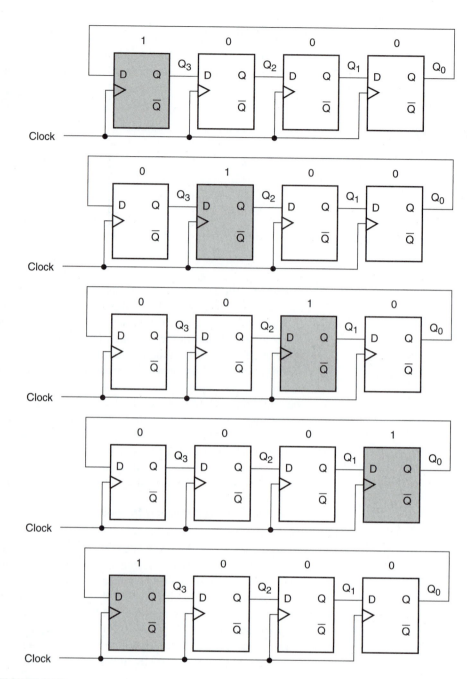

FIGURE 9.78
Circulating a 1 in a Ring Counter

The only required modification to the ring counter is one more flip-flop for each additional state. A 16-state ring counter needs 16 flip-flops and an 18-state ring counter must have 18 flip-flops. No decoding is required for either circuit.

Johnson Counters

Figure 9.81 shows a 4-bit Johnson counter constructed from D flip-flops. It is the same as a ring counter except for the inversion in the feedback loop where $\overline{Q_0}$ is connected to D_3. The circuit output is taken from flip-flop outputs Q_3 through Q_0. Since the feedback intro-

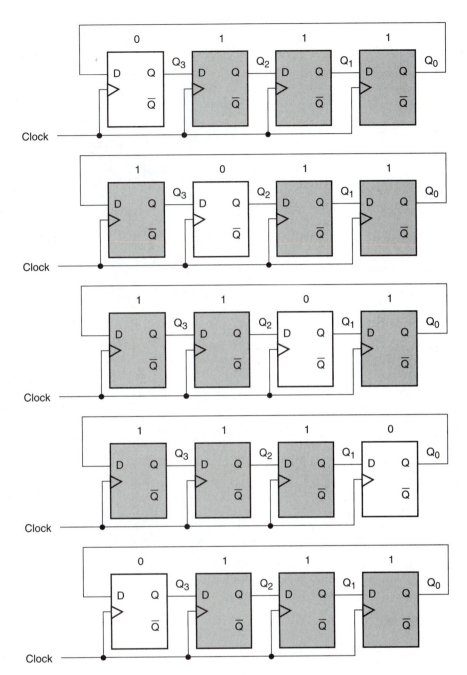

FIGURE 9.79
Circulating a 0 in a Ring Counter

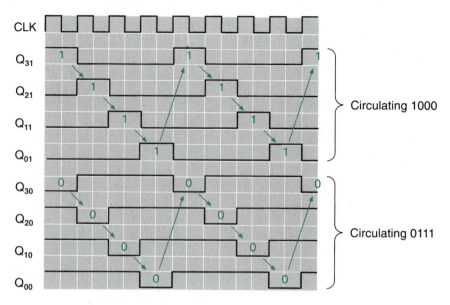

FIGURE 9.80
Timing Diagrams for Figures 9.78 and 9.79

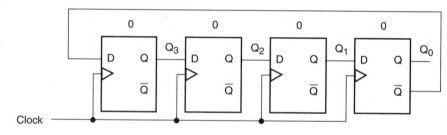

FIGURE 9.81
4-bit Johnson Counter

Table 9.17   Count Sequence of
a 4-bit Johnson Counter

Q_3	Q_2	Q_1	Q_0
0	0	0	0
1	0	0	0
1	1	0	0
1	1	1	0
1	1	1	1
0	1	1	1
0	0	1	1
0	0	0	1

duces a "twist" into the recirculating data, a Johnson counter is also called a "twisted ring counter."

Figure 9.82 shows the progress of data through a Johnson counter that starts cleared ($Q_3Q_2Q_1Q_0 = 0000$). The shaded flip-flops represents 1s and the unshaded flip-flops are 0s.

Every 0 at Q_0 is fed back to D_3 as a 1 and every 1 is fed back as a 0. The count sequence for this circuit is given in Table 9.17. There are 8 unique states in the count sequence table.

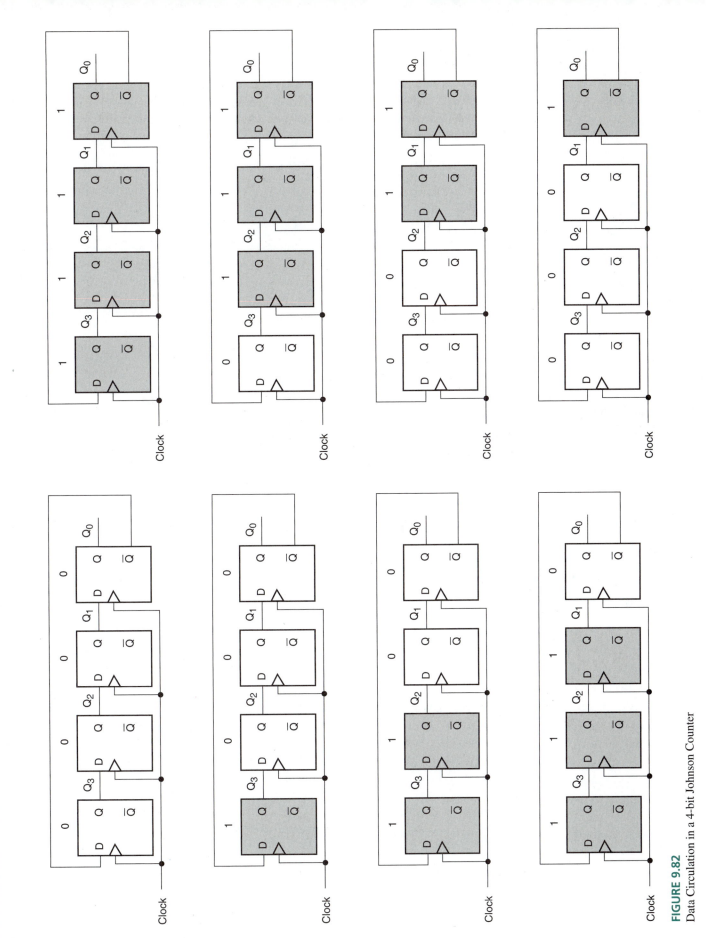

FIGURE 9.82
Data Circulation in a 4-bit Johnson Counter

443

▌▌ **EXAMPLE 9.19**

Write the VHDL code for a Johnson counter of generic width and instantiate it as an 8-bit counter. List the sequence of states in a table, assuming the counter is initially cleared, and create a simulation to verify the circuit's operation. Include a clear input (synchronous).

Solution   The VHDL design entities for the generic-width component and the 8-bit Johnson counter follow.

jnsn_ct.vhd
jnsn_ct8.vhd
jnsn_ct8.scf

```
-- jnsn_ct.vhd
-- Johnson counter of generic width
LIBRARY ieee;
USE ieee.std_logic_1164.ALL;

ENTITY jnsn_ct IS
  GENERIC (width : POSITIVE := 4);
  PORT (
       clk, clr : IN        STD_LOGIC;
       q          : BUFFER   STD_LOGIC_VECTOR(width-1 downto 0) );
END   jnsn_ct;

ARCHITECTURE johnson_counter of jnsn_ct IS
BEGIN
  PROCESS (clk)
  BEGIN
       IF (clk'EVENT and clk = '1') THEN
           IF clr = '0' THEN
                 q <= (others => '0'); -- n-bit clear function (n = width)
           ELSE
                 q(width-1 downto 0) <= (not q(0) ) & q(width-1 downto 1);
           END IF;
         END IF;
  END PROCESS;
END johnson_counter;

-- jnsn_ct8.vhd
-- 8-bit Johnson counter using component jnsn_ct
LIBRARY ieee;
USE ieee.std_logic_1164.ALL;

ENTITY jnsn_ct8 IS
 PORT(
       clock, clear : IN        STD_LOGIC;
       qo             : BUFFER  STD_LOGIC_VECTOR(7 downto 0));
END jnsn_ct8;

ARCHITECTURE johnson_counter of jnsn_ct8 IS
COMPONENT jnsn_ct GENERIC (width : POSITIVE);
 PORT(
       clk, clr : IN        STD_LOGIC;
       q          : BUFFER   STD_LOGIC_VECTOR(7 downto 0) );
END   COMPONENT;
BEGIN
  johnson: jnsn_ct
     GENERIC MAP (width=> 8)
     PORT MAP (clk => clock
              clr  => clear,
                q        => qo);
END johnson_counter;
```

Table 9.18   Count Sequence of an 8-bit Johnson Counter

$Q_7Q_6Q_5Q_4Q_3Q_2Q_1Q_0$
00000000
10000000
11000000
11100000
11110000
11111000
11111100
11111110
11111111
01111111
00111111
00011111
00001111
00000111
00000011
00000001

Note that in the component file (**jnsn_ct.vhd**), the counter is cleared synchronously by the statement (q <= (others => '0');). Recall that the clause (others => '0') can be used to set all bits of a signal aggregate to the value '0'. This is a simple way to clear a vector of unknown width without using a conversion function.

Table 9.18 shows the count sequence for the 8-bit Johnson counter.

The simulation of the Johnson counter, including one full cycle and a clear, is shown in Figure 9.83.

FIGURE 9.83
Example 9.19
Simulation of an 8-bit Johnson
Counter

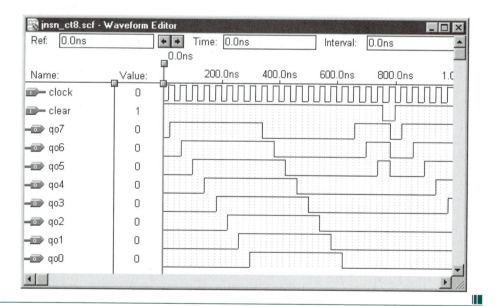

Johnson Counter Modulus and Decoding

> **NOTE**
>
> The maximum modulus of a Johnson counter is $2n$ for a circuit with n flip-flops.

The Johnson counter represents a compromise between binary and ring counters, whose maximum moduli are, respectively, 2^n and n for an n-bit counter.

If it is used for event sequencing, a Johnson counter must be decoded, unlike a ring counter. Its output states are such that each state can be decoded uniquely by a 2-input AND or NAND gate, depending on whether you need active-HIGH or active-LOW indication. This yields a simpler decoder than is required for a binary counter.

Table 9.19 shows the decoding of a 4-bit Johnson counter.

Table 9.19 Decoding a 4-bit Johnson Counter

Q_3	Q_2	Q_1	Q_0	Decoder Outputs	Comment
0	0	0	0	$\overline{Q_3}\,\overline{Q_0}$	MSB = LSB = 0
1	0	0	0	$Q_3\overline{Q_2}$	"1/0"
1	1	0	0	$Q_2\overline{Q_1}$	Pairs
1	1	1	0	$Q_1\overline{Q_0}$	
1	1	1	1	$Q_3 Q_0$	MSB = LSB = 1
0	1	1	1	$\overline{Q_3}Q_2$	"0/1"
0	0	1	1	$\overline{Q_2}Q_1$	Pairs
0	0	0	1	$\overline{Q_1}Q_0$	

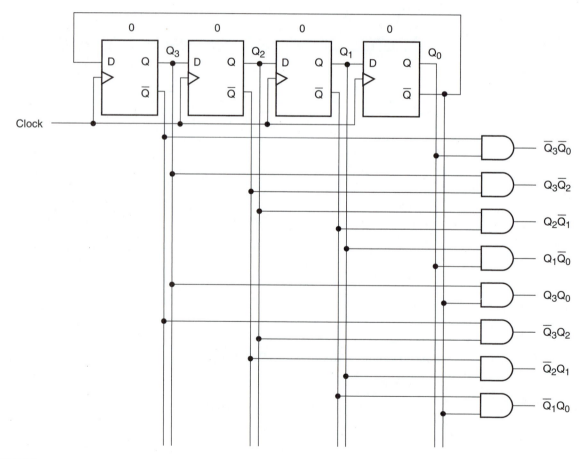

FIGURE 9.84
4-bit Johnson Counter with Output Decoding

Decoding a sequential circuit depends on the decoder responding uniquely to every possible state of the circuit outputs. If we want to use only 2-input gates in our decoder, it must recognize two variables for every state that are *both* active *only* in that state.

A Johnson counter decoder exploits what might be called the "1/0 interface" of the count sequence table. Careful examination of Tables 9.17 and 9.18 reveals that for every state, except where the outputs are all 1s or all 0s, there is a side-by-side 10 or 01 pair which exists only in that state.

Each of these pairs can be decoded to give unique indication of a particular state. For example, the pair $Q_3\overline{Q}_2$ uniquely indicates the second state since $Q_3 = 1$ AND $Q_2 = 0$ *only* in the second line of the count sequence table. (This is true for any size of Johnson counter; compare the second lines of Tables 9.17 and 9.18. In the second line of both tables, the MSB is 1 and the 2nd MSB is 0.)

For the states where the outputs are all 1s or all 0s, the most significant AND least significant bits can be decoded uniquely, these being the only states where MSB = LSB.

Figure 9.84 shows the decoder circuit for a 4-bit Johnson counter.

The output decoder of a Johnson counter does not increase in complexity as the modulus of the counter increases. The decoder will always consist of $2n$ 2-input AND or NAND gates for an n-bit counter. (For example, for an 8-bit Johnson counter, the decoder will consist of sixteen 2-input AND or NAND gates.)

▮▮ EXAMPLE 9.20

Draw the timing diagram of the Johnson counter decoder of Figure 9.84, assuming the counter is initially cleared.

Solution   Figure 9.85 shows the timing diagram of the Johnson counter and its decoder outputs.

FIGURE 9.85
Example 9.20
Johnson Counter Decoder
Outputs

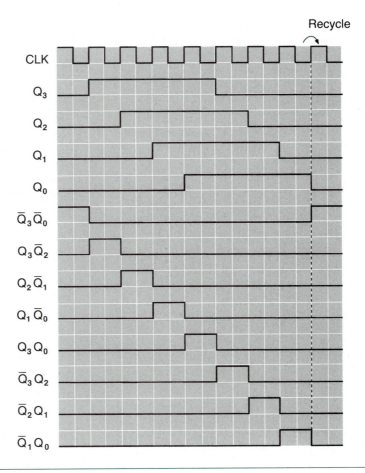

▓ SECTION 9.9 REVIEW PROBLEM

9.9 How many flip-flops are required to produce 24 unique states in each of the following types of counters: binary counter, ring counter, Johnson counter? How many and what type of decoding gates are required to produce an active-LOW decoder for each type of counter?

S U M M A R Y

1. A counter is a circuit that progresses in a defined sequence at the rate of one state per clock pulse.

2. The modulus of a counter is the number of states through which the counter output progresses before repeating.

3. A counter with an ascending sequence of states is called an UP counter. A counter with a descending sequence of states is called a DOWN counter.

4. In general, the maximum modulus of a counter is given by 2^n for an n-bit counter.

5. A counter whose modulus is 2^n is called a full-sequence counter. The count progresses from 0 to $2^n - 1$, which corresponds to a binary output of all 0s to all 1s.

6. A counter whose output is less than 2^n is called a truncated sequence counter.

7. The adjacent outputs of a full-sequence binary counter have a frequency ratio of $2:1$. The less significant of the two bits has the higher frequency.

8. The outputs of a truncated sequence counter do not necessarily have a simple frequency relationship.

9. A synchronous counter consists of a series of flip-flops, all clocked from the same source, that stores the present state of the counter and a combinational circuit that monitors the counter's present state and determines its next state.

10. A synchronous counter can be analyzed by a formal procedure that includes the following steps:

   a. Write the Boolean equations for the synchronous inputs of the counter flip-flops in terms of the present state of the flip-flip outputs.

   b. Evaluate each Boolean equation for an initial state to find the states of the synchronous inputs.

   c. Use flip-flop function tables to determine each flip-flop next state.

   d. Set the next state to the new present state.

   e. Continue until the sequence repeats.

11. The analysis procedure above should be applied to any unused states of the counter to ensure that they will enter the count sequence properly.

12. A synchronous counter can be designed using a formal method that relies on the excitation tables of the flip-flops used in the counter. An excitation table indicates the required logic levels on the flip-flop inputs to effect a particular transition.

13. The synchronous counter design procedure is based on the following steps:

   a. Draw the state diagram of the counter and use it to list the relationship between the counter's present and next states. The table should list the counter's present states in binary order.

   b. For the initial design, unused states can be set to a known destination, such as 0, or treated as don't care states.

   c. Use the flip-flop excitation table to determine the synchronous input levels for each present-to-next state transition.

   d. Use Boolean algebra or Karnaugh maps to find the simplest equations for the flip-flop inputs (*JK, D,* or *T*) in terms of Q.

   e. Unused states should be analyzed by substituting their values into the Boolean equations of the counter. This will verify whether or not an unused state will enter the count sequence properly.

14. If a counter must reset to 0 from an unused state, the flip-flops can be reset asynchronously to their initial states or the counter can be designed with the unused states always having 0 as their next state.

15. A counter can be designed in VHDL by using a behavioral description or a structural design that uses a component from the Library of Parameterized Modules (LPM).

16. A behavioral counter design requires a PROCESS statement that lists the clock signal and any asynchronous inputs in its sensitivity list. An IF statement inside the PROCESS can monitor the active clock edge by using the predefined EVENT attribute (e.g., clk'EVENT) and increment a count variable.

17. A variable is local to a PROCESS and is assigned with the := operator. A signal is global to the VHDL design entity and is assigned with the <= operator. (Recall that a signal is like an internal connecting wire and a variable is a piece of working memory.)

18. A structural counter design can use an LPM component (lpm_counter) and instantiate the component in a component instantiation statement. The statement's generic map specifies the component parameters, and its port map indicates the correspondence between the component port names and the user port, signal, or variable names.

19. Some of the most common control features available in synchronous counters include:

   a. Synchronous or asynchronous parallel load, which allows the count to be set to any value whenever a LOAD input is asserted

   b. Synchronous or asynchronous clear (reset), which sets all of the counter outputs to zero

   c. Count enable, which allows the count sequence to progress when asserted and inhibits the count when deasserted

   d. Bidirectional control, which determines whether the counter counts up or down

e. Output decoding, which activates one or more outputs when detecting particular states on the counter outputs

f. Ripple carry out or ripple clock out (RCO), a special case of output decoding that produces a pulse upon detecting the terminal count, or last state, of a count sequence

20. The parallel load function of a counter requires load *data* (the parallel input values) and a load *command* input, such as *LOAD,* that transfer the parallel data when asserted. If the load function is synchronous, a clock pulse is also required.

21. Synchronous load transfers data to the counter outputs on an active clock edge. Asynchronous load operates as soon as the load input activates, without waiting for the clock.

22. Synchronous load is implemented by a function select circuit that selects either the count logic or the direct parallel input to be applied to the synchronous input(s) of a flip-flop.

23. Asynchronous load is implemented by enabling or inhibiting a pair of NAND gates, one of which asserts a flip-flop clear input and the other of which asserts a preset input for the same flip-flop.

24. The count enable function enables or disables the count logic of a counter without affecting other functions, such as clock or clear. This can be done by ANDing the count logic with the count enable input signal.

25. A flip-flop in an UP counter toggles when all previous bits are HIGH. A flip-flop in a DOWN counter toggles when all previous bits are LOW. A circuit that selects one of these two conditions (a pair of AND-shaped gates, combined in an OR gate; essentially a 2-to-1 multiplexer) can implement a bidirectional count.

26. An output decoder asserts one output for each counter state. A special case is a terminal count decoder that detects the last state of a count sequence.

27. RCO (ripple clock out) generates one clock pulse upon terminal count, with its positive edge at the end of the count cycle.

28. Asynchronous inputs to a behaviorally defined counter in VHDL must be included in the sensitivity list of the process defining the counter. Asynchronous inputs must be checked inside the process before the clock is checked for an active edge.

29. Synchronous inputs to a behaviorally defined counter should not be included in the sensitivity list of the process defining the counter. Synchronous inputs must be checked inside the IF statement that checks the clock edge.

30. A shift register is a circuit for storing and moving data. Three basic movements in a shift register are: serial (from one flip-flop to another), parallel (into all flip-flops at once), and rotation (serial shift with a connection from the last flip-flop output to the first flip-flop input).

31. Serial shifting can be left (toward the MSB) or right (away from the MSB). This is the convention used by MAX+PLUS II. Some data sheets indicate the opposite relationship between right/left and LSB/MSB.

32. A function select circuit can implement several shift register variations: bidirectional serial shift, parallel load with serial shift, and universal shift (parallel/serial in/out and bidirectional in one device). The circuit directs data to the *D* inputs of each flip-flop from one of several sources, such as from the flip-flop immediately to the left or right or from an external parallel input.

33. A shift register can be created in VHDL by the structural, dataflow, or behavioral method.

34. A structural design instantiates components, such as D flip-flops, and connects them with internal signals.

35. A dataflow design uses internal Boolean relationships between inputs and outputs. It is similar to a structural model, except that it must contain a process to create the flip-flops.

36. A behavioral design method uses a description of the shift register function to generate the required hardware.

37. A VHDL component can be created with parameters (such as width) that are specified when the component is instantiated. The parameters are listed in a GENERIC clause in the component's entity declaration. Each parameter must be given a default value. The parameters are specified in a generic map in the design entity that instantiates the component.

38. A ring counter is a serial shift register with the serial output fed back to the serial input so that the internal data is continuously circulated. The initial value is generally set by asynchronous preset and clear functions.

39. The maximum modulus of a ring counter is n for a circuit with n flip-flops, as compared to 2^n for a binary counter. A ring counter output is self-decoding, whereas a binary counter requires $m \le 2^n$ AND or NAND gates with n inputs each.

40. A Johnson counter is a ring counter where the feedback is complemented. A Johnson counter has $2n$ states for an n-bit counter which can be uniquely decoded by $2n$ 2-input AND or NAND gates.

GLOSSARY

Attribute   A property associated with a named identifier in VHDL. (e.g., the attribute EVENT, when associated with the identifier clk (written clk'EVENT), indicates whether a transition has occurred on the input called clk.)

Behavioral design   A VHDL design technique that uses descriptions of required behavior to describe the design.

Bidirectional counter   A counter that can count up or down, depending on the state of a control input.

Bidirectional shift register   A shift register that can serially shift bits left or right according to the state of a direction control input.

Binary counter   A counter that generates a binary count sequence.

Clear   Reset (synchronous or asynchronous)

Command lines   Signals that connect the control section of a synchronous circuit to its memory section and direct the circuit from its present to its next state.

Conditional signal assignment statement   A signal assignment statement that is executed only when a Boolean condition is satisfied.

Control section   The combinational logic portion of a synchronous circuit that determines the next state of the circuit.

Count enable   A control function that allows a counter to progress through its count sequence when active and disables the counter when inactive.

Count sequence   The specific series of output states through which a counter progresses.

Counter   A sequential digital circuit whose output progresses in a predictable repeating pattern, advancing by one state for each clock pulse.

Count-sequence table   A list of counter states in the order of the count sequence.

Dataflow design   A VHDL design technique that uses Boolean equations to define relationships between inputs and outputs.

DOWN counter   A counter with a descending sequence.

Excitation table   A table showing the required input conditions for every possible transition of a flip-flop output.

Full-sequence counter   A counter whose modulus is the same as its maximum modulus ($m = 2^n$ for an n-bit counter).

GENERIC   A clause in the entity declaration of a VHDL component that lists the parameters that can be specified when the component is instantiated.

Johnson counter   A serial shift register with complemented feedback from the output of the last flip-flop to the input of the first. Also called a twisted ring counter

Left shift   A movement of data from the right to the left in a shift register. (Left is defined in MAX+PLUS II as toward the MSB.)

Maximum modulus (m_{max})   The largest number of counter states that can be represented by n bits ($m_{max} = 2^n$).

Memory section   A set of flip-flops in a synchronous circuit that hold its present state.

Modulo-n (or mod-n) counter   A counter with a modulus of n.

Modulus   The number of states through which a counter sequences before repeating.

Next state   The desired future state of flip-flop outputs in a synchronous sequential circuit after the next clock pulse is applied.

Parallel load   A function that allows simultaneous loading of binary values into all flip-flops of a synchronous circuit. Parallel loading can be synchronous or asynchronous.

Parallel-load shift register   A shift register that can be preset to any value by directly loading a binary number into its internal flip-flops.

Parallel transfer   Movement of data into all flip-flops of a shift register at the same time.

Present state   The current state of flip-flop outputs in a synchronous sequential circuit.

Presettable counter   A counter with a parallel load function.

Recycle   To make a transition from the last state of the count sequence to the first state.

Right shift   A movement of data from the left to the right in a shift register. (Right is defined in MAX+PLUS II as toward the LSB.)

Ring counter   A serial shift register with feedback from the output of the last flip-flop to the input of the first.

Ripple carry out or ripple clock out (RCO)   An output that produces one pulse with the same period as the clock upon terminal count.

Rotation   Serial shifting of data with the output(s) of the last flip-flop connected to the synchronous input(s) of the first flip-flop. The result is continuous circulation of the same data.

Serial shifting   Movement of data from one end of a shift register to the other at a rate of one bit per clock pulse.

Shift register   A synchronous sequential circuit that will store and move n-bit data, either serially or in parallel, in n flip-flops.

SRGn   Symbol for an n-bit shift register (e.g., SRG4 indicates a 4-bit shift register).

State diagram   A diagram showing the progression of states of a sequential circuit.

State machine   A synchronous sequential circuit.

Status lines   Signals that communicate the present state of a synchronous circuit from its memory section to its control section.

Structural design   A VHDL design technique that connects predesigned components using internal signals.

Synchronous counter   A counter whose flip-flops are all clocked by the same source and thus change in synchronization with each other.

Terminal count   The last state in a count sequence before the sequence repeats (e.g., 1111 is the terminal count of a 4-bit binary UP counter; 0000 is the terminal count of a 4-bit binary DOWN counter).

Truncated-sequence counter   A counter whose modulus is less than its maximum modulus ($m < 2^n$ for an n-bit counter)

Universal shift register   A shift register that can operate with any combination of serial and parallel inputs and outputs (i.e., serial in/serial out, serial in/parallel out, parallel in/serial out, parallel in/parallel out). A universal shift register is often bidirectional, as well.

UP counter   A counter with an ascending sequence.

PROBLEMS

Problem numbers set in color indicate more difficult problems; those with underlines indicate most difficult problems.

9.1   Basic Concepts of Digital Counters

9.1   A parking lot at a football stadium is monitored before a game to determine whether or not there is available space for more cars. When a car enters the lot, the driver takes a ticket from a dispenser which also produces a pulse for each ticket taken.

   The parking lot has space for 4095 cars. Draw a block diagram which shows how you can use a digital counter

to light a LOT FULL sign after 4095 cars have entered. (Assume no cars leave the lot until after the game, so you don't need to keep track of cars leaving the lot.) How many bits should the counter have?

9.2   Figure 9.86 shows a mod-16 which controls the operation of two digital sequential circuits, labeled Circuit 1 and Circuit 2. Circuit 1 is positive edge-triggered and clocked by counter output Q_1. Circuit 2 is negative edge-triggered and clocked by Q_3. (Q_3 is the MSB output of the counter.)

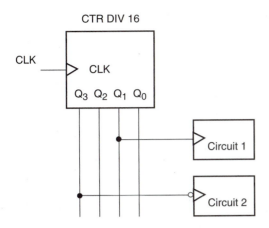

FIGURE 9.86

Problem 9.2

Mod-16 Counter Driving Two Sequential Circuits

    a. Draw the timing diagram for one complete cycle of the circuit operation. Draw arrows on the active edges of the waveforms that activate Circuit 1 and Circuit 2.

    b. State how many times Circuit 1 is clocked for each time that Circuit 2 is clocked.

9.3 Draw the timing diagram for one complete cycle of a mod-8 counter, including waveforms for *CLK*, Q_0, Q_1, and Q_2, where Q_0 is the LSB.

9.4 How many bits are required to make a counter with a modulus of 64? Why? What is the maximum count of such a counter?

9.5   a. Draw the state diagram of a mod-10 UP counter.

    b. Use the state diagram drawn in part **a** to answer the following questions:

       i. The counter is at state 0111. What is the count after 7 clock pulses are applied?

      ii. After 5 clock pulses, the counter output is at 0001. What was the counter state prior to the clock pulses?

    iii. The counter output is at 1000 after 15 clock pulses. What was the original output state?

9.6 What is the maximum modulus of a 6-bit counter? A 7-bit? 8-bit?

9.7 Draw the count sequence table and timing diagram of a mod-10 UP counter.

9.8 Draw the state diagram, count sequence table, and timing diagram of a mod-10 DOWN counter.

9.9 A mod-16 counter is clocked by a waveform having a frequency of 48 kHz. What is the frequency of each of the waveforms at Q_0, Q_1, Q_2, and Q_3?

9.10 A mod-10 counter is clocked by a waveform having a frequency of 48 kHz. What is the frequency of the Q_3 output waveform? The Q_0 waveform? Why is it difficult to determine the frequencies of Q_1 and Q_2?

9.2 Synchronous Counters

9.11 Draw the circuit for a synchronous mod-16 UP counter made from negative edge-triggered JK flip-flops.

9.12 Write the Boolean equations required to extend the counter drawn in Problem 9.11 to a mod-64 counter.

9.13 Write the *J* and *K* equations for the MSB of a synchronous mod-256 (8-bit) UP counter.

9.14 Analyze the operation of the synchronous counter in Figure 9.87 by drawing a state table showing all transitions, including unused states. Use this state table to draw a state diagram and a timing diagram. What is the counter's modulus?

9.15   a. Write the equations for the *J* and *K* inputs of each flip-flop of the synchronous counter represented in Figure 9.88.

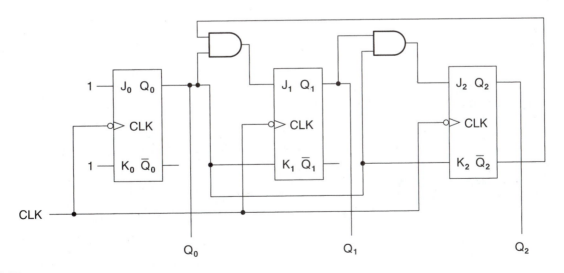

FIGURE 9.87

Problem 9.14

Synchronous Counter

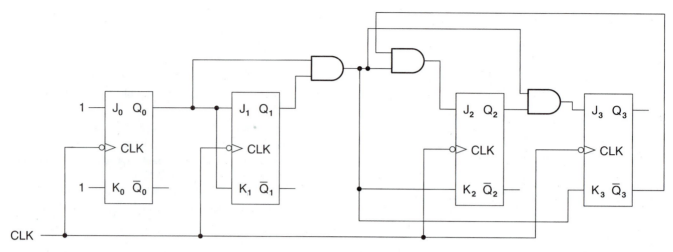

FIGURE 9.88
Problem 9.15
Synchronous Counter

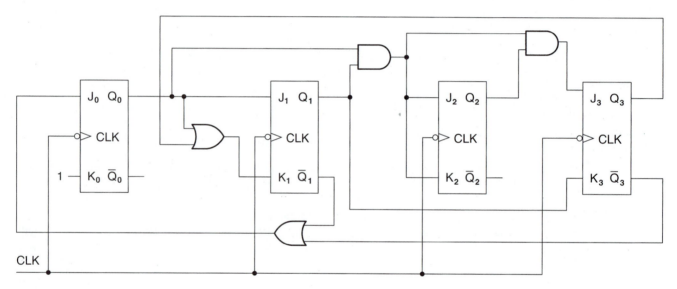

FIGURE 9.89
Problem 9.16
Counter

b. Assume that $Q_3Q_2Q_1Q_0 = 1010$ at some point in the count sequence. Use the equations from part **a** to predict the circuit outputs after each of three clock pulses.

9.16 Analyze the operation of the counter shown in Figure 9.89. Predict the count sequence by determining the J and K inputs and resulting transitions for each counter output state. Draw the state diagram and the timing diagram. Assume that all flip-flop outputs are initially 0.

9.3 Design of Synchronous Counters

9.17 Draw the timing diagram and state diagram of a synchronous mod-10 counter with a positive edge-triggered clock.

9.18 Design a synchronous mod-10 counter, using positive edge-triggered JK flip-flops. Check that unused states

properly enter the main sequence. Draw a state diagram showing the unused states.

9.19 Design a synchronous mod-10 counter, using positive edge-triggered D flip-flops. Check that unused states properly enter the main sequence. Draw a state diagram showing the unused states.

9.20 Design a synchronous 3-bit binary counter using T flip-flops.

9.21 Table 9.20 shows the count sequence for a **biquinary sequence** counter. The sequence has ten states, but does not progress in binary order. The advantage of the sequence is that its most significant bit has a divide-by-10 ratio, relative to a clock input, and a 50% duty cycle. Design the synchronous counter circuit for this sequence, using D flip-flops. *Hint:* When making the state table, list all *pre-*

Table 9.20 Biquinary
Sequence

$Q_3Q_2Q_1Q_0$
0000
0001
0010
0011
0100
1000
1001
1010
1011
1100

sent states in binary order. The next states *will not* be in binary order.

9.4 Programming Binary Counters in VHDL

9.22 Write the VHDL code for a behavioral description of a 6-bit binary counter with asynchronous clear.

9.23 Create a simulation file in MAX+PLUS II to verify the operation of the counter in Problem 9.22. (Use a 40 ns clock, which approximates the clock period of the oscillator on the Altera UP-1 board.) *Note: To make a useful simulation, you must include the recycle point, which may be beyond the default end time of the simulation (1 μs). To change the end time, select* **End Time** *from the MAX+PLUS II* **File** *menu in the Simulator menu. To change the clock period, select* **Grid Size** *from the MAX+PLUS II* **Options** *menu in the Simulator window. The default clock period is two grid spaces.*

9.24 Write a VHDL file that instantiates a counter from the Library of Parameterized Modules to make a 12-bit binary counter. Create a MAX+PLUS II simulation to verify the operation of the counter. (Refer to the note after Problem 9.23.)

9.5 Control Options for Synchronous Counters

9.25 Briefly explain the difference between asynchronous and synchronous parallel load in a synchronous counter. Draw a partial timing diagram that illustrates both functions for a 4-bit counter.

9.26 Refer to the 4-bit counter of Figure 9.26 (p. 391). The graphic design files for the counter are found on the CD accompanying this text as **4bit_sl.gdf** and **sl_count.gdf** in the folder *drive:\Student_Files\Chapter09*. Copy these files to a new folder and use the MAX+PLUS II graphic editor to expand the counter of Figure 9.26 to a 5-bit counter with synchronous load and asynchronous reset. Save and compile the file to make sure that there are no design errors.

9.27 Create a MAX+PLUS II simulation to verify the functions of the counter in Problem 9.26. The simulation must include the recycle point of the counter and show that the load is really synchronous and that the reset is really asynchronous.

9.28 Refer to the 4-bit counter of Figure 9.33 (p. 396). The graphic design files for the counter are found on the accompanying CD as **4bit_sle.gdf** and **sl_count.gdf** in the folder *drive:\Student Files\Chapter09*. Copy these files to a new folder and modify the synchronous count element **sl_count.gdf** so that it implements an active-HIGH synchronous load and an active-LOW synchronous clear function, as well as the binary count function. Create a default symbol for the new element and substitute it in **4bit_sle.gdf** for the existing counter elements **sl_count.** The load function should have priority over count enable, and clear (reset) should have priority over both. Save and compile the new file. *Hints:* (1) The clear function makes $Q = 0$ after a clock pulse. (2) Q follows D.

9.29 Create a MAX+PLUS II simulation to verify the functions of the counter in Problem 9.28. The simulation must include the recycle point of the counter and show that the load and clear really are synchronous and that load has priority over count enable and clear has priority over both.

9.30 Derive the Boolean equations for the synchronous DOWN-counter in Figure 9.35.

9.31 Write the Boolean equations for the count logic of the 4-bit bidirectional counter in Figure 9.38. Briefly explain how the logic works.

9.32 Draw a MAX+PLUS II Graphic Design File for a bidirectional counter, using T flip-flops. Create a simulation of the counter to verify its function

9.33 Use MAX+PLUS II to create a synchronous bidirectional counter with synchronous load, asynchronous reset, and count enable. The count enable should not affect the operation of the load and reset functions. The functions should have the following priority: (1) clear; (2) load; and (3) count. Create a MAX+PLUS II simulation to verify the operation of your design.

9.6 Programming Presettable and Bidirectional Counters in VHDL

9.34 Write the VHDL code for a counter that uses a behavioral description of the following functions: 12-bit binary UP count; active-LOW asynchronous clear, active-LOW synchronous load, active-LOW count enable, terminal count decoder. The clear function should have the highest priority, followed by load, then count enable. Create a simulation in MAX+PLUS II that verifies the functions of this counter.

9.35 Write the VHDL code for a behavioral description of a bidirectional counter with a modulus of 24. The counter should also have an active-LOW synchronous clear function that has priority over the count. Create a MAX+PLUS II simulation file to verify the counter operation.

9.36 Write the VHDL code for a 4-bit counter with two decoding outputs called **eq8** and **eq12**. Out **eq8** goes HIGH when the count equals 8 and **eq12** goes HIGH when the count equals 12 (decimal). The counter should also have an active-LOW asynchronous clear function that has priority over the count. Create a MAX+PLUS II simulation file to verify the counter operation.

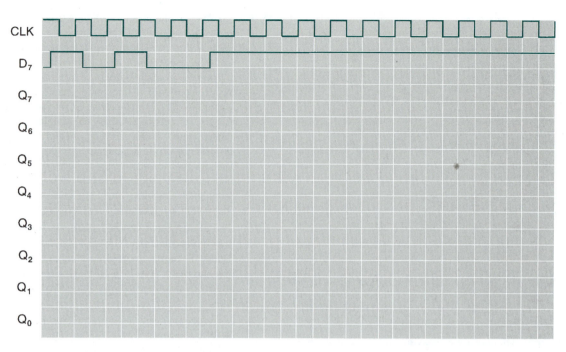

FIGURE 9.90
Problem 9.44
Timing Diagram

9.37 Modify the VHDL code in Example 9.10 (p. 412) so that the counter synchronously sets to all 1s (= 4095), rather than to 2047. Do not use SVALUE = 4095. Create a simulation in MAX+PLUS II that verifies the operation of the counter. State the main difference between the code for Example 9.10 and the solution to this problem.

9.38 Use a counter from the Library of Parameterized Modules to implement the counter described in Problem 9.35. Create a MAX+PLUS II simulation file to verify the operation of the counter.

9.39 Write a VHDL file that instantiates an 8-bit LPM count with synchronous load and clear, count enable, and directional control. Also include a terminal count decoder. (The LPM counter has no port for the terminal count function, so it must be done separately.) Create a MAX+PLUS II simulation to verify the operation of the counter.

9.7 Shift Registers

9.40 Use the MAX+PLUS II Graphic Editor to draw the circuit of a serial shift register constructed from JK flip-flops. Create a simulation to verify the operation of the shift register.

9.41 Use the MAX+PLUS II Graphic Editor to create the logic diagram of the 4-bit serial shift register based on JK

flip-flops that shifts left, rather than right. Create a simulation to verify the operation of the shift register.

9.42 The following bits are applied in sequence to the input of a 6-bit serial right-shift register: 0111111 (0 is applied first). Draw the timing diagram.

9.43 After the data in Problem 9.42 are applied to the 6-bit shift register, the serial input goes to 0 for the next 8 clock pulses and then returns to 1. Write the internal states, Q_5 through Q_0, of the shift register flip-flops after the first 2 clock pulses. Write the states after 6, 8, and 10 clock pulses.

9.44 Complete the timing diagram of Figure 9.90, which is for a serial shift register (right-shift). Assume the shift register is initially cleared. What happens to the state of the circuit if D_7 stays HIGH beyond the end of the diagram and the *CLK* input continues to pulse?

9.45 An 8-bit right-shift serial-in-serial-out shift register is initially cleared and has the following data clocked into its serial input: 1011001110. Draw a timing diagram of the circuit showing the *CLK, Serial Input,* and *Serial Output*. (Assume the individual flip-flop outputs are not accessible.)

9.46 Complete the logic circuit shown in Figure 9.91 to make a bidirectional shift register.

FIGURE 9.91
Problem 9.46
Logic Circuit

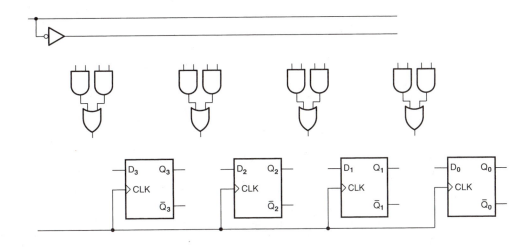

9.47 Complete the logic circuit shown in Figure 9.92 to make a parallel-in-serial-out shift register.

9.8 Programming Shift Registers in VHDL

9.48 Write the VHDL code for an 8-bit serial shift register using a structural design procedure. Use JK flip-flops. (MAX+PLUS II primitive: JKFF.) Create a MAX+PLUS II simulation file to verify the operation of your design.

9.49 Repeat Problem 9.48 using a dataflow design procedure.

9.50 Modify the VHDL code for the behaviorally designed shift register **srg4behv.vhd** so that the shift register moves the data left, not right. *Hint:* The statement

`q (3 downto 0) <= serial_in & q(3 downto 1);` is equivalent to the following two statements:

```
q(3) <= serial_in;
q(2 downto 0) <= q(3 downto 1);
```

Create a simulation file to verify the operation of this device.

9.51 Modify the VHDL code for the left-shift register Problem 9.50 to make a shift register of generic width. Use this

component in another VHDL file to make a 32-bit shift register that shifts left. Create a simulation file to verify the operation of this design.

9.52 Write the code for a VHDL design entity that implements a 4-bit universal shift register with asynchronous clear. Create a simulation that verifies the design function.

9.53 Use MAX+PLUS II to create simulations for the generic-width and the 16-bit universal shift registers in Example 9.16 (p. 432). What is the difference in width between the default value of the generic shift register and the instantiated component in the 16-bit file? Given this difference, why can the generic-width shift register be correctly used as a component in the 16-bit design entity?

9.54 Use an LPM shift register in a VHDL file to instantiate a 48-bit shift register with the following functions: serial input, parallel output, synchronous clear.

9.55 Use an LPM shift register in a VHDL file to instantiate a 10-bit shift register with the following functions: serial input and output whose internal value can be synchronously set to 960. Create a MAX+PLUS II simulation to verify the operation of the design.

FIGURE 9.92
Problem 9.47
Logic Circuit

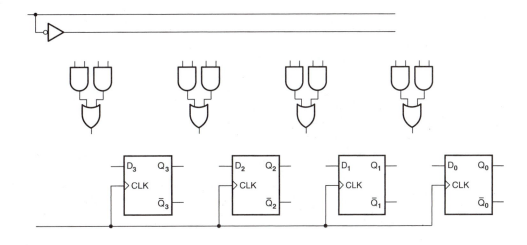

9.9   Shift Register Counters

9.56   Write the VHDL code for a ring counter of generic width and instantiate it as an 8-bit ring counter. List the sequence of states in a table, assuming the counter is initially cleared, and create a simulation to verify the circuit's operation. Include a clear input (synchronous).

9.57   Construct the count sequence table of a 5-bit Johnson counter, assuming the counter is initially cleared. What changes must be made to the decoder part of the circuit in Figure 9.84 (p. 446) if it is to decode the 5-bit Johnson counter?

9.58   A control sequence has ten steps, each activated by a logic HIGH. Use MAX+PLUS II to design a counter and decoder in each of the following configurations to produce the required sequence: binary counter, ring counter, and Johnson counter. You may use a Graphic Design File or VHDL. Create a simulation for each counter and decoder.

9.59   Use the MAX+PLUS II Graphic Editor to design a 4-bit ring counter that can be asynchronously initialized to $Q_3Q_2Q_1Q_0 = 1000$ by using only the clear inputs of its flip-flops. No presets allowed. *Hint:* use a circuit with a "double twist" in the data path.

ANSWERS TO SECTION REVIEW PROBLEMS

Section 9.1

9.1 A mod-24 UP counter goes from 00000 to 10111 (0 to 23). This requires 5 outputs. The counter is a truncated sequence since its modulus is less than $2^5 = 32$.

Section 9.2

9.2 1001, 0000

Section 9.3

9.3 JK flip-flops: $J_3K_3 = X0, J_2K_2 = 1X, J_1K_1 = X1, J_0K_0 = X1$
     D flip-flops: $D_3 = 1, D_2 = 1, D_1 = 0, D_0 = 0$

Section 9.4

9.4 `If (clock'EVENT AND clock = '0') THEN`
     `        count := count + 1;`
     `   END IF;`

Section 9.5

9.5 The completed timing diagram is shown in Figure 9.93.

Section 9.6

9.6 Asynchronous clear: PROCESS (clock, clear); Synchronous clear: PROCESS (clock)

Section 9.7

9.7 JK flip-flops can be used in the shift register of Figure 9.58. The Q output of any stage connects to the J input of the next stage and the $\overline{Q}$ output of any stage connects to the K input of the next. The **serial_in** input connects directly to the J input of the first flip-flop. **Serial_in** is applied to K of the first flip-flop through an inverter (NOT gate).

Section 9.8

9.8 A shift register output is defined as a port of mode BUFFER because this mode allows a signal to be fed back into the PLD matrix and reused as an input to another part of the circuit.

Section 9.9

Binary: 5 flip-flops, 24 5-inputs NANDs; Ring: 24 flip-flops, no decoding required;

Johnson: 12 flip-flops, 24 2-input NANDs.

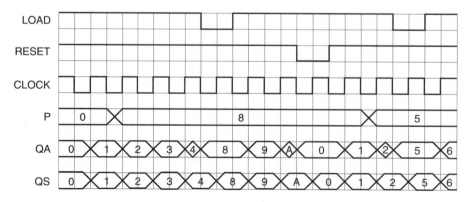

FIGURE 9.93
Answer to Section Review Problem 9.5

State Machine Design

C H A P T E R   O B J E C T I V E S

Upon successful completion of this chapter you will be able to:

- Describe the components of a state machine.
- Distinguish between Moore and Mealy implementations of state machines.
- Draw the state diagram of a state machine from a verbal description.
- Use the "classical" (state table) method of state machine design to determine the Boolean equations of the state machine.
- Translate the Boolean equations of a state machine into a Graphic Design File in Altera's MAX+PLUS II software.
- Write VHDL code to implement state machines.
- Create simulations in MAX+PLUS II to verify the function of a state machine design.
- Determine whether the output of a state machine is vulnerable to asynchronous changes of input.
- Design state machine applications, such as a switch debouncer, a single-pulse generator, and a traffic light controller.

10.1 State Machines

KEY TERMS

State machine   A synchronous sequential circuit, consisting of a sequential logic section and a combinational logic section, whose outputs and internal flip-flops progress through a predictable sequence of states in response to a clock and other input signals.

Moore machine   A state machine whose output is determined only by the sequential logic of the machine.

Mealy machine   A state machine whose output is determined by both the sequential logic and the combinational logic of the machine.

State variables   The variables held in the flip-flops of a state machine that determine its present state. *The number of state variables in a machine is equivalent to the number of flip-flops.*

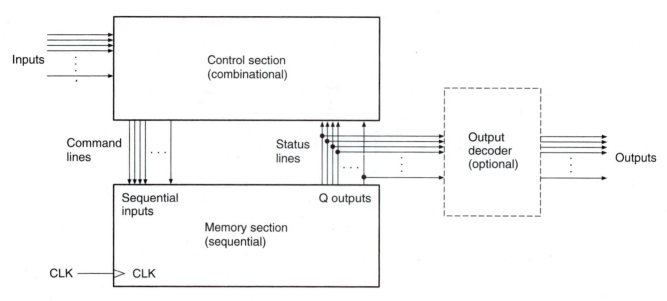

FIGURE 10.1
Moore-Type State Machine

The synchronous counters and shift registers we examined in Chapter 9 are examples of a larger class of circuits known as **state machines.** As described for synchronous counters in Section 9.2, a state machine consists of a memory section that holds the present state of the machine and a control section that determines the machine's next state. These sections communicate via a series of command and status lines. Depending on the type of machine, the outputs will either be functions of the present state only or of the present and next states.

Figure 10.1 shows the block diagram of a **Moore machine.** The outputs of a Moore machine are determined solely by the present state of the machine's memory section. The output may be directly connected to the Q outputs of the internal flip-flops, or the Q outputs might pass through a decoder circuit. The output of a Moore machine is synchronous to the system clock, since the output can only change when the machine's internal **state variables** change.

The block diagram of a **Mealy machine** is shown in Figure 10.2. The outputs of the Mealy machine are derived from the combinational (control) section of the machine, as

FIGURE 10.2
Mealy-Type State Machine

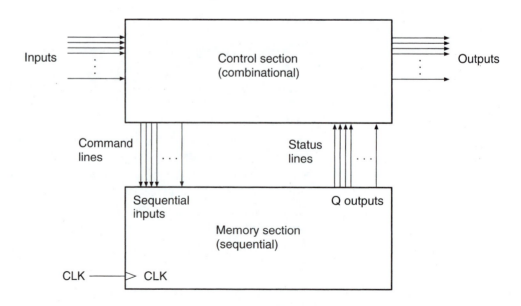

well as the sequential (memory) part of the machine. Therefore, the outputs can change asynchronously when the combinational circuit inputs change out of phase with the clock. (When we say that the outputs change asynchronously, we generally do not mean a change via a function such as asynchronous reset that directly affects the machine's flip-flops.)

■ SECTION 10.1 REVIEW PROBLEM

10.1 What is the main difference between a Moore-type state machine and a Mealy-type state machine?

10.2 State Machines with No Control Inputs

> **KEY TERMS**
>
> **Bubble** A circle in a state diagram containing the state name and values of the state variables.

A state machine can be designed using a classical technique, similar to that used to design a synchronous counter. We can also use a VHDL design method. We will design several state machines, using both classical and VHDL techniques.

As an example of these techniques, we will design a state machine whose output depends only on the clock input: a 3-bit counter with a Gray code count sequence. A 3-bit Gray code, shown in Table 10.1, changes only one bit between adjacent codes and is therefore not a binary-weighted sequence.

Table 10.1 3-bit Gray Code Sequence

$Q_2Q_1Q_0$
000
001
011
010
110
111
101
100

Gray code is often used in situations where it is important to minimize the effect of single-bit errors. For example, suppose the angle of a motor shaft is measured by a detected code on a Gray-coded shaft encoder, shown in Figure 10.3. The encoder indicates a 3-bit number for each of eight angular positions by having three concentric circular segments for each code. A dark band indicates a 1 and a transparent band indicates a 0, with the MSB as the outermost band. The dark or transparent bands are detected by three sensors that detect

FIGURE 10.3
Gray Code on a Shaft Encoder

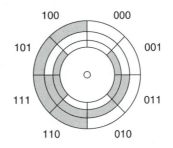

light shining through a transparent band. (A real shaft encoder has more bits to indicate an angle more precisely. For example, a shaft encoder that measures an angle of one degree would require nine bits, since there are 360 degrees in a circle and $2^8 \leq 360 \leq 2^9$.)

For most positions on the encoder, the error of a single bit results in a positional error of only one eighth of the circle. This is not true with binary coding, where single bit errors can give larger positional errors. For example if the positional decoder reads 100 instead of 000, this is a difference of 4 in binary. The same codes differ by only one position in Gray code.

Classical Design Techniques

We can summarize the classical design technique for a state machine, as follows:

1. Define the problem.

2. Draw a state diagram.

3. Make a state table that lists all possible present states and inputs and the next state and output state for each present state/input combination. *List the present states and inputs in **binary order.***

4. Use flip-flop excitation tables to determine at what states the flip-flop synchronous inputs must be to make the circuit go from each present state to its next state. *The next state variables are functions of the inputs and present state variables.*

5. Write the output value for each present state/input combination. *The output variables are functions of the inputs and present state variables.*

6. Simplify the Boolean expression for each output and synchronous input.

7. Use the Boolean expressions found in step 6 to draw the required logic circuit.

Let us follow this procedure to design a 3-bit Gray code counter. We will modify the procedure to account for the fact that there are no inputs other than the clock and no outputs that must be designed apart from the counter itself.

1. *Define the problem.* Design a counter whose outputs progress in the sequence defined in Table 10.1.

2. *Draw a state diagram.* The state diagram is shown in Figure 10.4. In addition to the values of state variables shown in each circle (or **bubble**), we also indicate a state name, such as s0, s1, s2, and so on. This name is independent of the value of state variables. We use numbered states (s0, s1, . . .) for convenience, but we could use any names we wanted to.

FIGURE 10.4
State Diagram for a 3-bit Gray Code Counter

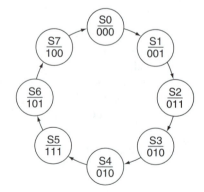

3. *Make a state table.* The state table, based on D flip-flops, is shown in Table 10.2. *Since there are eight unique states in the state diagram, we require three state variables ($2^3 = 8$), and hence three flip-flops.* Note that the present states are in binary-weighted order, even though the count does not progress in this order. In such a case, it is essential to have an accurate state diagram, from which we derive each next state. For example, if

Table 10.2 State Table for a 3-bit Gray Code Counter

Present State	Next State	Synchronous Inputs
$Q_2Q_1Q_0$	$Q_2Q_1Q_0$	$D_2D_1D_0$
000	001	001
001	011	011
010	110	110
011	010	010
100	000	000
101	100	100
110	111	111
111	101	101

the present state is 010, the next state is not 011, as we would expect, but 110, which we derive by examining the state diagram.

Why list the present states in binary order, rather than the same order as the output sequence? By doing so, we can easily simplify the equations for the D inputs of the flip-flops by using a series of Karnaugh maps. This is still possible, but harder to do, if we list the present states in order of the output sequence.

4. *Use flip-flop excitation tables to determine at what states the flip-flop synchronous inputs must be to make the circuit go from each present state to its next state.* This is not necessary if we use D flip-flops, since Q follows D. The D inputs are the same as the next state outputs. For JK or T flip-flops, we would follow the same procedure as for the design of synchronous counters outlined in Chapter 9.

5. *Simplify the Boolean expression for each synchronous input.* Figure 10.5 shows three Karnaugh maps, one for each D input of the circuit.

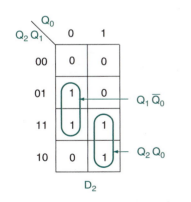
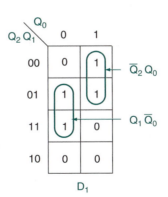
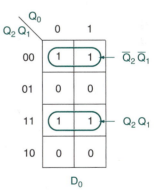

FIGURE 10.5
Karnaugh Maps for 3-bit Gray Code Counter

The K-maps yield three Boolean equations:

$$D_2 = Q_1\overline{Q_0} + Q_2Q_0$$
$$D_1 = Q_1\overline{Q_0} + \overline{Q_2}Q_0$$
$$D_0 = \overline{Q_2}\,\overline{Q_1} + Q_2Q_1$$

6. *Draw the logic circuit for the state machine.* Figure 10.6 shows the circuit for a 3-bit Gray code counter, drawn as a Graphic Design File in MAX+PLUS II. A simulation for this circuit is shown in Figure 10.7, with the outputs shown as individual waveforms and as a group with a binary value.

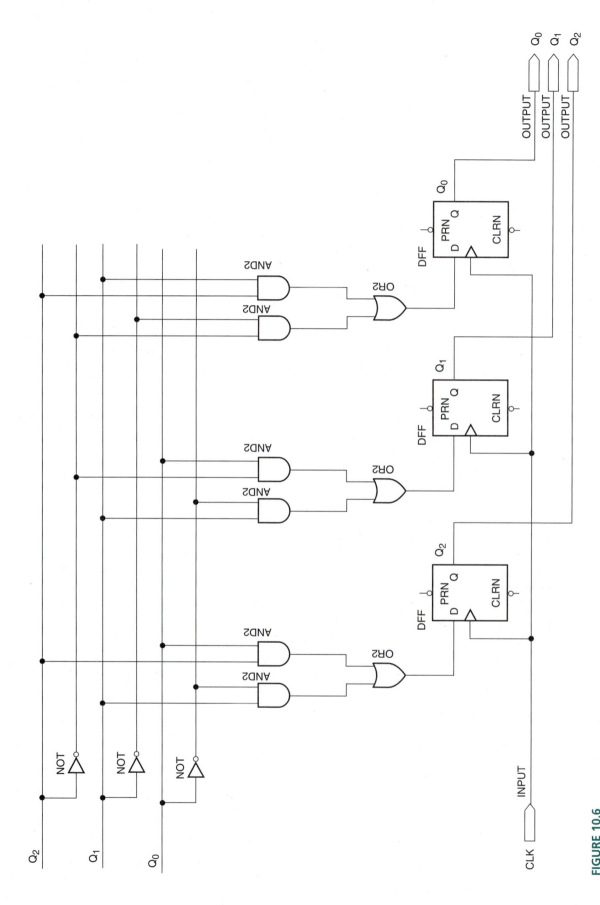

FIGURE 10.6
Logic Diagram of a 3-bit Gray Code Counter

gray_ct3.gof
gray_ct3.scf

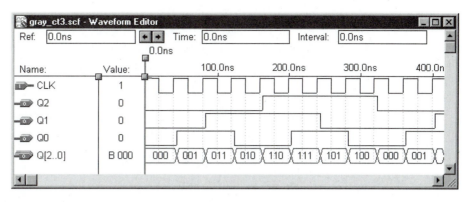

FIGURE 10.7
Simulation of a 3-bit Gray Code Counter (from Graphic Design File)

VHDL Design of State Machines

State machines can be defined in VHDL within a CASE statement. The VHDL code below illustrates the principle, using the 3-bit Gray code counter as an example.

```
-- gray_ct1.vhd
-- 3-bit Gray code counter
-- (state machine with decoded outputs)

LIBRARY ieee;
USE ieee.std_logic_1164.ALL;

ENTITY gray_ct1 IS
    PORT(
        clk : IN    STD_LOGIC;
        q   : OUT   STD_LOGIC_VECTOR(2 downto 0));
END gray_ct1;

ARCHITECTURE a OF gray_ct1 IS
    TYPE STATE_TYPE IS (s0, s1, s2, s3, s4, s5, s6, s7);
    SIGNAL state: STATE_TYPE;
BEGIN
    PROCESS (clk)
    BEGIN
        IF clk'EVENT AND clk = '1' THEN
            CASE state IS
                WHEN s0 =>
                    state <= s1;
                WHEN s1 =>
                    state <= s2;
                WHEN s2 =>
                    state <= s3;
                WHEN s3 =>
                    state <= s4;
                WHEN s4 =>
                    state <= s5;
```

gray_ct1.vhd

```
                    WHEN s5 =>
                        state <= s6;
                    WHEN s6=>
                        state <= s7;
                    WHEN s7 =>
                        state <= s0;
            END CASE;
        END IF;
END PROCESS;

WITH state SELECT
    q <= "000" WHEN s0,
         "001" WHEN s1,
         "011" WHEN s2,
         "010" WHEN s3,
         "110" WHEN s4,
         "111" WHEN s5,
         "101" WHEN s6,
         "100" WHEN s7;
END a;
```

Recall that the format of a CASE statement is:

```
CASE __expression IS
    WHEN__constant_value =>
        __statement;
        __statement;
    WHEN__constant_value =>
        __statement;
        __statement;
    WHEN OTHERS =>
        __statement;
        __statement;
END CASE;
```

The keyword **expression** in the CASE statement refers to a signal called **state** that we define to represent the state variables within the machine. For each possible value of **state,** we make an assignment indicating the next state of the machine. For example, the clause (WHEN s0 => (state <= s1)); indicates a transition from state s0 to state s1. The actual output values of the counter are assigned in a selected signal assignment statement after the PROCESS statement.

Notice that the signal **state** can have one of eight different values, from s0 to s7. Until now, we have seen signals with values such as '1' (BIT or STD_LOGIC types), "011" (BIT_VECTOR or STD_LOGIC_VECTOR types), or 7 (INTEGER types). The signal **state** is of type STATE_TYPE, which is a user-defined **enumerated type.** An enumerated type is simply a list of all values a signal, variable, or port of that type is allowed to have.

For example, we could define a type called DIRECTION with four values, with the statement:

```
TYPE DIRECTION IS (up, down, left, right);
```

We could then define a signal called **position** of type DIRECTION:

```
SIGNAL position: DIRECTION:
```

An IF statement or other construct could then assign one of the four defined values of type DIRECTION to the signal called **position:**

```
IF (x='0' and y='0') THEN
   position <= down;
ELSIF (x='0' and y='1') THEN
   position <= left;
ELSIF (x='1' and y='0') THEN
   position <= up;
ELSE
   position <= right;
END IF;
```

Thus the named identifier **position** of type DIRECTION can take on only the four values specified in the enumerated type definition.

An alternative way to encode the 3-bit counter is to include output assignments within the body of the CASE statement. Each case then has more than one statement, as indicated in the following VHDL code.

```
-- gray_ct2.vhd
-- 3-bit Gray code counter
-- (outputs defined within states)

LIBRARY ieee;
USE ieee.std_logic_1164.ALL;

ENTITY gray_ct2 IS
    PORT(
        clk : IN    STD_LOGIC;
        q   : OUT   STD_LOGIC_VECTOR(2 downto 0));
END gray_ct2;

ARCHITECTURE a OF gray_ct2 IS
    TYPE STATE_TYPE IS (s0, s1, s2, s3, s4, s5, s6, s7);
    SIGNAL state: STATE_TYPE;
BEGIN
    PROCESS (clk)
    BEGIN
        IF clk'EVENT AND clk = '1' THEN
            CASE state IS
                WHEN s0 =>
                    state <= s1;
                    q <= "001";
                WHEN s1 =>
                    state <= s2;
                    q <= "011";
                WHEN s2 =>
                    state <= s3;
                    q <= "010";
                WHEN s3 =>
                    state <= s4;
                    q <= "110";
                WHEN s4 =>
                    state <= s5;
                    q <= "111";
                WHEN s5 =>
                    state <= s6;
                    q <= "101";
                WHEN s6 =>
                    state <= s7;
                    q <= "100";
```

gray_ct2.vhd

```
            WHEN s7 =>
                state <= s0;
                q <= "000";
            END CASE;
         END IF;
      END PROCESS;
END a;
```

The above VHDL code is identical to that of the previous example, except for the way the outputs are assigned.

■■ SECTION 10.2 REVIEW PROBLEM

10.2 Write the Boolean equations for the J and K inputs of the flip-flops in a 3-bit Gray code counter based on JK flip-flops.

10.3 State Machines with Control Inputs

> ### KEY TERMS
>
> **Control input**   A state machine input that directs the machine from state to state.
>
> **Conditional transition**   A transition between states of a state machine that occurs only under specific conditions of one or more control inputs.
>
> **Unconditional transition**   A transition between states of a state machine that occurs regardless of the status of any control inputs.

As an extension of the techniques used in the previous section, we will examine the design of state machines that use **control inputs,** as well as the clock, to direct their operation. Outputs of these state machines will not necessarily be the same as the states of the machine's flip-flops. As a result, this type of state machine requires a more detailed state diagram notation, such as that shown in Figure 10.8.

FIGURE 10.8
State Diagram Notation

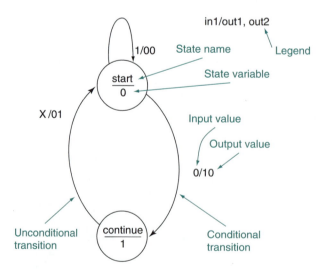

The state machine represented by the diagram in Figure 10.8 has two states, and thus requires only one state variable. Each state is represented by a bubble (circle) containing the state name and the value of the state variable. For example, the bubble containing the notation $\frac{\textbf{start}}{\textbf{0}}$ indicates that the state called **start** corresponds to a state variable with a value of 0. Each state must have a unique value for the state variable(s).

Transitions between states are marked with a combination of input and output values corresponding to the transition. The inputs and outputs are labeled **in1, in2, . . . , inx/out1, out2, . . . ,outx.** The inputs and outputs are sometimes simply indicated by the value of each variable for each transition. In this case, a legend indicates which variable corresponds to which position in the label.

For example, the legend in the state diagram of Figure 10.8 indicates that the inputs and outputs are labeled in the order **in1/out1, out2.** Thus if the machine is in the **start** state and the input **in1** goes to 0, there is a transition to the state **continue.** During this transition, **out1** goes to 1 and **out2** goes to 0. This is indicated by the notation 0/10 beside the transitional arrow. This is called a **conditional transition** because the transition depends on the state of **in1.** The other possibility from the **start** state is a no-change transition, with both outputs at 0, if **in1** = 1. This is shown as 1/00.

If the machine is in the state named **continue,** the notation X/01 indicates that the machine makes a transition back to the **start** state, regardless of the value of **in1,** and that **out1** = 0 and **out2** = 1 upon this transition. Since the transition always happens, it is called an **unconditional transition.**

What does this state machine do? We can determine its function by analyzing the state diagram, as follows.

1. There are two states, called **start** and **continue.** The machine begins in the **start** state and waits for a LOW input on **in1.** As long as **in1** is HIGH, the machine waits and the outputs **out1** and **out2** are both LOW.

2. When **in1** goes LOW, the machine makes a transition to **continue** in one clock pulse. Output **out1** goes HIGH.

3. On the next clock pulse, the machine goes back to **start.** The output **out2** goes HIGH and **out1** goes back LOW.

4. If **in1** is HIGH, the machine waits for a new LOW on **in1.** Both outputs are LOW again. If **in1** is LOW, the cycle repeats.

In summary, the machine waits for a LOW input on **in1,** then generates a pulse of one clock cycle duration on **out1,** then on **out2.** A timing diagram describing this operation is shown in Figure 10.9.

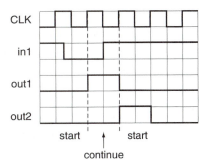

FIGURE 10.9
Ideal Operation of State Machine in Figure 10.8

Classical Design of State Machines with Control Inputs

We can use the classical design technique of the previous section to design a circuit that implements the state diagram of Figure 10.8.

1. *Define the problem.* Implement a digital circuit that generates a pulse on each of two outputs, as described above. For this implementation, let us use JK flip-flops for the state logic. If we so chose, we could also use D or T flip-flops.

2. *Draw a state diagram.* The state diagram is shown in Figure 10.8.

Table 10.3 State Table for State Diagram in Figure 10.8

Present State	Input	Next State	Sync. Inputs	Outputs	
Q	$in1$	Q	JK	$out1$	$out2$
0	0	1	1X	1	0
0	1	0	0X	0	0
1	0	0	X1	0	1
1	1	0	X1	0	1

3. *Make a state table.* The state table is shown in Table 10.3. The combination of present state and input are listed in binary order, thus making Table 10.3 into a truth table for the next state and output functions. Since there are two states, we require one state variable, Q. The next state of Q, a function of the present state and the input **in1**, is determined by examining the state diagram. (Thus, if you are in state 0, the next state is 1 if **in1** = 0 and 0 if **in1** = 1. If you are in state 1, the next state is always 0.)

Table 10.4 JK Flip-Flop Excitation Table

Transition	JK
0→0	0X
0→1	1X
1→0	X1
1→1	X0

4. *Use flip-flop excitation tables to determine at what states the flip-flop synchronous inputs must be to make the circuit go from each present state to its next state.* Table 10.4 shows the flip-flop excitation table for a JK flip-flop. The synchronous inputs are derived from the present-to-next state transitions in Table 10.4 and entered into Table 10.3. (Refer to the synchronous counter design process in Chapter 9 for more detail about using flip-flop excitation tables.)

5. *Write the output values for each present state/input combination.* These can be determined from the state diagram and are entered in the last two columns of Table 10.3.

6. *Simplify the Boolean expression for each output and synchronous input.* The following equations represent the next state and output logic of the state machine:

$$J = \overline{Q} \cdot \overline{in1} + Q \cdot \overline{in1} = \overline{in1}$$
$$K = 1$$
$$out1 = \overline{Q} \cdot \overline{in1}$$
$$out2 = Q \cdot \overline{in1} + Q \cdot in1 = Q$$

7. *Use the Boolean expressions found in step 6 to draw the required logic circuit.*

Figure 10.10 shows the circuit of the state machine drawn as a MAX+PLUS II Graphic Design File. Since **out1** is a function of the control section and the memory section of the machine, we can categorize the circuit as a Mealy machine. (All counter circuits that we have previously examined have been Moore machines since their outputs are derived solely from the flip-flop outputs of the circuit.)

Since the circuit is a Mealy machine, it is vulnerable to asynchronous changes of output due to asynchronous input changes. This is shown in the simulation waveforms of Figure 10.11.

state_x2a.gdf
state_x2a.scf

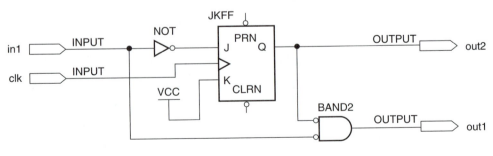

FIGURE 10.10
Implementation of State Machine of Figure 10.8

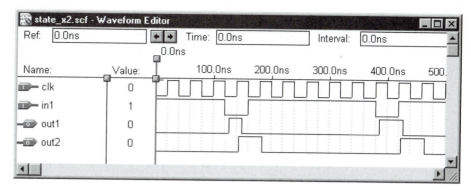

FIGURE 10.11
Simulation of State Machine Circuit of Figure 10.10

Ideally, **out1** should not change until the first positive clock edge after **in1** goes LOW. However, since **out1** is derived from a combinational output, it will change as soon as **in1** goes LOW, after allowing for a short propagation delay. Also, since **out2** is derived directly from a flip-flop and **out1** is derived from the same flip-flop via a gate, **out1** stays HIGH for a short time after **out2** goes HIGH. (The extra time represents the propagation delay of the gate.)

If output synchronization is a problem (and it may not be), it can be fixed by adding a synchronizing D flip-flop to each output, as shown in Figure 10.12.

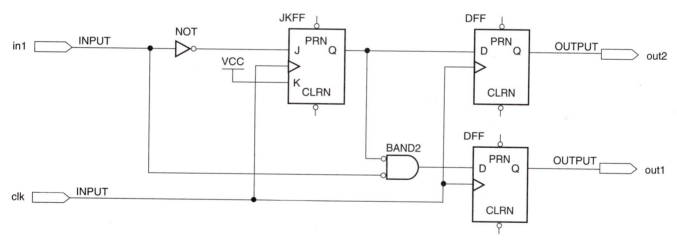

FIGURE 10.12
State Machine with Synchronous Outputs

state_x3a.gdf
state_x3a.scf

The state variable is stored as the state of the JK flip-flop. This state is clocked through a D flip-flop to generate **out2** and combined with **in1** to generate **out1** via another flip-flop. The simulation for this circuit, shown in Figure 10.13, indicates that the two outputs are synchronous with the clock, but delayed by one clock cycle after the state change.

VHDL Implementation of State Machines with Control Inputs

The VHDL code for a state machine with one or more control inputs is similar to that for a machine with no control inputs. The machine states are still defined using a CASE statement, but a case representing a conditional transition will contain an IF statement.

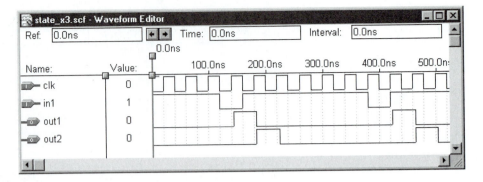

FIGURE 10.13
Simulation of the State Machine of Figure 10.12

The VHDL code for the state machine implemented above is as follows.

state_x1.vhd
state_x1.scf

```
-- state_x1.vhd
-- state machine example 1
-- Two states, one input, two outputs
-- Generates a pulse on one output, then the next
-- after receiving a LOW on the input

LIBRARY ieee;
USE ieee.std_logic_1164.ALL;

ENTITY state_x1 IS
    PORT(
        clk, in1   : IN   STD_LOGIC;
        out1, out2 : OUT  STD_LOGIC);
END state_x1;

ARCHITECTURE a OF state_x1 IS
    TYPE PULSER IS (start, continue);
    SIGNAL sequence: PULSER;
BEGIN
    PROCESS (clk)
    BEGIN
        IF clk'EVENT AND clk = '1' THEN
            CASE sequence IS
                WHEN start =>
                    IF in1 = '1' THEN
                        sequence <= start; -- no change if in1 = 1
                        out1 <= '0';
                        out2 <= '0';
                    ELSE
                        sequence <= continue; -- proceed if in1 = 0
                        out1 <= '1';           -- pulse on out1
                        out2 <= '0';
                    END IF;
                WHEN continue =>
                    sequence <= start;
                    out1 <= '0';
                    out2 <= '1';               -- pulse on out2
            END CASE;
        END IF;
    END PROCESS;
END a;
```

The transition from **start** is conditional, so the case for **start** contains an IF statement that defines the possible state transitions and their associated output states. The transition from **continue** is unconditional, so no IF statement is needed in the corresponding case.

Figure 10.14 shows the simulation for the VHDL design entity, **state_x1.vhd**. The values of the state variable, **sequence,** are also shown in the simulation. This gives us a ready indication of the machine's state (**start** or **continue**).

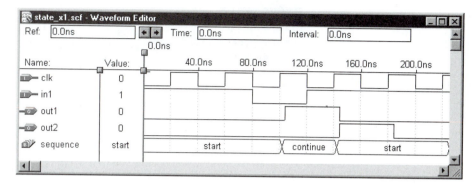

FIGURE 10.14

Simulation of the State Machine in VHDL Entity state_x1

The design of the state machine is such that if the input **in1** is held LOW beyond the end of one pulse cycle, the cycle will repeat, as shown in the simulation of Figure 10.15.

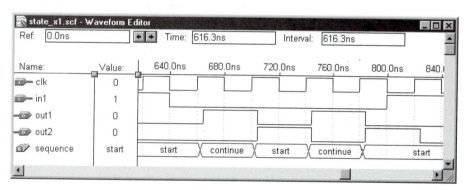

FIGURE 10.15

Simulation of VHDL State Machine Showing a Repeated Output Cycle

▌▎ EXAMPLE 10.1

Application

A state machine called a single-pulse generator operates as follows:

1. The circuit has two states: **seek** and **find,** an input called **sync** and an output called **pulse**.

2. The state machine resets to the state **seek.** If **sync** = 1, the machine remains in **seek** and the output, **pulse,** remains LOW.

3. When **sync** = 0, the machine makes a transition to **find.** In this transition, **pulse** goes HIGH.

4. When the machine is in state **find** and **sync** = 0, the machine remains in **find** and **pulse** goes LOW.

5. When the machine is in **find** and **sync** = 1, the machine goes back to **seek** and **pulse** remains LOW.

Use classical state machine design techniques to design the circuit for the single-pulse generator, using D flip-flops for the state logic. Use MAX+PLUS II to draw the state machine circuit. Create a simulation to verify the design operation. Briefly describe what this state machine does.

Solution   Figure 10.16 shows the state diagram derived from the description of the state machine. The state table is shown in Table 10.5. Since Q follows D, the D input is the same as the next state of Q.

FIGURE 10.16
Example 10.1
State Diagram for a Single-pulse
Generator

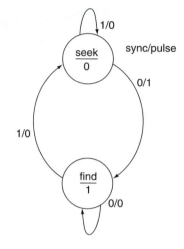

Table 10.5   State Table for Single-Pulse Generator

Present State	Input	Next State	Sync. Input	Output
Q	$sync$	Q	D	$pulse$
0	0	1	1	1
0	1	0	0	0
1	0	1	1	0
1	1	0	0	0

pulse1.gdf
pulse1.scf

The next-state and output equations are:

$$D = \overline{Q} \cdot \overline{sync} + Q \cdot \overline{sync} = \overline{sync}$$
$$pulse = \overline{Q} \cdot \overline{sync}$$

Figure 10.17 shows the state machine circuit derived from the above Boolean equations. The simulation for this circuit is shown in Figure 10.18. The simulation shows that the circuit generates one pulse when the input **sync** goes LOW, regardless of the length of time that **sync** is LOW. The circuit could be used in conjunction with a debounced pushbutton to produce exactly one pulse, regardless of how long the pushbutton was held down. Figure 10.19 shows such a circuit.

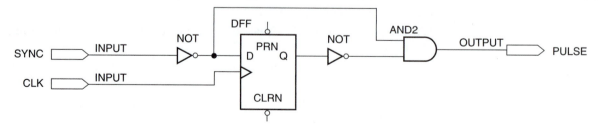

FIGURE 10.17
Example 10.1
Single-pulse Generator

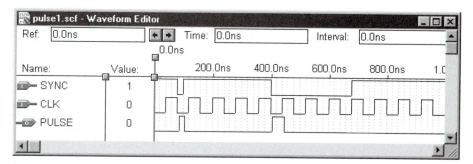

FIGURE 10.18
Example 10.1
Simulation of a Single-pulse Generator (from GDF)

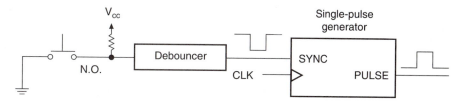

FIGURE 10.19
Example 10.1
Single-pulse Generator Used with a Debounced Pushbutton

▌▌ EXAMPLE 10.2

The state machine of Example 10.1 is vulnerable to asynchronous input changes. How do we know this from the circuit schematic and from the simulation waveform? Modify the circuit to eliminate the asynchronous behavior and show the effect of the change on a simulation of the design. How does this change improve the design?

Solution   The output, **pulse,** in the state machine of Figure 10.17 is derived from the state flip-flop and the combinational logic of the circuit. The output can be affected by a change that is purely combinational, thus making the output asynchronous. This is demonstrated on the first pulse of the simulation in Figure 10.18, where **pulse** momentarily goes HIGH between clock edges. Since no clock edge was present when either the input, **sync,** changed or when **pulse** changed, the output pulse must be due entirely to changes in the combinational part of the circuit.

The circuit output can be synchronized to the clock by adding an output flip-flop, as in Figure 10.20. A simulation of this circuit is shown in Figure 10.21. With the synchronized output, the output pulse is always the same width: one clock period. This gives a more predictable operation of the circuit.

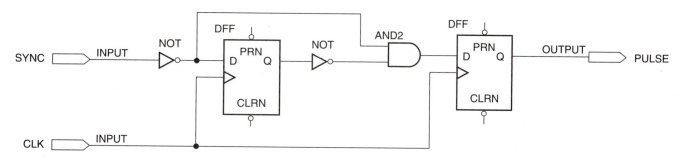

FIGURE 10.20
Example 10.2
Single-pulse Generator with Synchronous Output

pulse1a.gdf
pulse1a.scf

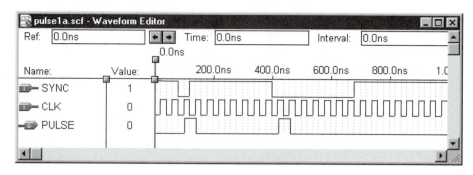

FIGURE 10.21
Example 10.2
Simulation of a Single-pulse Generator with Synchronous Output (from GDF)

■ EXAMPLE 10.3

Write the VHDL code for a design entity that implements the single-pulse generator, as described in Example 10.1. Create a simulation that verifies the operation of the design.

Solution   The required VHDL code is given here in the design entity **sngl_pls**.

```
LIBRARY ieee;
USE ieee.std_logic_1164.ALL;

ENTITY sngl_pls IS
    PORT(
        clk, sync : IN STD_LOGIC;
        pulse     : OUT STD_LOGIC);
END sngl_pls;

ARCHITECTURE pulser OF sngl_pls IS
    TYPE PULSE_STATE IS (seek, find);
    SIGNAL status: PULSE_STATE;
BEGIN
    PROCESS (clk, sync)
    BEGIN
        IF (clk'EVENT and clk = '1') THEN
            CASE status IS
                WHEN seek =>   IF (sync = '1') THEN
                                   status <= seek;
                                   pulse <= '0';
                               ELSE
                                   status <= find;
                                   pulse <= '1';
                               END IF;
                WHEN find =>   IF (sync = '1') THEN
                                   status <= seek;
                                   pulse <= '0';
                               ELSE
                                   status <= find;
                                   pulse <= '0';
                               END IF;
            END CASE;
        END IF;
    END PROCESS;
END pulser;
```

sngl_pls.vhd
sngl_pls.scf

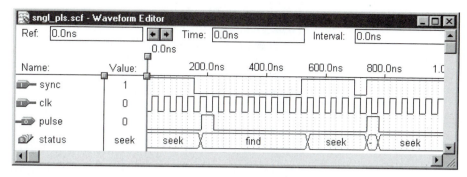

FIGURE 10.22
Example 10.3
Simulation of a Single-pulse Generator (VHDL)

The simulation of the VHDL design entity **sngl_pls** is shown in Figure 10.22

▮▮ SECTION 10.3 REVIEW PROBLEM

10.3 Briefly explain why the single-pulse circuit in Figure 10.20 has a flip-flop on its output.

www.electronictech.com

10.4 Switch Debouncer for a Normally Open Pushbutton Switch

> **KEY TERMS**
>
> **Form A contact**   A normally open contact on a switch or relay.
> **Form B contact**   A normally closed contact on a switch or relay.
> **Form C contact**   A pair of contacts, one normally open and one normally closed, that operate with a single action of a switch or relay.

A useful interface function is implemented by a digital circuit that removes the mechanical bounce from a pushbutton switch. The easiest way to debounce a pushbutton switch is with a NAND latch, as shown in Figure 10.23.

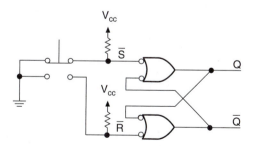

FIGURE 10.23
NAND Latch as a Switch Debouncer

The latch eliminates switch bounce by setting or resetting on the first bounce of a switch contact and ignoring further bounces. The limitation of this circuit is that the input switch must have **Form C contacts**. That is, the switch has normally open, normally closed, and common contacts. This is so that the switch resets the latch when pressed (i.e., when the normally open contact closes) and sets the latch when released (normally closed contact recloses). Each switch position activates an opposite latch function.

If the only available switch has a single set of contacts, such as the normally open (**Form A**) pushbuttons on the Altera UP-1 Education Board, a different debouncer circuit must be used. We will look at two solutions using VHDL: one based on an existing device (the Motorola MC14490 Contact Bounce Eliminator) and another that implements a state machine solution to the contact bounce problem.

Switch Debouncer Based on a 4-bit Shift Register

The circuit in Figure 10.24 is based on the same principle as the Motorola MC14490 Contact Bounce Eliminator, adapted for use in an Altera CPLD, such as the EPM7128S or the EPF10K20 on the Altera UP-1 Education Board.

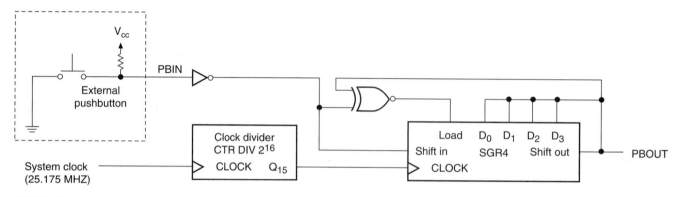

FIGURE 10.24
Switch Debouncer Based on a 4-bit Shift Register

The heart of the debouncer circuit in Figure 10.24 is a 2-bit comparator (an Exclusive NOR gate) and a 4-bit serial shift register, with active-HIGH synchronous LOAD. The XNOR gate compares the shift register serial input and output. When the shift register input and output are *different,* the input data are serially shifted through the register. When input and output of the shift register are *the same,* the binary value at the serial output is parallel-loaded back into all bits of the shift register.

Figure 10.25 shows the timing of the debouncer circuit with switch bounces on both make and break phases of the switch contact. The line labeled **4-bit delay** refers to the shift register flip-flop outputs. Pushbutton input is **pb_in**, debounced output is **pb_out** and **clk** is the UP-1 system clock, divided by 2^{16}. (Time values in Figure 10.25 are not to scale and should be disregarded.)

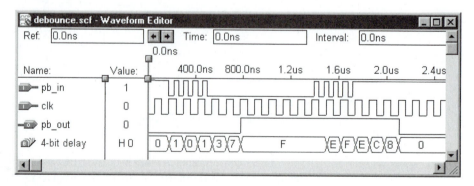

FIGURE 10.25
Simulation of the Shift Register-Based Debouncer

Assume the shift register is initially filled with 0s. The pushbutton rest state is HIGH. As shown in Figure 10.24, the pushbutton input value is inverted and applied to the shift register input. Therefore, before the switch is pressed, both input and output of the shift register are LOW. Since they are the same, the XNOR output is HIGH, which keeps the shift register in *LOAD* mode and the LOW at **pb_out** is reloaded to the register on every positive clock edge.

When the switch is pressed, it will bounce, as shown above the second, third, and fourth clock pulses on Figure 10.25. Just before the second clock pulse, **pb_in** is LOW. This makes the shift register input and output different, so a 1 is shifted in. (Recall that **pb_in** is at the opposite logic level to the shift register input.) On the next clock pulse, **pb_in** has bounced HIGH again. The shift register input and output are now the same, so the output value, 0, is loaded in parallel to all flip-flops of the shift register. On the fifth pulse, **pb_in** is stable at logic LOW. Since the shift register input is now HIGH and the output is LOW, the HIGH is shifted through the register. We see this by **4-bit delay** increasing in value: 0, 1, 3, 7, F, which in binary is equivalent to 0000, 0001, 0011, 0111, 1111. At this point, the input and output are now the same and the output value, 1, is parallel-loaded into the register on each clock pulse.

A similar process occurs when the waveform goes back to the HIGH state. When the input goes HIGH, a LOW is shifted into the shift register. If the input bounces back LOW, the shift register is parallel-loaded with HIGHs and the process starts over. When **pb_in** is stable at a HIGH level, a LOW is shifted through the register, resulting in the hexadecimal sequence F, E, C, 8, 0, which is equivalent to the binary values 1111, 1110, 1100, 1000, 0000.

To produce an output change, the shift register input and output must remain different for at least four clock pulses. This implies that the input is stable for that period of time. If the input and output are the same, this could mean one of two things. Either the input is stable and the shift register flip-flops should be kept at a constant state or the input has bounced back to its previous level and the shift register should be reinitialized. In either case, the output value should be parallel loaded back into the shift register. Serial shifting should only occur if there has been an input change.

The debouncer in Figure 10.24 is effective for removing bounce that lasts for no more than 4 clock periods. Since switch bounce is typically about 10 ms in duration, the clock should have a period of about 2.5 ms. At 25.175 MHz (a clock period of about 40 ns), the Altera UP-1 system clock is much too fast.

If we divide the oscillator frequency by 65536 ($= 2^{16}$) using a 16-bit counter, we obtain a clock waveform for the debouncer with a period of 2.6 ms. Four clock periods (10.2 ms) are sufficient to take care of switch bounce.

We can use VHDL to synthesize the switch debouncer by instantiating a counter and shift register from the Altera Library of Parameterized Modules and connecting them together with internal signals. The VHDL code is as follows.

debounce.vhd
debounce.scf

```
-- debounce.vhd
-- Switch Debouncer for a Form A contact, based on a 4-bit shift
-- register.  Function is similar to a Motorola MC14490 Contact
-- Bounce Eliminator.

-- Use modules from Library of Parameterized Modules (LPM):
--     LPM_SHIFTREG   (Shift Register)
--     LPM_COUNTER    (16-bit counter)

LIBRARY ieee;
USE ieee.std_logic_1164.ALL;
LIBRARY lpm;
USE lpm.lpm_components.ALL;
```

```
ENTITY debounce IS
    PORT(
        clk : IN STD_LOGIC;
        pb_in : IN STD_LOGIC;
        pb_out : OUT STD_LOGIC);
END debounce;

ARCHITECTURE debouncer OF debounce IS
-- Internal signals required to interconnect counter and shift
register
    SIGNAL srg_ser_out, srg_ser_in, srg_clk, srg_load : STD_LOGIC;
    SIGNAL srg_data    : STD_LOGIC_VECTOR(3 DOWNTO 0);
    SIGNAL ctr_q       : STD_LOGIC_VECTOR (15 DOWNTO 0);
BEGIN
-- Instantiate 16-bit counter
    clock_divider: lpm_counter
        GENERIC MAP (LPM_WIDTH    => 16)
        PORT MAP (clock      =>   clk,
                  q          =>   ctr_q(15 DOWNTO 0));

-- Instantiate 4-bit shift register
    four_bit_delay: lpm_shiftreg
        GENERIC MAP (LPM_WIDTH    => 4)
        PORT MAP (shiftin       => srg_ser_in,
                  clock         => srg_clk,
                  load          => srg_load,
                  data          => srg_data(3 downto 0),
                  shiftout      => srg_ser_out);

-- Shift register is clocked by counter output
-- (divides system clock by 2^16)
    srg_clk    <= ctr_q(15);

-- Undebounced pushbutton input to shift register
    srg_ser_in <=  not pb_in;

-- Shift register is parallel-loaded with output data if
-- shift register input and output are the same.
-- If input and output are different,
-- data are serial-shifted.
    srg_data(3)    <= srg_ser_out;
    srg_data(2)    <= srg_ser_out;
    srg_data(1)    <= srg_ser_out;
    srg_data(0)    <= srg_ser_out;
    pb_out         <= srg_ser_out;
    srg_load       <= not((not pb_in) xor srg_ser_out);
END debouncer;
```

Figure 10.26 shows a fairly easy way to test the switch debouncer. The debouncer output is used to clock an 8-bit counter whose outputs are decoded by two seven-segment decoders. (The decoders are VHDL files developed in a similar way to the seven-segment decoders in Chapter 5.)

Pin numbers are given for the EPM7128S CPLD on the Altera UP-1 circuit board. Since the clock and seven segment displays are hardwired on the Altera board, the only external connections required for the circuit are wires for the two pushbutton inputs, **reset** and **pb_in.**

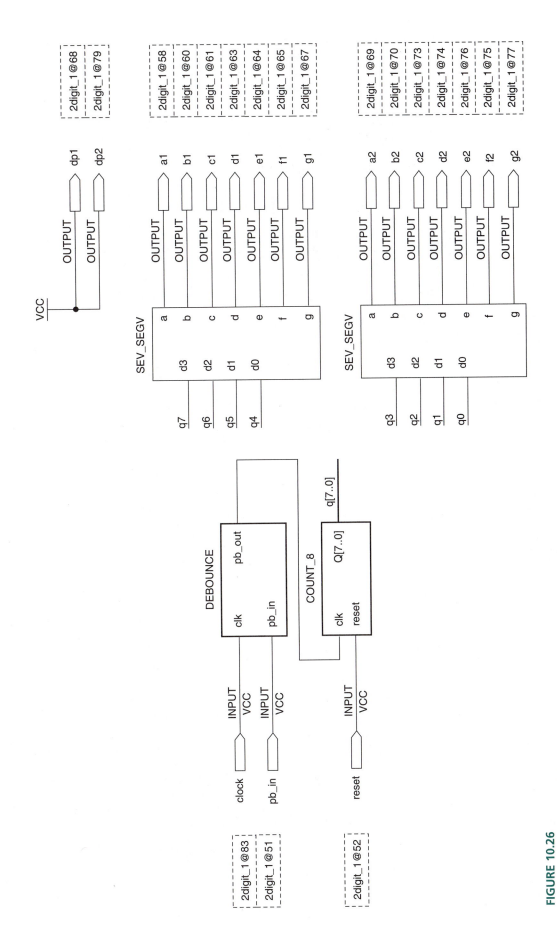

FIGURE 10.26
Test Circuit for a Switch Debouncer

479

2digit.gdf
count_8.vhd
sev_segv.vhd

If the debouncer is working properly, the seven-segment display should advance by one each time **pb_in** is pressed. If the debouncer is not working, the display will change by an unpredictable number with each switch press.

The component source files for the debouncer and test circuit components are supplied on the CD accompanying this book in the folder *drive:\Student Files\Chapter 10*. To use these files, create a symbol for each one (**File** menu; **Project**; **Set Project to Current File**; then **File** menu; **Create Default Symbol**) and draw the Graphic Design File of Figure 10.26.

Alternatively, you can instantiate each file as a component in a VHDL design entity (all components are designed in VHDL) and connect them together with internal signals.

Behaviorally Designed Switch Debouncer

We can also design a switch debouncer by using a behavioral state machine description in VHDL. In order to do so, we need to define the operation of the circuit with a state diagram, as in Figure 10.27.

FIGURE 10.27
State Diagram for a Behaviorally
Designed Switch Debouncer

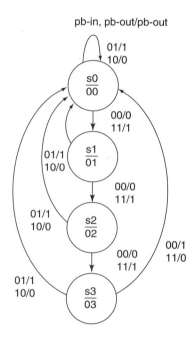

Transitions between states are determined by comparing **pb_in** and **pb_out**. If they are the same (00 or 11), the machine advances to the next state; if they are different (01 or 10), the machine reverts to the initial state, s0. At any point in the state diagram (including state s3, the last state), the machine will reset if **pb_in** and **pb_out** are different, indicating a bounce on the input.

If **pb_in** and **pb_out** are the same for four clock pulses, the input is deemed to be stable. Only at this point will the output change to its opposite state.

> **NOTE**
>
> In the shift register–based debouncer, the circuit advanced to the next state if the shift register input and output were different and reset if they were the same. This might appear to be opposite to our behavioral description, but it is not if you look carefully. The shift register debouncer circuit inverts **pb_in** before applying the signal to the serial input of the shift register. Therefore, viewed from the circuit input and output terminals, rather than at the shift register input and output, the description is the same in both cases.

The VHDL code corresponding to the behavioral description of the switch debouncer is given next. The only output change is specified on the transition from state s3 to s0 when **pb_in = pb_out**. Since no change is allowed at any other time, no other output state needs to be specified.

dbc_behv.vhd
dbc_behv.scf

```
-- dbc_behv.vhd
-- Behavioral definition of a switch debouncer
LIBRARY ieee;
USE ieee.std_logic_1164.ALL;

ENTITY dbc_behv IS
    PORT(
        clk, pb_in : IN STD_LOGIC;
        pb_out     : BUFFER STD_LOGIC);
END dbc_behv;

ARCHITECTURE debounce of dbc_behv IS
    TYPE sequence IS (s0, s1, s2, s3);
    SIGNAL state: sequence;
BEGIN
    PROCESS (clk, pb_in)
    BEGIN
        IF (clk'EVENT and clk='1') THEN
            CASE state IS
                WHEN s0=> IF (pb_in = pb_out) THEN
                                state <= s1;
                            ELSE
                                state <= s0;
                            END IF;
                WHEN s1=> IF (pb_in = pb_out) THEN
                                state <= s2;
                            ELSE
                                state <= s0;
                            END IF;
                WHEN S2=> IF (pb_in = pb_out) THEN
                                state <= s3;
                            ELSE
                                state <= s0;
                            END IF;
                WHEN s3=> IF (pb_in = pb_out) THEN
                                state <= s0;
                                pb_out <= not pb_out;
                            ELSE
                                state <= s0;
                            END IF;
                WHEN others => state <= s0;
            END CASE;
        END IF;
    END PROCESS;
END debounce;
```

Figure 10.28 shows a simulation of the behaviorally-designed switch debouncer. State s1 through s3 are of too short a duration to show properly on the simulation, so further details of the simulation are shown in Figures 10.29 and 10.30.

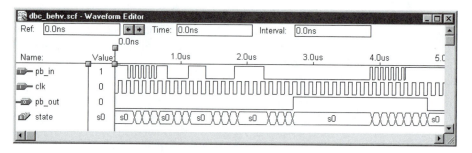

FIGURE 10.28
Simulation of a Behaviorally Designed Switch Debouncer

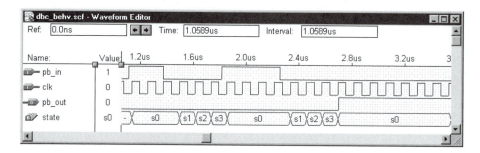

FIGURE 10.29
Simulation Detail (Behaviorally Designed Switch Debouncer)

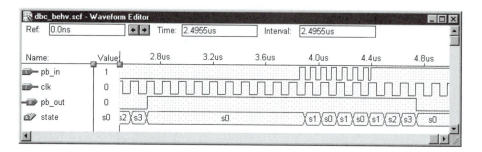

FIGURE 10.30
Simulation Detail (Behaviorally Designed Switch Debouncer)

Note that the behaviorally designed switch debouncer does not have a built-in clock divider. If we were to use the circuit on the Altera UP-1 board, we would need to include a divide-by-2^{16} counter to the circuit, as shown in Figure 10.31.

▌▌ SECTION 10.4 REVIEW PROBLEM

10.4 What is the fastest acceptable clock rate for the shift register portion of the debouncer in Figure 10.24 if the pushbutton switch bounces for 15ms?

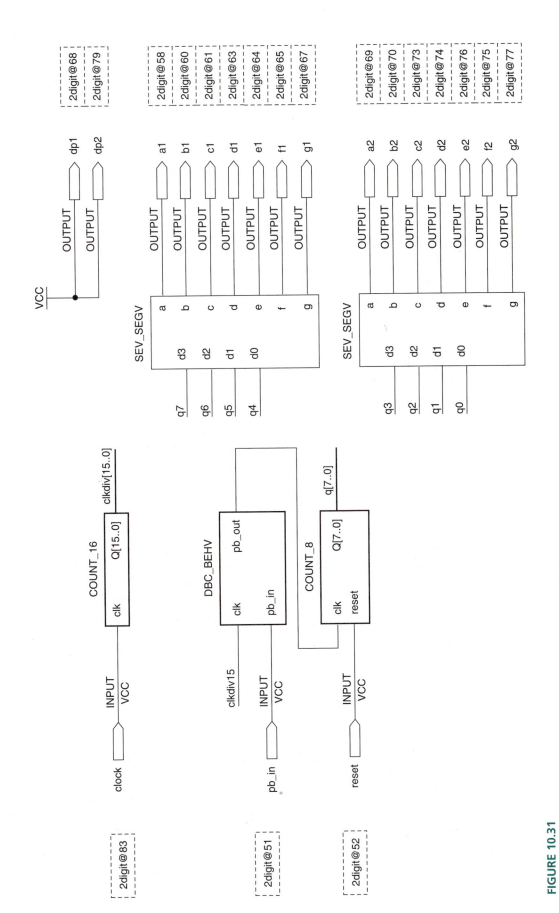

FIGURE 10.31

Using a Behaviorally Designed Debouncer with a 16-bit Clock Divider

483

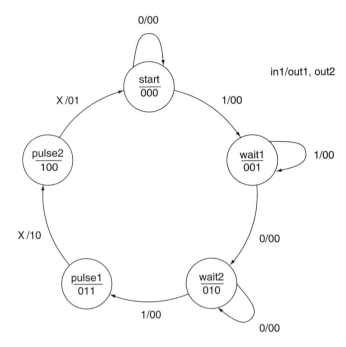

FIGURE 10.32
State Diagram for a Two-pulse Generator

10.5 Unused States in State Machines

In our study of counter circuits in Chapter 9, we found that when a counter modulus is not equal to a power of two there were unused states in the counter's sequence. For example, a mod-10 counter has six unused states, as the counter requires four bits to express ten states and the maximum number of 4-bit states is sixteen. The unused states (1010, 1011, 1100, 1101, 1110, and 1111) have to be accounted for in the design of a mod-10 counter.

The same is true of state machines whose number of states does not equal a power of two. For instance, a machine with five states requires three state variables. There are up to eight states available in a machine with three state variables, leaving three unused states. Figure 10.32 shows the state diagram of such a machine.

Unused states can be dealt with in two ways: they can be treated as don't care states, or they can be assigned specific destinations in the state diagram. In the latter case, the safest destination is the first state, in this case the state called **start.**

▌▌ EXAMPLE 10.4

Redraw the state diagram of Figure 10.32 to include the unused states of the machine's state variables. Set the unused states to have a destination state of **start.** Briefly describe the intended operation of the state machine.

Solution   Figure 10.33 shows the revised state diagram.

The machine begins in state **start** and waits for a HIGH on **in1**. The machine then makes a transition to **wait1** and stays there until **in1** goes LOW again. The machine goes to **wait2** and stays there until **in1** goes HIGH and then makes an unconditional transition to **pulse1** on the next clock pulse. Until this point, there is no change in either output.

The machine makes an unconditional transition to **pulse2** and makes **out1** go HIGH. The next transition, also unconditional, is to **start,** when **out1** goes LOW and **out2** goes HIGH. If **in1** is LOW, the machine stays in **start.** Otherwise, the cycle continues as above. In either case, **out2** goes LOW again.

FIGURE 10.33
Example 10.4
State Diagram for Two-pulse
Generator Showing Unused
States

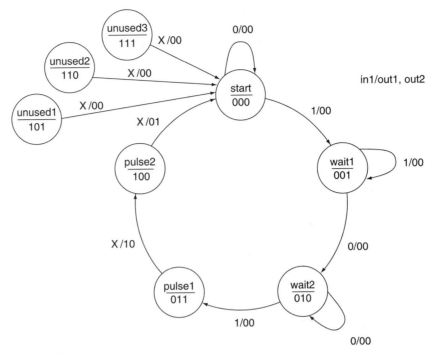

Thus the machine waits for a HIGH-LOW-HIGH input sequence and generates a pulse sequence on two outputs.

▌▌ EXAMPLE 10.5

Use classical state machine design techniques to implement the state machine described in the modified state diagram of Figure 10.33. Draw the state machine as a Graphic Design File in Max+PLUS II and create a simulation to verify its function.

Solution Table 10.6 shows the state table of the state machine represented by Figure 10.33.

Table 10.6 State Table for State Machine of Figure 10.33

Present State	Input	Next State	Outputs	
$Q_2Q_1Q_0$	$in1$	$Q_2Q_1Q_0$	$out1$	$out2$
000	0	000	0	0
000	1	001	0	0
001	0	010	0	0
001	1	001	0	0
010	0	010	0	0
010	1	011	0	0
011	0	100	1	0
011	1	100	1	0
100	0	000	0	1
100	1	000	0	1
101	0	000	0	0
101	1	000	0	0
110	0	000	0	0
110	1	000	0	0
111	0	000	0	0
111	1	000	0	0

Figure 10.34 shows the Karnaugh maps used to simplify the next-state equations for the state variable flip-flops. The output equations can be simplified by inspection.

The next-state and output equations for the state machine are:

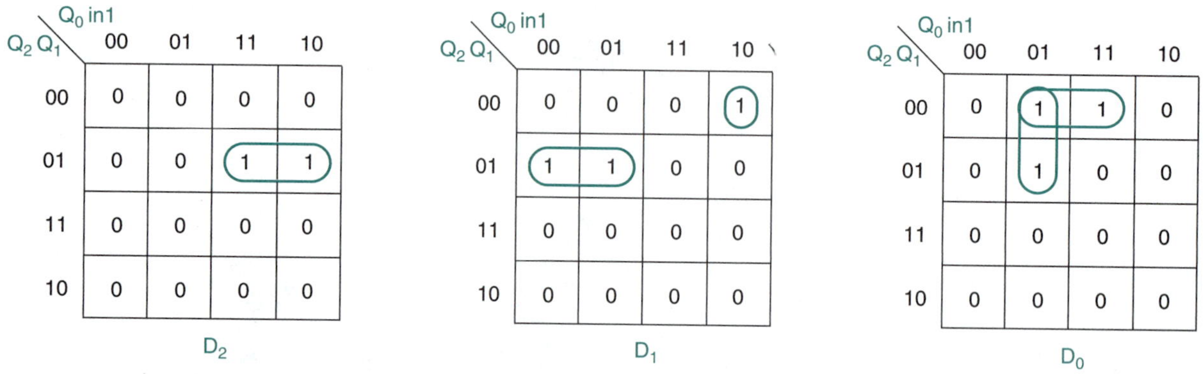

FIGURE 10.34
Example 10.5
K-Maps for Two-pulse Generator

$$D_2 = \overline{Q_2}Q_1Q_0$$
$$D_1 = \overline{Q_2}Q_1\overline{Q_0} + \overline{Q_2}\,\overline{Q_1}Q_0\overline{in1}$$
$$D_0 = \overline{Q_2}\,\overline{Q_0}in1 + \overline{Q_2}\,\overline{Q_1}in1$$
$$out1 = \overline{Q_2}Q_1Q_0$$
$$out2 = Q_2\overline{Q_1}\,\overline{Q_0}$$

Figure 10.35 shows the Graphic Design File schematic for the state machine. Figure 10.36 shows the MAX+PLUS II simulation waveforms.

We can monitor the state variables in the MAX+PLUS II simulation file by adding a group of waveforms for the buried nodes q2, q1, and q0. These are shown on the simulation as q[2..0].Q, meaning the Q outputs of the flip-flops named q2, q1, q0.

To add the buried nodes, select **Enter Node from SNF** from the **Node** menu in the simulator window. In the dialog box shown in Figure 10.37, check the box that says **All**, and click on **List**. Select the nodes q2.Q, q1.Q, and q0.Q from the **Available Nodes and Groups** and transfer them to the **Selected Nodes and Groups**. Click **OK**. Select the three new waveforms and from the **Node** menu, select **Group**. Click **OK** in the resulting dialog box.

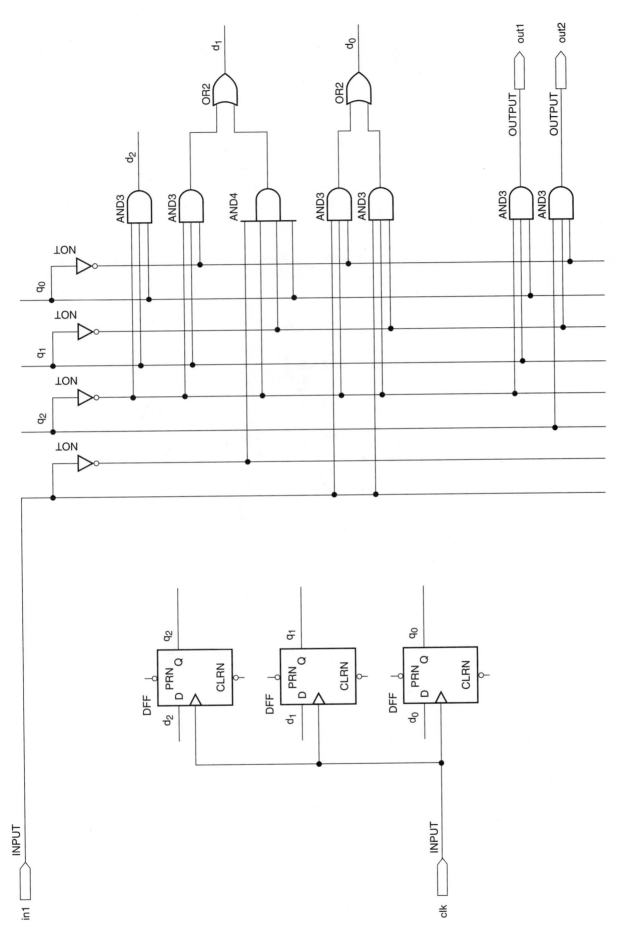

FIGURE 10.35
Example 10.5
Two-pulse Generator

487

FIGURE 10.36
Example 10.5
Simulation of a Two-pulse
Generator (GDF)

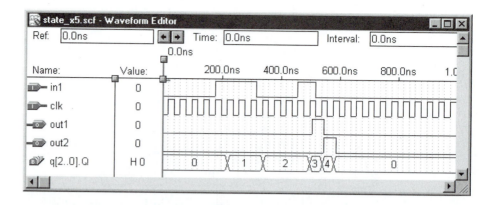

FIGURE 10.37
Adding Buried Nodes to a
Simulation

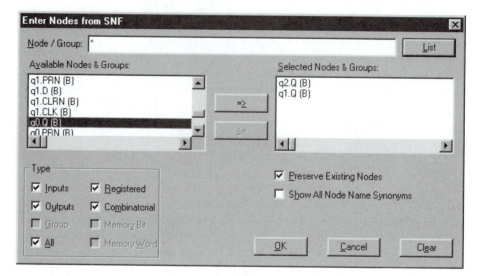

■▐ EXAMPLE 10.6

Write the VHDL code required to implement the two-pulse generator described in Examples 10.4 and 10.5. Create a MAX+PLUS II simulation to verify the operation of the design. Based on your examination of the simulations for the VHDL design and the GDF design of the previous example, how do the two designs differ in their operation? What is the reason for the difference?

Solution   The VHDL code for the state machine in design entity **two_pulse.vhd** follows. The unused states are accounted for in the **others** clause.

```
-- two_pulse.vhd
LIBRARY ieee;
USE ieee.std_logic_1164.ALL;

ENTITY two_pulse IS
   PORT(
       clk, in1  : IN    STD_LOGIC;
       output    : OUT   STD_LOGIC_VECTOR (1 to 2));
END two_pulse;

ARCHITECTURE a OF two_pulse IS
   TYPE SEQUENCE IS (start, wait1, wait2, pulse1, pulse2);
   SIGNAL pulse_state : SEQUENCE;
BEGIN
   PROCESS(clk)
   BEGIN
      IF (clk'EVENT and clk = '1') THEN
         CASE pulse_state IS
```

```
            WHEN start =>
                IF in1 = '0' THEN
                    pulse_state   <= start;
                    output        <= "00";
                ELSIF in1 = '1' THEN
                    pulse_state <= wait1;
                    output        <= "00";
                END IF;
            WHEN wait1 =>
                IF in1 = '0' THEN
                    pulse_state   <= wait2;
                    output        <= "00";
                ELSIF in1 = '1' THEN
                    pulse_state <= wait1;
                    output        <= "00";
                END IF;
            WHEN wait2 =>
                IF in1 = '0' THEN
                    pulse_state   <= wait2;
                    output        <= "00";
                ELSIF in1 = '1' THEN
                    pulse_state <= pulse1;
                    output        <= "00";
                END IF;
            WHEN pulse1 =>
                pulse_state   <= pulse2;
                output        <= "10";
            WHEN pulse2 =>
                pulse_state   <= start;
                output        <= "01";
            WHEN others =>
                pulse_state   <= start;
                output        <= "00";
        END CASE;
    END IF;
  END PROCESS;
END a;
```

Figure 10.38 shows the MAX+PLUS II simulation of the state machine.

If you closely examine the simulation waveforms in Figures 10.36 and 10.38, you will note that the pulse outputs in Figure 10.38 (VHDL design) occur one clock cycle later than they do in Figure 10.36 (graphical design). This is because the VHDL compiler has synthesized each output with a D flip-flop, as we did for the single-pulse circuit in Figure

FIGURE 10.38
Example 10.6
Simulation of a Two-pulse
Generator (VHDL)

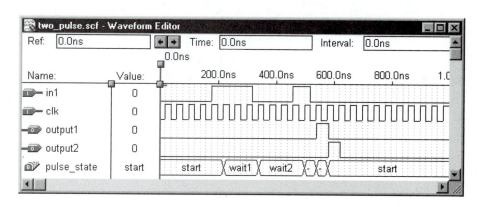

10.20, in order to ensure synchronous output operation.(We can verify this by examining the EQUATIONS section of the project report file, **two_pulse.rpt**.) Since the outputs are both derived entirely from flip-flop outputs, this synthesis step is not strictly necessary to ensure that the outputs are synchronous with the clock.

▌▌ SECTION 10.5 REVIEW PROBLEM

10.5 Is the state machine designed in Example 10.5 a Moore machine or a Mealy machine? Why?

10.6 Traffic Light Controller

A simple traffic light controller can be implemented by a state machine with a state diagram such as the one shown in Figure 10.39.

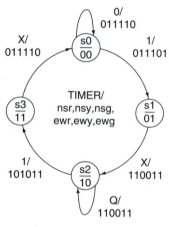

FIGURE 10.39

State Diagram of a Traffic Light Controller

The control scheme assumes control over a north-south road and an east-west road. The north-south lights are controlled by outputs called **nsr, nsy,** and **nsg** (north-south red, yellow, green). The east-west road is controlled by similar outputs called **ewr, ewy,** and **ewg.** A LOW controller output turns on a light. Thus an output 011110 corresponds to the north-south red and east-west green lights.

An input called *TIMER* controls the length of the two green-light cycles. When *TIMER* = 1, a transition from s0 to s1 or from s2 to s3 is possible (s0 represents the EW green; s2 the NS green). This transition accompanies a change from green to yellow on the active road. The light on the other road stays red. An unconditional transition follows, changing the yellow light to red on one road and the red light to green on the other.

The cycle can be set to any length by changing the signal on the *TIMER* input. (The yellow light will always be on for one clock pulse in this design.) For ease of observation, we will use a cycle of ten clock pulses. For either direction, the cycle consists of 4 clocks GREEN, 1 clock YELLOW, 5 clocks RED. This cycle can be generated by the MSB of a mod-5 counter, as shown in Figure 10.40. If we model the traffic controller using the Altera UP-1 board, we require a clock divider to slow down the 25.175 MHz clock to a rate of about 0.75 Hz, making it easy to observe the changes of lights. These blocks can all be instantiated in VHDL, which will be left as part of an exercise in the lab manual accompanying this book.

FIGURE 10.40

Traffic Control Demonstration Circuit for the Altera UP-1 Board

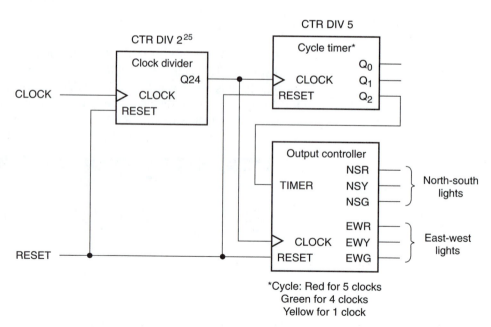

Figure 10.41 shows the simulation of the mod-5 counter that generates the *TIMER* control signal. The MSB goes HIGH for one clock period, then LOW for four. When applied to the *TIMER* input of the output controller, this signal directs the controller from state to state.

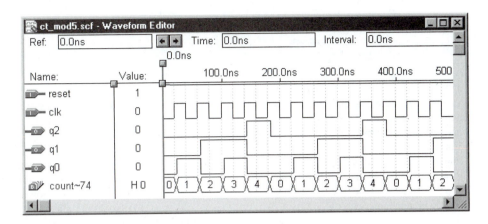

FIGURE 10.41
Simulation of a Mod-5 Counter

Figure 10.42 shows a simulation of the mod-5 counter and output controller. The north-south lights are red for five clock pulses (shown by 011 in the **north_south** waveform). At the same time, the east-west lights are green for four clock pulses (**east_west** = 110), followed by yellow for one clock pulse (**east_west** = 101). The cycle continues with an east-west red and north-south green and yellow.

According to the state diagram, the yellow light should happen on the transition where *TIMER* = 1. This corresponds to the point on the simulation waveforms where **count** = 4.

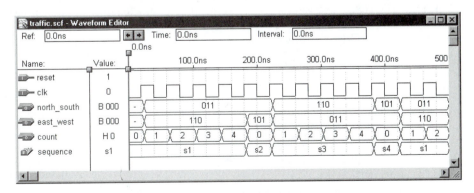

FIGURE 10.42
Simulation of a Traffic Light Controller

However, the yellow light does not come on until **count** = 0. This is because the MAX+PLUS II VHDL compiler synthesizes the controller outputs with synchronous outputs (flip-flops). As a result, the output states are delayed by one clock cycle. Since the relative lengths of the cycle proportions are preserved, this does not affect the operation of the controller.

SUMMARY

1. A state machine is a synchronous sequential circuit with a memory section (flip-flops) to hold the present state of the machine and a control section (gates) to determine the machine's next state.

2. The number of flip-flops in a state machine's memory section is the same as the number of state variables.

3. Two main types of state machine are the Moore machine and the Mealy machine.

4. The outputs of a Moore machine are entirely dependent on the states of the machine's flip-flops. Output changes will always be synchronous with the system clock.

5. The outputs of a Mealy machine depend on the states of the machine's flip-flops and the gates in the control section. A Mealy machine's outputs can change asynchronously, relative to the system clock.

6. A state machine can be designed in a classical fashion using the same method as in designing a synchronous counter, as follows:

    a. Define the problem and draw a state diagram.

    b. Construct a table of present and next states.

    c. Use flip-flop excitation tables to determine the flip-flop inputs for each state transition.

    d. Use Boolean algebra or K-maps to find the simplest Boolean expression for flip-flop inputs *(D, T, or JK)* in terms of outputs *(Q)*.

    e. Draw the logic diagram of the state machine.

7. The state names in a state machine can be named numerically (s0, s1, s2, . . .) or literally (start, idle, read, write), depending on the machine function. State names are independent of the values of the state variables.

8. A state machine can be defined in VHDL by using a CASE statement within a PROCESS to define the progression of states. The output values can be defined by a separate decoder construct or they can be assigned within each case of the CASE statement.

9. The possible values of the state variables of a machine are defined within an enumerated type definition. An enumerated type is a list of possible values that a port, variable, or signal of that type is allowed to have.

10. Notation for a state diagram includes a series of bubbles (circles) containing state names and values of state variables in the form $\dfrac{\text{state\_name}}{\text{state\_variable(s)}}$.

11. The inputs and outputs of a state machine are labeled **in1, in2,** . . . , **in***x***/out1, out2,** . . . ,**out***x*.

12. Transitions between states can be conditional or unconditional. A conditional transition happens only under certain conditions of a control input and is labeled with the relevant input condition. An unconditional transition happens under all conditions of input and is labeled with an X for each input variable.

13. Conditional transitions in a VHDL state machine are described by an IF statement within a particular case of the CASE statement that describes the machine.

14. Mealy machine outputs are susceptible to asynchronous output changes if a combinational input changes out of synchronization with the clock. This can be remedied by clocking each output through a separate synchronizing flip-flop.

15. A maximum of 2^n states can be assigned to a state machine that has n state variables. If the number of states is less than 2^n, the unused states must be accounted for. Either they can be treated as don't care states, or they can be assigned a specific destination state, usually the reset state.

16. In a VHDL implementation of a state machine, any unused states can be covered with an **others** clause in the CASE statement that defines the machine.

GLOSSARY

Conditional transition   A transition between states of a state machine that occurs only under specific conditions of one or more control inputs.

Control input   A state machine input that directs the operation of the machine from state to state.

Enumerated type   A user-defined type in VHDL in which all possible values of a named identifier are listed in a type definition statement.

Form A contact   A normally open contact on a switch or relay.

Form B contact   A normally closed contact on a switch or relay.

Form C contact   A pair of contacts, one normally open and one normally closed, that operate with a single action of a switch or relay.

Mealy machine   A state machine whose output is determined by both the sequential logic and the combinational logic of the machine.

Moore machine   A state machine whose output is determined only by the sequential logic of the machine.

State machine   A synchronous sequential circuit, consisting of a sequential logic section and a combinational logic section, whose outputs and internal flip-flops progress through a predictable sequence of states in response to a clock and other input signals.

State variables   The variables held in the flip-flops of a state machine that determine its present state.

Unconditional transition   A transition between states of a state machine that occurs regardless of the status of any control inputs.

PROBLEMS

Problem numbers set in color indicate more difficult problems: those with underlines indicate most difficult problems.

Section 10.1   State Machines

10.1   Is the state machine in Figure 10.43 a Moore machine or a Mealy machine? Explain your answer.

10.2   Is the state machine in Figure 10.44 a Moore machine or a Mealy machine? Explain your answer.

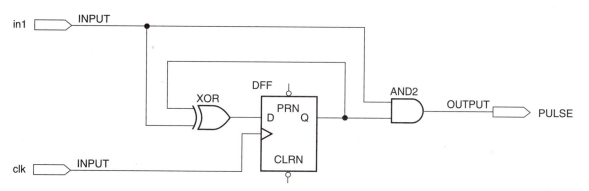

FIGURE 10.43
Problem 10.1
State Machine Circuit

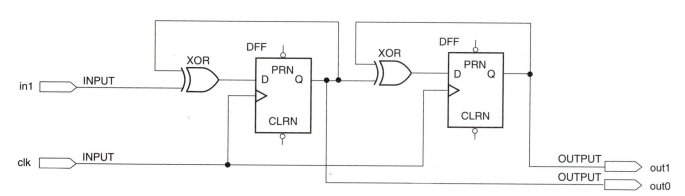

FIGURE 10.44
Problem 10.2
State Machine Circuit

Section 10.2   State Machines with No Control Inputs

10.3   A 4-bit Gray code sequence is shown in Table 10.7. Use classical design methods to design a counter with this sequence, using D flip-flops. Draw the resulting circuit diagram in a MAX+PLUS II Graphic Design File. Create a simulation to verify the circuit operation.

Table 10.7   4-bit Gray code sequence

$Q_3Q_2Q_1Q_0$
0000
0001
0011
0010
0110
0111
0101
0100
1100
1101
1111
1110
1010
1011
1001
1000

10.4   Use classical state machine design techniques to design a counter whose output sequence is shown in Table 10.8. (This is a divide-by-twelve counter in which the MSB output has a duty cycle of 50%.) Draw the state diagram, derive synchronous equations of the flip-flops, and draw the circuit implementation in MAX+PLUS II and create a simulation to verify the circuit's function.

Table 10.8   Counter Sequence for Problem 10.4

$Q_3Q_2Q_1Q_0$
0000
0001
0010
0011
0100
0101
1000
1001
1010
1011
1100
1101

10.5   Write the VHDL code required to implement a 4-bit Gray code counter. Create a simulation in MAX+PLUS II to verify the operation of the circuit.

10.6   Write the VHDL code required to implement a counter with the sequence shown in Table 10.8. Create a simulation in MAX+PLUS II to verify the operation of the circuit.

Section 10.3   State Machines with Control Inputs

10.7   Use classical state machine design techniques to find the Boolean next state and output equations for the state machine represented by the state diagram in Figure 10.45. Draw the state machine circuit as a Graphic Design File in MAX+PLUS II. Create a simulation file to verify the operation of the circuit. Briefly explain the intended function of the state machine.

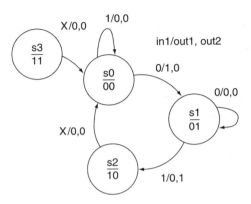

FIGURE 10.45
Problem 10.7
State Diagram

10.8   Referring to the simulation for the state machine in Problem 10.7, briefly explain why it is susceptible to asynchronous input changes. Modify the state machine circuit to eliminate the asynchronous behavior of the outputs. Create a MAX+PLUS II simulation to verify the function of the modified state machine.

10.9   Write the VHDL code required to implement the state machine in Problem 10.7. Create a simulation to verify the operation of the state machine.

10.10   A state machine is used to control an analog-to-digital converter, as shown in the block diagram of Figure 10.46.

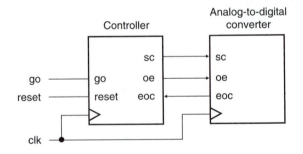

FIGURE 10.46
Problem 10.10
Analog-to-Digital Converter and Controller

The controller has four states, defined by state variables Q_1 and Q_0 as follows: **idle** (00), **start** (01), **waiting** (11), and **read** (10). There are two outputs: **sc** (Start Conversion; active-HIGH) and **oe** (Output Enable; active LOW). There are four inputs: clock, **go** (active-LOW) **eoc** (End of Conversion), and asynchronous reset (active LOW). The machine operates as follows:

a. In the **idle** state, the outputs are: sc = 0, oe = 1. The machine defaults to the **idle** state when the machine is reset.

b. Upon detecting a 0 at the **go** input, the machine makes a transition to the **start** state. In this transition, sc = 1, oe = 1.

c. The machine makes an unconditional transition to the **waiting** state; sc = 0, oe = 1. It remains in this state, with no output change, until input eoc = 1.

d. When eoc = 1, the machine goes to the **read** state; sc = 0, oe = 0.

e. The machine makes an unconditional transition to the **idle** state; sc = 0, oe = 1.

Use classical state machine design techniques to design the controller. Draw the required circuit in MAX+PLUS II and create a simulation to verify its operation. Is this machine vulnerable to asynchronous input change?

10.11 Use VHDL to implement the controller circuit of Problem 10.10. Create a simulation to verify its operation.

10.12 Write a VHDL file for a state machine that selects a 3-bit binary or Gray code count, depending on the state of an input called **gray**. If **gray** = **1**, count in Gray code. Otherwise count in binary. Create a simulation file that verifies the operation of the circuit, clearly showing the full Gray code count, binary count, and reset function.

Section 10.4 Switch Debouncer for a Normally Open Pushbutton Switch

10.13 Why is it not possible to debounce the pushbuttons on the Altera UP-1 board using a NAND latch?

10.14 Refer to the switch debouncer circuit in Figure 10.24 (p. 476). For how many clock periods must the input of the debouncer remain stable before the output can change?

10.15 What is the maximum switch bounce time that can be removed by the circuit of Figure 10.24 if the clock at the shift register is running at a rate of 480 Hz?

10.16 Briefly explain how the Exclusive NOR gate in the debounce circuit of Figure 10.24 determines if switch bounce has occurred.

10.17 Refer to the section on the behaviorally designed switch debouncer in Section 10.4. For how many clock periods must the input of the debouncer remain stable before the output can change? What is the maximum switch bounce time that can be removed by the circuit of Figure 10.24. if the state machine clock is running at a rate of 480 Hz?

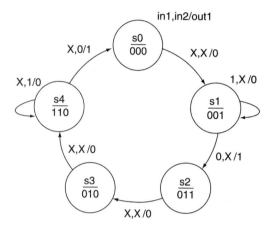

FIGURE 10.47
Problem 10.18
State Diagram

Section 10.5 Unused States in State Machines

10.18 Refer to the state diagram in Figure 10.47.

a. How many state variables are required to implement this state machine? Why?

b. How many unused states are there for this state machine? List the unused states.

c. Complete the partial timing diagram shown in Figure 10.48 to illustrate one complete cycle of the state machine represented by the state diagram of Figure 10.47.

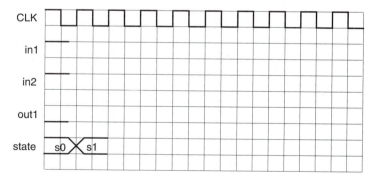

FIGURE 10.48
Problem 10.18
Partial Timing Diagram

10.19 Write the VHDL code required to implement the state machine described by the state diagram of Figure 10.47. Create a simulation file to verify the operation of the circuit.

10.20 Use classical state machine design techniques to design a state machine described by the state diagram of Figure 10.49. Briefly describe the intended operation of the circuit. Create a MAX+PLUS II simulation to verify the operation of the state machine design. Unused states may be treated as don't care states, but unspecified outputs should always be assigned to 0.

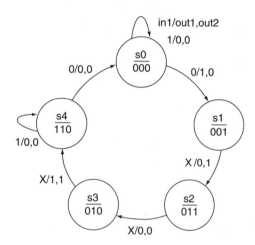

FIGURE 10.49
Problem 10.20
State Diagram

10.21 Determine the next state for each of the unused states of the state machine designed in Problem 10.20. Use this analysis to redraw the state diagram of Figure 10.49 so that it properly includes the unused states. (There is more than one right answer, depending on the result of the Boolean simplification process used in Problem 10.20.)

10.22 Write the VHDL code for the state machine described in Problem 10.20. Create a MAX+PLUS II simulation to verify the function of the state machine.

10.23 A state machine is used to control an analog-to-digital converter, as shown in the block diagram of Figure 10.46. (The following description is a modified version of the controller described in Problem 10.10.)

Five states are used: **idle, start, waiting1, waiting2,** and **read.** There are two outputs: **sc** (Start Conversion; active-HIGH) and **oe** (Output Enable; active HIGH). There are four inputs: clock, reset, **go,** and **eoc** (End of Conversion). The machine operates as follows:

a. In the **idle** state, the outputs are: sc = 0, oe = 0. The machine defaults to the **idle** state when asynchronously reset and remains there until go = 0.

b. When go = 0, the machine makes a transition to the **start** state. In this transition, sc = 1, oe = 0.

c. The machine makes an unconditional transition to the **waiting1** state; sc = 0, oe = 0. It remains in this state, with no output change, until input eoc = 0.

d. When eoc = 0, the machine goes to the **waiting2** state; sc = 0, oe = 0. It remains in this state, with no output change, until input eoc = 1.

e. The machine makes a transition to the **read** state when eoc = 1, sc = 0, oe = 1.

f. The machine makes an unconditional transition to the **idle** state; sc =, 0, oe = 0.

After reviewing the block diagram and the states just listed,

a. Draw the state diagram of the controller.

b. How many state variables are required for the controller described in this question?

10.24 Write the VHDL code for the state machine described in Problem 10.23. Create a simulation file to verify the function of the design.

ANSWERS TO SECTION REVIEW PROBLEMS

Section 10.1

10.1 A Moore state machine has outputs that depend only on the states of the flip-flops in the machine. A Mealy machine's outputs depend on the states of its flip-flops as well as the gates of the machine's control section. This can result in asynchronous output changes in the Mealy machine outputs.

Section 10.2

10.2

$$J_2 = Q_1\overline{Q_0}$$
$$K_2 = \overline{Q_1}Q_0$$
$$J_1 = Q_2Q_0$$
$$K_1 = Q_2Q_0$$
$$J_0 = \overline{Q_2}Q_1 + Q_2Q_1 = \overline{Q_2 \oplus Q_1}$$
$$K_0 = \overline{Q_2}Q_1 + Q_2\overline{Q_1} = Q_2 \oplus Q_1$$

Section 10.3

10.3 The output flip-flop synchronizes the output to the system clock, yielding the following advantages: (1) the output is always a known width of one clock cycle; and (2) the output is not vulnerable to change due to asynchronous changes of input.

Section 10.4

10.4 $T_c = 3.75$ ms; $f_c = 267$ Hz

Section 10.5

10.5 Moore machine. The outputs are derived entirely from the output states of the state machine and are not vulnerable to asynchronous changes of input.

Logic Gate Circuitry

CHAPTER OBJECTIVES

Upon successful completion of this chapter, you will be able to:

- Name the various logic families most commonly in use today and state several advantages and disadvantages of each.
- Define propagation delay.
- Calculate propagation delay of simple circuits, using data sheets.
- Define fanout and calculate its value, using data sheets.
- Calculate power dissipation of TTL and CMOS circuits.
- Calculate noise margin of a logic gate from data sheets.
- Draw circuits that will interface various CMOS and TTL gates.
- Explain how a bipolar junction transistor can be used as a logic inverter.
- Describe the function of a TTL input transistor in all possible input states: HIGH, LOW, and open-circuit.
- Explain the operation of a totem pole output.
- Illustrate how a totem pole output generates power line noise and describe how to remedy this problem.
- Illustrate why totem pole outputs cannot be tied together.
- Explain the difference between open-collector and totem pole outputs of a TTL gate.
- Illustrate the operation of TTL open-collector inverter, NAND, and NOR gates.
- Write the Boolean expression of a wired-AND circuit.
- Design a circuit that uses an open-collector gate to drive a high-current load.
- Calculate the value of a pull-up resistor at the output of an open-collector gate.
- Explain the operation of a tristate gate and name several of its advantages.
- Design a circuit using a tristate bus driver to direct the flow of data from one device to another.
- Describe the basic structure of a MOSFET and state its bias voltage requirements.
- Draw the circuit of an CMOS inverter and show how it works.

- Draw the circuits of CMOS NAND and NOR gates and explain the operation of each.
- Design a circuit using a CMOS transmission gate to enable and inhibit digital and analog signals.
- Interpret TTL data sheets to distinguish between the various TTL families.
- Describe the use of the Schottky barrier diode in TTL gates.
- Calculate speed-power products from data sheets.

Our study of logic gates and flip-flops in previous chapters has concentrated on digital *logic* and has largely ignored digital *electronics*. Digital logic devices are electronic circuits with their own characteristic voltages and currents. No serious study of digital circuitry is complete without some examination of this topic.

It is particularly important to understand the inputs and outputs of logic devices as electronic circuits. Knowing the input and output voltages and currents of these circuits is essential, since gate loading, power dissipation, noise voltages, and interfacing between logic families depend on them. The switching speed of device outputs is also fundamental and may be a consideration when choosing the logic family for a circuit design.

Input and output voltages of logic devices are specified in manufacturers' data sheets, which allows us to take a "black box" approach initially.

Later in the chapter, we will examine some basic digital circuits at a transistor level, since digital logic is based on transistor switching. Two major types of transistors, the bipolar junction transistor and the metal-oxide-semiconductor field effect transistor (MOSFET), form the basis of the major logic families in use today. Transistor-transistor logic (TTL) is based on the bipolar transistor. Complementary MOS (CMOS) is based on the MOSFET.

We will briefly study the operating characteristics of both bipolar transistors and MOSFETs and then see how these devices give rise to the electrical characteristics of simple logic gates.

11.1 Electrical Characteristics of Logic Gates

KEY TERMS

TTL Transistor-transistor logic. A logic family based on bipolar transistors.

CMOS Complementary metal-oxide semiconductor. A logic family based on metal-oxide-semiconductor field effect transistors (MOSFETs).

ECL Emitter coupled logic. A high-speed logic family based on bipolar transistors.

When we examine the electrical characteristics of logic circuits, we see them as practical, rather than ideal devices. We look at properties such as switching speed, power dissipation, noise immunity, and current-driving capability. There are several commonly available logic families in use today, each having a unique set of electrical characteristics that differentiates it from all the others. Each logic family gives superior performance in one or more of its electrical properties.

CMOS consumes very little power, has excellent noise immunity, and can be used with a wide range of power supply voltages.

TTL has a larger current-driving capability than CMOS. Its power consumption is higher than that of CMOS, and its power supply requirements are more rigid.

ECL is fast, making it the choice for high-speed applications. It is inferior to CMOS and TTL in terms of noise immunity and power consumption.

TTL and CMOS gates come in a wide range of subfamilies. Table 11.1 lists some of the TTL and CMOS variations of the quadruple 2-input NAND gate. All gates listed have

Table 11.1   Part Numbers for a Quad 2-input NAND Gate in Different Logic Families

	Part Number	Logic Family
TTL	74LS00	Low-power Schottky TTL
	74ALS00	Advanced low-power Schottky TTL
	74F00	Fast TTL
CMOS	74HC00	High-speed CMOS
	74HCT00	High-speed CMOS (TTL-compatible inputs)
	74LVX00	Low-voltage CMOS

the same logic function but different electrical characteristics. Other gates would be similarly designated, with the last two or three digits indicating the gate function (e.g., a quadruple 2-input NOR gate would be designated 74LS02, 74ALS02, 74F02, etc.).

We will examine four electrical characteristics of TTL and CMOS circuits: propagation delay, fanout, noise margin, and power dissipation. The first of these has to do with speed of output response to a change of input. The last three have to do with input and output voltages and currents. All four properties can be read directly from specifications given in a manufacturer's data sheet or derived from these specifications.

Figures 11.1 and 11.2 show how the input and output voltages and currents are defined in a 74XX00 NAND gate. This designation can be generalized to any logic gate input or output.

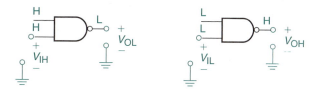

FIGURE 11.1
Input/Output Voltage Parameters

FIGURE 11.2
Input/Output Current Parameters

The voltages and currents are designated with two subscripts, one that designates an input or output and another that indicates the logic level. For example, V_{OL} is the voltage at the gate output when the output is in the logic LOW state. I_{IL} is the input current when the input is in the LOW state.

These voltages and currents are specified in manufacturers' published data sheets, which are usually available in print form in a data book or in an electronic format, such as Portable Document Format (**pdf**) on a CD or internet site.

Figure 11.3 shows a data sheet for a 74LS00 NAND gate, which also shows parameter values for a 54LS00 device. A 54-series device is manufactured to military specifications, which require a high range of environmental operating conditions. A 74-series device is suitable for general or commercial use. We will limit ourselves to the 74-series devices.

The voltage and current parameters indicated in Figures 11.1 and 11.2 are all shown in the 74LS00 data sheet. Some parameters are shown as typical values, as well as maximum or minimum. Typical values should be considered "information only" as device manufacturers

QUAD 2-INPUT NAND GATE

• ESD > 3500 Volts

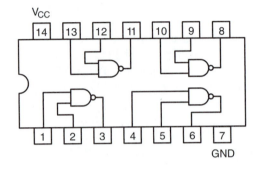

V_CC

| 14 | 13 | 12 | 11 | 10 | 9 | 8 |

| 1 | 2 | 3 | 4 | 5 | 6 | 7 |

GND

SN54/74LS00

QUAD 2-INPUT NAND GATE

LOW POWER SCHOTTKY

J SUFFIX
CERAMIC
CASE 632-08
14   1

N SUFFIX
PLASTIC
CASE 646-06
14   1

D SUFFIX
SOIC
CASE 751A-02
14   1

ORDERING INFORMATION

SN54LSXXJ	Ceramic
SN74LSXXN	Plastic
SN74LSXXD	SOIC

GUARANTEED OPERATING RANGES

Symbol	Parameter		Min	Typ	Max	Unit
V_{CC}	Supply Voltage	54	4.5	5.0	5.5	V
		74	4.75	5.0	5.25	
T_A	Operating Ambient Temperature Range	54	−55	25	125	°C
		74	0	25	70	
I_{OH}	Output Current — High	54, 74			−0.4	mA
I_{OL}	Output Current — Low	54			4.0	mA
		74			8.0	

FIGURE 11.3
74LS00 Data (1 of 2) Reprinted with permission of Motorola.

SN54/74LS00

DC CHARACTERISTICS OVER OPERATING TEMPERATURE RANGE (unless otherwise specified)

Symbol	Parameter		Min	Typ	Max	Unit	Test Conditions	
V_{IH}	Input HIGH Voltage		2.0			V	Guaranteed Input HIGH Voltage for All Inputs	
V_{IL}	Input LOW Voltage	54			0.7	V	Guaranteed Input LOW Voltage for All Inputs	
		74			0.8			
V_{IK}	Input Clamp Diode Voltage			−0.65	−1.5	V	V_{CC} = MIN, I_{IN} = −18 mA	
V_{OH}	Output HIGH Voltage	54	2.5	3.5		V	V_{CC} = MIN, I_{OH} = MAX, V_{IN} = V_{IH} or V_{IL} per Truth Table	
		74	2.7	3.5		V		
V_{OL}	Output LOW Voltage	54, 74		0.25	0.4	V	I_{OL} = 4.0 mA	V_{CC} = V_{CC} MIN, V_{IN} = V_{IL} or V_{IH} per Truth Table
		74		0.35	0.5	V	I_{OL} = 8.0 mA	
I_{IH}	Input HIGH Current				20	µA	V_{CC} = MAX, V_{IN} = 2.7 V	
					0.1	mA	V_{CC} = MAX, V_{IN} = 7.0 V	
I_{IL}	Input LOW Current				−0.4	mA	V_{CC} = MAX, V_{IN} = 0.4 V	
I_{OS}	Short Circuit Current (Note 1)		−20		−100	mA	V_{CC} = MAX	
I_{CC}	Power Supply Current Total, Output HIGH				1.6	mA	V_{CC} = MAX	
	Total, Output LOW				4.4			

Note 1: Not more than one output should be shorted at a time, nor for more than 1 second.

AC CHARACTERISTICS (T_A = 25°C)

Symbol	Parameter	Min	Typ	Max	Unit	Test Conditions
t_{PLH}	Turn-Off Delay, Input to Output		9.0	15	ns	V_{CC} = 5.0 V, C_L = 15 pF
t_{PHL}	Turn-On Delay, Input to Output		10	15	ns	

FIGURE 11.3
74LS00 Data (2 of 2) Reprinted with permission of Motorola.

do not guarantee these values. An exception to this would be the supply voltage, V_{CC}, whose typical value is simply indicated as the average of maximum and minimum values.

Note that I_{IH} and I_{IL} are shown in Figure 11.2 as flowing in opposite directions, as are I_{OH} and I_{OL}. On a data sheet, a current entering a gate is indicated as positive and a current leaving the gate is shown as having a negative value. The reason for these current directions will become apparent when we examine the internal circuits of the gates later in the chapter.

EXAMPLE 11.1

What is the maximum value of V_{OL} for a 74LS00 NAND gate when the output current is at its maximum value?

Solution When the output is in the LOW state, the output current is given by I_{OL}, which has a maximum value of 8 mA. The output voltage, V_{OL}, is specified for a value of 4 mA and for 8 mA. Since the output condition is specified for maximum I_{OL} (8 mA), then V_{OL} = 0.5 V.

The 74XX00 NAND gate data is sufficient to represent any logic functions having "normal" output current within its particular logic family. This data can be used for most gate or flip-flop circuits within the family. Some specialized devices with higher-current outputs (e.g., 74XX244 octal tristate buffers) have a different set of electrical characteristics within their family.

In the following sections of the chapter, we will use a NAND gate from each of three device families (74LS00, 74HC00A, and 74HCT00A) for illustrating the general principles of the various electrical characteristics. Devices from other families will also be used in examples and problems. Data sheets for the various devices are included in Appendix C.

▌▌ SECTION 11.1 REVIEW PROBLEM

11.1 What are the maximum values of voltage and current we can expect at the output of a 74LS00 NAND gate when both inputs are LOW?

11.2 Propagation Delay

KEY TERMS

t_{pHL} Propagation delay when the device output is changing from HIGH to LOW.

t_{pLH} Propagation delay when the device output is changing from LOW to HIGH.

Propagation delay occurs because the output of a logic gate or flip-flop cannot respond instantaneously to changes at its input. There is a short delay, on the order of several nanoseconds, between input change and output response. This is largely due to the charging and discharging of capacitances inherent in the switching transistors of the gate or flip-flop.

Figure 11.4 shows propagation delay in two gates: a 74XX00 NAND gate and a 74XX08 AND gate. Each gate has an identical input waveform, a LOW-HIGH-LOW pulse. After each input transition, the output changes after a short delay, t_p.

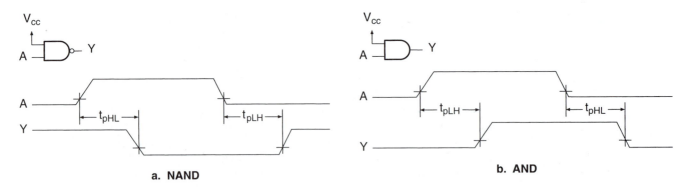

FIGURE 11.4
Propagation Delay in NAND and AND Gates

Two delays are shown for each gate: t_{pLH} and t_{pHL}. The *LH* and *HL* subscripts show the direction of change at the gate *output; LH* indicates that the output goes from LOW to HIGH, and *HL* shows the output changing from HIGH to LOW.

Propagation delay is the time between input and output voltages passing through a standard reference value. The reference voltage for standard TTL is 1.5 V. LSTTL and CMOS have different reference voltages, as follows.

> NOTE
>
> **Propagation Delay for Various Logic Families:**
>
> LSTTL: Time from 1.3 V at input to 1.3 V at output.
> Other TTL: Time from 1.5 V at input to 1.5 V at output.
> CMOS: Time from 50% of maximum input to 50% of maximum output.

EXAMPLE 11.2

Use the data sheet in Figure 11.3, as well as those in Appendix C, to find the maximum propagation delays for each of the following gates: 74LS00 (quadruple 2-input NAND), 74LS02 (quadruple 2-input NOR), 74LS08 (quadruple 2-input AND), and 74LS32 (quadruple 2-input OR).

Solution

Table 11.2 Propagation Delays of 74LS Gates

	74LS00	74LS02	74LS08	74LS32
t_{pLH}	15 ns	15 ns	15 ns	22 ns
t_{pHL}	15 ns	15 ns	20 ns	22 ns

Table 11.2 shows the variation of propagation delay among logic gates of the same family (74LS TTL). Since each logic function has a different circuit, its propagation delay will differ from those of gates with different functions.

EXAMPLE 11.3

Use data sheets to find the maximum propagation delays for each of the following logic gates: 74F00, 74AS00, 74ALS00, 74HC00, and 74HCT00.

Solution

Table 11.3 Propagation Delays of 74LS Gates

	74F00*	74AS00	74ALS00	74HC00**	74HCT00***
t_{pLH}	6 ns	4.5 ns	11 ns	15 ns	19 ns
t_{pHL}	5.3 ns	4 ns	8 ns	15 ns	19 ns

*Temperature range (74F00): 0°C to 70°C.
**$V_{CC} = 4.5$ V, temperature range (74HC00): -55°C to 25°C.
***$V_{CC} = 5$ V, temperature range (74HCT00): -55°C to 25°C.

As indicated by the notes for Table 11.3, propagation delay (and other parameters) vary with certain operating conditions, such as ambient temperature and power supply voltage. Always make sure that the operating conditions are correctly specified when looking up a data sheet parameter.

All gates in Example 11.3 have the same logic function (2-input NAND), but different propagation delay times. We might ask, "Why not always use the advanced Schottky TTL gate (74AS00), since it is the fastest?" The main reason is that it has the highest power dissipation of the gates shown. We wouldn't know this without looking up other specs on the data sheet. (We will learn how to do this later in the chapter.) Thus, it is important to make design decisions based on complete information, not just one parameter.

Propagation Delay in Logic Circuits

A circuit consisting of two or more gates or flip-flops has a propagation delay that is the sum of delays *in the input-to-output path*. Delays in gates that do not affect the circuit output are disregarded. Figure 11.5 shows how propagation delay works in a simple logic circuit consisting of a 74HC08 AND gate and a 74HC32 OR gate. Changes at inputs A and B must propagate through both gates to affect the output. The total delay in such a case is the sum of t_{p1} and t_{p2}. A change at input C must pass only through gate 2. The circuit delay resulting from this change is only t_{p2}.

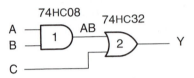

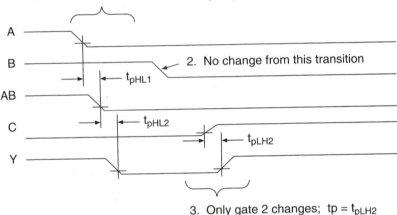

FIGURE 11.5
Propagation Delays in a Logic Gate Circuit

The timing diagram in Figure 11.5 shows the changes at inputs A, B, and C and the resulting transitions at all gate outputs.

Assume $V_{CC} = 4.5$ V and temperature range is $-55°C$ to $25°C$.

1. When A goes LOW, AB, the output of gate 1, also goes LOW after a maximum delay of $t_{pHL} = 15$ ns. This makes Y go LOW after a further delay of up to $t_{pHL} = 15$ ns. Total delay: $t_p = t_{pHL1} + t_{pHL2} = 15$ ns $+ 15$ ns $= 30$ ns, max.

2. The HIGH-to-LOW transition at input B has no effect, since there is no difference between $0 \cdot 1$ and $0 \cdot 0$. AB is already LOW.

3. The LOW-to-HIGH transition at input C makes Y go HIGH after a maximum delay of $t_{pLH2} = 15$ ns.

▌▌▌ SECTION 11.2 REVIEW PROBLEM

11.2 Assume the gates in Figure 11.5 are replaced by a 74LS08 AND gate and a 74LS32 OR gate. Repeat the calculations of for the propagation delays if the waveforms of Figure 11.5 are applied to the circuit. The data sheets for the 74LS08 and 74LS32 are found in Appendix C.

11.3 Fanout

We have assumed that logic gates are able to drive any number of other logic gates. Since gates are electrical devices with finite current-driving capabilities, this is obviously not the case. The number of gates ("loads") a logic gate can drive is referred to as its **fanout.**

NOTE

Fanout is simply an application of Kirchhoff's current law: The algebraic sum of currents at a node must be zero. Thus, the fanout of a logic gate is limited by:
a. The maximum current its output can supply safely in a given logic state (I_{OH} or I_{OL}), and
b. The current requirements of the load to which it is connected (I_{IH} or I_{IL}).

Figure 11.6 shows the fanout of an AND gate when its output is in the HIGH and LOW states. The AND gate, or **driving gate,** supplies current to the inputs of the other four gates, which are called the **load gates.**

Each load gate requires a fixed amount of input current, depending on which state it is in. The sum of these input currents equals the current supplied by the driving gate. The

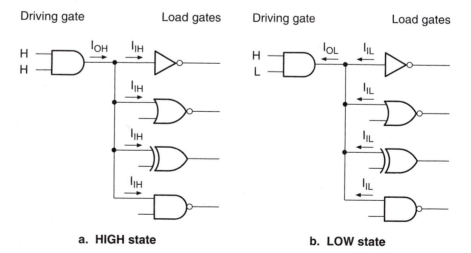

a. **HIGH state**   b. **LOW state**

FIGURE 11.6
Driving Gates and Load Gates

fanout is determined by the amount of current the driving gate can supply without damaging its output circuit.

The input and output currents of a gate are established by its internal circuitry. These values are usually the same for two gates in the same family, since the input and output circuitry of a gate is common to all members of the family. Exceptions may occur when the output of a particular gate, such as the 74XX244 octal three-state buffer, has additional output buffering or an input of a gate such as a 74LS86 Exclusive OR is equivalent to more than one input load.

III EXAMPLE 11.4

FIGURE 11.7
Example 11.4
Output Current due to One
Load Gate

The gates in Figure 11.7a and b are 74LS00 NAND gates. Determine the output current of the driving gate in each figure.

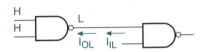

a. Low output on driving gate

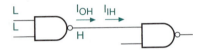

b. High output on driving gate

Solution   From the 74LS00 data sheet, $I_{IL} = -0.4$ mA and $I_{IH} = 20$ μA. (There are two values of I_{IH} given in the data sheet. Choose the one for the condition $V_{IN} = 2.7$ V, which is the minimum output voltage of a driving gate in the HIGH state (V_{OH}). The other value is not appropriate since a gate will never have a 7 V output, as specified in the condition, if its supply voltage is 5 V.)

Since the driving gate is driving one load, its output current is the same as the input current of the load gate. Therefore, the driving gate output currents are given by $I_{OL} = 0.4$ mA (positive, since it is entering the driving gate output) and $I_{OH} = -20$ μA (negative, since it is leaving the driving gate output).

III EXAMPLE 11.5

FIGURE 11.8
Example 11.5
Output Current due to Two Load
Gates

Determine the output current of the driving gate in each of Figures 11.8a and b if the gates are all 74LS00 NAND gates.

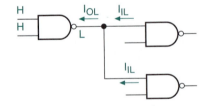

a. Low output on driving gate

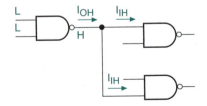

b. High output on driving gate

Solution   Since there are two identical load gates in the circuits of Figure 11.8, the driving gate output current will be twice the load gate input current.

$$I_{OL} = 2 \times 0.4 \text{ mA} = 0.8 \text{ mA}.$$
$$I_{OH} = 2 \times (-20 \text{ μA}) = -40 \text{ μA}.$$

Figure 11.9 shows the extension of the circuits in Figures 11.7 and 11.8, where the number of load gates is the maximum that can be driven by the driving gate. This is the condition used to calculate fanout.

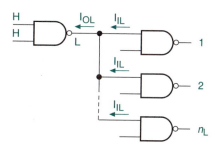

a. Low output on driving gate

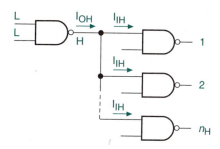

b. High output on driving gate

FIGURE 11.9
Output Current to Fanout Calculation

If the load gates each represent the same load, then by Kirchhoff's current law (KCL):

$$I_{OL} = I_{IL1} + I_{IL2} + \cdots I_{ILnL} = n_L \, I_{IL}$$
$$\text{and} \qquad I_{OH} = I_{IH1} + I_{IH2} + \cdots + I_{IHnH} = n_H \, I_{IH}$$

The fanout of the driving gate in the LOW and HIGH states can be calculated as:

$$n_L = \frac{I_{OL}}{I_{IL}}$$

$$\text{and} \quad n_H = \frac{I_{OH}}{I_{IH}}$$

By convention, current entering a gate (I_{IH}, I_{OL}) is denoted as positive, and current leaving a gate (I_{IL}, I_{OH}) is denoted as negative. When current is leaving a gate, we say the gate is **sourcing** current. When current is entering a gate, we say the gate is **sinking** current.

Note that the output of a gate does not always source current, nor does an input always sink current. The current direction changes for the HIGH and LOW states at the same terminal. The reason for this will become apparent when we study the circuitry of logic gate inputs and outputs.

▌▌ EXAMPLE 11.6

How many 74LS00 inputs can a 74LS00 NAND gate drive? (that is, what is the fanout of a 74LS00 NAND gate?)

Solution   We must consider the following cases:

a. When the output of the driving gate is LOW

b. When the output of the driving gate is HIGH

Output LOW:

$$I_{OL} = 8 \text{ mA (sinking)}$$
$$I_{IL} = -0.4 \text{ mA (sourcing)}$$
$$n_L = 8 \text{ mA}/0.4 \text{ mA} = 20$$

Output HIGH:

$$I_{OH} = -0.4 \text{ mA (sourcing)}$$
$$I_{IH} = 20 \text{ } \mu\text{A (sinking)}$$
$$n_H = 0.4 \text{ mA}/20 \text{ } \mu\text{A} = 20$$

Since $n_L = n_H$, fanout is 20.

We disregard the negative sign in our calculations, since the input current of the load gate and output current of the driving gate are actually in the same direction. For example, even though I_{OH} is leaving the driving gate (negative), I_{IH} is entering the load gates (positive). These currents flow in the same direction. If we include the minus sign in our calculation, we get a negative value of fanout, which is meaningless.

▌▌

The fanout in both HIGH and LOW states is the same in this case, but that is not always so. If the values of HIGH- and LOW-state fanout are different, the smallest value must be used. For example, if a gate can drive four loads in the HIGH state or eight in the LOW state, the fanout of the driving gate is four loads. If we attempt to drive eight loads, we can't guarantee enough driving current to supply all loads in both states.

If a gate from one logic family is used to drive gates from another logic family, we must use the output parameters (I_{OL}, I_{OH}) for the driving gate and the input parameters (I_{IL}, I_{IH}) for the load gates.

▌▌ EXAMPLE 11.7

Calculate the maximum number of Schottky TTL loads (74SXX series) that a 74LS86 XOR gate can drive.

Solution

Driving gate:	74LS86	$I_{OH} = -0.4 \text{ mA}$,
		$I_{OL} = 8 \text{ mA}$
Load gates:	74SXX	$I_{IH} = 50 \text{ } \mu\text{A}$,
		$I_{IL} = -2 \text{ mA}$

Output LOW:

$$I_{OL} = 8 \text{ mA (sinking)}$$
$$I_{IL} = -2 \text{ mA (sourcing)}$$
$$n_L = 8 \text{ mA}/2 \text{ mA} = 4$$

Output HIGH:

$$I_{OH} = -0.4 \text{ mA (sourcing)}$$
$$I_{IH} = 50 \text{ } \mu\text{A (sinking)}$$
$$n_H = 0.4 \text{ mA}/50 \text{ } \mu\text{A} = 8$$

Since $n_L < n_H$, fanout $= n_L = 4$.

▌▌

www.electronictech.com

What happens if we load a gate output beyond its rated fanout? Adding more load gates will do this by increasing the value of I_{OL} beyond its maximum rating. If enough load is added, the output of the driving gate might be destroyed by the heat generated by the excess current. More likely, the performance of the driving gate will be degraded.

Figure 11.10 shows the relationship between output voltage and current for a 74LS00 and a 74F00 NAND gate. Figure 11.10a shows that the output voltage (LOW state) increases with increasing sink current. Figure 11.10b indicates a decrease in HIGH state output voltage with an increase of source current.

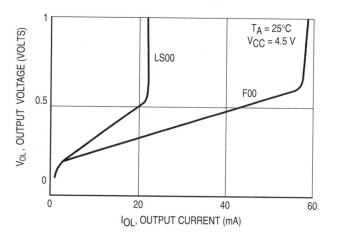

a. Output low characteristic

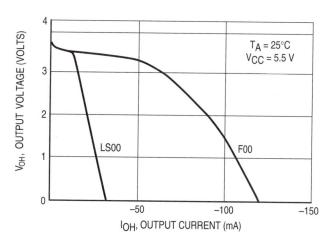

b. Output high characteristic

FIGURE 11.10

Output Characteristics of 74LS00 and 74F00 Gate. Reprinted with permission of Motorola

In other words, a greater load in either state takes the output voltage further away from its nominal value. This has an effect on other performance factors, such as noise margin, which we will examine in a later section of the chapter.

> **NOTE**
>
> The output voltage of a logic gate is defined in a datasheet for a particular value of output current.

We will examine the fanout of CMOS devices in a later section on interfacing between CMOS and TTL.

■■ SECTION 11.3 REVIEW PROBLEM

11.3 The input and output currents I_{OH}, I_{OL}, I_{IH}, and I_{IL} of a TTL device may be classified as source currents or sink currents. List each input or output current as a source or sink current.

14.4 Power Dissipation

> ### KEY TERMS
>
> **Power dissipation**   The electrical energy used by a logic circuit in a specified period of time. Abbreviation: P_D
>
> V_{CC} TTL or high-speed CMOS supply voltage.
>
> I_{CC} Total TTL or high-speed CMOS supply current.
>
> I_{CCH} TTL supply current with all outputs HIGH.
>
> I_{CCL} TTL supply current with all outputs LOW.
>
> I_T When referring to CMOS supply current, the sum of static and dynamic supply currents.
>
> C_{PD} Internal capacitance of a high-speed CMOS device used to calculate its power dissipation.

Electronic logic gates require a certain amount of electrical energy to operate. The measure of the energy used over time is called **power dissipation.** Each of the different families of logic has a characteristic range of values for the power it consumes.

For TTL and CMOS, the power dissipation is calculated as follows:

TTL: $P_D = V_{CC} I_{CC}$

High-Speed CMOS: $P_D = V_{CC} I_T$ (I_T = quiescent + dynamic supply current)

Figure 11.11 shows the supply voltage and current in a 74XX00 NAND gate.

FIGURE 11.11
Power Supply Voltage and Current in a 74XX00 NAND gate.

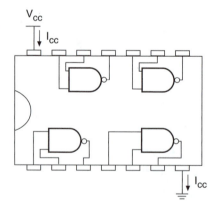

The main difference between the two families is the calculation of supply current.

The supply current in a TTL device is different when its outputs are HIGH than when they are LOW. Thus, supply current, I_{CC}, and therefore power dissipation, depends on the states of the device outputs. If the outputs are switching, I_{CC} is proportional to output duty cycle.

In a CMOS device, very little power is consumed when the device outputs are static. Much more current is drawn from the supply when the outputs switch from one state to another. Thus, the power dissipation of a device depends on the switching frequency of its outputs.

Power Dissipation in TTL Devices

Two values are given for supply current in a TTL data sheet. I_{CCL} is the current drawn from the power supply when all gate outputs are LOW. I_{CCH} is the current drawn from the supply when all outputs are HIGH. If the gate outputs are not all at the same level, the supply current is the sum of currents given by:

$$I_{CC} = \frac{n_H}{n} I_{CCH} + \frac{n_L}{n} I_{CCL}$$

where

n is the total number of gates in the package
n_H is the number of gates whose output is HIGH
n_L is the number of gates whose output is LOW

The power dissipation of a TTL chip also depends on the duty cycle of the gate outputs. That is, it depends on the fraction of time that the chip's outputs are HIGH.

If we assume that, on average, the outputs of a chip are switching with a duty cycle of 50%, the supply current can be calculated as follows:

$$I_{CC} = (I_{CCH} + I_{CCL})/2$$

If the output duty cycle is other than 50%, the supply current is given by:

$$I_{CC} = DC\, I_{CCH} + (1 - DC)\, I_{CCL}$$

where DC = duty cycle.

▌▌ EXAMPLE 11.8

Figure 11.12 shows a circuit constructed from the gates in a 74XX00 quadruple 2-input NAND gate package. Use the data sheet shown in Figure 11.3 to determine the maximum power dissipation of the circuit if the input is $DCBA = 1001$ and the gates are 74LS00 NANDs. Refer to the data sheets in Appendix C and repeat the calculation for 74ALS00 and 74AS00 gates.

FIGURE 11.12

Power Dissipation of 74XX00 NAND

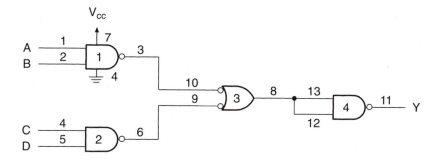

Solution

Gate 1: $\overline{AB} = 1$

Gate 2: $\overline{CD} = 1$

Gate 3: $\overline{AB + CD} = 0$

Gate 4: $\overline{\overline{AB + CD}} = 1$

Since three outputs are HIGH and one is LOW, the supply current is given by:

$$I_{CC} = \frac{n_H}{n} I_{CCH} + \frac{n_L}{n} I_{CCL}$$

$$= \frac{3}{4} I_{CCH} + \frac{1}{4} I_{CCL}$$

Maximum supply current for each device is:

$$74LS00: I_{CC} = 0.75(1.6 \text{ mA}) + 0.25(4.4 \text{ mA}) = 2.3 \text{ mA}$$
$$74ALS00: I_{CC} = 0.75(0.85 \text{ mA}) + 0.25(3 \text{ mA}) = 1.3875 \text{ mA}$$
$$74AS00: I_{CC} = 0.75(3.2 \text{ mA}) + 0.25(17.4 \text{ mA}) = 6.75 \text{ mA}$$

Maximum power dissipation for each device is:

$$74LS00: P_D = V_{CC} I_{CC} = (5 \text{ V})(2.3 \text{ mA}) = 11.5 \text{ mW}$$
$$74ALS00: P_D = V_{CC} I_{CC} = (5\text{V})(1.3875 \text{ mA}) = 6.94 \text{ mW}$$
$$74AS00: P_D = V_{CC} I_{CC} = (5\text{V})(6.75 \text{ mA}) = 33.75 \text{ mW}$$

$(1 \text{ mW} = 1 \text{ milliwatt} = 10^{-3} \text{ W.})$

III EXAMPLE 11.9

Find the maximum power dissipation of the circuit in Figure 11.12 if the gates are 74LS00 and the gate outputs are switching with an average duty cycle of 30%.

Solution

$$I_{CC} = 0.3 I_{CCH} + 0.7 I_{CCL}$$
$$I_{CC} = 0.3(1.6 \text{ mA}) + 0.7(4.4 \text{ mA})$$
$$= 3.56 \text{ mA}$$

$$P_D = V_{CC} I_{CC} = (5 \text{ V})(3.56 \text{ mA}) = 17.8 \text{ mW}$$

Power Dissipation in High-Speed CMOS Devices

CMOS gates draw the most power when their outputs are switching from one logic state to the other. When the outputs are static (not switching), the large internal impedances of the gate limit the supply current. A change of state requires the charging and discharging of internal gate capacitances, resulting in a greater demand on the power supply current. Thus, the faster a CMOS gate switches, the more current, and hence more power, it requires.

CMOS supply current has two components: a quiescent current that flows when the gate is in a steady state and a dynamic component that depends on frequency. For relatively high frequencies (about 1 MHz and up), the quiescent component is small compared to the dynamic component and can be neglected.

The quiescent current is usually specified for an entire chip package, regardless of the number of gates. It is given by $I_{CC} V_{CC}$. For a 74HC00A NAND gate, $I_{CC} = 1 \text{ μA}$ at room temperature for a supply voltage of $V_{CC} = 6.0$ V. The dynamic component calculation accounts for internal and load capacitance and is given, *per gate,* by:

$$(C_L + C_{PD}) V_{CC}^2 f$$

where C_L is the gate load capacitance
C_{PD} is the gate internal capacitance
V_{CC} is the supply voltage
f is the switching frequency of the gate output

III EXAMPLE 11.10

The circuit in Figure 11.12 is constructed from 74HC00A high-speed CMOS NAND gates. Calculate the power dissipation of the circuit:

a. When the gate inputs are steady at the state $DCBA = 1010$

b. When the outputs are switching at an average frequency of 10 kHz

c. When the outputs are switching at an average frequency of 1 MHz

Supply voltage is 5 V. Temperature range is 25°C to −55°C.

Solution Refer to the 74HC00A data sheet in Appendix C.

a. $P_D = V_{CC} I_{CC} = (5 \text{ V})(1 \text{ }\mu A) = 5 \text{ }\mu W$. This is the quiescent power dissipation of the circuit.

b. The 74HC00A data sheet indicates that each gate has a maximum input capacitance, C_{in} of 10 pF. Assume that this value represents the load capacitance of gates 1, 2, and 3 of the circuit in Figure 11.12. Further assume that gate 4 has a load capacitance of 0. The total power dissipation of the circuit is given by:

$$P_D = 3(22 \text{ pF} + 10 \text{ pF})(5 \text{ V})^2 (0.01 \text{ MHz})$$
$$+ (22 \text{ pF})(5 \text{ V})^2 (0.01 \text{ MHz}) + 5 \text{ }\mu W$$
$$= 3(8 \text{ }\mu W) + 5.5 \text{ }\mu W + 5 \text{ }\mu W$$
$$= 34.5 \text{ }\mu W$$

c. For $f = 1$ MHz, total power dissipation is given by:

$$P_D = 3(22 \text{ pF} + 10 \text{ pF})(5 \text{ V})^2 (1 \text{ MHz})$$
$$+ (22 \text{ pF})(5 \text{ V})^2 (1 \text{ MHz}) + 5 \text{ }\mu W$$
$$= 3(800 \text{ }\mu W) + 550 \text{ }\mu W + 5 \text{ }\mu W$$
$$= 2955 \text{ }\mu W = 2.95 \text{ mW}$$

▌▌ EXAMPLE 11.11

The circuit in Figure 11.12 is constructed using a 74LS00 quad 2-in NAND gate and again with a 74HC00 quad 2-in NAND. Both circuits have identical waveforms applied to their inputs that make all gate outputs switch with a duty cycle of 50%. Calculate the frequency at which the power dissipation of the 74HC00 circuit exceeds that of the 74LS00 circuit. Assume $V_{CC} = 5$ V and temperature = 25°C for both circuits.

Solution The power dissipation of the LSTTL circuit is:

$$P_D = V_{CC} I_{CC} = (V_{CC}) (I_{CCH} + I_{CCL})/2 = (5V) (1.6 \text{ mA} + 4.4 \text{ mA})/2$$
$$= (5 \text{ V}) (3.0 \text{ mA}) = 15 \text{ mW}$$

Neglect the quiescent current of the high-speed CMOS circuit.

Per gate: $P_D = (C_L + C_{PD})V_{CC}^2 f$
$C_{PD} = 22$ pF per gate
$C_L = 10$ pF for 3 gates and 0 pF for 1 gate

Total: $P_D = (3(10\text{pF} + 22 \text{ pF}) + 22 \text{ pF})(5 \text{ V})^2 f$
$= (3(32 \text{ pF}) + 22 \text{ pF}) (25 \text{ V}^2) f$
$= (96 \text{ pF} + 22 \text{ pF}) (25 \text{ V}^2) f = (118 \text{ pF}) (25 \text{ V}^2) f$

For $P_D = 15$ mW:

$$f = \frac{15\text{mW}}{(118\text{pF})(25\text{V}^2)} = 5.08\text{MHz}$$

The power dissipation of the 74HC00 circuit exceeds that of the 74LS00 circuit at 5.08 MHz.

▌▌

NOTE

The power saving in a high-speed CMOS circuit generally results from the fact that most device outputs are not switching at any given time. The power dissipation of a TTL circuit is independent of frequency and therefore draws some power at all times. This is not the case for CMOS, which draws the majority of its power when switching.

■■ SECTION 11.4 REVIEW PROBLEM

11.4 Why does CMOS power dissipation increase with frequency?

11.5 Noise Margin

KEY TERMS

Noise   Unwanted electrical signal, often resulting from electromagnetic radiation.

Noise margin   A measure of the ability of a logic circuit to tolerate noise.

V_{IH} Voltage level required to make the input of a logic circuit HIGH.

V_{IL} Voltage level required to make the input of a logic circuit LOW.

V_{OH} Voltage measured at a device output when the output is HIGH.

V_{OL} Voltage measured at a device output when the output is LOW.

Electrical circuits are susceptible to **noise**, or unwanted electrical signals. Such signals are often induced by electromagnetic fields of motors, fluorescent lighting, high-frequency electronic circuits, and cosmic rays. They can cause erroneous operation of a digital circuit. Since it is impossible to eliminate all noise from a circuit, it is desirable to build a certain amount of tolerance, or **noise margin**, into digital devices used in the circuit.

In all circuits studied so far, we have assumed that logic HIGH is +5 volts and logic LOW is 0 volts in devices with a 5-volt supply. In practice, there is a certain amount of tolerance on both the logic HIGH and LOW voltages; for TTL devices, a HIGH at a device input is anything above about +2 volts, and a LOW is any voltage below about +0.8 volts. Due to internal voltage drops, the HIGH output of a TTL gate is typically about +3.5 volts.

Figure 11.13 shows one inverter driving another. In Figure 11.13a, the output of the first inverter and the input of the second have the same logic threshold. That is, the input of the second gate recognizes any voltage above 2.7 volts as HIGH (V_{IH} = 2.7 V) and any voltage below 0.5 volts as LOW (V_{IL} = 0.5 V). The output of the first inverter produces at least 2.7 volts when HIGH (V_{OH} = 2.7 V) and no more than 0.5 volts as LOW (V_{OL} = 0.5 V).

If there is noise on the line connecting the two gates, it will likely cause the voltage of the second gate input to penetrate into the forbidden region between logic HIGH and LOW levels. This is shown on the graph of the waveform in Figure 11.13a. When the voltage enters the forbidden region, the gate will not operate reliably. Its output may switch states when it is not supposed to.

Figure 11.13b shows the same circuit with different logic thresholds at input and output. The output of the first inverter is guaranteed to be *at least 2.7 volts* when HIGH (V_{OH} = 2.7 V) and *no more than 0.5 volts* when LOW (V_{OL} = 0.5 V). The second gate recognizes any input voltage *greater than 2 volts* as a HIGH (V_{IH} = 2 V) and any input voltage *less than 0.8 volts* (V_{IL} = 0.8 V) a LOW.

The difference between logic thresholds allows for a small noise voltage, equal to or less than the difference, to be superimposed on the desired signal. It will not cause the input voltage of the second inverter to penetrate the forbidden region. This ensures reliable operation even in the presence of some noise.

For the 74LS04 inverter, the HIGH-state and LOW-state noise margins, V_{NH} and V_{NL}, are:

$$V_{NH} = V_{OH} - V_{IH} = 2.7 \text{ V} - 2.0 \text{ V} = 0.7 \text{ V}$$
$$V_{NL} = V_{IL} - V_{OL} = 0.8 \text{ V} - 0.5 \text{ V} = 0.3 \text{ V}$$

A device with these values of V_{IH} and V_{IL} is deemed to be **TTL compatible**.

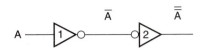

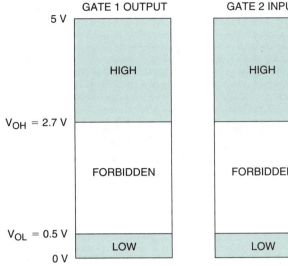

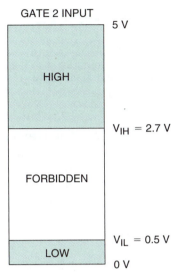

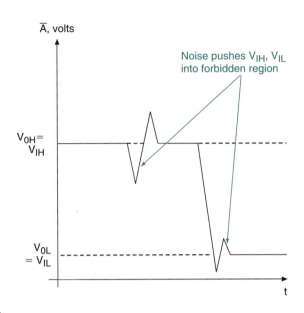

a. Zero noise margin

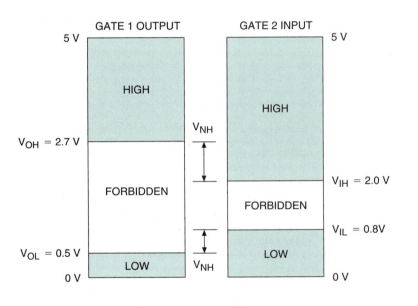

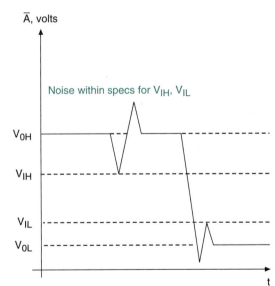

b. Nonzero noise margin

FIGURE 11.13
Noise Margins

▌▌ EXAMPLE 11.12

Use the 74HC00A data sheet in Appendix C to calculate the noise margins for this gate. Assume $V_{CC} = 4.5$ V, ambient temperature (T_A) is 25°C, and the driving gate is fully loaded ($I_{OUT} = \pm 4$ mA).

Solution

$$V_{NH} = V_{OH} - V_{IH} = 3.98 \text{ V} - 3.15 \text{ V} = 0.63 \text{ V}$$
$$V_{NL} = V_{IL} - V_{OL} = 1.35 \text{ V} - 0.26 \text{ V} = 1.09 \text{ V}$$

▐█

▮▮ SECTION 11.5 REVIEW PROBLEM

11.5 Calculate the noise margins of a 74HCT00A NAND gate from the data sheet in Appendix C. $V_{CC} = 4.5$ V, $T_A = 25°C$, $I_{OUT} = \pm 4$ mA

11.6 Interfacing TTL and CMOS Gates

> **KEY TERM**
>
> **TTL Compatible** Able to be driven directly by a TTL output. Usually implies voltage compatibility with TTL.

Interfacing different logic families is just an extension of the fanout and noise margin problems; you have to know what the load gates of a circuit require and what the driving gates can supply. In practice, this means you must know the specified values of input and output voltages and currents for the gates in question. Table 11.4, which is derived from the manufacturers' data sheets included in Appendix C, gives an overview of input and output parameters for a variety of TTL and CMOS families. Ambient temperature is assumed to be 25°C.

Table 11.4 TTL and CMOS Input and Output Parameters

	TTL				High-Speed CMOS				Low-Voltage CMOS	
	74LS	**74F**	**74AS**	**74ALS**	**74HC**	**74HCT**	**74VHC**	**74VHCT**	**74LVX**	**74LCX**
V_{CC} (V)	5.0	5.0	5.5	5.5	4.5	4.5	4.5	4.5	3.0	3.0
V_{OH} (V)	2.7	2.7	3.0	3.0	3.98	3.98	3.94	3.94	2.58	2.2
V_{OL} (V)	0.5	0.5	0.5	0.5	0.26	0.26	0.36	0.36	0.36	0.55
V_{IH} (V)	2.0	2.0	2.0	2.0	3.15	2.0	3.15	2.0	2.0	2.0
V_{IL} (V)	0.8	0.8	0.8	0.8	1.35	0.8	1.35	0.8	0.8	0.8
I_{OH} (mA)	−0.4	−1.0	−2.0	−0.4	−4.0	−4.0	−8.0	−8.0	−4.0	−24.0
I_{OL} (mA)	8.0	20.0	20.0	8.0	4.0	4.0	8.0	8.0	4.0	24.0
I_{IH} (mA)	0.02	0.1	0.02	0.02	0.0001	0.0001	0.0001	0.0001	0.0001	0.0001
I_{IL} (mA)	−0.4	−0.6	−0.5	−0.1	0.0001	0.0001	0.0001	0.0001	0.0001	0.0001

Table 11.4 is useful for comparison of logic families, but it is not a substitute for reading data sheets, as it gives parameters only under a restricted set of conditions. We can, however, make some observations based on the data in Table 11.4.

1. Input currents in a CMOS gate are very low, due to its high input impedance. As a result fanout is generally not a problem with CMOS loads. Interface problems to CMOS loads have to do with input voltage, not current.

2. CMOS devices, such as 74HCT, that have the same values of V_{IH} and V_{IL} as the TTL families in Table 11.4, are considered to be **TTL compatible**, since they can be driven directly by TTL drivers.

3. LSTTL is usually regarded as the benchmark for measuring TTL loading of a CMOS circuit. For example, a data sheet will claim that a device can drive 10 LSTTL loads. This claim depends on the values of I_{OH} and I_{OL} for the driving gate, which are not listed directly in CMOS data sheets, except as absolute maximum ratings. The values in Table 11.4 are the values of current for which the output voltages, V_{OH} and V_{OL}, are defined. (Recall from the section on fanout in this chapter that increasing output current causes output voltages to migrate away from their nominal values, thus reducing device noise margins.)

Let us examine four interfacing problems: high-speed CMOS driving 74LS, 74LS driving 74HC, 74LS driving 74HCT, and 74LS driving low-voltage CMOS.

High-Speed CMOS driving 74LS

To design an interface between any two logic families, we must examine the output voltages and currents of the driving gate and the input voltages and currents of the load gates.

Assume a 74HC00 NAND gate drives one or more 74LS00 NAND gates. From the 74HC00 data sheet, we determine that $V_{OH} = 3.98$ V and $V_{OL} = 0.26$ V for $V_{CC} = 4.5$ V. The 74LS00 requires at least 2.0 V at its input in the HIGH state and no more than 0.8 V in the LOW state. The 74HC00 therefore satisfies the input voltage requirement of the 74LS00.

For the defined output voltages, the 74HC00 gate can source or sink 4 mA. The fanout for the circuit is therefore calculated as follows:

$$n_H = \frac{I_{OH}}{I_{IH}} = \frac{4\text{mA}}{20\mu\text{A}} = 200$$

$$n_L = \frac{I_{OL}}{I_{IL}} = \frac{4\text{mA}}{0.4\text{A}} = 10$$

$$n = 10$$

Therefore a 74HC00 NAND can drive a 74LS00 directly, with a fanout of 10.

74LS Driving 74HC

As mentioned earlier, CMOS has a very small input current and therefore does not present a fanout problem to a 74LS driving gate. However, we must also examine the interface for voltage compatibility.

From data sheets, we see that a 74LS00 gate is guaranteed to provide at least 2.7 V in the HIGH state and no more than 0.5 V in the LOW state. A 74HC00 gate will recognize anything less than 1.35 V as a logic LOW and anything more than 3.15 V as a logic HIGH. The 74LS00 meets the LOW-state criterion, but it cannot guarantee sufficient output voltage in the HIGH state.

In order to properly drive a 74HC input with a 74LS output, we must provide a pull-up resistor to ensure sufficient HIGH-state voltage at the 74HC input. The circuit is illustrated in Figure 11.14. The pull-up resistor should be between 1 kΩ and 10 kΩ.

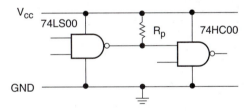

FIGURE 11.14
LSTTL driving 74HC CMOS

74LS Driving 74HCT

74HCT inputs are designed to be compatible with TTL outputs. As with 74HC devices, input currents are sufficiently low that fanout is not a problem with the 74LS-to-74HCT interface. 74HCT input voltages are the same as those for TTL ($V_{IH} = 2.0$ V and $V_{IL} = 0.8$ V). Therefore, 74HCT inputs can be driven directly by LSTTL outputs.

74LS Driving Low-voltage CMOS

CMOS families with supply voltages less than 5 V are rapidly becoming popular in new applications. Two of the reasons for their increasing prominence are reduced power dissipation (inversely proportional to the *square* of the supply voltage) and smaller feature size (i.e., size of the internal transistors) that allows more efficient packaging and faster operation. Low-voltage logic is particularly popular for battery-powered applications such as laptop computing or cell phones. Low voltage families typically operate at $V_{CC} = 3.3$ V or 2.5 V. Newer devices are available for $V_{CC} = 1.8$ V or 1.65 V.

Low-voltage CMOS families such as 74LVX or 74LCX can interface directly with TTL outputs if they are operated with a 3.0 V to 3.3 V power supply. These families are not really suitable for driving 5-volt TTL, as their noise margins are too small when they use a 3.0 V supply voltage.

If we wish to use a 74LS device to drive a 74HC device operating at a power supply voltage of less than 4.5 V, we can use a 74HC4049 or 74HC4050 buffer to translate the TTL logic level down to an appropriate value. The 74HC4049 is a package of six inverting buffers. The 74HC4050 has six noninverting buffers. These buffers can tolerate up to 15 V on their inputs. Their output voltages are determined by the value of their supply voltage.

Figure 11.15 shows an LSTTL-to-74HC interface circuit with a 74HC4050 buffer. Note that the interface buffer has the same power supply voltage as the load gate. Both sides of the interface are referenced to the same ground.

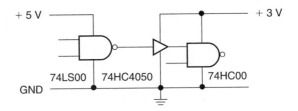

FIGURE 11.15
74LS-to-74HC Interface Using a 74HC4050 Buffer

▌▌ SECTION 11.6 REVIEW PROBLEM

11.6 A 74LS00 driving gate is to be interfaced to a 74HC00 load using a 74HC4050 non-inverting buffer. The 74HC00 has a power supply voltage of 2.5 V. What supply voltage should the 74HC4050 buffer have? Why?

11.7   Internal Circuitry of TTL Gates

> **K E Y   T E R M S**
>
> **Cutoff mode**   The operating mode of a bipolar transistor when there is no collector current flowing and the path from collector to emitter is effectively an open circuit. In a digital application, a transistor in cutoff mode is considered OFF.
>
> **Saturation mode**   The operating mode of a bipolar transistor when an increase in base current will not cause a further increase in the collector current and the path from collector to emitter is very nearly (but not quite) a short circuit. This is the ON state of a transistor in a digital circuit.

TTL has been around for a long time. The first transistor-transistor logic ICs were developed by Texas Instruments around 1965. Since then, there have been many improvements in the speed and power consumption of these devices, but the basic logic principles remain largely unchanged. Even though they are seldom used in modern designs, it makes sense to examine the internal circuitry of standard TTL gates such as the 7400 NAND, 7402 NOR, and 7404 inverter because the internal logic concepts are similar to the more advanced types of TTL.

The most important parts of the circuit, as far as a designer or technician is concerned, are the input and output circuits, because they are the only parts of the chip to which we have access. It is to these points that we interface other circuits and where we make diagnostic measurements. A basic understanding of the inputs and outputs of logic gate circuitry is helpful when we design or troubleshoot a digital circuit.

Bipolar Transistors as Logic Devices

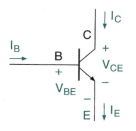

FIGURE 11.16

Currents and Voltages in an NPN Bipolar Transistor

The basic element of a TTL device is the bipolar junction transistor, illustrated in Figure 11.16. This is not the place to give a detailed analysis of the operation of a bipolar transistor, but a simplified summary of operating modes will be useful.

The bipolar transistor is a current amplifier having three terminals called the collector, emitter, and base. Current flowing into the base controls the amount of current flowing from the collector to the emitter. If base current is below a certain threshold, the transistor is in **cutoff mode** and no current flows in the collector. In this state, the base-emitter voltage is less than 0.6 V and the collector-emitter path acts like an open circuit. We can treat the collector-emitter path as an open switch, as shown in the lefthand diagram in Figure 11.17.

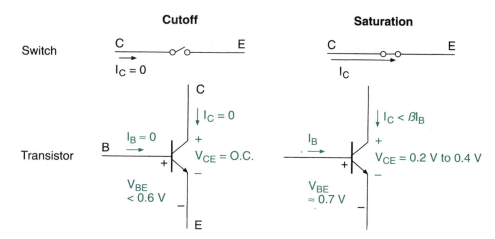

FIGURE 11.17

NPN Bipolar Transistor as a Switch

If the base current increases, the transistor enters the "active region," where the collector current is proportional to the base current by a current gain factor, β. This is the linear, or amplification, region of operation, used by analog amplifiers.

If the base current increases still further, collector current reaches a maximum value and will no longer increase with base current. This is called the **saturation mode** of the transistor. The saturated value of collector current, I_{CS}, is determined by (1) the resistance in the collector-emitter current path, (2) the voltage drop across the collector and emitter, V_{CE}, and (3) the collector supply voltage, V_{CC}. Base-emitter voltage is about 0.7 V and will not increase significantly with increasing base current. The voltage between collector and emitter is in the range from 0.2 V to 0.5 V. In this mode, we can treat the transistor as a closed switch, as shown in the righthand diagram of Figure 11.17.

Table 11.5 summarizes the voltages and currents in the cutoff, active, and saturation regions.

Table 11.5 Bipolar Transistor Characteristics

	Cutoff	Active	Saturation
I_C	0	$= \beta I_B$	$< \beta I_B$
V_{CE}	Open cct.	> 0.8 V	0.2 V–0.5 V
V_{BE}	< 0.6 V	0.6 V–0.7 V	≈ 0.7 V

▌▌ EXAMPLE 11.13

Figure 11.18 shows an NPN bipolar transistor connected in a common-emitter configuration. With the right choice of input voltages, this circuit acts as a digital inverter.

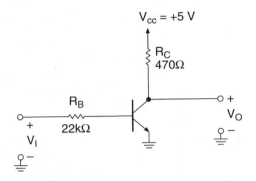

FIGURE 11.18
Example 11.13
Transistor as Inverter

Analyze the circuit to show that it acts as an inverter if a logic HIGH is defined as ≥3 V and a logic LOW is defined as ≤0.5 V. Assume that $\beta = 100$, and assume that $V_{BE} = 0.7$ V and $V_{CE} = 0.2$ V in saturation.

Solution   We will analyze the circuit with two input voltages: 3 V (logic HIGH) and 0.5 V (logic LOW). These two conditions are shown in Figure 11.19.

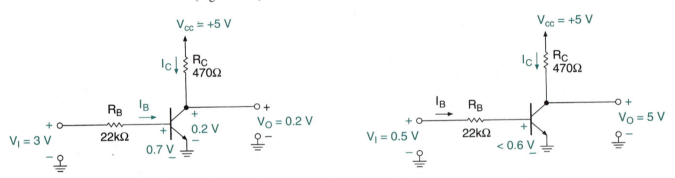

a. **HIGH input (saturation)**   b. **LOW input (cutoff)**

FIGURE 11.19
Example 11.13
Voltage and Current Analysis of Inverter

High input. We must prove that $V_I = 3$ V is sufficient to saturate the transistor. Let us assume that this is true and find out if calculations confirm our assumption.

Figure 11.19a shows the circuit with $V_I = 3$ V. By Kirchhoff's voltage law (KVL):

$$V_I = I_B R_B + V_{BE}, \text{ or}$$
$$I_B = (V_I - V_{BE})/R_B$$

If we assume that I_B is sufficient to saturate the transistor, then:

$$I_B = (3 \text{ V} - 0.7 \text{ V})/22 \text{ k}\Omega$$
$$= 105 \text{ } \mu\text{A}$$
$$\beta I_B = (100)(105 \text{ } \mu\text{A}) = 10.5 \text{ mA}$$

Collector current won't increase beyond its saturated value, even if base current increases. Therefore, if the transistor is saturated, βI_B will be larger than the current actually flowing in the collector-emitter path.

In saturation, the collector current can be calculated by KVL:

$$V_{CC} = I_C R_C + V_{CE}, \text{ or}$$
$$I_C = (V_{CC} - V_{CE})/R_C$$
$$I_C = (5 \text{ V} - 0.2 \text{ V})/470 \text{ } \Omega$$
$$= 10.2 \text{ mA}$$

Since $\beta I_B > I_C$, the transistor is saturated. Thus, an input voltage of 3 V will produce sufficient base current to saturate the transistor. The output is given by $V_O = V_{CE} = 0.2$ V, which is within the defined range of a logic LOW.

LOW input. Figure 11.19b shows the circuit with $V_I = 0.5$ V. By KVL:

$$V_I = I_B R_B + V_{BE}$$
$$V_{BE} = 0.5 \text{ V} - I_B R_B$$

Since V_{BE} must be <0.6 V, the transistor is in cutoff mode. Thus, in the collector circuit:

$$V_{CC} = I_C R_C + V_{CE}$$
$$5 \text{ V} = (0)(470 \text{ } \Omega) + V_{CE}$$
$$V_O = V_{CE} = 5 \text{ V (logic HIGH)}$$

Table 11.6 summarizes the operation of the circuit as an inverter.

Table 11.6 Input and Output of Single-Transistor Inverter

Input		Output	
V_I	Logic Level	V_O	Logic Level
0.5 V	LOW	5 V	HIGH
3 V	HIGH	0.2 V	LOW

TTL Open-Collector Inverter and NAND Gate

KEY TERM

Open-collector output   A TTL output where the collector of the LOW-state output transistor is brought out directly to the output pin. There is no built-in HIGH-state output circuitry, which allows two or more open-collector outputs to be connected without possible damage.

NOTE

The TTL gates (7405, 7401, 7404, 7400, and 7402) used in the following sections to illustrate TTL circuit principles are no longer in general use. They are from the original ("standard") TTL family, which has been superceded by faster and more efficient devices. However, the standard TTL devices are easier to understand than devices from the newer TTL subfamilies, since their circuit structure is simpler. The operating principles are similar in both the standard and newer families, so we will use the standard devices to illustrate the general principles of TTL operation.

Figure 11.20 shows the circuit of the simplest TTL gate: a 7405 inverter with **open-collector outputs.** This circuit performs the same function as the single-transistor inverter we examined in Example 11.13. These circuits differ most obviously in their input circuitry. The inverter circuit in Example 11.13 has a resistor as its input; the 7405 inverter has a transistor,

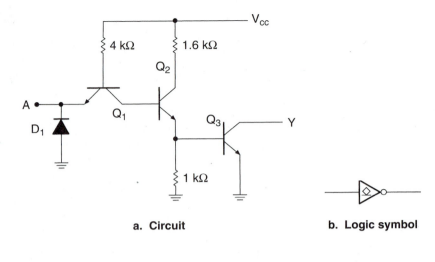

a. Circuit   b. Logic symbol

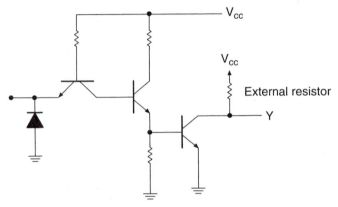

c. 7405 with HIGH-state pull-up

FIGURE 11.20
Open-Collector Inverter (7405)

Q_1, as its input. The input transistor allows faster switching of input states. This configuration is common to all standard TTL gates and will be examined in detail later in this section.

The logic function of the 7405 is performed by transistors Q_2 and Q_3. Output transistor Q_3 is switched ON and OFF by current flowing in the collector-emitter path of Q_2. When Q_3 is ON, Y is LOW.

However, when Q_3 is OFF, Y is floating. There is a high impedance between Y and ground, so the output is not LOW. But there is no connection to V_{CC} to make the output HIGH. In this condition, Y is neither HIGH nor LOW.

To enable the output to produce a HIGH state, we need to add an external pull-up resistor. The value of this resistor depends on the current sinking capability of Q_3, specified in the data sheet as I_{OL}. We will do such calculations in a later example.

TTL Inputs

Transistor Q_1 and diode D_1 make up the input circuit of the TTL inverter of Figure 11.20. The diode protects the input against small negative voltages. If the input goes more negative than about -0.7 V, the diode will conduct, effectively short-circuiting the input to ground plus one diode drop. This clamps the input to -0.7 V. D_1 has no logic function.

Q_1 can be treated as two back-to-back diodes, as shown in Figure 11.21. Figure 11.22 shows how the input responds to logic HIGH and LOW voltages.

LOW Input.   When a TTL input is made LOW, the base-emitter junction of Q_1 acts as a forward-biased diode, creating a current path from V_{CC} to ground via the input pin. This

FIGURE 11.21
Diode Equivalent of TTL Input Transistor

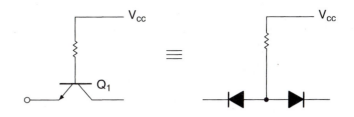

FIGURE 11.22
HIGH and LOW Inputs at a TTL Gate

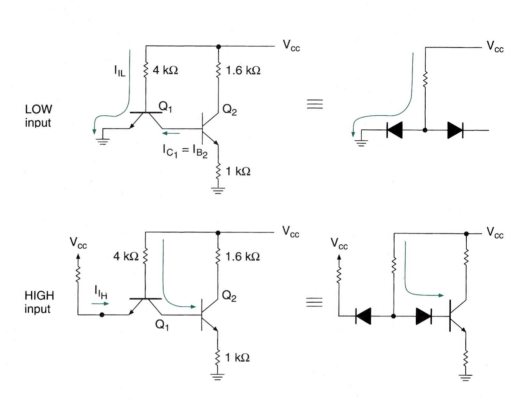

current makes up the majority of current I_{IL}, which has a maximum value of 1.6 mA in standard TTL (0.4mA in LSTTL).

At the moment the input is made LOW, the transistor action of Q_1 transports charge away from the base of Q_2, pulling it LOW and keeping it in cutoff mode. This current dies out when the base charge of Q_2 has been depleted, shortly after the LOW is applied to the input pin. The diode formed by the base-collector junction of Q_1 does not carry sufficient current to turn on Q_2, since the base-emitter path is of much lower impedance.

HIGH Input.   A HIGH at a TTL input reverse-biases the base-emitter junction of Q_1. Only a small leakage current, I_{IH}, flows. The maximum value of I_{IH} is 40 μA for standard TTL (20 μA for LSTTL).

Since the low-impedance current path to the input pin has not been established, current flows to the base of Q_2 via the forward-biased base-collector junction of Q_1. This current is sufficient to saturate Q_2.

Open (Floating) TTL Input.   An open-circuit TTL input acts as a logic HIGH, as illustrated by Figure 11.23. A TTL input relies on a logic LOW to establish a low-impedance current path from V_{CC} to the input pin. If the input is open, this LOW is not present and current flows in the base-collector junction of the transistor, by default. This is the same current that flows under the HIGH-input condition.

This HIGH is not stable; it can be converted to logic LOW by induced noise at the input pin. To avoid this uncertainty, an unused input should always be wired to a logic HIGH or LOW state.

FIGURE 11.23
LOW, HIGH, and Open
TTL Inputs

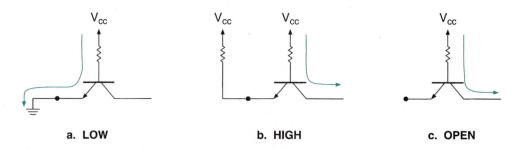

a. LOW   b. HIGH   c. OPEN

TTL Open-Collector Inverter

Figure 11.24 shows the operation of the 7405 open-collector inverter.

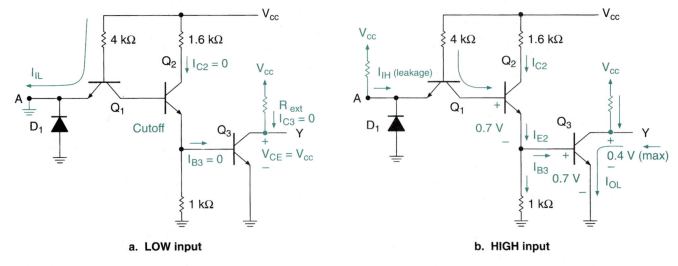

a. LOW input   b. HIGH input

FIGURE 11.24
7405 Operation

LOW Input. As was described above, a LOW input establishes a low-impedance path to ground, which draws current through the base-emitter junction of Q_1. This action also prevents base current from flowing in transistor Q_2, causing it to be in cutoff mode and making $I_{C2} = 0$.

Since I_{B3} is derived from I_{C2}, $I_{B3} = 0$ and Q_3 is cut off, making a high-impedance path between the collector and emitter of Q_3. As was the case with the single-transistor inverter in Example 11.13, when $I_{C3} = 0$, then $V_O = V_{CE} = V_{CC}$. (Since no current flows through the pull-up resistor, the voltage must be the same at both ends.) *Output Y is HIGH.*

HIGH Input. When input A is HIGH, the base-emitter junction of Q_1 does not have sufficient voltage across it to be forward-biased. Current flows through the base-collector junction of Q_1, saturating Q_2.

Since Q_2 is ON, current flows to the Q_2 emitter and splits through the 1-kΩ resistor and the base of Q_3. The output transistor, Q_3, turns ON, establishing a low-impedance current path from output Y to ground. Current is limited by the external pull-up resistor, which must be chosen to keep I_{OL} at or under its rated value of 16 mA. V_{CE3} is about 0.2 V to 0.4 V. *Output Y is LOW.*

TTL Open-Collector NAND

Figure 11.25 shows one gate of a 7401 quadruple 2-input NAND gate with open-collector outputs. The circuit is the same as that of the 7405 inverter, except that the input transistor has a second emitter. Multiple-emitter transistors of this type are common in TTL circuits and can be modeled by the diode equivalent in Figure 11.25b. Figure 11.26 shows the response of the multiple-emitter input transistor to various combinations of logic levels.

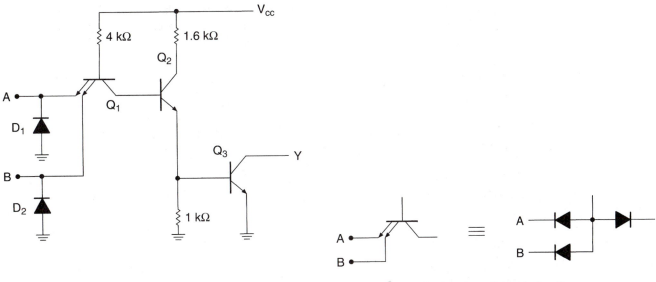

a. Circuit

b. Input equivalent circuit

FIGURE 11.25
TTL NAND with Open Collector Output

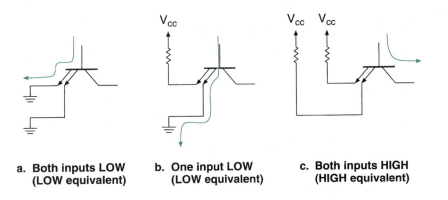

a. **Both inputs LOW**
   (LOW equivalent)

b. **One input LOW**
   (LOW equivalent)

c. **Both inputs HIGH**
   (HIGH equivalent)

FIGURE 11.26
Input Response of Multiple-Emitter Transistor

If both inputs are LOW, the NAND acts exactly the same as the 7405 inverter with a LOW input. (A low-impedance path is created through a base-emitter junction.) *Output Y is HIGH,* provided an external pull-up resistor is connected to output. A partial truth table for this condition is:

A	B	Y
0	0	1

If one input is LOW, the input acts the same as the inverter with a LOW input. The low-impedance current path through the one grounded emitter prevents sufficient base-collector current from flowing to forward-bias that junction. *Output Y is HIGH* if a pull-up resistor is connected to the output. A partial truth table is as follows:

A	B	Y
0	1	1
1	0	1

If both inputs are HIGH, the NAND circuit acts like the 7405 when its input is HIGH. (There is no base-emitter current path. A collector-emitter path is established by default.) *Output Y is LOW*. This condition can be represented by:

A	B	Y
1	1	0

Combining all these conditions, we get the standard NAND truth table:

A	B	Y
0	0	1
0	1	1
1	0	1
1	1	0

> **NOTE**
>
> If one or more emitters of a TTL multiple-emitter input transistor is LOW, the input is a LOW equivalent. All emitters must be HIGH to make the transistor input a HIGH equivalent.

These statements lead to the familiar NAND-gate descriptive sentences, illustrated by the gate symbols in Figure 11.27.

a. At least one input LOW makes the output HIGH.

b. Both inputs HIGH make the output LOW.

a. At least one input LOW
makes output HIGH

A	B	Y
0	0	1
0	1	1
1	0	1

b. Both inputs HIGH
make output LOW

A	B	Y
1	1	0

FIGURE 11.27
DeMorgan Equivalent Forms of a NAND Gate

▐▐▐ SECTION 11.7A REVIEW PROBLEM

11.7 What are the two main functions of the pull-up resistor on the output of an open-collector gate?

Open-Collector Applications

> **KEY TERM**
>
> **Wired-AND**   A connection where open-collector outputs of logic gates are wired together. The logical effect is the ANDing of connected functions.

A more common TTL output than the open collector is the totem pole output, which we will study later in this chapter. The totem pole output has its own internal pull-up circuit for HIGH outputs.

Gates with totem pole outputs cannot be used in all digital circuits. For example, open-collector gates are required when several outputs must be tied together, a connection called **wired-AND.** Totem pole outputs would be damaged by such a connection, since there is the possibility of conflict between an output HIGH and LOW state.

Open-collector outputs can also be used for applications requiring high current drive and for interfacing to circuits having supply voltages other than TTL levels.

A special symbol defined by IEEE/ANSI Standard 91-1984, an underlined square diamond, is shown in Figure 11.28. This symbol is added to a logic gate symbol to indicate that it has an open-collector output. Other symbols, such as a star (*), a dot (●), or the initials OC are also used.

Wired-AND

> **N O T E**
>
> A wired-AND connection combines the *outputs* of the connected gates in an AND function.

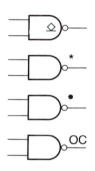

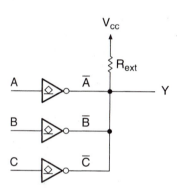

FIGURE 11.28

Open-Collector Symbols Shown
for a NAND Gate (e.g., 7401)

FIGURE 11.29

Three Inverters in a Wired-AND
Connection

Figure 11.29 shows three open-collector inverters connected in a wired-AND configuration. The output transistors of the inverters are shown in Figure 11.30, with different possible ON and OFF states. The only way output Y can remain HIGH is if all the transistors are in their OFF states, as in Figure 11.30c. This can happen only if the outputs of the inverters are all HIGH. This is the same as saying the outputs are ANDed together at Y.

The Boolean expression for Y is:

$$Y = \overline{A} \cdot \overline{B} \cdot \overline{C}$$
$$= \overline{A + B + C}$$

By DeMorgan's theorem, the wired-AND connection of inverter outputs is equivalent to a NOR function. Because of this DeMorgan equivalence, the connection is sometimes called "wired-OR."

Figure 11.31 shows three NAND gates in a wired-AND connection. Since the output functions are ANDed, the Boolean expression for Y is:

$$Y = \overline{AB} \cdot \overline{CD} \cdot \overline{EF}$$
$$= \overline{AB + CD + EF}$$

FIGURE 11.30
Output Transistors of Open-Collector Inverters in a Wired-AND Connection

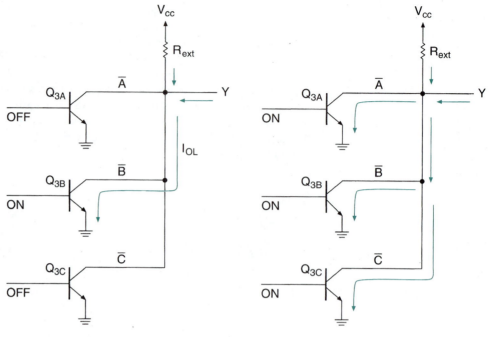

a. One inverter with LOW output

b. All inverter outputs LOW

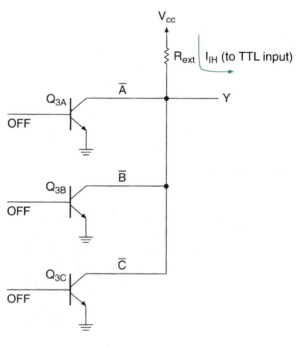

c. All inverter outputs HIGH

The resulting function is called AND-OR-INVERT. Normally this requires at least two types of logic gate—AND and NOR. The wired-AND configuration can synthesize any size of AND-OR-INVERT network using only NAND gates.

The wired-AND function is sometimes shown as an AND symbol around a soldered connection, as shown in Figure 11.31b.

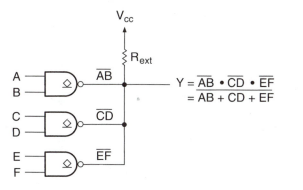

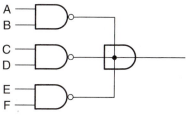

a. Pull-up resistor and open-collector gates

b. Wired-AND symbol

FIGURE 11.31
NAND Gates in Wired-AND Connection

High-Current Driver

Standard TTL outputs have higher current ratings in the LOW state than in the HIGH state. Thus, open-collector outputs are useful for driving loads that need more current than a standard TTL output can provide in the HIGH state. There are special TTL gates with higher ratings of I_{OL} to allow even larger loads to be driven. Typical loads would be LEDs, incandescent lamps, and relay coils, all of which require currents in the tens of milliamperes.

▌▌ EXAMPLE 11.14

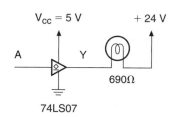

FIGURE 11.32
Example 11.14
74LS07 High-Current Driver

A 74LS07 hex buffer/driver contains six noninverting buffers whose outputs are open-collector, rated for $I_{OLmax} = 40$ mA and $V_{OHmax} = 30$ V. That is, even though there is no internal circuit to provide a logic HIGH at the output, the output transistor can withstand a voltage of up to 30 V without damage.

Figure 11.32 shows a 74LS07 buffer driving an incandescent lamp rated at 24 V, with a resistance of 690 Ω. Calculate the current that flows when the lamp is illuminated. What logic level at A turns the lamp on? Could the lamp be driven by a 74LS05 inverter? Why or why not?

Solution   From the 74LS07 data sheet in Appendix C, we see that $V_{OL} = 0.4$ V for $I_{OL} = 16$ mA and $V_{OL} = 0.7$ V for $I_{OL} = 40$ mA. Assume the latter value.

By KVL: 24 V − (I_{OL})(690 Ω) − V_{OL} = 0

Thus, I_{OL} = (24 V − 0.7 V)/690 Ω = 33.8 mA

Since the buffer is noninverting, and current flows when the output of the 74LS07 sinks current to ground (LOW), the lamp is on when A is LOW.

A 74LS05 open-collector inverter would not be a suitable driver for the circuit for two reasons: its output is only designed to withstand 5.5 V and it can only sink a maximum of 8 mA.

▌▌

Value of External Pull-up Resistor

The value of the pull-up resistor required by an open-collector circuit is calculated using manufacturer's specifications and the basic principles of circuit theory: Kirchhoff's voltage and current laws (KVL and KCL) and Ohm's law.

Figure 11.33 shows the circuit model for calculating the value of R_{ext}. It accounts for the current requirements of the loads, the LOW-state output voltage, and current-sinking capacity of the open-collector gate.

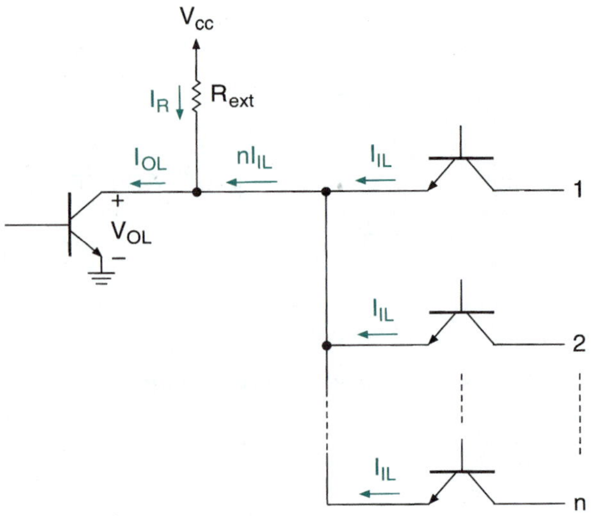

FIGURE 11.33
Circuit Model for Pull-up Resistor Calculation

> **NOTE**
>
> The main rule in resistor selection is to keep the sum of currents into the open-collector output to less than the maximum rated value of I_{OL}.
>
> $$I_{OL} = I_R + nI_{IL}$$
> $$I_R = (V_{CC} - V_{OL}/R_{ext}$$

▌▌ EXAMPLE 11.15

Calculate the minimum value of the pull-up resistor for a 74LS05 inverter if the circuit drives ten 74LS00 NAND gate inputs.

Solution

From 74LS00 specs: $I_{IL} = 0.4$ mA
For 10 gates: $nI_{IL} = 10I_{IL} = 4$ mA
From 74LS05 specs: $I_{OL} = 8$ mA

$$I_R = I_{OL} - nI_{IL}$$
$$= 8 \text{ mA} - 4 \text{ mA}$$
$$= 4 \text{ mA}$$

For $I_{OL} = 8$ mA, $V_{OL} = 0.5$ V

$$R_{ext} = (V_{CC} - V_{OL})/I_R$$
$$= (5 \text{ V} - 0.5 \text{ V})/4 \text{ mA}$$
$$= 4.5 \text{ V}/4 \text{ mA} = 1.125 \text{ k}\Omega$$

Use a 1.2-kΩ or 1.5-kΩ standard value resistor.

▌▌

▌▌ SECTION 11.7B REVIEW PROBLEM

11.8 Calculate the minimum value of pull-up resistor required for a 74LS05 inverter if it drives one input of a 74LS00 NAND gate. What is the minimum standard value of this resistor?

Totem Pole Outputs

Figure 11.34 shows one gate of a 7400 quadruple 2-input NAND with **totem pole outputs.** The circuit is the same as that for a 7401 open-collector NAND except for a transistor, resistor, and diode, which make up the HIGH-state output circuitry of the NAND gate.

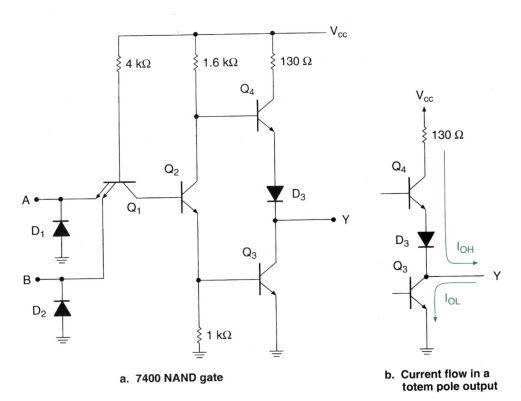

a. 7400 NAND gate

b. Current flow in a totem pole output

FIGURE 11.34
NAND Gate With Totem Pole Output

The totem pole output, shown in Figure 11.34b, has separate transistors to switch the output to the HIGH state (Q_4) and the LOW state (Q_3). These transistors are switched by Q_2, the **phase splitter.** Only one of them is ON at a time; the currents I_{OH} and I_{OL} never flow simultaneously.

The portion of the circuit consisting of Q_4, D_3, and the 130-Ω resistor replaces the external pull-up resistor required by the open-collector TTL output. Since the HIGH state is switched by its own transistor, we say that the circuit has an active pull-up.

The main advantage of the totem pole output over the open collector is that it can change states faster. The external pull-up resistance needed in an open-collector circuit slows down the output switching by contributing to the RC time constant of the output. The HIGH-state transistor circuit, with its relatively low output impedance, reduces this time constant and thus improves switching speed.

FIGURE 11.35
NAND Gate Operation

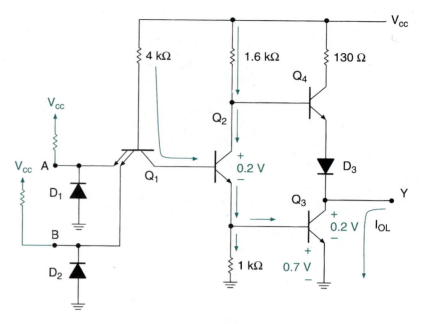

a. Both inputs HIGH

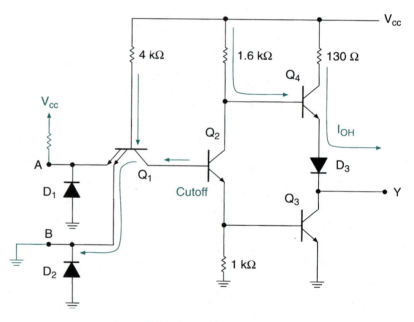

b. One input LOW

Figure 11.35 shows the operation of the 7400 NAND gate for HIGH and LOW input conditions.

HIGH Input. When both inputs are HIGH, there is no low-impedance base-emitter current path in Q_1. The base-collector junction of Q_1 acts as a forward-biased diode. Base current flows in Q_2, saturating the transistor. Sufficient current flows to Q_3 to saturate it. Y is connected to ground, via the collector-emitter path of Q_3. *The output is LOW.*

LOW Input. Figure 11.35b shows input B of a 7400 NAND gate pulled LOW. The circuit operates the same way if A or both A and B are LOW.

In this condition, a low-impedance path to ground is established through one of the base-emitter junctions of Q_1. This pulls the base of Q_2 LOW, causing it to be in cutoff mode. No current flows through the collector-emitter path of Q_2, so no base current flows in Q_3; it is also cut off.

Current flows through the 1.6-kΩ resistor to the base of Q_4, turning it ON. This connects the output, via Q_4, D_3, and the 130-Ω resistor, to V_{CC}. *The output is HIGH.*

Q_4 will not turn ON when Q_3 is ON. We can find out why by calculating $V_{BE4} + V_{D3}$. For Q_4 to conduct, two *pn* junctions (D_3 and the base-emitter junction of Q_4) must be forward-biased. Thus, ($V_{BE4} + V_{D3}$) must be greater than 0.6 V + 0.6 V = 1.2 V.

$$V_{BE4} + V_{D3} = V_{B4} - V_{CE3}$$

We can calculate V_{B4} by adding up voltage drops, as follows:

$$V_{B4} = V_{CE2} + V_{BE3} = 0.2 \text{ V} + 0.7 \text{ V} = 0.9 \text{ V}$$

Q_3 is saturated, thus:

$$V_{CE3} = 0.2 \text{ V}$$

The difference between these voltages is:

$$V_{B4} - V_{CE3} = 0.9 \text{ V} - 0.2 \text{ V} = 0.7 \text{ V}$$

This is insufficient to forward-bias BE_4 and D_3. Q_4 stays OFF.

NOTE

Without D_3 in the circuit,

$$V_{BE4} = (V_{CE2} + V_{BE3}) - V_{CE3}$$
$$= (0.2 \text{ V} + 0.7 \text{ V}) - 0.2 \text{ V}$$
$$= 0.7 \text{ V}$$

This is sufficient to saturate Q_4, even when Q_3 is ON. The diode is therefore necessary to keep Q_4 OFF when Q_3 is ON.

Switching Noise

KEY TERM

Storage time   Time required to transport stored charge away from the base region of a bipolar transistor before it can turn off.

A totem pole output is an inherently noisy circuit. Noise is generated on the supply voltage line when the output switches from LOW to HIGH.

When the output is in a steady HIGH or LOW state, Q_3 and Q_4 are always in opposite phase. The design of the totem pole output is such that when Q_3 is ON, it is saturated, but when Q_4 is ON, it operates in the transistor's active, or linear, region.

A saturated transistor takes longer to shut off than an unsaturated one due to **storage time,** the time required to transport stored charge away from the base region of the transistor. Thus, Q_3 takes longer to turn off than Q_4.

When a totem pole output is LOW, Q_3 is ON and Q_4 is OFF. When the output changes state, Q_4 turns ON before Q_3 can turn OFF, due to the storage time of Q_3. For a few nanoseconds, both transistors are ON. This condition momentarily shorts V_{CC} to ground, causing a surge of supply current, as shown in Figure 11.36.

The inductance of the power line produces a corresponding spike proportional to the instantaneous rate of change of the supply current ($v = L \, di/dt$, where L is the power line inductance and di/dt is the instantaneous rate of change of supply current).

These spikes on the supply voltage line can cause real problems, especially in synchronous circuits. They often cause erroneous switching that is nearly impossible to troubleshoot. The best cure for such problems is prevention.

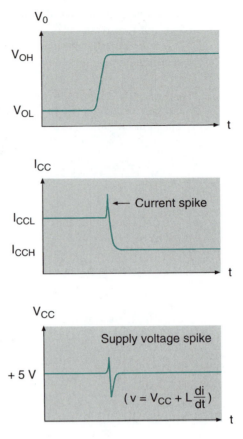

FIGURE 11.36
Spikes on Power Line During LOW-to-HIGH Transition of Totem Pole Output

Figure 11.37 shows the addition of a decoupling capacitor to a totem pole output to eliminate switching spikes. A low-inductance capacitor of about 0.1 μF is placed between the V_{CC} and ground pins of the chip to be decoupled. This capacitor offsets the power line inductance and acts as a low-impedance path to ground for high-frequency noise (i.e., spikes). Since a capacitor is an open circuit for low frequencies, the normal DC supply voltage is not shorted out.

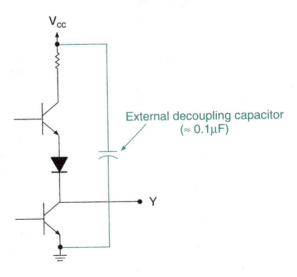

FIGURE 11.37
Decoupling the Power Supply

> **N O T E**
>
> It is important that the capacitor be placed *physically close* to the decoupled chip. Inductance of the power line accumulates with distance, and if the capacitor is far away from the chip (say, at the end of the circuit board), the decoupling effect of the capacitor is lost.

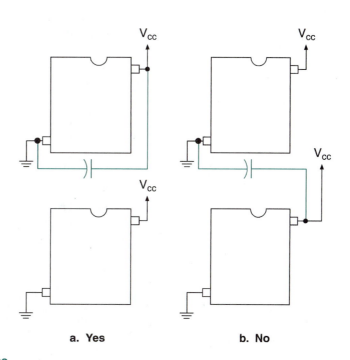

a. Yes   b. No

FIGURE 11.38
Placement of Decoupling Capacitor (Low-Frequency Designs)

It is not necessary to decouple every chip on a circuit board for designs operating at relatively low frequencies (≤ 1 MHz). In such cases, one capacitor for every two ICs is enough. The capacitor should be connected between V_{CC} and ground *of the same chip,* as shown in Figure 11.38.

For high-frequency designs, use one capacitor per IC, as shown in Figure 11.39. Connect directly to power and ground traces on a printed circuit board, *as close as possible* to the chip being decoupled.

Connection of Totem Pole Outputs

Totem pole outputs must never be connected together. As shown in Figure 11.40, the problem occurs when two connected outputs are in opposite states.

The active pull-up consisting of Q_4, D_3, and the 130-Ω resistor is designed to supply current to about 10 TTL inputs, each having a large input impedance. It will not withstand the current that flows when the output is forced to ground through the LOW output transistor of another gate.

Under this condition about 30 to 55 mA will flow through Q_{4A} and Q_{3B}. This exceeds the ratings of the outputs in both the HIGH and LOW state and will cause damage to the outputs over time. The outputs will probably withstand this sort of abuse for several minutes, but eventually will be damaged.

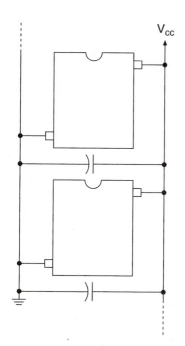

FIGURE 11.39
Placement of Decoupling Capacitors (High-Frequency Designs)

FIGURE 11.40
Totem Poles Connected Together

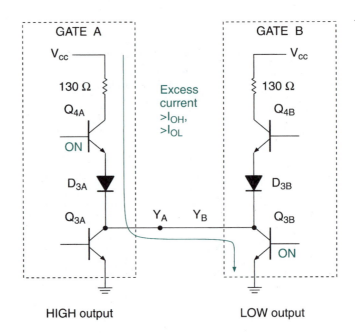

HIGH output   LOW output

▌▌ SECTION 11.7C REVIEW PROBLEM

11.9 A totem pole output is likely to be damaged when shorted to ground. Why?

Tristate Gates

KEY TERM

Tristate output   An output having three possible states: logic HIGH and LOW, and a high-impedance state, in which the output acts as an open circuit.

Figure 11.41 shows the circuits of two TTL inverters with **tristate outputs**. In addition to the usual binary states of HIGH and LOW, the output of the tristate inverter can also be in a "high-impedance" (Hi-Z) state. This state occurs when both Q_3 and Q_4 are OFF. The electrical effect is to produce an open circuit at the output, which is neither HIGH nor LOW.

The output of a tristate gate combines advantages of a totem pole output and an open-collector output. Like the totem pole output, it has an active pull-up with lower output impedance and faster switching than an open collector. Like the open collector, we can connect several outputs together, provided only one output is active at a time.

Input G, the "gating" or "enable" input, controls the gate. When G is active, the gate acts as an ordinary inverter. When inactive, the gate is in the high-impedance state. Table 11.6 summarizes the operation of the tristate inverters in Figure 11.41.

The tristate inverter in Figure 11.41a is enabled by a HIGH at the G input. The circuit is the same as a 7400 NAND gate with two exceptions: (1) an extra diode goes from the base of Q_4 to G, and (2) G connects directly to one of the emitters of Q_1.

When $G = 0$, Q_1 acts as though there was a LOW at a NAND gate input. In a 7400 NAND circuit, this causes Q_2 and Q_3 to be in cutoff mode.

Due to the opposite states of the emitter and collector in Q_2, Q_4 would normally be ON. Instead, the LOW at G pulls the base of Q_4 LOW through the extra diode. Thus, both Q_3 and Q_4 are OFF.

When $G = 1$, the G emitter of Q_1 acts like a HIGH NAND input. By the enable/inhibit rules of a NAND gate, $Y = \overline{A}$. The additional diode prevents the HIGH at G from activating Q_4.

The circuit in Figure 11.41b works the same way, except for the opposite sense of the activating input. This opposite active level is achieved by using an open-collector inverter, consisting of Q_5, Q_6, and Q_7, at input $\overline{G}$.

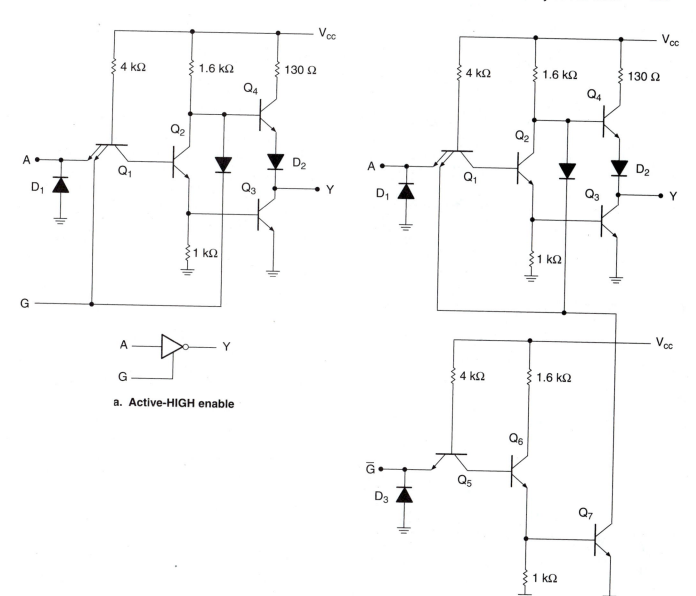

FIGURE 11.41
Tristate Inverters

Table 11.6 Truth Tables of Tristate Inverters

G	A	Y	$\overline{G}$	A	Y
0	0	Hi-Z	0	0	1
0	1	Hi-Z	0	1	0
1	0	1	1	0	Hi-Z
1	1	0	1	1	Hi-Z

▍▍ SECTION 11.7D REVIEW PROBLEM

11.10 Why is the diode from the base of Q_4 necessary in the tristate inverters in Figure 11.41?

Other Basic TTL Gates

Other TTL gates are similar to the NAND and inverter gates we have already examined. A significant variation is the OR/NOR circuit, which has a different input configuration than the AND/NAND/inverter type gates.

7402 NOR Gate

Figure 11.42 shows one gate of a 7402 quadruple 2-input NOR gate package. The difference between this gate and the 7400 NAND gate is the structure of the inputs. The NOR gate does not use the multiple-emitter transistor, but rather an individual transistor (Q_1 or Q_2) for each input. There are two phase splitters (Q_3 and Q_4), which are paralleled, emitter-to-emitter and collector-to-collector.

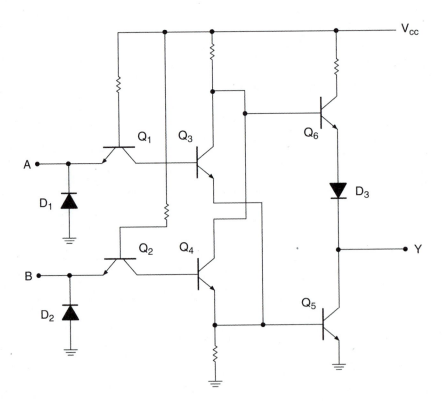

FIGURE 11.42
7402 NOR Gate Circuit

If either Q_3 or Q_4 is enabled by a HIGH at its corresponding input, it will turn on Q_5, making the output LOW.

If both gate inputs are LOW, both Q_3 and Q_4 are in cutoff mode, and so is Q_5. The output is HIGH through Q_6.

Table 11.7 shows the truth table and the states of the transistors for this gate. It is not strictly correct to refer to Q_1 and Q_2 as being ON or OFF, since there is current flowing in these transistors regardless of whether the inputs are HIGH or LOW. Let us define the ON

Table 11.7   7402 NOR Function and Truth Table

A	B	Q_1	Q_2	Q_3	Q_4	Q_5	Q_6	Y
0	0	ON	ON	OFF	OFF	OFF	ON	1
0	1	ON	OFF	OFF	ON	ON	OFF	0
1	0	OFF	ON	ON	OFF	ON	OFF	0
1	1	OFF	OFF	ON	ON	ON	OFF	0

state of an input transistor as the condition where the base-emitter junction is conducting (LOW input). If the base-collector junction conducts, we will consider the transistor OFF (HIGH input).

7408 AND Gate and 7432 OR Gate

It may not be obvious why we would choose to study NAND and NOR gates before AND and OR. After all, AND and OR are the more basic logic functions.

Electrically, it works the other way around. The simplest TTL circuit is the NAND/inverter, followed by the NOR. AND and OR gates are more complex since they are based on the NAND and NOR and require an extra inverter stage.

FIGURE 11.43

7408 AND Gate

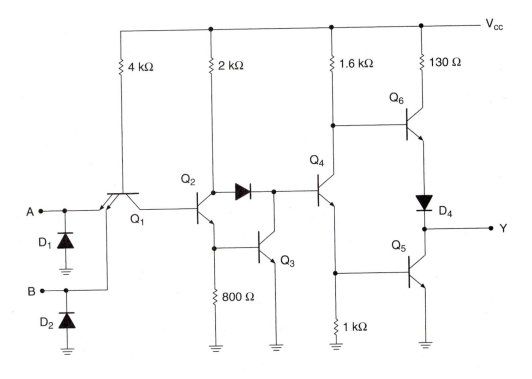

Figure 11.43 shows the circuit of a 7408 AND gate, and Figure 11.44 shows a 7432 TTL OR gate circuit. Each of these gates is like its NAND/NOR counterpart, except for an additional inverter, implemented by Q_3 in the AND gate and Q_5 in the OR gate.

Tables 11.8 and 11.9 show the transistor function and truth table for each gate. In keeping with the convention established for the NOR function table, an input transistor with a conducting base-emitter junction is considered ON.

FIGURE 11.44
7432 OR Gate

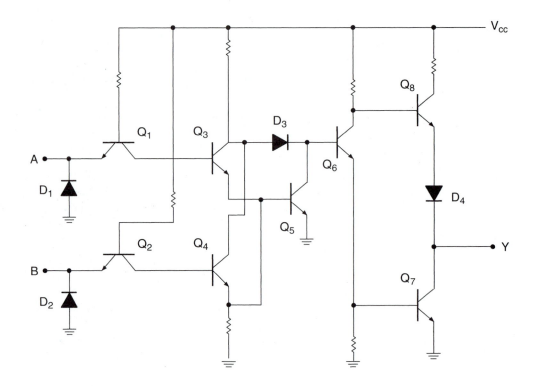

Table 11.8   7408 AND Function and Truth Table

A	B	Q_1	Q_2	Q_3	Q_4	Q_5	Q_6	Y
0	0	ON	OFF	OFF	ON	ON	OFF	0
0	1	ON	OFF	OFF	ON	ON	OFF	0
1	0	ON	OFF	OFF	ON	ON	OFF	0
1	1	OFF	ON	ON	OFF	OFF	ON	1

Table 11.9   7432 OR Function and Truth Table

A	B	Q_1	Q_2	Q_3	Q_4	Q_5	Q_6	Q_7	Q_8	Y
0	0	ON	ON	OFF	OFF	OFF	ON	ON	OFF	0
0	1	ON	OFF	OFF	ON	ON	OFF	OFF	ON	1
1	0	OFF	ON	ON	OFF	ON	OFF	OFF	ON	1
1	1	OFF	OFF	ON	ON	ON	OFF	OFF	ON	1

▌▌ SECTION 11.7E REVIEW PROBLEM

11.11 Why are noninverting gates more complex than inverting gates?

11.8   Internal Circuitry of MOS Gates

KEY TERMS

MOSFET   Metal-oxide-semiconductor field effect transistor. A MOSFET has three terminals—gate, source, and drain—which are analogous to the base, emitter, and collector of a bipolar junction transistor.

Enhancement-mode MOSFET   A MOSFET that creates a conduction path (a channel) between its drain and source terminals when the voltage between gate and source exceeds a specified threshold level.

Substrate   The foundation of *n*- or *p*-type silicon on which an integrated circuit is built.

***n*-channel enhancement-mode MOSFET**   A MOSFET built on a *p*-type substrate with *n*-type drain and source regions. An *n*-type channel is created in the p-substrate during conduction.

***p*-channel enhancement-mode MOSFET**   A MOSFET built on an *n*-type substrate with *p*-type drain and source regions. During conduction, a *p*-type channel is created in the *n*-substrate.

CMOS   A logic family based on the switching of *n*- and *p*-channel ("complementary") enhancement-mode MOSFETs.

All the logic circuits we have examined so far have been based on the switching of bipolar junction transistors. Another major logic family, **CMOS,** is based on the switching of metal-oxide-semiconductor field effect transistors, or **MOSFETS.**

There are two major types of MOSFETs, called depletion-mode and **enhancement-mode MOSFETs.** We will concentrate on the enhancement-mode devices, as they are the type used in the manufacture of digital ICs. Details of the differences between depletion- and enhancement-mode transistors can be found in any good textbook on electronic devices.

MOSFETs can be categorized in another way: as ***n*-channel** and ***p*-channel** devices, much as bipolar transistors are classified as NPN or PNP.

CMOS logic is constructed from both *n*- and *p*-channel MOSFETs. CMOS ("Complementary MOS") refers to the opposite, or complementary, operation of *n*- and *p*-channel transistors.

MOSFET Structure

Figure 11.45 shows the structure and symbol of an *n*-channel enhancement-mode MOSFET in an integrated circuit. The device is built on a **substrate** of *p*-type silicon, which has a deficiency of electrons in its structure. The drain and source regions are "wells" of *n*-type silicon, which has an excess of electrons. The drain and source are roughly equivalent to the emitter and collector of a bipolar transistor.

FIGURE 11.45
n-Channel MOSFET

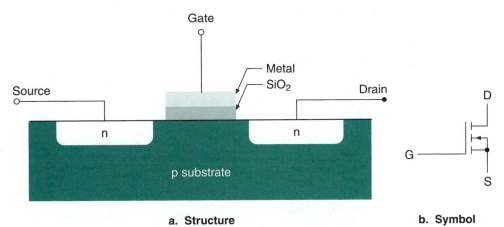

a. Structure   **b. Symbol**

The substrate is shown as a terminal with an arrow. The arrow points in for an *n*-channel device and out for a *p*-channel device. In nearly all cases, the substrate is shorted to the source terminal. (Some exceptions to this general rule will be examined when we look at circuits of CMOS gates.)

The gate terminal is similar to the base of a bipolar transistor in that it controls the flow of current between the drain and source. The difference is that a MOSFET uses gate *voltage* to control drain current, whereas a bipolar transistor uses base *current* to control collector current.

The gate consists of an insulating layer of silicon dioxide (SiO_2) and a layer of metal over the substrate between the drain and source. This gate structure is what gives the MOSFET its name (*metal-oxide-semiconductor* field effect transistor).

NOTE

The oxide layer of the gate structure is subject to damage if excessive voltage (greater than about 100 V) is applied. This especially includes static electricity, or electrostatic discharge (ESD). There are standard precautions for working with MOS devices that should be followed carefully.

Most important are ensuring that MOS devices are stored in antistatic or conducting material, that work surfaces are not likely to generate static, that unused inputs are not left open or floating, that you avoid touching the pins of a MOS device, and that if you must handle a MOS IC, you discharge any static on your person *before* touching it.

A conductive wrist strap with a high series resistance to ground (about 1 MΩ) is often worn to reduce static. The high resistance protects the operator from shock injury in the event of a short circuit.

A list of handling precautions is included in Appendix D.

▮▮ SECTION 11.8A REVIEW PROBLEM

11.12 Why are MOSFET circuits particularly susceptible to static damage?

Bias Requirement for MOS Transistors

KEY TERMS

Ohmic region   The MOSFET equivalent of saturation. When a MOSFET is biased ON, it acts like a relatively low resistance, or "ohmically."

n-type inversion layer   The conducting layer formed between drain and source when an enhancement-mode n-channel MOSFET is biased ON. Also referred to as the channel.

Threshold voltage $V_{GS(Th)}$ The minimum voltage between gate and source of a MOSFET for the formation of the conducting inversion layer (channel).

When we studied the operation of TTL gate circuits, we discovered that, for the most part, the bipolar transistors in the gates operated either in the saturation or the cutoff regions. In MOS-type gates, we make use of two similar operating regions in the constituent MOSFETs:

1. The cutoff region is the same as that for a bipolar transistor. Under this condition, there is a very high impedance between the drain and source terminals of the MOSFET.

2. The **ohmic region** is analogous to the saturation region of a bipolar transistor. In this state, there is a relatively low resistance between the MOSFET's drain and source.

The MOSFET switches between cutoff and ohmic regions when the voltage between gate and source, V_{GS}, is less than or greater than a value called the **threshold voltage.** The abbreviation for this voltage is $V_{GS(Th)}$; its value is between 1 and 5 volts, typically 1.5 V.

Figure 11.46 shows an n-channel MOSFET operating in the cutoff region. The gate-source voltage, V_{GS} is less than $V_{GS(Th)}$. There is no conduction between the drain and source. The resistance, $R_{DS(OFF)}$, between drain and source is very large, typically in the thousands of megohms.

FIGURE 11.46
n-Channel MOSFET in Cutoff
Region

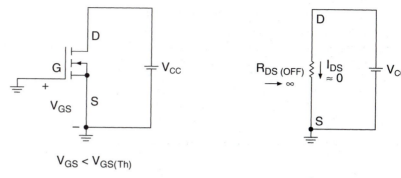

a. **Bias voltages**   b. **Equivalent circuit**

When the value of V_{GS} increases and exceeds the threshold voltage, the MOSFET enters the ohmic region. A conduction channel, called the ***n*-type inversion layer,** is created in the *p*-substrate of the transistor, as shown in Figure 11.47. This layer is like an artificially created region of *n*-type silicon, which allows conduction between the drain and source, provided there is sufficient potential difference between them.

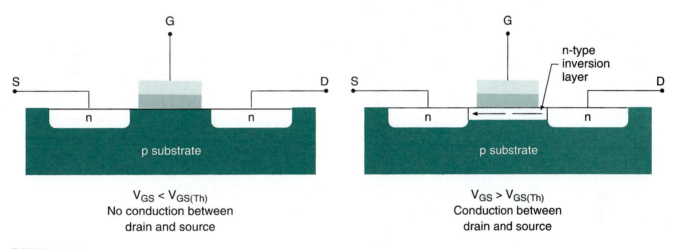

$V_{GS} < V_{GS(Th)}$
No conduction between
drain and source

$V_{GS} > V_{GS(Th)}$
Conduction between
drain and source

FIGURE 11.47
Channel Formation in an n-Channel MOSFET

Figure 11.48 shows a MOSFET operating in the ohmic region. $R_{DS(ON)}$, the equivalent resistance of a MOSFET in the ohmic region, is typically around 500 Ω to 2 kΩ. The drain-source current, I_{DS}, is determined by Ohm's law: $I_{DS} = V_{CC}/R_{DS(ON)}$.

FIGURE 11.48
n-Channel MOSFET in Ohmic
Region

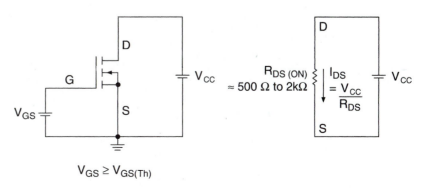

$V_{GS} \geq V_{GS(Th)}$

a. **Bias voltages**   b. **Equivalent circuit**

The operation of a p-channel MOSFET is similar, but with polarities reversed. If $V_{GS(Th)}$ is $+1.5$ V for an n-channel device, an equivalent p-channel MOSFET has a threshold voltage of -1.5 V. $V_{GS} > +1.5$ V turns ON an n-channel transistor; $V_{GS} < -1.5$ V turns ON a p-channel device.

Figure 11.49 summarizes the bias requirements for n- and p-channel enhancement-mode MOSFETs.

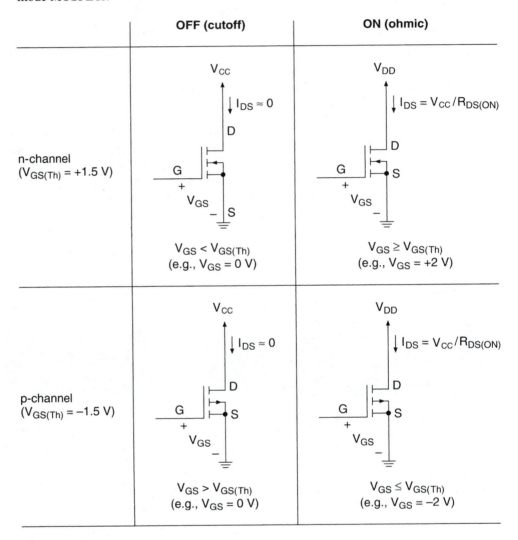

FIGURE 11.49
Bias Requirements of n- and p-Channel MOSFETs

CMOS Inverter

Figure 11.50 shows the circuit of a CMOS inverter, which consists of one n-channel and one p-channel MOSFET.

Recall the bias conditions of the two transistors:

n-channel: threshold voltage, $V_{GS(Th)} \approx +1.5$V

ON when $V_{GS} > V_{GS(Th)}$ (e.g., $V_{GS} = V_{CC}$)

OFF when $V_{GS} < V_{GS(Th)}$ (e.g., $V_{GS} = 0$ V)

p-channel: threshold voltage, $V_{GS(Th)} \approx -1.5$ V

ON when $V_{GS} < V_{GS(Th)}$ (e.g., $V_{GS} = -V_{CC}$)

OFF when $V_{GS} > V_{GS(Th)}$ (e.g., $V_{GS} = 0$ V)

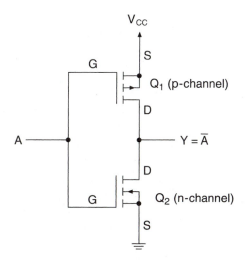

FIGURE 11.50
CMOS Inverter

The operation of the CMOS inverter, and any other CMOS gate, depends on arranging the bias conditions of each complementary pair of transistors so that they are always in opposite states. Whenever Q_1 is ON, Q_2 is OFF, and vice versa. Figure 11.51 shows how this is accomplished.

Assume that a LOW input is at ground potential and that a HIGH input is equal to V_{CC}.

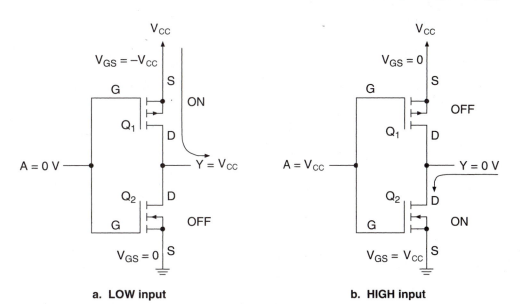

a. LOW input   **b. HIGH input**

FIGURE 11.51
Operation of CMOS Inverter

When *input A is LOW,* the gate voltage of Q_2 is the same as its source voltage; $V_{GS2} = 0$ and Q_2 is OFF. This places a high-impedance path between output Y and ground. At the same time, the gate voltage of Q_1 is 0 V and its source voltage is V_{CC}; $V_{GS1} = V_{G1} - V_{S1} = 0 - V_{CC} = -V_{CC}$. (The p-channel transistor, Q_1, is drawn "upside down" to make the complementary pair symmetrical.) Q_1 is ON, forming a low-impedance path from V_{DD} to the output Y. *Output Y is HIGH.*

When *input A is HIGH,* the gate-source voltage of the n-channel transistor is V_{CC}, causing Q_2 to turn ON. The gate of Q_1 is also at V_{CC}. Since the source of the p-channel transistor is at V_{CC}, $V_{GS1} = V_{G1} - V_{S1} = V_{CC} - V_{CC} = 0$ V; Q_1 is OFF. This combination creates a high impedance between V_{CC} and output Y and a low-impedance path from output Y to ground, as shown in Figure 15.51b. *Output Y is LOW.*

CMOS NAND/NOR Gates

CMOS NAND and NOR gates are constructed from complementary pairs of MOSFETs. Each MOSFET pair has an n-channel transistor that is turned ON by a HIGH input and a p-channel transistor that is turned ON by a LOW input. The n-channel devices switch the output to ground; the p-channel ones switch the output to V_{CC}. NAND and NOR functions are generated by arranging the MOSFET drain-source paths in series (AND) and parallel (OR) configurations.

In Figure 11.52, we see the DeMorgan equivalent forms of a NAND gate. Each form illustrates an aspect of NAND operation that can be described with a brief sentence and implemented by a MOSFET circuit. The combination of forms describes the complete operation of the device.

Figure 11.52a states that both NAND inputs must be HIGH to make the output LOW. A logic HIGH activates an n-channel MOSFET. The gate output is switched to ground by an n-channel MOSFET. Thus, the drain-source paths of two n-channel transistors must be connected in series to make the output LOW under the stated conditions.

Figure 11.52b shows that the NAND output is HIGH if either input is LOW. A p-channel transistor will turn ON when its gate is LOW and will switch a HIGH to the output. A parallel combination of p-channel MOSFETs will satisfy these conditions.

FIGURE 11.52
NAND Functions of MOSFETs

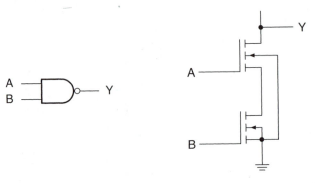

a. Both inputs HIGH make output LOW
(n-channels in series to ground)

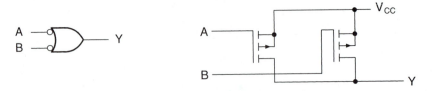

b. Either input LOW makes output HIGH
(p-channels in parallel to V_{CC})

The stated conditions are combined in the CMOS NAND circuit shown in Figure 11.53. Transistors Q_1 and Q_4 form a complementary pair, as do Q_2 and Q_3. When A and B are both HIGH, Q_1 and Q_2 are both OFF, cutting off the connection between V_{CC} and output Y. Q_3 and Q_4 are both ON, supplying a low-impedance path from output Y to ground, making the output LOW. This is shown in the partial truth table in Table 11.10.

When A is LOW and B is HIGH, Q_1 is ON. This creates a path from V_{CC} to output Y. At the same time, Q_4 is OFF. This cuts the Y-to-ground path through the n-channel

FIGURE 11.53
CMOS NAND Gate

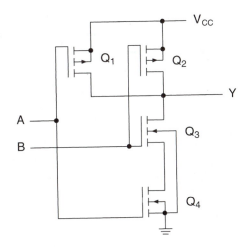

Table 11.10 Partial CMOS NAND Function and Truth Table

A	B	Q_1	Q_2	Q_3	Q_4	Y
1	1	OFF	OFF	ON	ON	0

Table 11.11 Partial CMOS NAND Function and Truth Table

A	B	Q_1	Q_2	Q_3	Q_4	Y
0	0	ON	ON	OFF	OFF	1
0	1	ON	OFF	ON	OFF	1
1	0	OFF	ON	OFF	ON	1

MOSFETs; the series path from output to ground is broken. One parallel path from V_{CC} to output has been established. Output Y is HIGH.

The remaining input combinations also make the output HIGH, as shown in Table 11.11. They do so by breaking the n-channel path from output to ground and enabling one or both p-channel paths from V_{CC} to output.

Each MOSFET in a logic circuit must have its own independent substrate bias. This ensures that the transistor will operate as expected when a logic HIGH or LOW is applied to its gate.

Normally, the substrate of a MOSFET is shorted to its source terminal. If a MOSFET source terminal is isolated from V_{CC} or ground, the substrate must be biased separately. For example, in the NAND gate in Figure 11.53, the substrate of Q_3 connects directly to ground.

NOR gates are similar to NANDs in construction. Figure 11.54 shows the DeMorgan equivalent forms of the NOR function and the MOSFET implementations of each aspect of the gate operation.

FIGURE 11.54
NOR Functions of MOSFETs

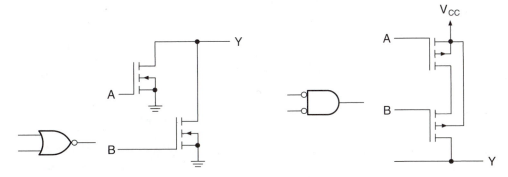

a. **Either input HIGH makes output LOW (n-channels in parallel to ground)**

b. **Both inputs LOW make output HIGH (p-channels in series to V_{CC})**

When either input is HIGH, the output is LOW. This function is implemented by two parallel n-channel MOSFETs. Both inputs must be LOW to make the output HIGH, which implies a series connection of two p-channel transistors. The complete NOR gate circuit is shown in Figure 11.55. (Note that the substrate of Q_2 is connected directly to V_{CC} to ensure that it has its own bias voltage.)

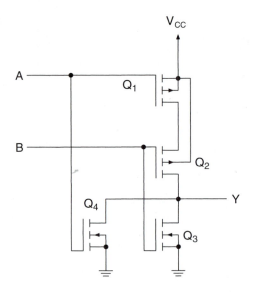

FIGURE 11.55
CMOS NOR Gate

As was the case with the NAND circuit, transistors Q_1 and Q_4 form a complementary MOSFET pair. Transistors Q_2 and Q_3 form the second pair.

When both inputs are LOW, both p-channel transistors are ON. This creates a low-impedance path from V_{CC} to output Y. The n-channel transistors, Q_3 and Q_4, are both OFF. This isolates the output from ground. Output Y is HIGH. Table 11.12 shows the MOSFET states under this condition.

Table 11.12 Partial CMOS NOR Function and Truth Table

A	B	Q_1	Q_2	Q_4	Q_4	Y
0	0	ON	ON	OFF	OFF	1

If either input is HIGH, one or both of the p-channel transistors will turn OFF. This action breaks the path from V_{CC} to output Y. The complementary n-channel transistor will turn ON. This creates a low-impedance path from output Y to ground. Output Y is LOW. Table 11.13 summarizes the possible input conditions and MOSFET states when the NOR output is LOW.

Table 11.13 Partial CMOS NOR Function and Truth Table

A	B	Q_1	Q_2	Q_3	Q_4	Y
0	1	ON	OFF	ON	OFF	0
1	0	OFF	ON	OFF	ON	0
1	1	OFF	OFF	ON	ON	0

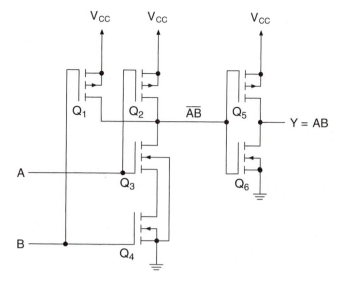

FIGURE 11.56
CMOS AND Gate

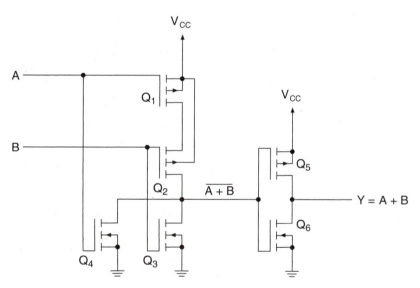

FIGURE 11.57
CMOS OR Gate

CMOS AND and OR Gates

Figures 11.56 and 11.57 show the circuits of CMOS AND and OR gates. The AND gate is the same as the NAND circuit, except for the output inverter section constructed from Q_5 and Q_6. The OR gate is the same as the NOR with an output inverter section.

▌▌ SECTION 11.8B REVIEW PROBLEM

11.13 Why is the source of a p-channel MOSFET connected to V_{CC} in a CMOS gate?

CMOS Transmission Gate

Figure 11.58 shows the circuit of a CMOS transmission gate. A CMOS transmission gate, or analog switch, conducts in both directions. This makes it possible to enable or inhibit

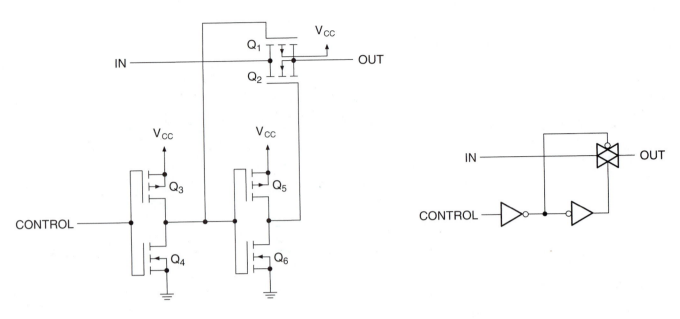

FIGURE 11.58
CMOS Transmission Gate

time-varying analog signals having both positive and negative values. Conduction takes place between the input and output terminals through MOSFETs Q_1 and Q_2. Positive current (left to right in the diagram) flows through Q_2, and negative current (right to left) flows through Q_1. Two inverters, consisting of the Q_3/Q_4 and Q_5/Q_6 pairs of MOSFETs, control the ON/OFF state of the circuit.

When $CONTROL = 1$, the inverters bias both Q_1 and Q_2 ON, allowing them to conduct. When $CONTROL = 0$, the circuit inhibits conduction between input and output.

The substrate terminal of Q_1 is connected, not to the source terminal of that transistor, but directly to V_{CC} thus providing the correct bias to Q_1 in the ON state.

A particular device with this function is the 74HC4066 quad analog switch, whose circuit symbol is shown in Figure 11.59. When the $CONTROL$ input is HIGH, analog and digital signals can pass between the bidirectional input terminals.

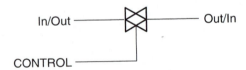

FIGURE 11.59
One of Four Analog Switches From 74HC4066

▐▌ EXAMPLE 11.16

Figure 11.60 shows a circuit where the analog switches in a 74HC4066 package are used to control the selection and muting of two pairs of speakers in a stereophonic audio system. Briefly explain the circuit operation.

Solution　The audio signal to each speaker is passed or blocked by a CMOS transmission gate. The speakers are paired into A and B groups. Each pair has a left and a right channel speaker. The same logic gate controls both speakers of each group.

The Select A switch enables the A speakers when it is open (logic HIGH). The Select B switch enables the B speakers when it is open. The Mute Toggle flip-flop mutes (disables) both sets of speakers when Q is LOW. This action inhibits both AND gates, making all transmission gate $CONTROL$ inputs LOW. The mute function toggles ON and OFF with each push of the Mute ON/OFF switch.

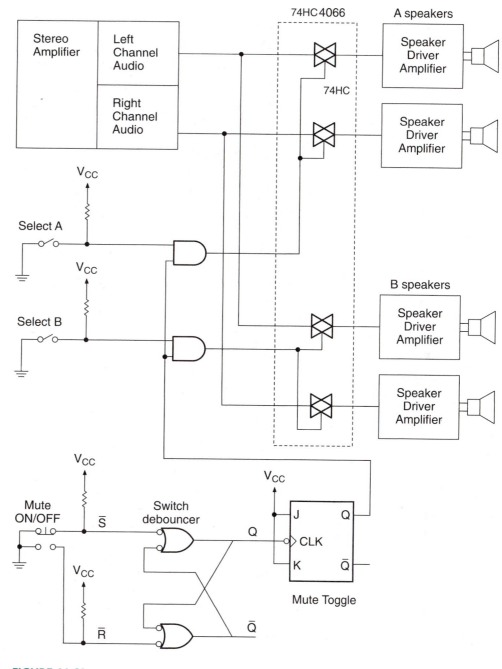

FIGURE 11.60
Example 11.16
74HC4066 Analog Switches as Audio Selectors

11.9 TTL and CMOS Variations

Standard (74NN) TTL and CMOS represented the two main standards of logic design for many years, and their influence is still visible in other, more advanced types of logic. The changes that have been made in newer logic families are not fundamental changes in the working concepts, but improvements to the specifications, particularly switching speed and power dissipation.

TTL Logic Families

Probably the most important development in TTL technology was the introduction, in the early 1970s, of the **Schottky barrier diode** into circuit designs. This made possible the first family of nonsaturated bipolar logic, with its resultant improvement in switching speed.

Figure 11.61 shows a bipolar transistor with a Schottky diode connected across its base and collector and the equivalent circuit symbol of this combination. We call this configuration a **Schottky transistor** and logic devices using such transistors **Schottky TTL**.

FIGURE 11.61
Schottky Transistor

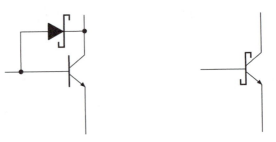

a. Schottky diode-clamped base-collector junction

b. Equivalent circuit symbol

Normally the base-collector junction of a saturated bipolar transistor has a drop of about 0.5 volts, as shown in Figure 11.62. The Schottky diode clamps this junction voltage to about 0.4 volts. This keeps the transistor out of deep saturation in its ON state. The base region of the Schottky-clamped transistor holds less charge than does a standard bipolar transistor. Its storage time, the time required to dissipate base charge upon turn-off, is substantially reduced. The transistor can switch faster with the Schottky diode than without.

FIGURE 11.62
ON-State Operating Voltages of Bipolar Transistors

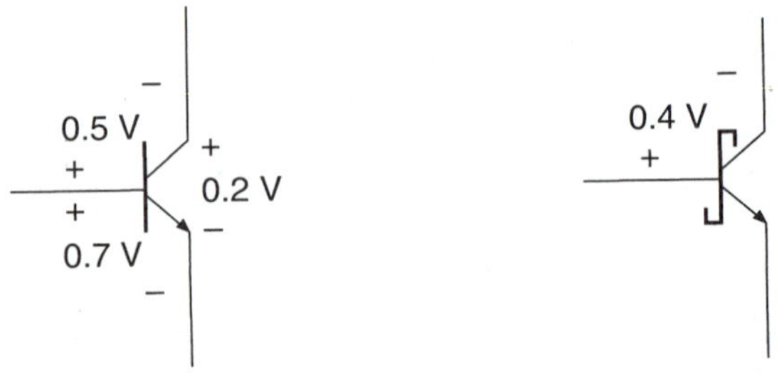

a. Saturated bipolar transistor

0.5 V
0.2 V
0.7 V

b. Base-collector voltage in Schottky transistor

0.4 V

Figure 11.63 shows the circuits of the 74S00 Schottky and 74LS00 low-power Schottky NAND gates. Compare these circuits to each other and to the 7400 standard TTL NAND gate in Figure 11.33.

FIGURE 11.63
Schottky TTL Circuits

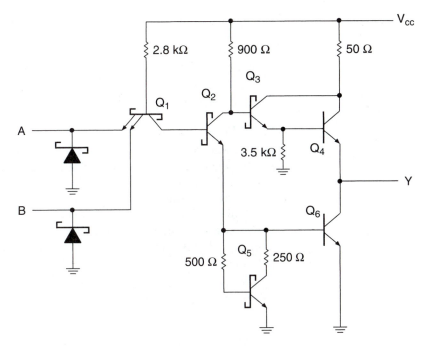

a. 74S00 Schottky NAND gate

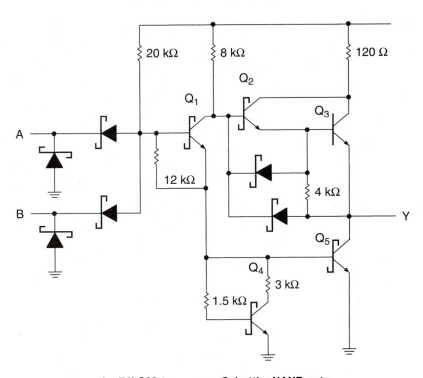

b. 74LS00 low-power Schottky NAND gate

In the 74S00 circuit, Q_1 acts as the input and Q_2 as the phase splitter, as in the 7400 gate. The HIGH output circuit consists of Q_3 and Q_4 connected as a modified Darlington pair. When Q_2 is OFF (at least one input is LOW), enough base current flows in Q_3 to turn it on. Collector-emitter current in Q_3 turns on Q_4, making the output HIGH.

When Q_2 is ON (both inputs are HIGH), the base of Q_3 is pulled LOW, turning it OFF. Sufficient current flows in the base of Q_5 to turn it ON. The resultant current through Q_5

will turn on Q_6, making the output LOW. A similar analysis can be made for the 74LS00 gate.

One difference between the 74S00 and 74LS00 circuits is the size of the resistors; the LS device has larger resistors. Less current flows in the gate circuit. This reduces power dissipation of the chip. The larger resistor values also slow down the switching times of the various transistors by increasing the RC time constants of the circuit elements.

Speed-Power Product

One measure of logic circuit efficiency is its **speed-power product,** calculated by multiplying switching speed and power dissipation, usually expressed in picojoules (pJ). (The joule is the SI unit of energy. Power is the rate of energy used per unit time.) A major goal of logic circuit design is the reduction of a device's speed-power product.

Table 11.14 shows the propagation delay, supply current, and speed-power product for a NAND gate in six TTL families: standard TTL (7400), Schottky (74S00), low-power Schottky (74LS00), fast TTL (74F00), advanced Schottky (74AS00), and advanced low-power Schottky (74ALS00).

Table 11.14   TTL Speed and Power Specifications

	7400	74LS00	74S00	74F00	74ALS00	74AS00
t_{pLH} **(max)**	22 ns	15 ns	4.5 ns	6 ns	11 ns	4.5 ns
t_{pHL} **(max)**	15 ns	15 ns	5 ns	5.3 ns	8 ns	4 ns
$I_{CCH}/4$ **(max)**	2 mA	0.4 mA	4 mA	0.7 mA	0.21 mA	0.8 mA
$I_{CCL}/4$ **(max)**	5.5 mA	1.1 mA	9 mA	2.6 mA	0.75 mA	4.35 mA
Speed-power product (per gate)	605 pJ	82.5 pJ	225 pJ	78.0 pJ	41.25 pJ	97.9 pJ

The speed-power product shown is the worst-case value. This is calculated by multiplying the largest value of $I_{CC}/4$ by the slowest switching speed by 5 volts for each family. We use $I_{CC}/4$ because I_{CC} is specified per chip (four gates).

A faster switching speed results in an overall increase in speed-power product, other factors being equal. For example, the speed-power product of either advanced Schottky family is lower than that of the LS and S families. However, the ALS series (the slower advanced Schottky family) has a lower speed-power product than the AS series.

The smaller resistors used to speed up output switching imply a proportional drop in propagation delay (higher speed) but an increased supply current. Power dissipation increases in proportion to the square of the supply current, thus offsetting the effect of the increased switching speed.

CMOS Logic Families

The CMOS gates we have looked at in this chapter are simpler than most gates actually in use. There are two main families of CMOS devices: metal-gate CMOS, and silicon-gate, or high-speed, CMOS.

Metal-Gate CMOS

There are two main variations on this type of circuit, designated B-series and UB-series CMOS. Most CMOS gates are B-series; UB-series is available in a limited number of inverting-type gates, such as inverters and 2-, 3-, and 4-input NAND and NOR gates. Figure 11.64 shows the difference in the two configurations.

Figure 11.64b shows one gate from a 4011UB quadruple 2-input NAND package. Its circuit is the same as the NAND configuration examined in Section 11.8. Power supply voltages in metal-gate CMOS are designated V_{DD} (power) and V_{SS} (ground). High-speed, or silicon-gate, CMOS uses the same power supply designations as TTL: V_{CC} and ground.

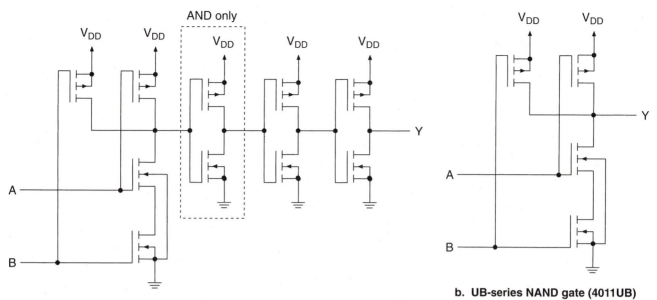

a. B-series AND/NAND gate (4081B/4011B)

b. UB-series NAND gate (4011UB)

FIGURE 11.64
Metal-Gate CMOS Circuits

The B-series configuration of this circuit has two additional inverter outputs in cascade with the NAND logic. (The same gate becomes an AND when we add a third output inverter.) The inverter configuration is actually an amplifier; extra inverter stages provide additional gain and increase noise margin by allowing the circuit to accept smaller input signals.

CMOS gates are sometimes used in analog applications, such as oscillators. The UB-series gates, with their lower gain, are more desirable for such applications. Due to its low switching speed, metal-gate CMOS is rarely used in new designs.

High-Speed CMOS

> **KEY TERM**
>
> **High-speed (silicon-gate) CMOS**   A CMOS logic family with a smaller device structure and thus higher speed than standard (metal-gate) CMOS.

Metal-gate CMOS has been considered a nearly ideal family for logic designs, with its high noise immunity, low power consumption, and flexible power supply requirements. Unfortunately, its propagation delay times, typically 10 to 20 times greater than those of equivalent TTL devices, are just not fast enough for use in modern microprocessor-based systems.

High-speed CMOS was developed to address the problem of switching speed, while striving to keep the other advantages of CMOS. This is achieved by using MOSFETs with a polysilicon material for the gate, rather than metal, as in standard CMOS. Because of advantages gained in this manufacturing process, each transistor is physically smaller and has a lower gate capacitance than metal-gate MOSFETs. Both these factors contribute to a lower propagation delay for the logic gate circuit.

Several subfamilies of high-speed CMOS are available for various logic and linear applications, designated by the labels 74HCNN, 74HC4NNN, 74HCTNN, and 74HCUNN.

The 74HCNN series duplicates equivalent LSTTL functions in packages having identical pinouts to LSTTL. The 74HC4NNN replaces CMOS functions pin for pin. Both these series have CMOS-equivalent input and output levels, within the power supply limits (2.0 V to 6.0 V) of high-speed CMOS.

Table 11.15   CMOS Speed and Power Specifications

	Metal-Gate CMOS		High-Speed CMOS			Advanced High-Speed CMOS		Low-Voltage CMOS	
	4011B	**4011UB**	**74HC00A**	**74HCT00A**	**74HCU04**	**74VHC00**	**74VHCT00**	**74LVX00**	**74LCX00**
t_{pLH}, t_{pHL}	250 ns	180 ns	15 ns	19 ns	14 ns	5.5 ns	6.9 ns	6.2 ns	5.2 ns
I_{DD} or I_{CC}	0.25 μA	0.25μA	0.25 μA	0.25 μA	0.17 μA	0.5 μA	0.5 μA	0.5 μA	0.25 μA
V_{DD} or V_{CC}	5.0 V	5.0 V	4.5 V	4.5 V	4.5 V	4.5 V	4.5 V	3.3 V	3.3 V
P_D (1 MHz)	1.5 mW	1.5 mW	446 μW	304 μW	303 μW	385 μW	385 μW	208 μW	272 μW
Speed-power product (quiescent)	0.31 pJ	0.23 pJ	0.017 pJ	0.021 pJ	0.011 pJ	0.012 pJ	0.015 pJ	0.010 pJ	0.043 pJ
Speed-power product (1 MHz)	375 pJ	270 pJ	6.68 pJ	5.77 pJ	4.25 pJ	2.12 pJ	2.65 pJ	1.29 pJ	1.42 pJ

The 74HCTNN devices are designed to be directly compatible with LSTTL devices, and thus have LSTTL-equivalent inputs and CMOS-equivalent outputs.

74HCUNN devices have no output buffers, like the 4000 UB-series standard CMOS devices. The 74HCU devices are used, as are the 4000UB devices, for linear applications such as oscillators and multivibrators.

Table 11.15 shows the relative performance of the various CMOS families. As in TTL, the 2-input NAND gate is used as the standard, except for the HCU family, where this gate is not available. The quiescent speed-power product of all CMOS families is much smaller than that of any TTL family. The high-speed CMOS families have propagation delays comparable to those of LSTTL.

The power dissipation of a CMOS device increases directly with frequency. The speed-power product also goes up with higher frequencies.

Table 11.15 shows CMOS speed-power product for a switching speed of 1 MHz. At these speeds, B-series CMOS has no advantage over the common TTL families in terms of its efficiency. It still has the edge on TTL with respect to noise immunity and power supply flexibility.

⦀ SECTION 11.9 REVIEW PROBLEM

11.14   Assuming that power dissipation of a 74HC00A NAND gate is directly proportional to its switching frequency, what is the speed-power product of the gate at 2 MHz, 5 MHz, and 10 MHz?

SUMMARY

1. TTL (transistor-transistor logic) and CMOS (complementary metal-oxide semiconductor) are two major logic families in use today. TTL is constructed from bipolar junction transistors. CMOS is made from metal-oxide-semiconductor field effect transistors (MOSFETs).

2. The main CMOS advantages include low power consumption, high noise immunity, and a flexibility in choosing a power supply voltage.

3. The main advantages of TTL include relatively high switching speed and an ability to drive loads with relatively high current requirements.

4. TTL and high-speed CMOS logic families are alphabetically designated by a part number having the form 74XXNN, where XX is the family and NN is a numeric logic function designator. (For example, 74HC00 and 74LS00 have the same logic function, but are from different logic families.)

Devices from earlier CMOS families are designated by a part number of the form 4NNNB or 4NNNUB.

5. Devices of the same logic family generally have the same electrical characteristics.

6. Data such as input/output voltages and currents are specified in manufacturers' datasheets. Only the maximum or minimum values of these parameters should be used as design information. "Typical" values should be regarded as "information only."

7. The time required for an a logic circuit output to change as a result of an input change is called propagation delay.

8. Propagation delay is specified as t_{pLH} when an output changes from LOW to HIGH and t_{pHL} when the output goes from HIGH to LOW.

9. Propagation delay in a circuit is the sum of all delays in the slowest input-to-output path. Gates whose outputs do not change are ignored in the calculation.

10. Fanout is the maximum number of device inputs that can be driven by the output of a logic device.

11. The actual value of output current in a driving gate is the sum of all load currents, which are the input currents of the load gates. For n loads,

$$I_{OL} = I_{IL1} + I_{IL2} + \cdots + I_{ILnL} = n_L\, I_{IL}$$

and $I_{OH} = I_{IH1} + I_{IH2} + \cdots + I_{IHnH} = n_H I_{IH}$

12. The fanout of the driving gate in the LOW and HIGH states can be calculated as:

$$n_L = \frac{I_{OL}}{I_{IL}}$$

$$\text{and } n_H = \frac{I_{OH}}{I_{IH}}$$

13. If the fanout is unequal for LOW and HIGH states, the smaller value must be used.

14. If the fanout of a gate is exceeded, the output voltage of the driving gate will drop if the output is HIGH and rise if the output is LOW. This move away from the nominal value degrades the general performance of the driving gate.

15. Power supply current (I_{CC}), and therefore power dissipation (P_D), of a TTL device depends on the number of outputs in the device that are HIGH or LOW. $P_D = V_{CC} I_{CC} = V_{CC}\left(\dfrac{n_H}{n} I_{CCH} + \dfrac{n_L}{n} I_{CCL}\right)$ for a device with n outputs, n_H of which are HIGH and n_L of which are LOW.

16. CMOS devices draw most current from the power supply when its outputs are switching and very little when they are static. Power dissipation of a high-speed CMOS device with n outputs has a static and a dynamic component, given by:

$$P_D = (C_L + C_{PD})V_{CC}^2 f + \frac{V_{CC} I_{CC}}{n}$$

At high frequencies (≥ 1 MHz), the quiescent current can be neglected.

17. Noise margin is a measure of the noise voltage that can be tolerated by a logic device input. In the HIGH state, it is given by $V_{NH} = V_{OH} - V_{IH}$. In the LOW state, it is given by $V_{NL} = V_{IL} - V_{OL}$. CMOS devices generally have higher noise margins than TTL.

18. When interfacing two devices from different logic families, the driving gate must satisfy the voltage and current requirements of the load gates.

19. Input current in a CMOS gate is very low, due to its high input impedance. Thus, fanout is generally not a problem with CMOS loads.

20. CMOS devices that have the same values of V_{IH} and V_{IL} as TTL are considered to be TTL compatible, since they can be driven directly by TTL drivers.

21. A 74HC or 74HCT device can drive 10 LSTTL loads directly. To calculate fanout, we use the output currents for which the driving gate output voltages are defined.

22. A 74LS device can drive one or more 74HC devices, provided each 74HC input has a pull-up resistor (about 1 kΩ to 10 kΩ) to supply sufficient voltage in the HIGH state.

23. A 74LS device can drive one or more 74HCT inputs directly.

24. Low-voltage CMOS (e.g., 74LVX or 74LCX) can be driven directly by a TTL device if the CMOS device is operated with a 3.3 V power supply. Noise margins are too small for a low-voltage CMOS driver to drive TTL loads.

25. 74HC or 74HCT gates can be operated at a low value of V_{CC} (e.g., 3 volts) and interfaced to a higher-voltage driver by an inverting or noninverting buffer, such as the 74HC4049 or 74HC4050. The interface buffer can tolerate relatively high input voltages (up to 15 V) and, if it shares the same supply voltage as the load gate, can provide correct input voltages to the load.

26. A bipolar transistor with a grounded emitter acts as an inverter or a digital switch. A HIGH at the base causes the transistor to conduct, pulling the collector to near-ground potential. If there is a pull-up resistor on the collector, there will be a HIGH state at the collector when the base is LOW.

27. The simplest TTL input is a transistor with its base connected to V_{CC} through a resistor. It can be treated as two diodes, back-to-back.

28. A TTL LOW input forward-biases the base-emitter junction of the input transistor, supplying a path to ground for input current.

29. A TTL HIGH input reverse-biases the base-emitter junction of the input transistor and forward-biases its base-collector junction. Input current in the HIGH state is restricted to reverse leakage current through the base-emitter junction.

30. An open TTL input is equivalent to a HIGH, as it provides no path to ground.

31. Some types of TTL gates, such as NAND, have multiple-emitter input transistors. Any one input LOW acts as a LOW for the whole circuit.

32. Other TTL gates, such as NOR, have separate transistors for each input. Any HIGH input acts as a HIGH for the whole circuit.

33. An open-collector output has one output transistor that switches on a path to ground (logic LOW) when it is turned on. There is no separate internal circuit for a HIGH output. This must be provided by an external pull-up resistor.

34. Open-collector outputs can be used to parallel outputs (wired-AND), drive high-current loads, or interface to a circuit with a different power supply voltage than the driving gate.

35. A totem pole output has a transistor that switches on for a LOW output and another that switches on for a HIGH output. These output transistors are always in opposite states, except briefly during times when the output is changing states.

36. Totem pole outputs generate noise spikes on the power line of a circuit when they switch between logic states. These

spikes can be amplified by inductance of the power line. Decoupling capacitors placed close to each device help minimize this problem.

37. TTL outputs should never be connected together, as they can be damaged when the outputs are in opposite states. (Too much output current flows.) The logic level under such conditions is not certain.

38. Gates with tristate outputs can generate logic LOW, logic HIGH, or high-impedance states. A high-impedance state is like an open circuit or electrical disconnection of the gate output from the circuit. In this state, both HIGH- and LOW-state output transistors are off.

39. The operation of a tristate output is controlled by the state of a control input. In one control state, the output is either HIGH or LOW. In the opposite control state, the output is in the high-impedance state.

40. CMOS (complementary MOS) devices are based on n-channel and p-channel MOSFETs (metal-oxide-semiconductor field effect transistors).

41. A MOSFET consists of a silicon substrate of a particular type of silicon (e.g., p-type), embedded with wells of the opposite type (e.g., n-type) that form the drain and source regions of the MOSFET. A gate electrode can bias the substrate to create a conduction channel between drain and source.

42. An n-channel enhancement mode MOSFET is biased on when its gate voltage exceeds its source voltage by a given amount called the threshold voltage.

43. A p-channel enhancement mode MOSFET is biased on when its gate voltage is less than its source voltage by a given amount called the threshold voltage.

44. An n-channel and p-channel MOSFET can be connected in such a way that one of the pair of MOSFETs is always on and one is always off. This connection is called a complementary pair and forms the basis for CMOS logic.

45. Logic functions, such as NAND and NOR, can be implemented with a complementary pair of MOSFETs for each input, with the MOSFETs in series or parallel to V_{CC} or ground, as required.

46. Many TTL families have been designed to incorporate Schottky barrier diodes, which limit the saturation of their transistors, allowing faster internal and output switching speeds.

47. Metal-gate CMOS has been superceded by high-speed (silicon-gate) CMOS, which has a smaller MOSFET size, resulting in faster switching and lower gate capacitance.

48. Speed-power product is a measure of the energy used by a gate. More advanced logic families have smaller values of speed-power product.

GLOSSARY

CMOS Complementary metal-oxide semiconductor. A logic family based on the switching of n- and p-channel metal-oxide-semiconductor field effect transistors (MOSFETs).

Cutoff mode The operating mode of a transistor when there is no collector or drain current flowing and the path from collector to emitter or drain to source is effectively an open circuit

Driving gate A gate whose output supplies current to the inputs of other gates.

ECL Emitter coupled logic. A high-speed logic family based on bipolar transistors.

Enhancement-mode MOSFET A MOSFET which creates a conduction path (a channel) between its drain and source terminals when the voltage between gate and source exceeds a specified threshold level.

Fanout The number of gate inputs that a gate output is capable of driving without possible logic errors.

Floating An undefined logic state, neither HIGH nor LOW.

High-speed (silicon-gate) CMOS A CMOS logic family with a smaller device structure and thus higher speed than standard (metal-gate) CMOS.

I_{CC} Total supply current in a TTL or high-speed CMOS device.

I_{CCH} TTL supply current with all outputs HIGH.

I_{CCL} TTL supply current with all outputs LOW.

I_{DD} CMOS supply current under static (nonswitching) conditions.

I_{IH} Current measured at a device input when the input is HIGH.

I_{IL} Current measured at a device input when the input is LOW.

I_{OH} Current measured at a device output when the output is HIGH.

I_{OL} Current measured at a device output when the output is LOW.

I_T When referring to CMOS supply current, the sum of static and dynamic supply currents.

Load gate A gate whose input current is supplied by the output of another gate.

MOSFET Metal-oxide-semiconductor field effect transistor. A MOSFET has three terminals—gate, source, and drain—which are analogous to the base, emitter, and collector of a bipolar junction transistor.

n-channel enhancement-mode MOSFET A MOSFET built on a p-type substrate with n-type drain and source regions. An n-type channel is created in the p-substrate during conduction.

Noise Unwanted electrical signal, often resulting from electromagnetic radiation.

Noise margin A measure of the ability of a logic circuit to tolerate noise.

n-type inversion layer The conducting layer formed between drain and source when an enhancement-mode n-channel MOSFET is biased ON. Also referred to as the channel.

Ohmic region The MOSFET equivalent of saturation. When a MOSFET is biased ON, it acts like a relatively low resistance, or "ohmically."

Open-collector output A TTL output where the collector of the LOW-state output transistor is brought out directly to the output pin. There is no built-in HIGH-state output circuitry which allows two or more open collector outputs to be connected without possible damage.

***p*-channel enhancement-mode MOSFET**   A MOSFET built on an *n*-type substrate with *p*-type drain and source regions. During conduction, a *p*-type channel is created in the *n*-substrate.

Phase splitter   A transistor in a TTL circuit which ensures that the LOW- and HIGH-state output transistors of a totem pole output are always in opposite phase (i.e., one ON, one OFF).

Power dissipation   The electrical energy used by a logic circuit in a specified period of time. Abbreviation: P_D

Propagation delay   The time required for the output of a digital circuit to change states after a change at one or more of its inputs.

Saturation mode   The operating mode of a bipolar transistor when an increase in base current will not cause a further increase in the collector current and the path from collector to emitter is very nearly (but not quite) a short circuit. This is the ON state of a transistor in a digital circuit.

Schottky barrier diode   A specialized diode with a forward drop of about +0.4 V.

Schottky transistor   A bipolar transistor with a Schottky diode across its base-collector junction, which prevents the transistor from going into deep saturation.

Schottky TTL   A series of unsaturated TTL logic families based on Schottky transistors. Schottky TTL switches faster than standard TTL due to decreased storage time in its transistors.

Sinking   A terminal on a gate or flip-flop is sinking current when the current flows into the terminal.

Sourcing   A terminal on a gate or flip-flop is sourcing current when the current flows out of the terminal.

Speed-power product   A measure of a logic circuit's efficiency, calculated by multiplying its propagation delay by its power dissipation. Unit: picojoule (pJ)

Storage time   Time required to transport stored charge away from the base region of a bipolar transistor before it can turn off.

Substrate   The foundation of *n*- or *p*-type silicon on which an integrated circuit is built.

Threshold voltage, $V_{GS(Th)}$ The minimum voltage between gate and source of a MOSFET for the formation of the conducting inversion layer (channel).

Totem pole output   A type of TTL output with a HIGH and a LOW output transistor, only one of which is active at any time.

t_{pHL}   Propagation delay when the device output is changing from HIGH to LOW.

t_{pLH}   Propagation delay when the device output is changing from LOW to HIGH.

Tristate output   An output having three possible states: logic HIGH, logic LOW, and a high-impedance state, in which the output acts as an open circuit.

TTL   Transistor-transistor logic. A logic family based on bipolar transistors.

TTL Compatible   Able to be driven directly by a TTL output. Usually implies voltage compatibility with TTL.

V_{CC}   Supply voltage for TTL and high-speed CMOS devices.

V_{DD}   Metal-gate CMOS supply voltage.

V_{IH}   Voltage level required to make the input of a logic circuit HIGH.

V_{IL}   Voltage level required to make the input of a logic circuit LOW.

V_{OH}   Voltage measured at a device output when the output is HIGH.

V_{OL}   Voltage measured at a device output when the output is LOW.

Wired-AND   A connection where open-collector outputs of logic gates are wired together. The logical effect is the ANDing of connected functions.

PROBLEMS

Problem numbers set in color indicate more difficult problems: those with underlines indicate most difficult problems.

Section 11.1   Electrical Characteristics of Logic Gates

11.1   Briefly list the advantages and disadvantages of TTL, CMOS, and ECL logic gates.

Section 11.2   Propagation Delay

11.2   Explain how propagation delay is measured in TTL devices and CMOS devices. How do these measurements differ?

11.3   Figure 11.65 shows the input and output waveforms of a logic gate. Use the graph to calculate t_{pHL} and t_{pLH}.

11.4   The inputs of the logic circuit in Figure 11.66 are in state 1 in the following table. The inputs change to state 2, then to state 3.

	A	B	C
State 1	1	0	1
State 2	0	0	1
State 3	0	0	0

a. Draw a timing diagram that uses the above changes of input state to illustrate the effect of propagation delay in the circuit.

b. Calculate the maximum time it takes for the output to change when the inputs change from state 1 to state 2.

c. Calculate the maximum time it takes for the output to change when the inputs change from state 2 to state 3.

FIGURE 11.65
Problem 11.3
Waveforms

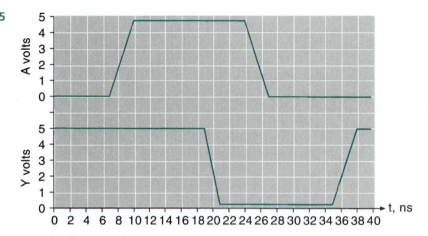

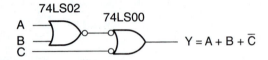

FIGURE 11.66
Problems 11.4 and 11.5
Logic Circuit

11.5   Repeat Problem 11.4 , parts b and c, for a 74HC00 NAND and a 74HC02 NOR gate.

Section 11.3   Fanout

11.6   Calculate the maximum number of low-power Schottky TTL loads (74LSNN series) that a 74S86 XOR gate can drive.

11.7   What is the maximum number of 74S32 OR gates that a 74LS00 NAND gate can drive?

11.8   What is the maximum number of 74LS00 NAND gates that a 74S32 OR gate can drive?

11.9   An LSTTL gate is driving seven LSTTL gate inputs, each equivalent to the load presented by a 74LS00 NAND input. Calculate the source and sink currents required from the driving gate.

11.10 Calculate the current values for the circuits shown in Figure 11.67. For each circuit, state the logic level at the output of gate 1.

FIGURE 11.67
Problem 11.10
Current Calculations

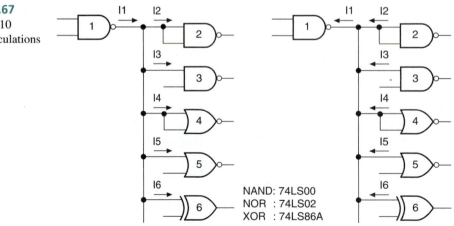

Section 11.4   Power Dissipation

11.11 The circuit in Figure 11.68 is constructed from the gates of a 74LS08 AND device. Calculate the power dissipation of the circuit for the following input logic levels:

	A	B	C	D	E
a.	0	0	0	0	0
b.	1	1	0	1	1
c.	1	1	1	1	0
d.	1	1	1	1	1

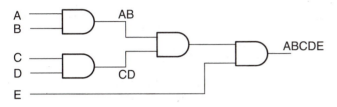

FIGURE 11.68
Problems 11.11 to 11.13
Logic Circuit

11.12 The gate outputs in Figure 11.68 are switching at an average frequency of 100 kHz, with an average duty cycle of 60%. Calculate the power dissipation if the gates are all 74S08 AND gates.

11.13 The gates in Figure 11.68 are 74HC08A high-speed CMOS gates.

a. Calculate the power dissipation of the circuit if the input state is $ABCDE = 010101$. ($V_{CC} = 4.5$ V, $T_A = 25°C$)

b. Calculate the circuit power dissipation if the outputs are switching at a frequency of 10 kHz, 50% duty cycle.

c. Repeat part b for a frequency of 2 MHz.

11.14 The circuit in Figure 11.69 consists of two 74LS00 NAND gates (gates 4 and 5) and three 74LS02 NOR gates (gates 1, 2, and 3). When this circuit is actually built, there will be two unused NAND gates and one unused NOR gate in the device packages.

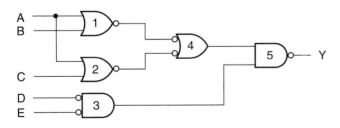

FIGURE 11.69
Problem 11.14
Logic Circuit

Calculate the maximum total power dissipation of the circuit when its input state is $ABCDE = 01100$. Include all unused gates. (Connect unused gate inputs so that they will dissipate the least amount of power.)

11.15 a. Calculate the no-load power dissipation of a single gate at 1 MHz for a 74HC00A quad 2-input NAND gate ($V_{CC} = 5$ V). (Neglect quiescent current.)

b. Calculate the percent change in power dissipation if the gate in part a of this question is operated with a new value of $V_{CC} = 3.3$ V. ($f = 1$ MHz)

Section 11.5   Noise Margin

11.16 Calculate the maximum noise margins, in both HIGH and LOW states, of:

a. A 74S00 NAND gate

b. A 74LS00 NAND gate

c. A 74AS00 NAND gate

d. A 74ALS00 NAND gate

e. A 74HC00 NAND gate ($V_{CC} = 5$ V)

f. A 74HCT00 NAND gate ($V_{CC} = 5$ V)

Section 11.6   Interfacing TTL and CMOS Gates

11.17 Why can an LSTTL gate drive a 74HCT gate directly, but not a 74HC? Show calculations.

11.18 Draw a circuit that allows an LSTTL gate to drive a 74HC gate. Explain briefly how it works.

11.19 How many LSTTL loads (e.g., 74LS00) can a 74HC00A NAND gate drive? Use data sheet parameters to support your answer. Assume $V_{CC} = 4.5$ V. Show all calculations.

Section 11.7   Internal Circuitry of TTL Gates

11.20 In what logic state is an open TTL input? Why?

11.21 Briefly describe the operation of the TTL open-collector inverter shown in Figure 11.20. What is the purpose of the diode?

11.22 Briefly explain the operation of a multiple-emitter input transistor used in a TTL NAND gate. Describe how the transistor responds to various combinations of HIGH and LOW inputs.

11.23 Draw a wired-AND circuit consisting of three open-collector NAND gates and an output pull-up resistor. The gate inputs are as follows:

Gate 1: Inputs A, B
Gate 2: Inputs C, D
Gate 3: Inputs E, F

Write the Boolean function of the circuit output.

11.24 Calculate the minimum value of the pull-up resistor if the circuit drawn in Problem 11.23 is to drive a logic gate having input current $I_{IL} = 0.8$ mA and the NAND gates can sink 12 mA in the LOW output state. (Assume that $V_{OL} = 0.4$ V.)

11.25 Draw a circuit consisting only of open-collector gates whose Boolean expression is the product-of-sums expression

$$(A + B)(C + D)(E + F)(G + H).$$

11.26 Is an open-collector TTL output likely to be damaged if shorted to ground? Why or why not?

11.27 Is an open-collector TTL output likely to be damaged if shorted to V_{CC}? Why or why not?

11.28 Draw the totem pole output of a standard TTL gate.

11.29 Refer to the TTL NAND gate in Figure 11.34.

    a. Why are Q_3 and Q_4 never on at the same time (ideally)?

    b. How does switching noise originate in a totem pole output? How can the problem be controlled?

11.30 Explain briefly why two totem pole outputs should not be connected together.

11.31 Two LED driver circuits are shown in Figure 11.70. For each circuit, calculate the current flowing when the LED is ON. Calculate the ratio between the LED ON current and I_{OL} or I_{OH} of the inverter, whichever is appropriate for each circuit. State which is the best connection for LED driving and explain why.

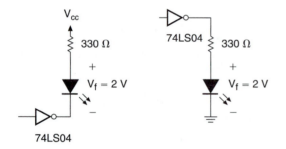

FIGURE 11.70
Problem 11.31
LED drivers

11.32 Calculate the current flowing when the lamp in Figure 11.71 is illuminated. Choose one of the following devices as a suitable driver: 74LS04, 74LS05 74LS06, 74LS16. Explain your choice. (Data sheets for these devices are found in Appendix C.)

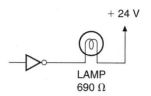

FIGURE 11.71
Problem 11.32
Lamp Driver

Section 11.8   Internal Circuitry of CMOS Gates

11.33 State several precautions that should be taken to prevent electrostatic damage to MOSFET circuits.

11.34   **a.** Draw the circuit symbols for an n-channel and a p-channel enhancement-mode MOSFET.

    b. Describe the required bias conditions for each type of MOSFET in the cutoff and ohmic regions.

    c. State the approximate channel resistance for a MOSFET in the cutoff and ohmic regions.

11.35 Draw the circuit diagram of a CMOS AND gate. Derive the truth table of the gate by analyzing the operation of all the transistors under all possible input conditions.

11.36 Repeat Problem 11.35 for a CMOS OR gate.

11.37 Figure 11.72 shows a circuit that can switch two analog signals to an automotive speedometer/tachometer. Each sensor produces an analog voltage proportional to its measured quantity. Briefly explain how these analog signals are switched to the display output circuitry.

Section 11.9   TTL and CMOS Variations

11.38 Briefly explain how a Schottky barrier diode can improve the performance of a transistor in a TTL circuit.

11.39 Is the speed-power product of a TTL gate affected by the switching frequency of its output? Explain.

11.40 Use data sheets to calculate the speed-power products of the following gates:

    a. 74LS00

    b. 74S00

    c. 74ALS00

    d. 74AS00

    e. 74HC00A (quiescent and 10 MHz)

    f. 74HCT00A (quiescent and 10 MHz)

    g. 74F00

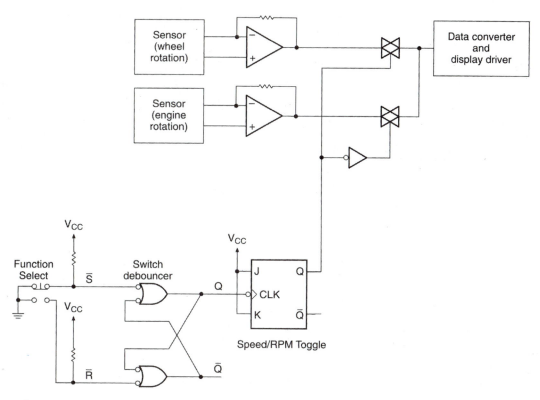

FIGURE 11.72
Problem 11.37
Speedometer/Tachometer Switching Circuit

11.41 Briefly explain the differences among the following high-speed CMOS logic families: 74HCNN, 74HC4NNN, 74HCTNN, and 74HCUNN.

11.42 Assume that the power dissipation of a metal-gate or high-speed CMOS gate increases in proportion to the switching frequency of its output. Calculate the speed-power product of the following gates at 2 MHz, 5 MHz, and 10 MHz:

a. 4011B

b. 74HCT04

c. 74HCU04

ANSWERS TO SECTION REVIEW PROBLEMS

Section 11.1

11.1 $V_{OH} = 2.7$ V min. (We cannot expect typical values for V_{OH}.) $I_{OH} = -0.4$ mA (The negative sign indicates that the current is leaving the gate. See Figure 11.2.)

Section 11.2

11.2 $t_{pHL1} + t_{pHL2} = 20$ ns $+ 22$ ns $= 42$ ns; $t_{pLH2} = 22$ ns

Section 11.3

11.3 Source currents: I_{OH}, I_{IL}; sink currents: I_{OL}, I_{IH}

Section 11.4

11.4 CMOS draws very little current when its outputs are not switching. Since the majority of current is drawn when the outputs switch, the more often the outputs switch, the more current is drawn from the supply. This is the same as saying that power dissipation increases with frequency.

Section 11.5

11.5 $V_{NH} = 1.98$ V, $V_{NL} = 0.66$ V

Section 11.6

11.6 2.5 V. The interface buffer and load should have the same supply voltage so that the output voltage of the buffer and input voltage of the load are compatible.

Section 11.7a

11.7 a. Provision of logic HIGH when output transistor is OFF
 b. Limitation of I_{OL} when output transistor is ON

Section 11.7b

11.8 $R_{ext} = 592$ Ω. Minimum standard value: 680 Ω

Section 11.7c

11.9 When the output is HIGH, current flows to ground through a low-impedance path, causing I_{OH} to exceed its rating.

Section 11.7d

11.10 The diode allows the base of Q_4 to be pulled LOW through *G*, but will not allow a HIGH at *G* to turn it on. This keeps both output transistors OFF in the high-impedance state and allows them to be in opposite states when the output is enabled.

Section 11.7e

11.11 Noninverting gates are actually double-inverting gates. They require an extra transistor stage to cancel the inversion introduced by NAND or NOR transistor logic.

Section 11.8a

11.12 The thin oxide layer in the gate region can be damaged by overvoltage, such as that caused by electrostatic dis-charge. If the oxide layer is damaged, it may no longer insulate the gate terminal from the MOSFET substrate, which causes the transistor to malfunction.

Section 11.8b

11.13 It allows complementary operation with an *n*-channel MOSFET. Specifically, a gate voltage of 0 V turns OFF an *n*-channel device having a grounded source. The same voltage turns ON the *p*-channel device whose source is tied to V_{CC}. It does so by making the *p*-channel gate-source voltage more negative than the required threshold.

Section 11.9

11.14 13.36 pJ, 33.4 pJ, and 66.8 pJ.

Interfacing Analog and Digital Circuits

CHAPTER OBJECTIVES

Upon successful completion of this chapter, you will be able to:

• Define the terms "analog" and "digital" and give examples of each.

• Explain the sampling of an analog signal and the effects of sampling frequency and quantization on the quality of the converted digital signal.

• Draw the block diagram of a generic digital-to-analog converter (DAC) and circuits of a weighted resistor DAC and an R-2R ladder DAC.

• Calculate analog output voltages of a DAC, given a reference voltage and a digital input code.

• Configure an MC1408 integrated circuit DAC for unipolar and bipolar output, and calculate output voltage from known component values, reference voltage, and digital inputs.

• Describe important performance specifications of a digital-to-analog converter.

• Draw the circuit for a flash analog-to-digital converter (ADC) and briefly explain its operation.

• Define "quantization error" and describe its effect on the output of an ADC.

• Explain the basis of the successive approximation ADC, draw its block diagram, and briefly describe its operation.

• Describe the operation of an integrator with constant input voltage.

• Draw the block diagram of a dual slope (integrating) ADC and briefly explain its operation.

• Explain the necessity of a sample and hold circuit in an ADC and its operation.

• State the Nyquist sampling theorem and do simple calculations of maximum analog frequencies that can be accurately sampled by an ADC system.

• Describe the phenomenon of aliasing and explain how it arises and how it can be remedied.

• Interface an ADC0808 analog-to-digital converter to a CPLD-based state machine.

• Design a 4-channel data acquisition system, including an ADC0808 analog-to-digital converter and a CPLD-based state machine.

Electronic circuits and signals can be divided into two main categories: analog and digital. Analog signals can vary continuously throughout a defined range. Digital signals take on specific values only, each usually described by a binary number.

Many phenomena in the world around us are analog in nature. Sound, light, heat, position, velocity, acceleration, time, weight, and volume are all analog quantities. Each of these can be represented by a voltage or current in an electronic circuit. This voltage or current is a copy, or analog, of the sound, velocity, or whatever.

We can also represent these physical properties digitally, that is, as a series of numbers, each describing an aspect of the property, such as its magnitude at a particular time. To translate between the physical world and a digital circuit, we must be able to convert analog signals to digital and vice versa.

We will begin by examining some of the factors involved in the conversion between analog and digital signals, including sampling rate, resolution, range, and quantization.

We will then examine circuits for converting digital signals to analog, since these have a fairly standard form. Analog-to-digital conversion has no standard method. We will study several of the most popular: simultaneous (flash) conversion, successive approximation, and dual slope (integrating) conversion.

12.1 Analog and Digital Signals

KEY TERMS

Continuous   Smoothly connected. An unbroken series of consecutive values with no instantaneous changes.

Discrete   Separated into distinct segments or pieces. A series of discontinuous values.

Analog   A way of representing some physical quantity, such as temperature or velocity, by a proportional continuous voltage or current. An analog voltage or current can have any value within a defined range.

Digital   A way of representing a physical quantity by a series of binary numbers. A digital representation can have only specific discrete values.

Analog-to-digital converter   A circuit that converts an analog signal at its input to a digital code. (Also called an A-to-D converter, A/D converter, or ADC.)

Digital-to-analog converter   A circuit that converts a digital code at its input to an analog voltage or current. (Also called a D-to-A converter, D/A converter, or DAC.)

Electronic circuits are tools to measure and change our environment. Measurement instruments tell us about the physical properties of objects around us. They answer questions such as "How hot is this water?", "How fast is this car going?", and "How many electrons are flowing past this point per second?" These data can correspond to voltages and currents in electronic instruments.

If the internal voltage of an instrument is directly proportional to the quantity being measured, with no breaks in the proportional function, we say that it is an **analog** voltage. Like the property being measured, the voltage can vary continuously throughout a defined range.

For example, sound waves are **continuous** movements in the air. We can plot these movements mathematically as a sum of sine waves of various frequencies. The patterns of magnetic domains on an audio tape are analogous to the sound waves that produce them and electromagnetically represent the same mathematical functions. When the tape is played, the playback head produces a voltage that is also proportional to the original sound waves. This analog audio voltage can be any value between the maximum and minimum voltages of the audio system amplifier.

If an instrument represents a measured quantity as a series of binary numbers, the representation is **digital**. Since the binary numbers in a circuit necessarily have a fixed number of bits, the instrument can represent the measured quantities only as having specific **discrete** values.

A compact disc stores a record of sound waves as a series of binary numbers. Each number represents the amplitude of the sound at a particular time. These numbers are decoded and translated into analog sound waves upon playback. The values of the stored numbers (the encoded sound information) are limited by the number of bits in each stored digital "word."

The main advantage of a digital representation is that it is not subject to the same distortions as an analog signal. Nonideal properties of analog circuits, such as stray inductance and capacitance, amplification limits, and unwanted phase shifts, all degrade an analog signal. Storage techniques, such as magnetic tape, can also introduce distortion due to the nonlinearity of the recording medium.

Digital signals, on the other hand, do not depend on the shape of a waveform to preserve the encoded information. All that is required is to maintain the integrity of the logic HIGHs and LOWs of the digital signal. Digital information can be easily moved around in a circuit and stored in a latch or on some magnetic or optical medium. When the information is required in analog form, the analog quantity is reproduced as a new copy every time it is needed. Each copy is as good as any previous one. Distortions are not introduced between copy generations, as is the case with analog copying techniques, unless the constituent bits themselves are changed.

Digital circuits give us a good way of measuring and evaluating the physical world, with many advantages over analog methods. However, most properties of the physical world are analog. How do we bridge the gap?

We can make these translations with two classes of circuits. An **analog-to-digital converter** accepts an analog voltage or current at its input and produces a corresponding digital code. A **digital-to-analog converter** generates a unique analog voltage or current for every combination of bits at its inputs.

Sampling an Analog Voltage

KEY TERMS

Sample An instantaneous measurement of an analog voltage, taken at regular intervals.

Sampling frequency The number of samples taken per unit time of an analog signal.

Quantization The number of bits used to represent an analog voltage as a digital number.

Resolution The difference in analog voltage corresponding to two adjacent digital codes. Analog step size.

Before we examine actual D/A and A/D converter circuits, we need to look at some of the theoretical issues behind the conversion process. We will look at the concept of **sampling** an analog signal and discover how the **sampling frequency** affects the accuracy of the digital representation. We will also examine **quantization,** or the number of bits in the digital representation of the analog sample, and its effect on the quality of a digital signal.

Figure 12.1 shows a circuit that converts an analog signal (a sine pulse) to a series of 4-bit digital codes, then back to an analog output. The analog input and output voltages are shown on the two graphs.

There are two main reasons why the output is not a very good copy of the input. First, the number of bits in the digital representation is too low. Second, the input signal is not

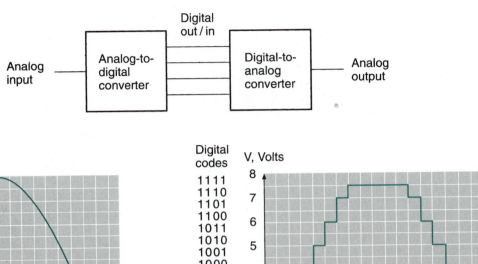

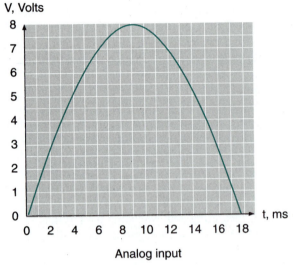

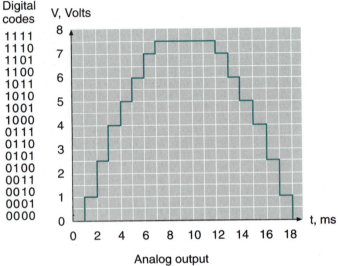

FIGURE 12.1
Analog Input and Output Signals

sampled frequently enough. To help us understand the effect of each of these factors, let us examine the conversion process in more detail.

The analog input signal varies between 0 and 8 volts. This is evenly divided into 16 ranges, each corresponding to a 4-bit digital code (0000 to 1111). We say that the signal is quantized into 4 bits. The **resolution,** or analog step size, for a 4-bit quantization is 8 V/16 steps = 0.5 V/step. Table 12.1 shows the codes for each analog range.

Table 12.1   4-bit Digital Codes
for 0 to 8 V Analog Range

Analog Voltage	Digital Code
0.00–0.25	0000
0.25–0.75	0001
0.75–1.25	0010
1.25–1.75	0011
1.75–2.25	0100
2.25–2.75	0101
2.75–3.25	0110
3.25–3.75	0111
3.75–4.25	1000
4.25–4.75	1001
4.75–5.25	1010
5.25–5.75	1011
5.75–6.25	1100
6.25–6.75	1101
6.75–7.25	1110
7.25–8.00	1111

The analog input is sampled and converted at the beginning of each time division on the graph. The 4-bit digital code does not change until the next conversion, 1 ms later. This is the same as saying that the system has a sampling frequency of 1 kHz ($f = 1/T = 1/(1 \text{ ms}) = 1$ kHz).

Table 12.2 shows the digital codes for samples taken from $t = 0$ to $t = 18$ ms. The analog voltages in Table 12.2 are calculated by the formula

$$V_{analog} = 8 \text{ V} \sin (t \times (10°/\text{ms}))$$

For example at $t = 2$ ms, $V_{analog} = 8$ V $\sin (2 \text{ ms} \times (10°/\text{ms})) = 8$ V $\sin (20°) = 2.736$ V.

The calculated analog values are compared to the voltage ranges in Table 12.1 and assigned the appropriate code. The value 2.736 V is between 2.25 V and 2.75 V and therefore is assigned the 4-bit value of 0101.

Table 12.2 4-bit Codes for a Sampled Analog Signal

Time (ms)	Analog Amplitude (volts)	Digital Code
0	0.000	0000
1	1.389	0011
2	2.736	0101
3	4.000	1000
4	5.142	1010
5	6.128	1100
6	6.928	1110
7	7.518	1111
8	7.878	1111
9	8.000	1111
10	7.878	1111
11	7.518	1111
12	6.928	1110
13	6.128	1100
14	5.142	1010
15	4.000	1000
16	2.736	0101
17	1.389	0011
18	0.000	0000

Table 12.3 8-bit Codes for a Sampled Analog Signal

Time (ms)	Analog Amplitude (volts)	Digital Code
0	0.000	00000000
1	1.389	00101100
2	2.736	01011100
3	4.000	10000000
4	5.142	10100101
5	6.128	11000010
6	6.928	11011110
7	7.518	11110001
8	7.878	11111100
9	8.000	11111111
10	7.878	11111100
11	7.518	11110001
12	6.928	11011110
13	6.128	11000010
14	5.142	10100101
15	4.000	10000000
16	2.736	01011100
17	1.389	00101100
18	0.000	00000000

The digital-to-analog converter in Figure 12.1 continuously converts the digital codes to their analog equivalents. Each code produces an analog voltage whose value is the midpoint of the range corresponding to that code.

For this particular analog waveform, the A/D converter introduces the greatest inaccuracy at the peak of the waveform, where the magnitude of the input voltage changes the least per unit time. There is not sufficient difference between the values of successive analog samples to map them into unique codes. As a result, the output waveform flattens out at the top.

This is the consequence of using a 4-bit quantization, which allows only 16 different analog ranges in the signal. By using more bits, we could divide the analog signal into a greater number of smaller ranges, allowing more accurate conversion of a signal having small changes in amplitude. For example, an 8-bit code would give us 256 steps (a resolution of 8 V/256 = 31.25 mV). This would yield the code assignments shown in Table 12.3. Note that for an 8-bit code, there is a unique value for every sampled voltage.

Figure 12.2 shows how different levels of quantization affect the accuracy of a digital representation of an analog signal. The analog input is a sine wave, converted to digital

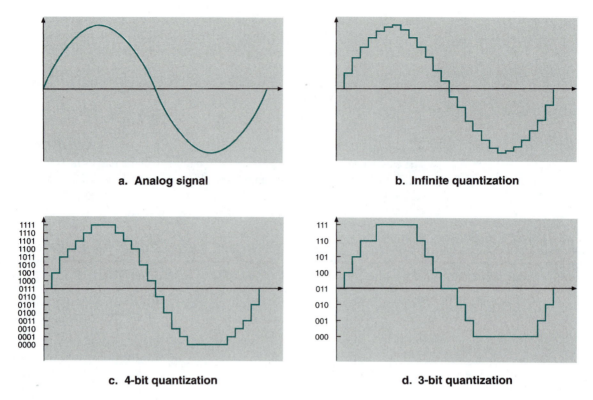

a. Analog signal

b. Infinite quantization

c. 4-bit quantization

d. 3-bit quantization

FIGURE 12.2
Effect of Quantization

codes and back to analog, as in Figure 12.1. The graphs show the analog input and three analog outputs, each of which has been sampled 28 times per cycle, but with different quantizations. The corresponding digital codes range from a maximum negative value of n 0s to a maximum positive value of n 1s for an n-bit quantization (e.g., for a 4-bit quantization, maximum negative = 0000, maximum positive = 1111).

The first output signal has an infinite number of bits in its quantization. Even the smallest analog change between samples has a unique code. This ideal case is not attainable, since a digital circuit always has a finite number of bits. We can see from the codes in Table 12.3 that an 8-bit quantization is sufficient to give unique codes for this waveform. An infinite quantization implies that the resolution is small enough that each sampled voltage can be represented, not only by a unique code, but as its exact value rather than a point within a range.

The 4-bit and 3-bit quantizations in the next two graphs show progressively worse representation of the original signal, especially at the peaks. The change in analog voltage is too small for each sample to have a unique code at these low quantizations.

Figure 12.3 shows how the digital representation of a signal can be improved by increasing its sampling frequency. It shows an analog signal and three analog waveforms resulting from an analog-digital-analog conversion. All waveforms have infinite quantization, but different numbers of samples in the analog-to-digital conversion. As the number of samples decreases, the output waveform becomes a poorer copy of the input.

In general, the sampling frequency affects the horizontal resolution of the digitized waveform and the quantization affects the vertical resolution.

FIGURE 12.3
Effect of Sampling Frequency

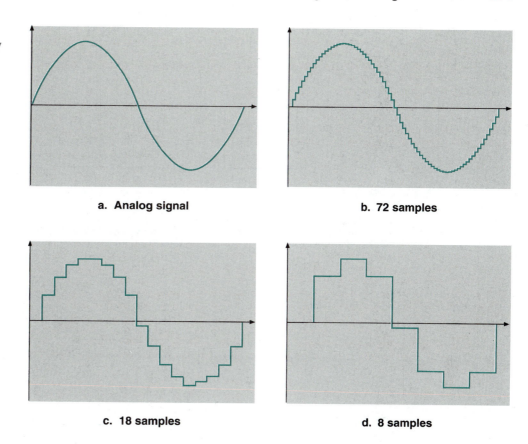

a. Analog signal   **b. 72 samples**

c. 18 samples   **d. 8 samples**

▐▐ SECTION 12.1 REVIEW PROBLEM

12.1 An analog signal has a range of 0 to 24 mV. The range is divided into 32 equal steps for conversion to a series of digital codes. How many bits are in the resultant digital codes? What is the resolution of the A/D converter?

12.2 Digital-to-Analog Conversion

KEY TERM

Full scale   The maximum analog reference voltage or current of a digital-to-analog converter.

Figure 12.4 shows the block diagram of a generalized digital-to-analog converter. Each digital input switches a proportionally weighted current on or off, with the current for the MSB being the largest. The second MSB produces a current half as large. The current generated by the third MSB is one quarter of the MSB current, and so on.

These currents all sum at the operational amplifier's (op amp's) inverting input. The total analog current for an n-bit circuit is given by:

$$I_a = \frac{b_{n-1}2^{n-1} + \cdots + b_2 2^2 + b_1 2^1 + b_0 2^0}{2^n} I_{ref}$$

The bit values $b_0, b_1, \ldots b_n$ can be only 0 or 1. The function of each bit is to include or exclude a term from the general expression.

FIGURE 12.4

Analysis of a Generalized
Digital-to-Analog Converter

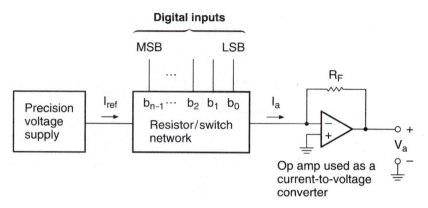

a. Generic DAC circuit

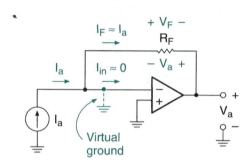

b. Op amp analysis

The op amp acts as a current-to-voltage converter. The analysis, illustrated in Figure 12.4b, is the same as for an inverting op amp circuit with a constant input current.

The input impedance of the op amp is the impedance between its inverting $(-)$ and noninverting $(+)$ terminals. This value is very large, on the order of 2 MΩ. If this is large compared to other circuit resistances, we can neglect the op amp input current, I_{in}.

This implies that the voltage drop across the input terminals is very small; the inverting and noninverting terminals are at approximately the same voltage. Since the noninverting input is grounded, we can say that the inverting input is "virtually grounded."

Current I_F flows in the feedback loop, through resistor R_F. Since $I_a - I_{in} - I_F = 0$ and $I_{in} \approx 0$, then $I_F \approx I_a$. By Ohm's law, the voltage across R_F is given by $V_F = I_a R_F$. The feedback resistor is connected to the output at one end and to virtual ground at the other. The op amp output voltage is measured with respect to ground. The two voltages are effectively in parallel. Thus, the output voltage is the same as the voltage across the feedback resistor, with a polarity opposite to V_F, calculated above.

$$V_a = -V_F = -I_a R_F$$
$$= \frac{-b_{n-1}2^{n-1} + \cdots + b_2 2^2 + b_1 2^1 + b_0 2^0}{2^n} I_{ref} R_F$$

The range of analog output voltage is set by choosing the appropriate value of R_F.

▌▌▌ EXAMPLE 12.1   Write the expression for analog current, I_a, of a 4-bit D/A converter. Calculate values of I_a for input codes $b_3b_2b_1b_0 = 0000, 0001, 1000, 1010$, and 1111, if $I_{ref} = 1$ mA.

Solution   The analog current of a 4-bit converter is:

$$I_a = \frac{b_3 2^3 + b_2 2^2 + b_1 2^1 + b_0 2^0}{2^4} I_{ref}$$

$$= \frac{8b_3 + 4b_2 + 2b_1 + b_0}{16} (1 \text{ mA})$$

$$b_3b_2b_1b_0 = 0000, I_a = \frac{(0 + 0 + 0 + 0)(1 \text{ mA})}{16} = 0$$

$$b_3b_2b_1b_0 = 0001, I_a = \frac{(0 + 0 + 0 + 1)(1 \text{ mA})}{16} = \frac{1 \text{ mA}}{16} = 62.5 \text{ μA}$$

$$b_3b_2b_1b_0 = 1000, I_a = \frac{(8 + 0 + 0 + 0)(1 \text{ mA})}{16} = \frac{8}{16} (1 \text{ mA}) = 0.5 \text{ mA}$$

$$b_3b_2b_1b_0 = 1010, I_a = \frac{(8 + 0 + 2 + 0)(1 \text{ mA})}{16} = \frac{10}{16} (1 \text{ mA}) = 0.625 \text{ mA}$$

$$b_3b_2b_1b_0 = 1111, I_a = \frac{(8 + 4 + 2 + 1)(1 \text{ mA})}{16} = \frac{15}{16} (1 \text{ mA}) = 0.9375 \text{ mA}$$

Example 12.1 suggests an easy way to calculate D/A analog current. I_a is a fraction of the reference current I_{ref}. The denominator of the fraction is 2^n for an n-bit converter. The numerator is the decimal equivalent of the binary input. For example, for input $b_3b_2b_1b_0 = 0111$, $I_a = (7/16)(I_{ref})$.

Note that when $b_3b_2b_1b_0 = 1111$, the analog current is not the full value of I_{ref}, but 15/16 of it. This is one least significant bit less than full scale.

This is true for any D/A converter, regardless of the number of bits. The maximum analog current for a 5-bit converter is 31/32 of full scale. In an 8-bit converter, I_a cannot exceed 255/256 of full scale. This is because the analog value 0 has its own code. An n-bit converter has 2^n input codes, ranging from 0 to $2^n - 1$.

The difference between the full scale *(FS)* of a digital-to-analog converter and its maximum output is the resolution of the converter. Since the resolution is the smallest change in output, equivalent to a change in the least significant bit, we can define the maximum output as $FS - 1 \text{ LSB}$. (As an example, in the case of an 8-bit converter $FS - 1 \text{ LSB} = 255/256 \, I_{ref}$.)

▌▌ SECTION 12.2A REVIEW PROBLEM

12.2 Calculate the range of analog voltage of a 4-bit D/A converter having values of $I_{ref} = 1 \text{ mA}$ and $R_F = 10 \text{ kΩ}$. Repeat the calculation for an 8-bit D/A converter.

Weighted Resistor D/A Converter

Figure 12.5 shows the circuit of a 4-bit weighted resistor D/A converter. The heart of this circuit is a parallel network of binary-weighted resistors. The MSB has a resistor value of R. Successive branches have resistor values that double with each bit: $2R$, $4R$, and $8R$. The branch currents decrease by halves with each descending bit value.

FIGURE 12.5
Weighted Resistor D-to-A Converter

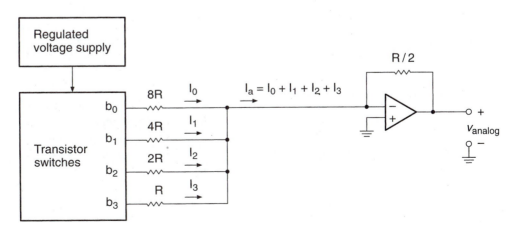

The bit inputs, b_3, b_2, b_1, and b_0, are either 0 V or V_{ref}. When the corresponding bits are HIGH, the branch currents are:

$$I_3 = V_{ref}/R$$
$$I_2 = V_{ref}/2R$$
$$I_1 = V_{ref}/4R$$
$$I_0 = V_{ref}/8R$$

The sum of branch currents gives us the analog current I_a.

$$I_a = \frac{b_3\,V_{ref}}{R} + \frac{b_2\,V_{ref}}{2R} + \frac{b_1\,V_{ref}}{4R} + \frac{b_0\,V_{ref}}{8R}$$

$$= \left[\frac{b_3}{1} + \frac{b_2}{2} + \frac{b_1}{4} + \frac{b_0}{8}\right]\frac{V_{ref}}{R}$$

We can calculate the analog voltage by Ohm's law:

$$V_2 = -I_a\,R_F = -I_a\,(R/2)$$

$$= -\left[\frac{b_3}{1} + \frac{b_2}{2} + \frac{b_1}{4} + \frac{b_0}{8}\right]\frac{V_{ref}}{R}\frac{R}{2}$$

$$= -\left[\frac{b_3}{1} + \frac{b_2}{2} + \frac{b_1}{4} + \frac{b_0}{8}\right]\frac{V_{ref}}{2}$$

$$= -\left[\frac{b_3}{2} + \frac{b_2}{4} + \frac{b_1}{8} + \frac{b_0}{16}\right]V_{ref}$$

The choice of $R_F = R/2$ makes the analog output a binary fraction of V_{ref}.

▌▍ EXAMPLE 12.2

Calculate the analog voltage of a weighted resistor D/A converter when the binary inputs have the following values: $b_3b_2b_1b_0$ = 0000, 1000, 1111. V_{ref} = 5 V.

Solution

$$b_3b_2b_1b_0 = 0000$$

$$V_a = -\left[\frac{0}{2} + \frac{0}{4} + \frac{0}{8} + \frac{0}{16}\right]V_{ref} = 0$$

$$b_3b_2b_1b_0 = 1000$$

$$V_a = -\left[\frac{1}{2} + \frac{0}{4} + \frac{0}{8} + \frac{0}{16}\right]V_{ref} = -\frac{1}{2}(5\text{ V}) = -2.5\text{ V}$$

$$b_3b_2b_1b_0 = 1111$$

$$V_a = -\left[\frac{1}{2} + \frac{1}{4} + \frac{1}{8} + \frac{1}{16}\right]V_{ref} = -\frac{15}{16}(5\text{ V}) = -4.69\text{ V}$$

▌▍

The weighted resistor DAC is seldom used in practice. One reason is the wide range of resistor values required for a large number of bits. Another reason is the difficulty in obtaining resistors whose values are sufficiently precise.

A 4-bit converter needs a range of resistors from R to $8R$. If $R = 1$ kΩ, then $8R = 8$ kΩ. An 8-bit DAC must have a range from 1 kΩ to 128 kΩ. Standard value resistors are specified to two significant figures; there is no standard 128-kΩ resistor. We would need to use relatively expensive precision resistors for any value having more than two significant figures.

Another DAC circuit, the R-2R ladder, is more commonly used. It requires only two values of resistance for any number of bits.

▌▌ SECTION 12.2B REVIEW PROBLEM

12.3 The resistor for the MSB of a 12-bit weighted resistor D/A converter is 1 kΩ. What is the resistor value for the LSB?

R-2R Ladder D/A Converter

Figure 12.6 shows the circuit of an R-2R ladder D/A converter. Like the weighted resistor DAC, this circuit produces an analog current that is the sum of binary-weighted currents. An operational amplifier converts the current to a proportional voltage.

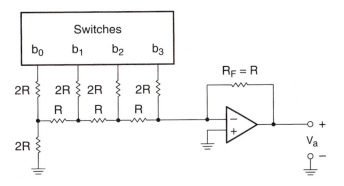

FIGURE 12.6
R-2R Ladder DAC

The circuit requires an operational amplifier with a high slew rate. Slew rate is the rate at which the output changes after a step change at the input. If a standard op amp (e.g., 741C) is used, the circuit will not accurately reproduce changes introduced by large changes in the digital input.

The method of generating the analog current for an R-2R ladder DAC is a little less obvious than for the weighted resistor DAC. As the name implies, the resistor network is a ladder that has two values of resistance, one of which is twice the other. This circuit is expandable to any number of bits simply by adding one resistor of each value for each bit.

The analog output is a function of the digital input and the value of the op amp feedback resistor. If logic HIGH = V_{ref}, logic LOW = 0 V, and $R_F = R$, the analog output is given by:

$$V_a = -\left[\frac{b_3}{2} + \frac{b_2}{4} + \frac{b_1}{8} + \frac{b_0}{16}\right] V_{ref}$$

One way to analyze this circuit is to replace the R-2R ladder with its Thévenin equivalent circuit and treat the circuit as an inverting amplifier. Figure 12.7 shows the equivalent circuit for the input code $b_3b_2b_1b_0 = 1000$.

Figure 12.8a shows the equivalent circuit of the R-2R ladder when $b_3b_2b_1b_0 = 1000$. All LOW bits are grounded, and the HIGH bit connects to V_{ref}. We can reduce the network to two resistors by using series and parallel combinations.

The two resistors at the far left of the ladder are in parallel: $2R \parallel 2R = R$. This equivalent resistance is in series with another: $R + R = 2R$. The new resistance is in parallel with yet another: $2R \parallel 2R = R$. We continue this process until we get the simplified circuit shown in Figure 12.8b.

FIGURE 12.7
Equivalent Circuit for
$b_3b_2b_1b_0 = 1000$

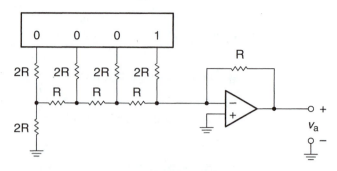

a. $b_3b_2b_1b_0 = 1000$

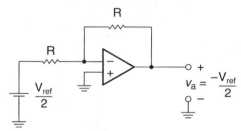

b. Equivalent circuit

FIGURE 12.8
R-2R Circuit Analysis for
$b_3b_2b_1b_0 = 1000$

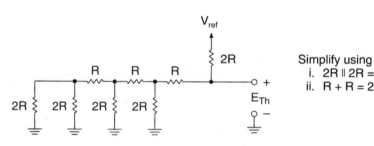

a. Equivalent circuit

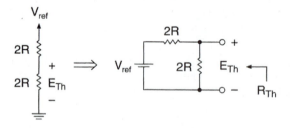

b. Simplified equivalent circuit   **c. Thévenin equivalent**

Next, we find the Thévenin equivalent of the simplified circuit. To find E_{Th}, calculate the terminal voltage of the circuit, using voltage division.

$$E_{Th} = \frac{2R}{2R + 2R} V_{ref} = V_{ref}/2$$

R_{Th} is the resistance of the circuit, as measured from the terminals, with the voltage source short-circuited. Its value is that of the two resistors in parallel: $R_{Th} = 2R \parallel 2R = R$.

NOTE

The value of the Thévenin resistance of the R-2R ladder will always be R, regardless of the digital input code. This is because we short-circuit any voltage sources when we make this calculation, which grounds the corresponding bit resistors. The other resistors are already grounded by logic LOWs. We reduce the circuit to a single resistor, R, by parallel and series combinations of R and $2R$. Figure 12.9 shows the equivalent circuit.

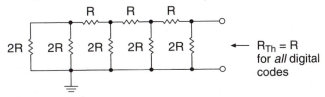

$$R_{Th} = (((((2R \parallel 2R) + R) \parallel 2R) + R) \parallel 2R) + R) \parallel 2R = R$$

FIGURE 12.9
Equivalent Circuit for Calculating R_{Th}

On the other hand, the value of E_{Th} will be different for each different binary input. It will be the sum of binary fractions of the full-scale output voltage, as previously calculated for the generic DAC.

Similar analysis of the R-2R ladder shows that when $b_3b_2b_1b_0 = 0100$, $V_a = -V_{ref}/4$, when $b_3b_2b_1b_0 = 0010$, $V_a = -V_{ref}/8$, and when $b_3b_2b_1b_0 = 0001$, $V_a = -V_{ref}/16$.

If two or more bits in the R-2R ladder are active, each bit acts as a separate voltage source. Analysis becomes much more complicated if we try to solve the network as we did for one active bit.

There is no need to go through a tedious circuit analysis to find the corresponding analog voltage. We can simplify the process greatly by applying the Superposition theorem. This theorem states that the effect of two or more sources in a network can be determined by calculating the effect of each source separately and adding the results.

The Superposition theorem suggests a generalized equivalent circuit of the R-2R ladder DAC. This is shown in Figure 12.10. A Thévenin equivalent source and resistance corresponds to each bit. The source and resistance are switched in and out of the circuit, depending on whether or not the corresponding bit is active.

FIGURE 12.10
Equivalent Circuit of R-2R DAC

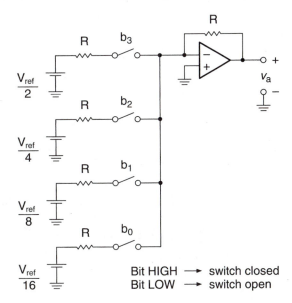

This model is easily expanded. The source for the most significant bit always has the value $V_{ref}/2$. Each source is half the value of the preceding bit. Thus, for a 5-bit circuit, the source for the least significant bit has a value of $V_{ref}/32$. An 8-bit circuit has an LSB equivalent source of $V_{ref}/256$.

EXAMPLE 12.3

A 4-bit DAC based on an R-2R ladder has a reference voltage of 10 volts. Calculate the analog output voltage, V_a, for the following input codes:

a. 0000

b. 1000

c. 0100

d. 1100

Solution

a. $V_a = -(0/16) V_{ref} = 0$ V

b. $V_a = -(8/16) V_{ref} = -(1/2) V_{ref} = -5$ V

c. $V_a = -(4/16) V_{ref} = -(1/4) V_{ref} = -2.5$ V

d. $V_a = -(12/16) V_{ref} = -(3/4) V_{ref} = -7.5$ V

EXAMPLE 12.4

Calculate the output voltage of an 8-bit DAC based on an R-2R ladder for the following input codes. What general conclusion can be drawn about each code when compared to the solutions in Example 12.3?

a. 00000000

b. 10000000

c. 01000000

d. 11000000

Solution

a. $V_a = -(0/256) V_{ref} = 0$ V

b. $V_a = -(128/256) V_{ref} = -(1/2) V_{ref} = -5$ V

c. $V_a = -(64/256) V_{ref} = -(1/4) V_{ref} = -2.5$ V

d. $V_a = -(192/256) V_{ref} = -(3/4) V_{ref} = -7.5$ V

In general, a DAC input code consisting of 1 followed by all 0s generates an output value of ½ full scale. A code of 01 followed by all 0s yields an output of ¼ full scale. An output of 11 followed by all 0s generates an output of ¾ full scale.

SECTION 12.2C REVIEW PROBLEM

12.4 Calculate V_a for an 8-bit R-2R ladder DAC when the input code is 10100001. Assume that V_{ref} is 10 V.

MC1408 Integrated Circuit D/A Converter

> **KEY TERM**
>
> **Multiplying DAC** A DAC whose output changes linearly with a change in DAC reference voltage.

A common and inexpensive DAC is the MC1408 8-bit multiplying digital-to-analog converter. This device also goes by the designation DAC0808. A logic symbol for this DAC is shown in Figure 12.11.

FIGURE 12.11
MC1408 DAC

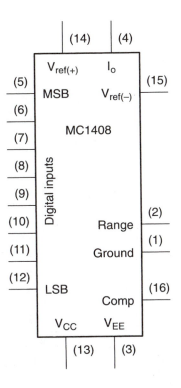

The output current, I_o, flows into pin 4. I_o is a binary fraction of the current flowing into pin 14, as specified by the states of the digital inputs. Other inputs select the range of output voltage and allow for phase compensation.

Figure 12.12 shows the MC1408 in a simple D/A configuration. R_{14} and R_{15} are approximately equal. Pin 14 is approximately at ground potential. This implies:

1. That the DAC reference current can be calculated using only V_{ref} (+) and R_{14} ($I_{\text{ref}} = V_{\text{ref}}$ (+)/R_{14})

2. That R_{15} is not strictly necessary in the circuit. (It is used primarily to stabilize the circuit against temperature drift.)

The reference voltage *must* be set up so that current flows into pin 14 and out of pin 15. Thus, V_{ref} (+) must be positive with respect to V_{ref} (−). (It is permissible to ground pin 14 if pin 15 is at a negative voltage.)

I_o is given by:

$$I_o = \left[\frac{b_7}{2} + \frac{b_6}{4} + \frac{b_5}{8} + \frac{b_4}{16} + \frac{b_3}{32} + \frac{b_2}{64} + \frac{b_1}{128} + \frac{b_0}{256}\right] \frac{V_{\text{ref}} (+)}{R_{14}}$$

Since the output is proportional to V_{ref} (+), we refer to the MC1408 as a **multiplying DAC.**

I_o should not exceed 2 mA. We calculate the output voltage by Ohm's law: $V_o = -I_o R_L$. The output voltage is negative because current flows from ground into pin 4.

The open pin on the Range input allows the output voltage dropped across R_L to range from +0.4 V to −5.0 V without damaging the output circuit of the DAC. If the Range input is grounded, the output can range from +0.4 to −0.55 V. The lower voltage range allows the output to switch about four times faster than it can in the higher range.

FIGURE 12.12
MC1408 Configured for
Unbuffered Analog Output

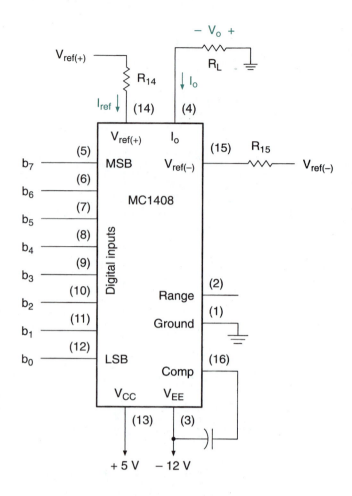

EXAMPLE 12.5

The DAC circuit in Figure 12.12 has the following component values: $R_{14} = R_{15} = 5.6$ kΩ; $R_L = 3.3$ kΩ. V_{ref} (+) is +8 V, and V_{ref} (−) is grounded.

Calculate the value of V_o for each of the following input codes: $b_7b_6b_5b_4b_3b_2b_1b_0 = 00000000, 00000001, 10000000, 10100000, 11111111$.

What is the resolution of this DAC?

Solution   First, calculate the value of I_{ref}.

$$I_{ref} = V_{ref} (+)/R_{14}$$
$$= +8 \text{ V}/5.6 \text{ k}\Omega = 1.43 \text{ mA}$$

Calculate the output current by using the binary fraction for each code. Multiply $-I_o$ by R_L to get the output voltage.

$b_7b_6b_5b_4b_3b_2b_1b_0 = 00000000$

$I_o = 0, V_o = 0$

$b_7b_6b_5b_4b_3b_2b_1b_0 = 00000001$

$I_o = (1/256) (1.43 \text{ mA}) = 5.58 \text{ }\mu\text{A}$

$V_o = -(5.58 \text{ }\mu\text{A})(3.3 \text{ k}\Omega) = -18.4 \text{ mV}$

$b_7b_6b_5b_4b_3b_2b_1b_0 = 10000000$

$I_o = (1/2) (1.43 \text{ mA}) = 714 \text{ }\mu\text{A}$

$V_o = -(714 \text{ }\mu\text{A})(3.3 \text{ k}\Omega) = -2.36 \text{ V}$

$$b_7b_6b_5b_4b_3b_2b_1b_0 = 10100000$$
$$I_o = = (1/2 + 1/8)(1.43 \text{ mA}) = (5/8)(1.43 \text{ mA}) = 893 \text{ } \mu A$$
$$V_o = -(893 \text{ } \mu A)(3.3 \text{ k}\Omega) = -2.95 \text{ V}$$
$$b_7b_6b_5b_4b_3b_2b_1b_0 = 11111111$$
$$I_o = (255/256) (1.43 \text{ mA}) = 1.42 \text{ mA}$$
$$V_o = -(1.42 \text{ mA})(3.3 \text{ k}\Omega) = -4.70 \text{ V}$$

Resolution is the same as the output resulting from the LSB: 18.4 mV/step

▌▌ SECTION 12.2D REVIEW PROBLEM

12.5 The output voltage range of an MC1408 DAC can be limited by grounding the Range pin. Why would we choose to do this?

Op Amp Buffering of MC1408

The MC1408 DAC will not drive much of a load on its own, particularly when the Range input is grounded. We can use an operational amplifier to increase the output voltage and current. This allows us to select the lower voltage range for faster switching while retaining the ability to drive a reasonable load. The output voltage is limited only by the op amp supply voltages. We use a 34071 high slew rate op amp for fast switching.

Figure 12.13 shows such a circuit. The 0.1-μF capacitor decouples the +5-V supply. (The manufacturer actually recommends that the +5-V logic supply not be used as a reference voltage. It doesn't matter for a demonstration circuit, but may introduce noise that is unacceptable in a commercial design.) The 75-pF capacitor is for phase compensation.

FIGURE 12.13

DAC With Op Amp Buffering

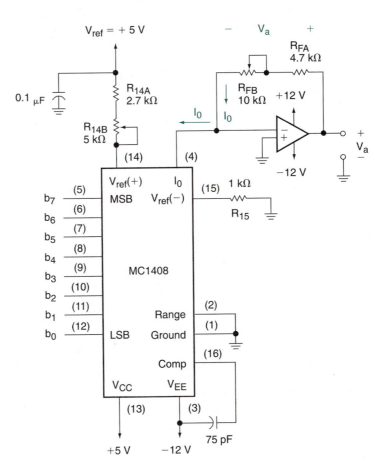

V_a is positive because the voltage drop across R_F is positive with respect to the virtual ground at the op amp ($-$) input. This feedback voltage is in parallel with (i.e., the same as) the output voltage, since both are measured from output to ground.

We can develop the formula for the analog voltage, V_a, in three stages:

1. Calculate the reference current:

$$I_{ref} = V_{ref}(+)/R_{14}$$

2. Determine the binary-weighted fraction of reference current to get DAC output current:

$$I_o = \left[\frac{b_7}{2} + \frac{b_6}{4} + \frac{b_5}{8} + \frac{b_4}{16} + \frac{b_3}{32} + \frac{b_2}{64} + \frac{b_1}{128} + \frac{b_0}{256}\right] I_{ref}$$

$$= \left[\frac{b_7}{2} + \frac{b_6}{4} + \frac{b_5}{8} + \frac{b_4}{16} + \frac{b_3}{32} + \frac{b_2}{64} + \frac{b_1}{128} + \frac{b_0}{256}\right] \frac{V_{ref}}{R_{14}}$$

$$= \left(\frac{\text{digital code}}{256}\right)\left(\frac{V_{ref}}{R_{14}}\right)$$

3. Use Ohm's law to calculate the op amp output voltage:

$$V_a = I_o R_F = \left(\frac{\text{digital code}}{256}\right)\left(\frac{R_F}{R_{14}}\right) V_{ref}$$

The resistor values in the above formulae are the total resistances for the corresponding part of the circuit. That is, $R_{14} = R_{14A} + R_{14B}$ and $R_F = R_{FA} + R_{FB}$. These both consist of a fixed and a variable resistor, which has two advantages: (a) The reference current and output voltage can be independently adjusted within a specified range by the variable resistors. (b) The resistances defining the reference and feedback currents cannot go below a specified minimum value, determined by the fixed resistance, ensuring that excessive current does not flow into the reference input or the DAC output terminal.

V_a can, in theory, be any positive value less than the op amp positive supply ($+12$ V in this case). Any attempt to exceed this voltage makes the op amp saturate. The actual maximum value, if not the same as the op amp's saturation voltage, depends on the values of R_F and R_{14}.

▌▌ EXAMPLE 12.6

Describe a step-by-step procedure that calibrates the DAC circuit in Figure 12.13 so that it has a reference current of 1 mA and a full-scale analog output voltage of 10 V, using only a series of measurements of the analog output voltage. When the procedure is complete, what are the resistance values in the circuit? What is the range of the DAC?

Solution   Since the maximum output of the DAC is 1 LSB less than full scale, we must indirectly measure the full scale value. We can do so by setting the digital input code to 10000000, which exactly represents the half-scale value of output current, and making appropriate adjustments.

Set the variable feedback resistor to zero so that the output voltage is due only to the fixed feedback resistor and the feedback current. Measure the output voltage of the circuit and adjust R_{14B} so that $V_a = 2.35$ volts. Ohm's law tells us that this sets the feedback current to $I_F = 2.35$ V/4.7 kΩ = 0.5 mA. Since the digital code is set for half scale, $I_{ref} = 2 I_F = 1$ mA.

Adjust R_{FB} so that the half-scale output voltage is 5.00 V.

After adjustment, $R_{14} = 2.7$ kΩ + 2.3 kΩ = 5 kΩ and $R_F = 4.7$ kΩ + 4.3 kΩ = 10 kΩ. In both cases the variable resistors were selected so that their final values are about half-way through their respective ranges.

The range of the DAC is 0 V to 9.961 V.

$$(FS - 1\ LSB = 10\ V - (10\ V/256) = 9.961\ V)$$

■▮ **EXAMPLE 12.7**

Figure 12.14 shows the circuit of an analog ramp (sawtooth) generator built from an MC1408 DAC, an op amp, and an 8-bit synchronous counter. (A ramp generator has numerous analog applications, such as sweep generation in an oscilloscope and frequency sweep in a spectrum analyzer.)

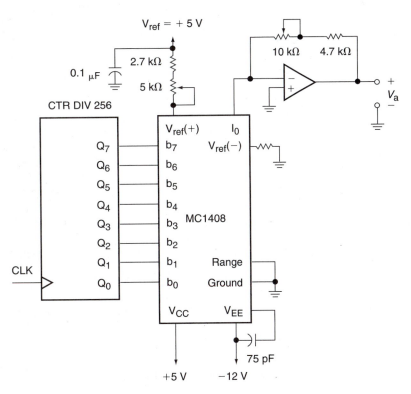

FIGURE 12.14
Example 16.5
DAC Ramp Generator

Briefly explain the operation of the circuit and sketch the output waveform. Calculate the step size between analog outputs resulting from adjacent codes. Assume that the DAC is set for +6-V output when the input code is 10000000.

Calculate the output sawtooth frequency when the clock is running at 1 MHz.

Solution   The 8-bit counter cycles from 00000000 to 11111111 and repeats continuously. This is a total of 256 states.

The DAC output is 0 V for an input code of 00000000 and (12 V − 1 LSB) for a code of 11111111. We know this because a code of 10000000 always gives an output voltage of half the full-scale value (6 V = 12 V/2), and the maximum code gives an output that is one step less than the full-scale voltage. The step size is 12 V/256 steps = 46.9 mV/step. The DAC output advances linearly from 0 to (12 V − 1 LSB) in 256 clock cycles.

Figure 12.15 shows the analog output plotted against the number of input clock cycles. The ramp looks smooth at the scale shown. A section enlarged 32 times shows the analog steps resulting from eight clock pulses.

One complete cycle of the sawtooth waveform requires 256 clock pulses. Thus, if $f_{CLK} = 1$ MHz, $f_o = 1$ MHz/256 = 3.9 kHz.

(Note that if we do not use a high slew rate op amp, the sawtooth waveform will not have vertical sides.)

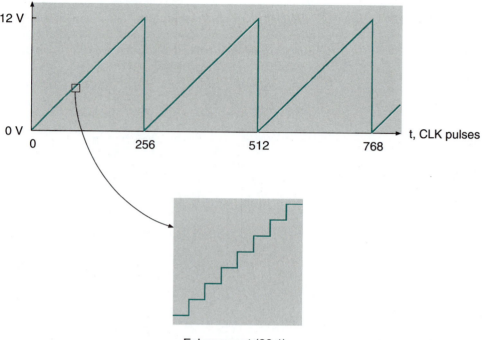

FIGURE 12.15
Example 12.7
Sawtooth Waveform Output of Circuit in Figure 12.14

Bipolar Operation of MC1408

Many analog signals are bipolar, that is, they have both positive and negative values. We can configure the MC1408 to produce a bipolar output voltage. Such a circuit is shown in Figure 12.16.

We can model the bipolar DAC as shown in Figure 12.16b. The amplitude of the constant-current sink, I_o, is set by V_{ref} (+), R_{14}, and the binary value of the digital inputs. I_s is determined by Ohm's law: $I_s = V_{ref}$ (+)$/R_4$.

The output voltage is set by the value of I_F:

$$V_a = I_F R_F = I_F (R_{FA} + R_{FB})$$

By Kirchhoff's current law:

$$I_s + I_F - I_o = 0$$

or

$$I_F = I_o - I_s$$

Thus, output voltage is given by:

$$V_a = (I_o - I_s)R_F = I_o R_F - I_s R_F$$

$$= \left[\frac{b_7}{2} + \frac{b_6}{4} + \frac{b_5}{8} + \frac{b_4}{16} + \frac{b_3}{32} + \frac{b_2}{64} + \frac{b_1}{128} + \frac{b_0}{256} \right] \frac{R_F}{R_{14}} V_{ref} - \frac{R_F}{R_4} V_{ref}$$

$$= \left(\frac{\text{digital code}}{256} \right) \left(\frac{R_F}{R_{14}} \right) V_{ref} - \frac{R_F}{R_4} V_{ref}$$

How do we understand the circuit operation from this mathematical analysis?

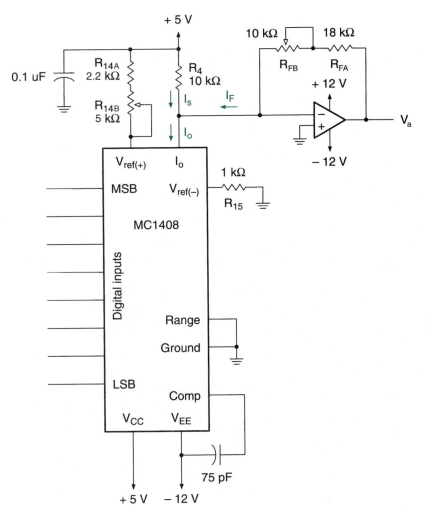

a. Bipolar DAC circuit

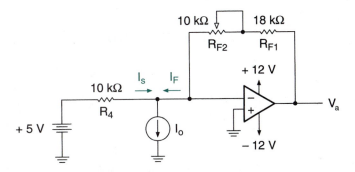

b. Equivalent circuit

FIGURE 12.16

MC1408 as a Bipolar D/A Converter

The current sink, I_o, is a variable element. The voltage source, V_{ref} (+), remains constant. To satisfy Kirchhoff's current law, the feedback current, I_F, must vary to the same degree as I_o. Depending on the value of I_o with respect to I_s, I_F can be positive or negative.

We can get some intuitive understanding of the circuit operation by examining several cases of the equation V_a.

Case 1: $I_o = 0$.   This corresponds to the digital input $b_7b_6b_5b_4b_3b_2b_1b_0 = 00000000$. The output voltage is:

$$V_a = (I_o - I_s)R_F = -I_s R_F = -\frac{R_F}{R_4}V_{ref}$$

This is the maximum negative output voltage.

Case 2: $0 < I_o < I_s$.   The term $(I_o - I_s)$ is negative, so output voltage is also a negative value.

Case 3: $I_o = I_s$.   The output is given by:

$$V_a = (I_o - I_s)R_F = 0$$

The digital code for this case could be any value, depending on the setting of R_{14}. To set the zero-crossing to half-scale, set the digital input to 10000000 and adjust R_{14} for 0 V.

Case 4: $I_o > I_s$.   Since the term $(I_o - I_s)$ is positive, output voltage is positive. The largest value of I_o (and thus the maximum positive output voltage) corresponds to the input code $b_7b_6b_5b_4b_3b_2b_1b_0 = 11111111$.

The magnitude of the maximum positive output voltage of this particular circuit is 2 LSB less than the magnitude of the maximum negative voltage. Specifically, $V_a = (127/128)(R_F/R_4)(V_{ref})$ if $R_4 = 2R_{14}$.

To summarize:

Input Code	Output Voltage
00000000	Maximum negative*
10000000	0 V**
11111111	Maximum positive

*As adjusted by R_{FB}
**As adjusted by R_{14B}

Negative Range:

00000000 to 01111111	(128 codes)

Positive Range:

10000001 to 11111111	(127 codes)

Zero:

00000000   (1 code)
                                        ─────────────
                                        256 codes

▌▌ EXAMPLE 12.8

Calculate the values to which R_{14} and R_F must be set to make the output of the bipolar DAC in Figure 12.16 range from -12 V to $(+12$ V $- 2$ LSB). Describe the procedure you would use to set the circuit output as specified.

Confirm that the calculated resistor settings generate the correct values of maximum and minimum output.

Solution   Set R_{14} so that the DAC circuit has an output of 0 V when input code is $b_7b_6b_5b_4b_3b_2b_1b_0 = 10000000$. We can calculate the value of R_{14} as follows:

$$\frac{R_F}{2R_{14}}V_{ref} - \frac{R_F}{R_4}V_{ref} = 0$$

The first term is set by the value of the input code. Solving for R_{14}, we get:

$$\left[\frac{1}{2R_{14}} - \frac{1}{R_4}\right]R_F \, V_{ref} = 0$$

$$\frac{1}{2R_{14}} - \frac{1}{R_4} = 0$$

$$\frac{1}{2R_{14}} = \frac{1}{R_4}$$

$$2R_{14} = R_4$$

$$R_{14} = R_4/2 = 10 \text{ k}\Omega/2 = 5 \text{ k}\Omega$$

To set the maximum negative value, set the input code to 00000000 and adjust R_{FB} for -12 V. $R_{FB} = R_F - R_{FA}$. Solve the following equation for R_F:

$$-\frac{R_F}{R_4} \, V_{ref} = -12 \text{ V}$$

$$-\frac{R_F}{10 \text{ k}\Omega} (5 \text{ V}) = -12 \text{ V}$$

$$R_F = (12 \text{ V})(10 \text{ k}\Omega)/5 \text{ V} = 24 \text{ k}\Omega$$

$$R_{FB} = 24 \text{ k}\Omega - 18 \text{ k}\Omega = 6 \text{ k}\Omega$$

Settings

$$R_{14} = R_4/2 = 5 \text{ k}\Omega \text{ for zero-crossing at half-scale.}$$

$$R_F = 24 \text{ k}\Omega \text{ for output of } \pm 12 \text{ V.}$$

Check Output Range

For $b_7b_6b_5b_4b_3b_2b_1b_0 = 00000000$:

$$V_a = \left[\frac{0}{R_{14}} - \frac{1}{R_4}\right]R_F \, V_{ref} = -\frac{(24 \text{ k}\Omega)(5 \text{ V})}{10 \text{ k}\Omega} = -12 \text{ V}$$

For $b_7b_6b_5b_4b_3b_2b_1b_0 = 11111111$:

$$V_a = \left[\frac{255}{256 \, R_{14}} - \frac{1}{R_4}\right]R_F \, V_{ref}$$

$$= \left[\frac{255}{(256)(5 \text{ k}\Omega)} - \frac{1}{10 \text{ k}\Omega}\right](24 \text{ k}\Omega)(5 \text{ V}) = 11.906 \text{ V}$$

(Note: 12 V $-$ 2 LSB = 12 V $-$ (12 V/128) = 12 V $-$ 94 mV = 11.906 V.)

▮▮ SECTION 12.2E REVIEW PROBLEM

12.6 Why is the actual maximum value of an 8-bit DAC less than its reference (i.e., its apparent maximum) voltage?

DAC Performance Specifications

A number of factors affect the performance of a digital-to-analog converter. The major factors are briefly described below.

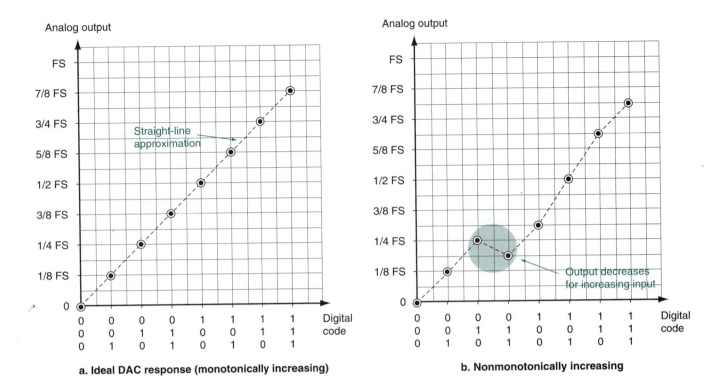

FIGURE 12.17
DAC Monotonicity

Monotonicity.   The output of a DAC is monotonic if the magnitude of the output voltage increases every time the input code increases. Figure 12.17 shows the output of a DAC that increases monotonically and the output of a DAC that does not.

We show the output response of a DAC as a series of data points joined by a straight-line approximation. One input code produces one voltage, so there is no value that corresponds to anything in between codes, but the straight-line approximation allows us to see a trend over the whole range of input codes.

Absolute accuracy.   This is a measure of DAC output voltage with respect to its expected value.

Relative accuracy.   Relative accuracy is a more frequently used measurement than absolute accuracy. It measures the deviation of the actual from the ideal output voltage as a fraction of the full-scale voltage. The MC1408 DAC has a relative accuracy of $\pm\frac{1}{2}$ LSB = $\pm0.195\%$ of full scale.

Settling time.   The time required for the outputs to switch and settle to within $\pm\frac{1}{2}$ LSB when the input code switches from all 0s to all 1s. The MC1408 has a settling time of 300 ns for 8-bit accuracy, limiting its output switching frequency to 1/300 ns = 3.33 MHz. Depending on the value of R_4, the output resistor, the settling time of the MC1408 may increase to as much as 1.2 μs when the Range input is open.

Gain error.   Gain error primarily affects the high end of the output voltage range. If the gain of a DAC is too high, the output saturates before reaching the maximum output code. Figure 12.18 shows the effect of gain error in a 3-bit DAC. In the high gain response, the last two input codes (110 and 111) produce the same output voltage.

FIGURE 12.18
DAC Gain Errors

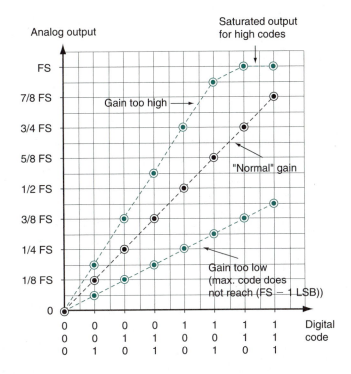

Linearity error. This error is present when the analog output does not follow a straight-line increase with increasing digital input codes. Figure 12.19 shows this error. A linearity error of more than $\pm\frac{1}{2}$ LSB can result in a nonmonotonic output. For example, in Figure 12.17b, the transition from 010 to 011 should result in an output change of $+1$ LSB. Instead, it results in a change of $-\frac{1}{2}$ LSB. This is an error of $-1\frac{1}{2}$ LSB, resulting in a nonmonotonic output.

In Figure 12.19, the code for 011 has a linearity error of $+\frac{1}{2}$ LSB and the adjacent code (100) has a linearity error of $-\frac{1}{2}$ LSB, yielding a flat output for the two codes. This makes it impossible to distinguish the value of input code for that analog output value.

FIGURE 12.19
DAC Linearity Error

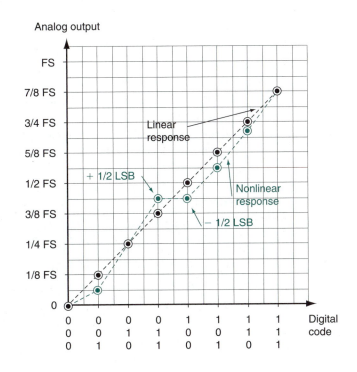

FIGURE 12.20
DAC Offset Error

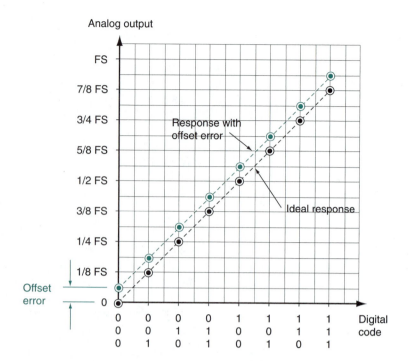

Differential nonlinearity. This specification measures the difference between actual and expected step size of a DAC when the input code is changed by 1 LSB. An actual step that is smaller than the expected step can result in a nonmonotonic output.

Offset error. This error occurs when the analog output of a positive-value DAC is not 0 V when the input code is all 0s. Figure 12.20 shows the effect of offset error.

▐▌ EXAMPLE 12.9

An 8-bit DAC has an output range of 0 to (+8 volts − 1 LSB). The hexadecimal value of the input is symbolized by x.

a. What is the value of 1 LSB?

b. Assuming an ideal DAC, what would the output be for a binary input $x = $ C0H?

c. If the DAC has an input of $x = $ 00H and the output voltage is 0.008 V, calculate the off-set error *(OE)* of the DAC in LSB and as a percentage of the full scale *(FS)*. Assume no other errors.

d. If the DAC has an input of $x = $ FFH and the output voltage is 7.98 V, calculate the gain error *(GE)* of the DAC in LSB and as a percentage of the full scale *(FS)*. Assume no other errors.

e. If the DAC output is 4 V for an input $x = $ 80H and the output is 0.015 V for an input of $x = $ 00H, calculate the linearity error *(LE)* and offset error *(OE)* of the DAC in LSB and as a percentage of the full scale *(FS)*.

Solution

a. 1 LSB $= FS/2^n = 8$ V$/2^8 = 8$ V$/256 = 31.25$ mV

b. C0H $= 192_{10}$

$V_a = $ (code/256) 8 V $= $ (192/256) 8 V $= 6$ V

alternatively: C0H $= 11000000$, which corresponds to $\frac{3}{4} FS = 6$ V

c. When $x = $ 00H, V_a should be 0 V. Therefore, $OE = 8$ mV.

OE[LSB] $= 8$ mV$/31.25$ mV $= 0.256$ LSB

OE[%FS] $= $ (8 mV/8 V) × 100% $FS = 0.1\%$ FS

d. When x = FFH, V_a should be (255/256) 8 V = 7.969 V. GE = 7.98 V − 7.96875 V = 11.25 mV.

$$GE[LSB] = 11.25 \text{ mV}/31.25 \text{ mV} = 0.36 \text{ LSB}$$

$$GE[\%FS] = (11.25 \text{ mV}/8 \text{ V}) \times 100\% \ FS = 0.14\% \ FS$$

e. Without accounting for other possible errors, the output value for an input of 80H appears to be correct. However, we find an offset error of 0.015 V that must be subtracted out of all measured values in the DAC output.

Adjusted value at 80H = 4 V − 0.015 V = 3.985 V. This error is exactly balanced by the offset error, so both have the same value.

$$LE[LSB] = OE[LSB] = 15 \text{ mV}/31.25 \text{ mV} = 0.48 \text{ LSB}$$

$$LE[\%FS] = OE[\%FS] = (15 \text{ mV}/8 \text{ V}) \times 100\% \ FS = 0.188\% \ FS$$

12.3 Analog-to-Digital Conversion

We saw in an earlier section of this chapter that all digital-to-analog converters can be described by a generic form. This is not true of analog-to-digital converters. There are many circuits for converting analog signals to digital codes, each with its own advantages. We will look at several of the most popular.

Flash A/D Converter

KEY TERMS

Flash converter (or simultaneous converter)   An analog-to-digital converter that uses comparators and a priority encoder to produce a digital code.

Priority encoder   An encoder that will produce a binary output corresponding to the subscript of the highest-priority active input. This is usually defined as the input with the largest subscript.

Figure 12.21 shows the circuit for a 3-bit **flash analog-to-digital converter.** The circuit consists of a resistive voltage divider, seven analog comparators, a **priority encoder,** and an output latch array.

The voltage divider has a total resistance of $8R$. The resistors are selected to produce seven equally spaced reference voltages ($V_{ref}/16, 3V_{ref}/16, 5V_{ref}/16, \ldots 15V_{ref}/16$; each is separated by $V_{ref}/8$). Each reference voltage is fed to the inverting input of a comparator.

A comparator output goes HIGH if the voltage at its noninverting (+) input is higher than the voltage at its inverting (−) input. If the (−) input voltage is greater than the (+) input voltage, the comparator output is LOW.

The analog voltage, V_a, is applied to the noninverting inputs of all comparators simultaneously. Thus, if the analog voltage exceeds the reference voltage of a particular comparator, that comparator switches its output to the HIGH state.

For most analog input values, more than one comparator will have a HIGH output. For example, the reference voltage of comparator 3 is ($5V_{ref}/16$). Comparator 4 has a reference voltage of ($7V_{ref}/16$). If the analog voltage is in the range ($5V_{ref}/16$) $\leq V_a <$ ($7V_{ref}/16$), comparators 3, 2, and 1 all have HIGH outputs and comparators 4, 5, 6, and 7 all have LOW outputs.

The priority encoder recognizes that input D_3 is the highest-priority active input and produces the digital code 011 at its outputs. The output latches store this value when the CLK input is pulsed.

FIGURE 12.21

Flash Converter (ADC)

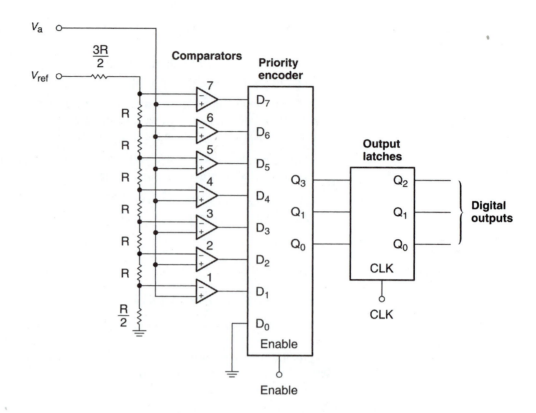

We can regularly sample an analog signal by applying a pulse waveform to the *CLK* input of the latch circuit. The sampling frequency is the same as the clock frequency.

The D_0 input of the priority encoder is grounded, rather than connected to a comparator output. No comparator is needed for this input; if $V_a < (V_{ref}/16)$, all comparator outputs are LOW and the resulting digital code is 000.

Figure 12.22 shows the transfer characteristic of the flash ADC with a reference voltage of 8 V. The digital steps are centered on the analog voltages that are whole-number fractions (1/8, 1/4, 3/8, . . . 7/8) of the reference voltage. The transitions are midway between these points. This is why the resistor for the least significant bit is $R/2$, rather than R.

FIGURE 12.22

Transfer Characteristic of Flash ADC

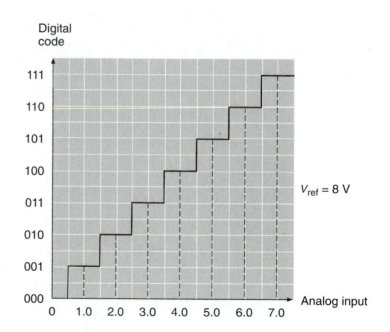

The general form of this circuit has $2^n - 1$ comparators for an n-bit output. For example, an 8-bit flash converter has $2^8 - 1 = 255$ comparators. For any large number of bits, the circuit becomes overly complex.

The main advantage of this circuit is its speed. Since the analog input is compared to the threshold values of all possible input codes at one time, conversion occurs in one clock cycle.

Successive Approximation A/D Converter

KEY TERMS

Successive approximation register   A state machine used to generate a sequence of closer and closer binary approximations to an analog signal.

Quantization error   Inaccuracy introduced into a digital signal by the inability of a fixed number of bits to represent the exact value of an analog signal.

Probably the most widely used type of analog-to-digital converter is the successive approximation ADC. The idea behind this type of converter is a technique a computer programmer would call "binary search."

The analog voltage to be converted is a number within a defined range. The search technique works by narrowing down progressively smaller binary fractions of the known range of numbers.

Suppose we know that the analog voltage is a number between 0 and 255, inclusive. We can find the binary value of any randomly chosen number in this range in no more than eight guesses, or approximations, since $2^8 = 256$. Each approximation adds one more bit to the estimated digital value.

The first approximation determines which half of the range the number is in. The second test finds which quarter of the range, the third test which eighth, the fourth test which sixteenth, and so on until we run out of bits.

❚❚ EXAMPLE 12.10

Use a binary search technique to find the value of a number in the range 0 to 255. (The number is 44.)

Solution

1. The number must be in the upper or lower half of the range. Cut the range in half:
   0–127, 128–125.
   Is $x \geq 128$? No. $0 \leq x < 128$.
2. Cut the remaining range in half: 0–63, 64–127.
   Is $x \geq 64$? No. $0 \leq x < 64$.
3. Cut the remaining range in half: 0–31, 32–63.
   Is $x \geq 32$? Yes. $32 \leq x < 64$.
4. Cut the remaining range in half: 32–47, 48–63.
   Is $x \geq 48$? No. $32 \leq x < 48$.
5. Cut the remaining range in half: 32–39, 40–47.
   Is $x \geq 40$? Yes. $40 \leq x < 48$.
6. Cut the remaining range in half: 40–43, 44–47.
   Is $x \geq 44$? Yes. $44 \leq x < 48$.
7. Cut the remaining range in half: 44–45, 46–47.
   Is $x \geq 46$? No. $44 \leq x < 46$.
8. Cut the remaining range in half: 44–45.
   Is $x \geq 45$? No. $x = 44$.

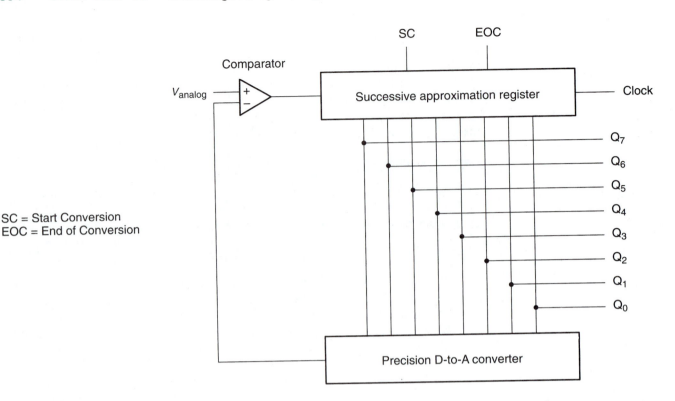

SC = Start Conversion
EOC = End of Conversion

FIGURE 12.23
Successive Approximation ADC

The test criteria for each step in Example 12.10 are phrased so that the answer is always yes or no. (For example, $x \geq 64$? can only be answered yes or no.) Assume that a 1 means yes and a 0 means no. The tests in Example 12.10 give the following sequence of results: 00101100. The decimal equivalent of this binary number is 44, our original value.

A successive approximation ADC such as the one shown in Figure 12.23 applies a similar technique. The circuit has three main components: an analog comparator, a digital-to-analog converter, and a state machine called a **successive approximation register** (SAR). The SAR is an 8-bit register whose bits can be set and cleared individually, according to a specific control sequence and the logic value at the output of the analog comparator.

When a pulse activates the Start Conversion input, bit Q_7 of the SAR is set. This makes the SAR output 10000000. The DAC converts the SAR output to an analog equivalent. When only the MSB is set, this is one half the reference voltage of the DAC.

The DAC output voltage is compared to an analog input voltage. (In effect, the SAR asks, "Is this approximation greater or less than the actual analog voltage?")

If $V_{analog} > V_{DAC}$, the comparator output is HIGH and the MSB remains set. Otherwise, the comparator output is LOW and the MSB is cleared. The process is repeated for all bits.

After all bits have been set or cleared, the End of Conversion (EOC) output changes state. This can be used to load the final digital value into an 8-bit latch.

■■ EXAMPLE 12.11

An 8-bit successive approximation ADC has an analog input voltage of 9.5 V. Describe the steps the circuit performs to generate an 8-bit digital equivalent value if the DAC in the circuit has a reference voltage of 12 V.

Solution Figure 12.24 shows the steps the converter performs to generate the 8-bit digital equivalent of 9.5 V. The conversion process is also summarized in Table 12.4.

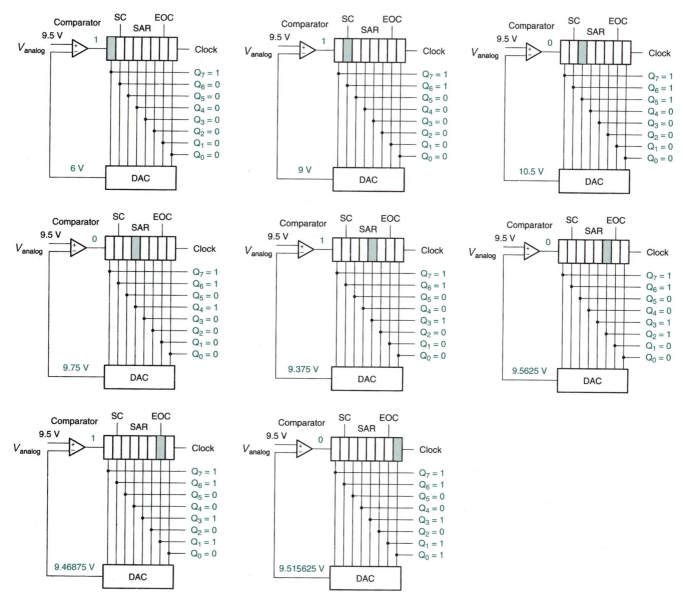

FIGURE 12.24
Example 12.11
Successive Approximation A/D Conversion

Table 12.4 8-Bit Successive Approximation Conversion

Bit	New Digital Value	Analog Equivalent	$V_{analog} \geq V_{DAC}$?	Comparator Output	Accumulated Digital Value
Q_7	10000000	6 V	Yes	1	10000000
Q_6	11000000	9 V	Yes	1	11000000
Q_5	11100000	10.5 V	No	0	11000000
Q_4	11010000	9.75 V	No	0	11000000
Q_3	11001000	9.375 V	Yes	1	11001000
Q_2	11001100	9.5625 V	No	0	11001000
Q_1	11001010	9.46875 V	Yes	1	11001010
Q_0	11001011	9.515625 V	No	0	11001010

The following steps occur for each bit:

1. The bit is set.

2. The digital output is converted to an analog voltage and compared to the actual analog input.

3. If the analog voltage is greater than the DAC output voltage, the bit remains set. Otherwise it is cleared.

∎∎

There is no exact 8-bit binary value for the analog voltage specified in Example 12.11 (9.5 V). The final answer is within 13 mV, out of 12 V, which is pretty close but not exact. This difference is called **quantization error.** The maximum value of quantization error is $\pm\frac{1}{2}$ LSB for any ADC, except on the lowest step, where the error is $+\frac{1}{2}$, -0 LSB, and on the highest step, where the error is $+1$, $-\frac{1}{2}$ LSB.

As more bits are added to the accumulated digital value, the analog equivalent of the approximation acquires more decimal places of accuracy. Note that once the analog value extends beyond the decimal point, the last decimal digit is always 5.

An advantage of a successive approximation ADC is that the conversion time is always the same, regardless of the analog input voltage. This is not true with all types of analog-to-digital converters. The constant conversion time allows the output to be synchronized so that it can be read at known intervals.

The conversion time can be as few as $(n + 1)$ clock pulses for an n-bit device, if a bit is set by a clock edge and cleared asynchronously or by the opposite clock edge. Some SARs require four or more clock pulses per bit.

Dual Slope A/D Converter

KEY TERMS

Integrator   A circuit whose output is the accumulated sum of all previous input values. The integrator's output changes linearly with time when the input voltage is constant.

Dual slope ADC   Also called an integrating ADC. An analog-to-digital converter based on an integrator. The name derives from the fact that during the conversion process the integrator output changes linearly over time, with two different slopes.

A **dual slope analog-to-digital converter** is based on an **integrator** circuit, such as the one shown in Figure 12.25. The circuit output is proportional to the integral of the input

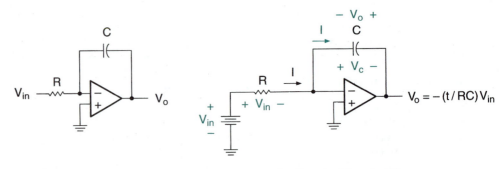

a. General circuit

b. Constant input voltage

FIGURE 12.25
Integrator

voltage as a function of time. Integration with respect to time is the summing of instantaneous values of a function over a specified period of time. In other words, the output of an integrator is the accumulated total of all previous values of input voltage.

We can analyze the circuit without calculus under special conditions, such as when the input voltage is constant. An integrator is similar to an inverting amplifier and can be analyzed using similar techniques. Since the input impedance of the op amp is large, there is very little current flowing into its ($-$) terminal. Ohm's law thus implies that there is very little voltage difference between the ($+$) and ($-$) terminals. Since they are at almost the same potential and the ($+$) terminal is grounded, we can say that the ($-$) terminal is "virtually grounded."

If the input voltage is constant, a DC current, I, flows in R. Since R is connected to the positive terminal of the input voltage source at one end and virtual ground at the other, the entire source voltage drops across the resistor. By Ohm's law,

$$I = V_{in}/R$$

Since the op amp input impedance is large, most current flows into the capacitor, causing it to charge over time. The current direction defines a polarity for V_c, the capacitor voltage.

The op amp output voltage is measured with respect to ground. The capacitor is connected from the op amp output to virtual ground. Therefore, the output voltage, V_o, is dropped across the capacitor. Notice that the polarities defined for V_o and V_c are opposite:

$$V_o = -V_c$$

The capacitor voltage is determined by the stored charge, Q, and the value of capacitance, C:

$$V_c = Q/C$$

The current I is the amount of charge flowing past a given point in a fixed time:

$$I = Q/t$$

Thus,

$$V_c = It/C$$

and

$$V_o = -It/C$$

Substitute the expression for I into this equation to get

$$V_o = -(t/RC)V_{in}$$

The output of an integrator with a constant input changes linearly with time, with a slope equal to $-\dfrac{V_{in}}{RC}$.

This equation describes the *change* in output voltage due to a constant input. When the input goes to 0 V, the capacitor holds its charge (ideally forever; in practice until it leaks away through circuit impedances) and maintains the output voltage at its final value. If a new input voltage is applied, we can use the integrator equation to calculate the change in output, which must then be added to the previous value.

▎▎ EXAMPLE 12.12

The integrator circuit of Figure 12.25 has the following component values:

$$C = 0.025 \ \mu F, \ R = 10 \ k\Omega$$

Sketch the graph of the output voltage if the waveform shown in the graph of Figure 12.26a is applied to the integrator input. The integrator output is originally at 0 V.

FIGURE 12.26
Example 12.12
Integrator Operation

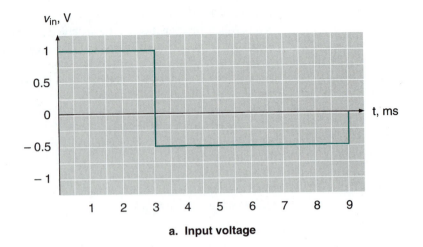

a. Input voltage

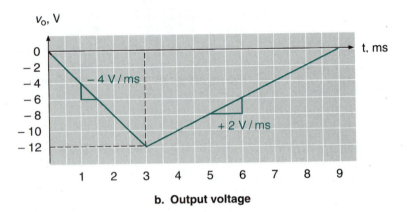

b. Output voltage

Solution   We must examine the graph in two sections:

1. From 0 to 3 ms
2. From 3 to 9 ms

A different constant input voltage is applied for each section of the graph.

0 to 3 ms:

The output at 3 ms is given by:

$$\text{slope} = \frac{V_{in}}{RC} = -\frac{1\text{V}}{(0.025\mu\text{F})(10\ \text{k}\Omega)}$$

$$= -4\ \text{V/ms}$$

$$V_o(3\ \text{ms}) = v_o(0) + (t/RC)\ V_{in}$$

$$= 0\text{V} + [(3\ \text{ms})(-4\ \text{V/ms})]$$

$$= -12\ \text{V}$$

The output changes at a rate of -4 V/ms for 3 ms.

3 to 9 ms:

The output at 9 ms is given by:

$$\text{slope} = \frac{V_{in}}{RC} = -\frac{(-0.5\ \text{V})}{(0.025\mu\text{F})(10\ \text{k}\Omega)}$$

$$= +2\ \text{V/ms}$$

$$V_o \,(9 \text{ ms}) = v_o(3 \text{ ms}) - (t/RC) \, V_{in}$$
$$= -12 \text{ V} + [(6 \text{ ms})(+2 \text{ V/ms})]$$
$$= -12 \text{ V} + (+12 \text{ V})$$
$$= 0 \text{ V}$$

The output changes at a rate of $+2$ V/ms for 6 ms. This cancels the effect of the original input.

Figure 12.27 shows the block diagram of an 8-bit dual slope analog-to-digital converter. Integrator output voltages for several input values are shown in Figure 12.28. Assume that the integrator has the same R and C values as in Figure 12.25.

FIGURE 12.27
Dual Slope ADC

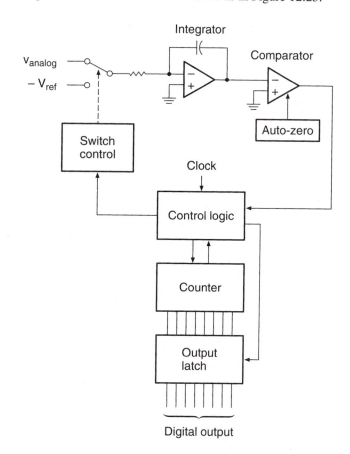

FIGURE 12.28
Integrator Outputs for Various
Input Voltages

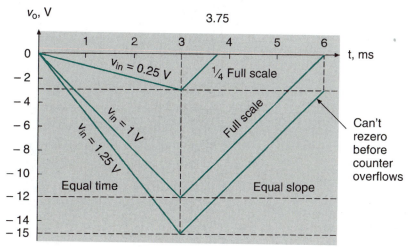

1. Before conversion starts, an auto-zero circuit sets the comparator output to 0 V by applying a compensating voltage to the comparator.

2. The input analog voltage causes the integrator output to increase in magnitude, as shown in the left half of Figure 12.28. As soon as this integrator voltage is nonzero, the comparator enables a counter via the control logic.

3. When the counter overflows (i.e., recycles to 00000000), the integrator input is switched from the analog input to $-V_{ref}$.

4. The reference voltage causes the integrator output to move toward 0 V at a known rate, as shown in the right half of Figure 12.28. During this rezeroing time, the counter continues to clock. When the integrator output voltage reaches 0 V, the comparator disables the counter. The digital equivalent of the analog voltage is now contained in the counter.

The reason this works is that in the initial integrating phase, the integrator output operates for a *known time,* producing a final output proportional to the input voltage. In the second phase, the output moves toward zero at a *known rate,* reaching zero in a time proportional to the final voltage of the first phase.

For example, assume that the components of the integrator and the clock rate of the counter are such that a 1-V input corresponds to the full-scale digital output *(FS).* The integrator output reaches a value of -12 V in 3 ms. The time required to rezero the integrator is the same as the initial integrating phase, 3 ms. The counter completes one cycle in the integrating phase and another cycle in the rezeroing phase, so that its final value is 00000000. (Note that this is the result obtained when 1 LSB is added to 11111111.)

If the input voltage is 0.25 volts, the integrator output is -3 V after 3 ms (one counter cycle). Since the integrator always rezeros at the same rate (4 V/ms), the rezeroing time is 0.75 ms, or one fourth of a counter cycle (since 12 V/4 = 3 V). The counter has time to reach state 01000000 or $\frac{1}{4}$ *FS.*

If we attempt to measure a voltage beyond that corresponding to full scale, the integrator output cannot rezero within the second counter cycle. Usually, an output pin on the ADC activates to show this condition. Some digital multimeters that use dual slope ADCs show an overvoltage or out-of-range condition by blanking the display, except for a leading digit 1.

One advantage of a dual slope ADC is its accuracy. One particular dual slope ADC is accurate to within $\pm0.05\% \pm 1$ count. This accuracy is balanced against a relatively slow conversion time, in the milliseconds, compared to microseconds for a successive approximation ADC and nanoseconds for a flash converter.

Another advantage is the ability of the integrator to reject noise. If we assume that noise voltage is random, then it will be positive about half the time and negative about half the time. Over time it should average out to zero.

As was alluded to above, a common application of this device is as a voltmeter circuit, where speed is less important than accuracy.

▉ SECTION 12.3 REVIEW PROBLEMS

12.7 Suppose that the dual slope ADC described above (same component values) has an input voltage of 0.375 V (3/8 full scale).

   a. What is the slope of the integrator voltage during the integrating phase?

   b. What is its slope during the rezeroing phase?

   c. How much time elapses during the rezeroing phase?

   d. What digital code is contained in the output latch after the conversion is complete?

Sample and Hold Circuit

For the sake of analysis, we have been assuming that the analog input voltage of any analog-to-digital converter is constant. This is an actual requirement. Most of these circuits will not produce a correct digital code if the analog voltage at the input changes during conversion time.

Unfortunately, most analog signals are not constant. Usually, we want to sample these signals at periodic intervals and generate a series of digital codes that tells us something about the way the input signal is changing over time. A circuit called a **sample and hold circuit** must be used to bridge the gap between a changing analog signal and a requirement for a constant ADC input voltage.

Figure 12.29 shows a basic sample and hold circuit. The voltage followers act as buffers with high input and low output impedances. The transmission gate is enabled during the sampling period, during which it charges the hold capacitor to the current value of the analog signal. During the hold period, the capacitor retains its charge, thus preserving the sampled analog voltage. The high input impedance of the second voltage follower prevents the capacitor from discharging significantly during the hold period.

FIGURE 12.29
Sample and Hold Circuit

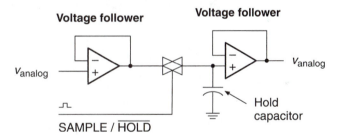

Figure 12.30 shows how a sample and hold circuit produces a steady series of constant analog voltages for an ADC input. Since these sampled values have yet to be converted to digital codes, they can take on any value within the analog range; they are not yet limited by the number of bits in the quantization.

FIGURE 12.30
Sample and Hold Output

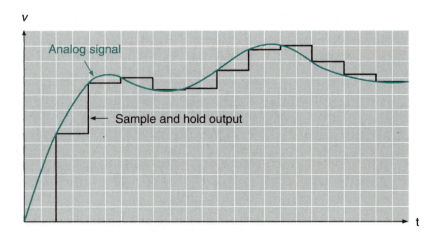

Ideally, a sample and hold circuit should charge quickly in sample mode and discharge slowly in hold mode. These characteristics are facilitated by the low output impedance and high input impedance of the voltage follower circuits.

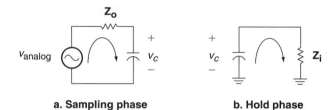

a. Sampling phase   **b. Hold phase**

FIGURE 12.31
Equivalent Circuits for Sample-and-Hold Circuit

Figure 12.31 shows the equivalent circuits of the sample and hold modes of the circuit in Figure 12.29. In sample mode, the capacitor charges through the output impedance, Z_o, of the first voltage follower. Since this is a very small value (about 75×10^{-5} Ω), the capacitor will charge quickly. In the hold mode, the capacitor discharges slowly through the very high input impedance of the second voltage follower (about 2×10^{11} Ω).

NOTE

The input and output impedances of the voltage follower are significantly different from the open-loop op amp values. This is because, in the voltage follower configuration, the input impedance is divided by the open loop gain (about 75 Ω/100,000) and the output impedance is multiplied by the open loop gain (about 2 MΩ × 100,000).

A variation of the sample and hold circuit is the track and hold circuit. The difference is not so much in the circuit as in the way it is operated. A sample and hold circuit is restricted by the charging speed of its hold capacitor. If there is a large change in signal level between samples, the hold capacitor may not be able to keep up with the change. A track and hold circuit samples the analog signal continuously, minimizing charging delays of the hold capacitor. When the analog signal needs to be converted, the track and hold circuit reverts to hold mode by closing the analog transmission gate. Many high-speed ADCs have a track and hold circuit as an integral part of the device.

Sampling Frequency and Aliasing

KEY TERMS

Nyquist sampling theorem   A theorem from information theory that states that, in order to preserve all information in a signal, it must be sampled at a rate of twice the highest-frequency component of the signal. ($f_s \geq 2f_{max}$)

Aliasing   A phenomenon that produces an unwanted low-frequency component in a sampled analog signal due to a sampling frequency that is too slow relative to the sampled analog signal.

Anti-aliasing filter   A low-pass filter with a corner frequency of twice the maximum frequency of a sampled signal, used to prevent aliasing in an ADC.

In the first section of this chapter, we saw that the sampling frequency of an ADC has a great effect on the quality of the digital representation of an analog signal. We may ask, what is the minimum value of the sampling frequency for any particular analog signal and what happens if this criterion is not met?

A theorem in information theory, called the **Nyquist sampling theorem**, states that a periodic signal must be sampled at least twice a cycle to preserve all its information. In practice, this means that the sampling frequency of a particular system must be twice the maximum frequency of any signal to be sampled by the system. (These frequencies might also include harmonics of a signal that add to the basic signal to give it its characteristic shape.) This can be expressed mathematically as $f_s \geq 2f_{max}$ for a sampling frequency f_s and a maximum-frequency component of f_{max}.

For example, the sampling frequency for compact disc audio is 44.1 kHz, which allows signals of up to 22.05 kHz to be sampled accurately. This fits in nicely with the statistical range of human hearing: 20 Hz–20 kHz. (People who have listened to any amount of rock music in their youth can probably only get up to 12 kHz.) Telephone-quality signals are sampled at 8 kHz, yielding a maximum frequency of 4 kHz, which is a bit more than the classical telephone-line bandwidth of 300 Hz–3300 Hz.

A sampling frequency of an ADC system that does not meet the criterion required by the Nyquist sampling theorem results in **aliasing,** a phenomenon that generates a false low-frequency component of the digital sample.

To get an idea of how aliasing works, let us examine a sine wave with a period of 12 μs ($f = 83.3$ kHz), shown in Figure 12.32. If we sampled the signal every 1 μs, we would capture the values listed in Table 12.5.

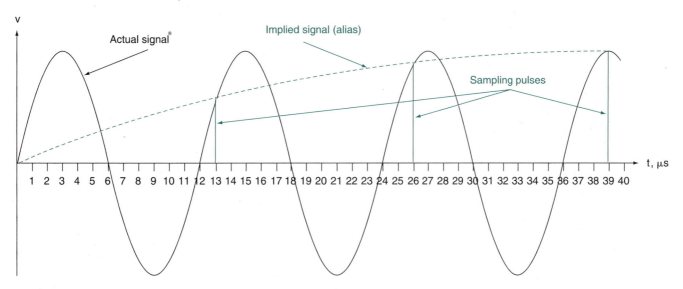

FIGURE 12.32
Effect of Sampling Too Slowly

The points in Table 12.5 can be used to accurately reconstruct the original sine wave. (The reconstructed output would need to be filtered to eliminate introduced high-frequency components, but the fundamental frequency would be correct.)

Suppose now that we sample the same sine wave at less than twice a cycle. Table 12.6 shows the samples captured by a series of sampling pulses that are spaced by 13 μs. The first four samples in the table are shown by vertical lines in Figure 12.32.

The samples in Table 12.6 have exactly the same amplitude as those taken in Table 12.5. However, the samples are spaced at 13 μs intervals, rather than 1 μs. For example, the sample at 13 μs measures the sine wave amplitude at 390°, which is the same as 30° of

Table 12.5 Sampled Values of an 83.3 kHz Sine Wave (1 μs Sampling)

Time (μs)	Degrees	Fraction of Peak
0	0°	0.000
1	30°	0.500
2	60°	0.866
3	90°	1.000
4	120°	0.866
5	150°	0.500
6	180°	0.000
7	210°	−0.500
8	240°	−0.866
9	270°	−1.000
10	300°	−0.866
11	330°	−0.500
12	360°	0.000

Table 12.6 Sampled Values of an 83.3 kHz Sine Wave (13 μs Sampling)

Time (μs)	Degrees	Fraction of Peak
0	0°	0.000
13	390°	0.500
26	780°	0.866
39	1170°	1.000
52	1560°	0.866
65	1950°	0.500
78	2340°	0.000
91	2730°	−0.500
104	3120°	−0.866
117	3510°	−1.000
130	3900°	−0.866
143	4290°	−0.500
156	4680°	0.000

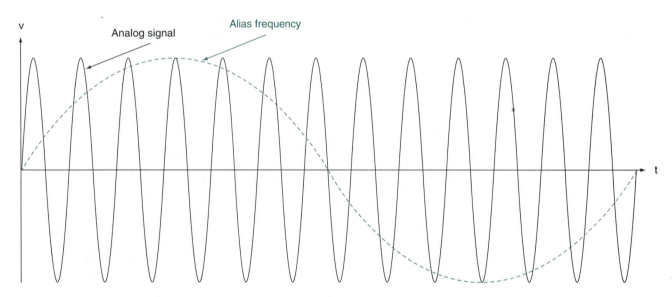

FIGURE 12.33
Aliasing

the *second* cycle. If these samples were used to reconstruct a sine wave, it would have a period of 156 μs, rather than 12 μs. This false low-frequency component is shown by the broken line connecting the first four samples in Figure 12.32, which represent measurements, not in a single cycle, but in four cycles of the original analog signal. Figure 12.33 shows one complete cycle of the alias frequency as a broken line and thirteen cycles of the sampled signal as a solid line.

Aliasing can be prevented by filtering the analog input to an ADC with an **anti-aliasing filter,** as shown in Figure 12.34.

The anti-aliasing filter is a low-pass filter with the corner frequency set to $2\,f_{max}$. Frequencies less than $2\,f_{max}$ are allowed to pass to the analog input of the ADC. Frequencies greater than $2\,f_{max}$ are attenuated. In this way, the ADC never converts any signal with a frequency greater than $2\,f_{max}$ and thus an alias frequency cannot develop.

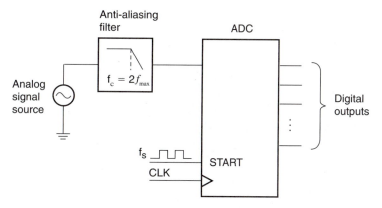

FIGURE 12.34
Anti-aliasing Filtering

12.4 Data Acquisition

>
> **Data acquisition network**   A circuit that gathers and digitizes data from several analog sources.

CPLD Interface for an ADC

Figure 12.35 shows the symbol for an ADC0808 analog-to-digital converter. This successive approximation ADC can form the basis of a **data acquisition network**, a system that can convert analog information from up to eight channels and store the converted values in a series of output latches.

www.electronictech.com

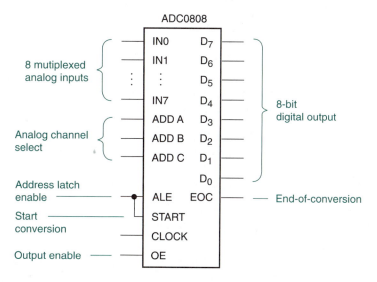

FIGURE 12.35
ADC0808 Analog-to-Digital Converter

The ADC0808 has a built-in 8-channel analog multiplexer with inputs *IN0* through *IN7*, which are selected by the states of three address inputs, *ADD C, ADD B,* and *ADD A,* where *ADD C* is the most significant bit. Before an analog input can be converted, its address must be stored in an internal address latch by a high-going pulse on *ALE* (Address Latch Enable).

The conversion process starts with a high-going pulse on the *START* input. (*START* and *ALE* can be tied together.) End-of-conversion is indicated by the *EOC* output. The conversion process is driven by the *CLOCK* input. After conversion is complete, the digital output can be read by making *OE* (Output Enable) HIGH. When *OE* is not active, the digital outputs are in the high-impedance state.

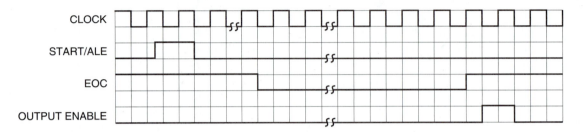

FIGURE 12.36
Timing Diagram for an ADC0808

Figure 12.36 shows a timing diagram relating the various control signals of the ADC0808. Figure 12.37 is an excerpt from the ADC0808 data sheet that shows relevant timing information. The ADC is reset on the rising edge of *START*. After *START/ALE* goes LOW, the ADC makes *EOC* go LOW within 8 clock cycles + 2 μs. *EOC* stays LOW until conversion is complete. The simplest way to operate the ADC on a stand-alone basis is to tie the *EOC* line to the *START/ALE* line so that the ADC starts as soon as *EOC* goes HIGH and the ADC continuously updates the value of the digital output.

We can design a state machine that controls the ADC and stores output values in an octal latch automatically. Figure 12.38 shows such a circuit. The controller is a state machine that will accept a LOW pulse from a pushbutton switch labeled **go,** perform one analog-to-digital conversion from one of eight analog channels and store the resulting 8-bit digital value in an octal latch. The analog channel is manually selected by DIP switches at the address select inputs. The entire state machine and latch portion of the circuit is contained in one CPLD, such as the Altera EPM7128SLC84-7 on the Altera UP-1 circuit board.

Figure 12.39 shows the state diagram of the controller, with two synchronous inputs called **go** and **eoc.** The asynchronous reset, which sets the machine to the **idle** state, is not shown on the state diagram. Outputs are **sc** (start conversion), **oe** (output enable), and **en** (latch enable). The states are as follows:

- **idle**—Wait for **go** = 0 (switch pressed). All outputs are LOW.
- **start**—Wait for **go** = 1 (switch released). Transition to **wait1** makes **sc** = 1 (*START/ALE* pulse). Other outputs are LOW.
- **wait1**—Wait for **eoc** = 0. (Wait for conversion to start. *EOC* is LOW during, but not before, conversion. Do not test for **eoc** = 1 until *after* conversion.) All outputs LOW.
- **wait2**—Wait for **eoc** = 1. (Conversion complete.) When complete, transition to **read.** At that time, **oe** = 1, **en** = 1.
- **read**—Enable ADC output (**oe** = 1) and make latch transparent (**en** = 1).
- **store**—Keep ADC output enabled, put latch in store mode (**en** = 0). ADC digital output is now stored in the output latch.

Figure 12.40 shows a simulation of the controller. Both the latch and controller can be implemented as VHDL design entities and instantiated as components in the top level of a VHDL hierarchy. This and later VHDL examples will be saved as exercises for the lab manual accompanying this book. The VHDL files are available to instructors in the Online Companion to this book.

ADC0808/ADC0809

Electrical Characteristics (Continued)

Digital Levels and DC Specifications: ADC0808CCN, ADC0808CCV, ADC0809CCN and ADC0809CCV, 4.75≤V$_{CC}$≤5.25V, −40 C≤T$_A$≤ +85° C unless otherwise noted

Symbol	Parameter	Conditions	Min	Typ	Max	Units
DATA OUTPUTS AND EOC (INTERRUPT)						
V$_{OUT(1)}$	Logical ™"1" Output Voltage	V$_{CC}$ = 4.75V I$_{OUT}$ = −360 μA I$_{OUT}$ = −10 μA		**2.4** **4.5**		V(min) V(min)
V$_{OUT(0)}$	Logical ™"0" Output Voltage	I$_O$=1.6 mA			0.45	V
V$_{OUT(0)}$	Logical ™"0" Output Voltage EOC	I$_O$=1.2 mA			0.45	V
I$_{OUT}$	TRI-STATE Output Current	V$_O$=5V V$_O$=0	−3		3	μA μA

Electrical Characteristics

Timing Specifications V$_{CC}$=V$_{REF(+)}$=5V, V$_{REF(_)}$=GND, t$_r$=t$_f$=20 ns and T$_A$=25°C unless otherwise noted.

Symbol	Parameter	Conditions		Min	Typ	Max	Units
t$_{WS}$	Minimum Start Pulse Width	(Figure 5)			100	200	ns
t$_{WALE}$	Minimum ALE Pulse Width	(Figure 5)			100	200	ns
t$_s$	Minimum Address Set-Up Time	(Figure 5)	2		25	50	ns
t$_H$	Minimum Address Hold Time	(Figure 5)	2		25	50	ns
t$_D$	Analog MUX Delay Time From ALE	R$_S$=0Ω (Figure 5)			1	2.5	μs
t$_{H1}$, t$_{H0}$	OE Control to Q Logic State	C$_L$=50 pF, R$_L$=10k (Figure 8)			125	250	ns
t$_{1H}$, t$_{0H}$	OE Control to Hi-Z	C$_L$=10 pF, R$_L$=10k (Figure 8)			125	250	ns
t$_c$	Conversion Time	f$_c$=640 kHz, (Figure 5) (Note 7)		90	100	116	μs
f$_c$	Clock Frequency			10	640	1280	kHz
t$_{EOC}$	EOC Delay Time	(Figure 5)		0		8+2μS	Clock Periods
C$_{IN}$	Input Capacitance	At Control Inputs			10	15	pF
C$_{OUT}$	TRI-STATE Output Capacitance	At TRI-STATE Outputs			10	15	pF

Note 1: Absolute Maximum Ratings indicate limits beyond which damage to the device may occur. DC and AC electrical specifications do not apply when operating the device beyond its specified operating conditions.

Note 2: All voltages are measured with respect to GND, unless othewise specified.

Note 3: A zener diode exists, internally, from V$_{CC}$ to GND and has a typical breakdown voltage of 7 V$_{DC}$.

Note 4: Two on-chip diodes are tied to each analog input which will forward conduct for analog input voltages one diode drop below ground or one diode drop greater than the V$_{CC}$n supply. The spec allows 100 mV forward bias of either diode. This means that as long as the analog V$_{IN}$ does not exceed the supply voltage by more than 100 mV, the output code will be correct. To achieve an absolute 0V$_{DC}$ to 5V$_{DC}$ input voltage range will therefore require a minimum supply voltage of 4.900 V$_{DC}$ over temperature variations, initial tolerance and loading.

Note 5: Total unadjusted error includes offset, full-scale, linearity, and multiplexer errors. See Figure 3. None of these A/Ds requires a zero or full-scale adjust. However, if an all zero code is desired for an analog input other than 0.0V, or if a narrow full-scale span exists (for example: 0.5V to 4.5V full-scale) the reference voltages can be adjusted to achieve this. See Figure 13.

Note 6: Comparator input current is a bias current into or out of the chopper stabilized comparator. The bias current varies directly with clock frequency and has little temperature dependence (Figure 6). See paragraph 4.0.

Note 7: The outputs of the data register are updated one clock cycle before the rising edge of EOC.

Note 8: Human body model, 100 pF discharged through a 1.5 kΩ resistor.

FIGURE 12.37

Extract from ADC0808 Datasheet (Reprinted with permission of National Semiconductor)

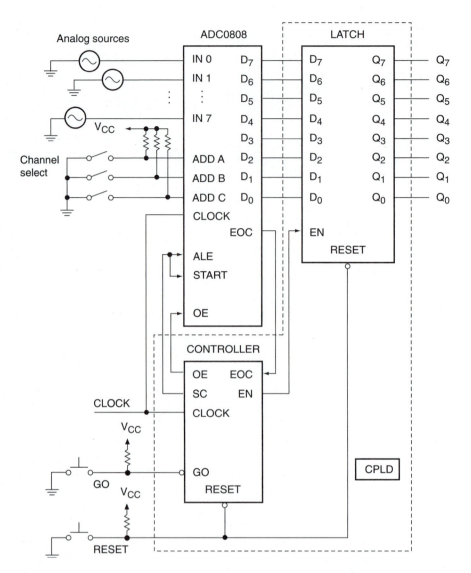

FIGURE 12.38
ADC Interface with One Output Channel and Manual Input Channel Selection

We should note that if the controller/latch circuit is to be implemented on the Altera UP-1 board, we must also include a clock divider circuit. The Altera UP-1 board has an on-board oscillator that runs at 25.175 MHz. The clock rate, as defined by the ADC0808 data sheet, must be in the range $10\text{kHz} \leq f_c \leq 1280 \text{ kHz}$.

A 5-bit counter can serve as a divide-by-32 circuit ($2^5 = 32$), as shown in Figure 12.41. If the UP-1 oscillator is applied to the counter clock, the Q_4 output frequency is given by 25.175 MHz/32 = 786.7 kHz, which is within the required range for the ADC clock. Q_4 should then be used to clock the state machine, as well as any other synchronous circuitry used in conjunction with the ADC0808.

We can make a few minor changes to the ADC interface in Figure 12.38 to make it run as a continuous-conversion circuit. First, we eliminate the **go** input and associated push-button. Second, we change the state diagram to eliminate the **idle** state, as shown in Figure 12.42. With no pushbutton to press and then release, we eliminate two wait transitions (previously associated with the **idle** and **start** states) from the state diagram. Otherwise, the circuit and controller remain the same. A simulation of the modified controller is shown in Figure 12.43.

FIGURE 12.39
State Diagram for an ADC Controller

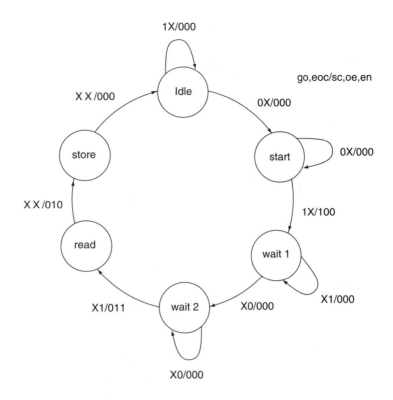

FIGURE 12.40
Simulation of State Machine ADC Controller

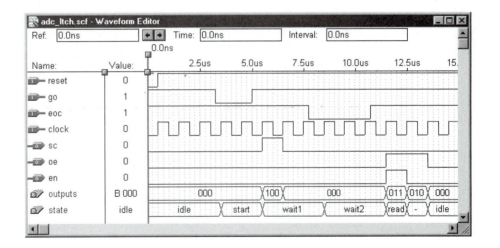

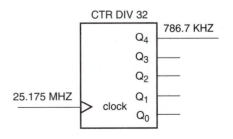

FIGURE 12.41
5-Bit Counter as Divide-by-32 Clock Divider

FIGURE 12.42
State Diagram for Continuous-
Convert ADC Controller

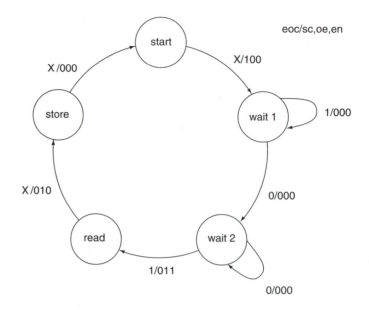

FIGURE 12.43
Simulation of Continuous-
Conversion ADC Controller

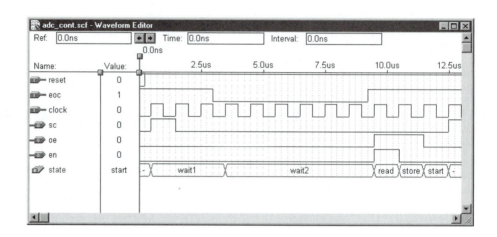

▌▌ EXAMPLE 12.13

From the ADC0808 data sheet extract in Figure 12.37, determine the number of clock cycles required for the conversion of an analog signal.

Solution For a clock frequency of 640 kHz, typical conversion time is given as 100 μs.

$$\frac{640 \times 10^3 \text{ clock cycles}}{\text{second}} \times (100 \times 10^{-6} \text{ seconds}) = 64 \text{ clock cycles}$$

▌▌ EXAMPLE 12.14

Calculate the highest-frequency analog input that can be accurately converted by an ADC0808 controlled by a state machine represented by the state diagram of Figure 12.42 if the system clock frequency is 787 kHz.

Solution One conversion cycle, T_s, requires 64 clock cycles for the ADC and an overhead of 13 clocks +2 μs for the state machine for a total of 77 clock cycles +2 μs. (**Start** to **wait1** requires one clock cycle. An additional 8 cycles +2 μs are needed before **eoc** goes LOW. According to Note 7 in Figure 12.37, the ADC conversion is complete one clock cycle before *EOC* goes HIGH. From this point back to **start** is 4 clocks.)

$$T_s = \frac{1}{787 \times 10^3 \text{ clock cycles/second}} \times 77 \text{ clock cycles} + 2\,\mu s = 99.8 \times 10^{-6} \text{ seconds}$$

$$f_s = \frac{1}{T_s} = \frac{1}{99.8 \times 10^{-6} \text{ s}} = 10.02 \text{ kHz}$$

According to the Nyquist sampling theorem, the maximum-frequency component of the sampled analog signal is $f_{max} = f_s/2 = 10.02 \text{ kHz}/2 = 5.01 \text{ kHz}$. This is of the same order of magnitude as a telephone-quality audio signal.

CPLD-Based Data Acquisition Network

Figure 12.44 shows a data acquisition system that continuously converts and stores data from four analog channels. All the circuitry within the broken line is contained within a single CPLD, such as the Altera EPM7128SLC84. The operation is similar to the system in

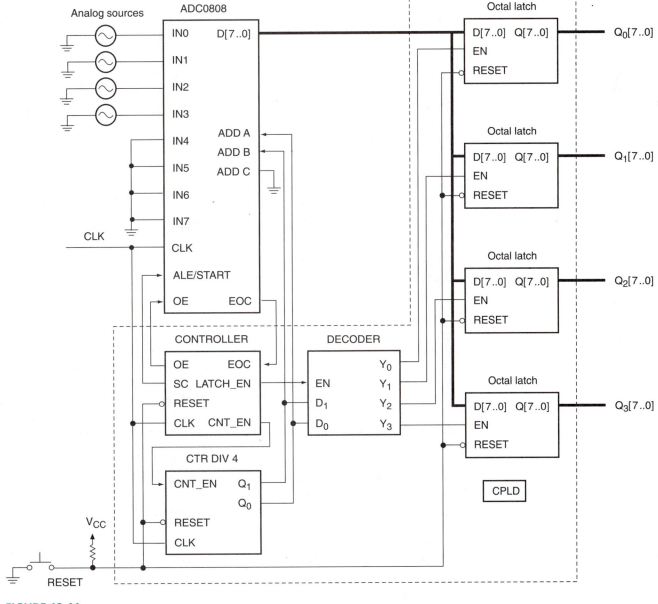

FIGURE 12.44
4-Channel Data Acquisition System

Figure 12.38, except that with multiple latches in the circuit, a counter and decoder are required to keep track of the selected channel.

The controller, whose state diagram is shown in Figure 12.45, generates the same control signals for the ADC as the system in Figure 12.38. When the conversion is complete and the controller detects a LOW on its **eoc** input, it reads the ADC output and transfers the contents to the selected 8-bit latch. The latch is selected, via the decoder, by the value of the counter (e.g., $Q_1Q_0 = 11$ selects analog input channel 3, decoder output Y_3, and latch 3). The selected latch input is enabled (i.e., made transparent) by the controller during the transition from **wait2** to **read.** At all other times all decoder outputs are LOW, disabling all latches, thus placing them in store mode. After the ADC data have been stored, the controller sets **cnt_en** HIGH, which allows the counter to be incremented on the next clock pulse. The next channel is now ready for a convert-and-store cycle. After all channels have been sampled, converted, and stored, the cycle begins again at channel 0 and continues indefinitely.

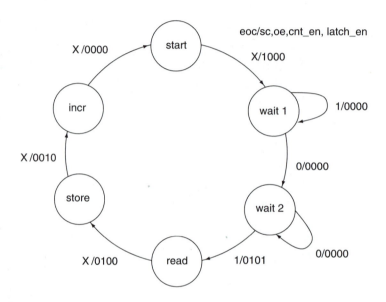

FIGURE 12.45
State Diagram for 4-channel Data Acquistion System

Figure 12.46 shows a simulation of the controller, counter, and decoder for the data acquisition system of Figure 12.44. During the read-and-store part of the cycle, only one of the latch enables, **y0** to **y3,** is active when **oe** is active. The number of the active latch enable is the same as the counter value on the second last waveform. The last line in the simulation (**controller|outputs3.Q**) is the **cnt_en** line from the controller to the counter. The counter is incremented on the first positive edge of the clock after this line goes HIGH. This point is indicated by the cursor line on the transition from channel 1 to channel 2.

The circuit in Figure 12.44 could be expanded to convert all eight analog channels from the ADC, but the chosen CPLD (EPM7128SLC84) does not have enough I/O pins. Eight 8-bit latch outputs require 64 pins; the CPLD only has 60 user I/Os. An 8-channel system could be implemented if it used eight external latches, such as eight 74HC373 octal latches, or internal latches on a different CPLD. Note that the CPLD has enough logic cells to implement the system, just not enough I/O pins. The identical device in a different package (EPM7128SQC100; 100-pin quad flat-pack) can accommodate the entire system.

For an 8-channel system, the counter would need to be expanded to 3 bits and the decoder to a 3-line-to-8-line device.

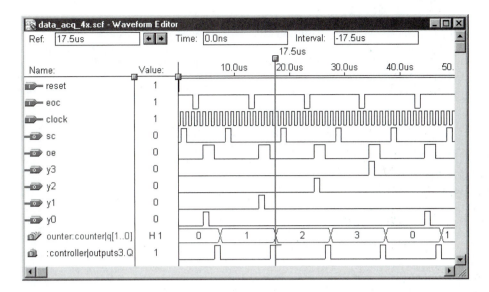

FIGURE 12.46
Simulation of 4-channel Data Acquisition System

▮▮ SECTION 12.4 REVIEW PROBLEM

12.8 Calculate the highest-frequency component of an analog signal that can be accurately converted by the 4-channel data acquisition system in Figure 12.44. Assume the system clock is running at 787 kHz.

SUMMARY

1. An analog system can represent a physical property (e.g., temperature, pressure, or velocity) by a proportional voltage or current. The mathematical function describing the analog voltage or current is continuous throughout a defined range.

2. A digital system can represent a physical property by a series of binary numbers of a fixed bit size.

3. Digital representations of data are not subject to the same distortions as analog representations. They are also easier to store and reproduce than analog.

4. The quality of a digital representation depends on the sampling frequency and quantization (number of bits) of the system that converts an analog input to a digital output.

5. The resolution of a system is a function of the number of bits in its digital representation. A greater number of bits implies that the sampled analog input can be broken up into more, smaller segments, allowing each segment to more closely approximate the original input value.

6. A digital-to-analog converter (DAC) uses electronic switches to sum binary-weighted currents to a total analog output current. Analog current can be calculated by:

$$I_a = \frac{b_{n-1}2^{n-1} + b_{n-2}\,2^{n-2} + \cdots + b_2 2^2 + b_1 2^1 + b_0 2^0}{2^n}\, I_{ref}$$

or, more simply:

$$I_a = \frac{\text{digital code}}{2^n}\, I_{ref}$$

for an n-bit DAC, where $b_{n-1}b_{n-2} \cdots b_2 b_1 b_0$ is the digital input code,
I_a is the analog output current, and
I_{ref} is the DAC reference (full scale) current.

7. The maximum output of a DAC is full scale *(FS)* minus the value represented by a change in the least significant bit of the input *(FS − 1 LSB)*. For example, for a 4-bit converter (1 LSB = 1/16 *FS*), the maximum output is *(FS − 1/16 FS)* = 15/16 *FS*. For an 8-bit converter (1 LSB = 1/256 *FS*), the maximum output is *(FS − 1/256 FS)* = 255/256 *FS*.

8. A weighted-resistor DAC derives its binary-weighted currents from binary-weighted resistors connected to the reference voltage supply.

9. An R-2R ladder DAC derives its binary weighted currents from a resistor ladder network that consists of resistors of two values only, one of which is twice the other. The R-2R ladder is more common than the weighted resistor DAC.

10. A DAC input code consisting of a 1 followed by all 0s represents an output of $\frac{1}{2}$ FS, regardless of the number of bits in the DAC input. A code of 01 followed by all 0s represents an output of $\frac{1}{4}$ FS. A code of 11 followed by all 0s is $\frac{3}{4}$ FS.

11. The MC1408 DAC is an example of a monolithic (single-chip) DAC. Output current at pin 4 is a binary-weighted fraction of the reference current at pin 14:

$$I_o = \left(\frac{\text{digital code}}{256}\right)\left(\frac{V_{\text{ref}}}{R_{14}}\right)$$

12. If the output of an MC1408 DAC is buffered by an noninverting op amp with a feedback resistance of R_F, the output voltage is given by:

$$V_a = I_o R_F = \left(\frac{\text{digital code}}{256}\right)\left(\frac{R_F}{R_{14}}\right)V_{\text{ref}}$$

13. An 8-bit DAC can be used as a ramp generator by connecting an 8-bit binary counter to the digital inputs.

14. An MC1408 DAC can be configured for bipolar output by connecting a pull-up resistor (R_4) from the output (pin 4) to the reference voltage supply. Output is given by:

$$V_a = I_o R_F - I_s R_F = \left(\frac{\text{digital code}}{256}\right)\left(\frac{R_F}{R_{14}}\right)V_{\text{ref}} - \frac{R_F}{R_4}V_{\text{ref}}$$

15. A DAC is monotonic if every increase in binary input results in an increase in analog output.

16. DAC errors include: offset error (nonzero output for zero input code), gain error (output falling above or below $FS -$ 1LSB for maximum input code due to an incorrect slope), linearity error (deviation from straight-line approximation between codes), and differential nonlinearity (deviation of step sizes from ideal of one step per LSB).

17. DAC linearity error of greater than $\pm\frac{1}{2}$ LSB can result in a nonmonotonic output.

18. Several popular types of analog-to-digital converters (ADC) are flash or simultaneous, successive approximation, and dual slope or integrating.

19. A flash ADC consists of a voltage divider with the same number of steps as output codes, a set of comparators (one for every output code), and a priority encoder. All comparators whose reference input is less than the analog input will fire, the priority encoder will detect the highest-value active comparator, and generate the corresponding output code. A flash ADC is fast, but requires 2^n comparators for an n-bit output code.

20. An ADC transfer characteristic is set up so that all codes are 1 LSB wide, except for the first and last codes. The code for 0 is $\frac{1}{2}$ LSB wide and the maximum code is $1\frac{1}{2}$ LSB wide. This offset places the nominal analog value of the code in the center of the code's range of analog input values.

21. A successive approximation ADC consists of a state machine called a successive approximation register (SAR) whose bits can be set and cleared individually in a specific sequence, a digital-to-analog converter, and an analog comparator.

22. A successive approximation ADC sets each bit of the SAR in turn as an approximation of the required digital code. For each bit, the approximation is converted back to analog form and compared with the incoming analog value. If the converted value is less than the actual analog value, the bit remains set and the next bit is tried. If the converted value is greater than the actual analog input, the bit is cleared and the next bit is tried.

23. A dual slope ADC consists of an integrator, comparator, counter, and control logic. The integrator output changes with a slope of $-V_{in}/RC$ for a constant input. This ADC allows the integrator to charge for the time required for the counter to complete one full cycle (known time). At that time, the integrator input is switched to a reference voltage of opposite polarity. The reference voltage discharges the integrator at a known rate. The time required to do this is stored in the counter and represents the fraction of full scale analog voltage applied to the converter.

24. A sample and hold circuit may be required to hold the input value of an ADC constant for the conversion time of the ADC. It samples an analog signal at periodic intervals and holds the sampled value in a capacitor until the next sample is taken. A track and hold circuit performs a similar function, but allows the capacitor to charge and discharge along with the changing analog signal, holding its value only during the conversion time of the ADC.

25. In order to preserve the information in an analog signal, it must be sampled at a frequency of at least twice the maximum-frequency component of the signal ($f_s \geq f_{\max}$). This criterion is called the Nyquist sampling theorem.

26. If the Nyquist sampling theorem is violated, an alias frequency, or false low-frequency component, will be added to the digital representation of the analog signal.

27. Alias frequencies can be eliminated with an anti-aliasing filter, a low-pass filter used to pass only frequencies less than $2f_s$ to the input of an ADC. This input frequency range automatically satisfies the Nyquist criterion at the ADC input.

28. An ADC0808 successive approximation ADC contains an 8-channel analog MUX and can be used as the basis for an 8-channel data acquisition system.

29. The conversion sequence for the ADC0808 is as follows:

   a. an analog input channel is selected by setting the appropriate address on lines *ADD C, ADD B,* and *ADD A.*

   b. *ALE* and *START* are pulsed HIGH.

   c. *EOC* (end-of-conversion) goes LOW no later than 8 clock cycles $+2$ μs after *START.*

   d. *EOC* goes HIGH when conversion is complete.

   e. *OE* (output enable) is set HIGH to read converted output.

This sequence can be controlled by a CPLD-based state machine.

30. A data acquisition system based on an ADC0808 requires an octal latch for each analog channel, a state-machine controller, and a counter/decoder circuit to select the active analog channel and latch.

GLOSSARY

Aliasing   A phenomenon that produces an unwanted low-frequency component in a sampled analog signal due to a sampling frequency that is too slow relative to the sampled analog signal.

Anti-aliasing filter   An low-pass filter with a corner frequency of twice the maximum frequency of a sampled signal, used to prevent aliasing in an ADC.

Analog   A way of representing some physical quantity, such as temperature or velocity, by a proportional continuous voltage or current. An analog voltage or current can have any value within a defined range.

Analog-to-digital converter   A circuit that converts an analog signal at its input to a digital code. (Also called an A-to-D converter, A/D converter, or ADC.)

Continuous   Smoothly connected. An unbroken series of consecutive values with no instantaneous changes.

Data acquisition network   A circuit that gathers and digitizes data from several analog sources.

Digital   A way of representing a physical quantity by a series of binary numbers. A digital representation can have only specific discrete values.

Digital-to-analog converter   A circuit that converts a digital code at its input to an analog voltage or current. (Also called a D-to-A converter, D/A converter, or DAC.)

Discrete   Separated into distinct segments or pieces. A series of discontinuous values.

Dual slope ADC   Also called an integrating ADC. An analog-to-digital converter based on an integrator. The name derives from the fact that during the conversion process the integrator output changes linearly over time, with two different slopes.

Flash converter (or simultaneous converter)   An analog-to-digital converter that uses comparators and a priority encoder to produce a digital code.

Full scale   The maximum analog reference voltage or current of a digital-to-analog converter.

Integrator   A circuit whose output is the accumulated sum of all previous input values. The integrator's output changes linearly with time when the input voltage is constant.

Multiplying DAC   A DAC whose output changes linearly with a change in DAC reference voltage.

Nyquist sampling theorem   A theorem from information theory that states that, in order to preserve all information in a signal, it must be sampled at a rate of twice the highest-frequency component of the signal. ($f_s \geq 2f_{max}$)

Priority encoder   An encoder that will produce a binary output corresponding to the subscript of the highest-priority active input. This is usually defined as the input with the largest subscript.

Quantization   The number of bits used to represent an analog voltage as a digital number.

Quantization error   Inaccuracy introduced into a digital signal by the inability of a fixed number of bits to represent the exact value of an analog signal.

Resolution   The difference in analog voltage corresponding to two adjacent digital codes. Analog step size.

Sample   An instantaneous measurement of an analog voltage, taken at regular intervals.

Sample and hold circuit   A circuit that samples an analog signal at periodic intervals and holds the sampled value long enough for an ADC to convert it to a digital code.

Sampling frequency   The number of samples taken per unit time of an analog signal.

Successive approximation register   A state machine used to generate a sequence of closer and closer binary approximations to an analog signal.

PROBLEMS

Problem numbers set in color indicate more difficult problems; those with underlines indicate most difficult problems.

Section 12.1   Analog and Digital Signals

12.1   An analog signal with a range of 0 to 12 V is converted to a series of 3-bit digital codes. Make a table similar to Table 12.1 showing the analog range for each digital code.

12.2   Sketch the positive half of a sine wave with a peak voltage of 12 V. Assume that this signal will be quantized according to the table constructed in Problem 12.1. Write the digital codes for the points 0, $T/8$, $T/4$, $3T/8$, . . . , T where T is the period of the half sine wave.

12.3   Repeat Problems 12.1 and 12.2 for a 4-bit quantization.

12.4   Write the 3-bit and 4-bit digital codes for the points 0, $T/16$, $T/8$, $3T/16$, . . . , T for the half sine wave described in Problem 12.2.

12.5   An analog-to-digital converter divides the range of an analog signal into 64 equal parts. The analog input has a range of 0 to 500 mV. How many bits are there in the resultant digital codes? What is the resolution of the A/D converter?

12.6   Repeat Problem 12.5 if the analog range is divided into 256 equal parts.

12.7   The analog range of a signal is divided into m equal parts, yielding a digital quantization of n bits. If the range is divided into $2m$ parts, how many bits are in the equivalent digital codes? (That is, how many extra bits do we get for each doubling of the number of codes?)

Section 12.2   Digital-to-Analog Conversion

12.8   **a.** Calculate the analog output voltage, V_a, for a 4-bit DAC when the input code is 1010.

   b. Calculate V_a for an 8-bit DAC when the input code is 10100000.

   c. Compare the results of parts a and b. What can you conclude from this comparison?

12.9 **a.** Calculate the analog output voltage, V_a, for a 4-bit DAC when the input code is 1100.

   b. Calculate V_a for an 8-bit DAC when the input code is 11001000.

   c. Compare the results of parts a and b. What can you conclude from this comparison? How does this differ from the comparison made in Problem 12.8?

12.10 Refer to the generalized D/A converter in Figure 12.4. For $I_{ref} = 500$ μA and $R_F = 22$ kΩ, calculate the range of analog output voltage, V_a, if the DAC is a 4-bit circuit. Repeat the calculation for an 8-bit DAC.

12.11 The resistor for the MSB of a 16-bit weighted resistor D/A converter is 1 kΩ. List the resistor values for all bits. What component problem do we encounter when we try to build this circuit?

12.12 Draw the circuit for an 8-bit R-2R ladder DAC.

12.13 Calculate the value of V_a of an R-2R ladder DAC when digital inputs are as follows. $V_{ref} = 12$ V.

   DCBA

   a. 1111

   b. 1011

   c. 0110

   d. 0011

12.14 An MC1408 DAC is configured as shown in Figure 12.12. $R_{14} = R_{15} = 6.8$ kΩ, $V_{ref}(+) = +12$ V, $V_{ref}(-) =$ ground, and $R_L = 2.2$ kΩ. Calculate the output voltage, V_a, for the following digital input codes: 00000000, 00000001, 10000000, 10101010, 11100010, 11111111.

12.15 Calculate the resolution of the DAC in Problem 12.14.

12.16 Refer to the op amp-buffered DAC in Figure 12.13. Assume the resistor values are changed as follows: $R_{14A} = 270$ Ω, $R_{14B} = 2$ kΩ (max), $R_{FA} = 1.2$ kΩ, $R_{FB} = 5$ kΩ (max). Describe a step-by-step procedure that calibrates the DAC so that it has a reference current of 4 mA and a full scale analog output voltage of 12 volts, using only a series of measurements of the analog output voltage. When the procedure is complete, what are the resistance values in the circuit? What is the range of the DAC?

12.17 The resistor networks shown in the DAC circuit of Figure 12.13 allow us to set our input reference current and output gain to values within a specified range. Using the values shown in Figure 12.13, fill in Table 12.7 for the cases when V_a is at minimum and maximum, and when the potentiometers are at their midpoint values. Assume the DAC input is set to 1111 1111. Show all calculations.

Table 12.7 DAC Output Range

	R_{14} (Ω)	R_F(Ω)	I_{ref} (mA)	I_o(mA)	V_a(V)
Minimum V_a					
Maximum V_a					
Pots at midpoint					

12.18 The waveform in Figure 12.47 is observed at the output of the DAC ramp generator of Figure 12.14. (Compare this to the proper waveform, found in Figure 12.15.) What is likely to be the problem with the circuit? Can it be easily fixed? How?

12.19 The waveform in Figure 12.48 is observed at the output of the DAC ramp generator in Figure 12.14. What is likely to be the problem with the circuit?

FIGURE 12.47
Problem 12.18
Waveform

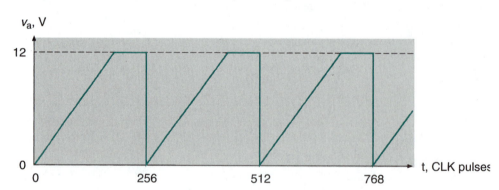

FIGURE 12.48
Problem 12.19
Waveform

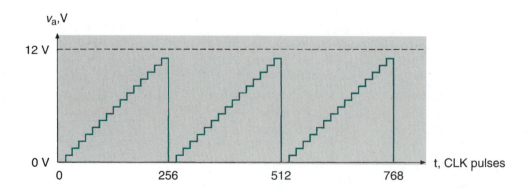

12.20 Refer to the bipolar DAC circuit in Figure 12.16. Describe how you would adjust the output for a range of -10 V to $(+10$ V $- 2$ LSB). Include values of variable components. Calculate the resolution of this circuit.

12.21 A 3-bit DAC has a reference voltage of 12 V and a transfer characteristic summarized in Table 12.8. Plot the data on a graph similar to those in Figures 12.18 through 12.20. From the data in Table 12.8, determine the offset error, gain error, and linearity error of the DAC, both in % of full scale and as a fraction of an LSB.

Table 12.8   DAC Transfer Characteristic for Problem 12.21

Digital Code	Analog Output (volts)
000	0.5
001	2.0
010	3.5
011	5.0
100	6.5
101	8.0
110	9.5
111	11.0

12.22 A 3-bit DAC has a reference voltage of 8 V and a transfer characteristic summarized in Table 12.9. Plot the data on a graph. From the data in Table 12.9, determine the offset error, gain error, linearity error, and differential nonlinearity of the DAC, both in % of full scale and as a fraction of an LSB.

Table 12.9   DAC Transfer Characteristic for Problem 12.22

Digital Code	Analog Output (volts)
000	0.000
001	1.036
010	2.071
011	3.107
100	4.143
101	5.179
110	6.214
111	7.250

12.23 A 3-bit DAC has a reference voltage of 4 V and a transfer characteristic summarized in Table 12.10. Plot the data on a graph. From the data in the Table 12.10, determine the offset error, gain error, and linearity error of the DAC, both in % of full scale and as a fraction of an LSB.

Table 12.10   DAC Transfer Characteristic for Problem 12.23

Digital Code	Analog Output (volts)
000	0.000
001	0.500
010	1.025
011	1.525
100	1.985
101	2.675
110	3.000
111	3.500

Section 12.3   Analog-to-Digital Conversion

12.24 How many comparators are needed to construct an 8-bit flash converter? Sketch the circuit of this converter. (It is only necessary to show a few of the comparators and indicate how many there are.)

12.25 Briefly explain the operation of a flash ADC. What is the purpose of the priority encoder? Explain how the latch can be used to synchronize the output to a particular sampling frequency.

12.26 Why do we choose a value of $R/2$ for the LSB resistor of a flash ADC?

12.27 An 8-bit successive approximation ADC has a reference voltage of $+16$ V. Describe the conversion sequence for the case where the analog input is 4.75 V. Summarize the steps in Table 12.11. (Refer to Example 12.11.)

12.28 What is displayed on the seven-segment display in Figure 12.49 when $v_{analog} = 5.25$ V? Assume that the reference voltage is 12 V and that the display can show hex digits.

12.29 Describe the operation of each part of the successive approximation ADC shown in Figure 12.49 when the analog input changes from 5.25 V to 8.0 V. What is the new number displayed on the seven-segment display?

Table 12.11   Table for Problem 16.23

Bit	New Digital Value	Analog Equivalent	$v_{analog} \geq v_{DAC}$?	Comparator Output	Accumulated Digital Value
Q_7					
Q_6					
Q_5					
Q_4					
Q_3					
Q_2					
Q_1					
Q_0					

FIGURE 12.49
Problem 12.28
Successive Approximation
ADC and Seven-Segment
Display

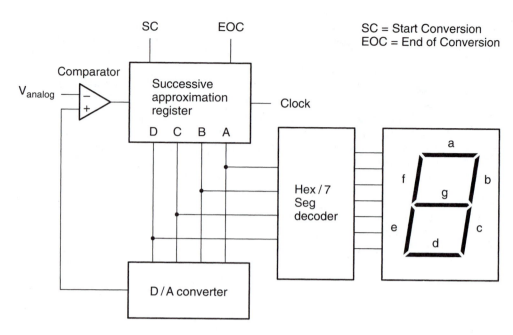

SC = Start Conversion
EOC = End of Conversion

12.30 a. An 8-bit successive approximation ADC has a reference voltage of 12 V. Calculate the resolution of this ADC.

   b. The analog input voltage to the ADC in part a is 8 V. Can this input voltage be represented exactly? What digital code represents the closest value to 8 V? What exact analog value does this represent? Calculate the percent error of this conversion.

12.31 What is the maximum quantization error of an ADC, relative to a fraction of 1 LSB?

12.32 An 8-bit dual slope analog-to-digital converter has a reference voltage of 16 V. The integrator component values are: R = 80 kΩ, C = 0.1 μF. The analog input voltage is 14 V.

   Calculate the slope of the integrator voltage during:

   a. the integrating phase, and

   b. the rezeroing phase.

   c. How much time elapses during the rezeroing phase? (Assume that (1) the integrating and rezeroing time are equal if the integrator output is at full scale, and (2) the reference voltage will rezero the integrator from full scale in exactly one counter cycle.)

   d. Sketch the integrator output waveform.

   e. What digital code is contained in the output latch after the conversion is complete?

12.33 Repeat Problem 12.32 if the analog input voltage is 3 V.

12.34 Repeat Problem 12.32 if the analog input voltage is 18 V.

12.35 make a sketch of a basic sample and hold circuit and briefly explain its operation.

12.36 Explain why a sample and hold circuit may be needed at the input of an analog-to-digital converter.

12.37 What is the highest-frequency component of an analog signal that can be accurately represented digitally if it is sampled at a rate of 100 kHz?

12.38 Calculate the minimum sampling frequency required to preserve all information when sampling a sine wave with a frequency of 130 kHz.

12.39 Suppose a sine wave with a period of 4.8 μs is sampled every 5.2 μs. What alias frequency will result? (Hint: see Figure 12.33.)

12.40 Calculate the corner frequency of an anti-aliasing filter for an ADC with a sampling frequency of 8 kHz. What type of filter (low-pass, high-pass, bandpass, etc.) is required?

Section 12.4   Data Acquisition

12.41 Refer to the data acquisition system in Figure 12.38. Write a VHDL file to implement the continuous-convert version of the ADC controller, as represented in the state diagram of Figure 12.42. Create a simulation in MAX+PLUS II to verify the operation of the controller.

12.42 Use the state machine controller from Problem 12.41 and an octal latch as components in a VHDL hierarchy that represents the ADC interface of Figure 12.38. Create a simulation in MAX+PLUS II to verify the operation of the design.

12.43 The data acquisition system in Figure 12.38 is designed with the controller from Problem 12.41. (The controller state diagram is shown in Figure 12.42.) Assume the controller and latch are interfaced with a different ADC that has a conversion time of 16 μs, which is equivalent to 64 clock cycles. Calculate the highest-frequency component that can be accurately converted with this system for a clock rate of 787 kHz.

12.44 Repeat Problem 12.43 for a 4-channel data acquisition system, assuming the same conversion rate for the ADC and the controller state diagram of Figure 12.45.

ANSWERS TO SECTION REVIEW PROBLEMS

Section 12.1

12.1 5 bits ($2^5 = 32$). Resolution = 24 mV/32 steps = 0.75 mV/step.

Section 12.2a

12.2 4-bit: $I_a = 0$ to $(15/16)(1 \text{ mA}) = 0$ to 0.9375 mA; $V_a = -I_a R_F = 0$ to -9.375 V 8-bit: $I_a = 0$ to $(255/256)(1 \text{ mA}) = 0$ to 0.9961 mA; $V_a = 0$ to -9.961 V

Section 12.2b

12.3 2.048 MΩ.

Section 12.2c

12.4 $V_a = (10 \text{ V}/2) + (10 \text{ V}/8) + (10 \text{ V}/256) = 6.29$ V

or $V_a = (161/256)10$ V = 6.29 V

Section 12.2d

12.5 The maximum switching speed is higher if we choose the lower range of output voltage.

Section 12.2e

12.6 The output 0 V requires its own code. This leaves 255, not 256, codes for the remaining output values. The maximum value of a positive-only output is 255/256 of the reference voltage. A bipolar DAC ranges from $-128/128$ to $+127/128$ of the reference voltage.

Section 12.3

12.7 a. -1.5 V/ms;   **b.** $+4$ V/ms;   **c.** 1.125 ms; **d.** 01100000.

Section 12.4

12.8 1.26 kHz

Memory Devices and Systems

CHAPTER OBJECTIVES

Upon successful completion of this chapter, you will be able to:

- Describe basic memory concepts of address and data.

- Understand how latches and flip-flops act as simple memory devices and sketch simple memory systems based on these devices.

- Distinguish between random access read/write memory (RAM) and read only memory (ROM).

- Describe the uses of tristate logic in data bussing.

- Sketch the circuits of static and dynamic RAM cells.

- Sketch a block diagram of a static or dynamic RAM chip.

- Describe various types of ROM cells and arrays: mask-programmed, UV erasable, and electrically erasable.

- Use various types of ROM in simple applications, such as digital function generation.

- Describe the basic configuration of flash memory.

- Describe the basic configuration and operation of two types of sequential memory: first-in-first-out (FIFO) and last-in-first-out (LIFO).

- Describe how dynamic RAM is configured into high capacity memory modules.

- Sketch a basic memory system, consisting of several memory devices, an address and a data bus, and address decoding circuitry.

- Represent the location of various memory device addresses on a system memory map.

- Recognize and eliminate conditions leading to bus contention in a memory system.

- Expand memory capacity by parallel bussing and CPLD-based decoding.

I n recent years, memory has become one of the most important topics in digital electronics. This is tied closely to the increasing prominence of cheap and readily available microprocessor chips. The simplest memory is a device we are already familiar with: the D flip-flop. This device stores a single bit of information as long as necessary. This simple concept is at the heart of all memory devices.

The other basic concept of memory is the organization of stored data. Bits are stored in locations specified by an "address," a unique number which tells a digital system how to find data that have been previously stored. (By analogy, think of your street address: a unique way to find you and anyone you live with.)

Some memory can be written to and read from in random order; this is called random access read/write memory (RAM). Other memory can be read only: read only memory (ROM). Yet another type of memory, sequential memory, can be read or written only in a specific sequence. There are several variations on all these basic classes.

Memory devices are usually part of a larger system, including a microprocessor, peripheral devices, and a system of tristate busses. If dynamic RAM is used in such a system, it is often in a memory module of some type. The capacity of a single memory chip is usually less than the memory capacity of the microprocessor system in which it is used. In order to use the full system capacity, it is necessary to use a method of memory address decoding to select a particular RAM device for a specified portion of system memory.

13.1   Basic Memory Concepts

KEY TERMS

Memory   A device for storing digital data in such a way that they can be recalled for later use in a digital system.

Data   Binary digits (0s and 1s) that contain some kind of information. The digital contents of a memory device.

Address   A number, represented by the binary states of a group of inputs or outputs, uniquely defining the location of data stored in a memory device.

Write   Store data in a memory device.

Read   Retrieve data from a memory device.

Byte   A group of 8 bits.

Nibble   Half a byte; 4 bits.

Address and Data

A **memory** is a digital device or circuit that can store one or more bits of **data.** The simplest memory device, a D-type latch, shown in Figure 13.1, can store 1 bit. A 0 or 1 is stored in the latch and remains there until changed.

A simple extension of the single D-type latch is an array of latches, shown in Figure 13.2, that can store 8 bits (1 **byte**) of data. Figure 13.3 shows this octal latch used as a component in a MAX+PLUS II graphic file and configured as an 8-bit memory.

When the *WRITEn* line goes LOW, then HIGH, data at the *DATA_IN* are stored in the eight latches. Data are available at the *DATA_OUT* pins when *READ* is HIGH. Note that although the *READ* and *WRITEn* inputs are separate in this design, their functions would often be implemented as opposite logic levels of the same pin.

Figure 13.4 shows a simulation of the 8-bit memory. The LOW pulses on *WRITEn* write the data, shown as two hexadecimal digits on the *DATA_IN* line, into the latches. To read the values stored in the eight latches, we set *READ* HIGH. In between read states, all *DATA_OUT* lines are in the high-impedance state, indicated by the notation ZZ.

octal_latch.gdf
1x8mem.gdf

1x8mem.scf

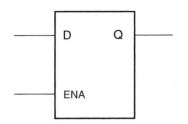

FIGURE 13.1
D-Type Latch

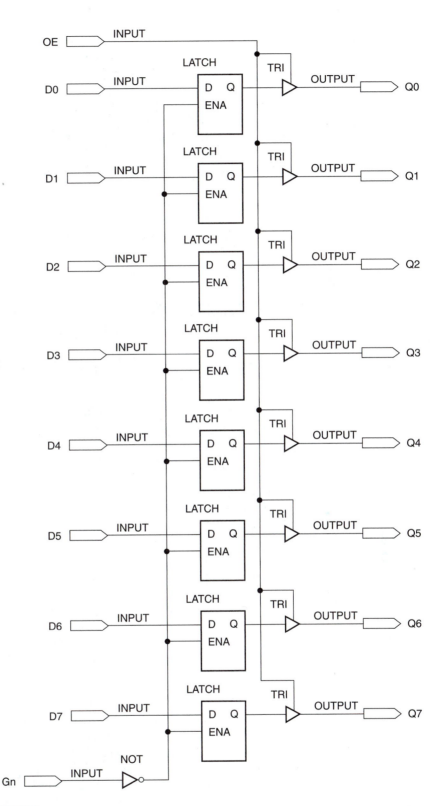

FIGURE13.2
Octal Latch

FIGURE13.3
Octal Latch as 8-bit Memory

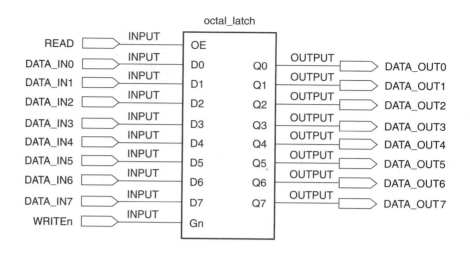

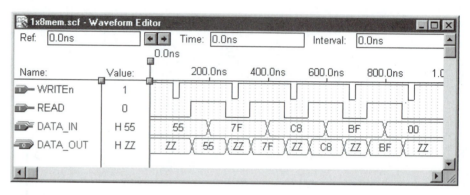

FIGURE 13.4
Simulation of 8-bit Memory

**4x8reg.gdf
ltch8lpm.vhd
dcdr2to4.vhd
oct4tol.vhd**

4x8reg.scf

Figure 13.5 shows an expanded version of the octal latch memory circuit. Four octal latches are configured to make a 4 × 8-bit memory that can store and recall four separate 8-bit words. The octal latches are based on 8-bit latches instantiated in VHDL from the Altera Library of Parameterized Modules (LPM). The remaining components of Figure 13.5 are behaviorally-designed VHDL components.

The 8-bit input data are applied to the inputs of all four octal latches simultaneously. Data are written to a particular latch when a 2-bit **address** and a LOW on *WRITEn* cause an output of a 2-line-to-4-line decoder to enable the selected latch. For example, when *ADDR[1..0]* = 01 AND *WRITEn* = 0, decoder output *Y1* goes HIGH, activating the *ENABLE* input on latch 1. The values at *DATA_IN[7..0]* are transferred to latch 1 and stored there when *WRITEn* goes HIGH.

The latch outputs are applied to the data inputs of an octal 4-to-1 multiplexer. Recall that this circuit will direct one of four 8-bit inputs to an 8-bit output. The selected set of inputs correspond to the binary value at the MUX select inputs, which is the same as the address applied to the decoder in the write phase. The MUX output is directed to the *DATA_OUT* lines by an octal tristate bus driver, which is enabled by the *READ* line. To read the contents of latch 1, we set the address to 01, as before, and make the *READ* line HIGH. If *READ* is LOW, the *DATA_OUT* lines are in the high-impedance state.

Figure 13.6 shows a simulation of the 4 × 8-bit memory. The address inputs change in a continuous binary sequence. For each address, a write pulse loads 8-bit data into the selected latch. After all four latches have been loaded, the latches are read in a rotating sequence. To read any new data from the memory, we would first have to write the new data into one or more of the latch locations.

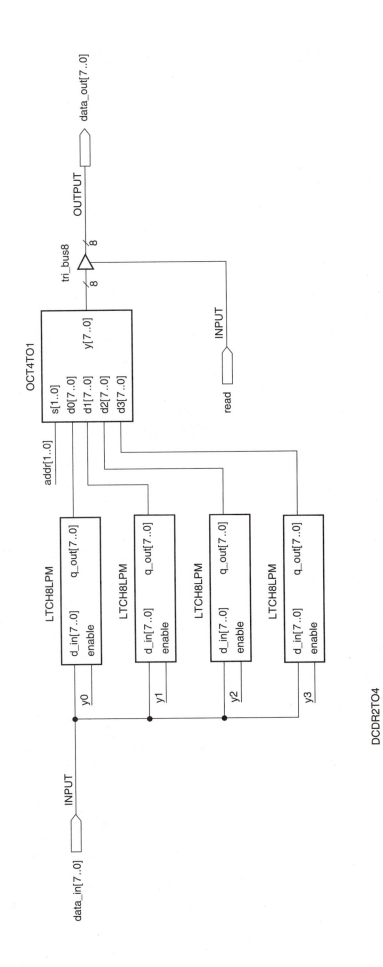

FIGURE 13.5

4 × 8-bit Memory from Octal Latches

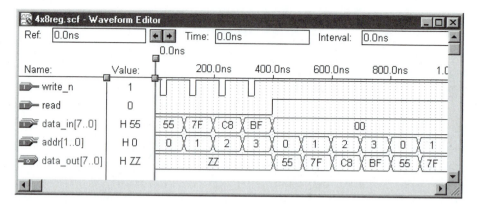

FIGURE 13.6
Simulation of 4 × 8 Memory

RAM and ROM

Random access memory (RAM)   A type of memory device where data can be accessed in any order, that is, randomly. The term usually refers to random access read/write memory.

Read only memory (ROM)   A type of memory where data are permanently stored and can only be read, not written.

The memory circuit in Figure 13.5 is one type of **random access memory,** or RAM. Data can be stored in or retrieved from any address at any time. The data can be accessed randomly, without the need to follow a sequence of addresses, as would be necessary in a sequential storage device such as magnetic tape.

RAM has come to mean random access read/write memory, memory that can have its data changed by a write operation, as well as have its data read. The data in another type of memory, called **read only memory,** or ROM, can also be accessed randomly, although it cannot be changed, or at least not changed as easily as RAM; there is no write function; hence the name "read only." Even though both types of memory are random access, we generally do not include ROM in this category.

Memory Capacity

b   Bit.

B   Byte.

K   1024 (= 2^{10}). Analogous to the metric prefix "k" (kilo-).

M   1,048,576 (= 2^{20}). Analogous to the metric prefix "M" (mega-).

The capacity of a memory device is specified by the address and data sizes. The circuit shown in Figure 13.5 has a capacity of 4 × 8 bits ("four-by-eight"). This tells us that the memory can store 32 bits, organized in groups of 8 bits at 4 different locations.

For large memories, with capacities of thousands or millions of bits, we use the shorthand designations **K** or **M** as prefixes for large binary numbers. The prefix K is analogous to, but not the same as, the metric prefix k (kilo). The metric kilo (lowercase k) indicates a multiplier of $10^3 = 1000$; the binary prefix K (uppercase) indicates a multiplier of $2^{10} = 1024$. Thus, one kilobit (Kb) is 1024 bits.

Similarly, the binary prefix M is analogous to the metric prefix M (mega). Both, unfortunately, are represented by uppercase M. The metric prefix represents a multiplier of $10^6 = 1,000,000$; the binary prefix M represents a value of $2^{20} = 1,048,576$. One megabit (Mb) is 1,048,576 bits. The next extension of this system is the multiplier G ($= 2^{30}$), which is analogous to the metric prefix G (giga; 10^9).

There is a move afoot to untangle all the inconsistencies in this notation and develop separate units for binary and metric applications, but to date, such new notation is not very widely used.

■■ EXAMPLE 13.1

A small microcontroller system (i.e., a stand-alone microcomputer system designed for a particular control application) has a memory with a capacity of 64 Kb, organized as 8K × 8. What is the total memory capacity of the system in bits? What is the memory capacity in bytes?

Solution   The total number of bits in the system memory is:

$$8K \times 8 = 8 \times 8 \times 1K = 64 \text{ Kb} = 64 \times 1024 \text{ bits} = 65,536 \text{ bits}$$

The number of bytes in system memory is:

$$\frac{64 \text{ Kb}}{8b/B} = 8 \text{ KB}$$

Usually, the range of numbers spanning 1K is expressed as the 1024 numbers from 0_{10} to 1023_{10} (0000000000_2 to 1111111111_2). This is the full range of numbers that can be expressed by 10 bits. In hexadecimal, the range of numbers spanning 1K is from 000H to 3FFH. The range of numbers in 1M is given as the full hexadecimal range of 20-bit numbers: 00000H to FFFFFH.

■■

The range of numbers spanning 8K can be written in 13 bits ($8 \times 1K = 2^3 \times 2^{10} = 2^{13}$). The addresses in an 8K × 8 memory range from 0000000000000 to 1111111111111, or 0000 to 1FFF in hexadecimal. Thus, a memory device that is organized as 8K × 8 has 13 address lines and 8 data lines.

Figure 13.7 shows the address and data lines of an 8K × 8 memory and a map of its contents. The addresses progress in binary order, but the contents of any location are the

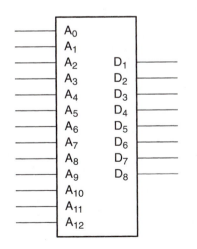

a. Address and data lines

D_8							D_1
1	0	1	1	0	1	0	1
0	0	0	1	1	0	1	1
1	1	0	1	0	0	1	1
0	0	0	0	0	1	1	1
0	1	1	1	0	1	1	1
1	0	0	0	1	0	1	0
0	1	0	1	1	1	1	1

1	0	1	0	1	0	1	0
0	0	0	1	1	1	1	1
1	1	0	0	1	0	1	1

	Addresses	
Binary	Hexadecimal
0 0000 0000 0000	0000
0 0000 0000 0001	0001
0 0000 0000 0010	0002
0 0000 0000 0011	0003
0 0000 0000 0100	0004
0 0000 0000 0101	0005
0 0000 0000 0110	0006
⋮	⋮
1 1111 1111 1101	1FFD
1 1111 1111 1110	1FFE
1 1111 1111 1111	1FFF

b. Contents (data) and location (address)

FIGURE 13.7
Address and Data in an 8K × 8 Memory

last data stored there. Since there is no way to predict what those data are, they are essentially random. For example, in Figure 13.7, the byte at address 0000000000100_2 (0004H) is 01110111_2 (77H). (One can readily see the advantage of using hexadecimal notation.)

▮▮ EXAMPLE 13.2

How many address lines are needed to access all addressable locations in a memory that is organized as 64K $\times$ 4? How many data lines are required?

Solution   Address lines: $2^n = 64K$

$$64K = 64 \times 1K = 2^6 \times 2^{10} = 2^{16}$$
$$n = 16 \text{ address lines}$$

Data lines: There are 4 data bits for each addressable location. Thus, the memory requires 4 data lines.

▮▮

Control Signals

Two memory devices are shown in Figure 13.8. The device in Figure 13.8a is a 1K $\times$ 4 random access read/write memory (RAM). Figure 13.8b shows 8K $\times$ 8 erasable programmable read only memory (EPROM). The address lines are designated by A and the data lines by DQ. The dual notation DQ indicates that these lines are used for both input (D) and output (Q) data, using the conventional designations of D-type latches. The input and output data are prevented from interfering with one another by a pair of opposite-direction tristate buffers on each input/output pin. One buffer goes to a memory cell input; the other comes from the memory cell output. The tristate outputs on the devices in Figure 13.8 allow the outputs to be electrically isolated from a system data bus that would connect several such devices to a microprocessor.

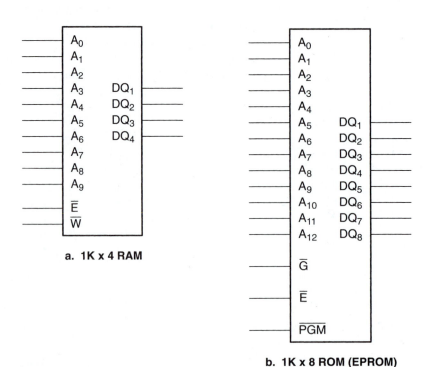

a. 1K x 4 RAM

b. 1K x 8 ROM (EPROM)

FIGURE 13.8
Address, Data, and Control Signals

In addition to the address and data lines, most memory devices, including those in Figure 13.8, have one or more of the following control signal inputs. (Different manufacturers use different notation, so several alternate designations for each function are listed.)

$\overline{E}$ (or $\overline{CE}$ or $\overline{CS}$). $\overline{\text{Enable}}$ (or $\overline{\text{Chip Enable}}$ or $\overline{\text{Chip Select}}$). The memory is enabled when this line is pulled LOW. If this line is HIGH, the memory cannot be written to or read from.

$\overline{W}$ (or $\overline{WE}$ or $R/\overline{W}$). $\overline{\text{Write}}$ (or $\overline{\text{Write Enable}}$ or Read/$\overline{\text{Write}}$). This input is used to select the read or write function when data input and output are on the same lines. When HIGH, this line selects the read (output) function if the chip is selected. When LOW, the write (input) function is selected.

$\overline{G}$ (or $\overline{OE}$). $\overline{\text{Gate}}$ (or $\overline{\text{Output Enable}}$). Some memory chips have a separate control to enable their tristate output buffers. When this line is LOW, the output buffers are enabled and the memory can be read. If this line is HIGH, the output buffers are in the high-impedance state. The chip select performs this function in devices without output enable pins.

The electrical functions of these control signals are illustrated in Figure 13.9.

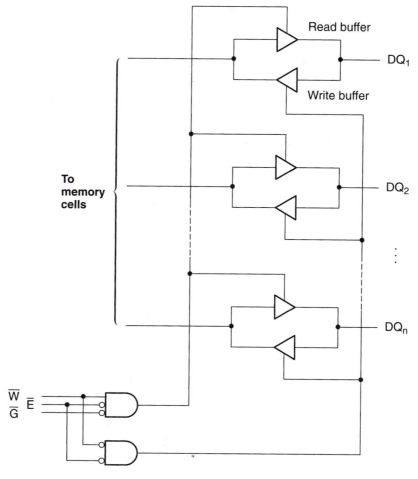

FIGURE 13.9
Memory Control Signals

13.2   Random Access Read/Write Memory (RAM)

Random access read/write memory (RAM) is used for temporary storage of large blocks of data. An important characteristic of RAM is that it is **volatile.** It can retain its stored data only as long as power is applied to the memory. When power is lost, so are the data. There are two main RAM configurations: static (SRAM) and dynamic (DRAM).

   Static RAM (SRAM) consists of arrays of memory cells that are essentially flip-flops. Data can be stored in a static RAM cell and left there indefinitely, as long as power is available to the RAM.

   A **dynamic RAM cell** stores a bit as the charged or discharged state of a small capacitor. Since the capacitor can hold its charge for only a few milliseconds, the charge must be restored ("refreshed") regularly. This makes a dynamic RAM (DRAM) system more complicated than SRAM, as it introduces a requirement for memory refresh circuitry.

   DRAMs have the advantage of large memory capacity over SRAMs. At the time of this writing, the largest SRAMs have a capacity of about 4 Mb, whereas the largest DRAMs have a capacity of 256 Mb. DRAM modules, that is, groups of DRAM chips on a small circuit board, have capacities of up to 1 GB. These figures are constantly increasing and are never up to date for very long. (The most famous estimate of the growth rate of semiconductor memory capacity, Moore's law, estimates that it doubles every 18 months. My casual observation is that this is accurate to within an order of magnitude.)

Static RAM Cells

The typical static RAM cell consists of at least two transistors that are cross-coupled in a flip-flop arrangement. Other parts of the cell include pull-up circuitry that can be active (transistor switches) or passive (resistors) and some decoding/switching logic. Figure 13.10 shows an SRAM cell in three technologies: bipolar, NMOS, and CMOS.

   Each of these cells can store 1 bit of data, a 0 or a 1, as the state of one of the transistors in the cell. The data are available in true or complement form, as the *BIT* and $\overline{BIT}$ outputs of the flip-flop.

   All types of SRAM cells operate in more or less the same way. We will analyze the operation of the NMOS cell (Figure 13.10b) and then compare it to the other types.

   Transistors Q_1 and Q_2 are permanently biased ON, making them into pull-up resistors. Channel width and length are chosen to give a resistance of about 1 kΩ. These NMOS load transistors are considered passive pull-ups, as they do not switch on and off.

   A bit is stored as V_{DS3}, the drain voltage of Q_3 with respect to its source. If this voltage is HIGH, the gate of Q_4 is HIGH with respect to its source and Q_4 is biased ON. This completes a conduction path from the drain of Q_4 to its source, making V_{DS4} logic LOW. This LOW is fed back to the gate of Q_3, turning it OFF. There is no conduction path between the drain and source of Q_3, so $V_{DS3} = V_{DD}$ or logic HIGH. The cell is storing a 1.

   This bit can be read by making the *ROW SELECT* line HIGH. This turns Q_5 and Q_6 ON, which puts the data onto the *BIT* and $\overline{BIT}$ lines where it can be read by other circuitry inside the RAM chip.

   To change the cell contents to a 0, we make the *BIT* line LOW and the *ROW SELECT* line HIGH. The *ROW SELECT* line gives access to the cell by turning on Q_5 and Q_6, com-

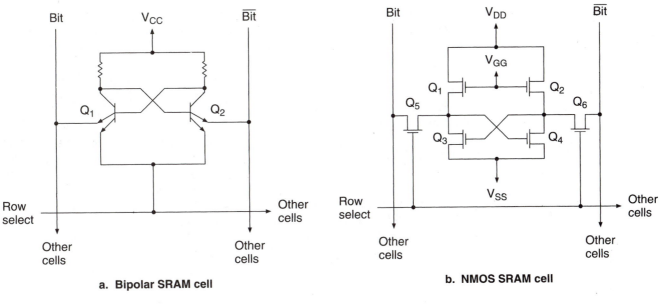

a. **Bipolar SRAM cell**

b. **NMOS SRAM cell**

c. **CMOS SRAM cell**

FIGURE 13.10
SRAM Cells

pleting the conduction path between the *BIT* lines and the flip-flop inputs. The LOW on the *BIT* line pulls the gate of Q_4 LOW, turning it OFF. This breaks the conduction path from Q_4 drain to source and makes $V_{DS4} = V_{DD}$, a logic HIGH. This HIGH is applied to the gate of Q_3, turning it ON. A conduction path is established between Q_3 drain and source, pulling the drain of Q_3 LOW. The cell now stores a logic 0.

NOTE

The contents of an SRAM cell must be changed by introducing a LOW on the *BIT* or the $\overline{BIT}$ line. The data cannot be changed by pulling an input HIGH without pulling the opposite input LOW. If a MOSFET gate is at the LOW state, a HIGH applied to that gate will be pulled down by the LOW level already existing there and will not cause the cell to change state.

The CMOS cell (Figure 13.10c) functions in the same way, except for the actions of Q_1 and Q_2. Q_1 and Q_3 are a complementary pair, as are transistors Q_2 and Q_4. For each of these pairs, when the p-channel transistor is ON, the n-channel is OFF, and vice versa. This arrangement is more energy efficient than the NMOS cell, since there is not the constant current drain associated with the load transistors. Power is consumed primarily during switching between states.

The main design goal of new memory technology is to increase speed and capacity while reducing power consumption and chip area. The NMOS cell has the advantage of being constructed from only one type of component. This makes it possible to manufacture more cells in the same chip area than can be done in either the CMOS or bipolar technologies. NMOS chips, however, are slower than bipolar. New advances in high-speed CMOS technologies have made possible CMOS memories that are as dense or denser than NMOS and faster. Because of this, NMOS will probably decline in importance over time.

Bipolar SRAMs can be either TTL, as shown in Figure 13.10a, or ECL, which is not shown. Of the two bipolar technologies, ECL is the faster. Historically, all bipolar SRAMs have had the advantage of speed over NMOS and CMOS chips. New CMOS devices however, have exceeded the speeds of TTL.

The bipolar SRAM cell is the least suitable for high-density memory. Both bipolar transistors and resistors are large components compared to a MOSFET. Thus, the bipolar cell is inherently larger than the CMOS or NMOS cell. Bipolar memories historically have been used when a small amount of high-speed memory is required.

The operation of the bipolar SRAM cell is similar to that of the MOSFET cells. In the quiescent state, the *ROW SELECT* line is LOW. In either the Read or the Write mode, the *ROW SELECT* line is HIGH. To change the data in the cell, pull one of the emitters LOW. When the emitter of Q_1 goes LOW, the cell contents become 0. When the emitter of Q_2 is pulled LOW, the cell contents are 1.

Static RAM Cell Arrays

KEY TERMS

Word-organized   A memory is word-organized if one address accesses one word of data.

Word   Data accessed at one addressable location.

Word length   Number of bits in a word.

Static RAM cell arrays are arranged in a square or rectangular format, accessible by groups in rows and columns. Each column corresponds to a complementary pair of *BIT* lines and each row to a *ROW SELECT* line, as shown in Figure 13.11.

The column lines have MOSFETs configured as pull-up resistors at one end and a circuit called a sense amplifier at the other. The sense amp is a large RAM cell that amplifies the charge of an active storage cell on the same *BIT* line. Having a larger RAM cell as a sense amp allows the storage cells to be smaller, since each individual cell need not carry the charge required for a logic level output.

Figure 13.12 shows the block diagram of a 4 megabit (Mb) SRAM array, including blocks for address decoding and output circuitry. The RAM cells are arrayed in a pattern of 512 rows and 8192 columns for efficient packaging. When a particular address is applied to address lines $A_{18} \ldots A_0$, the row and column decoders select an SRAM cell in the memory array for a read or write by activating the associated sense amps for the column and the row select line for the cell.

The columns are further subdivided into groups of eight, so that one column address selects eight bits (one byte) for a read or write operation. Thus, there are 512 separate row addresses (9 bits) and 1024 separate column addresses (10 bits) for every unique group of 8 data bits, requiring a total of 19 address lines and 8 data lines. The capacity of the SRAM can be written as $512 \times 1024 \times 8$.

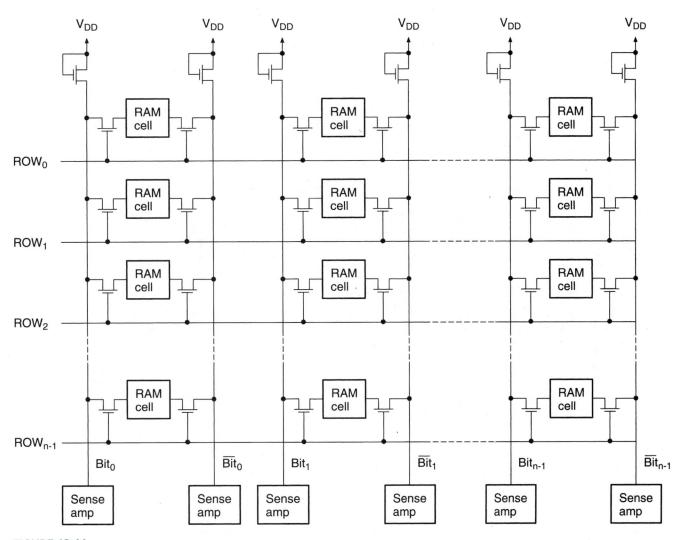

FIGURE 13.11
SRAM Cell Array

FIGURE 13.12
Block Diagram of a 4Mb (512 KB) SRAM

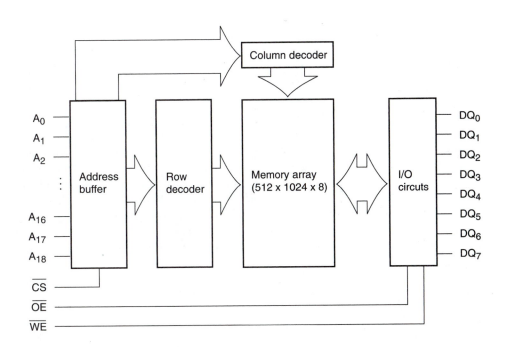

Since one address reads or writes 8 cells, we say that the SRAM in Figure 13.12 is **word-organized** and that the **word length** of the SRAM is 8 bits. Other popular word lengths for various memory arrays are 4, 16, 32, and 64 bits.

⦀ SECTION 13.2A REVIEW PROBLEM

13.1 If an SRAM array is organized as $512 \times 512 \times 16$, how many address and data lines are required? How does the bit capacity of this SRAM compare to that of Figure 13.12?

Dynamic RAM Cells

> ### KEY TERM
>
> **Refresh cycle** The process that periodically recharges the storage capacitors in a dynamic RAM.

A dynamic RAM (DRAM) cell consists of a capacitor and a pass transistor, as shown in Figure 13.13. A bit is stored in the cell as the charged or discharged state of the capacitor. The bit location is read from or written to by activating the cell MOSFET via the Word Select line, thus connecting the capacitor to the *BIT* line.

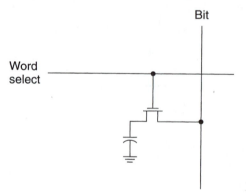

FIGURE 13.13
Dynamic RAM Cell

The major disadvantage of dynamic RAM is that the capacitor will eventually discharge by internal leakage current and must be recharged periodically to maintain integrity of the stored data. The recharging of the DRAM cell capacitors, known as refreshing the memory, must be done every 8 to 64 ms, depending on the device.

The **refresh cycle** adds an extra level of complication to the DRAM hardware and also to the timing of the read and write cycles, since the memory might have to be refreshed between read and write tasks. DRAM timing cycles are much more complicated than the equivalent SRAM cycles.

This inconvenience is offset by the high bit densities of DRAM, which are possible due to the simplicity of the DRAM cell. Up to 256 megabits of data can be stored on a single chip.

DRAM Cell Arrays

Dynamic RAM is sometimes **bit-organized** rather than word-organized. That is, one address will access one bit rather than one word of data. A bit-organized DRAM with a large capacity requires more address lines than a static RAM (e.g., 4 Mb $\times$ 1 DRAM requires 22 address lines (2^{22} = 4,194,304 = 4M) and 1 data line to access all cells).

In order to save pins on the IC package, a system of **address multiplexing** is used to specify the address of each cell. Each cell has a row address and a column address, which use the same input pins. Two negative-edge signals called **row address strobe** *(RAS)* and **column address strobe** *(CAS)* latch the row and column addresses into the DRAM's decoding circuitry. Figure 13.14 shows a simplified block diagram of the row and column addressing circuitry of a 1 Mb $\times$ 1 dynamic RAM.

Figure 13.15 shows the relative timing of the address inputs of a dynamic RAM. The first part of the address is applied to the address pins and latched into the row address buffers

FIGURE 13.14

Row and Column Decoding in a 1M $\times$ 1 Dynamic RAM

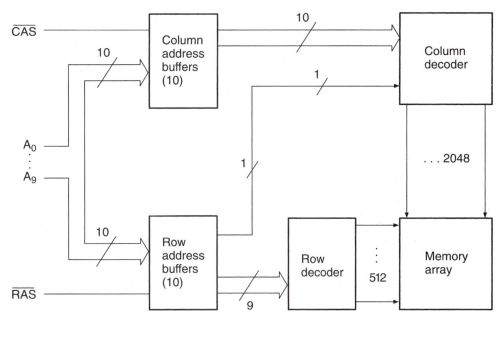

FIGURE 13.15

DRAM Address Latch Signals

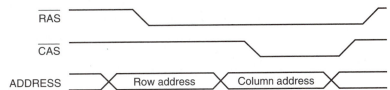

when $\overline{RAS}$ goes LOW. The second part of the address is then applied to the address pins and latched into the column address buffers by the $\overline{CAS}$ signal. This allows a 20-bit address to be implemented with 12 pins: 10 address and 2 control lines. Adding another address line effectively adds 2 bits to the address, allowing access to 4 times the number of cells.

The memory cell array in Figure 13.14 is rectangular, not square. One of the Row Address lines is connected internally to the Column Address decoder, resulting in a 512-row-by-2048-column memory array.

One advantage to the rectangular format shown is that it cuts the memory refresh time in half, since all the cells are refreshed by accessing the rows in sequence. Fewer rows means a faster refresh cycle. All cells in a row are also refreshed by normal read and write operations.

▮ SECTION 13.2B REVIEW PROBLEMS

13.2 How many address and data lines are required for the following sizes of dynamic RAM, assuming that each memory cell array is organized in a square format, with common Row and Column Address pins?

   a. $1M \times 1$

   b. $1M \times 4$

   c. $4M \times 1$

13.3 Read Only Memory (ROM)

> **KEY TERMS**
>
> **Hardware**   The electronic circuit of a digital or computer system.
>
> **Software**   Programming instructions required to make hardware perform specified tasks.
>
> **Firmware**   Software instructions permanently stored in ROM.

The main advantage of read only memory (ROM) over random access read/write memory (RAM) is that ROM is nonvolatile. It will retain data even when electrical power is lost to the ROM chip. The disadvantage of this is that stored data are difficult or impossible to change.

ROM is used for storing data required for tasks that never or rarely change, such as **software** instructions for a bootstrap loader in a personal computer or microcontroller. (The bootstrap loader—a term derived from the whimsical idea of pulling oneself up by one's bootstraps, that is, starting from nothing—is the software that gives the personal computer its minimum startup information. Generally, it contains the instructions needed to read a magnetic disk containing further operating instructions. This task is always the same for any given machine and is needed every time the machine is turned on, thus making it the ideal candidate for ROM storage.)

Software instructions stored in ROM are called **firmware.**

Mask-Programmed ROM

> **KEY TERM**
>
> **Mask-programmed ROM**   A type of read only memory (ROM) where the stored data are permanently encoded into the memory device during the manufacturing process.

The most permanent form of read only memory is the **mask-programmed ROM,** where the stored data are manufactured into the memory chip. Due to the inflexibility of this type of ROM and the relatively high cost of development, it is used only for well-developed high-volume applications. However, even though development cost of a mask-programmed ROM is high, volume production is cheaper than for some other types of ROM.

Examples of applications suitable to mask-programmed ROM include:

- Bootstrap loaders and BIOS (basic input/output system) for PCs.
- Character generators (decoders that convert ASCII codes into alphanumeric characters on a CRT display)
- Function lookup tables (tables corresponding to binary values of trigonometric, exponential, or other functions)
- Special software instructions that must be permanently stored and never changed (firmware)

Figure 13.16 shows a ROM based on a matrix of MOSFETs. Each cell is manufactured with a MOSFET and its gate and source connections. LOWs are programmed by

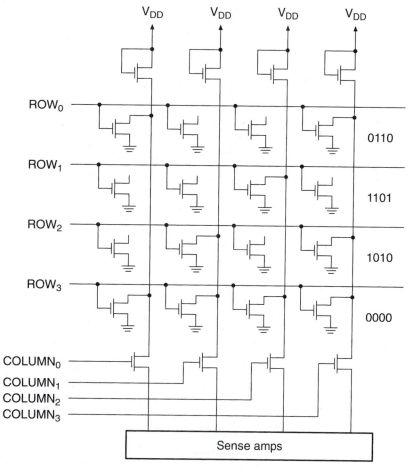

FIGURE 13.16
Mask-Programmed ROM

making a connection between the drain of the cell's MOSFET and the corresponding Bit line. When the appropriate Row Select goes HIGH, the MOSFET turns ON, providing a path to ground from the selected Bit line. Cells programmed HIGH have no connection between the MOSFET drain and the Bit line, which thus cannot be pulled LOW when the cell is selected.

These connections can be made by a custom overlay of connections (a mask) on top of the standard-cell layer. The standard-cell-plus-custom-overlay format is cheaper to manufacture than custom cells for each bit, even if many of the MOSFETs are never used.

EPROM

KEY TERMS

EPROM Erasable programmable read only memory. A type of ROM that can be programmed ("burned") by the user and erased later, if necessary, by exposing the chip to ultraviolet radiation.

FAMOS FET Floating-gate avalanche MOSFET. A MOSFET with a second, "floating" gate in which charge can be trapped to change the MOSFET's gate-source threshold voltage.

Mask-programmed ROM is useful because of its nonvolatility, but it is hard to program and impossible to erase. **Erasable programmable read only memory** (EPROM) combines the nonvolatility of ROM with the ability to change the internal data if necessary.

This erasability is particularly useful in the development of a ROM-based system. Anyone who has built a complex circuit or written a computer program knows that there is no such thing as getting it right the first time. Modifications can be made easily and cheaply to data stored in an EPROM. Later, when the design is complete, a mask ROM version can be prepared for mass production. Alternatively, if the design will be produced in small numbers, the ROM data can be stored in EPROMs, saving the cost of preparing a mask-programmed ROM.

The basis of the EPROM memory cell is the **FAMOS FET,** whose circuit symbol is shown in Figure 13.17. FAMOS stands for floating-gate avalanche metal-oxide-semiconductor. ("Avalanche" refers to electron behavior in a semiconductor under certain bias conditions.) This is a MOSFET with a second, or floating, gate that is insulated from the first by a thin oxide layer.

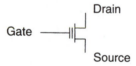

FIGURE 13.17
FAMOS FET

The floating gate has no electrical contact with either the first gate or the source and drain terminals. As is the case in a standard MOSFET, conduction between drain and source terminals is effected by the voltage of the gate terminal with respect to the source. If this voltage is above a certain threshold level, the transistor will turn ON, allowing current to flow between drain and source.

In the unprogrammed state the FAMOS transistor's threshold voltage is low enough for the transistor to be turned ON by a 5-V read signal on the Row Select line. During the programming operation, a relatively high voltage pulse (about 12 V to 25 V, depending on the device) on the Row Select line drives high-energy electrons into the floating gate and traps them there. This raises the threshold voltage of the programmed cell to a level where the cell won't turn ON when selected by a 5-V read.

The EPROM cells are configured so that an unprogrammed location contains a logic HIGH and the programming signal forces it LOW.

To erase an EPROM, the die (i.e., the silicon chip itself) must be exposed for about 20 to 45 minutes to high-intensity ultraviolet light of a specified wavelength (2537 angstroms) at a distance of 2.5 cm (1 inch). The high-energy photons that make up the UV radiation release the electrons trapped in the floating gate and restore the cell threshold voltages to their unprogrammed levels.

EPROMS are manufactured with a quartz window over the die to allow the UV radiation in. Since both sunlight and fluorescent light contain UV light of the right wavelength to erase the EPROM over time (several days to several years, depending on the intensity of the source), the quartz window should be covered by an opaque label after the EPROM has been programmed.

EPROM Application: Digital Function Generator

An EPROM can be used as the central component of a digital function generator. Other components in the system include a clock generator, a counter, a digital-to-analog converter, and an output op amp buffer. The portion of the circuit including the last three of these components is shown in Figure 13.18.

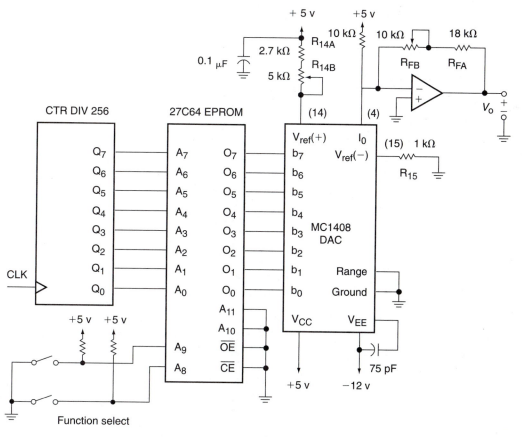

FIGURE 13.18
Digital Function Generator

The generator can produce the usual analog waveforms—sine, square, triangle, sawtooth—and any other waveforms that you wish to store in the EPROM. A single cycle of each waveform is stored as 256 consecutive 8-bit numbers. For example, the data for one cycle of the sine waveform are stored at addresses 0000H to FFFFH, as shown in hex form in Table 13.1. (FF is maximum positive, 80 is zero, and 00 is maximum negative.) The square wave data are stored at addresses 0100 to 01FF, also shown in Table 13.1. The data for other functions, stored in subsequent 256-byte blocks, are not shown. A full list of the function data and an ANSI C program to generate an EPROM record file (Intel format) are included in Appendix E.

Table 13.1 EPROM Sine and Square Wave Data

SINE

Base Address	0	1	2	3	4	5	6	7	8	9	A	B	C	D	E	F
0000	80	83	86	89	8C	8F	92	95	98	9C	9F	A2	A5	A8	AB	AE
0010	B0	B3	B6	B9	BC	BF	C1	C4	C7	C9	CC	CE	D1	D3	D5	D8
0020	DA	DC	DE	E0	E2	E4	E6	E8	EA	EC	ED	EF	F0	F2	F3	F5
0030	F6	F7	F8	F9	FA	FB	FC	FC	FD	FE	FE	FF	FF	FF	FF	FF
0040	FF	FF	FF	FF	FF	FF	FE	FE	FD	FC	FC	FB	FA	F9	F8	F7
0050	F6	F5	F3	F2	F0	EF	ED	EC	EA	E8	E6	E4	E2	E0	DE	DC
0060	DA	D8	D5	D3	D1	CE	CC	C9	C7	C4	C1	BF	BC	B9	B6	B3
0070	B0	AE	AB	A8	A5	A2	9F	9C	98	95	92	8F	8C	89	86	83
0080	7F	7C	79	76	73	70	6D	6A	67	63	60	5D	5A	57	54	51
0090	4F	4C	49	46	43	40	3E	3B	38	36	33	31	2E	2C	2A	27
00A0	25	23	21	1F	1D	1B	19	17	15	13	12	10	0F	0D	0C	0A
00B0	09	08	07	06	05	04	03	03	02	01	01	00	00	00	00	00
00C0	00	00	00	00	00	00	01	01	02	03	03	04	05	06	07	08
00D0	09	0A	0C	0D	0F	10	12	13	15	17	19	1B	1D	1F	21	23
00E0	25	27	2A	2C	2E	31	33	36	38	3B	3E	40	43	46	49	4C
00F0	4F	51	54	57	5A	5D	60	63	67	6A	6D	70	73	76	79	7C

SQUARE

Base Address	0	1	2	3	4	5	6	7	8	9	A	B	C	D	E	F
0100	FF	FF	FF	FF	FF	FF	FF	FF	FF	FF	FF	FF	FF	FF	FF	FF
0110	FF	FF	FF	FF	FF	FF	FF	FF	FF	FF	FF	FF	FF	FF	FF	FF
0120	FF	FF	FF	FF	FF	FF	FF	FF	FF	FF	FF	FF	FF	FF	FF	FF
0130	FF	FF	FF	FF	FF	FF	FF	FF	FF	FF	FF	FF	FF	FF	FF	FF
0140	FF	FF	FF	FF	FF	FF	FF	FF	FF	FF	FF	FF	FF	FF	FF	FF
0150	FF	FF	FF	FF	FF	FF	FF	FF	FF	FF	FF	FF	FF	FF	FF	FF
0160	FF	FF	FF	FF	FF	FF	FF	FF	FF	FF	FF	FF	FF	FF	FF	FF
0170	FF	FF	FF	FF	FF	FF	FF	FF	FF	FF	FF	FF	FF	FF	FF	FF
0180	00	00	00	00	00	00	00	00	00	00	00	00	00	00	00	00
0190	00	00	00	00	00	00	00	00	00	00	00	00	00	00	00	00
01A0	00	00	00	00	00	00	00	00	00	00	00	00	00	00	00	00
01B0	00	00	00	00	00	00	00	00	00	00	00	00	00	00	00	00
01C0	00	00	00	00	00	00	00	00	00	00	00	00	00	00	00	00
01D0	00	00	00	00	00	00	00	00	00	00	00	00	00	00	00	00
01E0	00	00	00	00	00	00	00	00	00	00	00	00	00	00	00	00
01F0	00	00	00	00	00	00	00	00	00	00	00	00	00	00	00	00

The counter and EPROM can also be implemented in a CPLD. Alternatively, a VHDL-designed state machine can replace the counter and EPROM, except for the sine function. These configurations are designed and built as exercises in the lab manual that accompanies this book.

The most significant bits of the EPROM address select the waveform function by selecting a block of 256 address. The 8 least significant bits of the EPROM address are connected to an 8-bit (mod-256) counter, which continuously cycles through the 256 selected addresses. A 27C64 EPROM (8K $\times$ 8) has 13 address lines. After the eight lower lines are accounted for, the remaining five lines can be used to select up to 32 digital functions. With the two binary Function Select switches, we can potentially select 4 functions.

For example, to select the Sine function, inputs A_9 and A_8, which comprise the most significant digit of the EPROM address, are set to 00. Thus, the 8-bit counter cycles through addresses 0000–00FF, the location of the sine data. The Square Wave function is selected by setting A_9 and A_8 to 01, thus selecting the address block 0100–01FF. Other functions can be similarly selected.

The data at each address are sent to the D/A converter (MC1408), which, in combination with the op amp, is configured to produce a bipolar (both positive and negative) output. (We use a high slew rate op amp so that the generated square waves will have vertical sides.) The circuit generates a continuous waveform by retracing the data points in one 256-byte section of the EPROM over and over.

The DAC/op amp combination produces a maximum negative voltage for a hex input of 00, a 0 V output for an input of 80, and a maximum positive voltage for an input of FF. (You might wish to refer to the section Bipolar Operation of MC1408 in Chapter 12 for details of the DAC operation.)

You can see from the Sine function data in Table 13.1 that 8 bits are not sufficient to represent each of the 256 steps of a digital sine function as a unique number. The peaks of the waveform are changing too slowly to be represented accurately by an 8-bit quantization, and as a result, the top of the sine wave is flat for several clock pulses. (Mathematically, a sine function is tangential to a horizontal line at its peak. However, since tangential means touching at one point, the flat top is a distortion.) A unique number for each of 256 steps of a sine function needs at least 13 bits,[1] but this requires additional bits on the D/A converter input, and therefore a different DAC and an expanded memory word length.

The output frequency of the function generator is 1/256 of the clock rate. Given that the settling time of the MC1408 DAC is about 300ns, the maximum clock rate of the circuit is 1/300 ns = 3.33 MHz. At this rate, the output frequency is 3.33 MHz/256 = 13 kHz.

EEPROM

> **KEY TERM**
>
> **EEPROM (or E²PROM)**   Electrically erasable programmable read only memory. A type of read only memory that can be field-programmed and selectively erased while still in a circuit.

---

[1]Bits required:
- 360°/256 steps = 1.40625°/step.
- Sine function changes most slowly at peak, so calculate $A \sin(90° \pm 1.40625°)$ to find smallest amplitude change.
- The smallest power-of-2 amplitude, A, for which $A \sin(90°) - A \sin(90° \pm 1.40625°) \geq 1$ is 4096.
- The amplitude range $-4096 \leq A \leq 4095$ can be represented by a 13-bit number.

As was discussed in the previous section, EPROMs have the useful property of being erasable. The problem is that they must be removed from the circuit for erasure, and bits cannot be selectively erased; the whole memory cell array is erased as a unit.

Electrically erasable programmable read only memory (EEPROM or E²PROM) provides the advantages of EPROM along with the additional benefit of allowing erasure of selected bits while the chip is in the circuit; it combines the read/write properties of RAM with the nonvolatility of ROM. EEPROM is useful for storage of data that need to be changed occasionally, but that must be retained when power is lost to the EEPROM chip. One example is the memory circuit in an electronically tuned car radio that stores the channel numbers of local stations.

Like the UV-erasable EPROM, the memory cell of the EEPROM is based on the FAMOS transistor. Unlike the EPROM, the FAMOS FET is coupled with a standard MOS-FET, as shown in Figure 13.19.

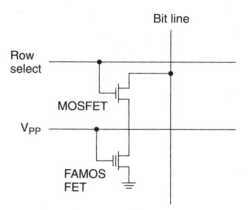

FIGURE 13.19
EEPROM Cell

The FAMOS FET is programmed in the same way as UV-erasable EPROM: a programming voltage pulse (V_{PP}) drives high-energy electrons into the floating gate of the FAMOS transistor, where they remain trapped and change the threshold voltage of the transistor. The cell is read by keeping the programming line at 5 V and making the cell's Row Select line HIGH. The FAMOS transistor will or will not turn on, depending on its programmed state.

The FAMOS transistors used in EPROM and EEPROM differ in one important respect. The EEPROM transistor is manufactured with a very thin oxide layer between the drain and the upper (nonfloating) gate. This construction allows trapped electrons in the floating gate to be forced out electrically, thus erasing the cell contents.

Given the obvious advantages of EEPROM, why doesn't it replace all other types of memory? There are several reasons:

1. EEPROM has a much slower access time than RAM and is thus not good for high-speed applications.

2. The currently available EEPROMs have significantly smaller bit capacities than commercially available RAM (especially dynamic RAM) and EPROM.

3. EEPROM has a fixed number of write/erase cycles, typically 100,000. After that, new data cannot be programmed into the device.

Flash Memory

> ### KEY TERMS
>
> **Flash memory** A nonvolatile type of memory that can be programmed and erased in sectors, rather than byte-at-a-time.
>
> **Sector** A segment of flash memory that forms the smallest amount that can be erased and reprogrammed at one time.
>
> **Boot block** A sector in a flash memory reserved for primary firmware.
>
> **Top boot block** A boot block sector in a flash memory placed at the highest address in the memory.
>
> **Bottom boot block** A boot block sector in flash memory paced at the lowest address in the memory.

A popular variation on EEPROM is **flash memory.** This type of nonvolatile memory generally has a larger byte capacity (e.g., 8 Mb) than EEPROM devices and thus can be used to store fairly large amounts of firmware, such as the BIOS (basic input/output system) of a PC.

A flash memory is divided into **sectors,** groups of bytes that are programmed and erased at one time. One sector is designated as the **boot block,** which is either the sector with the highest (**top boot block**) or lowest (**bottom boot block**) address. The primary firmware is usually stored in the boot block, with the idea that the system using the flash memory is configured to look there first for firmware instructions. The boot block can also be protected from unauthorized erasure or modification (e.g., by a virus), thus adding a security feature to the device.

Figure 13.20 shows the arrangement of sectors of a 512K × 8-bit (4 Mb) flash memory with a bottom boot block architecture. The range of addresses are shown alongside the blocks. For example, sector S0 (the boot block) has a 16 KB address range of 00000H to 03FFFH. Sector S1 has an 8 KB address range from 04000H to 05FFFH. The first 64 KB of the memory are divided into one 16 KB, two 8 KB, and one 32 KB sectors. The remainder of the memory is divided into equal 64 KB sectors. Note that even though the boot block is drawn at the top of Figure 13.20, it is a bottom boot block because it is the sector with the lowest address.

A flash memory with a top boot block would have the same proportions given over to its sectors, but mirror-image to the diagram in Figure 13.20. That is, S10 (boot block) would be a 16 KB sector from 7C000H to 7FFFFH. The other sectors would be identical to the bottom boot block architecture, but in reverse order.

As with other EEPROM devices, a flash memory can be erased and reprogrammed while installed in a circuit. The memory cells in a flash device have a limited number of program/erase cycles, like other EEPROMs. The sector architecture of the flash memory makes it faster to erase and program than other EEPROM-based memories which must erase or program bytes one at a time. This same characteristic makes it unsuitable for use as system RAM, which must be able to program single bytes.

▮▮ SECTION 13.3 REVIEW PROBLEM

13.3 A flash memory has a capacity of 8 Mb, organized as 1M × 8-bit. List the address range for the 32 KB boot block sector of the memory if the device has a bottom boot block architecture and if it has a top boot block architecture.

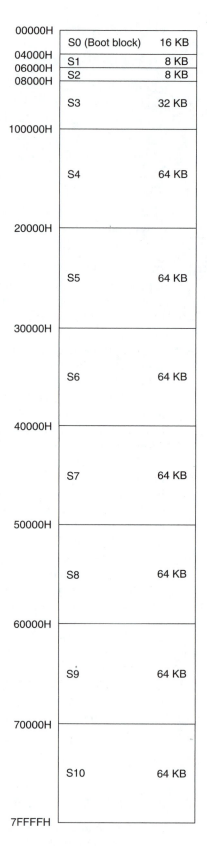

FIGURE 13.20

Sectors in a 512K × 8b Flash Memory (Bottom Boot Block)

13.4 Sequential Memory: FIFO and LIFO

The RAM and ROM devices we have examined up until now have all been random access devices. That is, any data could be read from or written to any sequence of addresses in any order. There is another class of memory in which the data must be accessed in a particular order. Such devices are called **sequential memory**.

There are two main ways of organizing a sequential memory—as a **queue** or as a **stack**. Figure 13.21 shows the arrangement of data in each of these types of memory.

A queue is a **first-in first-out** (FIFO) memory, meaning that the data can be read only in the same order they are written, much as railway cars always come out of a tunnel in the same order they go in.

One common use for FIFO memory is to connect two devices that have different data rates. For instance, a computer can send data to a printer much faster than the printer can use it. To keep the computer from either waiting for the printer to print everything or periodically interrupting the computer's operation to continue the print task, data can be sent in a burst to a FIFO, where the printer can read them as needed. The only proviso is that there

FIGURE 13.21
Sequential Memory

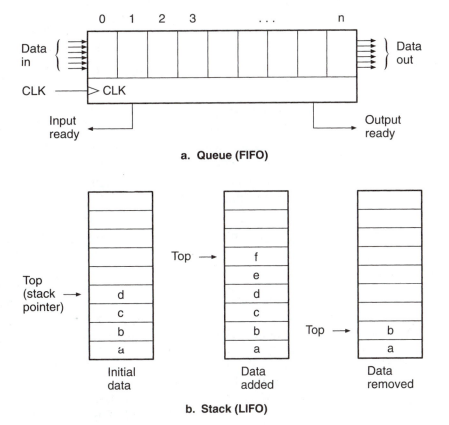

a. Queue (FIFO)

b. Stack (LIFO)

must be some logic signal to the computer telling it when the queue is full and not to send more data and another signal to the printer letting it know that there are some data to read from the queue.

The **last-in-first-out** (LIFO), or stack, memory configuration, also shown in Figure 13.21, is not available as a special chip, but rather is a way of organizing RAM in a memory system.

The term "stack" is analogous to the idea of a spring-loaded stack of plates in a cafeteria line. When you put a bunch of plates on the stack, they settle into the recessed storage area. When a plate is removed, the stack springs back slightly and brings the second plate to the top level. (The other plates, of course, all move up a notch.) The top plate is the only one available for removal from the stack, and plates are always removed in reverse order from that in which they were loaded.

Figure 13.21b shows how data are transferred to and from a LIFO memory. A block of addresses in a RAM is designated as a stack, and one or two bytes of data in the RAM store a number called the stack pointer, which is the current address of the top of the stack.

In Figure 13.21, the value of the stack pointer changes with every change of data in the stack, pointing to the last-in data in every case. When data are removed from the stack, the stack pointer is used to locate the data that must be read first. After the read, the stack pointer is modified to point to the next-out data. Some stack configurations have the stack pointer painting to the next empty location on the stack.

The most common application for LIFO memory is in a computer system. If a program is interrupted during its execution by a demand from the program or some piece of hardware that needs attention, the status of various registers within the computer are stored on a stack and the computer can pay attention to the new demand, which will certainly change its operating state. After the interrupting task is finished, the original operating state of the computer can be taken from the top of the stack and reloaded into the appropriate registers, and the program can resume where it left off.

▏▎▏ SECTION 13.4 REVIEW PROBLEM

13.4 State the main difference between a stack and a queue.

13.5 Dynamic RAM Modules

> **KEY TERMS**
>
> **Memory module**   A small circuit board containing several dynamic RAM chips.
>
> **Single in-line memory module (SIMM)**   A memory module with DRAMs and connector pins on one side of the board only.
>
> **Dual in-line memory module (DIMM)**   A memory module with DRAMs and connector pins on both sides of the board.

Dynamic RAM chips are often combined on a small circuit board to make a **memory module**. This is because the data bus widths of systems requiring the DRAMs are not always the same as the DRAMs themselves. For example, Figure 13.22 shows how four 64M × 8 DRAMs are combined to make a 64M × 32 memory module. The block diagram of the module is shown in Figure 13.22, and the mechanical outline is shown in Figure 13.23. The data input/output lines are separate from one another so that there are 32 data I/Os *(DQ)*. The address lines (ADDR[12..0]) for the module are parallel on all chips. With address multiplexing, this 13-bit address bus yields a 26-bit address, giving a 64M address range. Chip selects (CS) for all devices are connected together so that selecting the module selects all chips on the module.

This particular memory module is configured as a **single in-line memory module (SIMM)**, which has the DRAM chips and pin connections on one side of the board only. A

FIGURE 13.22
SIMM Block Diagram

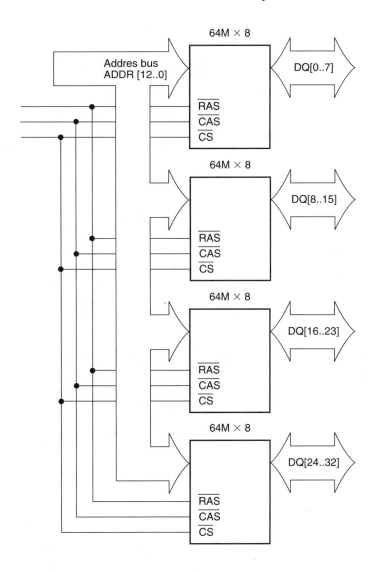

FIGURE 13.23
SIMM Layout

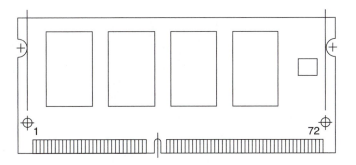

dual in-line memory module (DIMM) has the DRAMs mounted on both sides of the circuit board and pin connections on both sides of the board as well.

▍▍ SECTION 13.5 REVIEW PROBLEM

13.5 A SIMM has a capacity of 16M × 32. How many 16M × 8 DRAMs are required to make this SIMM? How many address lines does the SIMM require? How should the DRAMs be connected?

13.6   Memory Systems

www.electronictech.com

In the section on memory modules, we saw how multiple memory devices can be combined to make a system that has the same number of addressable locations as the individual devices making up the system, but with a wider data bus. We can also create memory systems where the data I/O width of the system is the same as the individual chips, but where the system has more addressable locations than any chip within the system.

In such a system, the data I/O and control lines from the individual memory chips are connected in parallel, as are the lower bits of an address bus connecting the chips. However, it is important that only one memory device be enabled at any given time, in order to avoid **bus contention**, the condition that results when more than one output attempts to drive a common bus line. To avoid bus contention, one or more additional address lines must be decoded by an **address decoder** that allows only one chip to be selected at a time.

Figure 13.24 shows two 32K $\times$ 8 SRAMs connected to make a 64K $\times$ 8 memory system. A single 32K $\times$ 8 SRAM, as shown in Figure 13.24a, requires 15 address lines, 8 data lines, a write enable (WE), and chip select (CS) line. To make a 64K $\times$ 8 SRAM system, all of these lines are connected in parallel, except the CS lines. In order to enable only one at a time, we use one more address line, A_{15}, and enable the top SRAM when $A_{15} = 0$ and the bottom SRAM when $A_{15} = 1$.

The address range of one 32K $\times$ 8 SRAM is given by the range of states of the address lines A[14..0]:

Lowest single-chip address:	000 0000 0000 0000 = 0000H
Highest single-chip address:	111 1111 1111 1111 = 7FFFH

The address range of the whole system must also account for the A_{15} bit:

Lowest system address:	0000 0000 0000 0000 = 0000H
Highest system address:	1111 1111 1111 1111 = FFFFH

Within the context of the system, each individual SRAM chip has a range of addresses, depending on the state of A_{15}. Assume $SRAM_0$ is selected when $A_{15} = 0$ and $SRAM_1$ is selected when $A_{15} = 1$.

Lowest $SRAM_0$ address:	0000 0000 0000 0000 = 0000H
Highest $SRAM_0$ address:	0111 1111 1111 1111 = 7FFFH
Lowest $SRAM_1$ address:	1000 0000 0000 0000 = 8000H
Highest $SRAM_1$ address:	1111 1111 1111 1111 = FFFFH

Figure 13.25 shows a **memory map** of the 64K $\times$ 8 SRAM system, indicating the range of addresses for each device in the system. The total range of addresses in the system is called the **address space**.

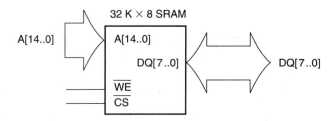

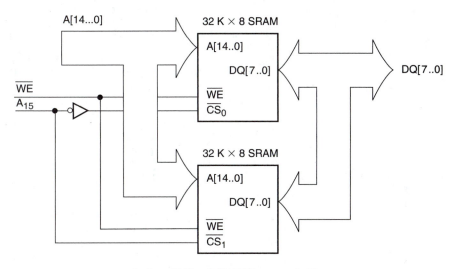

FIGURE 13.24
Expanding Memory Space

FIGURE 13.25
Memory Map

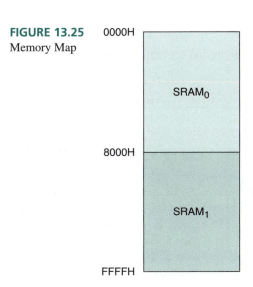

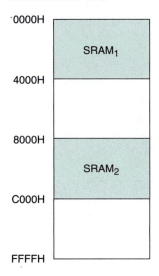

FIGURE 13.26
Memory Map Showing Non-contiguous Decoded Blocks.

Figure 13.26 shows a memory map for a system with an address space of 64K (16 address lines). Two 16K × 8 blocks of SRAM are located at start addresses of 0000H and 8000H, respectively. Sketch a memory system that implements the memory map of Figure 13.26.

Solution A 16K address block requires 14 address lines, since

$$16K = 16 \times 1024 = 2^4 \times 2^{10} = 2^{14}$$

The entire 64K address space requires 16 address lines, since

$$64K = 64 \times 1024 = 2^6 \times 2^{10} = 2^{16}$$

The highest address in a block is the start address plus the block size.

	16K block size:	11 1111 1111 1111 = 3FFFH
$SRAM_0$:	Lowest address:	0000 0000 0000 0000 = 0000H
	Highest address:	0011 1111 1111 1111 = 3FFFH
$SRAM_2$:	Lowest Address:	1000 0000 0000 0000 = 8000H
	Highest Address:	1011 1111 1111 1111 = BFFFH

$A_{15}A_{14} = 00$ for the entire range of the $SRAM_0$ block. $A_{15}A_{14} = 10$ for the entire $SRAM_2$ range. These can be decoded by the gates shown in Figure 13.27.

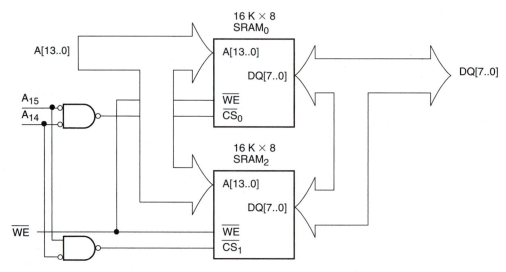

FIGURE 13.27
Example 13.3
32K × 8 SRAM with non-contiguous blocks.

Address Decoding with *n*-line-to-*m*-line Decoders

Figure 13.28 shows a 64K memory system with four 16K chips: one EPROM at 0000H and three SRAMs at 4000H, 8000H, and C000H, respectively. In this circuit, the address decoding is done by a 2-line-to-4-line decoder, which can be an off-the-shelf MSI decoder, such as a 74HC139 decoder or a PLD-based design.

Table 13.2 shows the address ranges decoded by each decoder output. The first two address bits are the same throughout any given address range. Figure 13.29 shows the memory map for the system.

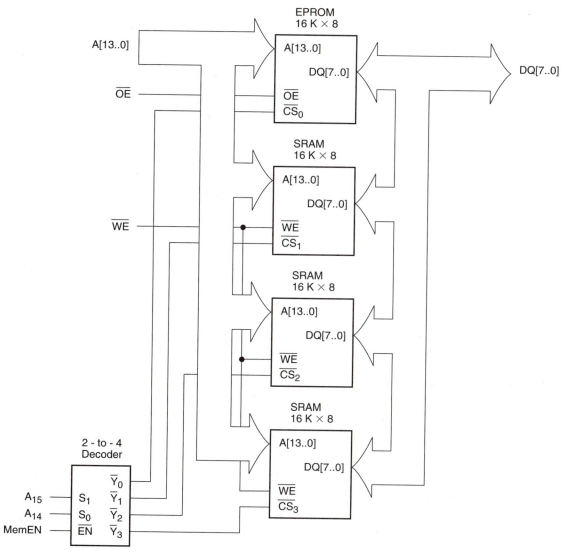

FIGURE 13.28
64K Memory System

Table 13.2 Address Decoding for Figure 13.28

A_{15}	A_{14}	Active Decoder Output	Device	Address Range
0	0	Y_0	EPROM	0000 0000 0000 0000 = 0000H
				0011 1111 1111 1111 = 3FFFH
0	1	Y_1	SRAM$_1$	0100 0000 0000 0000 = 4000H
				0111 1111 1111 1111 = 7FFFH
1	0	Y_2	SRAM$_2$	1000 0000 0000 0000 = 8000H
				1011 1111 1111 1111 = BFFFH
1	1	Y_3	SRAM$_3$	1100 0000 0000 0000 = C000H
				1111 1111 1111 1111 = FFFFH

FIGURE 13.29
Memory Map for Figure 13.28

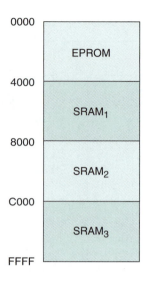

```
0000 ┌─────────────┐
     │             │
     │    EPROM    │
     │             │
4000 ├─────────────┤
     │             │
     │   SRAM₁     │
     │             │
8000 ├─────────────┤
     │             │
     │   SRAM₂     │
     │             │
C000 ├─────────────┤
     │             │
     │   SRAM₃     │
     │             │
FFFF └─────────────┘
```

▌▐ SECTION 13.6 REVIEW PROBLEM

13.6 Calculate the number of 128K memory blocks will fit into a 1M address space. Write the start addresses for the blocks.

S U M M A R Y

1. A memory is a device that can accept data and store them for later recall.

2. Data are located in a memory by an address, a binary number at a set of address inputs that uniquely locates the block of data.

3. The operation that stores data in a memory is called the write function. The operation that recalls the stored data is the read function. These functions are controlled by functions such as write enable ($\overline{\text{WE}}$), chip select ($\overline{\text{CS}}$), and output enable ($\overline{\text{OE}}$).

4. RAM is random access memory. RAM can be written to and read from in any order of addresses. RAM is volatile. That is, it loses its data when power is removed from the device.

5. ROM is read only memory. Original ROM devices could not be written to at all, except at the time of manufacture. Modern variations can also be written to, but not as easily as RAM. ROM is nonvolatile; it retains its data when power is removed from the device.

6. Memory capacity is given as $m \times n$ for m addressable locations and an n-bit data bus. For example, a 64K $\times$ 8 memory has 65,536 addressable locations, each with 8-bit data.

7. Large blocks of memory are designated with the binary prefixes K ($2^{10} = 1024$), M ($2^{20} = 1,048,576$), and G ($2^{30} = 1,073,741,824$).

8. RAM can be divided into two major classes: static RAM (SRAM) and dynamic RAM (DRAM). SRAM retains its data as long as power is applied to the device. DRAM requires its data to be refreshed periodically.

9. Typically DRAM capacity is larger than SRAM because DRAM cells are smaller than SRAM cells. An SRAM cell is essentially a flip-flop consisting of several transistors. A DRAM cell has only one transistor and a capacitor.

10. RAM cells are arranged in rectangular arrays for efficient packaging. Internal circuitry locates each cell at the intersection of a row and column within the array.

11. For packaging efficiency, DRAM addresses are often multiplexed so that the device receives half its address as a row address, latched in to the device by a $\overline{\text{RAS}}$ (row address strobe) signal and the second half as a column address, latched in by a $\overline{\text{CAS}}$ (column address strobe) signal.

12. Read only memory (ROM) is used where it is important to retain data after power is removed.

13. Mask-programmed ROM is programmed at the time of manufacture. Programming is done by making a custom overlay of connections onto a standard cell array. Data cannot be changed. This is suitable for mature designs in high-volume production.

14. Erasable programmable read only memory (EPROM) can be programmed by the user and erased by exposure to ultraviolet light of a specified frequency and intensity. An EPROM must be removed from its circuit for erasing and reprogramming.

15. Electrically erasable read only memory (EEPROM or E²PROM) can be programmed and erased in-circuit. It is nonvolatile, but unsuitable for use as system RAM due to

its long programming/erase times and finite number of program/erase cycles.

16. Flash memory is a type of EEPROM that is organized into sectors that are erased all at once. This is faster than other EEPROM, which must be erased byte-by-byte.

17. Flash memory is often configured with one sector as a boot block, where primary firmware is stored. A bottom boot block architecture has the boot block at the lowest chip address. A top boot block architecture has the boot block at the highest chip address.

18. Sequential memory must have its data accessed in sequence. Two major classes are first-in first-out (FIFO) and last-in first-out (LIFO). FIFO is also called a queue and LIFO is called a stack.

19. Dynamic RAM chips are often configured as memory modules, small circuit boards with multiple DRAMs. The modules usually have the same number of address locations as the individual chips on the module, but a wider data bus.

20. Memory systems can be configured to have the same data width as individual memory devices comprising the system, but with more addressable locations than any chip in the system. The additional addresses require additional system address lines, which are decoded to enable one chip at a time within the system.

GLOSSARY

Address   A number, represented by the binary states of a group of inputs or outputs, uniquely defining the location of data stored in a memory device.

Address decoder   A circuit enabling a particular memory device to be selected by the address bus of a larger memory system.

Address multiplexing   A technique of addressing storage cells in a dynamic RAM which sequentially uses the same inputs for row address and column address of the cell.

Address Space   A block of addresses in a memory system.

b   Bit.

B   Byte.

Bit-organized   A memory is bit-organized if one address accesses one bit of data.

Boot block   A sector in a flash memory reserved for primary firmware.

Bottom boot block   A boot block sector in flash memory paced at the lowest address in the memory.

Bus   A group of parallel conductors carrying related logic signals, such as multi-bit data or addresses.

Bus contention   The condition that results when two or more devices try to send data to a bus at the same time. Bus contention can damage the output buffers of the devices involved.

Byte   A group of 8 bits.

CAS   Column address strobe. A signal used to latch the column address into the decoding circuitry of a dynamic RAM with multiplexed addressing.

Data   Binary digits (0s and 1s) which contain some kind of information. In the context of memory, the digital contents of a memory device.

Dual in-line memory module (DIMM)   A memory module with DRAMs and connector pins on both sides of the board.

Dynamic RAM   A random access memory which cannot retain data for more than a few (e.g., 64) milliseconds without being "refreshed."

EEPROM (or E²PROM)   Electrically erasable programmable read only memory. A type of read only memory that can be field-programmed and selectively erased while still in a circuit.

EPROM   Erasable programmable read only memory. A type of ROM that can be programmed ("burned") by the user and erased later, if necessary, by exposing the chip to ultraviolet radiation.

FAMOS FET   Floating-gate avalanche. MOSFET. A MOSFET with a second, "floating" gate in which charge can be trapped to change the MOSFET's gate-source threshold voltage. A FAMOS transistor is the memory element in an EPROM cell.

FIFO   First-in first-out. A sequential memory in which the stored data can only be read in the order in which it was written.

Firmware   Software instructions permanently stored in ROM.

Flash memory   A nonvolatile type of memory that can be programmed and erased in sectors, rather than byte-at-a-time.

Hardware   The electronic circuit of a digital or computer system.

I/O   Input/output.

K $1024 (=2^{10})$ Analogous to the metric prefix "k" (kilo).

LIFO   Last-In first-out. A sequential memory in which the last data written is the first data read.

M $1,048,576 (=2^{20})$ Analogous to the metric prefix "M" (mega).

Mask-programmed ROM   A type of read only memory (ROM) where the stored data are permanently encoded into the memory device during the manufacturing process.

Memory   A device for storing digital data in such a way that it can be recalled for later use in a digital system.

Memory map   A diagram showing the total address space of a memory system and the placement of various memory devices within that space.

Memory module   A small circuit board containing several dynamic RAM chips.

Nibble   Half a byte; 4 bits.

PROM   Programmable read only memory. A type of ROM whose data need not be manufactured into the chip, but can be programmed by the user.

Queue   A FIFO memory.

RAM cell   The smallest storage unit of a RAM, capable of storing one bit.

Random access memory (RAM)   A type of memory device where data at any address can be accessed in any order, that is, randomly. The term usually refers to random access read/write memory.

$\overline{\text{RAS}}$ Row address strobe. A signal used to latch the row address into the decoding circuitry of a dynamic RAM with multiplexed addressing.

Read   Retrieve data from a memory device.

Read only memory (ROM)   A type of memory where data is permanently stored and can only be read, not written.

Refresh cycle   The process which periodically recharges the storage capacitors in a dynamic RAM.

Sector   A segment of flash memory that forms the smallest amount that can be erased and reprogrammed at one time.

Sequential memory   Memory in which the stored data cannot be read or written in random order, but must be addressed in a specific sequence.

Single in-line memory module (SIMM)   A memory module with DRAMs and connector pins on one side of the board only.

Software   Programming instructions required to make hardware perform specified tasks.

Stack   A LIFO memory.

Static RAM   A random access memory which can retain data indefinitely as long as electrical power is available to the chip.

Top boot block   A boot block sector in a flash memory placed at the highest address in the memory.

Volatile   A memory is volatile if its stored data is lost when electrical power is lost.

Word   Data accessed at one addressable location.

Word length   Number of bits in a word.

Word-organized   A memory is word-organized if one address accesses one word of data.

Write   Store data in a memory device.

PROBLEMS

Section 13.2   Basic Memory Concepts

13.1   How many address lines are necessary to make an 8×8 memory similar to the 4×8 memory in Figure 13.5? How many address lines are necessary to make a 16×8 memory?

13.2   Briefly explain the difference between RAM and ROM.

13.3   Calculate the number of address lines and data lines needed to access all stored data in each of the following sizes of memory:

    a. $64K \times 8$

    b. $128K \times 16$

    c. $128K \times 32$

    d. $256K \times 16$

    Calculate the total bit capacity of each memory.

13.4   Explain the difference between the chip enable $(\overline{E})$ and the output enable $(\overline{G})$ control functions in a RAM.

13.5   Refer to Figure 13.9. Briefly explain the operation of the $\overline{W}, \overline{E},$ and $\overline{G}$ functions of the RAM shown.

Section 13.2   Random Access Read/Write Memory (RAM)

13.6   Draw the circuit for an NMOS static RAM cell. Label one output *BIT* and the other $\overline{BIT}$.

13.7   Refer to the NMOS static RAM cell drawn in Problem 13.6. Assume that *BIT* = 1. Describe the operation required to change *BIT* to 0.

13.8   Describe the main difference between a CMOS and an NMOS static RAM cell.

13.9   Explain how a particular RAM cell is selected from a group of many cells.

13.10   How many address lines are required to access all elements in a $1M \times 1$ dynamic RAM with address multiplexing?

13.11   What is the capacity of an address-multiplexed DRAM with one more address line than the DRAM referred to in Problem 13.10? With two more address lines?

13.12   How many address lines are required to access all elements in a $256M \times 16$ DRAM with address multiplexing?

Section 13.3   Read Only Memory (ROM)

13.13   Briefly list some of the differences between maskprogrammed ROM, UV-erasable EPROM, EEPROM, and flash memory.

13.14   Briefly describe the programming and erasing process of a UV-EPROM.

13.15   Briefly explain the difference between flash memory and other EEPROM. What is the advantage of each configuration?

13.16   A flash memory has a capacity of 8 Mb, organized as $512K \times 16$-bit. List the address range for the 16 KB boot block sector of the memory if the device has a bottom boot block architecture and if it has a top boot block architecture.

13.17   Briefly state why EEPROM is not suitable for use as system RAM.

13.18   Briefly state why flash memory is unsuitable for use as system RAM.

Section 13.4   Sequential Memory

13.19   State one possible application for a FIFO and for a LIFO memory.

Section 13.5   Memory Modules

13.20   A SIMM has a capacity of $32M \times 64$. How many $32M \times 8$ DRAMs are required to make this SIMM? How many address lines does the SIMM require? How should the DRAMs be connected?

Section 13.6   Memory Systems

13.21 A microcontroller system with a 16-bit address bus is connected to a 4K × 8 RAM chip and an 8K × 8 RAM chip. The 8K address begins at 6000H. The 4K address block starts at 2000H.

    Calculate the end address for each block and show address blocks for both memory chips on a 64K memory map.

13.22 Draw the memory system of Problem 13.21.

13.23 A microcontroller system with a 16-bit address bus has the following memory assignments:

Memory	Size	Start Address
RAM$_0$	16K	4000H
RAM$_1$	8K	8000H
RAM$_2$	8K	A000H

Show the blocks on a 64K memory map.

13.24 Draw the memory system described in Problem 13.23.

13.25 The memory map of a microcontroller system with a 16-bit address bus is shown in Figure 13.30. Make a table of start and end addresses for each of the blocks shown. Indicate the size of each block.

13.26 Sketch the memory system described in Problem 13.25.

13.27 How many 16M × 32 DIMMs are required to make a 256M × 32 memory system? Make a table showing the start and end addresses of each block.

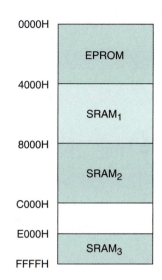

FIGURE 13.30
Problem 13.25

ANSWERS TO SECTION REVIEW PROBLEMS

Section 13.2a

13.1 18 address lines, 16 data lines; capacity = 4Mb, same as Figure 3.12

Section 13.2b

13.2 a. 10 address, 1 data; **b.** 10 address, 4 data; **c.** 11 address, 1 data.

Section 13.3

13.3 Bottom boot block: 00000H to 07FFFH; top boot block: F8000H to FFFFFH.

Section 13.4

13.4 A stack is a last-in first-out (LIFO) memory and a queue is a first-in first-out (FIFO) memory.

Section 13.5

13.5 Four DRAMs. 12 address lines. Address and control lines are in parallel with all DRAMs. Data I/O lines are separate.

Section 13.6

13.6 Eight blocks. Start addresses: 00000H, 20000H, 40000H, 60000H, 80000H, A0000H, C0000H, E0000H.

Altera UP-1 User Guide

Current versions of the Altera UP-1 board are shipped with version 9.23 of MAX+PLUS II software. See the file SE_READ.txt on the accompanying CD for installation instructions.

University Program
Design Laboratory Package

August 1997, ver. 1	User Guide

Introduction

The University Program Design Laboratory Package was designed to meet the needs of universities teaching digital logic design with state-of-the-art development tools and programmable logic devices (PLDs). The package provides all of the necessary tools for creating and implementing digital logic designs, including the following features:

- MAX+PLUS® II version 7.21 Student Edition development software
- UP 1 Education Board
    - EPM7128S device in an 84-pin plastic J-lead chip carrier (PLCC) package
    - EPF10K20 device in a 240-pin power quad flat pack (RQFP) package
- ByteBlaster™ Parallel Port Download Cable

MAX+PLUS II Version 7.21 Student Edition Software

The MAX+PLUS II version 7.21 Student Edition software contains many of the features available in the commercial version of MAX+PLUS II, including a completely integrated design flow and an intuitive graphical user interface. This software supports schematic capture and text-based hardware description language design entry, including the Altera® Hardware Description Language (AHDL™) and VHDL. It also provides design programming, compilation, and verification support for the EPM7128S and EPF10K20 devices.

To maximize learning, the MAX+PLUS II software includes complete and instantly accessible on-line help. The student version of the MAX+PLUS II software can be freely distributed to students for installation on their personal computers.

 For information on how to install the MAX+PLUS II version 7.21 Student Edition software on your computer, see "Software Installation" on page 17.

UP 1 Education Board

The UP 1 Education Board is a stand-alone experiment board based on two of Altera's leading device families: MAX® 7000 and FLEX® 10K. Its simple design, when used with the MAX+PLUS II software, provides a superior platform for learning digital logic design using high-level development tools and PLDs.

The UP 1 Education Board was designed to meet the needs of the educator and the design laboratory environment. The UP 1 Education Board supports both product-term based and look-up table (LUT)-based architectures and includes two PLDs. The EPM7128S device can be programmed in-system with the ByteBlaster download cable. The EPF10K20 device can be configured in-system with either the ByteBlaster download cable or an EPC1 Configuration EPROM (not included).

EPM7128S Device

The EPM7128S device, a mid-density member of the high-density, high-performance MAX 7000S family, is based on EEPROM elements. The EPM7128S device comes in a socket-mounted 84-pin PLCC package and has 128 macrocells. Each macrocell has a programmable-AND/fixed-OR array as well as a configurable register with independently programmable clock, clock enable, clear, and preset functions. With a capacity of 2,500 gates and a simple architecture, the EPM7128S device is ideal for introductory designs as well as larger combinatorial and sequential logic functions.

 For more information on MAX 7000 devices, go to the *MAX 7000 Programmable Logic Device Family Data Sheet*.

EPF10K20 Device

The EPF10K20 device, a member of Altera's high-density FLEX 10K family, is based on reconfigurable SRAM elements. The EPF10K20 device comes in a 240-pin RQFP package and has 1,152 logic elements (LEs) and 6 embedded array blocks (EABs). Each LE consists of a 4-input look-up table (LUT), a programmable flipflop, and dedicated signal paths for carry and cascade functions. Each EAB provides 2,048 bits of memory, which can be used to create RAM, ROM, or first-in first-out (FIFO) functions. The EABs can also be used to implement logic functions, such as multipliers, microcontrollers, state machines, and digital signal processing (DSP) functions. With a typical gate count of 20,000 gates, the EPF10K20 device is ideal for advanced designs, including computer architecture, communications, and DSP applications.

 For more information on FLEX 10K devices, go to the *FLEX 10K Embedded Programmable Logic Family Data Sheet*.

ByteBlaster Parallel Port Download Cable

Designs can be easily and quickly downloaded into the UP 1 Education Board using the ByteBlaster download cable, which is a hardware interface to a standard parallel port. This cable channels programming or configuration data between the MAX+PLUS II software and the UP 1 Education Board. Because design changes are downloaded directly to the devices on the board, prototyping is easy and multiple design iterations can be accomplished in quick succession.

For more information on the ByteBlaster download cable, go to the *ByteBlaster Parallel Port Download Cable Data Sheet*.

UP 1 Education Board Description

The UP 1 Education Board contains the elements described in this section. Figure 1 shows a block diagram of the UP 1 Education Board.

Figure 1. UP 1 Education Board Block Diagram

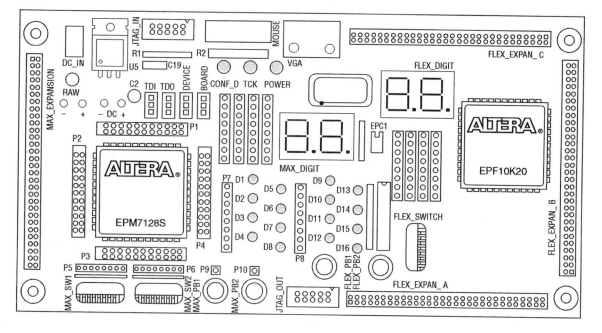

DC_IN & RAW Power Input

The DC_IN power input accepts a 2.5 mm × 5.55 mm female connector. The acceptable DC input is 7 to 12 V DC at a minimum of 250 mA. The RAW power input consists of two holes for connecting an unregulated power source. The hole marked with a plus sign (+) is the positive input; the hole marked with a minus sign (–) is board common.

On-Board Voltage Regulator

The on-board voltage regulator, an LM340T, regulates the DC positive input at 5 V. The DC input consist of two holes for connecting a 5-V DC regulated power source. The hole marked with a plus sign (+) is the positive input; the hole marked with a minus sign (–) is board common. A green light-emitting diode (LED) labeled POWER is illuminated when current is flowing from the 5-V DC regulated power source.

Oscillator

The UP 1 Education Board contains a 25.175-MHz crystal oscillator. The output of the oscillator drives the global clock input on the EPM7128S device (pin 83) and the global clock input on the EPF10K20 device (pin 91).

JTAG_IN Header

The 10-pin female plug on the ByteBlaster download cable connects with the JTAG_IN 10-pin male header on the UP 1 Education Board. The UP 1 Education Board provides power and ground to the ByteBlaster download cable. Data is shifted into the devices via the TDI pin and shifted out of the devices via the TDO pin. Table 1 identifies the JTAG_IN pin names when the ByteBlaster is operating in JTAG mode.

Table 1. JTAG_IN 10-Pin Header Pin-Outs

Pin	JTAG Signal
1	TCK
2	GND
3	TDO
4	VCC
5	TMS
6	No Connect
7	No Connect
8	No Connect
9	TDI
10	GND

Jumpers

The UP 1 Education Board contains four three-pin jumpers (TDI, TDO, DEVICE, and BOARD) that set the JTAG configuration. You can set the JTAG chain for a variety of configurations (i.e., to program only the EPM7128S device, to configure only the EPF10K20 device, to configure/program both devices, or to connect multiple UP 1 Education Boards together). Figure 1 shows the positions of the three connectors (C1, C2, and C3) on each of the four jumpers.

Figure 2. Position of C1, C2 & C3 Connectors

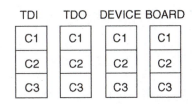

Table 2 defines the settings for each configuration.

University Program Design Laboratory Package User Guide

Table 2. JTAG Jumper Settings				
Desired Action	**TDI**	**TDO**	**DEVICE**	**BOARD**
Program EPM7128S device only	C1 & C2	C1 & C2	C1 & C2	C1 & C2
Configure EPF10K20 device only	C2 & C3	C2 & C3	C1 & C2	C1 & C2
Program/configure both devices, *Note (1)*	C2 & C3	C1 & C2	C2 & C3	C1 & C2
Connect multiple boards together, *Note (2)*	C2 & C3	OPEN	C2 & C3	C2 & C3

Notes:
(1) The first device in the JTAG chain is the EPF10K20, and the second device is the EPM7128S.
(2) The first device in the JTAG chain is the EPF10K20, and the second device is the EPM7128S. The last board in the chain must be set for a single board configuration (i.e., for programming only the EPM7128S device, configuring only the EPF10K20 device, or configuring/programming both devices). The last board cannot be set for connecting multiple boards together.

During configuration, the green CONF_D LED will turn off and the green TCK LED will modulate to indicate that data is transferring. After the device has successfully configured, the CONF_D LED will illuminate.

 For information on how to program or configure EPM7128S and EPF10K20 devices, see "Programming or Configuring Devices" on page 18.

EPM7128S Device

The UP 1 Education Board provides the following resources for the EPM7128S device.

- Socket-mounted 84-pin PLCC package
- Signal pins that are accessible via female headers
- JTAG chain connection for the ByteBlaster
- 2 momentary push-button switches
- 2 octal dipswitches
- 16 LEDs
- Dual-digit 7-segment display
- On-board oscillator (25.175 MHz)
- Expansion port with 42 I/O pins and the dedicated global CLR, OE1, and OE2/GCLK2 pins

The pins from the EPM7128S device are not pre-assigned to switches and LEDs on the board, but are instead connected to female headers. With direct access to the pins, students can concentrate on design fundamentals and learn about the programmability of I/O pins and PLDs. After successfully compiling and verifying a design with MAX+PLUS II, students can easily connect the assigned I/O pins to the switches and LEDs using common hook-up wire. Students can then download their design into the device and compare their design's simulation to the actual hardware implementation.

EPM7128S Prototyping Headers

The EPM7128S prototyping headers are female headers that surround the device and provide access to the device's signal pins. The 21 pins on each side of the 84-pin PLCC package connect to one of the 22-pin, dual-row 0.1-inch female headers. The pin numbers for the EPM7128S device are printed on the UP 1 Education Board; an X indicates an unassigned pin. Table 3 lists the pin numbers for the four female headers: P1, P2, P3, and P4. The power, ground, and JTAG signal pins are not accessible through these female headers.

Table 3. Pin Numbers for Each Prototyping Header Note (1)							
P1		**P2**		**P3**		**P4**	
Outside	**Inside**	**Outside**	**Inside**	**Outside**	**Inside**	**Outside**	**Inside**
75	76	12	13	33	34	54	55
77	78	14	15	35	36	56	57
79	80	16	17	37	38	58	59
81	82	18	19	39	40	60	61
83	84	20	21	41	42	62	63
1	2	22	23	43	44	64	65
3	4	24	25	45	46	66	67
5	6	26	27	47	48	68	69
7	8	28	29	49	50	70	71
9	10	30	31	51	52	72	73
11	X	32	X	53	X	74	X

Note:
(1) Inside refers to the row of female headers closest to the device; outside refers to the row of female headers furthest from the device.

University Program Design Laboratory Package User Guide

MAX_PB1 & MAX_PB2 Push-Buttons

MAX_PB1 and MAX_PB2 are two push-buttons that provide active-low signals and are pulled-up through 10-KΩ resistors. Connections to these signals are easily made by inserting one end of the hook-up wire into the push-button female header. The other end of the hook-up wire should be inserted into the appropriate female header assigned to the I/O pin of the EPM7128S device.

MAX_SW1 & MAX_SW2 Switches

MAX_SW1 and MAX_SW2 each contain eight switches that provide logic-level signals. These switches are pulled-up through 10-KΩ resistors. Connections to these signals are easily made by inserting one end of the hook-up wire into the female header aligned with the appropriate switch. The other end of the hook-up wire should be inserted into the appropriate female header assigned to the I/O pin of the EPM7128S device. The switch output is set to logic 1 when the switch is open and set to logic 0 when the switch is closed.

D1 through D16 LEDs

The UP 1 Education Board contains 16 LEDs that are pulled-up with a 330-Ω resistor. An LED is illuminated when a logic 0 is applied to the female header associated with the LED. LEDs D1 through D8 are connected in the same sequence to the female headers (i.e., D1 is connected to position 1, and D2 is connected to position 2). LEDs D9 through D16 are connected in the same sequence to the female headers (i.e., D9 is connected to position 1, and D10 is connected to position 2). See Figure 3.

Figure 3. LED Positions

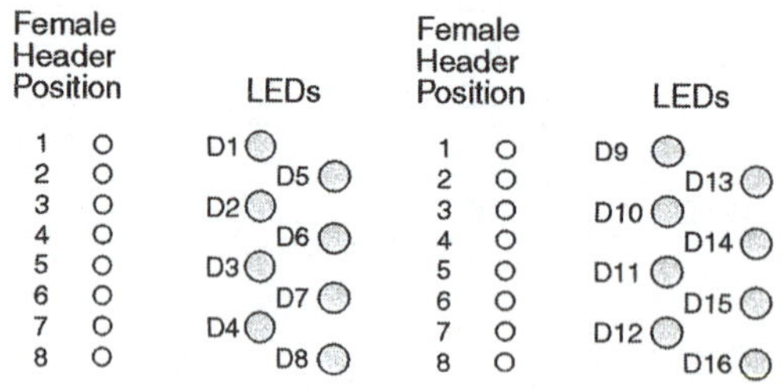

MAX_DIGIT Display

MAX_DIGIT is a dual-digit seven-segment display connected directly to the EPM7128S device. Each LED segment of the display can be illuminated by driving the connected EPM7128S device I/O pin with a logic 0. Figure 4 shows the name of each segment.

Figure 4. Display Segment Name

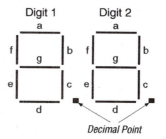

Table 4 lists the pin assignments for each segment.

Table 4. MAX_DIGIT Segment I/O Connections		
Display Segment	**Pin for Digit 1**	**Pin for Digit 2**
a	58	69
b	60	70
c	61	73
d	63	74
e	64	76
f	65	75
g	67	77
Decimal point	68	79

MAX_EXPANSION

MAX_EXPANSION is a dual row of 0.1-inch spaced holes for accessing signal I/O pins and global signals on the EPM7128S device, power, and ground. Figure 5 shows the numbering convention for the holes.

University Program Design Laboratory Package User Guide

Figure 5. MAX_EXPANSION Numbering Convention

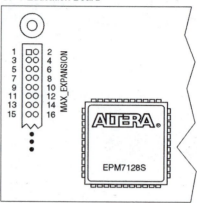

UP 1 Education Board

Table 5 lists the signal names and the EPM7128S device pins connected to each hole.

Table 5. MAX_EXPANSION Signal Names & Device Connections (Part 1 of 2)			
Hole Number	**Signal/Pin**	**Hole Number**	**Signal/Pin**
1	RAW	2	GND
3	VCC	4	GND
5	VCC	6	GND
7	No Connect	8	No Connect
9	No Connect	10	No Connect
11	No Connect	12	GCLRn/1
13	OE1/84	14	OE2/GCLK2/2
15	4	16	5
17	6	18	8
19	9	20	10
21	11	22	12
23	15	24	16
25	17	26	18
27	20	28	21
29	22	30	24
31	25	32	27
33	28	34	29
35	30	36	31
37	33	38	34

University Program Design Laboratory Package User Guide

Table 5. MAX_EXPANSION Signal Names & Device Connections (Part 2 of 2)			
Hole Number	Signal/Pin	Hole Number	Signal/Pin
39	35	40	36
41	37	42	39
43	40	44	41
45	44	46	45
47	46	48	48
49	49	50	50
51	51	52	52
53	54	54	55
55	56	56	57
57	VCC	58	GND
59	VCC	60	GND

EPF10K20 Device

The UP 1 Education Board provides the following resources for the EPF10K20 device. The pins from the EPF10K20 device are pre-assigned to switches and LEDs on the board.

- JTAG chain connection for the ByteBlaster
- Socket for an EPC1 Configuration EPROM
- 2 momentary push-button switches
- 1 octal dipswitch
- Dual-digit 7-segment display
- On-board oscillator (25.175 MHz)
- VGA port
- Mouse port
- 3 expansion ports, each with 42 I/O pins and 7 global pins   .

FLEX_PB1 & FLEX_PB2 Push-Buttons

FLEX_PB1 and FLEX_PB2 are two push-buttons that provide active-low signals to two general-purpose I/O pins on the EPF10K20 device. FLEX_PB1 is connected to pin 28, and FLEX_PB2 is connected to pin 29. Each push-button is pulled-up through a 10-KΩ resistor.

FLEX_SW1 Switches

FLEX_SW1 contains eight switches that provide logic-level signals to eight general-purpose I/O pins on the EPF10K20 device. An input pin is set to logic 1 when the switch is open and set to logic 0 when the switch is closed. Table 6 lists the pin assignment for each switch.

Table 6. FLEX_SW1 Pin Assignments

Switch	EPF10K20 Pin
FLEX_SWITCH-1	41
FLEX_SWITCH-2	40
FLEX_SWITCH-3	39
FLEX_SWITCH-4	38
FLEX_SWITCH-5	36
FLEX_SWITCH-6	35
FLEX_SWITCH-7	34
FLEX_SWITCH-8	33

FLEX_DIGIT Display

FLEX_DIGIT is a dual-digit seven-segment display connected directly to the EPF10K20 device. Each LED segment on the display can be illuminated by driving the connected EPF10K20 device I/O pin with a logic 0. See Figure 4 on page 9 for the name of each segment. Table 7 lists the pin assignment for each segment.

Table 7. FLEX_DIGIT Segment I/O Connections

Display Segment	Pin for Digit 1	Pin for Digit 2
a	6	17
b	7	18
c	8	19
d	9	20
e	11	21
f	12	23
g	13	24
Decimal point	14	25

VGA Interface

The VGA interface allows the EPF10K20 device to control an external video monitor. This interface is composed of a simple diode-resistor network and a 15-pin D-sub connector (labeled VGA), where the monitor can plug into the UP 1 Education Board. The diode-resistor network and D-sub connector are designed to generate voltages that conform to the VGA standard.

Information about the color of the screen, and the row and column indexing of the screen, are sent from the EPF10K20 device to the monitor via five signals. Three VGA signals are red, green, and blue, while the other two signals are horizontal and vertical synchronization. Manipulating these signals allows images to be written to the monitor's screen.

 See "VGA Driver Operation" on page 25 for details on how the VGA interface operates.

Table 8 lists the D-sub connector and the EPF10K20 device connections.

Table 8. D-Sub Connections

Signal	D-Sub Connector Pin	EPF10K20 Pin
RED	1	236
GREEN	2	237
BLUE	3	238
GND	6, 7, 8, 10, 11	–
HORIZ_SYNC	13	240
VERT_SYNC	14	239
No Connect	4, 5, 9, 15	–

MOUSE Connector

The MOUSE interface, which consists of a 6-pin mini-DIN connector, allows the EPF10K20 device to receive data from a PS/2 mouse or a PS/2 keyboard. The UP 1 Education Board provides power and ground to the attached mouse or keyboard. The EPF10K20 device outputs the DATA_CLOCK signal to the mouse and inputs the data signal from the mouse. Table 9 lists the signal names and the mini-DIN and EPF10K20 pin connections.

 See "MOUSE Interface Operation" on page 27 for details on how the MOUSE interface operates.

Table 9. MOUSE Connections

Mouse Signal	Mini-DIN Pin	EPF10K20 Pin
MOUSE_CLK	1	30
MOUSE_DATA	3	31
VCC	5	–
GND	2	–

University Program Design Laboratory Package User Guide

FLEX_EXPAN_A, FLEX_EXPAN_B & FLEX_EXPAN_C

FLEX_EXPAN_A, FLEX_EXPAN_B, and FLEX_EXPAN_C are dual rows of 0.1-inch spaced holes for accessing signal I/O pins and global signals on the EPF10K20 device, power, and ground. Figure 6 shows the numbering convention for these holes.

Figure 6. FLEX_EXPAN_A, FLEX_EXPAN_B & FLEX_EXPAN_C Numbering Convention

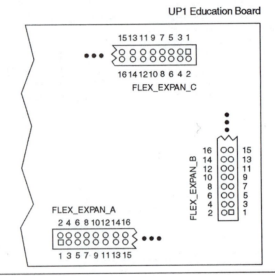

Tables 10, 11, and 12 list the signal name and the EPF10K20 device pin connected to each hole.

Table 10. FLEX_EXPAN_A Signal Names & Device Connections (Part 1 of 2)			
Hole Number	**Signal/Pin**	**Hole Number**	**Signal/Pin**
1	RAW	2	GND
3	VCC	4	GND
5	VCC	6	GND
7	No Connect	8	DI1/99
9	DI2/92	10	DI3/210
11	DI4/212	12	DEV_CLR/209
13	DEV_OE/213	14	DEV_CLK2/211
15	45	16	46
17	48	18	49
19	50	20	51
21	53	22	54

University Program Design Laboratory Package User Guide

Table 10. FLEX_EXPAN_A Signal Names & Device Connections (Part 2 of 2)

Hole Number	Signal/Pin	Hole Number	Signal/Pin
23	55	24	56
25	61	26	62
27	63	28	64
29	65	30	66
31	67	32	68
33	70	34	71
35	72	36	73
37	74	38	75
39	76	40	78
41	79	42	80
43	81	44	82
45	83	46	84
47	86	48	87
49	88	50	94
51	95	52	97
53	98	54	99
55	100	56	101
57	VCC	58	GND
59	VCC	60	GND

Table 11. FLEX_EXPAN_B Signal Names & Device Connections (Part 1 of 2)

Hole Number	Signal/Pin	Hole Number	Signal/Pin
1	RAW	2	GND
3	VCC	4	GND
5	VCC	6	GND
7	No Connect	8	DI1/99
9	DI2/92	10	DI3/210
11	DI4/212	12	DEV_CLR/209
13	DEV_OE/213	14	DEV_CLK2/211
15	109	16	110
17	111	18	113
19	114	20	115
21	116	22	117
23	118	24	119
25	120	26	126

University Program Design Laboratory Package User Guide

Table 11. FLEX_EXPAN_B Signal Names & Device Connections (Part 2 of 2)

Hole Number	Signal/Pin	Hole Number	Signal/Pin
27	127	28	128
29	129	30	131
31	132	32	133
33	134	34	136
35	137	36	138
37	139	38	141
39	142	40	143
41	144	42	146
43	147	44	148
45	149	46	151
47	152	48	153
49	154	50	156
51	157	52	158
53	159	54	161
55	162	56	163
57	VCC	58	GND
59	VCC	60	GND

Table 12. FLEX_EXPAN_C Signal Names & Device Connections (Part 1 of 2)

Hole Number	Signal/Pin	Hole Number	Signal/Pin
1	RAW	2	GND
3	VCC	4	GND
5	VCC	6	GND
7	No Connect	8	DI1/99
9	DI2/92	10	DI3/210
11	DI4/212	12	DEV_CLR/209
13	DEV_OE/213	14	DEV_CLK2/211
15	175	16	181
17	182	18	183
19	184	20	185
21	186	22	187
23	188	24	190
25	191	26	192
27	193	28	194
29	195	30	196

University Program Design Laboratory Package User Guide

Table 12. FLEX_EXPAN_C Signal Names & Device Connections (Part 2 of 2)			
Hole Number	**Signal/Pin**	**Hole Number**	**Signal/Pin**
31	198	32	198
33	200	34	201
35	202	36	203
37	204	38	206
39	207	40	208
41	214	42	215
43	217	44	218
45	219	46	220
47	221	48	222
49	223	50	225
51	226	52	227
53	228	54	229
55	230	56	231
57	VCC	58	GND
59	VCC	60	GND

Software Installation

This section describes how to install the MAX+PLUS II version 7.21 Student Edition software for the following operating systems:

- Windows 3.11 and Windows NT 3.51
- Windows 95 and Windows NT 4.0

After installation, students can register to obtain an authorization code via the Altera world-wide web site at the following URL: **http://www.altera.com/maxplus2-student**.

For complete installation instructions, refer to the **read.me** file on the *MAX+PLUS II 7.21 Student Edition CD-ROM* or go to the *MAX+PLUS II Getting Started* manual.

Windows 3.1 & Windows NT 3.51

Follow the steps shown below to install the MAX+PLUS II version 7.21 Student Edition software on your PC.

1. Insert the *MAX+PLUS II 7.21 Student Edition CD-ROM* into your CD-ROM drive.

2. In the Windows Program Manager, choose **Run** (File menu).

3. Type: <*CD-ROM drive*>`:\pc\maxplus2\install` and choose **OK**. You are guided through the installation procedure.

Windows 95 & Windows NT 4.0

Follow the steps shown below to install the MAX+PLUS II version 7.21 Student Edition software on your PC.

1. Insert the *MAX+PLUS II 7.21 Student Edition CD-ROM* into your CD-ROM drive.

2. Choose **Run** (Start menu).

3. Type: <*CD-ROM drive*>`:\pc\maxplus2\install` and choose **OK**. You are guided through the installation procedure.

Programming or Configuring Devices

Programming or configuring the devices on the UP 1 Education Board requires setting the on-board jumpers and the JTAG programming options in MAX+PLUS II, and connecting the ByteBlaster download cable to the PC's parallel port and to the JTAG_IN connector on the UP 1 Education Board. This section describes how to set these options to perform the following actions:

- Program only the EPM7128S device
- Configure only the EPF10K20 device
- Configure/program both devices
- Connect multiple UP 1 Education Boards together in a chain

EPM7128S Programming

This section describes the procedures for programming only the EPM7128S device, (i.e., how to set the on-board jumpers, connect the ByteBlaster download cable, and set options in the MAX+PLUS II software).

Setting the On-Board Jumpers for EPM7128S Programming

To program only the EPM7128S device in a JTAG chain, set the jumpers TDI, TDO, DEVICE, and BOARD as shown in Figure 7.

Figure 7. Jumper Settings for Programming Only the EPM7128S Device

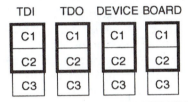

Connecting the ByteBlaster Download Cable for EPM7128S Programming

Attach the ByteBlaster directly to the PC's parallel port and to the JTAG_IN connector on the UP 1 Education Board. For more information on setting up the ByteBlaster, go to the *ByteBlaster Parallel Port Download Cable Data Sheet*.

Setting the JTAG Options in MAX+PLUS II for EPM7128S Programming

The following steps describe how to use the MAX+PLUS II software to program the EPM7128S device in a JTAG chain. For more information on how to use MAX+PLUS II, see MAX+PLUS II Help.

1. To program more than one EPM7128S device, turn on the *Multi-Device JTAG Chain* command (JTAG menu) in the MAX+PLUS II Programmer.

2. Choose **Multi-Device JTAG Chain Setup** (JTAG menu).

3. In the **Multi-Device JTAG Chain Setup** dialog box, select *EPM7128S* in the *Device Name* drop-down list box.

4. Type the name of the programming file for the EPM7128S device in the *Programming File Name* box. The **Select Programming File** button can also be used to browse your computer's directory structure to locate the appropriate programming file.

5. Choose **Add** to add the device and associated programming file to the *Device Names & Programming File Names* box. The number to the left of the device name shows the order of the device in the JTAG chain. The device's associated programming file is displayed on the same line as the device name. If no programming file is associated with a device, "<none>" is displayed next to the device name.

6. Choose **Detect JTAG Chain Info** to have the ByteBlaster check the device count, JTAG ID code, and total instruction length of the JTAG chain. A message just above the **Detect JTAG Chain Info** button reports the information detected by the ByteBlaster. You must manually verify that this message matches the information in the *Device Names & Programming File Names* box.

7. To save the current settings to a JTAG Chain File (**.jcf**) for future use, choose **Save JCF**. In the **Save JCF** dialog box, type the name of the file in the *File Name* box and then select the desired directory in the *Directories* box. Choose **OK**.

8. Choose **OK** to save your changes.

9. In the MAX+PLUS II Programmer, choose **Program**.

EPF10K20 Configuration

This section describes the procedures for configuring only the EPF10K20 device, (i.e., how to set the on-board jumpers, connect the ByteBlaster download cable, and set options in the MAX+PLUS II software).

Setting the On-Board Jumpers for EPF10K20 Configuration

To configure only the EPF10K20 device in a JTAG chain, set the jumpers TDI, TDO, DEVICE, and BOARD as shown in Figure 8.

Figure 8. Jumper Settings for Configuring Only the EPF10K20 Device

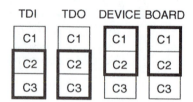

University Program Design Laboratory Package User Guide

Connecting the ByteBlaster Download Cable for EPF10K20 Configuration

Attach the ByteBlaster directly to the PC's parallel port and to the JTAG_IN connector on the UP 1 Education Board. For more information on setting up the ByteBlaster, go to the *ByteBlaster Parallel Port Download Cable Data Sheet.*

Setting the JTAG Options in MAX+PLUS II for EPF10K20 Configuration

The following steps describe how to use MAX+PLUS II to configure the EPF10K20 device in a JTAG chain. For more information on how to configure a device, see MAX+PLUS II Help.

1. To configure more than one EPF10K20 device, turn on the *Multi-Device JTAG Chain* command (JTAG menu) in the MAX+PLUS II Programmer.

2. Choose **Multi-Device JTAG Chain Setup** (JTAG menu).

3. In the **Multi-Device JTAG Chain Setup** dialog box, select *EPF10K20* in the *Device Name* drop-down list box.

4. Type the name of the programming file for the EPF10K20 device in the *Programming File Name* box. The **Select Programming File** button can also be used to browse your computer's directory structure to locate the appropriate programming file.

5. Choose **Add** to add the device and associated programming file to the *Device Names & Programming File Names* box. The number to the left of the device name shows the order of the device in the JTAG chain. The device's associated programming file is displayed on the same line as the device name. If no programming file is associated with a device, "<none>" is displayed next to the device name.

6. Choose **Detect JTAG Chain Info** to have the ByteBlaster check the device count, JTAG ID code, and total instruction length of the JTAG chain. A message just above the **Detect JTAG Chain Info** button reports the information detected by the ByteBlaster. You must manually verify that this message matches the information in the *Device Names & Programming File Names* box.

7. To save the current settings to a JCF for future use, choose **Save JCF**. In the **Save JCF** dialog box, type the name of the file in the *File Name* box and then select the desired directory in the *Directories* box. Choose **OK**.

8.   Choose **OK** to save your changes.

9.   In the MAX+PLUS II Programmer, choose **Configure**.

Configure/Program Both Devices

This section describes the procedures for configuring/programming both the EPF10K20 and EPM7128S devices in a JTAG chain, (i.e., how to set the on-board jumpers, connect the ByteBlaster download cable, and set options in the MAX+PLUS II software).

Setting the On-Board Jumpers for Configuring/Programming Both Devices

To configure/program both the EPF10K20 and EPM7128S devices in a multi-device JTAG chain, set the jumpers TDI, TDO, DEVICE, and BOARD as shown in Figure 9.

Figure 9. Jumper Settings for Configuring/Programming Both Devices

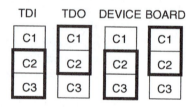

Connecting the ByteBlaster Download Cable for Configuring/Programming Both Devices

Attach the ByteBlaster directly to the PC's parallel port and to the JTAG_IN connector on the UP 1 Education Board. For more information on setting up the ByteBlaster, go to the *ByteBlaster Parallel Port Download Cable Data Sheet*.

Setting the JTAG Options in MAX+PLUS II for Configuring/Programming Both Devices

The following steps describe how to use MAX+PLUS II to configure/program both devices in a multi-device JTAG chain. For more information on how to program or configure a device, see MAX+PLUS II Help.

1. Turn on the *Multi-Device JTAG Chain* command (JTAG menu).

2. Choose **Multi-Device JTAG Chain Setup** (JTAG menu).

3. In the **Multi-Device JTAG Chain Setup** dialog box, select the first target device name in the *Device Name* drop-down list box.

4. In the *Programming File Name* box, type the name of the programming file for the device listed in the *Device Name* box. The **Select Programming File** button can also be used to browse your computer's directory structure to locate the appropriate programming file.

5. Choose **Add** to add the device and associated programming file to the *Device Names & Programming File Name* box. The number to the left of the device name shows the device's order in the JTAG chain. The device's associated programming file is displayed on the same line as the device name. If no programming file is associated with a device, "<none>" is displayed next to the device name.

6. Repeat steps 3 through 5 to add information for each device in the JTAG chain.

7. Choose **Detect JTAG Chain Info** to have the ByteBlaster check the device count, JTAG ID code, and total instruction length of the multi-device JTAG chain. A message just above the **Detect JTAG Chain Info** button reports the information detected by the ByteBlaster. You must manually verify that this message matches the information in the *Device Names & Programming File Names* box.

8. To save the current settings to a JCF for future use, choose **Save JCF**. In the **Save JCF** dialog box, type the name of the file in the *File Name* box and then select the desired directory in the *Directories* box. Choose **OK**.

9. Choose **OK** to save the changes.

10. In the MAX+PLUS II Programmer, choose **Configure** to configure all the EPF10K20 devices in the JTAG chain. Then, choose **Program** to program all the EPM7128S devices in the JTAG chain.

Connect Multiple UP 1 Education Boards Together in a Chain

This section describes the procedures for connecting multiple UP 1 Education Boards together, (i.e., how to set the on-board jumpers, connect the ByteBlaster download cable, and set options in the MAX+PLUS II software).

University Program Design Laboratory Package User Guide

Setting the On-Board Jumpers for Connecting Multiple UP 1 Education Boards Together

To configure/program EPM7128S and EPF10K20 devices on multiple UP 1 Education Boards connected in a multi-device JTAG chain, set the jumpers TDI, TDO, DEVICE, and BOARD for all boards except the last board in the chain as shown in Figure 10.

Figure 10. Jumper Settings for All Boards Except the Last Board in the Chain

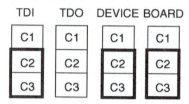

The last UP 1 Education Board in the chain can configure/program one or both devices. However, the BOARD jumper must be set as shown in Figure 11.

Figure 11. Jumper Settings for the Last Board in the Chain

The TDI, TDO, and DEVICE settings depend on which configuration you use.

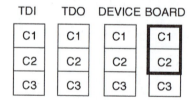

Connecting the ByteBlaster Download Cable for Connecting Multiple UP 1 Education Boards Together

Attach the ByteBlaster directly to your PC's parallel port and to the JTAG_IN connector on the UP 1 Education Board. For more information on setting up the ByteBlaster, go to the *ByteBlaster Parallel Port Download Cable Data Sheet*.

Setting the JTAG Options in MAX+PLUS II for Connecting Multiple UP 1 Education Boards Together

For information on how to set the JTAG Options in MAX+PLUS II, see "Setting the JTAG Options in MAX+PLUS II for Configuring/Programming Both Devices" on page 22.

VGA Driver Operation

A standard VGA monitor consists of a grid of pixels that can be divided into rows and columns. A VGA monitor typically contains 480 rows, with 640 pixels per row, as shown in Figure 12. Each pixel can display various colors, depending on the state of the red, green, and blue signals.

Figure 12. VGA Monitor

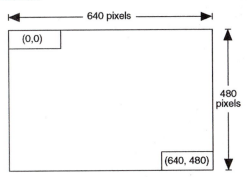

Each VGA monitor has an internal clock that determines when each pixel is updated. This clock operates at the VGA-specified frequency of 25.175 MHz. The monitor refreshes the screen in a prescribed manner that is partially controlled by the horizontal and vertical synchronization signals. The monitor starts each refresh cycle by updating the pixel in the top left-hand corner of the screen, which can be treated as the origin of an X–Y plane (see Figure 12). After the first pixel is refreshed, the monitor refreshes the remaining pixels in the row. When the monitor receives a pulse on the horizontal synchronization, it refreshes the next row of pixels. This process is repeated until the monitor reaches the bottom of the screen. When the monitor reaches the bottom of the screen, the vertical synchronization pulses, causing the monitor to begin refreshing pixels at the top of the screen (i.e., at [0,0]).

VGA Timing

For the VGA monitor to work properly, it must receive data at specific times with specific pulses. Horizontal and vertical synchronization pulses must occur at specified times to synchronize the monitor while it is receiving color data. Figures 13 and 14 show the timing waveforms for the color information with respect to the horizontal and vertical synchronization signals.

University Program Design Laboratory Package User Guide

Figure 13. Horizontal Refresh Cycle

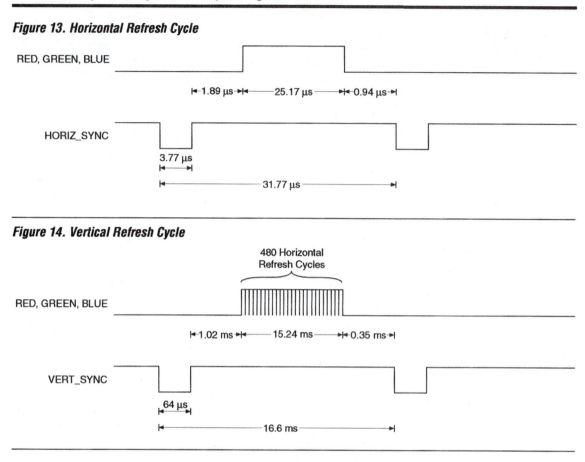

Figure 14. Vertical Refresh Cycle

The frequency of operation and the number of pixels that the monitor must update determines the time required to update each pixel, and the time required to update the whole screen. The following equations roughly calculate the time required for the monitor to perform all of its functions.

University Program Design Laboratory Package User Guide

$$T_{pixel} = 1/f_{CLK}$$
$$= 40 \text{ ns}$$

$$T_{ROW} = A$$
$$= (T_{pixel} \times 640 \text{ pixels})/(row + guard\ bands)$$
$$= 25\ \mu s + B + C + E$$
$$= 31.77\ \mu s$$

$$T_{screen} = (T_{ROW} \times 480 \text{ rows}) + guard\ bands$$
$$= 15.5\ ms + P + Q + S$$
$$= 16.6\ ms$$

$$f_{RR} = 1/T_{ROW}$$
$$= 31.5\ KHz$$

$$f_{SR} = 1/T_{screen}$$
$$= 60\ Hz$$

Where:
T_{pixel} = Time required to update a pixel
f_{CLK} = 25.175 MHz
T_{ROW} = Time required to update one row
T_{screen} = Time required to update the screen
f_{RR} = Row refresh frequency
f_{SR} = Screen refresh frequency

The monitor writes to the screen by sending red, green, blue, horizontal sync, and vertical synchronization signals when the screen is at the expected location. Once the timing of the horizontal and vertical synchronization signals is accurate, the monitor only needs to keep track of the current location, so it can send the correct color data to the pixel.

MOUSE Interface Operation

You can connect a mouse to the UP 1 Education Board via the 6-pin mini-DIN connector. The data is sent using a synchronous serial protocol, and the transmission is controlled by the CLK and DATA signals. During non-transmission, CLK is at logic 1 and DATA can be either logic 0 or logic 1.

Each transmission contains one start bit, eight data bits, odd parity, and one stop bit. Data transmission starts from the least significant bit (LSB), i.e., the sequence of transmission is start bit, DATA0 through DATA7, parity, stop bit. Start bits are logic 0, and stop bits are logic 1. Each clock period is 30 to 50 μsec; the data transition to the falling edge of the clock is 5 to 25 μsec. Table 13 shows the data packet format.

University Program Design Laboratory Package User Guide

Table 13. Data Packet Format						Note (1)		
Packet Number	D7	D6	D5	D4	D3	D2	D1	D0
1	YV	XV	YS	XS	1	0	R	L
2	X7	X6	X5	X4	X3	X2	X1	X0
3	Y7	Y6	Y5	Y4	Y3	Y2	Y1	Y0

Note:
(1)   where: L   = Left button state (1 = left mouse button is pressed down)
              R   = Right button state (1 = right mouse button is pressed down)
              X0 – X7   = Movement in X direction
              Y0 – Y7   = Movement in Y direction
              XS, YS   = Movement data sign (1 = negative)
              XV, YV   = Movement data overflow (1 = overflow has occurred)

The mouse operates on a Cartesian coordinate system (i.e., moving to the right is positive, moving to the left is negative, moving up is positive, and moving down is negative). The magnitude of the movement is a function of the mouse's rate of movement. The faster the mouse moves, the greater the magnitude.

101 Innovation Drive
San Jose, CA 95134
(408) 544-7000
http://www.altera.com
University Program:
university@altera.com
Literature Services:
(888) 3-ALTERA
lit_req@altera.com

nsai

I.S. EN ISO 9001

University Program
Design Laboratory Package

November 1997, ver. 1.1   **User Guide Supplement**

This user guide supplement provides updated pin-out and timing information for the UP 1 Education Board. This supplement should be used together with the *University Program Design Laboratory Package User Guide.*

Pin-Out Information

Table 1 provides updated pin-out information for the MAX_EXPANSION port on the UP 1 Education Board.

Table 1. MAX_EXPANSION Signal Names & Device Connections (Part 1 of 2)			
Hole Number	**Signal/Pin,** *Note (1)*	**Hole Number**	**Signal/Pin,** *Note (1)*
1	RAW	2	GND
3	VCC	4	GND
5	VCC	6	GND
7	No Connect	8	No Connect
9	No Connect	10	No Connect
11	No Connect	12	GCLRn/1
13	OE1/84	14	OE2/GCLK2/2
15	4	16	5
17	6	18	8
19	9	20	10
21	11	22	12
23	15	24	16
25	17	26	18
27	20	28	21
29	22	30	25
31	24	32	27
33	29	34	28
35	31	36	30
37	33	38	34
39	35	40	36
41	37	42	40
43	39	44	41
45	44	46	46

University Program Design Laboratory Package User Guide Supplement

Table 1. MAX_EXPANSION Signal Names & Device Connections (Part 2 of 2)

Hole Number	Signal/Pin, *Note (1)*	Hole Number	Signal/Pin, *Note (1)*
47	45	48	48
49	50	50	49
51	52	52	51
53	54	54	55
55	56	56	57
57	VCC	58	GND
59	VCC	60	GND

Note:
(1)   The updated pin numbers are highlighted in gray.

Table 2 shows updated pin-out information for the FLEX_EXPAN_C port on the UP 1 Education Board.

Table 2. FLEX_EXPAN_C Signal Names & Device Connections (Part 1 of 2)

Hole Number	Signal/Pin, *Note (1)*	Hole Number	Signal/Pin, *Note (1)*
1	RAW	2	GND
3	VCC	4	GND
5	VCC	6	GND
7	No Connect	8	DI1/99
9	DI2/92	10	DI3/210
11	DI4/212	12	DEV_CLR/209
13	DEV_OE/213	14	DEV_CLK2/211
15	175	16	181
17	182	18	183
19	184	20	185
21	186	22	187
23	188	24	190
25	191	26	192
27	193	28	194
29	195	30	196
31	198	32	199
33	200	34	201
35	202	36	203
37	204	38	206
39	207	40	208
41	214	42	215
43	217	44	218

University Program Design Laboratory Package User Guide Supplement

Table 2. FLEX_EXPAN_C Signal Names & Device Connections (Part 2 of 2)

Hole Number	Signal/Pin, *Note (1)*	Hole Number	Signal/Pin, *Note (1)*
45	219	46	220
47	221	48	222
49	223	50	225
51	226	52	227
53	228	54	229
55	230	56	231
57	VCC	58	GND
59	VCC	60	GND

Note:
(1)   The updated pin numbers are highlighted in gray.

VGA Timing Information

For the VGA monitor to work properly, it must receive data at specific times with specific pulses. Horizontal and vertical synchronization pulses must occur at specified times to synchronize the monitor while it is receiving color data. Figures 1 and 2 show updated timing waveforms for color information with respect to horizontal and vertical synchronization signals.

Figure 1. Horizontal Refresh Cycle

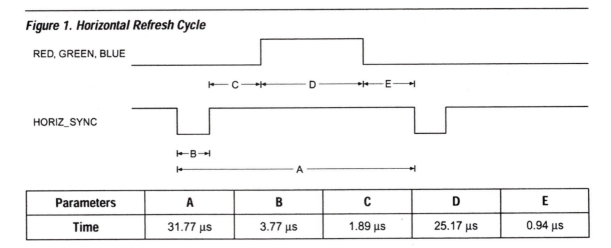

Parameters	A	B	C	D	E
Time	31.77 µs	3.77 µs	1.89 µs	25.17 µs	0.94 µs

University Program Design Laboratory Package User Guide Supplement

Figure 2. Vertical Refresh Cycle

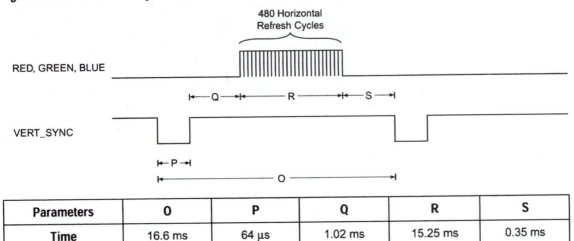

Parameters	O	P	Q	R	S
Time	16.6 ms	64 μs	1.02 ms	15.25 ms	0.35 ms

The following updated equations determine the time required for a monitor to update each pixel and to update a whole screen.

$$T_{pixel} = 1/f_{CLK} = 40 \text{ ns}$$

$$
\begin{aligned}
T_{ROW} &= A = B + C + D + E \\
&= (T_{pixel} \times 640 \text{ pixels}) + \text{row} + \text{guard bands} = 31.77 \text{ μs}
\end{aligned}
$$

$$
\begin{aligned}
T_{screen} &= O = P + Q + R + S \\
&= (T_{ROW} \times 480 \text{ rows}) + \text{guard bands} = 16.6 \text{ ms}
\end{aligned}
$$

Where:
T_{pixel} = Time required to update a pixel
f_{CLK} = 25.175 MHz
T_{ROW} = Time required to update one row
T_{screen} = Time required to update the screen
B, C, E, P, Q, S = Guard bands

APPENDIX B

VHDL Language Reference

1.1 Valid Names

A valid name in VHDL consists of a letter followed by any number of letters or numbers, without spaces. VHDL is not case sensitive. An underscore may be used within a name, but may not begin or end the name. Two consecutive underscores are not permitted.

III EXAMPLES

Valid names:
```
decode4
just_in_time
What_4
```
Invalid names:
```
4decode        (begins with a digit)
in__time       (two consecutive underscores)
_What_4        (begins with underscore)
my design      (space inside name)
your_words?    (special character ? not allowed)
```

1.2 Comments

A comment is explanatory text that is ignored by the VHDL compiler. It is indicated by two consecutive hyphens.

III EXAMPLE

```
-- This is a comment.
```

1.3 Entity and Architecture

All VHDL files require an entity declaration and an architecture body. The entity declaration indicates the input and output ports of the design. The architecture body details the internal relationship between inputs and outputs. The VHDL file name must be the same as the entity name.

Syntax:

```
LIBRARY ieee;
USE ieee.std_logic_1164.ALL;

ENTITY __entity_name IS
  GENERIC(define parameters);
  PORT(define inputs and outputs);
END __entity_name;

ARCHITECTURE a OF __entity_name IS
  SIGNAL and COMPONENT declarations;
BEGIN
  statements;
END a;
```

▌ EXAMPLES:

```
--Majority vote circuit (majority.vhd)
LIBRARY ieee;
USE ieee.std_logic_1164.ALL;

ENTITY majority IS
  PORT(
      a, b, c: IN   STD_LOGIC;
      y      : OUT  STD_LOGIC);
END majority;

ARCHITECTURE a OF majority IS
BEGIN
  y <= (a and b) or (b and c) or (a and c);
END a;

-- 2-line-to-4-line decoder with active-HIGH outputs (decoder.vhd)
LIBRARY ieee;
USE ieee.std_logic_1164.ALL;

ENTITY decoder IS
  PORT(
      d : IN  STD_LOGIC_VECTOR (1 downto 0);
      y : OUT STD_LOGIC_VECTOR (3 downto 0));
END decoder;

ARCHITECTURE a OF decoder IS
BEGIN
  WITH d SELECT
      y <=   "0001" WHEN "00",
             "0010" WHEN "01",
             "0100" WHEN "10",
             "1000" WHEN "11",
             "0000" WHEN others;
END a;
```

1.4 Ports

A port in VHDL is a connection from a VHDL design entity to the outside world. The direction or directions in which a port may operate is called its mode. A VHDL port may have one of four modes: IN (input only), OUT (output only), INOUT (bidirectional), and BUFFER (output, with feedback from the output back into the design entity). The mode of a port is declared in the port statement of an entity declaration or component declaration.

EXAMPLES:

```
ENTITY mux IS
PORT(
    s1, s0      : IN STD_LOGIC;
    y0, y1, y2, y3 : OUT STD_LOGIC);
END mux;


ENTITY srg8 IS
PORT(
    clock, reset  : IN        STD_LOGIC;
    q             : BUFFER    STD_LOGIC_VECTOR (7 downto 0));
END srg8;
```

1.5 Signals and Variables

A signal is like an internal wire connecting two or more points inside an architecture body. It is declared before the BEGIN statement of an *architecture body* and is global to the architecture. Its value is assigned with the <= operator.

A variable is an piece of working memory, local to a specific process. It is declared before the BEGIN statement of a *process* and is assigned using the := operator.

EXAMPLE:

```
ARCHITECTURE a OF design4 IS
    SIGNAL connect : STD_LOGIC_VECTOR ( 7 downto 0);
BEGIN
    PROCESS check IS
        VARIABLE count : INTEGER RANGE 0 TO 255;
    BEGIN
        IF (clock'EVENT and clock = '1') THEN
            count := count + 1;     -- Variable assignment statement
        END IF;
    END PROCESS;
    connect <= a and b;             -- Signal assignment statement
END a;
```

1.6 Type

The type of a port, signal, or variable determines the values it can have. For example, a signal of type BIT can only have values '0' and '1'. A signal of type INTEGER can have any integer value, up to the limits of the bit size of the particular computer system for which the VHDL compiler is designed. Some common types are:

Type	Values	How written
BIT	'0', '1'	Single quotes
STD_LOGIC	'U', 'X', '0', '1', 'Z', 'W',	Single quotes
(see **Section 1.6.1**)	'L', 'H', '-'	
INTEGER	Integer values	No quotes
BIT_VECTOR	Multiple instances of '0' and '1'	Double quotes (e.g., "00101")
STD_LOGIC_VECTOR	Multiple instances of 'U', 'X',	Double quotes (e.g., "11ZZ00")
	'0', '1', 'Z', 'W', 'L', 'H', '-'	

1.6.1   STD_LOGIC

The STD_LOGIC (standard logic) type, also called IEEE Std.1164 Multi-Valued Logic, gives a broader range of output values than just '0' and '1'. Any port, signal, or variable of type STD_LOGIC or STD_LOGIC_VECTOR can have any of the following values.

```
'U',  -- Uninitialized
'X',  -- Forcing  Unknown
'0',  -- Forcing  0
'1',  -- Forcing  1
'Z',  -- High Impedance
'W',  -- Weak     Unknown
'L',  -- Weak     0
'H',  -- Weak     1
'-',  -- Don't care
```

"Forcing" levels are deemed to be the equivalent of a gate output. "Weak" levels are specified by a pull-up or pull-down resistor. The 'Z' state is used as the high-impedance state of a tristate buffer.

The majority of applications can be handled by 'X', '0', '1', and 'Z' values.

To use STD_LOGIC in a VHDL file, you must include the following reference to the VHDL **library** called **ieee** and the **std_logic_1164** package before the entity declaration.

```
LIBRARY ieee;
USE ieee.std_logic_1164.ALL;
```

1.6.2   Enumerated Type

An enumerated type is a user-defined type that lists all possible values for a port, signal, or variable. One use of an enumerated type is to list the states of a state machine.

▌▌ EXAMPLE:

```
TYPE STATE_TYPE IS (idle, start, pulse, read);
SIGNAL state: STATE_TYPE;
```

1.7   Libraries and Packages

A library is a collection of previously compiled VHDL constructs that can be used in a design entity. A package is an uncompiled collection of VHDL constructs that can be used in multiple design entities. Library names must be included at the beginning of a VHDL file, before the entity declaration, to use certain types or functions. The most obvious is the library **ieee**, which in the package **std_logic_1164**, defines the STD_LOGIC (standard logic) type.

Syntax:

```
LIBRARY __ library_name;
USE __library_name.__package_name.ALL;
```

EXAMPLES:

```
LIBRARY ieee;
USE ieee.std_logic_1164.ALL;      -- Defines STD_LOGIC type
USE ieee.std_logic_arith.ALL;     -- Defines arithmetic functions

LIBRARY lpm;                      -- Component declarations for the
USE lpm.lpm_components.ALL;       -- Library of Parameterized Modules

LIBRARY altera;                   -- Component declarations for
USE altera.maxplus2.ALL;          -- MAX+PLUS II primitives
```

2. Concurrent Structures

2.1. Concurrent Signal Assignment Statement

2.2. Selected Signal Assignment Statement

2.3. Conditional Signal Assignment Statements

2.4. Components

2.4.1. Component Declaration

2.4.2. Component Instantiation

2.4.3. Generic Clause

2.5. Generate Statement

2.6. Process Statement

A concurrent structure in VHDL acts as a separate component. A change applied to multiple concurrent structures acts on all affected structures at the same time. This is similar to a signal applied to multiple components in a circuit; a change in the signal will operate on all the components simultaneously.

2.1 Concurrent Signal Assignment Statement

A concurrent signal assignment statement assigns a port or signal the value of a Boolean expression or constant. This statement is useful for encoding a Boolean equation. Since the operators and, or, not, and xor have equal precedence in VHDL, the order of precedence must be made explicit by parentheses.

Syntax:

```
__signal <= __expression;
```

EXAMPLES:

```
sum <= (a xor b) xor c;
c_out <= ((a xor b) and c_in) or (a and b);
```

2.2 Selected Signal Assignment Statement

A selected signal assignment statement assigns one of several alternative values to a port or signal, based on the value of a selecting signal. It can be used to implement a truth table or a selecting circuit like a multiplexer.

Syntax:

```
label: WITH  __expression SELECT
__signal <= __expression WHEN __constant_value,
                __expression WHEN __constant_value,
                __expression WHEN __constant_value,
                __expression WHEN __constant_value;
```

▌▌ EXAMPLES:

```
-- decoder implemented as a truth table (2 inputs, 4 outputs)
-- d has been defined as STD_LOGIC_VECTOR (1 downto 0)
-- y has been defined as STD_LOGIC_VECTOR (3 downto 0)
WITH d SELECT
      y <=    "0001" WHEN "00",
              "0010" WHEN "01",
              "0100" WHEN "10",
              "1000" WHEN "11",
              "0000" WHEN others;
-- multiplexer
-- input signal assigned to y, depending on states of s1, s0
M:    WITH s SELECT
      y <=    d0 WHEN "00",
              d1 WHEN "01",
              d2 WHEN "10",
              d3 WHEN "11";
```

2.3 Conditional Signal Assignment Statement

A conditional signal assignment statement assigns a value to a port or signal based on a series of linked conditions. The basic structure assigns a value if the first condition is true. If not, another value is assigned if a second condition is true, and so on, until a default condition is reached. This is an ideal structure for a priority encoder.

Syntax:

```
__label:
__signal <=     __expression WHEN __boolean_expression ELSE
                __expression WHEN __boolean_expression ELSE
                __expression;
```

▌▌ EXAMPLE:

```
-- priority encoder
-- q defined as INTEGER RANGE 0 TO 7
-- d defined as STD_LOGIC_VECTOR (7 downto 0)
encoder:
  q   <= 7 WHEN d(7)='1' ELSE
         6 WHEN d(6)='1' ELSE
         5 WHEN d(5)='1' ELSE
         4 WHEN d(4)='1' ELSE
         3 WHEN d(3)='1' ELSE
         2 WHEN d(2)='1' ELSE
         1 WHEN d(1)='1' ELSE
         0;
```

2.4   Components

A VHDL file can use another VHDL file as a component. The general form of a design entity using components is:

```
ENTITY entity_name IS
    PORT ( input and output definitions);
END entity_name;

ARCHITECTURE arch_name OF entity_name IS
    component declaration(s);
    signal declaration(s);
BEGIN
    Component instantiation(s);
    Other statements;
END arch_name;
```

2.4.1   Component Declaration

A component declaration is similar in form to an entity declaration, in that it includes the required ports and parameters of the component. The difference is that it refers to a design described in a separate VHDL file. The ports and parameters in the component declaration may be a subset of those in the component file, but they must have the same names.

Syntax:

```
COMPONENT __component_name
GENERIC(__parameter_name : string := __default_value;
        __parameter_name : integer := __default_value);
PORT(
    __input  name, __input_name      : IN  STD_LOGIC;
    __bidir  name, __bidir_name      : INOUT STD_LOGIC;
    __output name, __output_name     : OUT  STD_LOGIC);
END COMPONENT;
```

EXAMPLE:

```
ARCHITECTURE adder OF add4pa IS
    COMPONENT full_add
        PORT(
            a, b, c_in : IN BIT;
            c_out, sum : OUT BIT);
    END COMPONENT;

    SIGNAL c : BIT_VECTOR (3 downto 1);
BEGIN
    statements
END adder;
```

2.4.2   Component Instantiation

Each instance of a component requires a component instantiation statement. Ports can be assigned explicitly with the => operator, or implicitly by inserting the user port name in the position of the corresponding port name within the component declaration.

Syntax:

```
__instance_name: __component_name
 GENERIC MAP (__parameter_name => __parameter_value ,
              __parameter_name => __parameter_value)
   PORT MAP (__component_port => __connect_port,
              __component_port => __connect_port);
```

▍▍ EXAMPLES:

```
-- Four Component Instantiation Statements
-- Explicit port assignments
adder1: full_add
    PORT MAP ( a     => a(1),
               b     => b(1),
               c_in  => c0,
               c_out => c(1),
               sum   => sum(1));
adder2: full_add
    PORT MAP ( a     => a(2),
               b     => b(2),
               c_in  => c(1),
               c_out => c(2),
               sum   => sum(2));
adder3: full_add
    PORT MAP ( a     => a(3),
               b     => b(3),
               c_in  => c(2),
               c_out => c(3),
               sum   => sum(3));
adder4: full_add
    PORT MAP ( a     => a(4),
               b     => b(4),
               c_in  => c(3),
               c_out => c4,
               sum   => sum (4));

-- Four component instantiations
-- Implicit port assignments
adder1: full_add PORT MAP (a(1), b(1), c0,   c(1), sum(1));
adder2: full_add PORT MAP (a(2), b(2), c(1), c(2), sum(2));
adder3: full_add PORT MAP (a(3), b(3), c(2), c(3), sum(3));
adder4: full_add PORT MAP (a(4), b(4), c(3), c4,   sum(4));
```

2.4.3 Generic Clause

A generic clause allows a component to be designed with one or more unspecified properties ("parameters") that are specified when the component is instantiated. A parameter specified in a generic clause must be given a default value with the := operator.

Syntax:

```
-- parameters defined in entity declaration of component file
ENTITY entity_name IS
    GENERIC(__parameter_name : type := __default_value;
            __parameter_name : type := __default_value);
    PORT (port declarations);
END entity  name;
```

```
                    -- Component declaration in top-level file also has generic clause.
                    -- Default values of parameters not specified.
                    COMPONENT component_name IS
                        GENERIC(__parameter_name : type;
                                __parameter_name : type);
                        PORT (port declarations);
                    END COMPONENT;

                    -- Parameters specified in generic map in component instantiation
                        __instance_name: __component_name
                        GENERIC MAP  (__parameter_name => __parameter_value,
                                      __parameter_name => __parameter_value)
                        PORT MAP (port instantiations);
```

█ EXAMPLE:

```
-- Component: behaviorally defined shift register
-- with default width of 4.
ENTITY srt_bhv IS
    GENERIC (width : POSITIVE := 4);
    PORT(
        serial_in, clk  : IN        STD_LOGIC;
        q               : BUFFER    STD_LOGIC_VECTOR(width-1 downto 0));
END srt_bhv;

ARCHITECTURE right_shift of srt_bhv IS
BEGIN
  PROCESS (clk)
  BEGIN
     IF (clk'EVENT and clk = '1') THEN
          q(width-1 downto 0) <= serial  in & q(width-1 downto 1);
     END IF;
  END PROCESS;
END right_shift;


-- srt8_bhv.vhd
-- 8-bit shift register that instantiates srt_bhv
LIBRARY ieee;
USE ieee.std_logic_1164.ALL;

ENTITY srt8_bhv IS
 PORT(
     data_in, clock : IN        STD_LOGIC;
        qo          : BUFFER    STD_LOGIC_VECTOR(7 downto 0));
END srt8_bhv;

ARCHITECTURE right  shift of srt8_bhv IS
-- component declaration
COMPONENT srt_bhv
 GENERIC (width : POSITIVE);
 PORT(
     serial_in, clk : IN  STD_LOGIC;
     q              : OUT STD_LOGIC_VECTOR(7 downto 0));
END COMPONENT;

(example continues)
```

```
BEGIN
-- component instantiation
Shift_right_8: srt_bhv
    GENERIC MAP  (width=> 8)
     PORT MAP  (serial_in => data_in,
                clk       => clock,
                q         => qo);
END right_shift;
```

2.5 Generate Statement

A generate statement is used to create multiple instances of a particular hardware structure. It relies on the value of one or more index variables to create the required number of repetitions.

Syntax:

```
   __generate_label:
FOR __index_variable IN __range GENERATE
    __statement;
    __statement;
END GENERATE;
```

▌▌ EXAMPLES:

```
-- Instantiate four full adders
adders:
FOR i IN 1 to 4 GENERATE
    adder: full_add PORT MAP (a(i), b(i), c(i-1), c(i), sum(i));
END GENERATE;

-- Instantiate four latches from MAX+PLUS II primitives
-- Requires the statements LIBRARY altera; and
-- USE altera.maxplus.ALL;
latch4:
FOR i IN 3 downto 0 GENERATE
latch_primitive: latch
    PORT MAP (d => d_in(i), ena => enable, q => q_out (i));
END GENERATE;
```

2.6 Process Statement

A process is a concurrent statement, but the statements inside the process are sequential. For example, a process can define a flip-flop, a separate component whose ports are affected concurrently, but the inside of the flip-flop acts sequentially. A process executes all statements inside it when there is a change of a signal in its sensitivity list. The process label is optional.

Syntax:

```
__process_label:
PROCESS (sensitivity list)
   variable declarations
BEGIN
   sequential statements
END PROCESS __process_label;
```

▌ EXAMPLE:

```
-- D latch
PROCESS (en)
BEGIN
    IF (en = '1') THEN
        q <= d;
    END IF;
END PROCESS;
```

3. Sequential Structures

3.1. If Statement

3.1.1. Evaluating Clock Functions

3.2. Case Statement

A sequential structure in VHDL is one in which the order of statements affects the operation of the circuit. It can be used to implement combinational circuits, but is primarily used to implement sequential circuits such as latches, counters, shift registers, and state machines. Sequential statements must be contained within a process.

3.1 If Statement

An IF statement executes one or more statements if a Boolean condition is satisfied.

Syntax:

```
IF __expression THEN
    __statement;
    __statement;
ELSIF __expression THEN
    __statement;
    __statement;
ELSE
    __statement;
    __statement;
END IF;
```

▌ EXAMPLE:

```
PROCESS (reset, load, clock)
    VARIABLE count INTEGER RANGE 0 TO 255;
BEGIN
    IF (reset = '0') THEN
        q <= 0;
    ELSIF (reset = '1' and load = '0') THEN
        q <= p;
    ELSIF (clock'EVENT and clock = '1') THEN
        count := count + 1;
        q <= count;
    END IF;
END PROCESS;
```

3.1.1   Evaluating Clock Functions

As implied in previous examples, the state of a system clock can be checked with an IF statement using the predefined attribute called EVENT. The clause clock'EVENT ("clock tick EVENT") is true if there has been activity on the signal called clock. Thus (clock'EVENT and clock = '1') is true just after a positive edge on clock.

3.2   Case Statement

A case statement is used to execute one of several sets of statements, based on the evaluation of a signal.

Syntax:

```
CASE __expression IS
    WHEN __constant_value =>
        __statement;
        __statement;
    WHEN __constant_value =>
        __statement;
        __statement;
    WHEN OTHERS =>
        __statement;
        __statement;
END CASE;
```

EXAMPLES:

```
-- Case evaluates 2-bit value of s and assigns
-- 4-bit values of x and y accordingly
-- Default case (others) required if using STD_LOGIC
CASE s IS
    WHEN "00" =>
        y <= "0001";
        x <= "1110";
    WHEN "01" =>
        y <= "0010";
        x <= "1101";
    WHEN "10" =>
        y <= "0100";
        x <= "1011";
    WHEN "11" =>
        y <= "1000";
        x <= "0111";
    WHEN others =>
        y <= "0000";
        x <= "1111";
END CASE;
```

```
-- This case evaluates the state variable "sequence"
-- that can have two possible values: "start" and "continue"
-- Values of out1 and out2 are also assigned for each case.
CASE sequence IS
    WHEN start =>
        IF in1 = '1' THEN
            sequence <= start;
            out1 <= '0';
            out2 <= '0';
        ELSE
            sequence <= continue;
            out1 <= '1';
            out2 <= '0';
END IF;
    WHEN continue =>
            sequence <= start;
            out1 <= '0';
            out2 <= '1';
END CASE;
```

Manufacturers' Data Sheets

Data Sheet List

Device	Description	Source/File Name	Pages
74LS00	Quad 2-input NAND Gate	Motorola/sn74ls00rev6.pdf	703
74LS02	Quad 2-input NOR Gate	Motorola/sn74ls02rev5.pdf	705
74LS04	Hex Inverter	Motorola/sn74ls04rev6.pdf	707
74LS05	Hex Inverter (Open Collector)	Motorola/sn74ls05rev6.pdf	709
74LS06/16	Hex Inverting Buffer (Open Collector)	Texas Instruments/sdls020a.pdf	711
75LS07	Hex Noninverting Buffer (Open Collector)	Texas Instruments/sdls021a.pdf	714
74LS08	Quad 2-input AND Gate	Motorola/sn74ls08rev6.pdf	717
74LS32	Quard 2-input OR Gate	Motorola/sn74ls32rev6.pdf	719
74LS86	Quard 2-input XOR Gate	Motorola/sn74ls86rev6.pdf	721
74F00	Quad 2-input NAND Gate	Texas Instruments/sdfs035a.pdf	723
74AS/ALS00	Quad 2-input NAND Gate	Texas Instruments/sdas187a.pdf	726
74HC00	Quad 2-input NAND Gate	Motorola/mc74hc00arev7a.pdf	731
74HCT00	Quad 2-input NAND Gate (TTL Input Levels)	Motorola/mc74hct00arev6.pdf	735
74VHC00	Quad 2-input NAND Gate	Motorola/mc74vhc00arev0.pdf	738
74VHCT00	Quad 2-input NAND Gate (TTL Input Levels)	Motorola/mc74vhct00arev0.pdf	741
74HCU04	Hex Inverter (Unbuffered)	Motorola/mc74hcu04arev1.pdf	744
74HC4049/4050	Hex Buffer	Motorola/mc74hc4049rev6.pdf	749
74LVX00	Quad 2-input NAND Gate	Motorola/mc74lvx00rev0b.pdf	753
74LCX00	Quad 2-input NAND Gate	Motorola/mc74lcx00rev1.pdf	756
MC14XXXB	4000B-series CMOS Gates	Motorola/mc14001brev3.pdf	759

SN74LS00

Quad 2-Input NAND Gate

- ESD > 3500 Volts

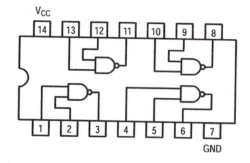

ON Semiconductor
Formerly a Division of Motorola
http://onsemi.com

**LOW
POWER
SCHOTTKY**

**PLASTIC
N SUFFIX
CASE 646**

**SOIC
D SUFFIX
CASE 751A**

GUARANTEED OPERATING RANGES

Symbol	Parameter	Min	Typ	Max	Unit
V_{CC}	Supply Voltage	4.75	5.0	5.25	V
T_A	Operating Ambient Temperature Range	0	25	70	°C
I_{OH}	Output Current – High			−0.4	mA
I_{OL}	Output Current – Low			8.0	mA

ORDERING INFORMATION

Device	Package	Shipping
SN74LS00N	14 Pin DIP	2000 Units/Box
SN74LS00D	14 Pin	2500/Tape & Reel

SN74LS00

DC CHARACTERISTICS OVER OPERATING TEMPERATURE RANGE (unless otherwise specified)

Symbol	Parameter	Limits			Unit	Test Conditions
		Min	Typ	Max		
V_{IH}	Input HIGH Voltage	2.0			V	Guaranteed Input HIGH Voltage for All Inputs
V_{IL}	Input LOW Voltage			0.8	V	Guaranteed Input LOW Voltage for All Inputs
V_{IK}	Input Clamp Diode Voltage		−0.65	−1.5	V	V_{CC} = MIN, I_{IN} = −18 mA
V_{OH}	Output HIGH Voltage	2.7	3.5		V	V_{CC} = MIN, I_{OH} = MAX, V_{IN} = V_{IH} or V_{IL} per Truth Table
V_{OL}	Output LOW Voltage		0.25	0.4	V	I_{OL} = 4.0 mA / V_{CC} = V_{CC} MIN, V_{IN} = V_{IL} or V_{IH} per Truth Table
			0.35	0.5	V	I_{OL} = 8.0 mA
I_{IH}	Input HIGH Current			20	μA	V_{CC} = MAX, V_{IN} = 2.7 V
				0.1	mA	V_{CC} = MAX, V_{IN} = 7.0 V
I_{IL}	Input LOW Current			−0.4	mA	V_{CC} = MAX, V_{IN} = 0.4 V
I_{OS}	Short Circuit Current (Note 1)	−20		−100	mA	V_{CC} = MAX
I_{CC}	Power Supply Current Total, Output HIGH			1.6	mA	V_{CC} = MAX
	Total, Output LOW			4.4		

Note 1: Not more than one output should be shorted at a time, nor for more than 1 second.

AC CHARACTERISTICS (T_A = 25°C)

Symbol	Parameter	Limits			Unit	Test Conditions
		Min	Typ	Max		
t_{PLH}	Turn–Off Delay, Input to Output		9.0	15	ns	V_{CC} = 5.0 V C_L = 15 pF
t_{PHL}	Turn–On Delay, Input to Output		10	15	ns	

 **MOTOROLA**

QUAD 2-INPUT NOR GATE

SN54/74LS02

QUAD 2-INPUT NOR GATE

LOW POWER SCHOTTKY

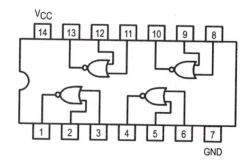

V_{CC}

14 13 12 11 10 9 8

1 2 3 4 5 6 7

GND

J SUFFIX
CERAMIC
CASE 632-08

N SUFFIX
PLASTIC
CASE 646-06

D SUFFIX
SOIC
CASE 751A-02

ORDERING INFORMATION

SN54LSXXJ   Ceramic
SN74LSXXN   Plastic
SN74LSXXD   SOIC

GUARANTEED OPERATING RANGES

Symbol	Parameter		Min	Typ	Max	Unit
V_{CC}	Supply Voltage	54	4.5	5.0	5.5	V
		74	4.75	5.0	5.25	
T_A	Operating Ambient Temperature Range	54	−55	25	125	°C
		74	0	25	70	
I_{OH}	Output Current — High	54, 74			−0.4	mA
I_{OL}	Output Current — Low	54			4.0	mA
		74			8.0	

SN54/74LS02

DC CHARACTERISTICS OVER OPERATING TEMPERATURE RANGE (unless otherwise specified)

Symbol	Parameter		Limits			Unit	Test Conditions	
			Min	Typ	Max			
V_{IH}	Input HIGH Voltage		2.0			V	Guaranteed Input HIGH Voltage for All Inputs	
V_{IL}	Input LOW Voltage	54			0.7	V	Guaranteed Input LOW Voltage for All Inputs	
		74			0.8			
V_{IK}	Input Clamp Diode Voltage			−0.65	−1.5	V	V_{CC} = MIN, I_{IN} = −18 mA	
V_{OH}	Output HIGH Voltage	54	2.5	3.5		V	V_{CC} = MIN, I_{OH} = MAX, V_{IN} = V_{IH} or V_{IL} per Truth Table	
		74	2.7	3.5		V		
V_{OL}	Output LOW Voltage	54, 74		0.25	0.4	V	I_{OL} = 4.0 mA	V_{CC} = V_{CC} MIN, V_{IN} = V_{IL} or V_{IH} per Truth Table
		74		0.35	0.5	V	I_{OL} = 8.0 mA	
I_{IH}	Input HIGH Current				20	μA	V_{CC} = MAX, V_{IN} = 2.7 V	
					0.1	mA	V_{CC} = MAX, V_{IN} = 7.0 V	
I_{IL}	Input LOW Current				−0.4	mA	V_{CC} = MAX, V_{IN} = 0.4 V	
I_{OS}	Short Circuit Current (Note 1)		−20		−100	mA	V_{CC} = MAX	
I_{CC}	Power Supply Current Total, Output HIGH				3.2	mA	V_{CC} = MAX	
	Total, Output LOW				5.4			

Note 1: Not more than one output should be shorted at a time, nor for more than 1 second.

AC CHARACTERISTICS (T_A = 25°C)

Symbol	Parameter	Limits			Unit	Test Conditions
		Min	Typ	Max		
t_{PLH}	Turn-Off Delay, Input to Output		10	15	ns	V_{CC} = 5.0 V C_L = 15 pF
t_{PHL}	Turn-On Delay, Input to Output		10	15	ns	

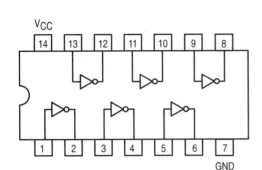

ON Semiconductor
Formerly a Division of Motorola
http://onsemi.com

**LOW
POWER
SCHOTTKY**

GUARANTEED OPERATING RANGES

Symbol	Parameter	Min	Typ	Max	Unit
V_{CC}	Supply Voltage	4.75	5.0	5.25	V
T_A	Operating Ambient Temperature Range	0	25	70	°C
I_{OH}	Output Current ± High			±0.4	mA
I_{OL}	Output Current ± Low			8.0	mA

**PLASTIC
N SUFFIX
CASE 646**

**SOIC
D SUFFIX
CASE 751A**

ORDERING INFORMATION

Device	Package	Shipping
SN74LS04N	14 Pin DIP	2000 Units/Box
SN74LS04D	14 Pin	2500/Tape & Reel

SN74LS04

DC CHARACTERISTICS OVER OPERATING TEMPERATURE RANGE (unless otherwise specified)

Symbol	Parameter	Min	Typ	Max	Unit	Test Conditions	
V_{IH}	Input HIGH Voltage	2.0			V	Guaranteed Input HIGH Voltage for All Inputs	
V_{IL}	Input LOW Voltage			0.8	V	Guaranteed Input LOW Voltage for All Inputs	
V_{IK}	Input Clamp Diode Voltage		±0.65	±1.5	V	V_{CC} = MIN, I_{IN} = ±18 mA	
V_{OH}	Output HIGH Voltage	2.7	3.5		V	V_{CC} = MIN, I_{OH} = MAX, V_{IN} = V_{IH} or V_{IL} per Truth Table	
V_{OL}	Output LOW Voltage		0.25	0.4	V	I_{OL} = 4.0 mA	V_{CC} = V_{CC} MIN, V_{IN} = V_{IL} or V_{IH} per Truth Table
			0.35	0.5	V	I_{OL} = 8.0 mA	
I_{IH}	Input HIGH Current			20	µA	V_{CC} = MAX, V_{IN} = 2.7 V	
				0.1	mA	V_{CC} = MAX, V_{IN} = 7.0 V	
I_{IL}	Input LOW Current			±0.4	mA	V_{CC} = MAX, V_{IN} = 0.4 V	
I_{OS}	Short Circuit Current (Note 1)	±20		±100	mA	V_{CC} = MAX	
I_{CC}	Power Supply Current Total, Output HIGH			2.4	mA	V_{CC} = MAX	
	Total, Output LOW			6.6			

Note 1: Not more than one output should be shorted at a time, nor for more than 1 second.

AC CHARACTERISTICS (T_A = 255C)

Symbol	Parameter	Min	Typ	Max	Unit	Test Conditions
t_{PLH}	Turn±Off Delay, Input to Output		9.0	15	ns	V_{CC} = 5.0 V C_L = 15 pF
t_{PHL}	Turn±On Delay, Input to Output		10	15	ns	

SN74LS05

Hex Inverter

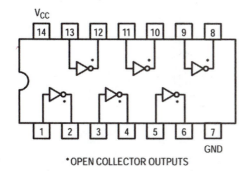

V_{CC}

14 | 13 | 12 | 11 | 10 | 9 | 8

1 | 2 | 3 | 4 | 5 | 6 | 7

GND

*OPEN COLLECTOR OUTPUTS

ON Semiconductor
Formerly a Division of Motorola
http://onsemi.com

**LOW
POWER
SCHOTTKY**

GUARANTEED OPERATING RANGES

Symbol	Parameter	Min	Typ	Max	Unit
V_{CC}	Supply Voltage	4.75	5.0	5.25	V
T_A	Operating Ambient Temperature Range	0	25	70	°C
V_{OH}	Output Voltage – High			5.5	V
I_{OL}	Output Current – Low			8.0	mA

14
1

**PLASTIC
N SUFFIX
CASE 646**

14
1

**SOIC
D SUFFIX
CASE 751A**

ORDERING INFORMATION

Device	Package	Shipping
SN74LS05N	14 Pin DIP	2000 Units/Box
SN74LS05D	14 Pin	2500/Tape & Reel

SN74LS05

DC CHARACTERISTICS OVER OPERATING TEMPERATURE RANGE (unless otherwise specified)

Symbol	Parameter	Limits			Unit	Test Conditions	
		Min	Typ	Max			
V_{IH}	Input HIGH Voltage	2.0			V	Guaranteed Input HIGH Voltage for All Inputs	
V_{IL}	Input LOW Voltage			0.8	V	Guaranteed Input LOW Voltage for All Inputs	
V_{IK}	Input Clamp Diode Voltage		−0.65	−1.5	V	V_{CC} = MIN, I_{IN} = −18 mA	
I_{OH}	Output HIGH Current			100	μA	V_{CC} = MIN, V_{OH} = MAX	
V_{OL}	Output LOW Voltage		0.25	0.4	V	I_{OL} = 4.0 mA	V_{CC} = V_{CC} MIN, V_{IN} = V_{IL} or V_{IH} per Truth Table
			0.35	0.5	V	I_{OL} = 8.0 mA	
I_{IH}	Input HIGH Current			20	μA	V_{CC} = MAX, V_{IN} = 2.7 V	
				0.1	mA	V_{CC} = MAX, V_{IN} = 7.0 V	
I_{IL}	Input LOW Current			−0.4	mA	V_{CC} = MAX, V_{IN} = 0.4 V	
I_{CC}	Power Supply Current Total, Output HIGH			2.4	mA	V_{CC} = MAX	
	Total, Output LOW			6.6			

AC CHARACTERISTICS (T_A = 25°C)

Symbol	Parameter	Limits			Unit	Test Conditions
		Min	Typ	Max		
t_{PLH}	Turn–Off Delay, Input to Output		17	32	ns	V_{CC} = 5.0 V C_L = 15 pF, R_L = 2.0 kΩ
t_{PHL}	Turn–On Delay, Input to Output		15	28	ns	

SN54LS06, SN54LS16, SN74LS06, SN74LS16
HEX INVERTER BUFFERS/DRIVERS WITH
OPEN-COLLECTOR HIGH-VOLTAGE OUTPUTS

SDLS020A – MAY 1990

- **Converts TTL Voltage Levels to MOS Levels**
- **High Sink-Current Capability**
- **Input Clamping Diodes Simplify System Design**
- **Open-Collector Driver for Indicator Lamps and Relays**
- **Package Options Include "Small Outline" Packages, Ceramic Chip Carriers, and Standard Plastic and Ceramic 300-mil DIPs**

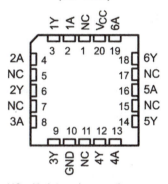

SN54LS06, SN54LS16 . . . J PACKAGE
SN74LS06, SN74LS16 . . . D OR N PACKAGE
(TOP VIEW)

1A	1	14	V$_{CC}$
1Y	2	13	6A
2A	3	12	6Y
2Y	4	11	5A
3A	5	10	5Y
3Y	6	9	4A
GND	7	8	4Y

SN54LS06, SN54LS16 . . . FK PACKAGE
(TOP VIEW)

NC – No internal connection

description

These monolithic hex inverter buffers/drivers feature high-voltage open-collector outputs to interface with high-level circuits (such as MOS), or for driving high-current loads, and are also characterized for use as inverter buffers for driving TTL inputs. The 'LS06 has a rated output voltage of 30 V and the 'LS16 has a rated output voltage of 15 V. The maximum sink current for the SN54LS06 and SN54LS16 is 30 mA and the SN74LS06 and SN74LS16 is 40 mA.

These circuits are compatible with most TTL families. Inputs are diode-clamped to minimize transmission-effects, which simplifies design. Typical power dissipation is 175 mW and average propagation delay time is 8 ns.

The SN54LS06 and SN54LS16 are characterized over the full military temperature range of –55°C to 125°C. The SN74LS06 and SN74LS16 are characterized for operation from 0°C to 70°C.

logic symbol†

1A	1	2	1Y
2A	3	4	2Y
3A	5	6	3Y
4A	9	8	4Y
5A	11	10	5Y
6A	13	12	6Y

† This symbol is in accordance with ANSI/IEEE Std 91-1984 and IEC Publication 617-12.
Pin numbers shown are for D, J, and N packages.

logic diagram (positive logic)

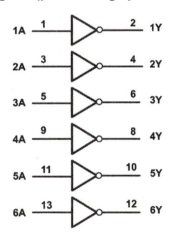

1A	1	2	1Y
2A	3	4	2Y
3A	5	6	3Y
4A	9	8	4Y
5A	11	10	5Y
6A	13	12	6Y

SN54LS06, SN54LS16, SN74LS06, SN74LS16 HEX INVERTER BUFFERS/DRIVERS WITH OPEN-COLLECTOR HIGH-VOLTAGE OUTPUTS

SDLS020A – MAY 1990

schematic (each gate)

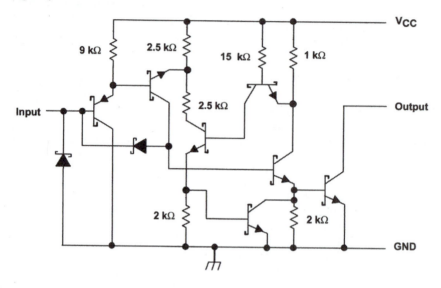

absolute maximum ratings over operating free-air temperature range (unless otherwise noted)†

Supply voltage, V_{CC} .... 7 V

Input voltage, V_I (see Note 1) .... 5.5 V

Output voltage, V_O (see Notes 1 and 2): SN54LS06, SN74LS06 .... 30 V

SN54LS16, SN74LS16 .... 15 V

Operating free-air temperature range: SN54LS06, SN54LS16 .... −55°C to 125°C

SN74LS06, SN74LS16 .... 0°C to 70°C

Storage temperature range .... −65°C to 150°C

† Stresses beyond those listed under "absolute maximum ratings" may cause permanent damage to the device. This are stress ratings only, and functional operation of the device at these or any other conditions beyond those indicated under "recommended operating conditions" is not implied. Exposure to absolute-maximum-rated conditions for extended periods may affect device reliability.

NOTES: 1. Voltage values are with respect to network ground terminal.

2. This is the maximum voltage that should be applied to any output when it is in the off state.

recommended operating conditions

			SN54LS06 SN54LS16			SN74LS06 SN74LS16			UNIT
			MIN	NOM	MAX	MIN	NOM	MAX	
V_{CC}	Supply voltage		4.5	5	5.5	4.75	5	5.25	V
V_{IH}	High-level input voltage		2			2			V
V_{IL}	Low-level input voltage				0.8			0.8	V
V_{OH}	High-level output voltage	'LS06			30			30	V
		'LS16			15			15	
I_{OL}	Low-level output current				30			40	mA
T_A	Operating free-air temperature		−55		125	0		70	°C

SN54LS06, SN54LS16, SN74LS06, SN74LS16
HEX INVERTER BUFFERS/DRIVERS WITH
OPEN-COLLECTOR HIGH-VOLTAGE OUTPUTS

SDLS020A – MAY 1990

electrical characteristics over recommended operating free-air temperature range (unless otherwise noted)

PARAMETER	TEST CONDITIONS†		SN54LS06 SN54LS16			SN74LS06 SN74LS16			UNIT
			MIN	TYP‡	MAX	MIN	TYP‡	MAX	
V_{IK}	V_{CC} = MIN,	I_I = −12 mA			−1.5			−1.5	V
I_{OH}	V_{CC} = MIN, V_{IL} = 0.8 V	'LS06, V_{OH} = 30 V			0.25			0.25	mA
		'LS16, V_{OH} = 15 V			0.25			0.25	
V_{OL}	V_{CC} = MIN, V_{IH} = 2 V	I_{OL} = 16 mA		0.25	0.4		0.25	0.4	V
		I_{OL} = 30 mA			0.7				
		I_{OL} = 40 mA						0.7	
I_I	V_{CC} = MAX,	V_I = 7 V			1			1	mA
I_{IH}	V_{CC} = MAX,	V_I = 2.4 V			20			20	μA
I_{IL}	V_{CC} = MAX,	V_I = 0.4 V			−0.2			−0.2	mA
I_{CCH}	V_{CC} = MAX				18			18	mA
I_{CCL}	V_{CC} = MAX				60			60	mA

† For conditions shown as MIN or MAX, use the appropriate value specified under recommended operating conditions.
‡ All typical values are at V_{CC} = 5 V, and T_A = 25°C.

switching characteristics, V_{CC} = 5 V, T_A = 25°C (see Note 3)

PARAMETER	FROM (INPUT)	TO (OUTPUT)	TEST CONDITIONS	MIN	TYP	MAX	UNIT
t_{PLH}	A	Y	R_L = 110 Ω, C_L = 15 pF		7	15	ns
t_{PHL}					10	20	

NOTE 3: Load circuit and voltage waveforms are shown in Section 1 of *TTL Logic Data Book*, 1988.

SN54LS07, SN74LS07, SN74LS17
HEX BUFFERS/DRIVERS WITH
OPEN-COLLECTOR HIGH-VOLTAGE OUTPUTS
SDLS021A, D3517, MAY 1990–REVISED AUGUST 1991

- Converts TTL-Voltage Levels to MOS Levels

- High Sink-Current Capability

- Input Clamping Diodes Simplify System Design

- Open-Collector Driver for Indicator Lamps and Relays

- Package Options Include "Small Outline" Packages, Ceramic Chip Carriers, and Standard and Ceramic 300-mil DIPs

description

These monolithic hex buffers/drivers feature high-voltage open-collector outputs to interface with high-level circuits or for driving high-current loads. They are also characterized for use as buffers for driving TTL inputs. The 'LS07 has a rated output voltage of 30 V and the 'LS17 has a rated output voltage of 15 V. The maximum sink current is 30 mA for the SN54LS07 and 40 mA for the SN74LS07 and SN74LS17.

These circuits are compatible with most TTL families. Inputs are diode-clamped to minimize transmission-line effects, which simplifies design. Typical power dissipation is 140 mW and average propagation delay time is 12 ns.

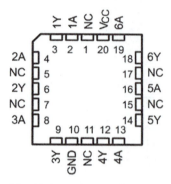

SN54LS07 . . . J PACKAGE
SN74LS07, SN74LS17 . . . D OR N PACKAGE
(TOP VIEW)

SN54LS07 . . . FK PACKAGE
(TOP VIEW)

NC – No internal connection

The SN54LS07 is characterized over the full military temperature range of –55°C to 125°C. The SN74LS07 and SN74LS17 are characterized for operation from 0°C to 70°C.

logic symbol†

† This symbol is in accordance with ANSI/IEEE Std 91-1984 and IEC Publication 617-12.
Pin numbers shown are for D, J, and N packages.

logic diagram (positive logic)

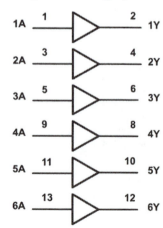

SN54LS07, SN74LS07, SN74LS17
HEX BUFFERS/DRIVERS WITH
OPEN-COLLECTOR HIGH-VOLTAGE OUTPUTS

schematic (each gate)

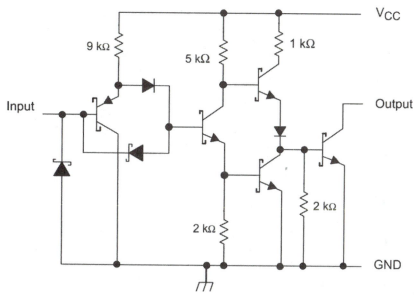

Resistor values shown are nominal.

absolute maximum ratings over operating free-air temperature range (unless otherwise noted)†

Supply voltage, V_{CC} . . . . 7 V
Input voltage, V_I (see Note 1) . . . . 5.5 V
Output voltage, V_O (see Notes 1 and 2): SN54LS07, SN74LS07 . . . . 30 V
                                     SN74LS17 . . . . 15 V
Operating free-air temperature range: SN54LS07 . . . . −55°C to 125°C
                                SN54LS07, SN74LS17 . . . . 0°C to 70°C
Storage temperature range . . . . −65°C to 150°C

† Stresses beyond those listed under "absolute maximum ratings" may cause permanent damage to the device. These are stress ratings only, and
  functional operation of the device at these or any other conditions beyond those indicated under "recommended operating conditions" is not
  implied. Exposure to absolute-maximum-rated conditions for extended periods may affect device reliability.
NOTES: 1. Voltage values are with respect to network ground terminal.
         2. This is the maximum voltage that should be applied to any output when it is in the off state.

recommended operating conditions

			SN54LS07			SN74LS07 SN74LS17			UNIT
			MIN	NOM	MAX	MIN	NOM	MAX	
V_{CC}	Supply voltage		4.5	5	5.5	4.75	5	5.25	V
V_{IH}	High-level input voltage		2			2			V
V_{IL}	Low-level input voltage				0.8			0.8	V
V_{OH}	High-level output voltage	'LS07			30			30	V
		'LS17						15	
I_{OL}	Low-level output current				30			40	mA
T_A	Operating free-air temperature		−55		125	0		70	°C

SN54LS07, SN74LS07, SN74LS17
HEX BUFFERS/DRIVERS WITH
OPEN-COLLECTOR HIGH-VOLTAGE OUTPUTS

electrical characteristics over recommended operating free-air temperature range (unless otherwise noted)

PARAMETER	TEST CONDITIONS†		SN54LS07			SN74LS07 SN74LS17			UNIT
			MIN	TYP‡	MAX	MIN	TYP‡	MAX	
V_{IK}	V_{CC} = MIN,	I_I = −12 mA			−1.5			−1.5	V
I_{OH}	V_{CC} = MIN, V_{IH} = 2 V	'LS07, V_{OH} = 30 V			0.25			0.25	mA
		'LS17, V_{OH} = 15 V			0.25			0.25	
V_{OL}	V_{CC} = MIN, V_{IL} = 0.8 V	I_{OL} = 16 mA			0.4			0.4	V
		I_{OL} = MAX§			0.7			0.7	
I_I	V_{CC} = MAX,	V_I = 7 V			1			1	mA
I_{IH}	V_{CC} = MAX,	V_I = 2.4 V			20			20	μA
I_{IL}	V_{CC} = MAX,	V_I = 0.4 V			−0.2			−0.2	mA
I_{CCH}	V_{CC} = MAX				14			14	mA
I_{CCL}	V_{CC} = MAX				45			45	mA

† For conditions shown as MIN or MAX, use the appropriate value specified under recommended operating conditions.
‡ All typical values are at V_{CC} = 5 V, T_A = 25°C.
§ I_{OL} = 30 mA for SN54 series parts and 40 mA for SN74 series parts.

switching characteristics, V_{CC} = 5 V, T_A = 25°C (see Note 3)

PARAMETER	FROM (INPUT)	TO (OUTPUT)	TEST CONDITIONS	MIN	TYP	MAX	UNIT
t_{PLH}	A	Y	R_L = 110 Ω, C_L = 15 pF		6	10	ns
t_{PHL}					19	30	

NOTE 3: Load circuit and voltage waveforms are shown in Section 1 of *TTL Logic Data Book*, 1988.

SN74LS08

Quad 2-Input AND Gate

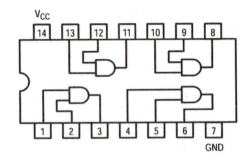

ON Semiconductor
Formerly a Division of Motorola
http://onsemi.com

**LOW
POWER
SCHOTTKY**

GUARANTEED OPERATING RANGES

Symbol	Parameter	Min	Typ	Max	Unit
V_{CC}	Supply Voltage	4.75	5.0	5.25	V
T_A	Operating Ambient Temperature Range	0	25	70	°C
I_{OH}	Output Current – High			−0.4	mA
I_{OL}	Output Current – Low			8.0	mA

**PLASTIC
N SUFFIX
CASE 646**

**SOIC
D SUFFIX
CASE 751A**

ORDERING INFORMATION

Device	Package	Shipping
SN74LS08N	14 Pin DIP	2000 Units/Box
SN74LS08D	14 Pin	2500/Tape & Reel

SN74LS08

DC CHARACTERISTICS OVER OPERATING TEMPERATURE RANGE (unless otherwise specified)

Symbol	Parameter	Limits			Unit	Test Conditions	
		Min	Typ	Max			
V_{IH}	Input HIGH Voltage	2.0			V	Guaranteed Input HIGH Voltage for All Inputs	
V_{IL}	Input LOW Voltage			0.8	V	Guaranteed Input LOW Voltage for All Inputs	
V_{IK}	Input Clamp Diode Voltage		−0.65	−1.5	V	V_{CC} = MIN, I_{IN} = −18 mA	
V_{OH}	Output HIGH Voltage	2.7	3.5		V	V_{CC} = MIN, I_{OH} = MAX, V_{IN} = V_{IH} or V_{IL} per Truth Table	
V_{OL}	Output LOW Voltage		0.25	0.4	V	I_{OL} = 4.0 mA	V_{CC} = V_{CC} MIN, V_{IN} = V_{IL} or V_{IH} per Truth Table
			0.35	0.5	V	I_{OL} = 8.0 mA	
I_{IH}	Input HIGH Current			20	μA	V_{CC} = MAX, V_{IN} = 2.7 V	
				0.1	mA	V_{CC} = MAX, V_{IN} = 7.0 V	
I_{IL}	Input LOW Current			−0.4	mA	V_{CC} = MAX, V_{IN} = 0.4 V	
I_{OS}	Short Circuit Current (Note 1)	−20		−100	mA	V_{CC} = MAX	
I_{CC}	Power Supply Current Total, Output HIGH			4.8	mA	V_{CC} = MAX	
	Total, Output LOW			8.8			

Note 1: Not more than one output should be shorted at a time, nor for more than 1 second.

AC CHARACTERISTICS (T_A = 25°C)

Symbol	Parameter	Limits			Unit	Test Conditions
		Min	Typ	Max		
t_{PLH}	Turn–Off Delay, Input to Output		8.0	15	ns	V_{CC} = 5.0 V C_L = 15 pF
t_{PHL}	Turn–On Delay, Input to Output		10	20	ns	

SN74LS32

Quad 2-Input OR Gate

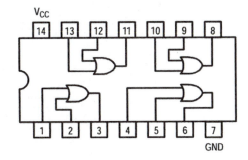

ON Semiconductor
Formerly a Division of Motorola
http://onsemi.com

**LOW
POWER
SCHOTTKY**

GUARANTEED OPERATING RANGES

Symbol	Parameter	Min	Typ	Max	Unit
V_{CC}	Supply Voltage	4.75	5.0	5.25	V
T_A	Operating Ambient Temperature Range	0	25	70	°C
I_{OH}	Output Current – High			−0.4	mA
I_{OL}	Output Current – Low			8.0	mA

**PLASTIC
N SUFFIX
CASE 646**

**SOIC
D SUFFIX
CASE 751A**

ORDERING INFORMATION

Device	Package	Shipping
SN74LS32N	14 Pin DIP	2000 Units/Box
SN74LS32D	14 Pin	2500/Tape & Reel

SN74LS32

DC CHARACTERISTICS OVER OPERATING TEMPERATURE RANGE (unless otherwise specified)

Symbol	Parameter	Limits			Unit	Test Conditions
		Min	Typ	Max		
V_{IH}	Input HIGH Voltage	2.0			V	Guaranteed Input HIGH Voltage for All Inputs
V_{IL}	Input LOW Voltage			0.8	V	Guaranteed Input LOW Voltage for All Inputs
V_{IK}	Input Clamp Diode Voltage		−0.65	−1.5	V	V_{CC} = MIN, I_{IN} = −18 mA
V_{OH}	Output HIGH Voltage	2.7	3.5		V	V_{CC} = MIN, I_{OH} = MAX, V_{IN} = V_{IH} or V_{IL} per Truth Table
V_{OL}	Output LOW Voltage		0.25	0.4	V	I_{OL} = 4.0 mA	V_{CC} = V_{CC} MIN, V_{IN} = V_{IL} or V_{IH} per Truth Table
			0.35	0.5	V	I_{OL} = 8.0 mA	
I_{IH}	Input HIGH Current			20	μA	V_{CC} = MAX, V_{IN} = 2.7 V
				0.1	mA	V_{CC} = MAX, V_{IN} = 7.0 V
I_{IL}	Input LOW Current			−0.4	mA	V_{CC} = MAX, V_{IN} = 0.4 V
I_{OS}	Short Circuit Current (Note 1)	−20		−100	mA	V_{CC} = MAX
I_{CC}	Power Supply Current Total, Output HIGH			6.2	mA	V_{CC} = MAX
	Total, Output LOW			9.8		

Note 1: Not more than one output should be shorted at a time, nor for more than 1 second.

AC CHARACTERISTICS (T_A = 25°C)

Symbol	Parameter	Limits			Unit	Test Conditions
		Min	Typ	Max		
t_{PLH}	Turn-Off Delay, Input to Output		14	22	ns	V_{CC} = 5.0 V C_L = 15 pF
t_{PHL}	Turn-On Delay, Input to Output		14	22	ns	

SN74LS86

Quad 2-Input Exclusive OR Gate

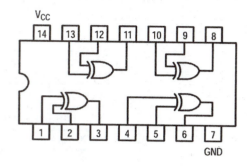

ON Semiconductor
Formerly a Division of Motorola
http://onsemi.com

**LOW
POWER
SCHOTTKY**

**PLASTIC
N SUFFIX
CASE 646**

**SOIC
D SUFFIX
CASE 751A**

TRUTH TABLE

IN		OUT
A	B	Z
L	L	L
L	H	H
H	L	H
H	H	L

GUARANTEED OPERATING RANGES

Symbol	Parameter	Min	Typ	Max	Unit
V_{CC}	Supply Voltage	4.75	5.0	5.25	V
T_A	Operating Ambient Temperature Range	0	25	70	°C
I_{OH}	Output Current – High			−0.4	mA
I_{OL}	Output Current – Low			8.0	mA

ORDERING INFORMATION

Device	Package	Shipping
SN74LS86N	14 Pin DIP	2000 Units/Box
SN74LS86D	14 Pin	2500/Tape & Reel

SN74LS86

DC CHARACTERISTICS OVER OPERATING TEMPERATURE RANGE (unless otherwise specified)

Symbol	Parameter	Limits			Unit	Test Conditions	
		Min	Typ	Max			
V_{IH}	Input HIGH Voltage	2.0			V	Guaranteed Input HIGH Voltage for All Inputs	
V_{IL}	Input LOW Voltage			0.8	V	Guaranteed Input LOW Voltage for All Inputs	
V_{IK}	Input Clamp Diode Voltage		−0.65	−1.5	V	V_{CC} = MIN, I_{IN} = −18 mA	
V_{OH}	Output HIGH Voltage	2.7	3.5		V	V_{CC} = MIN, I_{OH} = MAX, V_{IN} = V_{IH} or V_{IL} per Truth Table	
V_{OL}	Output LOW Voltage		0.25	0.4	V	I_{OL} = 4.0 mA	V_{CC} = V_{CC} MIN, V_{IN} = V_{IL} or V_{IH} per Truth Table
			0.35	0.5	V	I_{OL} = 8.0 mA	
I_{IH}	Input HIGH Current			40	µA	V_{CC} = MAX, V_{IN} = 2.7 V	
				0.2	mA	V_{CC} = MAX, V_{IN} = 7.0 V	
I_{IL}	Input LOW Current			−0.8	mA	V_{CC} = MAX, V_{IN} = 0.4 V	
I_{OS}	Short Circuit Current (Note 1)	−20		−100	mA	V_{CC} = MAX	
I_{CC}	Power Supply Current			10	mA	V_{CC} = MAX	

Note 1: Not more than one output should be shorted at a time, nor for more than 1 second.

AC CHARACTERISTICS (T_A = 25°C)

Symbol	Parameter	Limits			Unit	Test Conditions
		Min	Typ	Max		
t_{PLH} t_{PHL}	Propagation Delay, Other Input LOW		12 10	23 17	ns	V_{CC} = 5.0 V C_L = 15 pF
t_{PLH} t_{PHL}	Propagation Delay, Other Input HIGH		20 13	30 22	ns	

SN54F00, SN74F00
QUADRUPLE 2-INPUT POSITIVE-NAND GATES

SDFS035A – MARCH 1987 – REVISED OCTOBER 1993

- **Package Options Include Plastic Small-Outline Packages, Ceramic Chip Carriers, and Standard Plastic and Ceramic 300-mil DIPs**

description

These devices contain four independent 2-input NAND gates. They perform the Boolean functions $Y = \overline{A \cdot B}$ or $Y = \overline{A} + \overline{B}$ in positive logic.

The SN54F00 is characterized for operation over the full military temperature range of −55°C to 125°C. The SN74F00 is characterized for operation from 0°C to 70°C.

SN54F00 . . . J PACKAGE
SN74F00 . . . D OR N PACKAGE
(TOP VIEW)

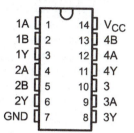

FUNCTION TABLE
(each gate)

INPUTS		OUTPUT
A	B	Y
H	H	L
L	X	H
X	L	H

SN54F00 . . . FK PACKAGE
(TOP VIEW)

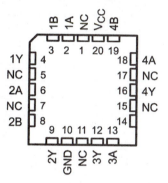

NC – No internal connection

logic symbol†

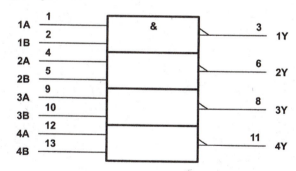

† This symbol is in accordance with ANSI/IEEE Std 91-1984 and IEC Publication 617-12.

logic diagram (positive logic)

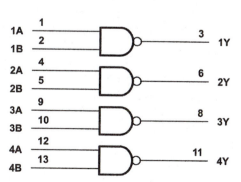

Pin numbers shown are for the D, J, and N packages.

SN54F00, SN74F00
QUADRUPLE 2-INPUT POSITIVE-NAND GATES

SDFS035A – MARCH 1987 – REVISED OCTOBER 1993

absolute maximum ratings over operating free-air temperature range (unless otherwise noted)†

Supply voltage range, V_{CC}	−0.5 V to 7 V
Input voltage range, V_I (see Note 1)	−1.2 V to 7 V
Input current range	−30 mA to 5 mA
Voltage range applied to any output in the high state	−0.5 V to V_{CC}
Current into any output in the low state	40 mA
Operating free-air temperature range: SN54F00	−55°C to 125°C
SN74F00	0°C to 70°C
Storage temperature range	−65°C to 150°C

† Stresses beyond those listed under "absolute maximum ratings" may cause permanent damage to the device. These are stress ratings only and functional operation of the device at these or any other conditions beyond those indicated under "recommended operating conditions" is not implied. Exposure to absolute-maximum-rated conditions for extended periods may affect device reliability.

NOTE 1:   The input voltage ratings may be exceeded provided the input current ratings are observed.

recommended operating conditions

		SN54F00			SN74F00			UNIT
		MIN	NOM	MAX	MIN	NOM	MAX	
V_{CC}	Supply voltage	4.5	5	5.5	4.5	5	5.5	V
V_{IH}	High-level input voltage	2			2			V
V_{IL}	Low-level input voltage			0.8			0.8	V
I_{IK}	Input clamp current			−18			−18	mA
I_{OH}	High-level output current			−1			−1	mA
I_{OL}	Low-level output current			20			20	mA
T_A	Operating free-air temperature	−55		125	0		70	°C

electrical characteristics over recommended operating free-air temperature range (unless otherwise noted)

PARAMETER	TEST CONDITIONS		SN54F00			SN74F00			UNIT
			MIN	TYP‡	MAX	MIN	TYP‡	MAX	
V_{IK}	$V_{CC} = 4.5$ V,	$I_I = -18$ mA			−1.2			−1.2	V
V_{OH}	$V_{CC} = 4.5$ V,	$I_{OH} = -1$ mA	2.5	3.4		2.5	3.4		V
	$V_{CC} = 4.75$ V,	$I_{OH} = -1$ mA				2.7			
V_{OL}	$V_{CC} = 4.5$ V,	$I_{OL} = 20$ mA		0.3	0.5		0.3	0.5	V
I_I	$V_{CC} = 5.5$ V,	$V_I = 7$ V			0.1			0.1	mA
I_{IH}	$V_{CC} = 5.5$ V,	$V_I = 2.7$ V			20			20	µA
I_{IL}	$V_{CC} = 5.5$ V,	$V_I = 0.5$ V			−0.6			−0.6	mA
I_{OS}§	$V_{CC} = 5.5$ V,	$V_O = 0$	−60		−150	−60		−150	mA
I_{CCH}	$V_{CC} = 5.5$ V,	$V_I = 0$		1.9	2.8		1.9	2.8	mA
I_{CCL}	$V_{CC} = 5.5$ V,	$V_I = 4.5$ V		6.8	10.2		6.8	10.2	mA

‡ All typical values are at $V_{CC} = 5$ V, $T_A = 25$°C.

§ Not more than one output should be shorted at a time, and the duration of the short circuit should not exceed one second.

SN54F00, SN74F00
QUADRUPLE 2-INPUT POSITIVE-NAND GATES

SDFS035A – MARCH 1987 – REVISED OCTOBER 1993

switching characteristics (see Note 2)

PARAMETER	FROM (INPUT)	TO (OUTPUT)	V_{CC} = 5 V, C_L = 50 pF, R_L = 500 Ω, T_A = 25°C			V_{CC} = 4.5 V to 5.5 V, C_L = 50 pF, R_L = 500 Ω, T_A = MIN to MAX†				UNIT
			'F00			SN54F00		SN74F00		
			MIN	TYP	MAX	MIN	MAX	MIN	MAX	
t_{PLH}	A or B	Y	1.6	3.3	5	2	7	1.6	6	ns
t_{PHL}			1	2.8	4.3	1.5	6.5	1	5.3	

† For conditions shown as MIN or MAX, use the appropriate value specified under recommended operating conditions.
NOTE 2:   Load circuits and waveforms are shown in Section 1.

SN54ALS00A, SN54AS00, SN74ALS00A, SN74AS00
QUADRUPLE 2-INPUT POSITIVE-NAND GATES

SDAS187A – APRIL 1982 – REVISED DECEMBER 1994

● **Package Options Include Plastic Small-Outline (D) Packages, Ceramic Chip Carriers (FK), and Standard Plastic (N) and Ceramic (J) 300-mil DIPs**

description

These devices contain four independent 2-input positive-NAND gates. They perform the Boolean functions $Y = \overline{A \bullet B}$ or $Y = \overline{A} + \overline{B}$ in positive logic.

The SN54ALS00A and SN54AS00 are characterized for operation over the full military temperature range of −55°C to 125°C. The SN74ALS00A and SN74AS00 are characterized for operation from 0°C to 70°C.

SN54ALS00A, SN54AS00 . . . J PACKAGE
SN74ALS00A, SN74AS00 . . . D OR N PACKAGE
(TOP VIEW)

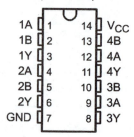

SN54ALS00A, SN54AS00 . . . FK PACKAGE
(TOP VIEW)

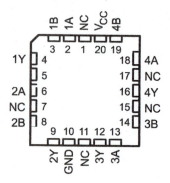

NC – No internal connection

FUNCTION TABLE
(each gate)

INPUTS		OUTPUT
A	**B**	**Y**
H	H	L
L	X	H
X	L	H

logic symbol†

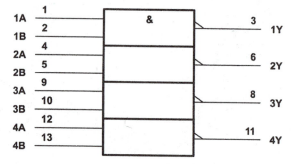

† This symbol is in accordance with ANSI/IEEE Std 91-1984 and IEC Publication 617-12.
Pin numbers shown are for the D, J, and N packages.

SN54ALS00A, SN54AS00, SN74ALS00A, SN74AS00
QUADRUPLE 2-INPUT POSITIVE-NAND GATES

SDAS187A – APRIL 1982 – REVISED DECEMBER 1994

logic diagram (positive logic)

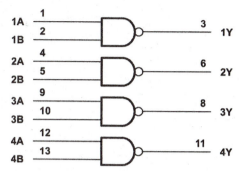

Pin numbers shown are for the D, J, and N packages.

absolute maximum ratings over operating free-air temperature range (unless otherwise noted)†

Supply voltage, V_{CC} .... 7 V
Input voltage, V_I .... 7 V
Operating free-air temperature range, T_A: SN54ALS00A .... −55°C to 125°C
SN74ALS00A .... 0°C to 70°C
Storage temperature range .... −65°C to 150°C

† Stresses beyond those listed under "absolute maximum ratings" may cause permanent damage to the device. These are stress ratings only, and functional operation of the device at these or any other conditions beyond those indicated under "recommended operating conditions" is not implied. Exposure to absolute-maximum-rated conditions for extended periods may affect device reliability.

recommended operating conditions

		SN54ALS00A			SN74ALS00A			UNIT
		MIN	NOM	MAX	MIN	NOM	MAX	
V_{CC}	Supply voltage	4.5	5	5.5	4.5	5	5.5	V
V_{IH}	High-level input voltage	2			2			V
V_{IL}	Low-level input voltage			0.8‡			0.8	V
				0.7§				
I_{OH}	High-level output current			−0.4			−0.4	mA
I_{OL}	Low-level output current			4			8	mA
T_A	Operating free-air temperature	−55		125	0		70	°C

‡ Applies over temperature range −55°C to 70°C
§ Applies over temperature range 70°C to 125°C

SN54ALS00A, SN54AS00, SN74ALS00A, SN74AS00
QUADRUPLE 2-INPUT POSITIVE-NAND GATES

SDAS187A – APRIL 1982 – REVISED DECEMBER 1994

electrical characteristics over recommended operating free-air temperature range (unless otherwise noted)

PARAMETER	TEST CONDITIONS		SN54ALS00A MIN	SN54ALS00A TYP†	SN54ALS00A MAX	SN74ALS00A MIN	SN74ALS00A TYP†	SN74ALS00A MAX	UNIT
V_{IK}	$V_{CC} = 4.5$ V,	$I_I = -18$ mA			-1.2			-1.5	V
V_{OH}	$V_{CC} = 4.5$ V to 5.5 V,	$I_{OH} = -0.4$ mA	$V_{CC} -2$			$V_{CC} -2$			V
V_{OL}	$V_{CC} = 4.5$ V	$I_{OL} = 4$ mA		0.25	0.4		0.25	0.4	V
		$I_{OL} = 8$ mA					0.35	0.5	
I_I	$V_{CC} = 5.5$ V,	$V_I = 7$ V			0.1			0.1	mA
I_{IH}	$V_{CC} = 5.5$ V,	$V_I = 2.7$ V			20			20	µA
I_{IL}	$V_{CC} = 5.5$ V,	$V_I = 0.4$ V			-0.1			-0.1	mA
I_O‡	$V_{CC} = 5.5$ V,	$V_O = 2.25$ V	-20		-112	-30		-112	mA
I_{CCH}	$V_{CC} = 5.5$ V,	$V_I = 0$		0.5	0.85		0.5	0.85	mA
I_{CCL}	$V_{CC} = 5.5$ V,	$V_I = 4.5$ V		1.5	3		1.5	3	mA

† All typical values are at $V_{CC} = 5$ V, $T_A = 25°C$.
‡ The output conditions have been chosen to produce a current that closely approximates one half of the true short-circuit output current, I_{OS}.

switching characteristics (see Figure 1)

PARAMETER	FROM (INPUT)	TO (OUTPUT)	$V_{CC} = 4.5$ V to 5.5 V, $C_L = 50$ pF, $R_L = 500$ Ω, $T_A = $ MIN to MAX§				UNIT
			SN54ALS00A MIN	SN54ALS00A MAX	SN74ALS00A MIN	SN74ALS00A MAX	
t_{PLH}	A or B	Y	3	15	3	11	ns
t_{PHL}			2	9	2	8	

§ For conditions shown as MIN or MAX, use the appropriate value specified under recommended operating conditions.

SN54ALS00A, SN54AS00, SN74ALS00A, SN74AS00
QUADRUPLE 2-INPUT POSITIVE-NAND GATES

SDAS187A – APRIL 1982 – REVISED DECEMBER 1994

absolute maximum ratings over operating free-air temperature range (unless otherwise noted)[†]

Supply voltage, V_{CC} .... 7 V
Input voltage, V_I .... 7 V
Operating free-air temperature range, T_A: SN54AS00 .... −55°C to 125°C
SN74AS00 .... 0°C to 70°C
Storage temperature range .... −65°C to 150°C

[†] Stresses beyond those listed under "absolute maximum ratings" may cause permanent damage to the device. These are stress ratings only, and functional operation of the device at these or any other conditions beyond those indicated under "recommended operating conditions" is not implied. Exposure to absolute-maximum-rated conditions for extended periods may affect device reliability.

recommended operating conditions

		SN54AS00			SN74AS00			UNIT
		MIN	NOM	MAX	MIN	NOM	MAX	
V_{CC}	Supply voltage	4.5	5	5.5	4.5	5	5.5	V
V_{IH}	High-level input voltage	2			2			V
V_{IL}	Low-level input voltage			0.8			0.8	V
I_{OH}	High-level output current			−2			−2	mA
I_{OL}	Low-level output current			20			20	mA
T_A	Operating free-air temperature	−55		125	0		70	°C

electrical characteristics over recommended operating free-air temperature range (unless otherwise noted)

PARAMETER	TEST CONDITIONS		SN54AS00			SN74AS00			UNIT
			MIN	TYP[‡]	MAX	MIN	TYP[‡]	MAX	
V_{IK}	V_{CC} = 4.5 V,	I_I = −18 mA			−1.2			−1.2	V
V_{OH}	V_{CC} = 4.5 V to 5.5 V,	I_{OH} = −2 mA	V_{CC} −2			V_{CC} −2			V
V_{OL}	V_{CC} = 4.5 V,	I_{OL} = 20 mA		0.35	0.5		0.35	0.5	V
I_I	V_{CC} = 5.5 V,	V_I = 7 V			0.1			0.1	mA
I_{IH}	V_{CC} = 5.5 V,	V_I = 2.7 V			20			20	µA
I_{IL}	V_{CC} = 5.5 V,	V_I = 0.4 V			−0.5			−0.5	mA
I_O[§]	V_{CC} = 5.5 V,	V_O = 2.25 V	−30		−112	−30		−112	mA
I_{CCH}	V_{CC} = 5.5 V,	V_I = 0		2	3.2		2	3.2	mA
I_{CCL}	V_{CC} = 5.5 V,	V_I = 4.5 V		10.8	17.4		10.8	17.4	mA

[‡] All typical values are at V_{CC} = 5 V, T_A = 25°C.
[§] The output conditions have been chosen to produce a current that closely approximates one half of the true short-circuit output current, I_{OS}.

switching characteristics (see Figure 1)

PARAMETER	FROM (INPUT)	TO (OUTPUT)	V_{CC} = 4.5 V to 5.5 V, C_L = 50 pF, R_L = 500 Ω, T_A = MIN to MAX[¶]				UNIT
			SN54AS00		SN74AS00		
			MIN	MAX	MIN	MAX	
t_{PLH}	A or B	Y	1	5	1	4.5	ns
t_{PHL}			1	5	1	4	

[¶] For conditions shown as MIN or MAX, use the appropriate value specified under recommended operating conditions.

SN54ALS00A, SN54AS00, SN74ALS00A, SN74AS00
QUADRUPLE 2-INPUT POSITIVE-NAND GATES

SDAS187A – APRIL 1982 – REVISED DECEMBER 1994

PARAMETER MEASUREMENT INFORMATION
SERIES 54ALS/74ALS AND 54AS/74AS DEVICES

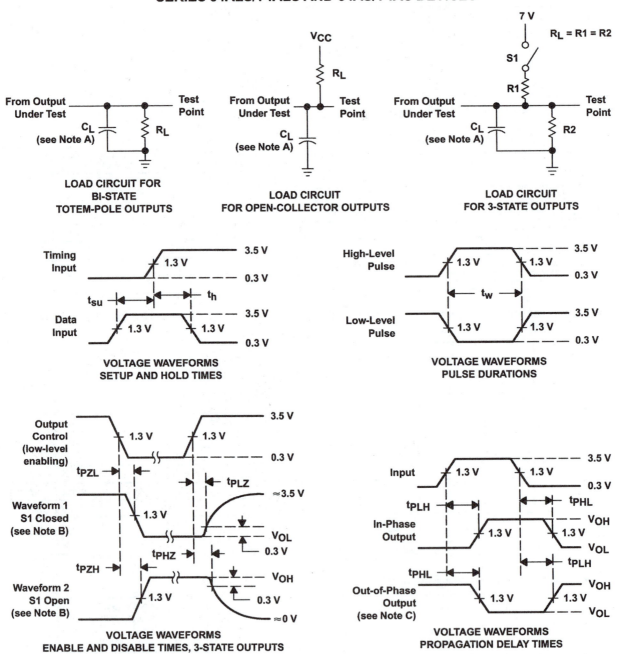

NOTES: A. C_L includes probe and jig capacitance.
B. Waveform 1 is for an output with internal conditions such that the output is low except when disabled by the output control.
Waveform 2 is for an output with internal conditions such that the output is high except when disabled by the output control.
C. When measuring propagation delay items of 3-state outputs, switch S1 is open.
D. All input pulses have the following characteristics: PRR ≤ 1 MHz, $t_r = t_f = 2$ ns, duty cycle = 50%.
E. The outputs are measured one at a time with one transition per measurement.

Figure 1. Load Circuits and Voltage Waveforms

MOTOROLA
SEMICONDUCTOR TECHNICAL DATA

Quad 2-Input NAND Gate
High–Performance Silicon–Gate CMOS

The MC54/74HC00A is identical in pinout to the LS00. The device inputs are compatible with Standard CMOS outputs; with pullup resistors, they are compatible with LSTTL outputs.

- Output Drive Capability: 10 LSTTL Loads
- Outputs Directly Interface to CMOS, NMOS and TTL
- Operating Voltage Range: 2 to 6V
- Low Input Current: 1μA
- High Noise Immunity Characteristic of CMOS Devices
- In Compliance With the JEDEC Standard No. 7A Requirements
- Chip Complexity: 32 FETs or 8 Equivalent Gates

MC54/74HC00A

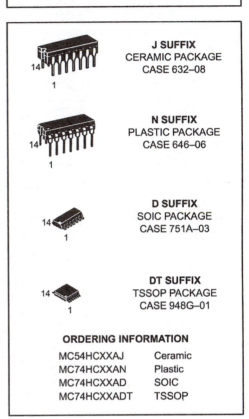

J SUFFIX
CERAMIC PACKAGE
CASE 632–08

N SUFFIX
PLASTIC PACKAGE
CASE 646–06

D SUFFIX
SOIC PACKAGE
CASE 751A–03

DT SUFFIX
TSSOP PACKAGE
CASE 948G–01

ORDERING INFORMATION

MC54HCXXAJ	Ceramic
MC74HCXXAN	Plastic
MC74HCXXAD	SOIC
MC74HCXXADT	TSSOP

FUNCTION TABLE

Inputs		Output
A	**B**	**Y**
L	L	H
L	H	H
H	L	H
H	H	L

LOGIC DIAGRAM

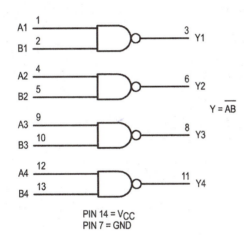

$Y = \overline{AB}$

PIN 14 = V$_{CC}$
PIN 7 = GND

Pinout: 14–Lead Packages (Top View)

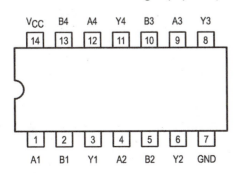

MC54/74HC00A

MAXIMUM RATINGS*

Symbol	Parameter	Value	Unit
V_{CC}	DC Supply Voltage (Referenced to GND)	-0.5 to $+7.0$	V
V_{in}	DC Input Voltage (Referenced to GND)	-0.5 to $V_{CC} + 0.5$	V
V_{out}	DC Output Voltage (Referenced to GND)	-0.5 to $V_{CC} + 0.5$	V
I_{in}	DC Input Current, per Pin	± 20	mA
I_{out}	DC Output Current, per Pin	± 25	mA
I_{CC}	DC Supply Current, V_{CC} and GND Pins	± 50	mA
P_D	Power Dissipation in Still Air, Plastic or Ceramic DIP† SOIC Package† TSSOP Package†	750 500 450	mW
T_{stg}	Storage Temperature	-65 to $+150$	°C
T_L	Lead Temperature, 1 mm from Case for 10 Seconds Plastic DIP, SOIC or TSSOP Package Ceramic DIP	 260 300	°C

* Maximum Ratings are those values beyond which damage to the device may occur.
 Functional operation should be restricted to the Recommended Operating Conditions.
†Derivating — Plastic DIP: -10 mW/°C from 65° to 125°C
       Ceramic DIP: -10 mW/°C from 100° to 125°C
       SOIC Package: -7 mW/°C from 65° to 125°C
       TSSOP Package: -6.1 mW/°C from 65° to 125°C
For high frequency or heavy load considerations, see Chapter 2 of the Motorola High–Speed CMOS Data Book (DL129/D).

This device contains protection circuitry to guard against damage due to high static voltages or electric fields. However, precautions must be taken to avoid applications of any voltage higher than maximum rated voltages to this high–impedance circuit. For proper operation, V_{in} and V_{out} should be constrained to the range GND $\leq$ (V_{in} or V_{out}) $\leq$ V_{CC}.

Unused inputs must always be tied to an appropriate logic voltage level (e.g., either GND or V_{CC}). Unused outputs must be left open.

RECOMMENDED OPERATING CONDITIONS

Symbol	Parameter		Min	Max	Unit
V_{CC}	DC Supply Voltage (Referenced to GND)		2.0	6.0	V
V_{in}, V_{out}	DC Input Voltage, Output Voltage (Referenced to GND)		0	V_{CC}	V
T_A	Operating Temperature, All Package Types		-55	$+125$	°C
t_r, t_f	Input Rise and Fall Time (Figure 1)	$V_{CC} = 2.0$ V $V_{CC} = 4.5$ V $V_{CC} = 6.0$ V	0 0 0	1000 500 400	ns

DC CHARACTERISTICS (Voltages Referenced to GND)

Symbol	Parameter	Condition	V_{CC} V	Guaranteed Limit			Unit
				−55 to 25°C	≤85°C	≤125°C	
V_{IH}	Minimum High–Level Input Voltage	V_{out} = 0.1V or V_{CC} −0.1V $|I_{out}|$ ≤ 20µA	2.0 3.0 4.5 6.0	1.50 2.10 3.15 4.20	1.50 2.10 3.15 4.20	1.50 2.10 3.15 4.20	V
V_{IL}	Maximum Low–Level Input Voltage	V_{out} = 0.1V or V_{CC} − 0.1V $|I_{out}|$ ≤ 20µA	2.0 3.0 4.5 6.0	0.50 0.90 1.35 1.80	0.50 0.90 1.35 1.80	0.50 0.90 1.35 1.80	V
V_{OH}	Minimum High–Level Output Voltage	V_{in} = V_{IH} or V_{IL} $|I_{out}|$ ≤ 20µA	2.0 4.5 6.0	1.9 4.4 5.9	1.9 4.4 5.9	1.9 4.4 5.9	V
		V_{in} = V_{IH} or V_{IL} $|I_{out}|$ ≤ 2.4mA $|I_{out}|$ ≤ 4.0mA $|I_{out}|$ ≤ 5.2mA	3.0 4.5 6.0	2.48 3.98 5.48	2.34 3.84 5.34	2.20 3.70 5.20	
V_{OL}	Maximum Low–Level Output Voltage	V_{in} = V_{IH} or V_{IL} $|I_{out}|$ ≤ 20µA	2.0 4.5 6.0	0.1 0.1 0.1	0.1 0.1 0.1	0.1 0.1 0.1	V
		V_{in} = V_{IH} or V_{IL} $|I_{out}|$ ≤ 2.4mA $|I_{out}|$ ≤ 4.0mA $|I_{out}|$ ≤ 5.2mA	3.0 4.5 6.0	0.26 0.26 0.26	0.33 0.33 0.33	0.40 0.40 0.40	
I_{in}	Maximum Input Leakage Current	V_{in} = V_{CC} or GND	6.0	±0.1	±1.0	±1.0	µA
I_{CC}	Maximum Quiescent Supply Current (per Package)	V_{in} = V_{CC} or GND I_{out} = 0µA	6.0	1.0	10	40	µA

NOTE: Information on typical parametric values can be found in Chapter 2 of the Motorola High–Speed CMOS Data Book (DL129/D).

AC CHARACTERISTICS (C_L = 50 pF, Input t_r = t_f = 6 ns)

Symbol	Parameter	V_{CC} V	Guaranteed Limit			Unit
			−55 to 25°C	≤85°C	≤125°C	
t_{PLH}, t_{PHL}	Maximum Propagation Delay, Input A or B to Output Y (Figures 1 and 2)	2.0 3.0 4.5 6.0	75 30 15 13	95 40 19 16	110 55 22 19	ns
t_{TLH}, t_{THL}	Maximum Output Transition Time, Any Output (Figures 1 and 2)	2.0 3.0 4.5 6.0	75 27 15 13	95 32 19 16	110 36 22 19	ns
C_{in}	Maximum Input Capacitance		10	10	10	pF

NOTE: For propagation delays with loads other than 50 pF, and information on typical parametric values, see Chapter 2 of the Motorola High–Speed CMOS Data Book (DL129/D).

		Typical @ 25°C, V_{CC} = 5.0 V, V_{EE} = 0 V	
C_{PD}	Power Dissipation Capacitance (Per Buffer)*	22	pF

* Used to determine the no–load dynamic power consumption: $P_D = C_{PD} V_{CC}^2 f + I_{CC} V_{CC}$. For load considerations, see Chapter 2 of the Motorola High–Speed CMOS Data Book (DL129/D).

MC54/74HC00A

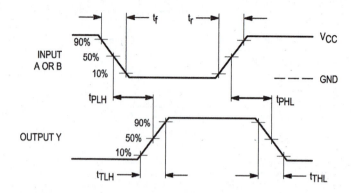

Figure 1. Switching Waveforms

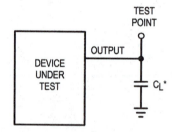

*Includes all probe and jig capacitance

Figure 2. Test Circuit

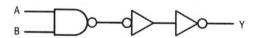

**Figure 3. Expanded Logic Diagram
(1/4 of the Device)**

MOTOROLA
SEMICONDUCTOR TECHNICAL DATA

Quad 2-Input NAND Gate with LSTTL-Compatible Inputs
High–Performance Silicon–Gate CMOS

The MC54/74HCT00A may be used as a level converter for interfacing TTL or NMOS outputs to high–speed CMOS inputs.
The HCT00A is identical in pinout to the LS00.

- Output Drive Capability: 10 LSTTL Loads
- TTL/NMOS–Compatible Input Levels
- Outputs Directly Interface to CMOS, NMOS and TTL
- Operating Voltage Range: 4.5 to 5.5 V
- Low Input Current: 1.0 µA
- In Compliance with the Requirements Defined by JEDEC Standard No. 7A
- Chip Complexity: 48 FETs or 12 Equivalent Gates

LOGIC DIAGRAM

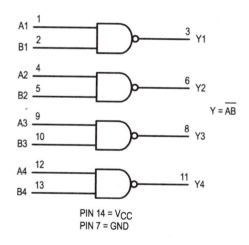

$$Y = \overline{AB}$$

PIN 14 = V_{CC}
PIN 7 = GND

MC54/74HCT00A

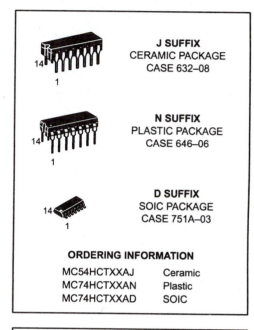

J SUFFIX
CERAMIC PACKAGE
CASE 632–08

N SUFFIX
PLASTIC PACKAGE
CASE 646–06

D SUFFIX
SOIC PACKAGE
CASE 751A–03

ORDERING INFORMATION

MC54HCTXXAJ	Ceramic
MC74HCTXXAN	Plastic
MC74HCTXXAD	SOIC

PIN ASSIGNMENT

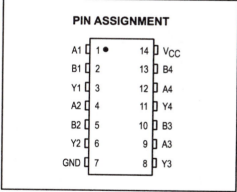

A1	1		14	V_{CC}
B1	2		13	B4
Y1	3		12	A4
A2	4		11	Y4
B2	5		10	B3
Y2	6		9	A3
GND	7		8	Y3

FUNCTION TABLE

Inputs		Output
A	B	Y
L	L	H
L	H	H
H	L	H
H	H	L

MC54/74HCT00A

MAXIMUM RATINGS*

Symbol	Parameter	Value	Unit
V_{CC}	DC Supply Voltage (Referenced to GND)	− 0.5 to + 7.0	V
V_{in}	DC Input Voltage (Referenced to GND)	− 0.5 to V_{CC} + 0.5	V
V_{out}	DC Output Voltage (Referenced to GND)	− 0.5 to V_{CC} + 0.5	V
I_{in}	DC Input Current, per Pin	± 20	mA
I_{out}	DC Output Current, per Pin	± 25	mA
I_{CC}	DC Supply Current, V_{CC} and GND Pins	± 50	mA
P_D	Power Dissipation in Still Air, Plastic or Ceramic DIP† SOIC Package†	750 500	mW
T_{stg}	Storage Temperature	− 65 to + 150	°C
T_L	Lead Temperature, 1 mm from Case for 10 Seconds SOIC or Plastic Package Ceramic Dip	260 300	°C

This device contains protection circuitry to guard against damage due to high static voltages or electric fields. However, precautions must be taken to avoid applications of any voltage higher than maximum rated voltages to this high–impedance circuit. For proper operation, V_{in} and V_{out} should be constrained to the range GND $\leq$ (V_{in} or V_{out}) $\leq$ V_{CC}.

Unused inputs must always be tied to an appropriate logic voltage level (e.g., either GND or V_{CC}). Unused outputs must be left open.

* Maximum Ratings are those values beyond which damage to the device may occur.
  Functional operation should be restricted to the Recommended Operating Conditions.
†Derating — Plastic DIP: − 10 mW/°C from 65° to 125°C
  Ceramic DIP: − 10 mW/°C from 100° to 125°C
  SOIC Package: − 7 mW/°C from 65° to 125°C
For high frequency or heavy load considerations, see Chapter 2 of the Motorola High–Speed CMOS Data Book (DL129/D).

RECOMMENDED OPERATING CONDITIONS

Symbol	Parameter	Min	Max	Unit
V_{CC}	DC Supply Voltage (Referenced to GND)	2.0	6.0	V
V_{in}, V_{out}	DC Input Voltage, Output Voltage (Referenced to GND)	0	V_{CC}	V
T_A	Operating Temperature, All Package Types	− 55	+ 125	°C
t_r, t_f	Input Rise and Fall Time (Figure 1)	0	500	ns

DC CHARACTERISTICS FOR THE MC54/74HCT00A (Voltages Referenced to GND)

Symbol	Parameter	Test Conditions	V_{CC} V	− 55 to 25°C Min	Max	$\leq$ 85°C Min	Max	$\leq$ 125°C Min	Max	Unit
V_{IH}	Minimum High–Level Input Voltage	V_{out} = 0.1 or V_{CC} − 0.1 V $|I_{out}| \leq$ 20 µA	4.5 5.5	2.00 2.00		2.00 2.00		2.00 2.00		V
V_{IL}	Maximum Low–Level Input Voltage	V_{out} = 0.1 or V_{CC} − 0.1 V $|I_{out}| \leq$ 20 µA	4.5 5.5		0.80 0.80		0.80 0.80		0.80 0.80	V
V_{OH}	Minimum High–Level Output Voltage	V_{in} = V_{IH} or V_{IL} $|I_{out}| \leq$ 20 µA	4.5 5.5	4.40 5.40		4.40 5.40		4.40 5.40		V
		V_{in} = V_{IH} or V_{IL} $|I_{out}| \leq$ 4.0 mA	4.5	3.98		3.84		3.70		
V_{OL}	Maximum Low–Level Output Voltage	V_{in} = V_{IH} or V_{IL} $|I_{out}| \leq$ 20 µA	4.5 5.5		0.10 0.10		0.10 0.10		0.10 0.10	V
		V_{in} = V_{IH} or V_{IL} $|I_{out}|$ = 4.0 mA	4.5		0.26		0.33		0.40	
I_{in}	Maximum Input Leakage Current	V_{in} = V_{CC} or GND	5.5		± 0.10		± 1.00		± 1.00	µA
I_{CC}	Maximum Quiescent Supply Current (per Package)	V_{in} = V_{CC} or GND $|I_{out}| \leq$ 0 µA	5.5		1		10		40	µA
ΔI_{CC}	Additional Quiescent Supply Current	V_{in} = 2.4 V, Any One Input V_{in} = V_{CC} or GND, Other Inputs I_{out} = 0 µA		$\geq$ − 55°C			25 to 125°C			
			5.5	2.9			2.4			mA

NOTE: Information on typical parametric values can be found in Chapter 2 of the Motorola High–Speed CMOS Data Book (DL129/D).

MC54/74HCT00A

AC CHARACTERISTICS FOR THE MC54/74HCT00A (V_{CC} = 5.0 V ± 10%, CL = 50 pF, Input t_r = t_f = 6.0 ns)

Symbol	Parameter	Fig.	Guaranteed Limits						Unit
			− 55 to 25°C		≤85°C		≤125°C		
			Min	Max	Min	Max	Min	Max	
t_{PLH}, t_{PHL}	Maximum Propagation Delay, Input A or B to Output Y	1, 2		19		24		28	ns
t_{TLH}, t_{THL}	Maximum Output Transition Time, Any Output	1, 2		15		19		22	ns
C_{in}	Maximum Input Capacitance	—		10		10		10	pF

NOTE: For propagation delays with loads other than 50 pF, and information on typical parametric values, see Chapter 2 of the Motorola High–Speed CMOS Data Book (DL129/D).

		Typical @ 25°C, V_{CC} = 5.0 V	
C_{PD}	Power Dissipation Capacitance (Per Gate)*	15	pF

* Used to determine the no–load dynamic power consumption: $P_D = C_{PD} V_{CC}^2 f + I_{CC} V_{CC}$. For load considerations, see Chapter 2 of the Motorola High–Speed CMOS Data Book (DL129/D).

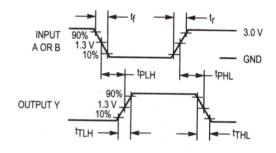

Figure 1. Switching Waveforms

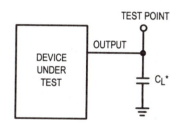

* Includes all probe and jig capacitance

Figure 2. Test Circuit

3

EXPANDED LOGIC DIAGRAM
(1/4 OF THE DEVICE)

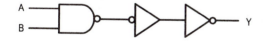

MOTOROLA
SEMICONDUCTOR TECHNICAL DATA

Quad 2-Input NAND Gate

The MC74VHC00 is an advanced high speed CMOS 2–input NAND gate fabricated with silicon gate CMOS technology. It achieves high speed operation similar to equivalent Bipolar Schottky TTL while maintaining CMOS low power dissipation.

The internal circuit is composed of three stages, including a buffer output which provides high noise immunity and stable output. The inputs tolerate voltages up to 7V, allowing the interface of 5V systems to 3V systems.

- High Speed: t_{PD} = 3.7ns (Typ) at V_{CC} = 5V
- Low Power Dissipation: I_{CC} = 2µA (Max) at T_A = 25°C
- High Noise Immunity: V_{NIH} = V_{NIL} = 28% V_{CC}
- Power Down Protection Provided on Inputs
- Balanced Propagation Delays
- Designed for 2V to 5.5V Operating Range
- Low Noise: V_{OLP} = 0.8V (Max)
- Pin and Function Compatible with Other Standard Logic Families
- Latchup Performance Exceeds 300mA
- ESD Performance: HBM > 2000V; Machine Model > 200V
- Chip Complexity: 32 FETs or 8 Equivalent Gates

MC74VHC00

D SUFFIX
14–LEAD SOIC PACKAGE
CASE 751A–03

DT SUFFIX
14–LEAD TSSOP PACKAGE
CASE 948G–01

M SUFFIX
14–LEAD SOIC EIAJ PACKAGE
CASE 965–01

ORDERING INFORMATION

MC74VHCXXD   SOIC
MC74VHCXXDT TSSOP
MC74VHCXXM   SOIC EIAJ

FUNCTION TABLE

Inputs		Output
A	**B**	**$\overline{Y}$**
L	L	H
L	H	H
H	L	H
H	H	L

LOGIC DIAGRAM

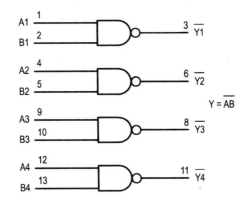

$Y = \overline{AB}$

Pinout: 14–Lead Packages (Top View)

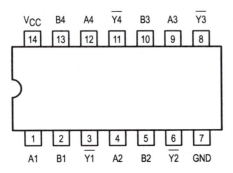

MC74VHC00

MAXIMUM RATINGS*

Symbol	Parameter	Value	Unit
V_{CC}	DC Supply Voltage	− 0.5 to + 7.0	V
V_{in}	DC Input Voltage	− 0.5 to + 7.0	V
V_{out}	DC Output Voltage	− 0.5 to V_{CC} + 0.5	V
I_{IK}	Input Diode Current	− 20	mA
I_{OK}	Output Diode Current	± 20	mA
I_{out}	DC Output Current, per Pin	± 25	mA
I_{CC}	DC Supply Current, V_{CC} and GND Pins	± 50	mA
P_D	Power Dissipation in Still Air, SOIC Packages† TSSOP Package†	500 450	mW
T_{stg}	Storage Temperature	− 65 to + 150	°C

> This device contains protection circuitry to guard against damage due to high static voltages or electric fields. However, precautions must be taken to avoid applications of any voltage higher than maximum rated voltages to this high–impedance circuit. For proper operation, V_{in} and V_{out} should be constrained to the range GND ≤ (V_{in} or V_{out}) ≤ V_{CC}.
>
> Unused inputs must always be tied to an appropriate logic voltage level (e.g., either GND or V_{CC}). Unused outputs must be left open.

* Absolute maximum continuous ratings are those values beyond which damage to the device may occur. Exposure to these conditions or conditions beyond those indicated may adversely affect device reliability. Functional operation under absolute–maximum–rated conditions is not implied.

†Derating — SOIC Packages: − 7 mW/°C from 65° to 125°C
TSSOP Package: − 6.1 mW/°C from 65° to 125°C

RECOMMENDED OPERATING CONDITIONS

Symbol	Parameter		Min	Max	Unit
V_{CC}	DC Supply Voltage		2.0	5.5	V
V_{in}	DC Input Voltage		0	5.5	V
V_{out}	DC Output Voltage		0	V_{CC}	V
T_A	Operating Temperature, All Package Types		− 40	+ 85	°C
t_r, t_f	Input Rise and Fall Time	V_{CC} = 3.3V ±0.3V V_{CC} =5.0V ±0.5V	0 0	100 20	ns/V

DC ELECTRICAL CHARACTERISTICS

Symbol	Parameter	Test Conditions	V_{CC} V	T_A = 25°C Min	T_A = 25°C Typ	T_A = 25°C Max	T_A = − 40 to 85°C Min	T_A = − 40 to 85°C Max	Unit
V_{IH}	High–Level Input Voltage		2.0 3.0 to 5.5	1.50 V_{CC} x 0.7			1.50 V_{CC} x 0.7		V
V_{IL}	Low–Level Input Voltage		2.0 3.0 to 5.5			0.50 V_{CC} x 0.3		0.50 V_{CC} x 0.3	V
V_{OH}	High–Level Output Voltage	V_{in} = V_{IH} or V_{IL} I_{OH} = − 50μA	2.0 3.0 4.5	1.9 2.9 4.4	2.0 3.0 4.5		1.9 2.9 4.4		V
		V_{in} = V_{IH} or V_{IL} I_{OH} = − 4mA I_{OH} = − 8mA	3.0 4.5	2.58 3.94			2.48 3.80		
V_{OL}	Low–Level Output Voltage	V_{in} = V_{IH} or V_{IL} I_{OL} = 50μA	2.0 3.0 4.5		0.0 0.0 0.0	0.1 0.1 0.1		0.1 0.1 0.1	V
		V_{in} = V_{IH} or V_{IL} I_{OL} = 4mA I_{OL} = 8mA	3.0 4.5			0.36 0.36		0.44 0.44	

DC ELECTRICAL CHARACTERISTICS

Symbol	Parameter	Test Conditions	V_{CC} V	$T_A = 25°C$			$T_A = -40$ to $85°C$		Unit
				Min	Typ	Max	Min	Max	
I_{in}	Input Leakage Current	V_{in} = 5.5V or GND	0 to 5.5			± 0.1		± 1.0	μA
I_{CC}	Quiescent Supply Current	V_{in} = V_{CC} or GND	5.5			2.0		20.0	μA

AC ELECTRICAL CHARACTERISTICS (Input $t_r = t_f$ = 3.0ns)

Symbol	Parameter	Test Conditions		$T_A = 25°C$			$T_A = -40$ to $85°C$		Unit
				Min	Typ	Max	Min	Max	
t_{PLH}, t_{PHL}	Propagation Delay, A or B to Y	V_{CC} = 3.3 ± 0.3V	C_L = 15pF		5.5	7.9	1.0	9.5	ns
			C_L = 50pF		8.0	11.4	1.0	13.0	
		V_{CC} = 5.0 ± 0.5V	C_L = 15pF		3.7	5.5	1.0	6.5	
			C_L = 50pF		5.2	7.5	1.0	8.5	
C_{in}	Input Capacitance				4	10		10	pF

		Typical @ 25°C, V_{CC} = 5.0V	
C_{PD}	Power Dissipation Capacitance (Note 1.)	19	pF

1. C_{PD} is defined as the value of the internal equivalent capacitance which is calculated from the operating current consumption without load. Average operating current can be obtained by the equation: $I_{CC(OPR)} = C_{PD} \bullet V_{CC} \bullet f_{in} + I_{CC}/4$ (per gate). C_{PD} is used to determine the no–load dynamic power consumption; $P_D = C_{PD} \bullet V_{CC}^2 \bullet f_{in} + I_{CC} \bullet V_{CC}$.

NOISE CHARACTERISTICS (Input $t_r = t_f$ = 3.0ns, C_L = 50pF, V_{CC} = 5.0V, Measured in SOIC Package)

Symbol	Characteristic	$T_A = 25°C$		Unit
		Typ	Max	
V_{OLP}	Quiet Output Maximum Dynamic V_{OL}	0.3	0.8	V
V_{OLV}	Quiet Output Minimum Dynamic V_{OL}	− 0.3	− 0.8	V
V_{IHD}	Minimum High Level Dynamic Input Voltage		3.5	V
V_{ILD}	Maximum Low Level Dynamic Input Voltage		1.5	V

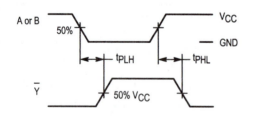

Figure 1. Switching Waveforms

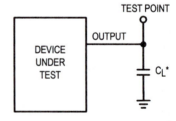

* Includes all probe and jig capacitance

Figure 2. Test Circuit

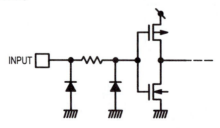

Figure 3. Input Equivalent Circuit

MOTOROLA
SEMICONDUCTOR TECHNICAL DATA

Quad 2-Input NAND Gate

The MC74VHCT00A is an advanced high speed CMOS 2–input NAND gate fabricated with silicon gate CMOS technology. It achieves high speed operation similar to equivalent Bipolar Schottky TTL while maintaining CMOS low power dissipation.

The VHCT inputs are compatible with TTL levels. This device can be used as a level converter for interfacing 3.3V to 5.0V, because it has full 5V CMOS level output swings.

The VHCT00A input structures provide protection when voltages between 0V and 5.5V are applied, regardless of the supply voltage. The output structures also provide protection when V_{CC} = 0V. These input and output structures help prevent device destruction caused by supply voltage – input/output voltage mismatch, battery backup, hot insertion, etc.

- High Speed: t_{PD} = 5.0ns (Typ) at V_{CC} = 5V
- Low Power Dissipation: I_{CC} = 2μA (Max) at T_A = 25°C
- TTL–Compatible Inputs: V_{IL} = 0.8V; V_{IH} = 2.0V
- Power Down Protection Provided on Inputs and Outputs
- Balanced Propagation Delays
- Designed for 4.5V to 5.5V Operating Range
- Low Noise: V_{OLP} = 0.8V (Max)
- Pin and Function Compatible with Other Standard Logic Families
- Latchup Performance Exceeds 300mA
- ESD Performance: HBM > 2000V; Machine Model > 200V
- Chip Complexity: 48 FETs or 12 Equivalent Gates

MC74VHCT00A

D SUFFIX
14–LEAD SOIC PACKAGE
CASE 751A–03

DT SUFFIX
14–LEAD TSSOP PACKAGE
CASE 948G–01

M SUFFIX
14–LEAD SOIC EIAJ PACKAGE
CASE 965–01

ORDERING INFORMATION

MC74VHCTXXAD	SOIC
MC74VHCTXXADT	TSSOP
MC74VHCTXXAM	SOIC EIAJ

FUNCTION TABLE

Inputs		Output
A	**B**	**$\overline{Y}$**
L	L	H
L	H	H
H	L	H
H	H	L

LOGIC DIAGRAM

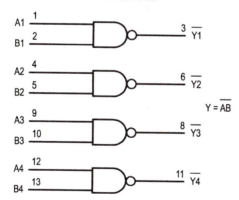

$Y = \overline{AB}$

Pinout: 14–Lead Packages (Top View)

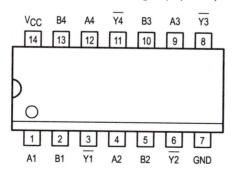

MC74VHCT00A

MAXIMUM RATINGS*

Symbol	Parameter		Value	Unit
V_{CC}	DC Supply Voltage		-0.5 to $+7.0$	V
V_{in}	DC Input Voltage		-0.5 to $+7.0$	V
V_{out}	DC Output Voltage	$V_{CC} = 0$ High or Low State	-0.5 to $+7.0$ -0.5 to $V_{CC} + 0.5$	V
I_{IK}	Input Diode Current		-20	mA
I_{OK}	Output Diode Current ($V_{OUT} <$ GND; $V_{OUT} > V_{CC}$)		± 20	mA
I_{out}	DC Output Current, per Pin		± 25	mA
I_{CC}	DC Supply Current, V_{CC} and GND Pins		± 50	mA
P_D	Power Dissipation in Still Air, SOIC Packages† TSSOP Package†		500 450	mW
T_{stg}	Storage Temperature		-65 to $+150$	°C

> This device contains protection circuitry to guard against damage due to high static voltages or electric fields. However, precautions must be taken to avoid applications of any voltage higher than maximum rated voltages to this high–impedance circuit. For proper operation, V_{in} and V_{out} should be constrained to the range GND $\leq$ (V_{in} or V_{out}) $\leq V_{CC}$.
>
> Unused inputs must always be tied to an appropriate logic voltage level (e.g., either GND or V_{CC}). Unused outputs must be left open.

* Absolute maximum continuous ratings are those values beyond which damage to the device may occur. Exposure to these conditions or conditions beyond those indicated may adversely affect device reliability. Functional operation under absolute–maximum–rated conditions is not implied.

†Derating — SOIC Packages: -7 mW/°C from 65° to 125°C
TSSOP Package: -6.1 mW/°C from 65° to 125°C

RECOMMENDED OPERATING CONDITIONS

Symbol	Parameter		Min	Max	Unit
V_{CC}	DC Supply Voltage		4.5	5.5	V
V_{in}	DC Input Voltage		0	5.5	V
V_{out}	DC Output Voltage	$V_{CC} = 0$ High or Low State	0 0	5.5 V_{CC}	V
T_A	Operating Temperature		-40	$+85$	°C
t_r, t_f	Input Rise and Fall Time	$V_{CC} = 5.0$V ± 0.5V	0	20	ns/V

DC ELECTRICAL CHARACTERISTICS

Symbol	Parameter	Test Conditions	V_{CC} V	$T_A = 25$°C Min	$T_A = 25$°C Typ	$T_A = 25$°C Max	$T_A = -40$ to 85°C Min	$T_A = -40$ to 85°C Max	Unit
V_{IH}	Minimum High–Level Input Voltage		4.5 to 5.5	2.0			2.0		V
V_{IL}	Maximum Low–Level Input Voltage		4.5 to 5.5			0.8		0.8	V
V_{OH}	Minimum High–Level Output Voltage $V_{in} = V_{IH}$ or V_{IL}	$I_{OH} = -50\mu$A	4.5	4.4	4.5		4.4		V
		$I_{OH} = -8$mA	4.5	3.94			3.80		
V_{OL}	Maximum Low–Level Output Voltage $V_{in} = V_{IH}$ or V_{IL}	$I_{OL} = 50\mu$A	4.5		0.0	0.1		0.1	V
		$I_{OL} = 8$mA	4.5			0.36		0.44	
I_{in}	Maximum Input Leakage Current	$V_{in} = 5.5$ V or GND	0 to 5.5			± 0.1		± 1.0	μA
I_{CC}	Maximum Quiescent Supply Current	$V_{in} = V_{CC}$ or GND	5.5			2.0		20.0	μA
I_{CCT}	Quiescent Supply Current	Per Input: $V_{IN} = 3.4$V Other Input: V_{CC} or GND	5.5			1.35		1.50	mA
I_{OPD}	Output Leakage Current	$V_{OUT} = 5.5$V	0			0.5		5.0	μA

MC74VHCT00A

AC ELECTRICAL CHARACTERISTICS (Input $t_r = t_f = 3.0$ns)

Symbol	Parameter	Test Conditions		$T_A = 25°C$			$T_A = -40$ to $85°C$		Unit
			Min	Typ	Max	Min	Max		
t_{PLH}, t_{PHL}	Propagation Delay, A or B to Y	$V_{CC} = 5.0 \pm 0.5$V $\quad C_L = 15$pF $\quad\quad\quad\quad\quad\quad\quad C_L = 50$pF		5.0 5.5	6.9 7.9	1.0 1.0	8.0 9.0		
C_{in}	Input Capacitance			4	10		10		pF

		Typical @ 25°C, $V_{CC} = 5.0$V	
C_{PD}	Power Dissipation Capacitance (Note NO TAG)	17	pF

1. C_{PD} is defined as the value of the internal equivalent capacitance which is calculated from the operating current consumption without load. Average operating current can be obtained by the equation: $I_{CC(OPR)} = C_{PD} \cdot V_{CC} \cdot f_{in} + I_{CC}/4$ (per gate). C_{PD} is used to determine the no-load dynamic power consumption; $P_D = C_{PD} \cdot V_{CC}^2 \cdot f_{in} + I_{CC} \cdot V_{CC}$.

NOISE CHARACTERISTICS (Input $t_r = t_f = 3.0$ns, $C_L = 50$pF, $V_{CC} = 5.0$V, Measured in SOIC Package)

Symbol	Characteristic	$T_A = 25°C$		Unit
		Typ	Max	
V_{OLP}	Quiet Output Maximum Dynamic V_{OL}	0.4	0.8	V
V_{OLV}	Quiet Output Minimum Dynamic V_{OL}	-0.4	-0.8	V
V_{IHD}	Minimum High Level Dynamic Input Voltage		2.0	V
V_{ILD}	Maximum Low Level Dynamic Input Voltage		0.8	V

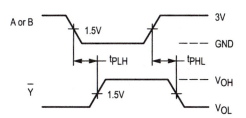

Figure 1. Switching Waveforms

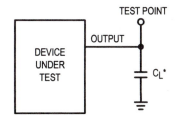

* Includes all probe and jig capacitance

Figure 2. Test Circuit

MOTOROLA
SEMICONDUCTOR TECHNICAL DATA

Hex Unbuffered Inverter
High–Performance Silicon–Gate CMOS

The MC74HCU04A is identical in pinout to the LS04 and the MC14069UB. The device inputs are compatible with standard CMOS outputs; with pullup resistors, they are compatible with LSTTL outputs.

This device consists of six single–stage inverters. These inverters are well suited for use as oscillators, pulse shapers, and in many other applications requiring a high–input impedance amplifier. For digital applications, the HC04A is recommended.

- Output Drive Capability: 10 LSTTL Loads
- Outputs Directly Interface to CMOS, NMOS, and TTL
- Operating Voltage Range: 2 to 6 V; 2.5 to 6 V in Oscillator Configurations
- Low Input Current: 1 µA
- High Noise Immunity Characteristic of CMOS Devices
- In Compliance with the Requirements Defined by JEDEC Standard No. 7A
- Chip Complexity: 12 FETs or 3 Equivalent Gates

MC74HCU04A

N SUFFIX
PLASTIC PACKAGE
CASE 646–06

D SUFFIX
SOIC PACKAGE
CASE 751A–03

DT SUFFIX
TSSOP PACKAGE
CASE 948G–01

ORDERING INFORMATION

MC74HCUXXAN	Plastic
MC74HCUXXAD	SOIC
MC74HCUXXADT	TSSOP

LOGIC DIAGRAM

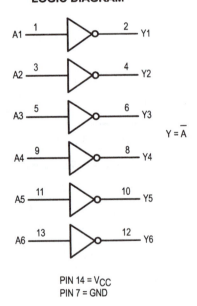

$Y = \overline{A}$

PIN 14 = V$_{CC}$
PIN 7 = GND

PIN ASSIGNMENT

A1	1	14	V$_{CC}$
Y1	2	13	A6
A2	3	12	Y6
Y2	4	11	A5
A3	5	10	Y5
Y3	6	9	A4
GND	7	8	Y4

FUNCTION TABLE

Inputs A	Outputs Y
L	H
H	L

MC74HCU04A

MAXIMUM RATINGS*

Symbol	Parameter		Value	Unit
V_{CC}	DC Supply Voltage (Referenced to GND)		− 0.5 to + 7.0	V
V_{in}	DC Input Voltage (Referenced to GND)		− 0.5 to V_{CC} + 0.5	V
V_{out}	DC Output Voltage (Referenced to GND)		− 0.5 to V_{CC} + 0.5	V
I_{in}	DC Input Current, per Pin		± 20	mA
I_{out}	DC Output Current, per Pin		± 25	mA
I_{CC}	DC Supply Current, V_{CC} and GND Pins		± 50	mA
P_D	Power Dissipation in Still Air	Plastic DIP† SOIC Package† TSSOP Package†	750 500 450	mW
T_{stg}	Storage Temperature		− 65 to + 150	°C
T_L	Lead Temperature, 1 mm from case for 10 Seconds Plastic DIP, SOIC or TSSOP Package		260	°C

> This device contains protection circuitry to guard against damage due to high static voltages or electric fields. However, precautions must be taken to avoid applications of any voltage higher than maximum rated voltages to this high–impedance circuit. For proper operation, V_{in} and V_{out} should be constrained to the range GND ≤ (V_{in} or V_{out}) ≤ V_{CC}.
>
> Unused inputs must always be tied to an appropriate logic voltage level (e.g., either GND or V_{CC}). Unused outputs must be left open.

* Maximum Ratings are those values beyond which damage to the device may occur.
 Functional operation should be restricted to the Recommended Operating Conditions.
†Derating — Plastic DIP: –10mW/°C from 65° to 125°C
     SOIC Package: –7mW/°C from 65° to 125°C
     TSSOP Package: – 6.1 mW/°C from 65° to 125°C
For high frequency or heavy load considerations, see Chapter 2 of the Motorola High–Speed CMOS Data Book (DL129/D).

RECOMMENDED OPERATING CONDITIONS

Symbol	Parameter	Min	Max	Unit
V_{CC}	DC Supply Voltage (Referenced to GND)	2.0	6.0	V
V_{in}, V_{out}	DC Input Voltage, Output Voltage (Referenced to GND)	0	V_{CC}	V
T_A	Operating Temperature, All Package Types	− 55	+ 125	°C
t_r, t_f	Input Rise and Fall Time (Figure 1)	—	No Limit	ns

DC ELECTRICAL CHARACTERISTICS (Voltages Referenced to GND)

Symbol	Parameter	Test Conditions		V_{CC} V	Guaranteed Limit			Unit
					− 55 to 25°C	≤ 85°C	≤ 125°C	
V_{IH}	Minimum High–Level Input Voltage	V_{out} = 0.5 V* $\|I_{out}\|$ ≤ 20 μA		2.0 3.0 4.5 6.0	1.7 2.5 3.6 4.8	1.7 2.5 3.6 4.8	1.7 2.5 3.6 4.8	V
V_{IL}	Maximum Low–Level Input Voltage	V_{out} = V_{CC} − 0.5 V* $\|I_{out}\|$ ≤ 20 μA		2.0 3.0 4.5 6.0	0.3 0.5 0.8 1.1	0.3 0.5 0.8 1.1	0.3 0.5 0.8 1.1	V
V_{OH}	Minimum High–Level Output Voltage	V_{in} = GND $\|I_{out}\|$ ≤ 20 μA		2.0 4.5 6.0	1.8 4.0 5.5	1.8 4.0 5.5	1.8 4.0 5.5	V
		V_{in} = GND	$\|I_{out}\|$ ≤ 2.4 mA $\|I_{out}\|$ ≤ 4.0 mA $\|I_{out}\|$ ≤ 5.2 mA	3.0 4.5 6.0	2.36 3.86 5.36	2.26 3.76 5.26	2.20 3.70 5.20	
V_{OL}	Maximum Low–Level Output Voltage	V_{in} = V_{CC} $\|I_{out}\|$ ≤ 20 μA		2.0 4.5 6.0	0.2 0.5 0.5	0.2 0.5 0.5	0.2 0.5 0.5	V
		V_{in} = V_{CC}	$\|I_{out}\|$ ≤ 2.4 mA $\|I_{out}\|$ ≤ 4.0 mA $\|I_{out}\|$ ≤ 5.2 mA	3.0 4.5 6.0	0.32 0.32 0.32	0.32 0.37 0.37	0.32 0.40 0.40	

MC74HCU04A

DC ELECTRICAL CHARACTERISTICS (Voltages Referenced to GND)

Symbol	Parameter	Test Conditions	V_{CC} V	Guaranteed Limit			Unit
				− 55 to 25°C	≤ 85°C	≤ 125°C	
I_{in}	Maximum Input Leakage Current	$V_{in} = V_{CC}$ or GND	6.0	± 0.1	± 1.0	± 1.0	μA
I_{CC}	Maximum Quiescent Supply Current (per Package)	$V_{in} = V_{CC}$ or GND $I_{out} = 0$ μA	6.0	1	10	40	μA

NOTE: Information on typical parametric values can be found in Chapter 2 of the Motorola High–Speed CMOS Data Book (DL129/D).
* For $V_{CC} = 2.0$ V, $V_{out} = 0.2$ V or $V_{CC} - 0.2$ V.

AC ELECTRICAL CHARACTERISTICS ($C_L = 50$ pF, Input $t_r = t_f = 6$ ns)

Symbol	Parameter	V_{CC} V	Guaranteed Limit			Unit
			− 55 to 25°C	≤ 85°C	≤ 125°C	
t_{PLH}, t_{PHL}	Maximum Propagation Delay, Input A to Output Y (Figures 1 and 2)	2.0	70	90	105	ns
		3.0	40	45	50	
		4.5	14	18	21	
		6.0	12	15	18	
t_{TLH}, t_{THL}	Maximum Output Transition Time, Any Output (Figures 1 and 2)	2.0	75	95	110	ns
		3.0	27	32	36	
		4.5	15	19	22	
		6.0	13	16	19	
C_{in}	Maximum Input Capacitance	—	10	10	10	pF

NOTES:
  1. For propagation delays with loads other than 50 pF, see Chapter 2 of the Motorola High–Speed CMOS Data Book (DL129/D).
  2. Information on typical parametric values can be found in Chapter 2 of the Motorola High–Speed CMOS Data Book (DL129/D).

		Typical @ 25°C, $V_{CC} = 5.0$ V	
C_{PD}	Power Dissipation Capacitance (Per Inverter)*	15	pF

* Used to determine the no–load dynamic power consumption: $P_D = C_{PD} V_{CC}^2 f + I_{CC} V_{CC}$. For load considerations, see Chapter 2 of the Motorola High–Speed CMOS Data Book (DL129/D).

MC74HCU04A

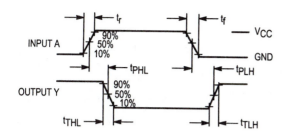

Figure 1. Switching Waveforms

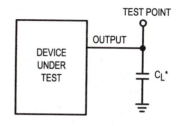

* Includes all probe and jig capacitance

Figure 2. Test Circuit

LOGIC DETAIL
(1/6 of Device Shown)

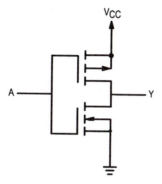

MC74HCU04A

TYPICAL APPLICATIONS

Crystal Oscillator

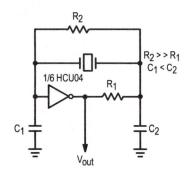

$R_2 >> R_1$
$C_1 < C_2$

Stable RC Oscillator

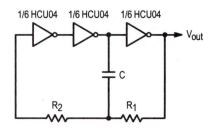

Schmitt Trigger

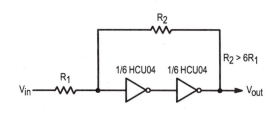

$R_2 > 6R_1$

High Input Impedance Single–Stage Amplifier with a 2 to 6 V Supply Range

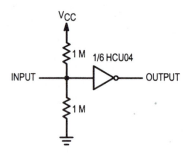

Multi–Stage Amplifier

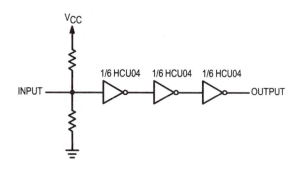

LED Driver

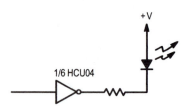

For reduced power supply current, use high–efficiency LEDs
such as the Hewlett–Packard HLMP series or equivalent.

MOTOROLA
SEMICONDUCTOR TECHNICAL DATA

Hex Buffers/Logic-Level Down Converters
High–Performance Silicon–Gate CMOS

The MC54/74HC4049 consists of six inverting buffers, and the MC54/74HC4050 consists of six noninverting buffers. They are identical in pinout to the MC14049UB and MC14050B metal–gate CMOS buffers. The device inputs are compatible with standard CMOS outputs; with pullup resistors, they are compatible with LSTTL outputs.

The input protection circuitry on these devices has been modified by eliminating the V_{CC} diodes to allow the use of input voltages up to 15 volts. Thus, the devices may be used as logic–level translators that convert from a high voltage to a low voltage while operating at the low–voltage power supply. They allow MC14000–series CMOS operating up to 15 volts to be interfaced with High–Speed CMOS at 2 to 6 volts. The protection diodes to GND are Zener diodes, which protect the inputs from both positive and negative voltage transients.

- Output Drive Capability: 10 LSTTL Loads
- Outputs Directly Interface to CMOS, NMOS, and TTL
- Operating Voltage Range: 2 to 6 V
- Low Input Current: 5 µA
- High Noise Immunity Characteristic of CMOS Devices
- In Compliance with the Requirements Defined by JEDEC Standard No. 7A
- Chip Complexity: 36 FETs or 9 Equivalent Gates (4049)
  24 FETs or 6 Equivalent Gates (4050)

MC54/74HC4049
MC54/74HC4050

J SUFFIX
CERAMIC PACKAGE
CASE 620–10

N SUFFIX
PLASTIC PACKAGE
CASE 648–08

D SUFFIX
SOIC PACKAGE
CASE 751B–05

ORDERING INFORMATION

MC54HCXXXXJ	Ceramic
MC74HCXXXXN	Plastic
MC74HCXXXXD	SOIC

PIN ASSIGNMENT

V_{CC}	1●	16	NC
Y0	2	15	Y5
A0	3	14	A5
Y1	4	13	NC
A1	5	12	Y4
Y2	6	11	A4
A2	7	10	Y3
GND	8	9	A3

NC = NO CONNECTION

FUNCTION TABLE

A	Y Outputs	
Input	HC4049	HC4060
L	H	L
H	L	H

LOGIC DIAGRAMS

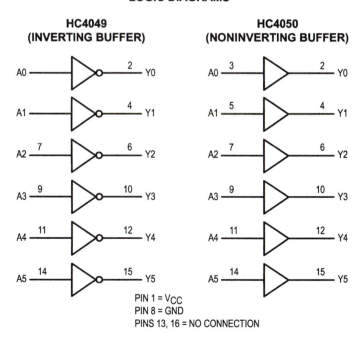

HC4049
(INVERTING BUFFER)

HC4050
(NONINVERTING BUFFER)

PIN 1 = V_{CC}
PIN 8 = GND
PINS 13, 16 = NO CONNECTION

MC54/74HC4049 MC54/74HC4050

MAXIMUM RATINGS*

Symbol	Parameter	Value	Unit
V_{CC}	DC Supply Voltage (Referenced to GND)	-0.5 to $+7.0$	V
V_{in}	DC Input Voltage (Referenced to GND)	-1.5 to $+18$	V
V_{out}	DC Output Voltage (Referenced to GND)	-0.5 to $V_{CC} + 0.5$	V
I_{in}	DC Input Current, per Pin	± 20	mA
I_{out}	DC Output Current, per Pin	± 25	mA
I_{CC}	DC Supply Current, V_{CC} and GND Pins	± 50	mA
P_D	Power Dissipation in Still Air, Plastic or Ceramic DIP† SOIC Package†	750 500	mW
T_{stg}	Storage Temperature	-65 to $+150$	°C
T_L	Lead Temperature, 1 mm from Case for 10 Seconds (Plastic DIP or SOIC Package) (Ceramic DIP)	 260 300	°C

This device contains circuitry to protect the inputs against damage due to high static voltages or electric fields referenced to the GND pin, only. Extra precautions must be taken to avoid applications of any voltage higher than maximum rated voltages to this high–impedance circuit. For proper operation, the ranges GND $\leq$ V_{in} $\leq$ 15 V and GND $\leq$ V_{out} $\leq$ V_{CC} are recommended.

Unused inputs must always be tied to an appropriate logic voltage level (e.g., either GND or V_{CC}).

* Maximum Ratings are those values beyond which damage to the device may occur.
  Functional operation should be restricted to the Recommended Operating Conditions.
†Derating — Plastic DIP: -10 mW/°C from 65° to 125°C
                      Ceramic DIP: -10 mW/°C from 100° to 125°C
                      SOIC Package: -7 mW/°C from 65° to 125°C
For high frequency or heavy load considerations, see Chapter 2 of the Motorola High–Speed CMOS Data Book (DL129/D).

RECOMMENDED OPERATING CONDITIONS

Symbol	Parameter		Min	Max	Unit
V_{CC}	DC Supply Voltage (Referenced to GND)		2.0	6.0	V
V_{in}	DC Input Voltage (Referenced to GND)		0	V_{CC} to 15	V
V_{out}	DC Output Voltage (Referenced to GND)		0	V_{CC}	V
T_A	Operating Temperature, All Package Types		-55	$+125$	°C
t_r, t_f	Input Rise and Fall Time (Figure 1)	$V_{CC} = 2.0$ V $V_{CC} = 4.5$ V $V_{CC} = 6.0$ V	0 0 0	1000 500 400	ns

DC ELECTRICAL CHARACTERISTICS (Voltages Referenced to GND)

Symbol	Parameter	Test Conditions	V_{CC} V	Guaranteed Limit			Unit
				-55 to 25°C	≤ 85°C	≤ 125°C	
V_{IH}	Minimum High–Level Input Voltage	$V_{out} = V_{CC} - 0.1$ V $\lvert I_{out} \rvert \leq 20$ μA	2.0 4.5 6.0	1.5 3.15 4.2	1.5 3.15 4.2	1.5 3.15 4.2	V
V_{IL}	Maximum Low–Level Input Voltage	$V_{out} = 0.1$ V or $V_{CC} - 0.1$ V $\lvert I_{out} \rvert \leq 20$ μA	2.0 4.5 6.0	0.3 0.9 1.2	0.3 0.9 1.2	0.3 0.9 1.2	V
V_{OH}	Minimum High–Level Output Voltage	$V_{in} = V_{IH}$ $\lvert I_{out} \rvert \leq 20$ μA	2.0 4.5 6.0	1.9 4.4 5.9	1.9 4.4 5.9	1.9 4.4 5.9	V
		$V_{in} = V_{IH}$ or V_{IL} $\lvert I_{out} \rvert \leq 4.0$ mA $\lvert I_{out} \rvert \leq 5.2$ mA	4.5 6.0	3.98 5.48	3.84 5.34	3.70 5.20	
V_{OL}	Maximum Low–Level Output Voltage	$V_{in} = V_{IH}$ or V_{IL} $\lvert I_{out} \rvert \leq 20$ μA	2.0 4.5 6.0	0.1 0.1 0.1	0.1 0.1 0.1	0.1 0.1 0.1	V
		$V_{in} = V_{IH}$ or V_{IL} $\lvert I_{out} \rvert \leq 4.0$ mA $\lvert I_{out} \rvert \leq 5.2$ mA	4.5 6.0	0.26 0.26	0.33 0.33	0.40 0.40	
I_{in}	Maximum Input Leakage Current	$V_{in} = V_{CC}$ or GND $V_{in} = 15$ V	6.0 6.0	± 0.1 0.5	± 1.0 5.0	± 1.0 5.0	μA
I_{CC}	Maximum Quiescent Supply Current (per Package)	$V_{in} = 15$ V or GND $I_{out} = 0$ μA	6.0	2	20	40	μA

NOTE: Information on typical parametric values can be found in Chapter 2 of the Motorola High–Speed CMOS Data Book (DL129/D).

AC ELECTRICAL CHARACTERISTICS (C_L = 50 pF, Input $t_r = t_f$ = 6 ns)

Symbol	Parameter	V_{CC} V	Guaranteed Limit			Unit
			− 55 to 25°C	≤ 85°C	≤ 125°C	
t_{PLH}, t_{PHL}	Maximum Propagation Delay, Input A to Output Y (Figures 1 and 2)	2.0 4.5 6.0	85 17 14	105 21 18	130 26 22	ns
t_{TLH}, t_{THL}	Maximum Output Transition Time, Any Output (Figures 1 and 2)	2.0 4.5 6.0	75 15 13	95 19 16	110 22 19	ns
C_{in}	Maximum Input Capacitance		10	10	10	pF

NOTES:
1. For propagation delays with loads other than 50 pF, see Chapter 2 of the Motorola High–Speed CMOS Data Book (DL129/D).
2. Information on typical parametric values can be found in Chapter 2 of the Motorola High–Speed CMOS Data Book (DL129/D).

		Typical @ 25°C, V_{CC} = 5.0 V	
C_{PD}	Power Dissipation Capacitance (Per Buffer)*	27	pF

* Used to determine the no–load dynamic power consumption: $P_D = C_{PD} V_{CC}^2 f + I_{CC} V_{CC}$. For load considerations, see Chapter 2 of the Motorola High–Speed CMOS Data Book (DL129/D).

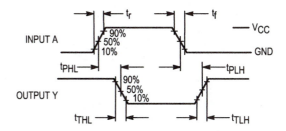

Figure 1a. Switching Waveforms (HC4049)

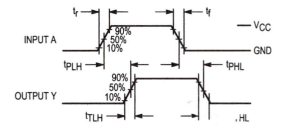

Figure 1b. Switching Waveforms (HC4050)

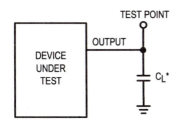

* Includes all probe and jig capacitance

Figure 2. Test Circuit

MC54/74HC4049 MC54/74HC4050

LOGIC DETAIL

HC4049
(1/6 of the Device)

HC4050
(1/6 of the Device)

TYPICAL APPLICATIONS

LSTTL to Low–Voltalge HSCMOS

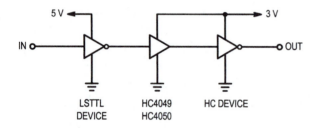

LSTTL   HC4049   HC DEVICE
DEVICE   HC4050

High–Voltage CMOS to HSCMOS

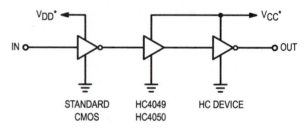

STANDARD   HC4049   HC DEVICE
CMOS   HC4050

NOTE: To determine the noise immunity for the LSTTL to low–voltage
  configuration, use Eq. 1 and Eq. 2:
  (TTL) V_{OH} – (CMOS) V_{IH} Eq. 1
  (TTL) V_{OL} – (CMOS) V_{IL} Eq. 2

  For the supply levels shown:
  $2.4 - 3 (75\%) = 2.4 - 2.25 = 0.15$ V
  $0.4 - 3 (15\%) = 0.4 - 0.45 = 0.05$ V

  Therefore, worst case noise immunity is 50 mV.
   For supply levels greater than 4.5 volts use
  the 74HCT04A for direct interface to TTL outputs.

*Table 1. Supply Examples

V_{DD}	V_{CC}
15 V	2 V
12 V	5 V
12 V	3 V

MOTOROLA
SEMICONDUCTOR TECHNICAL DATA

Quad 2-Input NAND Gate
With 5V-Tolerant Inputs

The MC74LVX00 is an advanced high speed CMOS 2–input NAND gate. The inputs tolerate voltages up to 7V, allowing the interface of 5V systems to 3V systems.

- High Speed: t_{PD} = 4.1ns (Typ) at V_{CC} = 3.3V
- Low Power Dissipation: I_{CC} = 2μA (Max) at T_A = 25°C
- Power Down Protection Provided on Inputs
- Balanced Propagation Delays
- Low Noise: V_{OLP} = 0.5V (Max)
- Pin and Function Compatible with Other Standard Logic Families
- Latchup Performance Exceeds 300mA
- ESD Performance: HBM > 2000V; Machine Model > 200V

MC74LVX00

LVX

LOW–VOLTAGE CMOS

D SUFFIX
14–LEAD SOIC PACKAGE
CASE 751A–03

DT SUFFIX
14–LEAD TSSOP PACKAGE
CASE 948G–01

M SUFFIX
14–LEAD SOIC EIAJ PACKAGE
CASE 965–01

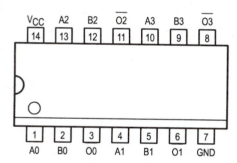

Figure 1. 14–Lead Pinout (Top View)

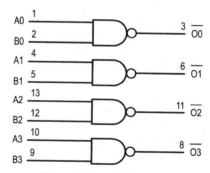

Figure 2. Logic Diagram

PIN NAMES

Pins	Function
An, Bn	Data Inputs
On	Outputs

FUNCTION TABLE

Inputs		Outputs
An	**Bn**	**On**
L	L	H
L	H	H
H	L	H
H	H	L

MC74LVX00

MAXIMUM RATINGS*

Symbol	Parameter	Value	Unit
V_{CC}	DC Supply Voltage	− 0.5 to + 7.0	V
V_{in}	DC Input Voltage	− 0.5 to + 7.0	V
V_{out}	DC Output Voltage	− 0.5 to V_{CC} + 0.5	V
I_{IK}	Input Diode Current	− 20	mA
I_{OK}	Output Diode Current	± 20	mA
I_{out}	DC Output Current, per Pin	± 25	mA
I_{CC}	DC Supply Current, V_{CC} and GND Pins	± 50	mA
P_D	Power Dissipation	180	mW
T_{stg}	Storage Temperature	− 65 to + 150	°C

* Absolute maximum continuous ratings are those values beyond which damage to the device may occur. Exposure to these conditions or conditions beyond those indicated may adversely affect device reliability. Functional operation under absolute–maximum–rated conditions is not implied.

RECOMMENDED OPERATING CONDITIONS

Symbol	Parameter	Min	Max	Unit
V_{CC}	DC Supply Voltage	2.0	3.6	V
V_{in}	DC Input Voltage	0	5.5	V
V_{out}	DC Output Voltage	0	V_{CC}	V
T_A	Operating Temperature, All Package Types	− 40	+ 85	°C
$\Delta t/\Delta V$	Input Rise and Fall Time	0	100	ns/V

DC ELECTRICAL CHARACTERISTICS

Symbol	Parameter	Test Conditions	V_{CC} V	T_A = 25°C Min	T_A = 25°C Typ	T_A = 25°C Max	T_A = − 40 to 85°C Min	T_A = − 40 to 85°C Max	Unit
V_{IH}	High–Level Input Voltage		2.0 3.0 3.6	1.5 2.0 2.4			1.5 2.0 2.4		V
V_{IL}	Low–Level Input Voltage		2.0 3.0 3.6			0.5 0.8 0.8		0.5 0.8 0.8	V
V_{OH}	High–Level Output Voltage (V_{in} = V_{IH} or V_{IL})	I_{OH} = −50µA I_{OH} = −50µA I_{OH} = −4mA	2.0 3.0 3.0	1.9 2.9 2.58	2.0 3.0		1.9 2.9 2.48		V
V_{OL}	Low–Level Output Voltage (V_{in} = V_{IH} or V_{IL})	I_{OL} = 50µA I_{OL} = 50µA I_{OL} = 4mA	2.0 3.0 3.0		0.0 0.0	0.1 0.1 0.36		0.1 0.1 0.44	V
I_{in}	Input Leakage Current	V_{in} = 5.5V or GND	3.6			± 0.1		± 1.0	µA
I_{CC}	Quiescent Supply Current	V_{in} = V_{CC} or GND	3.6			2.0		20.0	µA

MC74LVX00

AC ELECTRICAL CHARACTERISTICS (Input $t_r = t_f = 3.0$ns)

Symbol	Parameter	Test Conditions		$T_A = 25°C$			$T_A = -40$ to $85°C$		Unit
				Min	Typ	Max	Min	Max	
t_{PLH}, t_{PHL}	Propagation Delay, Input to Output	$V_{CC} = 2.7V$	$C_L = 15$pF $C_L = 50$pF		5.4 7.9	10.1 13.6	1.0 1.0	12.5 16.0	ns
		$V_{CC} = 3.3 \pm 0.3V$	$C_L = 15$pF $C_L = 50$pF		4.1 6.6	6.2 9.7	1.0 1.0	7.5 11.0	
t_{OSHL}, t_{OSLH}	Output–to–Output Skew (Note 1.)	$V_{CC} = 2.7V$ $V_{CC} = 3.3 \pm 0.3V$	$C_L = 50$pF $C_L = 50$pF			1.5 1.5		1.5 1.5	ns

1. Skew is defined as the absolute value of the difference between the actual propagation delay for any two separate outputs of the same device. The specification applies to any outputs switching in the same direction, either HIGH–to–LOW (t_{OSHL}) or LOW–to–HIGH (t_{OSLH}); parameter guaranteed by design.

CAPACITIVE CHARACTERISTICS

Symbol	Parameter		$T_A = 25°C$			$T_A = -40$ to $85°C$		Unit
		Min	Typ	Max	Min	Max		
C_{in}	Input Capacitance		4	10		10		pF
C_{PD}	Power Dissipation Capacitance (Note 2.)		19					pF

2. C_{PD} is defined as the value of the internal equivalent capacitance which is calculated from the operating current consumption without load. Average operating current can be obtained by the equation: $I_{CC(OPR)} = C_{PD} \cdot V_{CC} \cdot f_{in} + I_{CC}/4$ (per gate). C_{PD} is used to determine the no–load dynamic power consumption; $P_D = C_{PD} \cdot V_{CC}^2 \cdot f_{in} + I_{CC} \cdot V_{CC}$.

NOISE CHARACTERISTICS (Input $t_r = t_f = 3.0$ns, $C_L = 50$pF, $V_{CC} = 3.3V$, Measured in SOIC Package)

Symbol	Characteristic	$T_A = 25°C$		Unit
		Typ	Max	
V_{OLP}	Quiet Output Maximum Dynamic V_{OL}	0.3	0.5	V
V_{OLV}	Quiet Output Minimum Dynamic V_{OL}	-0.3	-0.5	V
V_{IHD}	Minimum High Level Dynamic Input Voltage		2.0	V
V_{ILD}	Maximum Low Level Dynamic Input Voltage		0.8	V

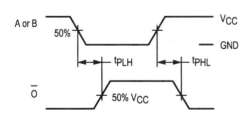

Figure 3. Switching Waveforms

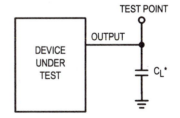

* Includes all probe and jig capacitance

Figure 4. Test Circuit

MOTOROLA
SEMICONDUCTOR TECHNICAL DATA

Low-Voltage CMOS Quad 2-Input NAND Gate
With 5V-Tolerant Inputs

The MC74LCX00 is a high performance, quad 2–input NAND gate operating from a 2.7 to 3.6V supply. High impedance TTL compatible inputs significantly reduce current loading to input drivers while TTL compatible outputs offer improved switching noise performance. A V_I specification of 5.5V allows MC74LCX00 inputs to be safely driven from 5V devices.

Current drive capability is 24mA at the outputs.

- Designed for 2.7 to 3.6V V_{CC} Operation
- 5V Tolerant Inputs — Interface Capability With 5V TTL Logic
- LVTTL Compatible
- LVCMOS Compatible
- 24mA Balanced Output Sink and Source Capability
- Near Zero Static Supply Current (10μA) Substantially Reduces System Power Requirements
- Latchup Performance Exceeds 500mA
- ESD Performance: Human Body Model >2000V; Machine Model >200V

MC74LCX00

LCX

**LOW–VOLTAGE CMOS
QUAD 2–INPUT NAND GATE**

D SUFFIX
PLASTIC SOIC
CASE 751A–03

M SUFFIX
PLASTIC SOIC EIAJ
CASE 965–01

SD SUFFIX
PLASTIC SSOP
CASE 940A–03

DT SUFFIX
PLASTIC TSSOP
CASE 948G–01

Pinout: 14–Lead (Top View)

V_{CC}	A2	B2	$\overline{O2}$	A3	B3	$\overline{O3}$
14	13	12	11	10	9	8

1	2	3	4	5	6	7
A0	B0	$\overline{O0}$	A1	B1	$\overline{O1}$	GND

LOGIC DIAGRAM

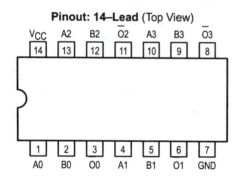

PIN NAMES

Pins	Function
An, Bn	Data Inputs
On	Outputs

FUNCTION TABLE

Inputs		Outputs
An	Bn	$\overline{On}$
L	L	H
L	H	H
H	L	H
H	H	L

MC74LCX00

ABSOLUTE MAXIMUM RATINGS*

Symbol	Parameter	Value	Condition	Unit
V_{CC}	DC Supply Voltage	−0.5 to +7.0		V
V_I	DC Input Voltage	$−0.5 \leq V_I \leq +7.0$		V
V_O	DC Output Voltage	$−0.5 \leq V_O \leq V_{CC} + 0.5$	Note 1.	V
I_{IK}	DC Input Diode Current	−50	$V_I <$ GND	mA
I_{OK}	DC Output Diode Current	−50	$V_O <$ GND	mA
		+50	$V_O > V_{CC}$	mA
I_O	DC Output Source/Sink Current	±50		mA
I_{CC}	DC Supply Current Per Supply Pin	±100		mA
I_{GND}	DC Ground Current Per Ground Pin	±100		mA
T_{STG}	Storage Temperature Range	−65 to +150		°C

* Absolute maximum continuous ratings are those values beyond which damage to the device may occur. Exposure to these conditions or conditions beyond those indicated may adversely affect device reliability. Functional operation under absolute–maximum–rated conditions is not implied.

1. Output in HIGH or LOW State. I_O absolute maximum rating must be observed.

RECOMMENDED OPERATING CONDITIONS

Symbol	Parameter		Min	Typ	Max	Unit
V_{CC}	Supply Voltage	Operating	2.0	3.3	3.6	V
		Data Retention Only	1.5	3.3	3.6	
V_I	Input Voltage		0		5.5	V
V_O	Output Voltage	(HIGH or LOW State)	0		V_{CC}	V
I_{OH}	HIGH Level Output Current, V_{CC} = 3.0V – 3.6V				−24	mA
I_{OL}	LOW Level Output Current, V_{CC} = 3.0V – 3.6V				24	mA
I_{OH}	HIGH Level Output Current, V_{CC} = 2.7V – 3.0V				−12	mA
I_{OL}	LOW Level Output Current, V_{CC} = 2.7V – 3.0V				12	mA
T_A	Operating Free–Air Temperature		−40		+85	°C
$\Delta t/\Delta V$	Input Transition Rise or Fall Rate, V_{IN} from 0.8V to 2.0V, V_{CC} = 3.0V		0		10	ns/V

DC ELECTRICAL CHARACTERISTICS

Symbol	Characteristic	Condition	T_A = −40°C to +85°C Min	Max	Unit
V_{IH}	HIGH Level Input Voltage (Note 2.)	$2.7V \leq V_{CC} \leq 3.6V$	2.0		V
V_{IL}	LOW Level Input Voltage (Note 2.)	$2.7V \leq V_{CC} \leq 3.6V$		0.8	V
V_{OH}	HIGH Level Output Voltage	$2.7V \leq V_{CC} \leq 3.6V$; I_{OH} = −100µA	V_{CC} − 0.2		V
		V_{CC} = 2.7V; I_{OH} = −12mA	2.2		
		V_{CC} = 3.0V; I_{OH} = −18mA	2.4		
		V_{CC} = 3.0V; I_{OH} = −24mA	2.2		
V_{OL}	LOW Level Output Voltage	$2.7V \leq V_{CC} \leq 3.6V$; I_{OL} = 100µA		0.2	V
		V_{CC} = 2.7V; I_{OL} = 12mA		0.4	
		V_{CC} = 3.0V; I_{OL} = 16mA		0.4	
		V_{CC} = 3.0V; I_{OL} = 24mA		0.55	

2. These values of V_I are used to test DC electrical characteristics only.

DC ELECTRICAL CHARACTERISTICS (continued)

Symbol	Characteristic	Condition	T_A = −40°C to +85°C		Unit
			Min	Max	
I_I	Input Leakage Current	$2.7V \leq V_{CC} \leq 3.6V$; $0V \leq V_I \leq 5.5V$		±5.0	µA
I_{CC}	Quiescent Supply Current	$2.7 \leq V_{CC} \leq 3.6V$; V_I = GND or V_{CC}		10	µA
		$2.7 \leq V_{CC} \leq 3.6V$; $3.6 \leq V_I \leq 5.5V$		±10	µA
ΔI_{CC}	Increase in I_{CC} per Input	$2.7 \leq V_{CC} \leq 3.6V$; $V_{IH} = V_{CC} - 0.6V$		500	µA

AC CHARACTERISTICS ($t_R = t_F$ = 2.5ns; C_L = 50pF; R_L = 500Ω)

Symbol	Parameter	Waveform	Limits			Unit
			T_A = −40°C to +85°C			
			V_{CC} = 3.0V to 3.6V		V_{CC} = 2.7V	
			Min	Max	Max	
t_{PLH} t_{PHL}	Propagation Delay Input to Output	1	1.5 1.5	5.2 5.2	6.0 6.0	ns
t_{OSHL} t_{OSLH}	Output–to–Output Skew (Note 3.)			1.0 1.0		ns

3. Skew is defined as the absolute value of the difference between the actual propagation delay for any two separate outputs of the same device. The specification applies to any outputs switching in the same direction, either HIGH–to–LOW (t_{OSHL}) or LOW–to–HIGH (t_{OSLH}); parameter guaranteed by design.

DYNAMIC SWITCHING CHARACTERISTICS

Symbol	Characteristic	Condition	T_A = +25°C			Unit
			Min	Typ	Max	
V_{OLP}	Dynamic LOW Peak Voltage (Note 4.)	V_{CC} = 3.3V, C_L = 50pF, V_{IH} = 3.3V, V_{IL} = 0V		0.8		V
V_{OLV}	Dynamic LOW Valley Voltage (Note 4.)	V_{CC} = 3.3V, C_L = 50pF, V_{IH} = 3.3V, V_{IL} = 0V		0.8		V

4. Number of outputs defined as "n". Measured with "n–1" outputs switching from HIGH–to–LOW or LOW–to–HIGH. The remaining output is measured in the LOW state.

CAPACITIVE CHARACTERISTICS

Symbol	Parameter	Condition	Typical	Unit
C_{IN}	Input Capacitance	V_{CC} = 3.3V, V_I = 0V or V_{CC}	7	pF
C_{OUT}	Output Capacitance	V_{CC} = 3.3V, V_I = 0V or V_{CC}	8	pF
C_{PD}	Power Dissipation Capacitance	10MHz, V_{CC} = 3.3V, V_I = 0V or V_{CC}	25	pF

MOTOROLA
SEMICONDUCTOR TECHNICAL DATA

B-Suffix Series CMOS Gates

The B Series logic gates are constructed with P and N channel enhancement mode devices in a single monolithic structure (Complementary MOS). Their primary use is where low power dissipation and/or high noise immunity is desired.

- Supply Voltage Range = 3.0 Vdc to 18 Vdc
- All Outputs Buffered
- Capable of Driving Two Low–power TTL Loads or One Low–power Schottky TTL Load Over the Rated Temperature Range.
- Double Diode Protection on All Inputs Except: Triple Diode Protection on MC14011B and MC14081B
- Pin–for–Pin Replacements for Corresponding CD4000 Series B Suffix Devices (Exceptions: MC14068B and MC14078B)

L SUFFIX	P SUFFIX	D SUFFIX
CERAMIC	PLASTIC	SOIC
CASE 632	CASE 646	CASE 751A

ORDERING INFORMATION

MC14XXXBCP   Plastic
MC14XXXBCL   Ceramic
MC14XXXBD   SOIC

T_A = – 55° to 125°C for all packages.

MAXIMUM RATINGS* (Voltages Referenced to V_{SS})

Symbol	Parameter	Value	Unit
V_{DD}	DC Supply Voltage	– 0.5 to + 18.0	V
V_{in}, V_{out}	Input or Output Voltage (DC or Transient)	– 0.5 to V_{DD} + 0.5	V
I_{in}, I_{out}	Input or Output Current (DC or Transient), per Pin	± 10	mA
P_D	Power Dissipation, per Package†	500	mW
T_{stg}	Storage Temperature	– 65 to + 150	°C
T_L	Lead Temperature (8–Second Soldering)	260	°C

* Maximum Ratings are those values beyond which damage to the device may occur.
†Temperature Derating:
   Plastic "P and D/DW" Packages: – 7.0 mW/°C From 65°C To 125°C
   Ceramic "L" Packages: – 12 mW/°C From 100°C To 125°C

MC14001B
Quad 2-Input NOR Gate

MC14002B
Dual 4-Input NOR Gate

MC14011B
Quad 2-Input NAND Gate

MC14012B
Dual 4-Input NAND Gate

MC14023B
Triple 3-Input NAND Gate

MC14025B
Triple 3-Input NOR Gate

MC14068B
8-Input NAND Gate

MC14071B
Quad 2-Input OR Gate

MC14072B
Dual 4-Input OR Gate

MC14073B
Triple 3-Input AND Gate

MC14075B
Triple 3-Input OR Gate

MC14078B
8-Input NOR Gate

MC14081B
Quad 2-Input AND Gate

MC14082B
Dual 4-Input AND Gate

LOGIC DIAGRAMS

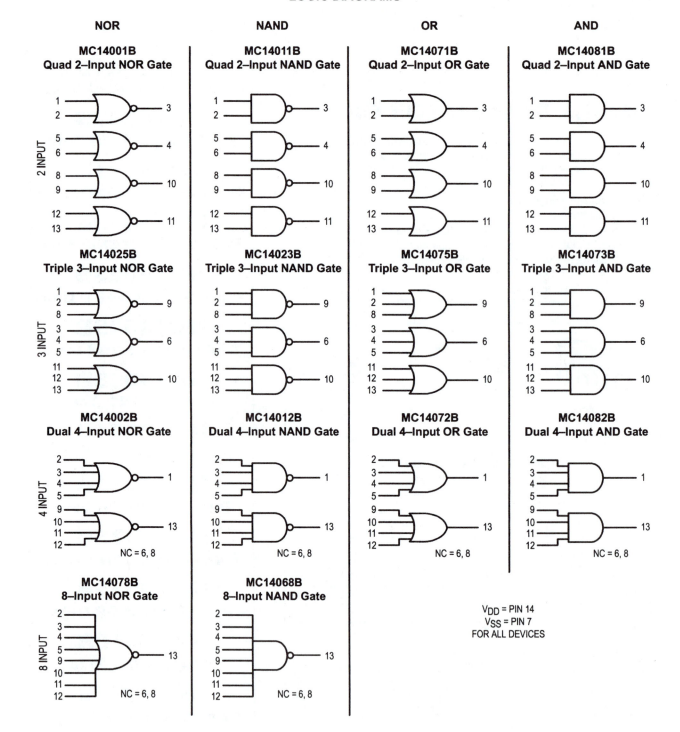

V_{DD} = PIN 14
V_{SS} = PIN 7
FOR ALL DEVICES

PIN ASSIGNMENTS

MC14001B
Quad 2–Input NOR Gate

IN 1$_A$	1	14	V$_{DD}$
IN 2$_A$	2	13	IN 2$_D$
OUT$_A$	3	12	IN 1$_D$
OUT$_B$	4	11	OUT$_D$
IN 1$_B$	5	10	OUT$_C$
IN 2$_B$	6	9	IN 2$_C$
V$_{SS}$	7	8	IN 1$_C$

MC14002B
Dual 4–Input NOR Gate

OUT$_A$	1	14	V$_{DD}$
IN 1$_A$	2	13	OUT$_B$
IN 2$_A$	3	12	IN 4$_B$
IN 3$_A$	4	11	IN 3$_B$
IN 4$_A$	5	10	IN 2$_B$
NC	6	9	IN 1$_B$
V$_{SS}$	7	8	NC

MC14011B
Quad 2–Input NAND Gate

IN 1$_A$	1	14	V$_{DD}$
IN 2$_A$	2	13	IN 2$_D$
OUT$_A$	3	12	IN 1$_D$
OUT$_B$	4	11	OUT$_D$
IN 1$_B$	5	10	OUT$_C$
IN 2$_B$	6	9	IN 2$_C$
V$_{SS}$	7	8	IN 1$_C$

MC14012B
Dual 4–Input NAND Gate

OUT$_A$	1	14	V$_{DD}$
IN 1$_A$	2	13	OUT$_B$
IN 2$_A$	3	12	IN 4$_B$
IN 3$_A$	4	11	IN 3$_B$
IN 4$_A$	5	10	IN 2$_B$
NC	6	9	IN 1$_B$
V$_{SS}$	7	8	NC

MC14023B
Triple 3–Input NAND Gate

IN 1$_A$	1	14	V$_{DD}$
IN 2$_A$	2	13	IN 3$_C$
IN 1$_B$	3	12	IN 2$_C$
IN 2$_B$	4	11	IN 1$_C$
IN 3$_B$	5	10	OUT$_C$
OUT$_B$	6	9	OUT$_A$
V$_{SS}$	7	8	IN 3$_A$

MC14025B
Triple 3–Input NOR Gate

IN 1$_A$	1	14	V$_{DD}$
IN 2$_A$	2	13	IN 3$_C$
IN 1$_B$	3	12	IN 2$_C$
IN 2$_B$	4	11	IN 1$_C$
IN 3$_B$	5	10	OUT$_C$
OUT$_B$	6	9	OUT$_A$
V$_{SS}$	7	8	IN 3$_A$

MC14068B
8–Input NAND Gate

NC	1	14	V$_{DD}$
IN 1	2	13	OUT
IN 2	3	12	IN 8
IN 3	4	11	IN 7
IN 4	5	10	IN 6
NC	6	9	IN 5
V$_{SS}$	7	8	NC

MC14071B
Quad 2–Input OR Gate

IN 1$_A$	1	14	V$_{DD}$
IN 2$_A$	2	13	IN 2$_D$
OUT$_A$	3	12	IN 1$_D$
OUT$_B$	4	11	OUT$_D$
IN 1$_B$	5	10	OUT$_C$
IN 2$_B$	6	9	IN 2$_C$
V$_{SS}$	7	8	IN 1$_C$

MC14072B
Dual 4–Input OR Gate

OUT$_A$	1	14	V$_{DD}$
IN 1$_A$	2	13	OUT$_B$
IN 2$_A$	3	12	IN 4$_B$
IN 3$_A$	4	11	IN 3$_B$
IN 4$_A$	5	10	IN 2$_B$
NC	6	9	IN 1$_B$
V$_{SS}$	7	8	NC

MC14073B
Triple 3–Input AND Gate

IN 1$_A$	1	14	V$_{DD}$
IN 2$_A$	2	13	IN 3$_C$
IN 1$_B$	3	12	IN 2$_C$
IN 2$_B$	4	11	IN 1$_C$
IN 3$_B$	5	10	OUT$_C$
OUT$_B$	6	9	OUT$_A$
V$_{SS}$	7	8	IN 3$_A$

MC14075B
Triple 3–Input OR Gate

IN 1$_A$	1	14	V$_{DD}$
IN 2$_A$	2	13	IN 3$_C$
IN 1$_B$	3	12	IN 2$_C$
IN 2$_B$	4	11	IN 1$_C$
IN 3$_B$	5	10	OUT$_C$
OUT$_B$	6	9	OUT$_A$
V$_{SS}$	7	8	IN 3$_A$

MC14078B
8–Input NOR Gate

NC	1	14	V$_{DD}$
IN 1	2	13	OUT
IN 2	3	12	IN 8
IN 3	4	11	IN 7
IN 4	5	10	IN 6
NC	6	9	IN 5
V$_{SS}$	7	8	NC

MC14081B
Quad 2–Input AND Gate

IN 1$_A$	1	14	V$_{DD}$
IN 2$_A$	2	13	IN 2$_D$
OUT$_A$	3	12	IN 1$_D$
OUT$_B$	4	11	OUT$_D$
IN 1$_B$	5	10	OUT$_C$
IN 2$_B$	6	9	IN 2$_C$
V$_{SS}$	7	8	IN 1$_C$

MC14082B
Dual 4–Input AND Gate

OUT$_A$	1	14	V$_{DD}$
IN 1$_A$	2	13	OUT$_B$
IN 2$_A$	3	12	IN 4$_B$
IN 3$_A$	4	11	IN 3$_B$
IN 4$_A$	5	10	IN 2$_B$
NC	6	9	IN 1$_B$
V$_{SS}$	7	8	NC

NC = NO CONNECTION

ELECTRICAL CHARACTERISTICS (Voltages Referenced to V_{SS})

Characteristic		Symbol	V_{DD} Vdc	-55°C Min	-55°C Max	25°C Min	25°C Typ #	25°C Max	125°C Min	125°C Max	Unit
Output Voltage "0" Level V_{in} = V_{DD} or 0		V_{OL}	5.0 10 15	— — —	0.05 0.05 0.05	— — —	0 0 0	0.05 0.05 0.05	— — —	0.05 0.05 0.05	Vdc
"1" Level V_{in} = 0 or V_{DD}		V_{OH}	5.0 10 15	4.95 9.95 14.95	— — —	4.95 9.95 14.95	5.0 10 15	— — —	4.95 9.95 14.95	— — —	Vdc
Input Voltage "0" Level (V_O = 4.5 or 0.5 Vdc) (V_O = 9.0 or 1.0 Vdc) (V_O = 13.5 or 1.5 Vdc)		V_{IL}	5.0 10 15	— — —	1.5 3.0 4.0	— — —	2.25 4.50 6.75	1.5 3.0 4.0	— — —	1.5 3.0 4.0	Vdc
"1" Level (V_O = 0.5 or 4.5 Vdc) (V_O = 1.0 or 9.0 Vdc) (V_O = 1.5 or 13.5 Vdc)		V_{IH}	5.0 10 15	3.5 7.0 11	— — —	3.5 7.0 11	2.75 5.50 8.25	— — —	3.5 7.0 11	— — —	Vdc
Output Drive Current (V_{OH} = 2.5 Vdc) (V_{OH} = 4.6 Vdc) (V_{OH} = 9.5 Vdc) (V_{OH} = 13.5 Vdc)	Source	I_{OH}	5.0 5.0 10 15	-3.0 -0.64 -1.6 -4.2	— — — —	-2.4 -0.51 -1.3 -3.4	-4.2 -0.88 -2.25 -8.8	— — — —	-1.7 -0.36 -0.9 -2.4	— — — —	mAdc
(V_{OL} = 0.4 Vdc) (V_{OL} = 0.5 Vdc) (V_{OL} = 1.5 Vdc)	Sink	I_{OL}	5.0 10 15	0.64 1.6 4.2	— — —	0.51 1.3 3.4	0.88 2.25 8.8	— — —	0.36 0.9 2.4	— — —	mAdc
Input Current		I_{in}	15	—	±0.1	—	±0.00001	±0.1	—	±1.0	µAdc
Input Capacitance (V_{in} = 0)		C_{in}	—	—	—	—	5.0	7.5	—	—	pF
Quiescent Current (Per Package)		I_{DD}	5.0 10 15	— — —	0.25 0.5 1.0	— — —	0.0005 0.0010 0.0015	0.25 0.5 1.0	— — —	7.5 15 30	µAdc
Total Supply Current**† (Dynamic plus Quiescent, Per Gate, C_L = 50 pF)		I_T	5.0 10 15	$I_T = (0.3\ \mu A/kHz)\ f + I_{DD}/N$ $I_T = (0.6\ \mu A/kHz)\ f + I_{DD}/N$ $I_T = (0.9\ \mu A/kHz)\ f + I_{DD}/N$							µAdc

\#Data labelled "Typ" is not to be used for design purposes but is intended as an indication of the IC's potential performance.

** The formulas given are for the typical characteristics only at 25°C.

†To calculate total supply current at loads other than 50 pF:

$$I_T(C_L) = I_T(50\ pF) + (C_L - 50)\ Vfk$$

where: I_T is in µA (per package), C_L in pF, $V = (V_{DD} - V_{SS})$ in volts, f in kHz is input frequency, and k = 0.001 x the number of exercised gates per package.

B–SERIES GATE SWITCHING TIMES

SWITCHING CHARACTERISTICS* (C_L = 50 pF, T_A = 25°C)

Characteristic	Symbol	V_{DD} Vdc	Min	Typ #	Max	Unit
Output Rise Time, All B–Series Gates	t_{TLH}					ns
$\quad t_{TLH}$ = (1.35 ns/pF) C_L + 33 ns		5.0	—	100	200	
$\quad t_{TLH}$ = (0.60 ns/pF) C_L + 20 ns		10	—	50	100	
$\quad t_{TLH}$ = (0.40 ns/PF) C_L + 20 ns		15	—	40	80	
Output Fall Time, All B–Series Gates	t_{THL}					ns
$\quad t_{THL}$ = (1.35 ns/pF) C_L + 33 ns		5.0	—	100	200	
$\quad t_{THL}$ = (0.60 ns/pF) C_L + 20 ns		10	—	50	100	
$\quad t_{THL}$ = (0.40 ns/pF) C_L + 20 ns		15	—	40	80	
Propagation Delay Time	t_{PLH}, t_{PHL}					ns
$\quad$ MC14001B, MC14011B only						
$\quad\quad t_{PLH}$, t_{PHL} = (0.90 ns/pF) C_L + 80 ns		5.0	—	125	250	
$\quad\quad t_{PLH}$, t_{PHL} = (0.36 ns/pF) C_L + 32 ns		10	—	50	100	
$\quad\quad t_{PLH}$, t_{PHL} = (0.26 ns/pF) C_L + 27 ns		15	—	40	80	
$\quad$ All Other 2, 3, and 4 Input Gates						
$\quad\quad t_{PLH}$, t_{PHL} = (0.90 ns/pF) C_L + 115 ns		5.0	—	160	300	
$\quad\quad t_{PLH}$, t_{PHL} = (0.36 ns/pF) C_L + 47 ns		10	—	65	130	
$\quad\quad t_{PLH}$, t_{PHL} = (0.26 ns/pF) C_L + 37 ns		15	—	50	100	
$\quad$ 8–Input Gates (MC14068B, MC14078B)						
$\quad\quad t_{PLH}$, t_{PHL} = (0.90 ns/pF) C_L + 155 ns		5.0	—	200	350	
$\quad\quad t_{PLH}$, t_{PHL} = (0.36 ns/pF) C_L + 62 ns		10	—	80	150	
$\quad\quad t_{PLH}$, t_{PHL} = (0.26 ns/pF) C_L + 47 ns		15	—	60	110	

* The formulas given are for the typical characteristics only at 25°C.

#Data labelled "Typ" is not to be used for design purposes but is intended as an indication of the IC's potential performance.

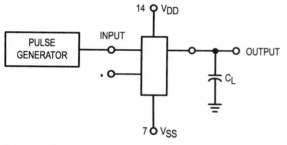

* All unused inputs of AND, NAND gates must be connected to V_{DD}.
  All unused inputs of OR, NOR gates must be connected to V_{SS}.

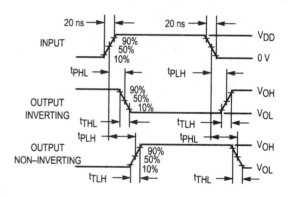

Figure 1. Switching Time Test Circuit and Waveforms

CIRCUIT SCHEMATIC
NOR, OR GATES

MC14001B, MC14071B
One of Four Gates Shown

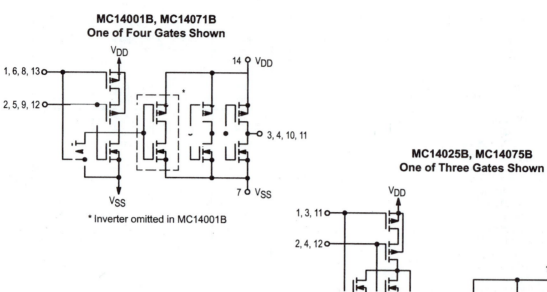

* Inverter omitted in MC14001B

MC14025B, MC14075B
One of Three Gates Shown

* Inverter omitted in MC14025B

MC14002B, MC14072B
One of Two Gates Shown

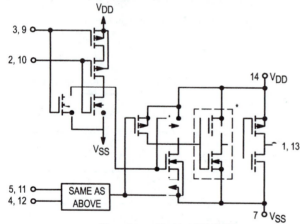

* Inverter omitted in MC14002B

MC14078B
Eight Input Gate

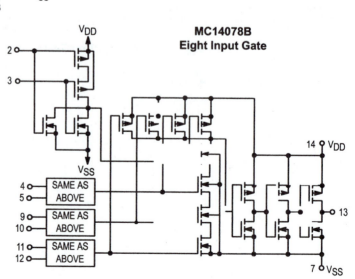

CIRCUIT SCHEMATIC
NAND, AND GATES

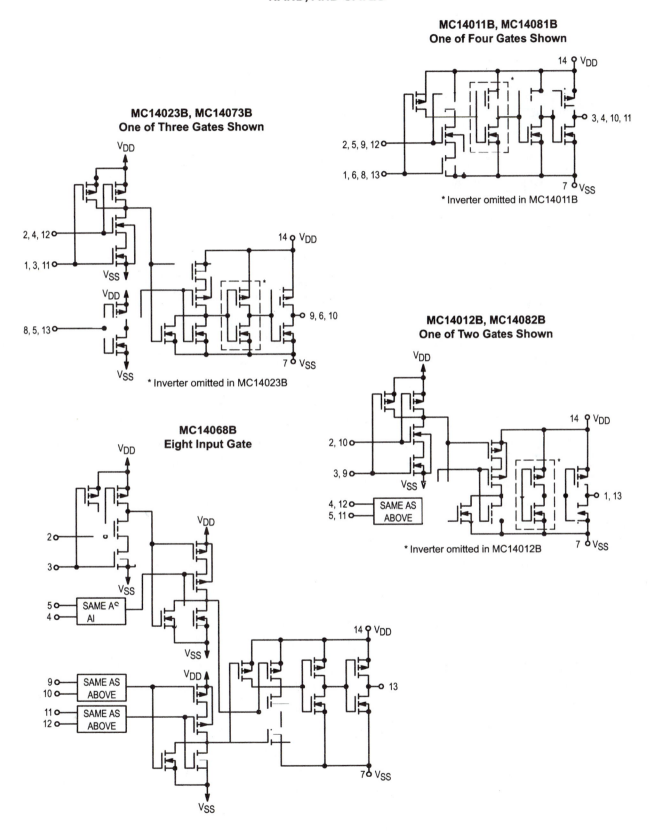

MC14011B, MC14081B
One of Four Gates Shown

* Inverter omitted in MC14011B

MC14023B, MC14073B
One of Three Gates Shown

* Inverter omitted in MC14023B

MC14012B, MC14082B
One of Two Gates Shown

* Inverter omitted in MC14012B

MC14068B
Eight Input Gate

APPENDIX D

Handling Precautions for CMOS

HANDLING AND DESIGN GUIDELINES

HANDLING PRECAUTIONS

All MOS devices have insulated gates that are subject to voltage breakdown. The gate oxide for Motorola CMOS devices is about 800 Å thick and breaks down at a gate-source potential of about 100 volts. To guard against such a breakdown from static discharge or other voltage transients, the protection network shown in Figure 1 is used on each input to the CMOS device.

Static damaged devices behave in various ways, depending on the severity of the damage. The most severely damaged inputs are the easiest to detect because the input has been completely destroyed and is either shorted to V_{DD}, shorted to V_{SS}, or open-circuited. The effect is that the device no longer responds to signals present at the damaged input. Less severe cases are more difficult to detect because they show up as intermittent failures or as degraded performance. Another effect of static damage is that the inputs generally have increased leakage currents.

Although the input protection network does provide a great deal of protection, CMOS devices are not immune to large static voltage discharges that can be generated during handling. For example, static voltages generated by a person walking across a waxed floor have been measured in the 4-15 kV range (depending on humidity, surface conditions, etc.). Therefore, the following precautions should be observed:

1. Do not exceed the Maximum Ratings specified by the data sheet.
2. All unused device inputs should be connected to V_{DD} or V_{SS}.
3. All low-impedance equipment (pulse generators, etc.) should be connected to CMOS inputs only after the device is powered up. Similarly, this type of equipment should be disconnected before power is turned off.
4. Circuit boards containing CMOS devices are merely extensions of the devices, and the same handling precautions apply. Contacting edge connectors wired directly to device inputs can cause damage. Plastic wrapping should be avoided. When external connections to a PC board are connected to an input of a CMOS device, a resistor should be used in series with the input. This resistor helps limit accidental damage if the PC board is removed and brought into contact with static generating materials. The limiting factor for the series resistor is the added delay. This is caused by the time constant formed by the series resistor and

input capacitance. Note that the maximum input rise and fall times should not be exceeded. In Figure 2, two possible networks are shown using a series resistor to reduce ESD (Electrostatic Discharge) damage. For convenience, an equation for added propagation delay and rise time effects due to series resistance size is given.

5. All CMOS devices should be stored or transported in materials that are antistatic. CMOS devices must not be inserted into conventional plastic "snow", styrofoam, or plastic trays, but should be left in their original container until ready for use.
6. All CMOS devices should be placed on a grounded bench surface and operators should ground themselves prior to handling devices, since a worker can be statically charged with respect to the bench surface. Wrist straps in contact with skin are strongly recommended. See Figure 3 for an example of a typical work station.
7. Nylon or other static generating materials should not come in contact with CMOS devices.
8. If automatic handlers are being used, high levels of static electricity may be generated by the movement of the device, the belts, or the boards. Reduce static build-up by using ionized air blowers or room humidifiers. All parts of machines which come into contact with the top, bottom, or sides of IC packages must be grounded to metal or other conductive material.
9. Cold chambers using CO_2 for cooling should be equipped with baffles, and the CMOS devices must be contained on or in conductive material.
10. When lead-straightening or hand-soldering is necessary, provide ground straps for the apparatus used and be sure that soldering ties are grounded.
11. The following steps should be observed during wave solder operations:
    a. The solder pot and conductive conveyor system of the wave soldering machine must be grounded to an earth ground.
    b. The loading and unloading work benches should have conductive tops which are grounded to an earth ground.
    c. Operators must comply with precautions previously explained.
    d. Completed assemblies should be placed in antistatic containers prior to being moved to subsequent stations.

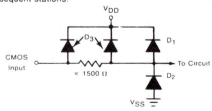

FIGURE 1 — INPUT PROTECTION NETWORK

12. The following steps should be observed during board-cleaning operations:
    a. Vapor degreasers and baskets must be grounded to an earth ground.
    b. Brush or spray cleaning should not be used.
    c. Assemblies should be placed into the vapor degreaser immediately upon removal from the antistatic container.
    d. Cleaned assemblies should be placed in antistatic containers immediately after removal from the cleaning basket.
    e. High velocity air movement or application of solvents and coatings should be employed only when assembled printed circuit boards are grounded and a static eliminator is directed at the board.
13. The use of static detection meters for production line surveillance is highly recommended.
14. Equipment specifications should alert users to the presence of CMOS devices and require familiarization with this specification prior to performing any kind of maintenance or replacement of devices or modules.

15. Do not insert or remove CMOS devices from test sockets with power applied. Check all power supplies to be used for testing devices to be certain there are no voltage transients present.
16. Double check test equipment setup for proper polarity of V_{DD} and V_{SS} before conducting parametric or functional testing.
17. Do not recycle shipping rails or trays. Repeated use causes deterioration of their antistatic coating.

RECOMMENDED FOR READING:

"Total Control of the Static in Your Business"

Available by writing to:
3M Company
Static Control Systems
P.O. Box 2963
Austin, Texas 78769-2963
Or by Calling:
1-800-328-1368

FIGURE 2 — NETWORKS FOR MINIMIZING ESD AND REDUCING CMOS LATCH UP SUSCEPTIBILITY

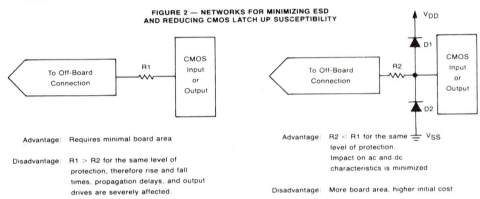

Advantage: Requires minimal board area

Disadvantage: R1 > R2 for the same level of protection, therefore rise and fall times, propagation delays, and output drives are severely affected.

Advantage: R2 < R1 for the same level of protection. Impact on ac and dc characteristics is minimized

Disadvantage: More board area, higher initial cost

Note: These networks are useful for protecting the following
A digital inputs and outputs   C 3-state outputs
B analog inputs and outputs   D bidirectional (I/O) ports

PROPAGATION DELAY AND RISE TIME vs. SERIES RESISTANCE

$$R \approx \frac{t}{C \cdot k}$$

where:
R = the maximum allowable series resistance in ohms
t = the maximum tolerable propagation delay or rise time in seconds
C = the board capacitance plus the driven device's input capacitance in farads
k = 0.7 for propagation delay calculations
k = 2.3 for rise time calculations

FIGURE 3 — TYPICAL MANUFACTURING WORK STATION

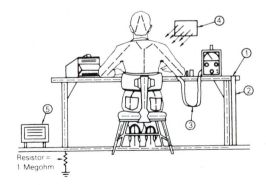

Resistor = 1 Megohm

NOTES: 1. 1/16 inch conductive sheet stock covering bench top work area.
2. Ground strap.
3. Wrist strap in contact with skin.
4. Static neutralizer. (Ionized air blower directed at work.) Primarily for use in areas where direct grounding is impractical.
5. Room humidifier. Primarily for use in areas where the relative humidity is less than 45%. Caution: building heating and cooling systems usually dry the air causing the relative humidity inside of buildings to be less than outside humidity.

EPROM Data For A Digital Function Generator

Included in Appendix E:

E.1 A complete set of EPROM data for the EPROM-based digital function generator described in Section 13.4. The file can be used to program a standard EPROM or as an initialization file for an LPM_ROM component in MAX+PLUS II. (LPM_ROM can only be used with the FLEX 10K device on the Altera UP-1 board.)

E.2 A program written in ANSI C to generate an EPROM record file (Intel format).

E.3 A copy of the generated record file.

E.1 EPROM Data

00 is the maximum negative voltage of a waveform, FF is maximum positive, and 80 is the zero-crossing point.

SINE

Base Address	_Byte Addresses_															
	0	1	2	3	4	5	6	7	8	9	A	B	C	D	E	F
0000	80	83	86	89	8C	8F	92	95	98	9C	9F	A2	A5	A8	AB	AE
0010	B0	B3	B6	B9	BC	BF	C1	C4	C7	C9	CC	CE	D1	D3	D5	D8
0020	DA	DC	DE	E0	E2	E4	E6	E8	EA	EC	ED	EF	F0	F2	F3	F5
0030	F6	F7	F8	F9	FA	FB	FC	FC	FD	FE	FE	FF	FF	FF	FF	FF
0040	FF	FF	FF	FF	FF	FF	FE	FE	FD	FC	FC	FB	FA	F9	F8	F7
0050	F6	F5	F3	F2	F0	EF	ED	EC	EA	E8	E6	E4	E2	E0	DE	DC
0060	DA	D8	D5	D3	D1	CE	CC	C9	C7	C4	C1	BF	BC	B9	B6	B3
0070	B0	AE	AB	A8	A5	A2	9F	9C	98	95	92	8F	8C	89	86	83
0080	7F	7C	79	76	73	70	6D	6A	67	63	60	5D	5A	57	54	51
0090	4F	4C	49	46	43	40	3E	3B	38	36	33	31	2E	2C	2A	27
00A0	25	23	21	1F	1D	1B	19	17	15	13	12	10	0F	0D	0C	0A
00B0	09	08	07	06	05	04	03	03	02	01	01	00	00	00	00	00
00C0	00	00	00	00	00	00	01	01	02	03	03	04	05	06	07	08
00D0	09	0A	0C	0D	0F	10	12	13	15	17	19	1B	1D	1F	21	23
00E0	25	27	2A	2C	2E	31	33	36	38	3B	3E	40	43	46	49	4C
00F0	4F	51	54	57	5A	5D	60	63	67	6A	6D	70	73	76	79	7C

SQUARE

Base Address	\ Byte Addresses															
	0	1	2	3	4	5	6	7	8	9	A	B	C	D	E	F
0100	FF	FF	FF	FF	FF	FF	FF	FF	FF	FF	FF	FF	FF	FF	FF	FF
0110	FF	FF	FF	FF	FF	FF	FF	FF	FF	FF	FF	FF	FF	FF	FF	FF
0120	FF	FF	FF	FF	FF	FF	FF	FF	FF	FF	FF	FF	FF	FF	FF	FF
0130	FF	FF	FF	FF	FF	FF	FF	FF	FF	FF	FF	FF	FF	FF	FF	FF
0140	FF	FF	FF	FF	FF	FF	FF	FF	FF	FF	FF	FF	FF	FF	FF	FF
0150	FF	FF	FF	FF	FF	FF	FF	FF	FF	FF	FF	FF	FF	FF	FF	FF
0160	FF	FF	FF	FF	FF	FF	FF	FF	FF	FF	FF	FF	FF	FF	FF	FF
0170	FF	FF	FF	FF	FF	FF	FF	FF	FF	FF	FF	FF	FF	FF	FF	FF
0180	00	00	00	00	00	00	00	00	00	00	00	00	00	00	00	00
0190	00	00	00	00	00	00	00	00	00	00	00	00	00	00	00	00
01A0	00	00	00	00	00	00	00	00	00	00	00	00	00	00	00	00
01B0	00	00	00	00	00	00	00	00	00	00	00	00	00	00	00	00
01C0	00	00	00	00	00	00	00	00	00	00	00	00	00	00	00	00
01D0	00	00	00	00	00	00	00	00	00	00	00	00	00	00	00	00
01E0	00	00	00	00	00	00	00	00	00	00	00	00	00	00	00	00
01F0	00	00	00	00	00	00	00	00	00	00	00	00	00	00	00	00

TRIANGLE

Base Address	Byte Addresses															
	0	1	2	3	4	5	6	7	8	9	A	B	C	D	E	F
0200	80	82	84	86	88	8A	8C	8E	90	92	94	96	98	9A	9C	9E
0210	A0	A2	A4	A6	A8	AA	AC	AE	B0	B2	B4	B6	B8	BA	BC	BE
0220	C0	C2	C4	C6	C8	CA	CC	CE	D0	D2	D4	D6	D8	DA	DC	DE
0230	E0	E2	E4	E6	E8	EA	EC	F0	F2	F4	F6	F8	FA	FC	FE	
0240	FE	FC	FA	F8	F6	F4	F2	F0	EE	EC	EA	E8	E6	E4	E2	E0
0250	DE	DC	DA	D8	D6	D4	D2	D0	CE	CC	CA	C8	C6	C4	C2	C0
0260	BE	BC	BA	B8	B6	B4	B2	B0	AE	AC	AA	A8	A6	A4	A2	A0
0270	9E	9C	9A	98	96	94	92	90	8E	8C	8A	88	86	84	82	80
0280	7E	7C	7A	78	76	74	72	70	6E	6C	6A	68	66	64	62	60
0290	5E	5C	5A	58	56	54	52	50	4E	4C	4A	48	46	44	42	40
02A0	3E	3C	3A	38	36	34	32	30	2E	2C	2A	28	26	24	22	20
02B0	1E	1C	1A	18	16	14	12	10	0E	0C	0A	08	06	04	02	00
02C0	02	04	06	08	0A	0C	0E	10	12	14	16	18	1A	1C	1E	20
02D0	22	24	26	28	2A	2C	2E	30	32	34	36	38	3A	3C	3E	40
02E0	42	44	46	48	4A	4C	4E	50	52	54	56	58	5A	5C	5E	60
02F0	62	64	66	68	6A	6C	6E	70	72	74	76	78	7A	7C	7E	80

SAWTOOTH

Base Address	Byte Addresses															
	0	1	2	3	4	5	6	7	8	9	A	B	C	D	E	F
0300	00	01	02	03	04	05	06	07	08	09	0A	0B	0C	0D	0E	0F
0310	10	11	12	13	14	15	16	17	18	19	1A	1B	1C	1D	1E	1F
0320	20	21	22	23	24	25	26	27	28	29	2A	2B	2C	2D	2E	2F
0330	30	31	32	33	34	35	36	37	38	39	3A	3B	3C	3D	3E	3F
0340	40	41	42	43	44	45	46	47	48	49	4A	4B	4C	4D	4E	4F
0350	50	51	52	53	54	55	56	57	58	59	5A	5B	5C	5D	5E	5F
0360	60	61	62	63	64	65	66	67	68	69	6A	6B	6C	6D	6E	6F
0370	70	71	72	73	74	75	76	77	78	79	7A	7B	7C	7D	7E	7F
0380	80	81	82	83	84	85	86	87	88	89	8A	8B	8C	8D	8E	8F
0390	90	91	92	93	94	95	96	97	98	99	9A	9B	9C	9D	9E	9F
03A0	A0	A1	A2	A3	A4	A5	A6	A7	A8	A9	AA	AB	AC	AD	AE	AF
03B0	B0	B1	B2	B3	B4	B5	B6	B7	B8	B9	BA	BB	BC	BD	BE	BF
03C0	C0	C1	C2	C3	C4	C5	C6	C7	C8	C9	CA	CB	CC	CD	CE	CF
03D0	D0	D1	D2	D3	D4	D5	D6	D7	D8	D9	DA	DB	DC	DD	DE	DF
03E0	E0	E1	E2	E3	E4	E5	E6	E7	E8	E9	EA	EB	EC	ED	EE	EF
03F0	F0	F1	F2	F3	F4	F5	F6	F7	F8	F9	FA	FB	FC	FD	FD	FF

E.2 C Program

This program is also on the accompanying CD in the file
\Student_Files\EPROM\EPROM.C.

```c
/*     Hex File Generator
*      Written by: Ronan Capina and Robert Dueck
*
*      This program is to create a hex file in a specific EPROM record format.
* The EPROM is addressed in blocks of 256 bytes (8 address lines) by an 8 bit
* counter to create a digital image of one of several waveform outputs.
* Two additional address bits select one of four output functions.
*      When the EPROM data are run through a D/A Converter, they create an
* analog waveform running at the frequency of the counter divided by 256.
*      The waveforms are sine, square, triangle, and sawtooth.
*      The record format is as follows. (Spaces are inserted only for clarity.
* The actual record must have NO spaces.)
*
*      : 10 0080 00 AF5F67F0602703E0322CFA92007780C3 61
*
*      (:Record Length = 10hex = 16dec)
*      (Address = 0080hex; location in EPROM of first data byte in record)
*      (Record type = 00 = data)
*      (16 data bytes = 32 hex digits)
*      (Checksum; Record Length + Address High byte + Address Low byte
*      + Record type + data bytes + Checksum = 00, after discarding carry)
*
*      An END record is also required, having a similar format, except that
* Record Type = 01 = END and there are no data bytes. Checksum is still
* required.
*      eg. :00000001FF
*/
#include <stdio.h>
#include <stdlib.h>
#include <math.h>
#include <string.h>
#include <conio.h>

void Sawtooth(int);
void triangle(int);
void Square(int);
int AddrByte(int Addr);
int Chksum(int sum);
void sine(int);
char *HexString(int value);

char *Hex = HULL;
int Ampl;

const double Pi = 3.141592654;

int main(void)
{
  FILE *fp;
      int Fcn =1, Linenum, sum, Byte, Addr = 0;
  char Record[256];
  clrscr();
```

```
/*      Hex file is written to c:\eprom\eprom.hex
 *      To change the file name, modify the following line to:
 *              if (!(fp = fopen("c:\\YourDirectoryName\\YourFile.hex", "w+t")))
{
 */
  if (!(fp = fopen("c:\\eprom\\eprom.hex", "w+t"))) {
        printf("Error opening output file.\r\n");
        exit(1);
  } else
        printf("File opened successfully. \r\n");

  while (Fcn != 5) {      /* Create records for 4 functions */
                          /* Each function has 16 lines of data */
      for (Linenum = 1; Linenum <= 16; Linenum++) {
          /* Create record and address information */
          if (Addr < 16)
            sprintf(Record, ":10000%x", Addr);
          else
            if (Addr < 256)
                sprintf(Record, ":1000%x", Addr);
            else
                sprintf(Record, ":100%x", Addr);
          strcat(Record, "00");

          /* Accumulate sum for calculating checksum */
          sum = 16 + AddrByte(Addr);

          /* Calculate byte values for selected function */
          for (Byte = 1; Byte <= 16; Byte++) {
            if (Fcn == 1)
                sine(Addr % 256);
            else if (Fcn == 2)
                Square(Addr % 256);
            else if (Fcn == 3)
                triangle(Addr % 256);
            else
                Sawtooth(Addr % 256);

          /* Append calculated byte value (amplitude) to the record
                and update checksum accumulator */
          strcat (Record, HexString(Ampl));
          sum = sum + Ampl;
          Addr++;
          }

          strcat (Record, HexStrIng(Chksum(sum)));
          fprintf(fp, "%s\r\n", Record);
      }
      Fcn++;
  }
      fprintf(fp, ":00000001FF\r\n");
  fclose(fp);

  return(0);
}
```

```c
int AddrByte(int Addr)
{
  if (Addr < 256)
      return Addr;
  else
      return (((int)(Addr / 256)) + (Addr % 256));
}
int Chksum(int sum)
{
  int IntRem = sum % 256;
  if (IntRem == 0)
      return 0;
  else
      return (256 - IntRem);
}


char *HexString(int value)
{
  if (value < 16) {
      sprintf(Hex, "0%x", value);
      return Hex;
  } else if (value > 255)
      return "FF";
  else {
      sprintf(Hex, "%x", value);
      return Hex;
  }
}


void Sawtooth(int Addr)
{
      Ampl = Addr;
      return;
}


void sine(int Addr)
{
      double angle = ((((float)Addr) * 2 * Pi) / 256);
      Ampl = ((int)((sin(angle) * 128) + 128));
      if (Ampl > 255)
       Ampl = 255;
       return;
}
void Square(int Addr)
{
      if (Addr < 128)
       Ampl = 255;
      else
       Ampl = 0;
      return;
}
```

```
void triangle(int Addr)
{
     if (Addr < 64)
      Ampl = 128 + (2 * Addr);
     else if ((Addr >= 64) && (Addr < 192))
      Ampl = 256 - 2 * (Addr - 63);
     else
      Ampl = 2 * (Addr - 191);
     return;
}
```

E.3   Resultant Record File

```
:10000000808386898C8F9295989C9FA2A5A8ABAE81
:10001000B0B3B6B9BCBFC1C4C7C9CCCED1D3D5D893
:10002000DADCDEE0E2E4E6E8EAECEDEFF0F2F3F54C
:10003000F6F7F8F9FAFBFCFDFEFEFFFFFFFFFFFF01
:10004000FFFFFFFFFFFFFFEFEFDFCFCFBFAF9F8F7E8
:10005000F6F5F3F2F0EFEDECEAE8E6E4E2E0DEDC00
:10006000DAD8D5D3D1CECCC9C7C4C1BFBCB9B6B319
:10007000B0AEABA8A5A29F9C9895928F8C898683E1
:100080007F7C797673706D6A6763605D5A575451EF
:100090004F4C494643403E3B383633312E2C2A27BD
:1000A0002523211F1D1B1917151312100F0D0C0AE4
:1000B000090807060504030302010100000000000F
:1000C000000000000000000000010203040506070808
:1000D000090A0C0D0F1012131517191B1D1F2123D0
:1000E00025272A2C2E313336383B3E404346494C97
:1000F0004F5154575A5D6063676A6D707376797CAF
:10010000FFFFFFFFFFFFFFFFFFFFFFFFFFFFFFFFFF
:10011000FFFFFFFFFFFFFFFFFFFFFFFFFFFFFFFFEF
:10012000FFFFFFFFFFFFFFFFFFFFFFFFFFFFFFFFDF
:10013000FFFFFFFFFFFFFFFFFFFFFFFFFFFFFFFFCF
:10014000FFFFFFFFFFFFFFFFFFFFFFFFFFFFFFFFBF
:10015000FFFFFFFFFFFFFFFFFFFFFFFFFFFFFFFFAF
:10016000FFFFFFFFFFFFFFFFFFFFFFFFFFFFFFFF9F
:10017000FFFFFFFFFFFFFFFFFFFFFFFFFFFFFFFF8F
:10018000000000000000000000000000000000006F
:10019000000000000000000000000000000000005F
:1001A000000000000000000000000000000000004F
:1001B000000000000000000000000000000000003F
:1001C000000000000000000000000000000000002F
:1001D000000000000000000000000000000000001F
:1001E000000000000000000000000000000000000F
:1001F00000000000000000000000000000000000FF
:1002000080828486888A8C8E90929496989A9C9EFE
:10021000A0A2A4A6A8AAACAEB0B2B4B6B8BABCBEEE
:10022000C0C2C4C6C8CACCCED0D2D4D6D8DADCDEDE
:10023000E0E2E4E6E8EAECEEF0F2F4F6F8FAFCFECE
:10024000FEFCFAF8F6F4F2F0EEECEAE8E6E4E2E0BE
:10025000DEDCDAD8D6D4D2D0CECCCAC8C6C4C2C0AE
:10026000BEBCBAB8B6B4B2B0AEACAAA8A6A4A2A09E
:100270009E9C9A98969492908E8C8A88868482808E
:100280007E7C7A78767472706E6C6A68666462607E
```

```
:100290005E5C5A58565452504E4C4A48464442406E
:1002A0003E3C3A38363432302E2C2A28262422205E
:1002B0001E1C1A18161412100E0C0A08060402004E
:1002C000020406080A0C0E10121416181A1C1E201E
:1002D000222426282A2C2E30323436383A3C3E400E
:1002E0004244464 84A4C4E50525456585A5C5E60FE
:1002F000626466686A6C6E70727476787A7C7E80EE
:10030000000102030405060708090A0B0C0D0E0F75
:1003100010111213141516171819 1A1B1C1D1E1F65
:100320002021222324252627282 92A2B2C2D2E2F55
:100330003031323334353637383 93A3B3C3D3E3F45
:100340004041424344454647484 94A4B4C4D4E4F35
:100350005051525354555657585 95A5B5C5D5E5F25
:100360006061626364656667686 96A6B6C6D6E6F15
:100370007071727374757677787 97A7B7C7D7E7F05
:100380008081828384858687888 98A8B8C8D8E8FF5
:100390009091929394959697989 99A9B9C9D9E9FE5
:1003A000A0A1A2A3A4A5A6A7A8A9AAABACADAEAFD5
:1003B000B0B1B2B3B4B5B6B7B8B9BABBBCBDBEBFC5
:1003C000C0C1C2C3C4C5C6C7C8C9CACBCCCDCECFB5
:1003D000D0D1D2D3D4D5D6D7D8D9DADBDCDDDEDFA5
:1003E000E0E1E2E3E4E5E6E7E8E9EAEBECEDEEEF95
:1003F000F0F1F2F3F4F5F6F7F8F9FAFBFCFDFEFF85
:00000001FF
```

Answers to Selected Odd-Numbered Problems

Chapter 1

1.1 Analog quantities:
   a. Water temperature at the beach;
   b. weight of a bucket of sand;
   e. height of a wave;
   Digital quantities:
   c. grains of sand in a bucket;
   d. waves hitting the beach in one hour;
   f. people in a square mile.
   Generally, any quantity that can be expressed as "the number of. . ." is digital.

1.3 **a.** 4;   **b.** 8;   **c.** 25;
   d. 6;   **e.** 21;   **f.** 29;
   g. 59;   **h.** 93;   **i.** 33;
   j. 185

1.5 101, 110, 111, 1000

1.7 16

1.9 **a.** 0.625;   **b.** 0.375;   **c.** 0.8125

1.11 1/3

1.13 **a.** 0.11;
   b. 0.101;
   c. 0.0011;
   d. 0.10$\overline{1001}$;
   e. 1.11;
   f. 11.11$\overline{1100}$;
   g. 1000011.1101011100001. . . (nonrepeating)

1.15 9F7, 9F8, 9F9, 9FA, 9FB, 9FC, 9FD, 9FE, 9FF, A00, A01, A02, A03

1.17 **a.** 2C5;   **b.** 761;   **c.** FFF;
   d. 1000;   **e.** 2790;   **f.** 7D00;
   g. 8000

1.19 **a.** 5E86;   **b.** B6A;   **c.** C5B;
   d. 6BC4;   **e.** 15785;   **f.** 198B7;
   g. 28000

1.21 **Periodic: b., c., e.** Each of these waveforms repeats itself in a fixed period of time. (Note that waveform **b.** may not immediately appear to be periodic. However, if we count the sequence of short pulse, short space, medium pulse, medium space, short pulse, long space, we will find that each repetition of this sequence takes the same time.)

   Aperiodic: a., d. Neither of these waveforms repeats in a fixed period of time. Waveform **a.** has three equally-spaced pulses of equal width, but this pattern does not repeat in the time shown. Waveform **d.** has pulses of equal duration, spaced at increasing (i.e. unequal) intervals.

1.23 From the graph in Figure 1.14, read the times corresponding to the 10%, 50%, and 90% values of the pulse on both leading and trailing edges.

   Leading edge: 10%: 5 µs   **Trailing edge:** 90%: 40 µs
                50%: 7.5 µs   50%: 45 µs
                90%: 10 µs   10%: 50 µs

   Pulse width: 50% of leading edge to 50% of trailing edge.

$$t_w = 45\,\mu s - 7.5\,\mu s = 37.5\,\mu s$$

   Rise time: 10% of rising edge to 90% of rising edge.

$$t_r = 10\,\mu s - 5\,\mu s = 5\,\mu s$$

   Fall time: 90% of falling edge to 10% of falling edge.
$$t_f = 50\,\mu s - 40\,\mu s = 10\,\mu s$$

Chapter 2

2.1 See Figure ANS2.1.

a. Distinctive Shape   **b. Rectangular Outline**

FIGURE ANS2.1

2.3 See Figure ANS2.3.

A
B
C
Y = A + B + C

a. Distinctive Shape

A
B
C
≥1
Y = A + B + C

b. Rectangular Outline

FIGURE ANS2.3

2.5 *N* is HIGH if *J* OR *K* OR *L* OR *M* IS HIGH. See Table ANS2.5.

Table ANS2.5   4-input OR Truth Table

J	K	L	M	N
0	0	0	0	0
0	0	0	1	1 *
0	0	1	0	1 *
0	0	1	1	1 *
0	1	0	0	1 *
0	1	0	1	1 *
0	1	1	0	1 *
0	1	1	1	1 *
1	0	0	0	1 *
1	0	0	1	1 *
1	0	1	0	1 *
1	0	1	1	1 *
1	1	0	0	1 *
1	1	0	1	1 *
1	1	1	0	1 *
1	1	1	1	1 *

2.7 The switches must be connected in parallel. See Figure ANS2.7.

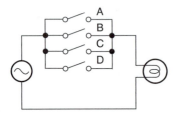

FIGURE ANS2.7

2.9 Active LOW. When the switch is pressed, it generates a logic LOW.

2.11 The anode must be at a higher voltage than the cathode by a specified amount.

2.13 See Figure ANS2.13.

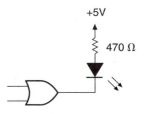

+5V

470 Ω

FIGURE ANS2.13

2.15 **a.** Output *Y* is LOW when *A* OR *B* OR *C* OR *D* are HIGH. The truth table is shown in Table ANS2.15.

Table ANS2.15   4-input NOR Truth Table

A	B	C	D	Y
0	0	0	0	1
0	0	0	1	0
0	0	1	0	0
0	0	1	1	0
0	1	0	0	0
0	1	0	1	0
0	1	1	0	0
0	1	1	1	0
1	0	0	0	0
1	0	0	1	0
1	0	1	0	0
1	0	1	1	0
1	1	0	0	0
1	1	0	1	0
1	1	1	0	0
1	1	1	1	0

b. $Y = \overline{A + B + C + D}$

c. See Figure ANS2.15

A
B
C
D
$Y = \overline{A + B + C + D}$

a. Distinctive Shape

A
B
C
D
≥1
Y

b. Rectangular Outline

FIGURE ANS2.15

2.17 Output *Y* is LOW if inputs *A* AND *B* AND *C* AND *D* AND *E* are all HIGH.

2.19 Required gate is a 2-input AND.

2.21 XNOR.

2.23 Output is HIGH if an odd number of inputs is HIGH.

2.25 **a.** and **c.** The attributes of shape, input level, and output level are all different between these two symbols.

2.27 See Figure ANS2.27.

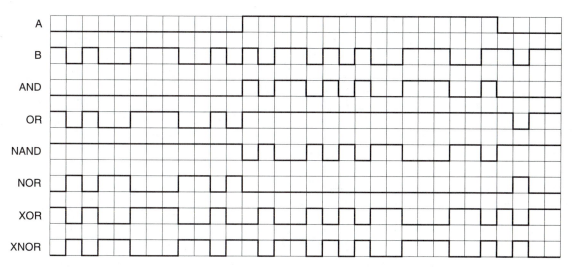

FIGURE ANS2.27

2.29 See Figure ANS2.29.

2.31 A HIGH is required to enable the AND gate. This allows the lamp to flash.

2.33 No. An XOR gate has no inhibit state. The lamp always flashes.

2.35 Transistor-Transistor Logic (TTL) and Complementary Metal-Oxide-Semiconductor (CMOS). Typically, TTL can drive higher-current loads. CMOS has more flexible power supply requirements and uses less power.

2.37 Low power Schottky TTL: 74LS02; CMOS: 4001B; High-speed CMOS: 74HC02. NANDs and NORs are differentiated by the last two digits in their part numbers.

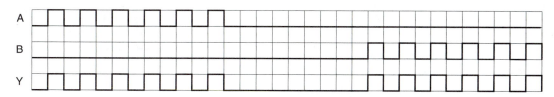

FIGURE ANS2.29

Chapter 3

3.1 **a.** $Y = ABC$;

  b. $X = PQ + RS$;

  c. $M = HJKL$;

  d. $A = W + X + Y + Z$;

  e. $Y = (A + B)(C + D)$;

  f. $Y = \overline{(A + B)(C + D)}$;

  g. $Y = (\overline{A} + \overline{B})(\overline{C} + \overline{D})$;

  h. $X = \overline{P}\,\overline{Q} + \overline{R}\,\overline{S}$;

  i. $X = \overline{\overline{P}\,\overline{Q} + \overline{R}\,\overline{S}}$

FIGURE ANS3.3

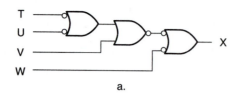

a.

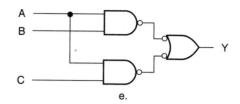

e.

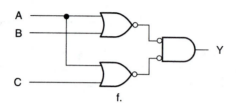

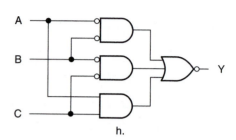

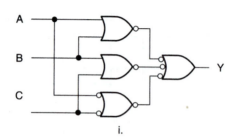

f.

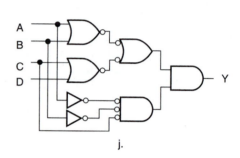

h.

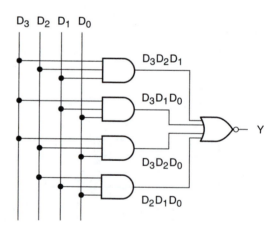

i.

j.

3.3 See Figure ANS3.3.

Boolean expressions:

a. $X = \overline{T} + \overline{U} + V + \overline{W}$;

e. $Y = AB + AC$;

f. $Y = (A+B)(A+C)$;

h. $Y = \overline{A}\,\overline{B} + \overline{B}\,\overline{C} + A\,C$;

i. $Y = (A + B) + (B + C) + (\overline{A} + \overline{C}) = 1$;

j. $Y = (A + B + C + D)AB\overline{C} = AB\overline{C}$

3.5 $Y = \overline{D_3}D_2D_1D_0 + D_3\overline{D_2}D_1D_0 + D_3D_2\overline{D_1}D_0 + D_3D_2D_1\overline{D_0}$ for a circuit that indicates that *exactly* three inputs are HIGH. If *at least* three inputs are HIGH, the equation simplifies to $Y = D_2D_1D_0 + D_3D_1D_0 + D_3D_2D_0 + D_3D_2D_1$. The latter circuit is shown in Figure ANS3.5.

FIGURE ANS3.5

$D_3D_2D_1$

$D_3D_1D_0$

$D_3D_2D_0$

$D_2D_1D_0$

Y

3.7   **e.** $Y = (\overline{A} + \overline{C}) + \overline{B}\,\overline{C}$;

f. $Y = A\overline{B}C + C$;

g. $(\overline{ABD})(B + \overline{C}) + \overline{A}\,\overline{C}$;

h. $Y = (\overline{AB})(\overline{AC})(BC)$;

i. $Y = (A + \overline{B}) + (\overline{A}\,C)(BC)$

All of the above equations could be simplified further with Boolean algebra.

3.9 a.

T	U	V	W	X
0	0	0	0	1
0	0	0	1	1
0	0	1	0	1
0	0	1	1	1
0	1	0	0	1
0	1	0	1	1
0	1	1	0	1
0	1	1	1	1
1	0	0	0	1
1	0	0	1	1
1	0	1	0	1
1	0	1	1	1
1	1	0	0	1
1	1	0	1	1
1	1	1	0	1
1	1	1	1	0

j.

A	B	C	D	Y
0	0	0	0	0
0	0	0	1	0
0	0	1	0	0
0	0	1	1	0
0	1	0	0	0
0	1	0	1	0
0	1	1	0	0
0	1	1	1	0
1	0	0	0	0
1	0	0	1	0
1	0	1	0	0
1	0	1	1	0
1	1	0	0	1
1	1	0	1	1
1	1	1	0	0
1	1	1	1	0

h.

A	B	C	Y
0	0	0	0
0	0	1	0
0	1	0	1
0	1	1	1
1	0	0	0
1	0	1	0
1	1	0	1
1	1	1	0

i.

A	B	C	Y
0	0	0	1
0	0	1	1
0	1	0	1
0	1	1	1
1	0	0	1
1	0	1	1
1	1	0	1
1	1	1	1

3.11 SOP: $Y = \overline{A}\,\overline{B}\,\overline{C} + \overline{A}\,\overline{B}\,C + \overline{A}\,B\,\overline{C} + \overline{A}\,B\,C$

POS: $Y = (\overline{A} + B + C)(\overline{A} + B + \overline{C})(\overline{A} + \overline{B} + C)$
$(\overline{A} + \overline{B} + \overline{C})$

See Figure ANS3.11

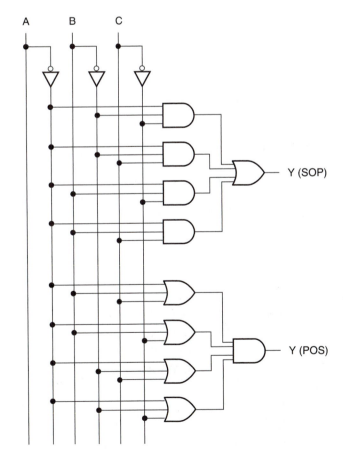

FIGURE ANS3.11

3.13   SOP: $Y = \overline{A}\,\overline{B}\,C + \overline{A}\,B\,\overline{C} + A\,\overline{B}\,C + A\,B\,\overline{C} + A\,B\,C$

POS: $Y = (A + B + C)(A + \overline{B} + \overline{C})(\overline{A} + B + C)$

See Figure ANS3.13

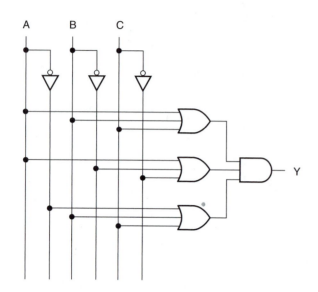

FIGURE ANS3.13

3.15 $Y = (A + B)(\overline{A} + \overline{B})$ See Figure ANS3.15.

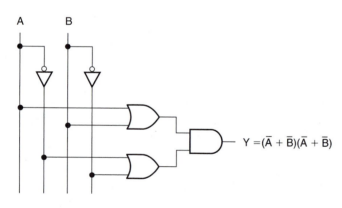

FIGURE ANS3.15

3.17 $Y = (A + B + C)\overline{D} = A\overline{D} + B\overline{D} + C\overline{D}$

3.19   **a.** $Y = AB + C;$ **b.** $Y = C;$

   c. $J = K;$ **d.** $S = 0;$

   e. $S = T;$ **f.** $Y = B\,\overline{C}\,\overline{D} + A\,\overline{B}\,F + \overline{C}\,F$

3.21   **a.** $Y = \overline{A} + \overline{B};$

   b. $Y = C\,D + \overline{C}\,\overline{D} + A\,B;$

   c. $K = M\overline{N} + ML$

3.23   SOP: $Y = \overline{A}\,\overline{C} + B\,\overline{C};$ POS: $Y = (\overline{A} + B)\overline{C}$

3.25 $Y = AD + B\overline{C}$

3.27 $Y = \overline{A}D + \overline{C}D + BC\overline{D}$

3.29 $Y = \overline{A}\,\overline{B}\,\overline{C}\,D + A\,\overline{B}\,\overline{C}\,D + BC$

3.31 $Y = \overline{A}\,\overline{B}\,\overline{C} + A\,B\,\overline{C} + \overline{A}\,\overline{B}\,D + A\,\overline{D}$

See Figure ANS3.31.

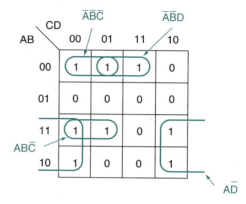

FIGURE ANS3.31

3.33 $Y = \overline{A}\,B + CD$ See Figure ANS3.33.

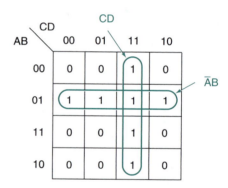

FIGURE ANS3.33

3.35 $Y = AD + \overline{B}C$ See Figure ANS3.35.

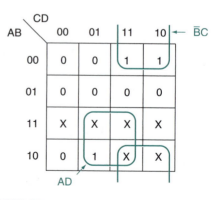

FIGURE ANS3.35

3.37 $Y = AB\overline{C}\overline{D} + \overline{A}D + CD$. See Figure ANS3.37.

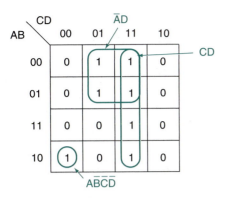

FIGURE ANS3.37.

3.39 $Y = \overline{AB}\,\overline{C}D + \overline{A}B + B\overline{D}$. See Figure ANS3.39.

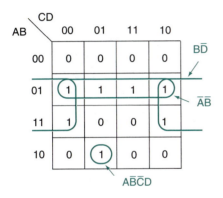

FIGURE ANS3.39

3.41 $Y = D$. See Figure ANS3.41.

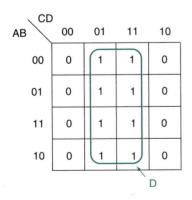

FIGURE ANS3.41

3.43 $Y = A\overline{C} + A\overline{B} + BCD$. See Figure ANS3.43.

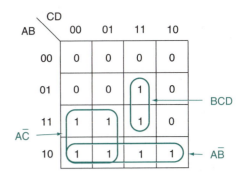

FIGURE ANS3.43

3.45 $Y = (\overline{A} + C)(A + \overline{C})(A + B + \overline{D})$. See Figure ANS3.45.

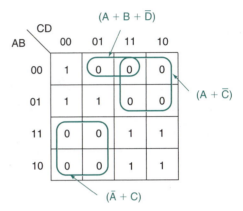

a. K-map

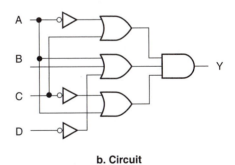

b. Circuit

FIGURE ANS3.45

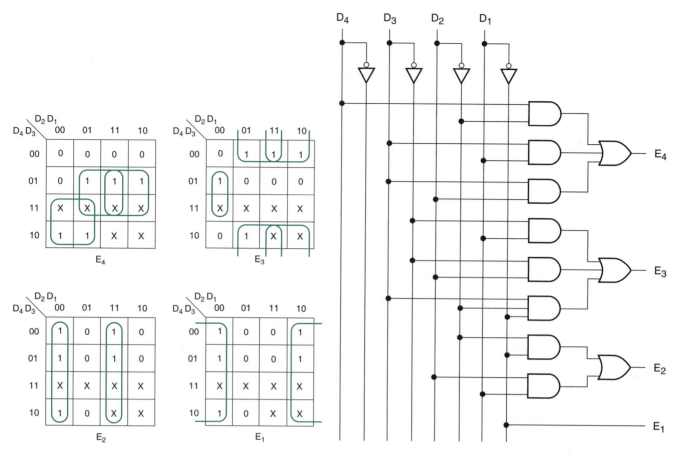

FIGURE ANS3.47

3.47 $E_4 = D_4\overline{D}_2 + D_3D_1 + D_3D_2$

$E_3 = \overline{D}_3D_2 + \overline{D}_3D_1 + D_3\overline{D}_2\overline{D}_1$

$E_2 = \overline{D}_2\overline{D}_1 + D_2D_1$

$E_1 = \overline{D}_1$

See Figure ANS3.47.

Chapter 4

4.1 Advantages of programmable logic: User is not restricted to standard digital functions from a device manufacturer; only required functions need be implemented; package count can be reduced; design can be reprogrammed or reconfigured without changing the circuit board.

4.3 PAL (Programmable Array Logic); GAL (Generic Array Logic); EPLD (Erasable Programmable Logic Device); FPGA (Field-Programmable Gate Array)

4.5 A design file in MAX+PLUS II is a single file with descriptive information, such as a schematic or text in a hardware description language. A project is a collection of files associated with a design entered in MAX+PLUS II.

4.7 **Primitives**—Basic functional blocks, such as logic gates, used in PLD design files.

Instance—A single copy of a component in a PLD design file.

4.9 The **gdf** for the 4-channel demultiplexer circuit is shown in Figure ANS4.9.

4.11 The **gdf** for the half adder is shown in Figure ANS4.11a. The default symbol for the half adder is shown in Figure ANS4.11b.

4.13 The **gdf** for the full adder (hierarchical design) is shown in Figure ANS4.13

4.15 **AHDL**—Altera Hardware Description Language

VHDL—VHSIC Hardware Description Language

VHSIC—Very High Speed Integrated Circuit

4.17 The two minimum VHDL structures are an **entity declaration** and an **architecture body.** The entity describes the input and output terminals of the design. The architecture defines the relationship between the inputs, outputs, and internal signals of the design.

4.19 A VHDL port of mode OUT can be used as an output only. A port of mode BUFFER is an output that can also be fed back into the design entity for use by other functions within the entity.

FIGURE ANS4.9

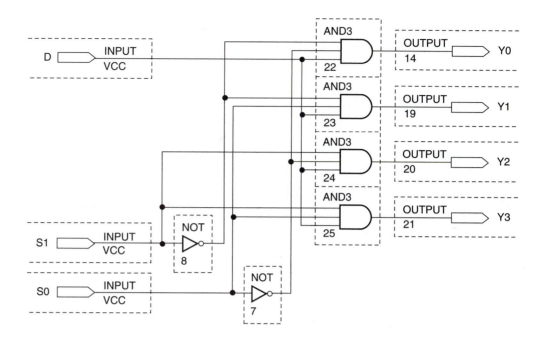

FIGURE ANS4.11

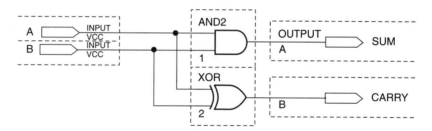

a. Half Adder Circuit

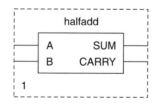

b. Symbol

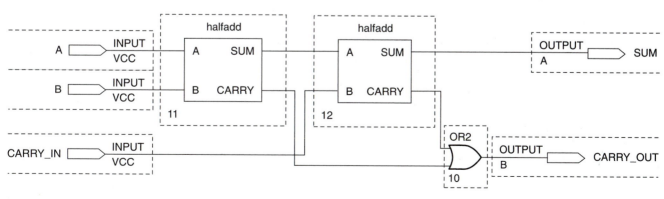

FIGURE ANS4.13

4.21
```
-- mux4.vhd
-- 4-to-1 multiplexer
-- Directs one of four input signals (d0 to d3) to output,
-- depending on status of select bits (s1, s0).

-- STD_LOGIC types
LIBRARY ieee;
USE ieee.std_logic_1164.ALL;

-- Define inputs and outputs
ENTITY mux4 IS
PORT(    d0, d1, d2, d3    : IN STD_LOGIC;    -- data inputs
         s: IN STD_LOGIC_VECTOR (1 downto 0); -- select inputs
         y: OUT STD_LOGIC);
END mux4;

-- Define i/o relationship
ARCHITECTURE mux4to1 OF mux4 IS
BEGIN
         -- Choose a signal assignment for y
         -- based on binary value of d
         -- Default case: output LOW
         WITH s SELECT
             y <= d0 WHEN "00",
                  d1 WHEN "01",
                  d2 WHEN "10",
                  d3 WHEN "11",
                  '0' WHEN others;
END mux4to1;
```

4.23
```
-- dmux4.vhd
-- 4-channel demultiplexer
-- Directs input to one of four outputs,
-- depending on state of select inputs (s1, s0)

-- Standard VHDL models
LIBRARY ieee;
USE ieee.std_logic_1164.ALL;

-- Define inputs and outputs
ENTITY dmux4 IS
    PORT(
           d, s1, s0         : IN  STD_LOGIC;
           y0, y1, y2, y3    : OUT STD_LOGIC);
END dmux4;

-- Define i/o relationship
ARCHITECTURE four_ch_dmux OF dmux4 IS
BEGIN
    --  Concurrent Signal Assignment
    y0    <=    (not s1) and (not s0) and d;
    y1    <=    (not s1) and (    s0) and d;
    y2    <=    (    s1) and (not s0) and d;
    y3    <=    (    s1) and (    s0) and d;
END four_ch_dmux;
```

4.25
```
-- half add.vhd
-- Half Adder
-- Adds two bits, A and B and produces SUM and CARRY outputs
```

```
-- Standard VHDL models
LIBRARY ieee;
USE ieee.std_logic_1164.ALL;

-- Define inputs and outputs
ENTITY half_add IS
   PORT(
          a, b        : IN STD_LOGIC;
          sum, carry : OUT STD_LOGIC);
END half_add;

-- Define relationship between A, B and SUM, CARRY
ARCHITECTURE half_adder OF half_add IS
BEGIN
   -- Concurrent Signal Assignment
   sum    <= a xor b;
   carry <= a and b;
END half_adder;
```

4.27   The **gdf** for the full adder with VHDL half adder components is the same as Figure ANS4.13.

Chapter 5

5.1   1100, 0001, 1111; $Y = D_3 D_2 \overline{D_1} \overline{D_0}$; $Y = \overline{D_3} \overline{D_2} \overline{D_1} D_0$; $Y = D_3 D_2 D_1 D_0$

5.3   See Figure ANS5.3.

FIGURE ANS5.3

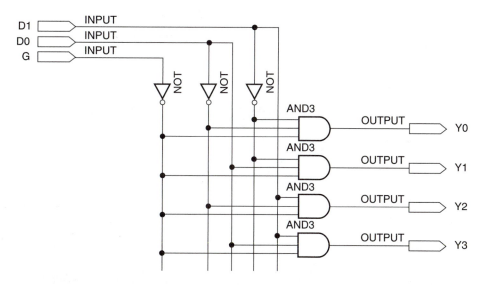

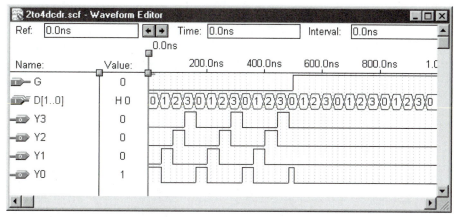

5.5 **a** 32;   **b.** 64;   **c.** 256; $m = 2^n$.

5.7 A selected signal assignment assigns an output value based on alternative input values. Each choice is independent of the others. A conditional signal assignment evaluates one input choice and assigns a value to an output if true. Otherwise, a second choice is evaluated, then a third, and so on. Low-priority choices are assigned only if higher-priority alternatives are false. This linked conditional structure tends to generate a more "serial" hardware, as opposed to the more "parallel" structure of the selected signal assignment. The selected signal assignment is preferable because it is generally results in a better use of chip resources and is more efficient.

5.9 See Figures ANS5.9a and ANS5.9c.

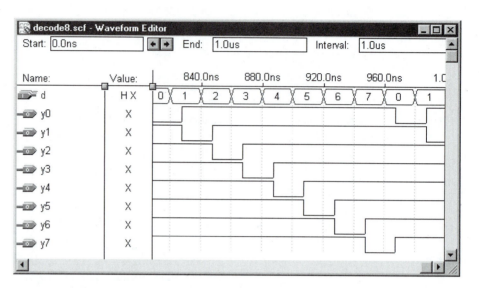

FIGURE ANS5.9A

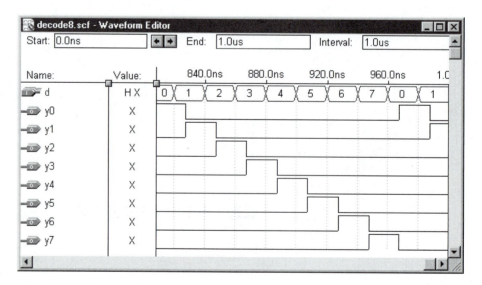

FIGURE ANS5.9C

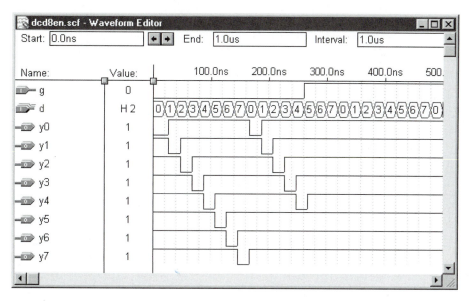

FIGURE ANS5.11

5.11 See Figure ANS5.11.

5.13 $a = \overline{D_3}\overline{D_2}\overline{D_1}D_0 + \overline{D_3}D_2\overline{D_1}\overline{D_0} + D_3\overline{D_2}D_1D_0 +$
$\quad\quad D_3D_2\overline{D_1}D_0$

$b = \overline{D_3}D_2\overline{D_1}D_0 + D_3D_2D_1 + D_3D_2\overline{D_0} + D_3D_1D_0 +$
$\quad\quad D_2D_1D_0$

$c = \overline{D_3}\overline{D_2}D_1\overline{D_0} + D_3D_2\overline{D_1}\overline{D_0} + D_3D_2D_1$

$d = \overline{D_3}\overline{D_2}\overline{D_1}D_0 + \overline{D_3}D_2\overline{D_1}\overline{D_0} + D_3\overline{D_2}D_1\overline{D_0} + D_2D_1D_0$

$e = \overline{D_3}D_0 + \overline{D_3}D_2\overline{D_1} + \overline{D_2}\overline{D_1}D_0$

$f = \overline{D_3}\overline{D_2}D_0 + \overline{D_3}\overline{D_2}D_1 + \overline{D_3}D_1D_0 + D_3D_2\overline{D_1}D_0$

$g = \overline{D_3}\overline{D_2}\overline{D_1} + \overline{D_3}D_2D_1D_0 + D_3D_2\overline{D_1}\overline{D_0}$

5.17   **a.** 1000;   **b.** 1001;   **c.** 1001

5.19 See Figures ANS5.19a and b.

5.21 See Figure ANS5.21.

Truth Table for an 8-to-1 MUX

S_2	S_1	S_0	Y
0	0	0	D_0
0	0	1	D_1
0	1	0	D_2
0	1	1	D_3
1	0	0	D_4
1	0	1	D_5
1	1	0	D_6
1	1	1	D_7

Truth Table for a 16-to-1 MUX

S_3	S_2	S_1	S_0	Y
0	0	0	0	D_0
0	0	0	1	D_1
0	0	1	0	D_2
0	0	1	1	D_3
0	1	0	0	D_4
0	1	0	1	D_5
0	1	1	0	D_6
0	1	1	1	D_7
1	0	0	0	D_8
1	0	0	1	D_9
1	0	1	0	D_{10}
1	0	1	1	D_{11}
1	1	0	0	D_{12}
1	1	0	1	D_{13}
1	1	1	0	D_{14}
1	1	1	1	D_{15}

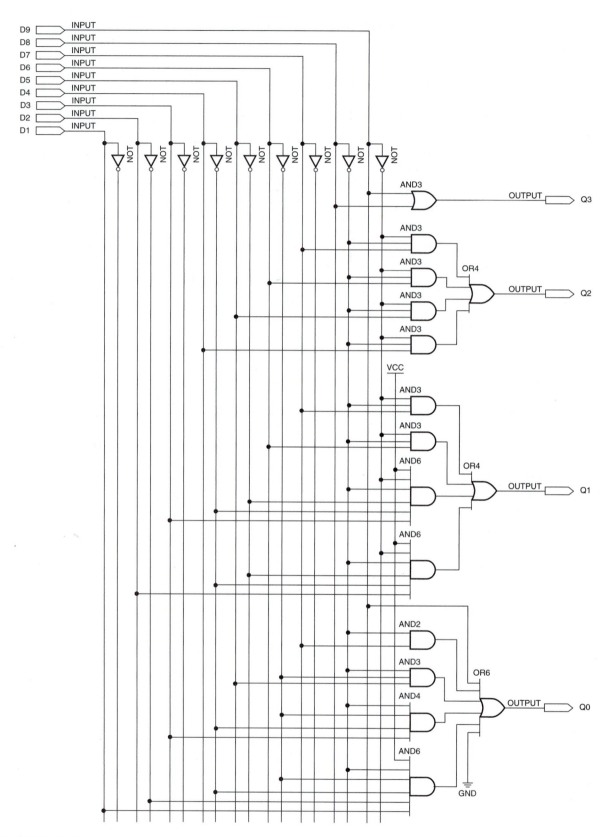

FIGURE ANS5.19A

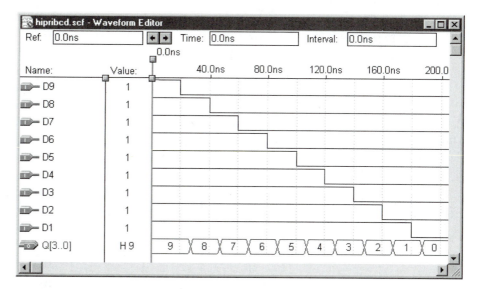

FIGURE ANS5.19B

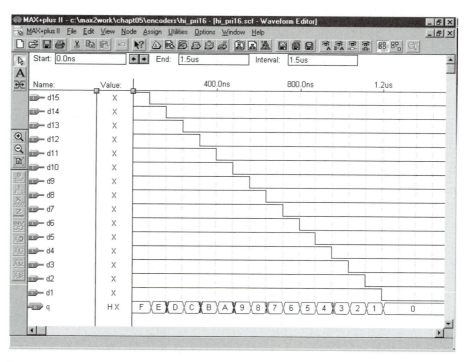

FIGURE ANS5.21

5.23 See Figure ANS5.23

See page 787 for Truth Tables.

5.25 $Y = \bar{S}_2\bar{S}_1\bar{S}_0 D_0 + \bar{S}_2\bar{S}_1 S_0 D_1 + \bar{S}_2 S_1 \bar{S}_0 D_2 + \bar{S}_2 S_1 S_0 D_3 + S_2\bar{S}_1\bar{S}_0 D_4 + S_2\bar{S}_1 S_0 D_5 + S_2 S_1\bar{S}_0 D_6 + S_2 S_1 S_0 D_7$

$= \bar{1}\cdot\bar{0}\cdot\bar{1}\cdot D_0 + \bar{1}\cdot\bar{0}\cdot 1\cdot D_1 + \bar{1}\cdot 0\cdot\bar{1}\cdot D_2 + \bar{1}\cdot 0\cdot 1\cdot D_3 + 1\cdot\bar{0}\cdot\bar{1}\cdot D_4 + 1\cdot\bar{0}\cdot 1\cdot D_5 + 1\cdot 0\cdot\bar{1}\cdot D_6 + 1\cdot 0\cdot 1\cdot D_7$

$= 0\cdot D_0 + 0\cdot D_1 + 0\cdot D_2 + 0\cdot D_3 + \cdot D_4 + 1\cdot D_5 + 0\cdot D_6 + 0\cdot D_7$

$= D_5$

FIGURE ANS5.23

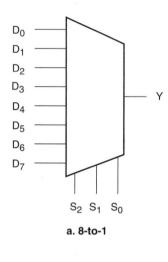

a. 8-to-1

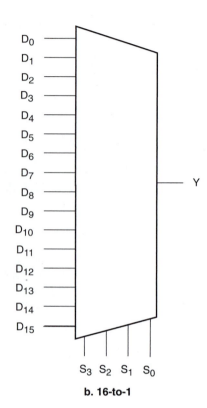

b. 16-to-1

5.27

```
-- quad8to1.vhd

-- Eight-channel 4-bit multiplexer
-- One of eight sets four inputs
-- (d03..d00), (d13..d10), (d23..d20), (d33..d30),
-- (d43..d40), (d53..d50), (d63..d60), (d73..d70)
-- is directed to an output (y), based on the status of three
-- select inputs (s2, s1, s0).

ENTITY quad8to1 IS
      PORT(
            s       : IN   INTEGER RANGE 0 to 7;
            d0      : IN   BIT_VECTOR (3 downto 0);
            d1      : IN   BIT_VECTOR (3 downto 0);
            d2      : IN   BIT_VECTOR (3 downto 0);
            d3      : IN   BIT_VECTOR (3 downto 0);
            d4      : IN   BIT_VECTOR (3 downto 0);
            d5      : IN   BIT_VECTOR (3 downto 0);
            d6      : IN   BIT_VECTOR (3 downto 0);
            d7      : IN   BIT_VECTOR (3 downto 0);
            y       : OUT  BIT_VECTOR (3 downto 0));
END quad8to1;
```

```
ARCHITECTURE mux8 OF quad8to1 IS
BEGIN

      -- Selected Signal Assignment
MUX4: WITH s SELECT
            y <=        d0 WHEN 0,
                        d1 WHEN 1,
                        d2 WHEN 2,
                        d3 WHEN 3,
                        d4 WHEN 4,
                        d5 WHEN 5,
                        d6 WHEN 6,
                        d7 WHEN 7;

   END mux8;
```

The simulation of this circuit is shown in Figure ANS5.27.

FIGURE ANS5.27

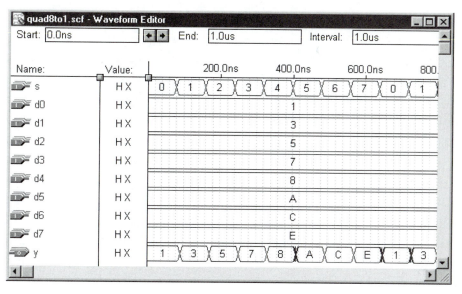

```
5.29  -- oct4to1.vhd
      -- Four-channel 8-bit multiplexer
      -- One of four sets eight inputs
      -- (d07..d00), (d17..d10), (d27..d20), or (d37..d30)
      -- is directed to a an output (y), based on the status of two
      -- select inputs (s1, s0).

ENTITY oct4to1 IS
      PORT(
            s      : IN  INTEGER RANGE 0 to 3;
            d0     : IN  BIT_VECTOR (7 downto 0);
            d1     : IN  BIT_VECTOR (7 downto 0);
            d2     : IN  BIT_VECTOR (7 downto 0);
            d3     : IN  BIT_VECTOR (7 downto 0);
            y      : OUT BIT_VECTOR (7 downto 0));
END oct4to1;
```

```
ARCHITECTURE mux4 OF oct4to1 IS
BEGIN
        -- Selected Signal Assignment
MUX8: WITH s SELECT
             y      <=      d0 WHEN 0,
                            d1 WHEN 1,
                            d2 WHEN 2,
                            d3 WHEN 3;

     END mux4;
```

The simulation is shown in Figure ANS5.29.

FIGURE ANS5.29

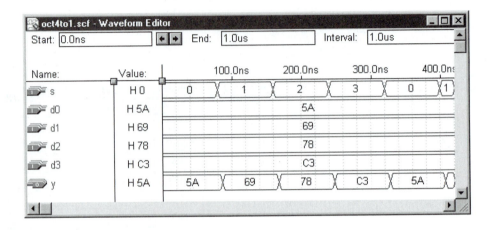

5.31
```
ENTITY mux_8ch IS
        PORT(
             sel    : IN  BIT_VECTOR (2 downto 0);
             d      : IN  BIT_VECTOR (7 downto 0);
             y      : OUT BIT);
     END mux_8ch;


ARCHITECTURE a OF mux_8ch IS
BEGIN
        -- Selected Signal Assignment
MUX8: WITH sel SELECT
             y      <=      d(0) WHEN "000",
                            d(1) WHEN "001",
                            d(2) WHEN "010",
                            d(3) WHEN "011",
                            d(4) WHEN "100",
                            d(5) WHEN "101",
                            d(6) WHEN "110",
                            d(7) WHEN "111";

     END a;
```

An 8-to-1 MUX can be easily extended to a 16-bit device by adding one select input, eight data inputs, and eight lines to the selected signal assignment statement.

5.35 00110011; 00001111

5.37 See Figure ANS5.37.

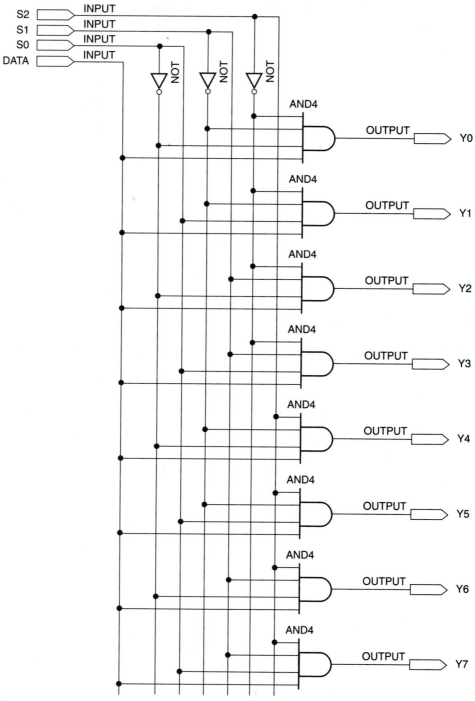

FIGURE ANS5.37A

FIGURE ANS5.37B

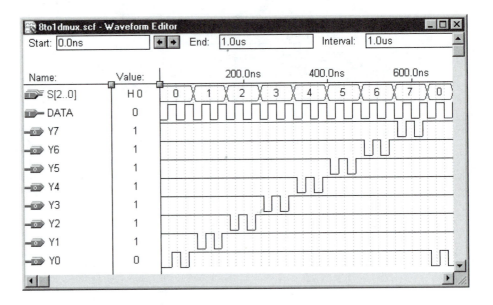

5.39 An analog switch can transmit a range of positive and negative voltages, not just 0V and 5V.

5.41 See Figure ANS5.41

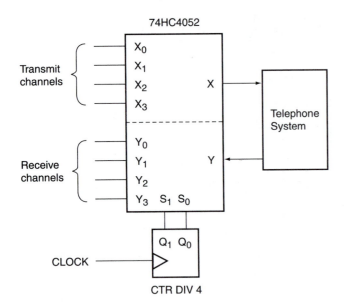

FIGURE ANS5.41

5.43 See Figure ANS5.43. The glitch in the simulation is caused by propagation delay.

5.45 $\text{AEQB} = \overline{(A_5 \oplus B_5)}\,\overline{(A_4 \oplus B_4)}\,\overline{(A_3 \oplus B_3)}\,\overline{(A_2 \oplus B_2)}$
$\overline{(A_1 \oplus B_1)}\,\overline{(A_0 \oplus B_0)}$

$\text{AGTB} = A_5\overline{B_5} + A_4\overline{B_4}\,\overline{(A_5 \oplus B_5)} +$
$A_3\overline{B_3}\,\overline{(A_5 \oplus B_5)}\,\overline{(A_4 \oplus B_4)}$
$+ A_2\overline{B_2}\,\overline{(A_5 \oplus B_5)}\,\overline{(A_4 \oplus B_4)}\,\overline{(A_3 \oplus B_3)}$
$+ A_1\overline{B_1}\,\overline{(A_5 \oplus B_5)}\,\overline{(A_4 \oplus B_4)}\,\overline{(A_3 \oplus B_3)}$
$\overline{(A_2 \oplus B_2)}$
$+ A_0\overline{B_0}\,\overline{(A_5 \oplus B_5)}\,\overline{(A_4 \oplus B_4)}\,\overline{(A_3 \oplus B_3)}$
$\overline{(A_2 \oplus B_2)}\,\overline{(A_1 \oplus B_1)}$

$\text{ALTB} = \overline{A_5}B_5 + \overline{A_4}B_4\,\overline{(A_5 \oplus B_5)} + \overline{A_3}B_3$
$\overline{(A_5 \oplus B_5)}\,\overline{(A_4 \oplus B_4)}$
$+ \overline{A_2}B_2\,\overline{(A_5 \oplus B_5)}\,\overline{(A_4 \oplus B_4)}\,\overline{(A_3 \oplus B_3)}$
$+ \overline{A_1}B_1\,\overline{(A_5 \oplus B_5)}\,\overline{(A_4 \oplus B_4)}\,\overline{(A_3 \oplus B_3)}$
$\overline{(A_2 \oplus B_2)}$
$+ \overline{A_0}B_0\,\overline{(A_5 \oplus B_5)}\,\overline{(A_4 \oplus B_4)}\,\overline{(A_3 \oplus B_3)}$
$\overline{(A_2 \oplus B_2)}\,\overline{(A_1 \oplus B_1)}$

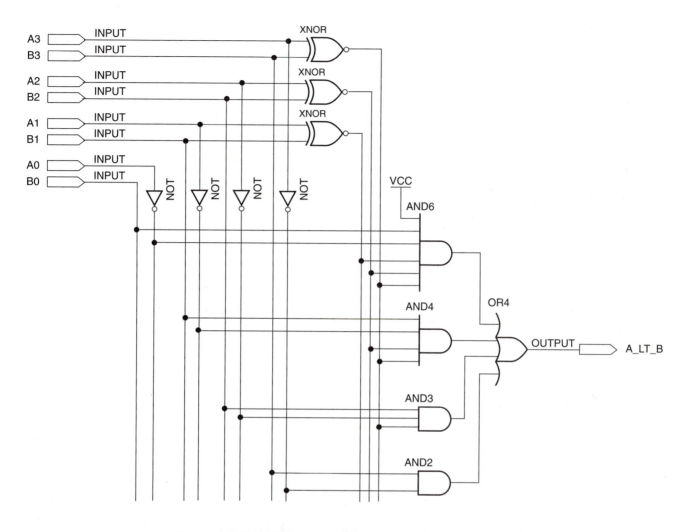

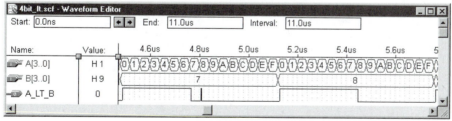

FIGURE ANS5.43

5.47 -- cmp4x6.vhd

```
LIBRARY ieee;
USE ieee.std_logic_1164.ALL;

ENTITY cmp4x6 IS
     PORT(
            a, b  : IN INTEGER RANGE 0 TO 15;
            altb, aleb, aeqb, aneb, ageb, agtb : OUT STD_LOGIC);
END cmp4x6;
```

```
ARCHITECTURE a OF cmp4x6 IS
      SIGNAL compare    : STD_LOGIC_VECTOR (5 downto 0);
BEGIN
      PROCESS (a,b)
      BEGIN
            IF a<b THEN
                  compare    <=    "001011";
            ELSIF a=b THEN
                  compare    <=    "100101";
            ELSIF a>b THEN
                  compare    <=    "111000";
            ELSE
                  compare    <=    "111111";
            END IF;
            altb  <=    compare (5); -- a is less than b
            aleb  <=    compare (4); -- a is less than or equal to b
            aeqb  <=    compare (3); -- a equals b
            aneb  <=    compare (2); -- a is not equal to b
            ageb  <=    compare (1); -- a is greater than or equal to b
            agtb  <=    compare (0); -- a is greater than b
      END PROCESS;
END a;
```

5.49 **a.** 1111100; five 1s; $P_E = 1$; $P_O = 0$;
  b. 1010110; four 1s; $P_E = 0$; $P_O = 1$;
  c. 0001101; three 1s; $P_E = 1$; $P_O = 0$

5.51 **a.** ABCDEFGHP = 110101100; $P' = 1$; Error in bit D.
  b. ABCDEFGHP = 110001101; $P' = 1$; Error in parity bit.

  c. ABCDEFGHP = 110001100; $P' = 0$; Data received correctly.

  d. ABCDEFGHP = 110010100; $P' = 0$; Errors in bits E and F undetected

5.53 See Figure ANS5.53.

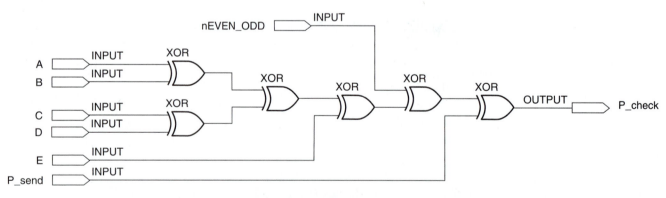

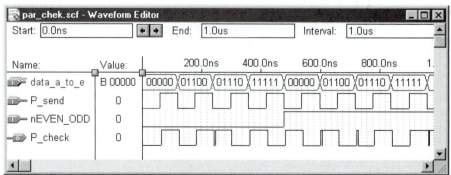

FIGURE ANS5.53

Chapter 6

6.1 **a.** 11111;   **b.** 100000;   **c.** 11110;
         d. 101010; **e.** 101100; **f.** 1100100

6.3

	Decimal	True Magnitude	1's Complement	2's Complement
a.	−110	11101110	10010001	10010010
b.	67	01000011	01000011	01000011
c.	−54	10110110	11001001	11001010
d.	−93	11011101	10100010	10100011
e.	0	00000000	00000000	00000000
f.	−1	10000001	11111110	11111111
g.	127	01111111	01111111	01111111
h.	−127	11111111	10000000	10000001

6.5 Largest: $01111111_2 = +127_{10}$;
         smallest: $10000000_2 = -128_{10}$

6.7 Overflow in an 8-bit signed addition results if the sum is outside the range $-128 \leq \text{sum} \leq +127$. The sums in parts **a.** and **f.** do not generate an overflow. The sums in parts **b.**, **c.**, **d.**, and **e.** do.

6.9 **a.** 3D;
         b. 120;
         c. B1A;
         d. FFF;
         e. 2A7F

6.11

Decimal	True Binary
709	1011000101
1889	11101100001
2395	100101011011
1259	10011101011
3972	111110000100
7730	1111000110010

8421 BCD	Excess-3
0111 0000 1001	1010 0011 1100
0001 1000 1000 1001	0100 1011 1011 1100
0010 0011 1001 0101	0101 0110 1100 1000
0001 0010 0101 1001	0100 0101 1000 1100
0011 1001 0111 0010	0110 1100 1010 0101
0111 0111 0011 0000	1010 1010 0110 0011

6.15 The sequence of codes yields the following text:

```
57 41 52 4E 49 4E 47 21 20 54 68 69 73 20
W  A  R  N  I  N  G  !  SP T  h  i  s  SP
```

```
63 6F 6D 6D 61 6E 64 20 65 72 61 73 65 73
c  o  m  m  a  n  d  SP e  r  a  s  e  s
```

```
20 36 34 30 4D 20 6F 66 20 6D 65 6D 6F 72 79 2E
SP 6  4  0  M  SP o  f  SP m  e  m  o  r  y  .
```

6.21 A fast carry circuit is "flatter", but "wider" than a ripple carry circuit. There are more gate levels for an input change to propagate through in a ripple carry circuit. The ripple carry is thus slower. The limitation on a fast carry circuit is its width, both in the number of gates and on the number of inputs on the gates. Both factors increase with adder bit size.

6.23 A carry is generated if the MSB of either A or B is HIGH AND the second bit of either A or B is HIGH AND the third bits of both A and B are HIGH.

6.25 To generate all possible combinations of input for an 8-bit adder requires $2^{16} = 65{,}536$ combinations. (A simulation with one change every 40 ns would have an end time of 2.62144 ms.)

6.27 See Figure ANS6.27

The transition from the sum FFF+000=FFF to FFF+001=000 (plus a carry) is given in the following table:

Time	Sum (Hex)	Sum (Binary)	From a1 to:
0	FFF	1111 1111 1111	
7.5 ns	FFC	1111 1111 1100	sum1, sum2
12.5 ns	FC0	1111 1100 0000	sum3-sum6
17.5 ns	F00	1111 0000 0000	sum7, sum8
22.5 ns	E00	1110 0000 0000	sum9
26.5 ns	C00	1100 0000 0000	sum10
31.5 ns	000	0000 0000 0000	sum11, sum12

FIGURE ANS6.27

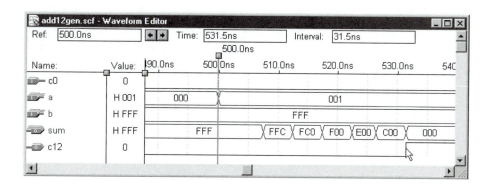

6.29 The 4-bit parallel adder/subtractor is shown in Figure ANS6.29a. The component **add4**, a parallel binary adder, is shown in Figure ANS6.29b.

SUB = 1: Input carry is forced HIGH, automatically adding 1 to the output sum; the XOR gates act as inverters, making the inputs to the adder equal to the one's complement of B; the output is A + (one's complement of B) + 1 = A − B.

SUB = 0: Input carry is forced LOW, adding 0 to the output sum; the XOR gates act as noninverting buffers, making the inputs to the adder equal the true binary value of B; the output is A + B + 0 = A + B.

FIGURE ANS6.29A

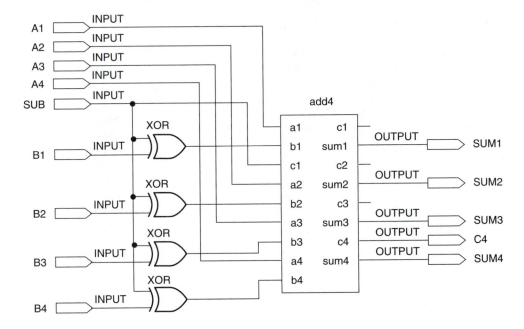

FIGURE ANS6.29B

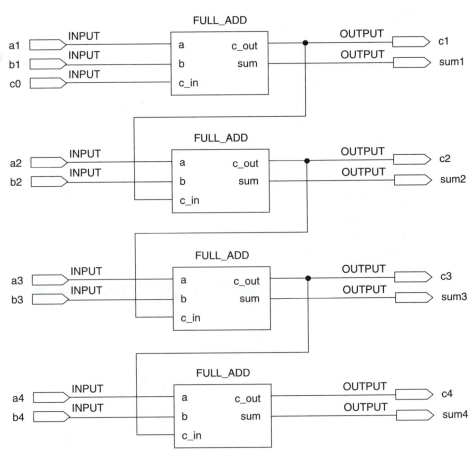

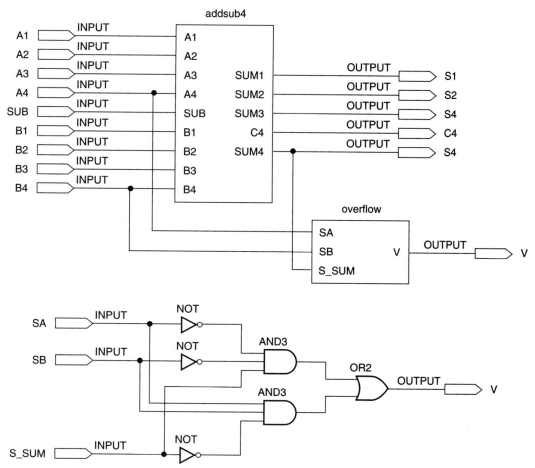

FIGURE ANS6.31

6.31 See Figure ANS6.31.

6.33 `-- addsubv1.vhd`

```
-- 4-bit parallel adder with overflow detection,
-- using a generate statement and components
-- overflow: SOP network

ENTITY addsubv1 IS
     PORT(
            sub           : IN         BIT;
            a, b          : IN         BIT_VECTOR(4 downto 1);
            c4, v         : OUT        BIT;
            sum           : BUFFER     BIT_VECTOR(4 downto 1));
END addsubv1;

ARCHITECTURE adder OF addsubv1 IS
     -- Component declaration
     COMPONENT full_add
         PORT(
                a, b, c_in  : IN  BIT;
                c_out, sum  : OUT BIT);
     END COMPONENT;
     -- Define a signal for internal carry bits
     SIGNAL c        : BIT_VECTOR (4 downto 0);
     SIGNAL b_comp   : BIT_VECTOR (4 downto 1);
```

```
      BEGIN
            -- Carry input depends on add or subtract (sub=1 for subtract)
            c(0)   <=     sub;
            adders:
            FOR I IN 1 to 4 GENERATE
                  -- invert b for subtract function (b(i) xor 1)
                  -- do not invert b for add function (b(i) xor 0)
                  b_comp(i) <= b(i) xor sub;
            adder: full_add PORT MAP (a(i), b_comp(i), c(i-1), c(i), sum(i));
            END GENERATE;
            c4     <=     c(4);
            v <= (a(4) and b(4) and (not sum(4)))
                  or ((not a(4)) and (not b(4)) and sum(4));
      END adder;


-- addsubv2.vhd
-- 4-bit parallel adder with overflow detection,
-- using a generate statement and components
-- overflow: xor gate

ENTITY addsubv2 IS
      PORT(
            sub           : IN  BIT;
            a, b          : IN  BIT_VECTOR(4 downto 1);
            c4, v         : OUT BIT;
            sum           : OUT BIT_VECTOR(4 downto 1));
END addsubv2;


ARCHITECTURE adder OF addsubv2 IS
      -- Component declaration
      COMPONENT full_add
            PORT(
                  a, b, c_in   : IN  BIT;
                  c_out, sum   : OUT BIT);
      END COMPONENT;
      -- Define  a signal for internal carry bits
      SIGNAL c          : BIT_VECTOR (4 downto 0);
      SIGNAL b_comp     : BIT_VECTOR (4 downto 1);
BEGIN
      -- Carry input depends on add or subtract (sub=1 for subtract)
      c(0)   <=     sub;
      adders:
      FOR i IN 1 to 4 GENERATE
            -- invert b for subtract function (b(i) xor 1)
            -- do not invert b for add function (b(i) xor 0)
            b_comp(i) <= b(i) xor sub;
            adder: full_add PORT MAP (a(i), b_comp(i), c(i-1), c(i), sum(i));
            END GENERATE;
            c4     <=     c(4);
            v <= c(4) xor c(3);
      END adder;
```

6.35 1999; 3½ digits

6.37 1 followed by n 9s.

6.39 See Figure 6.26 in text.

6.41
```
-- add4bcd.vhd
-- 4-bit bcd adder
LIBRARY ieee;
USE ieee.std_logic_1164.ALL;
ENTITY add4bcd IS
     PORT(
          c0                    : IN  STD_LOGIC:
          a_bcd, b_bcd          : IN  STD_LOGIC_VECTOR(4 down to 1);
          sum_bcd               : OUT STD_LOGIC_VECTOR(5 downto 1));
END add4bcd;

ARCHITECTURE adder OF add4bcd IS
     -- Component declaration
     COMPONENT add4gen
     PORT(
          c0          : IN  STD_LOGIC;
          a, b        : IN  STD_LOGIC_VECTOR(4 downto 1);
          c4          : OUT STD_LOGIC;
          sum         : OUT STD_LOGIC_VECTOR(4 downto 1));
     END COMPONENT;

     COMPONENT bin2bcd
     PORT(
          bin : IN STD_LOGIC_VECTOR(5 downto 1);
          bcd : OUT STD_LOGIC_VECTOR(5 downto 1);
     END COMPONENT;
     -- Define a signal for internal carry bits
     SIGNAL connect : STD_LOGIC_VECTOR (5 downto 1);
BEGIN
          adder: add4gen PORT MAP (c0, a_bcd, b_bcd, connect(5),
               connect (4 downto 1));
          converter: bin2bcd PORT MAP (connect, sum_bcd);
END adder;
```

6.43 The circuit will be like Figure 6.27 in the text, minus the thousands digit. It will generate a 3½ digit output.

Chapter 7

7.1 See Figure ANS7.1.

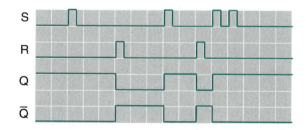

FIGURE ANS7.1

7.3 See Figure ANS7.3.

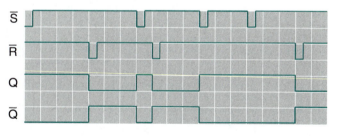

FIGURE ANS7.3

7.5   See Figure ANS7.5.

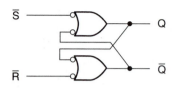

FIGURE ANS7.5

$\overline{S}$	$\overline{R}$	
0	0	Latch tries to set and reset at the same time. Forbidden state.
0	1	Set input active. Q = 1.
1	0	Reset input active. Q = 0.
1	1	Neither set nor reset active. No change.

7.7   See Figure ANS7.7.

7.9   See Figure ANS7.9.

7.11   **a.** See Figure ANS7.11.

b. i. R is last input active. Latch resets;   **ii.** S is last input active. Latch resets;   **iii.** S and R go from both active to no change state. The latch cannot predictably resolve this transition. Output unknown.

c. Both set and reset are active at the same time.

7.13   See Figure ANS7.13.

FIGURE ANS7.7

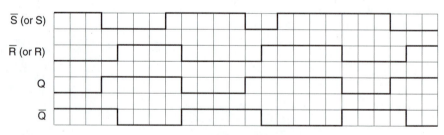

FIGURE ANS7.9

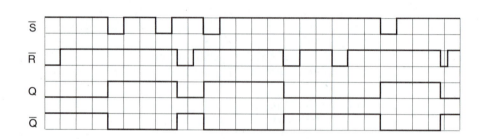

NAND waveforms

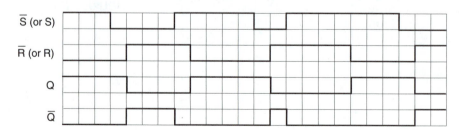

NOR waveforms

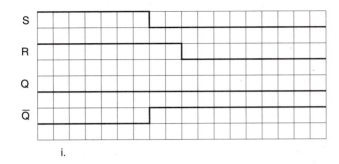

i.

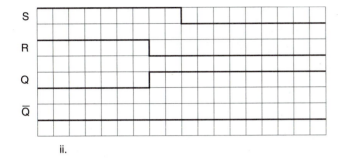

ii.

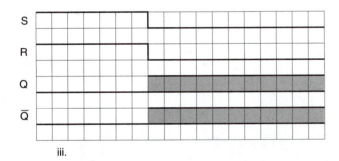

iii.

FIGURE ANS7.11

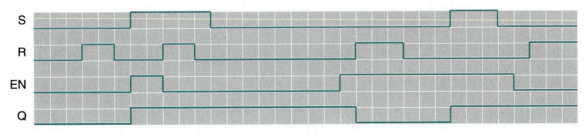

FIGURE ANS7.13

7.15 See Figure ANS7.15.

7.17 See Figure ANS7.17.

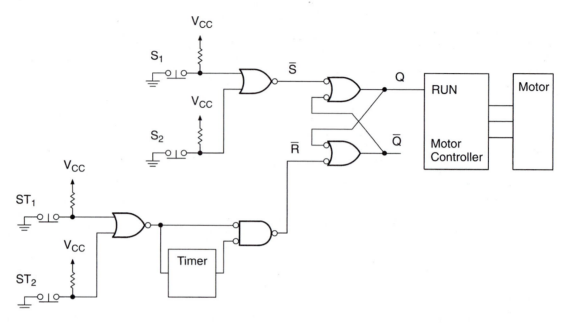

FIGURE ANS7.15

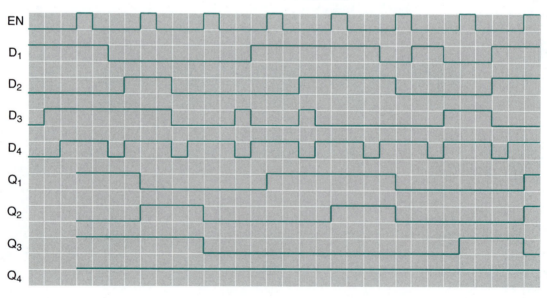

FIGURE ANS7.17

7.19 -- ltch8prm.vhd
-- D latch with active-HIGH level-sensitive enable

```
LIBRARY ieee;
USE ieee.std_logic_1164.ALL;
LIBRARY altera;
USE altera.maxplus2.ALL;

ENTITY ltch8prm IS
      PORT(d_in    : IN  STD_LOGIC_VECTOR(7 downto 0);
            enable : IN STD_LOGIC;
            q_out  : OUT STD_LOGIC_VECTOR(7 downto 0));
END ltch8prm;

ARCHITECTURE a OF ltch8prm IS
BEGIN
      -- Instantiate a latch from a MAX+PLUS II primitive
      latch8:
      FOR i IN 7 downto 0 GENERATE
      latch_primitive: latch
            PORT MAP (d => d_in(i), ena => enable, q => q_out(i));
      END GENERATE;
END a;
```

See Figure ANS7.19.

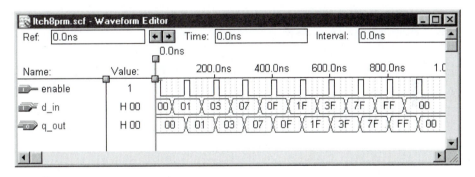

FIGURE ANS7.19

7.21 See Figure ANS7.21.

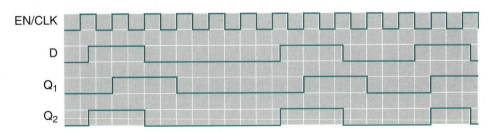

FIGURE ANS7.21

FIGURE ANS7.23

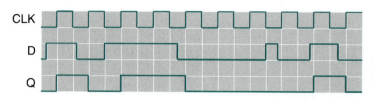

7.23 See Figure ANS7.23.

7.25 See Figure ANS7.25.

FIGURE ANS7.25

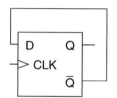

7.27
```
-- dff12lpm.vhd
-- 12-BIT D flip-flop
-- Uses a flip-flop component from the Library of Parameterized Modules (LPM)

LIBRARY ieee;
USE ieee.std_logic_1164.ALL;
LIBRARY lpm;
USE lpm.lpm_components.ALL;
ENTITY dff12lpm IS
      PORT(d_in  : IN  STD_LOGIC_VECTOR(11 downto 0);
           clk   : IN  STD_LOGIC;
           q_out : OUT STD_LOGIC_VECTOR(11 downto 0));
END dff12lpm;

ARCHITECTURE a OF dff12lpm IS
BEGIN
      -- Instantiate flip-flop from an LPM component
      dff12: lpm_ff
            GENERIC MAP (LPM_WIDTH => 12)
            PORT MAP (data  => d_in,
                      clock => clk,
                      q     => q_out);
END a;
```

FIGURE ANS7.29

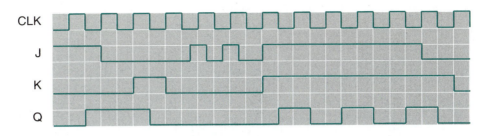

7.29 See Figure ANS7.29.

7.31 See Figure ANS7.31. The circuit generates the following repeating pattern: 111, 110, 101, 100, 011, 010, 001, 000. This is a 3-bit binary down-count sequence.

7.33 The circuit generates a 4-bit binary sequence from 0000 to 1111, then repeats indefinitely.

7.35 See Figure 7.35.

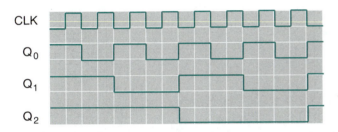

FIGURE ANS7.31

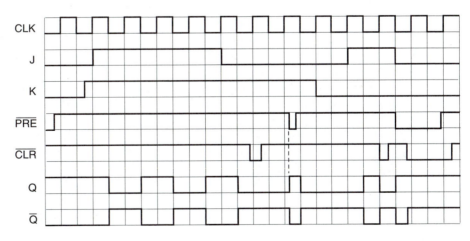

FIGURE 7.35

7.37 See Figure 7.37

7.39 Similarity: an asynchronous circuit and an asynchronous input cause outputs to change out of synchronization with a system clock. Difference: an asynchronous circuit may

be clocked, but at different times throughout the circuit; an asynchronous input is independent of the clock function altogether.

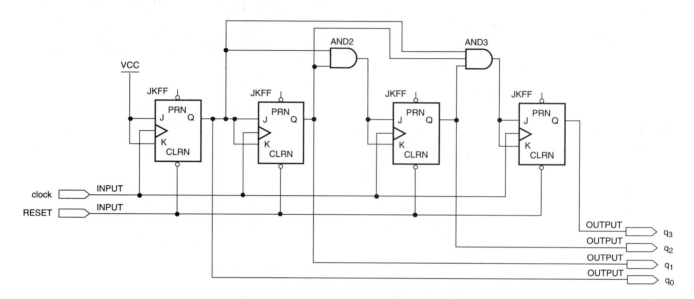

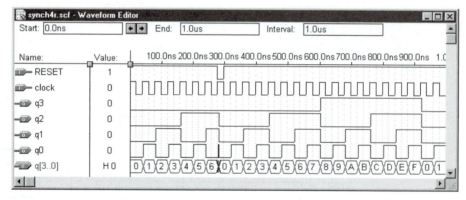

FIGURE 7.37

7.41
```
-- d12lpmcl.vhd
-- 4-BIT D latch with active-HIGH level-sensitive enable
-- Uses a latch component from the Library of Parameterized Modules

LIBRARY ieee;
USE ieee.std_logic_1164.ALL;
LIBRARY lpm;
USE lpm.lpm_components.ALL;

ENTITY d12lpmcl IS
      PORT(d_in                : IN  STD_LOGIC_VECTOR(11 downto 0);
            clk, set, reset    : IN  STD_LOGIC;
            q_out              : OUT STD_LOGIC_VECTOR(11 downto 0));
END d12lpmcl;

ARCHITECTURE a OF d12lpmcl IS
      SIGNAL clrn : STD_LOGIC;
      SIGNAL prn  : STD_LOGIC;
```

```
BEGIN
      -- Instantiate flip-flop from an LPM component
      dff12: lpm_ff
            GENERIC MAP (LPM_WIDTH => 12)
            PORT MAP (  data  => d_in,
                        clock => clk,
                        aclr  => clrn,
                        aset  => prn,
                        q     => q_out);
      -- Make set and reset active-LOW
      clrn <= not reset;
      prn <= not set;
END a;
```

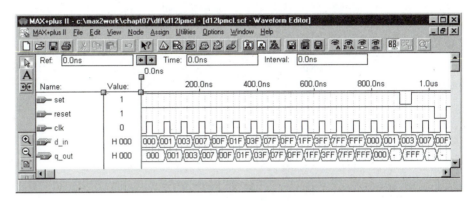

FIGURE ANS7.41

7.43   See Figure ANS7.43.

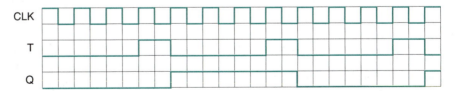

FIGURE ANS7.43

7.45   `-- syn4tprm.vhd`
       `-- 4-bit sync counter (TFF primitives)`

```
LIBRARY ieee;
USE ieee.std_logic_1164.ALL;
LIBRARY altera;
USE altera.maxplus2.ALL;

ENTITY syn4tprm   IS
      PORT (clock, reset : IN  STD_LOGIC;
                  q           : OUT STD_LOGIC_VECTOR(3 downto 0));
END syn4tprm;
```

```
ARCHITECTURE a OF syn4tprm IS
        -- Declare component only with ports actually used
        COMPONENT TFF
            PORT ( t : IN      STD_LOGIC;
                   clk : IN    STD_LOGIC;
                   clrn: IN    STD_LOGIC;
                   q  : OUT    STD_LOGIC);
        END COMPONENT;
        SIGNAL q_int : STD_LOGIC_VECTOR(2 downto 0);
        SIGNAL t_int : STD_LOGIC_VECTOR(3 downto 0);
BEGIN
        -- Instantiate 4 T flip-flops.
        ff0: tff
              PORT MAP (t_int(0), clock, reset, q_int(0));
        ff1: tff
              PORT MAP (t_int(1), clock, reset, q_int(1));
        ff2: tff
              PORT MAP (t_int(2), clock, reset, q_int(2));
        ff3: tff
              PORT MAP (t_int(3), clock, reset, q(3));

        -- Connect flip-flops internally
        t_int(0)    <=  '1';
        t_int(1)    <=  q_int(0);
        t_int(2)    <=  q_int(0) and q_int(1);
        t_int(3)    <=  q_int(0) and q_int(1) and q_int(2);

        q(0)        <=  q_int(0);
        q(1)        <=  q_int(1);
        q(2)        <=  q_int(2);
END a;
```

7.47 $t_{su} = 20$ ns, $t_h = 0$

7.49 clock pulse width: $t_w = 12$ ns; setup time: $t_{su} = 10$ ns; hold time: $t_h = 5$ ns

Chapter 8

8.1 See Figure 8.2.

8.7 **a.** 4;   **b.** 6;   **c.** 8

8.9 A global architecture cell configures all macrocells in the PLD. A local architecture cell works only on the macrocell of which it is a part.

8.11 Registered/active LOW; registered/active HIGH; combinatorial/active LOW; combinatorial/active HIGH

8.13 No. Global clock only.

8.15 Global. These functions operate simultaneously on all macrocells.

8.17 **a.** 32;
        b. 64;
        c. 128;
        d. 160

8.19 $n/16$ Logic Array Blocks for n macrocells. (e.g. 128/16 = 8 LABs for an EPM7128S)

8.21 Macrocells without pin connections can be used for internal logic.

8.23 A MAX7000S macrocell can be reset from a global clear pin (GCLRn) or locally from a product term.

8.25 5 dedicated product terms; by using terms from shared logic expanders and parallel logic expanders; 5 dedicated, up to 15 from parallel logic expanders; up to 16 from shared logic expanders.

8.27 A sum-of-products network constructs Boolean expressions by switching signals into an OR-gate output via a programmable matrix of AND gates. A look-up table network stores the output values of the network in a small memory whose storage locations are selected by combinations of the input signals.

8.29 A carry chain allows for efficient fast-carry implementation of adders, comparators, and other circuits whose inputs become wider with higher-order bits.

8.31 2048

Chapter 9

9.1 See Figure ANS9.1. The 12-bit counter recycles to 0 after 4096 cars have entered the parking lot. The last car causes all bits to go LOW. The negative edge on the MSB clocks a flip flop whose output enables the LOT FULL sign. Every car out of the gate resets the flip-flop and turns off the sign.

A better circuit would have the exit gate make the counter output decrease by 1 with every vehicle exiting.

9.3 See Figure ANS9.3.

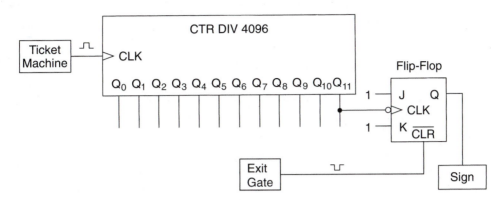

FIGURE ANS9.1

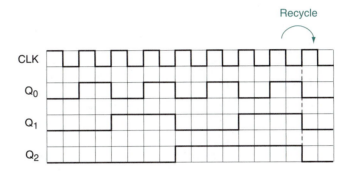

FIGURE ANS9.3

9.5 **a.** See Figure ANS9.5
b. i. 0100;
ii. 0110;
iii. 0011

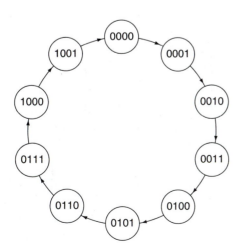

FIGURE ANS9.5

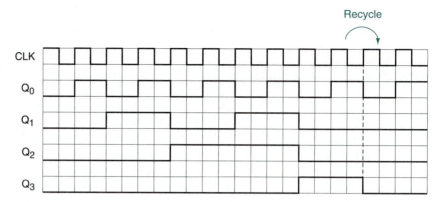

FIGURE ANS9.7

9.7 Figure ANS9.7 shows the timing diagram of a mod-10 counter.

Q_3	Q_2	Q_1	Q_0
0	0	0	0
0	0	0	1
0	0	1	0
0	0	1	1
0	1	0	0
0	1	0	1
0	1	1	0
0	1	1	1
1	0	0	0
1	0	0	1

9.9 Q_0: 24 kHz; Q_1: 12 kHz; Q_2: 6 kHz; Q_3: 3 kHz

9.11   See Figure ANS9.11

9.13 $J_0 = K_0 = 1$
$J_1 = K_1 = Q_0$
$J_2 = K_2 = Q_1 Q_0$
$J_3 = K_3 = Q_2 Q_1 Q_0$
$J_4 = K_4 = Q_3 Q_2 Q_1 Q_0$
$J_5 = K_5 = Q_4 Q_3 Q_2 Q_1 Q_0$
$J_6 = K_6 = Q_5 Q_4 Q_3 Q_2 Q_1 Q_0$
$J_7 = K_7 = Q_6 Q_5 Q_4 Q_3 Q_2 Q_1 Q_0$

9.15 **a.** $J_3 = Q_2 Q_1 Q_0$
$K_3 = Q_1 Q_0$
$J_2 = \overline{Q_3} Q_1 Q_0$
$K_2 = Q_1 Q_0$
$J_1 = Q_0$
$K_1 = Q_0$
$J_0 = 1$
$K_0 = 1$

b. 1011, 0000, 0001

9.19 See Figure ANS9.19

FIGURE ANS9.11

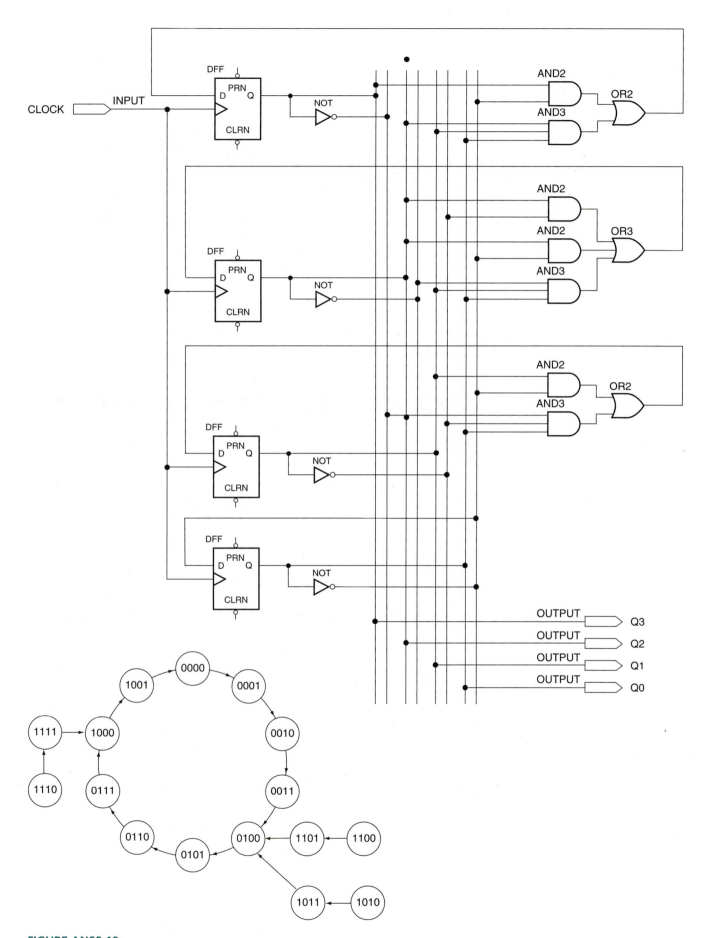

FIGURE ANS9.19

9.21 Boolean equations:

$D_3 = \overline{Q}_3 Q_2 + Q_3 \overline{Q}_2$

$D_2 = Q_1 Q_0$

$D_1 = \overline{Q}_1 Q_0 + Q_1 \overline{Q}_0$

$D_0 = \overline{Q}_2 \overline{Q}_0$

9.23 See Figure ANS9.23a for a simulation of the clear function and Figure ANS9.23b for the recycle point of the counter.

9.25 See Figure 9.22 in the text. Asynchronous load transfers data directly to the flip-flops of a counter as soon as the load input is asserted; it does not wait for a clock edge. Synchronous load waits for an active clock edge to load a value into the counter flip-flops.

9.27 Figure ANS9.27 shows the part of the simulation where the value 1AH is synchronously loaded into the counter.

9.29 See Figure ANS9.29

9.31 $D_0 = \overline{Q}_0$

$D_1 = Q_0\text{DIR} + \overline{Q}_0\overline{\text{DIR}}$

$D_2 = Q_1 Q_0\text{DIR} + \overline{Q}_1\overline{Q}_0\overline{\text{DIR}}$

$D_3 = Q_2 Q_1 Q_0\text{DIR} + \overline{Q}_2\overline{Q}_1\overline{Q}_0\overline{\text{DIR}}$

The right-hand product term of each equation represents the down-count logic, which is enabled whenever DIR = 0. The left-hand product term is the up-count logic, enabled when DIR = 1. D_0 is always the opposite of Q_0, regardless of whether the count is up or down.

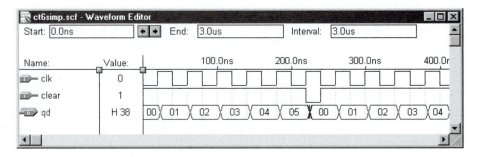

FIGURE ANS9.23A

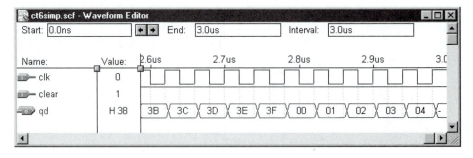

FIGURE ANS9.23B

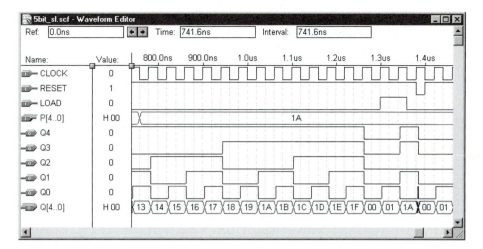

FIGURE ANS9.27

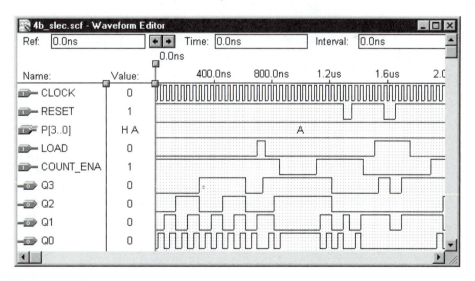

FIGURE ANS 9.29

9.33   The circuit is shown in Figure ANS9.33a. The counter
module **sl_count** is shown is Figure 9.25 in the text. The
simulation is shown in Figure ANS9.33b.

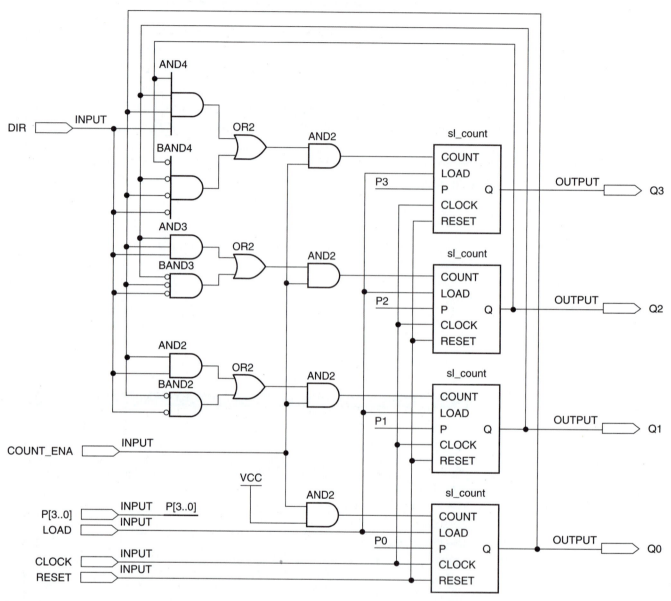

FIGURE ANS9.33A

FIGURE ANS9.33B

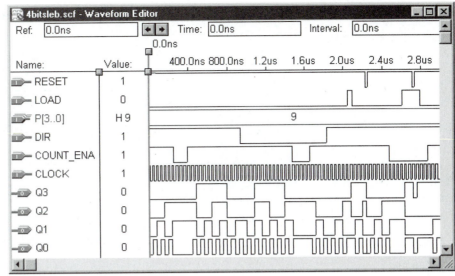

9.35
```
-- ct_mod24
-- Presettable counter with synchronous clear and load
-- and a modulus of 24

ENTITY ct_mod24 IS
      PORT(
        clk                    : IN    BIT;
        clear, direction       : IN    BIT;
        q                      : OUT INTEGER RANGE 0 TO 23);
END ct_mod24;

ARCHITECTURE a OF ct_mod24 IS
      BEGIN
      PROCESS (clk)
          VARIABLE    cnt        : INTEGER RANGE 0 TO 23;
            BEGIN
                IF (clk'EVENT AND clk = '1') THEN
                    IF (clear = '0')  THEN    -- Synchronous clear
                        cnt    :=    0;
                    ELSIF (direction = '0') THEN
                        IF cnt = 0 THEN
                            cnt := 23;
                        ELSE
                            cnt := cnt - 1;
                        END IF;
                    ELSIF (direction = '1') THEN
                        IF cnt = 23 THEN
                            cnt := 0;
                        ELSE
                            cnt := cnt + 1;
                        END IF;
                    END IF;
                END IF;
                q      <=    cnt;
            END PROCESS;
        END a;
```

See Figure ANS9.35 for simulation.

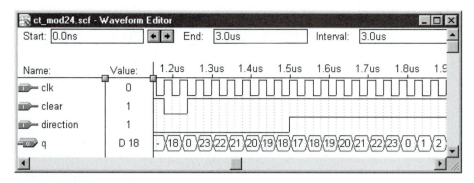

FIGURE ANS9.35

9.37 -- sst1_lpm.vhd
-- 12-bit LPM counter with sst1 and aclr (Chapter problem)

```
LIBRARY ieee;
USE ieee.std_logic_1164.ALL;
LIBRARY lpm;
USE lpm.lpm_components.ALL;

ENTITY sst1_lpm IS
      PORT(
         clk              : IN    STD_LOGIC;
         clear, set       : IN STD_LOGIC;
         q                : OUT   STD_LOGIC_VECTOR (11 downto 0));
END sst1_lpm;

ARCHITECTURE a OF sst1_lpm IS
BEGIN
      counter1: lpm_counter
            GENERIC MAP (LPM_WIDTH => 12)
            PORT MAP ( clock => clk,
                       sset  => set,
                       aclr  => clear,
                       q     => q);
END a;
```

The counter in this problem sets to all 1s (1111 1111 1111 = FFFH), rather than 0111 1111 1111 (= 7FFH). See Figure ANS9.37 for the simulation of the counter in problem 9.37.

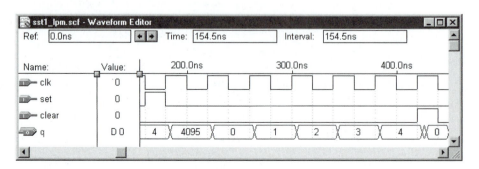

FIGURE ANS9.37

9.39
```
-- lpm8term
-- 8-bit presettable counter with synchronous clear and load,
-- count enable, a directional control port,
-- and terminal count decoding

LIBRARY ieee;
USE ieee.std_logic_1164.ALL;
LIBRARY lpm;
USE lpm.lpm_components.ALL;

ENTITY lpm8term IS
      PORT(
          clk, count_ena          : IN      STD_LOGIC;
          clear, load, direction  : IN      STD_LOGIC;
          p                       : IN STD_LOGIC_VECTOR(7 downto 0);
          max_min                 : OUT STD_LOGIC;
q                                 : OUT    STD_LOGIC_VECTOR(7 downto 0));
END lpm8term;

ARCHITECTURE a OF lpm8term IS
      SIGNAL cnt : STD_LOGIC_VECTOR(7 downto 0);
BEGIN
      counter1: lpm_counter
          GENERIC MAP (LPM_WIDTH => 8)
          PORT MAP (  clock  => clk,
                      updown => direction,
                      cnt_en => count_ena,
                      data   => p,
                      sload  => load,
                      sclr   => clear,
                      q      => cnt);
      q <= cnt;
      PROCESS (clk, cnt)
      BEGIN
                  -- Terminal count decoder
                  IF (cnt = "00000000" and direction = '0') THEN
                      max_min      <= '1';
                  ELSIF (cnt = "11111111" and direction = '1') THEN
                      max_min <=   '1';
                  ELSE
                      max_min <=   '0';
                  END IF;
      END PROCESS;
END a;
```

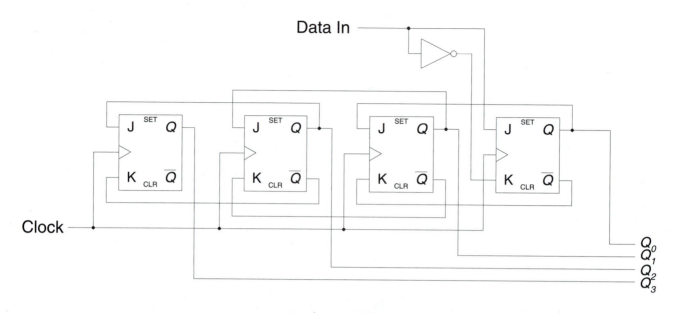

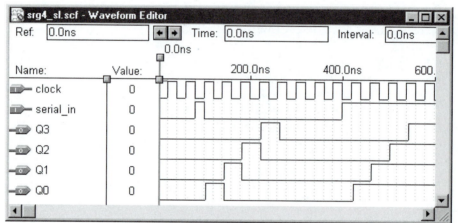

FIGURE ANS9.41

9.41 See Figure ANS9.41.

9.43 001111, 000000, 000000, 110000

9.45 See Figure ANS9.45. The serial output is the same as the serial input, only delayed by eight clock pulses and synchronized to the positive edge of the clock.

9.47 See Figure ANS9.47.

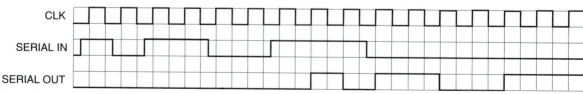

FIGURE ANS9.45

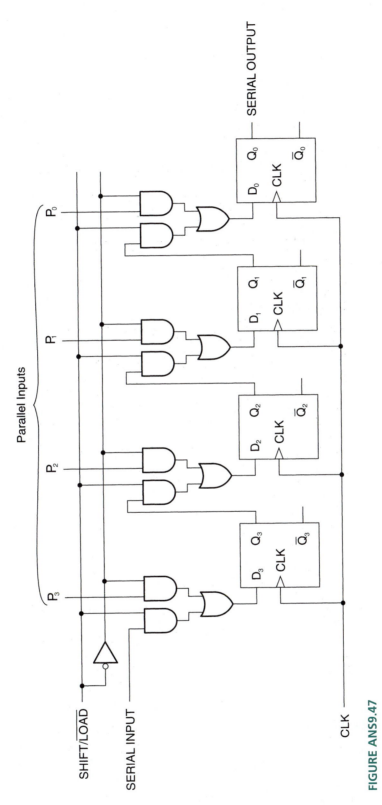

FIGURE ANS9.47

9.51
```
-- Left-shift register of generic width
LIBRARY ieee;
USE ieee.std_logic_1164.ALL;

ENTITY slt_bhv IS
      GENERIC (width : POSITIVE);
      PORT(
            serial_in, clk    : IN        STD_LOGIC;
            q                 : BUFFER    STD_LOGIC_VECTOR(width-1 downto 0));
END slt_bhv;

ARCHITECTURE left_shift of slt_bhv IS
BEGIN
      PROCESS (clk)
      BEGIN
            IF (clk'EVENT and clk = '1') THEN
                  q(width-1 downto 0) <= q(width-2 downto 0) & serial_in;
            END IF;
      END PROCESS;
END left_shift;

-- 32-bit left-shift register
LIBRARY ieee;
USE ieee.std_logic_1164.ALL;

ENTITY slt32_bhv IS
      PORT(
            data_in, clock    : IN        STD_LOGIC;
            qo                : BUFFER    STD_LOGIC_VECTOR(31 downto 0));
END slt32_bhv;

ARCHITECTURE left_shift of slt32_bhv IS
COMPONENT slt_bhv
      GENERIC (width : POSITIVE);
      PORT(
            serial_in, clk    : IN  STD_LOGIC;
            q                 : OUT STD_LOGIC_VECTOR(31 downto 0));
END COMPONENT;
BEGIN
      Shift_left_32: slt_bhv
            GENERIC MAP (width=> 32)
            PORT MAP (serial_in => data_in,
                      clk        => clock,
                      q          => qo);
END left_shift;
```

Figure ANS9.51 shows a partial simulation of the shift
register.

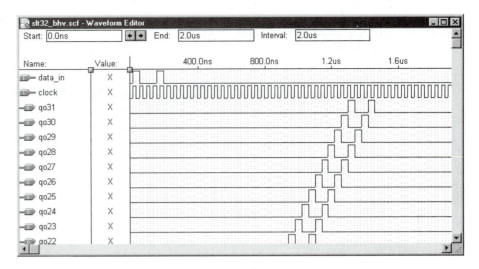

FIGURE ANS9.51

9.53 The generic component has a default width of 8 bits. The instantiated component has an assigned width of 16 bits. The generic map in the instantiated component overrides the default parameter.

9.55
```
-- srg10lpm.vhd
-- 10-bit serial shift register (shift right)
LIBRARY ieee;
USE ieee.std_logic_1164.ALL;
LIBRARY lpm;
USE lpm.lpm_components.ALL;

ENTITY srg10lpm IS
      PORT(
            clk        : IN  STD_LOGIC;
            serial_in  : IN  STD_LOGIC;
            sync_set   : IN  STD_LOGIC;
            serial_out : OUT STD_LOGIC);
END srg10lpm;

ARCHITECTURE lpm-shift of srg10lpm IS
COMPONENT lpm_shiftreg
      GENERIC(LPM_WIDTH: POSITIVE; LPM_SVALUE: STRING);
      PORT(
            clock, shiftin : IN  STD_LOGIC;
            sset           : IN  STD_LOGIC;
            shiftout       : OUT STD_LOGIC);
END COMPONENT;
BEGIN
      Shift_10: lpm_shifreg
            GENERIC MAP (LPM_WIDTH=> 10, LPM_SVALUE => "960")
            PORT MAP (clk, serial_in, sync_set, serial_out);
END lpm_shift;
```

The parameter LPM_SVALUE is set to 960, the decimal equivalent of H"3C0". Figure ANS9.55 shows the simulation of the shift register.

FIGURE ANS9.55

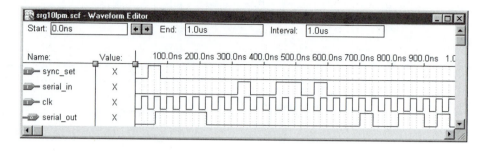

9.57

Q_4	Q_3	Q_2	Q_1	Q_0
0	0	0	0	0
1	0	0	0	0
1	1	0	0	0
1	1	1	0	0
1	1	1	1	0
1	1	1	1	1
0	1	1	1	1
0	0	1	1	1
0	0	0	1	1
0	0	0	0	1

All gates used in the decoder of Figure 9.84 remain unchanged except those decoding the MSB/LSB pairs (Q_3Q_0 and $\overline{Q_3}\,\overline{Q_0}$). Change these to decode Q_4Q_0 and $\overline{Q_4}\,\overline{Q_0}$. Add two new gates to decode $Q_4\overline{Q_3}$ (2nd state) and $\overline{Q_4}Q_3$ (7th state).

9.59   See Figure ANS9.59

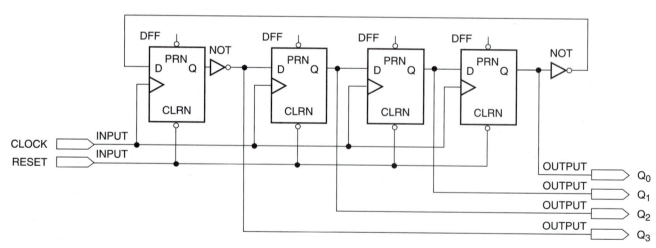

FIGURE ANS9.59

Chapter 10

10.1   Mealy machine. The output is fed by combinational, as well as sequential, logic.

10.3 $D_3 = Q_2\overline{Q_1}\,\overline{Q_0} + Q_3Q_1 + Q_3Q_0$
$D_2 = \overline{Q_3}Q_1\overline{Q_0} + Q_2\overline{Q_1} + Q_2Q_0$

$D_1 = \overline{Q_3}\,\overline{Q_2}Q_0 + Q_3Q_2Q_0 + Q_1\overline{Q_0}$
$D_0 = \overline{Q_3}\,\overline{Q_2}\,\overline{Q_1} + Q_3Q_2\overline{Q_1} + \overline{Q_3}Q_2Q_1 + Q_3\overline{Q_2}Q_1$

See Figure ANS10.3.

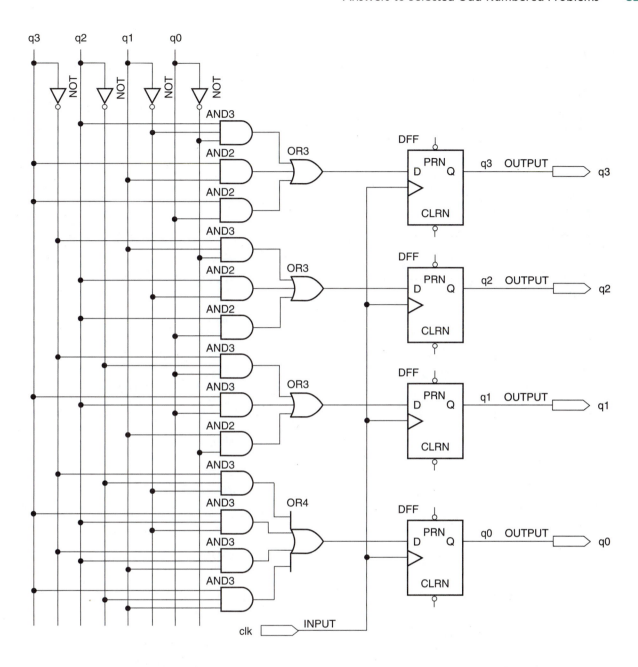

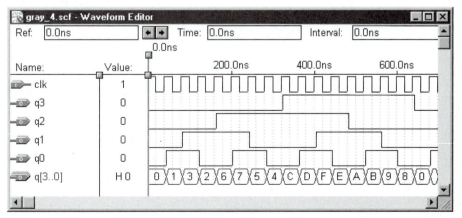

FIGURE ANS10.3

10.5 $J_2 = Q_1\overline{Q_0}$
$K_2 = \overline{Q_1}Q_0$
$J_1 = \overline{Q_2}Q_0$
$K_1 = Q_2Q_0$
$J_0 = \overline{Q_2}\overline{Q_1} + Q_2Q_1$
$K_0 = \overline{Q_2}Q_1 + Q_2\overline{Q_1}$

See Figure ANS10.5

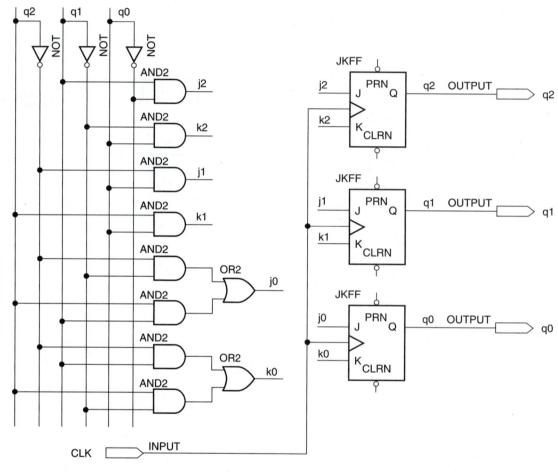

FIGURE ANS10.5

10.7 $D_1 = \overline{Q_1}Q_0\text{in}1$
$D_0 = \overline{Q_1}\overline{\text{in}1}$
$\text{out}1 = \overline{Q_1}Q_0\overline{\text{in}1}$
$\text{out}2 = \overline{Q_1}Q_0\text{in}1$

See Figure ANS10.7. The circuit generates a HIGH pulse on **out1** when **in1** goes LOW and a HIGH pulse on **out2** when the input goes back HIGH.

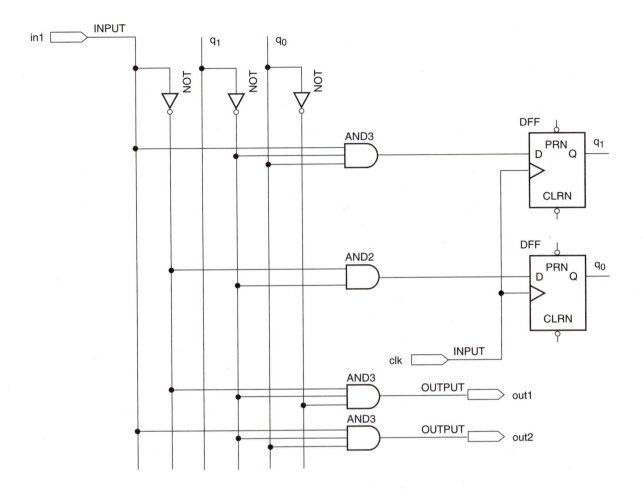

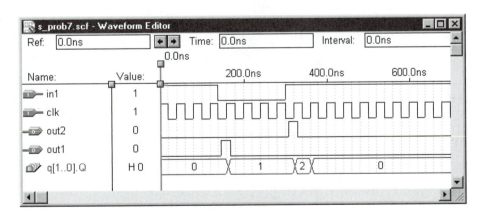

FIGURE ANS10.7

10.9 -- prob10_9.vhd

```vhdl
LIBRARY ieee;
USE ieee.std_logic_1164.ALL;

ENTITY prob10_9 IS
        PORT(
                clk, in1    : IN STD_LOGIC;
                out1, out2  : OUT STD_LOGIC);
END prob10_9;

ARCHITECTURE a OF prob10_9 IS
        TYPE PULSER IS (s0, s1, s2, s3);
        SIGNAL sequence: PULSER;
BEGIN
        PROCESS (clk)
        BEGIN
                IF clk'EVENT AND clk = '1' THEN
                        CASE sequence IS
                                WHEN s0 =>
                                        IF in1 = '1' THEN
                                                sequence <= s0;   -- no change if in1 = 1
                                                out1 <= '0';
                                                out2 <= '0';
                                        ELSE
                                                sequence <= s1;   -- proceed if in1 = 0
                                                out1 <= '1';          -- pulse on out1
                                                out2 <= '0';
                                        END IF;
                                WHEN s1 =>
                                        IF in1 = '0' THEN
                                                sequence <= s1;   -- outputs LOW
                                                out1 <= '0';
                                                out2 <= '0';
                                        ELSE
                                                sequence <= s2;
                                                out1 <= '0';
                                                out2 <= '1';          -- pulse on out2
                                        END IF;
                                WHEN s2 =>
                                                sequence <= s0;
                                                out1 <= '0';
                                                out2 <= '0';
                                WHEN others =>
                                                sequence <= s0;
                                                out1 <= '0';
                                                out2 <= '0';
                        END CASE;
                END IF;
        END PROCESS;
END a;
```

See Figure ANS10.9.

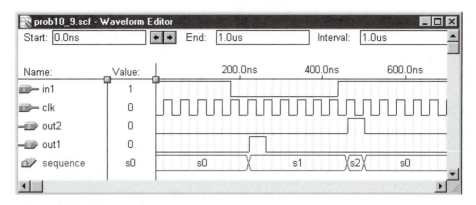

FIGURE ANS10.9

10.11
```
LIBRARY ieee;
USE ieee.std_logic_1164.ALL;
ENTITY prob10_11 IS
      PORT(
              clk, go, reset, eoc   : IN STD_LOGIC;
              sc, oe    : OUT STD_LOGIC);
END prob10_11;

ARCHITECTURE a OF prob10_11 IS
      TYPE ADC IS (idle, start, waiting, read);
      SIGNAL state: ADC;
      SIGNAL outputs: STD_LOGIC_VECTOR(1 downto 0);
BEGIN
      sc <= outputs(1);
      oe <= outputs(0);
      PROCESS (clk)
      BEGIN
            IF clk'EVENT AND clk = '1' THEN
                  IF reset = '0' THEN
                        state <= idle;
                        outputs <= "01";
                  ELSE
                        CASE state IS
                              WHEN idle =>
                                    IF go = '0' THEN
                                          state <= idle;
                                          outputs <= "01";
                                    ELSIF go = '1' THEN
                                          state <= start;
                                          outputs <= "11";
                                    END IF;
                              WHEN start =>
                                    state <= waiting;
                                    outputs <= "01";
                              WHEN waiting =>
                                    IF eoc = '0' THEN
                                          state <= waiting;
                                          outputs <= "01";
                                    ELSIF eoc = '1' THEN
                                          state <= read;
                                          outputs <= "00";
                                    END IF;
```

```
                        WHEN read =>
                            state <= idle;
                            outputs <= "01";
                END CASE;
            END IF;
        END IF;
    END PROCESS;
END a;
```

See Figure ANS 10.11.

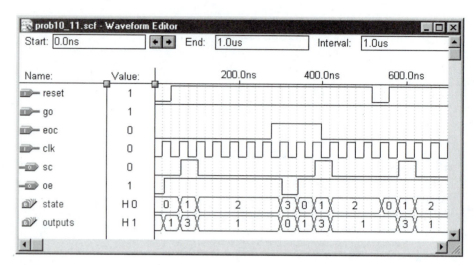

FIGURE ANS10.11

10.13 A NAND latch can only debounce a switch with a normally open and a normally closed contact: one to set and the other to reset the latch. The pushbutton on the Altera UP-1 board has only a normally open contact.

10.15 8.33 ms

10.17 Four clock periods. 8.33 ms

10.19 -- prob10_19.vhd

```
LIBRARY ieee;
USE ieee.std_logic_1164.ALL;

ENTITY prob10_19 IS
    PORT(
        clk, in1, in2    : IN  STD_LOGIC;
        out1             : OUT STD_LOGIC);
END prob10_19;

ARCHITECTURE a OF prob10_19 IS
    TYPE STATE_TYPE IS (s0, s1, s2, s3, s4);
    SIGNAL state: STATE_TYPE;
```

```
BEGIN
    PROCESS (clk)
    BEGIN
        IF clk'EVENT AND clk = '1' THEN
            CASE state IS
                WHEN s0 =>
                    state <= s1;
                    out1 <= '0';
                WHEN s1 =>
                    IF in1 = '1' THEN
                        state <= s1;
                        out1 <= '0';
                    ELSIF in1 = '0' THEN
                        state <= s2;
                        out1 <= '1';
                    END IF;
                WHEN s2 =>
                    state <= s3;
                    out1 <= '0';
                WHEN s3 =>
                    state <= s4;
                    out1 <= '0';
                WHEN s4 =>
                    IF in2 = '1' THEN
                        state <= s4;
                        out1 <= '0';
                    ELSIF in2 = '0' THEN
                        state <= s0;
                        out1 <= '1';
                    END IF;
            END CASE;
        END IF;
    END PROCESS;
END a;
```

See Figure ANS10.19

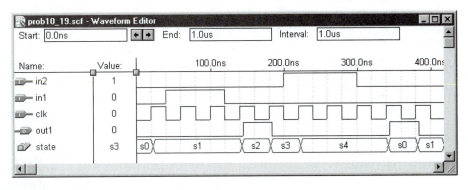

FIGURE ANS10.19

10.23 See Figure ANS10.23.

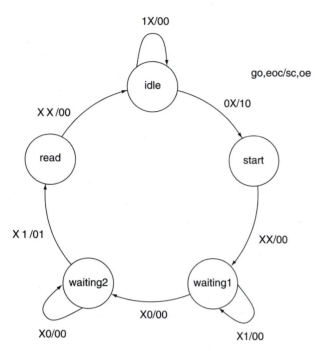

FIGURE ANS10.23

Chapter 11

11.1 **TTL:** advantages—relatively high speed, high current driving capability; disadvantages—high power consumption, rigid power supply requirements. **CMOS:** advantages—low power consumption, high noise immunity, flexible power supply requirements; disadvantages—low output current **ECL:** advantages—high speed; disadvantages—high susceptibility to noise, high power consumption.

11.3 $t_{pHL} = 12$ ns, $t_{pLH} = 10$ ns

11.5 Transition from state 1 to state 2: $t_p = t_{pLH02} + t_{pHL00} = 16$ ns + 15 ns = 31 ns. Transition from state 2 to state 3: $t_p = t_{pLH00} = 15$ ns (Assume $V_{CC} = 4.5$ volts; T = 25°C to −55°C)

11.7 Driving gate (74LS00): $I_{OL} = 8$ mA; $I_{OH} = -0.4$ mA
Load gate (74S32): $I_{IL} = -2$ mA; $I_{IH} = 20$ μA
$n_L = I_{OL}/I_{IL} = 8$ mA/2 mA = 4
$n_H = I_{OH}/I_{IH} = 0.4$ mA/0.02 mA = 20
$n = n_L = 4$

11.9 Source: $I_{OH} = -0.14$ mA; sink: $I_{OL} = 2.8$ mA

11.11 **a.** 44 mW;

b. 39 mW;

c. 29 mW;

d. 20 mW

11.13 **a.** 4.5 μW;

b. 28.3 μW;

c. 4.46 mW

11.15 **a.** 550 μW;

b. −56.4%

11.17 The outputs of a 74LS00 gates are guaranteed to produce output voltages of $V_{OH} \geq 2.7$ V and $V_{OL} \leq 0.8$ V. The inputs of a 74HCT series gate are voltage compatible with LSTTL outputs since $V_{IH} \geq 2$ V and $V_{IL} \leq 0.5$V. This is not the case for 74HC series gates, where $V_{IH} \geq 3.15$V and $V_{IL} \leq 1.35$V. The 74LS gate is not guaranteed to drive the 74HC gate in the HIGH state.

11.19 10 loads, since the 74HC output voltages are defined for an output current of ±4mA.

11.21 HIGH input: The base-emitter junction of transistor Q_1 is reverse-biased. Current flows, by default, through the base-collector junction of Q_1, supplying base current to Q_2, saturating it. This in turn, saturates Q_3, making its collector LOW.

LOW input: Current has a path to ground via the base-emitter junction of Q_1. This transports charge away from the base region of Q_2, making it cutoff. Since Q_2 is cutoff, no base current flows in Q_3. If an external pull-up resistor is connected to the collector of Q_3, the output will be HIGH. Otherwise, it is floating.

11.23 See Figure 11.31 in text. $Y = \overline{AB} \cdot \overline{CD} \cdot \overline{EF} = AB + CD + EF$

11.27 Yes. When the output transistor saturates (output LOW), there will be a direct connection from the output to V_{CC}. Since there is relatively little resistance in the current path, the current will likely exceed the rated output current, I_{OL}.

11.29 **a.** Q_3 and Q_4 are never on at the same time because the phase splitter, Q_2, keeps them in opposite states. The voltage in the circuit is divided such that when Q_2 is on, it pulls the base of Q_4 into the cutoff region for that transistor. At the same time, Q_3 is supplied with base current and thus saturated. When Q_2 is off, there is no base current in Q_3, making it cutoff. The base voltage at Q_4 is now such that it is on.

b. Switching noise originates in a totem pole output because the HIGH output transistor, Q_4, can switch on faster than the LOW output transistor, Q_3, can switch off. For a brief time, both transistors are on, causing a supply current spike. This can be counteracted by connecting a small capacitor between the supply voltage, V_{CC}, and ground.

11.31 7.58 mA, 95% of I_{OL}; 2.12 mA, 530% of I_{OH} The first circuit is more suitable, as it can drive a higher current to the LED and still remain within the output specification of the inverter.

11.33 Store MOS devices in antistatic or conducting material. Work only on an antistatic work surface and wear a conductive wrist strap. Connect all unused device inputs to power or ground. Do not touch the pins of the MOS device.

11.35 See Figure 11.56 in text.

A	B	Q_1	Q_2	Q_3	Q_4	Q_5	Q_6	Y
0	0	ON	ON	OFF	OFF	OFF	ON	0
0	1	OFF	ON	OFF	ON	OFF	ON	0
1	0	ON	OFF	ON	OFF	OFF	ON	0
1	1	OFF	OFF	ON	ON	ON	ONN	1

11.37 The state of flip-flop output Q selects which signal is switched to the data converter/display driver by enabling one of the CMOS transmission gates. When Q=1, the wheel rotation sensor is selected. Q=0 selects the engine rotation sensor.

11.39 No. TTL power dissipation, and therefore the speed-power product, depends on the logic states of the device outputs, not on frequency.

11.41 74HCNN: pin replacement for TTL device; CMOS-compatible inputs; TTL-compatible outputs. 74HC4NNN: pin replacement for CMOS device; CMOS-compatible inputs; TTL-compatible outputs. 74HCTNN: pin replacement for TTL device; TTL-compatible inputs; TTL-compatible outputs. 74HCUNN: unbuffered CMOS outputs.

Chapter 12

12.1

Analog Voltage	Code	Analog Voltage	Code
0 − 0.75	000	0.000 − 0.375	0000
0.75 − 2.25	001	0.375 − 1.125	0001
2.25 − 3.75	010	1.125 − 1.875	0010
3.75 − 5.25	011	1.875 − 2.625	0011
5.25 − 6.75	100	2.625 − 3.375	0100
6.75 − 8.25	101	3.375 − 4.125	0101
8.25 − 9.75	110	4.125 − 4.875	0110
9.75 − 12.00	111	4.875 − 5.625	0111
		5.625 − 6.375	1000
		6.375 − 7.125	1001
		7.125 − 7.875	1010
		7.875 − 8.625	1011
		8.625 − 9.375	1100
		9.375 − 10.125	1101
		10.125 − 10.875	1110
		10.875 − 12.000	1111

12.3

Fraction of T	Sine Voltage	Digital Code
0	0 V	0000
T/8	4.59 V	0110
T/4	8.48 V	1011
3T/8	11.09 V	1111
T/2	12.00 V	1111
5T/8	11.09 V	1111
3T/4	8.48 V	1011
7T/8	4.59 V	0110
T	0 V	0000

12.5 6 bits, since $64 = 2^6$. Resolution = 500 mV/64 = 7.8125 mV.

12.7 n+1 (One extra bit for each doubling of the number of codes.)

12.9 **a.** $V_a = (code/2^4) V_{ref} = (12/16) V_{ref} = 0.75 V_{ref}$;

b. $V_a = (code/2^8) V_{ref} = (200/256) V_{ref} = 0.78125 V_{ref}$;

c. A 4-bit and 8-bit quantization of the same analog voltage are the same in the first four bits. The additional bit in the lower 4 bits adds an extra voltage to the analog output.

12.11 From most to least significant bits: 1 kΩ, 2 kΩ, 4 kΩ, 8 kΩ, 16 kΩ, 32 kΩ, 64 kΩ, 128 kΩ, 256 kΩ, 512 kΩ, 1024 kΩ, 2048 kΩ, 4096 kΩ, 8192 kΩ, 16,384 kΩ, 32,768 kΩ. All resistors greater than 64 kΩ are specified to three or more significant figures. These values, which are necessary to maintain conversion accuracy, are not available as commercial components.

12.13 **a.** $V_a = (15/16) \times 12 V = 11.25 V$

b. $V_a = (11/16) \times 12 V = 8.25 V$

c. $V_a = (6/16) \times 12 V = 4.8 V$

d. $V_a = (3/16) \times 12 V = 2.25 V$

12.15 Resolution = $(1/256)(2.2 kΩ \times 12 V)/6.8 kΩ = 15.16 mV$

12.19 There are only 16 steps in the waveform and they reach to 15/16 of the reference value. Therefore, the four least significant bits are stuck at logic LOW.

12.21 Offset error (OE) = 0.5 V; OE = 0.333 LSB; OE = 4.167% FS. Gain error = 0; Linearity error = 0

12.23 Linearity error (LE) = 0.175 V; LE = 0.35 LSB; LE = 4.375%. Gain error = 0; Offset error = 0

12.25 The priority encoder converts the highest active comparator voltage to a digital code. The enable input of the latch can be pulsed with a waveform having the same frequency as the sampling frequency.

12.27

Bit	New Digital Value	Analog Equivalent from DAC	$V_{analog} \geq V_{DAC}$	Comparator Output	Accumulated Digital Value
Q_7	10000000	8 V	No	0	00000000
Q_6	01000000	4 V	Yes	1	01000000
Q_5	01100000	6 V	No	0	01000000
Q_4	01010000	5 V	No	0	01000000
Q_3	01001000	4.5 V	Yes	1	01001000
Q_2	01001100	4.75 V	Yes	1	01001100
Q_1	01001110	4.875 V	No	0	01001100
Q_0	01001101	4.8125 V	No	0	01001100

12.29 (8 V/12 V) $\times$ 16 = 10.667. Since the SAR method of A/D conversion truncates a result, the new code value will be 1010. The new hex digit is A.

12.31 $\pm\frac{1}{2}$ LSB

12.33 a. Integrating phase: The slope for a Full Scale input is given by:

$$-(v_{in})/RC = -16 \text{ V}/(80 \text{ k}\Omega)(0.1 \text{ }\mu\text{F}) = -2 \text{ V/ms}.$$

Since the slope is proportional to the input voltage, the slope for a 3 V input is:

$$(3/16) \times (-2 \text{ V/ms}) = -0.375 \text{ V/ms}$$

b. Rezeroing phase: At +2 V/ms, the integrator would take 8 seconds to rezero from Full Scale. This is always the slope when the circuit rezeros.

c. It would take (3/16) $\times$ (8 s) = 1.5 s to rezero for an input of 3 V.

d. The integrator waveform is similar to that for the input of 1/4 Full Scale shown in Figure 12.28 in the text.

e. Code = (3/16) $\times$ 256 = 48_{10} = 00110000_2.

12.35 See Figure 12.29 in text.

12.37 200 kHz

12.39 The sampling frequency is 13/12 times the period of the sampled analog waveform. This is 1-1/12 periods, or 30 degrees greater than the sampled waveform. Thus one full cycle of the alias frequency is 5.2 μs $\times$ 12 = 62.4 μs. The alias frequency is approximately 16 kHz.

12.41
```
-- adc_cont.vhd
-- State machine interface to ADC0808
-- Continuous conversion, single latch,
-- analog channel selected externally

LIBRARY ieee;
USE ieee.std_logic_1164.ALL;

ENTITY adc_cont IS
PORT(
      clock, reset, eoc : IN STD_LOGIC;
      sc, oe, en : OUT STD_LOGIC);
END adc_cont;
```

```
ARCHITECTURE adc OF adc_cont IS
      TYPE state_type IS (start, wait1, wait2, read, store);
      SIGNAL state: state_type;
      SIGNAL outputs: STD_LOGIC_VECTOR (1 to 3);
BEGIN
      PROCESS (clock, reset)
      BEGIN
          IF (reset = '0') THEN
                state <= start;
                outputs <= "000";
          ELSIF (clock'EVENT and clock = '1') THEN
                CASE state IS
                      WHEN start =>
                            state <= wait1;
                            outputs <= "100";
                      WHEN wait1 =>
                            IF (eoc = '1') THEN
                                  state <= wait1;
                                  outputs <= "000";
                            ELSIF (eoc = '0') THEN
                                  state <= wait2;
                                  outputs <= "000";
                            END IF;
                      WHEN wait2 =>
                            IF (eoc = '0') THEN
                                  state <= wait2;
                                  outputs <= "000";
                            ELSIF (eoc = '1') THEN
                                  state <= read;
                                  outputs <= "011";
                            END IF;
                      WHEN read =>
                            state <= store;
                            outputs <= "010";
                      WHEN store =>
                            state <= start;
                            outputs <= "000";
                END CASE;
          END IF;
          sc          <= outputs(1);
          oe          <= outputs(2);
          en          <= outputs(3);
      END PROCESS;
END adc;
```

See Figure ANS12.41.

FIGURE ANS12.41

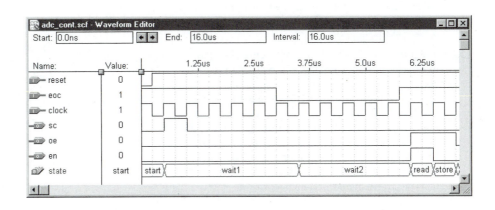

12.43 23.5 kHz

Chapter 13

13.1 The number of address lines is n for 2^n memory locations. Thus, an 8×8 memory requires 3 address lines ($2^3 = 8$). A 16×8 memory requires 4 address lines ($2^4 = 16$).

13.3 **a.** $64K = 2^6 \times 2^{10} = 2^{16}$; 16 address lines, 8 data lines

b. $128K = 2^7 \times 2^{10} = 2^{17}$; 17 address lines, 16 data lines

c. $128K = 2^7 \times 2^{10} = 2^{17}$; 17 address lines, 32 data lines

d. $256K = 2^8 \times 2^{10} = 2^{18}$; 18 address lines, 16 data lines

13.5 The inputs $\overline{W}$ (Write), $\overline{G}$ (Gate), and $\overline{E}$ (Enable) control the flow of data into or out of the RAM shown by enabling or disabling the two tristate buffers on each pin. There is an output (read) buffer and an input (write) buffer for each pin.

The read buffers are enabled when $\overline{W} = 1$, $\overline{E} = 0$, and $\overline{G} = 0$. The write buffers are enabled when $\overline{W} = 0$ and $\overline{E} = 0$. $\overline{G}$ is not required for the write buffer. Thus $\overline{W}$ controls the direction of the data (read or write), $\overline{E}$ enables the tristate buffers in either direction, and $\overline{G}$ enables the output buffers only.

13.7 To change the cell contents to a 0, we make the BIT line LOW and the ROW SELECT line HIGH. The ROW SELECT line gives access to the cell by turning on Q_5 and Q_6, completing the conduction path between the BIT lines and the flip flop inputs. The LOW on the BIT line pulls the gate of Q_4 LOW, turning it OFF. This breaks the conduction path from Q_4 drain to source and makes $V_{DS4} = V_{DD}$, a logic HIGH. This HIGH is applied to the gate of Q_3, turning it ON. A conduction path is established between Q_3 drain and source, pulling the drain of Q_3 LOW. The cell now stores a logic 0.

13.9 A selected RAM cell is at the junction of an active ROW line and an active COLUMN line in a rectangular matrix of cells.

13.11 The DRAM in Problem 13.10 has 10 multiplexed ROW/COLUMN address lines. Adding one more line makes 11 lines, each of which are used for a ROW address and also a column address. This make a total of 22 lines, giving an address capacity of $2^{22} = 4M$ locations. Adding another address line gives 12 multiplexed lines, each used for ROW and COLUMN, giving a total of $2^{24} = 16M$ locations.

13.13 The primary difference between the different types of ROM is how easy each type is to program and erase.

Mask-programmed ROM has the data manufactured into the device, making it difficult to program and impossible to erase. It is relatively cheap to mass-produce and is useful for storing unchanging data that must always be retained, including after power failure. An example is the "boot ROM" in a personal computer that contains data for minimal start-up instructions.

UV-erasable EPROM is fairly expensive because of the specialized packaging it requires. It is user-programmable and can be easily erased by exposure to ultraviolet light when removed from the circuit. It is useful for unfinished designs, since stored data can be changed as development of a product proceeds.

EEPROM can be used for applications which require data to be stored after power is removed from a device, but which require periodic in-circuit changes of data. One example might be an EEPROM which stores the numbers of several local channels in a digitally-programmed car radio.

13.15 Unlike EEPROM, flash memory is organized into sectors that can be erased all at one time. One sector, called the boot block, can be protected against unauthorized erasure or modification, thus adding a level of security to the memory.

13.17 EEPROM has slower access time and smaller bit capacity than RAM. It also has a finite number of program/erase cycles.

13.19 FIFO: buffer for serial data transmission; LIFO: memory stack in a microcomputer

13.21 $4K = 2^{12}$. Range = 0000 0000 0000 to 1111 1111 1111 (000H to FFFH); End address = Start + Maximum = 2000H + FFFH = 2FFFH.

$8K = 2^{13}$. Range = 0 0000 0000 0000 to 1 1111 1111 1111 (0000H to 1FFFH);

End = Start + Maximum = 6000H + 1FFFH = 7FFFH

See Figure ANS13.21.

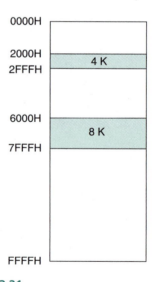

FIGURE ANS13.21

13.23 See Figure ANS13.23

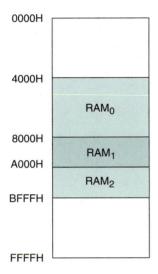

0000H
4000H
RAM_0
8000H
RAM_1
A000H
RAM_2
BFFFH
FFFFH

FIGURE ANS13.23

13.25

Device	Start Address	End Address	Size
EPROM	0000H	3FFFH	16K
$SRAM_1$	4000H	7FFFH	16K
$SRAM_2$	8000H	BFFFH	16K
$SRAM_3$	E000H	FFFFH	8K

13.27   16 DIMMs

Device	Start Address	End Address
0	0000000H	0FFFFFFH
1	1000000H	1FFFFFFH
2	2000000H	2FFFFFFH
3	3000000H	3FFFFFFH
4	4000000H	4FFFFFFH
5	5000000H	5FFFFFFH
6	6000000H	6FFFFFFH
7	7000000H	7FFFFFFH
8	8000000H	8FFFFFFH
9	9000000H	9FFFFFFH
10	A000000H	AFFFFFFH
11	B000000H	BFFFFFFH
12	C000000H	CFFFFFFH
13	D000000H	DFFFFFFH
14	E000000H	EFFFFFFH
15	F000000H	FFFFFFFH

Index